The Theory of Chaotic Attractors

Springer
New York
Berlin
Heidelberg
Hong Kong
London
Milan
Paris
Tokyo

Brian R. Hunt
Judy A. Kennedy
Tien-Yien Li
Helena E. Nusse
Editors

The Theory of Chaotic Attractors

Springer

Brian R. Hunt
Institute for Physical Science and Technology
University of Maryland
College Park, MD 20742
USA
bhunt@ipst.umd.edu

Judy A. Kennedy
Department of Mathematics
University of Delaware
Newark, Delaware 19716
USA
kennedy@math.udel.edu

Tien-Yien Li
Department of Mathematics
Michigan State University
East Lansing, Michigan 48824
USA
li@math.msu.edu

Helena E. Nusse
Department of Econometrics
University of Groningen
NL-9700 AV Groningen
The Netherlands
h.e.nusse@eco.rug.nl

Mathematics Subject Classification (2000): 37D45, 37A40, 37C40, 37Exx, 28D05, 01A75

Library of Congress Cataloging-in-Publication Data
The theory of chaotic attractors / [edited by] Brian R. Hunt, Judy A. Kennedy, Tien-Yien Li, Helena E. Nusse.
p. cm.
Includes bibliographical references
ISBN 0-387-40349-3 (alk. paper)
1. Chaotic behavior in systems. I. Nusse, Helena Engelina, 1952— II. Hunt, Brian R.
Q172.5.C45T44 2003
033'.857—dc21 2003053003

ISBN 0-387-40349-3 Printed on acid-free paper.

Printed in the United States of America. (EB)

9 8 7 6 5 4 3 2 1 SPIN 10936931

Springer-Verlag is a part of *Springer Science+Business Media*

springeronline.com

This volume is dedicated to James A. Yorke in commemoration of his 60th birthday.

Preface

The editors felt that the time was right for a book on an important topic, the history and development of the notions of chaotic attractors and their "natural" invariant measures. We wanted to bring together a coherent collection of readable, interesting, outstanding papers for detailed study and comparison. We hope that this book will allow serious graduate students to hold seminars to study how the research in this field developed. Limitation of space forced us painfully to exclude many excellent, relevant papers, and the resulting choice reflects the interests of the editors.

Since James Alan Yorke was born August 3, 1941, we chose to have this book commemorate his sixtieth birthday, honoring his research in this field. The editors are four of his collaborators.

We would particularly like to thank Achi Dosanjh (senior editor mathematics), Elizabeth Young (assistant editor mathematics), Joel Ariaratnam (mathematics editorial), and Yong-Soon Hwang (book production editor) from Springer Verlag in New York for their efforts in publishing this book.

College Park, MD
June 2003

Brian R. Hunt
Judy A. Kennedy
Tien-Yien Li
Helena E. Nusse

Contents

Introduction

Part I: Who is James Yorke?

Recently, the Science and Technology Foundation of Japan announced that James A. Yorke had been named a winner of the 2003 Japan Prize for his work in the field of chaos theory. This foundation has awarded Japan Prizes since 1985 under the auspices of the Japanese prime minister. It is a Japanese version of the Nobel Prize. One of the most esteemed of science and technology prizes, Japan Prizes are given to scientists whose "original and outstanding achievements in science and technology are recognized as having advanced the frontiers of knowledge and served the cause of peace and prosperity for mankind." The Science and Technology Foundation of Japan announced that Yorke and Benoit Mandelbrot, of Yale University, will share $415,000 (50 million yen) for the "creation of universal concepts in complex systems: chaos and fractals". The other 2003 Japan Prize was awarded for discovering the principle behind magnetic resonance imaging (MRI). Past Japan prizes have been awarded for achievements such as the discovery of the HIV virus, the invention of the World Wide Web, the development of artificial intelligence, and the eradication of smallpox. "For a scientist to be awarded the Japan Prize is a distinction as great as any in the world," said University of Maryland President C.D. Mote, Jr. "Jim Yorke has now been officially recognized for his original achievements in nonlinear dynamics that have monumentally advanced the frontiers of science and technology and served the cause of peace and prosperity for mankind. I am so pleased for him and for the inspiration his recognition will provide for others who by pursuing their passions can hope to have an impact as great as Jim's." The presentation ceremony for the 2003 Japan Prize laureates is scheduled to be held in Tokyo in April 2003.

James A. Yorke came to the University of Maryland for graduate studies in 1963 at the age of 22, in part because of the interdisciplinary opportunities offered by the faculty of an interdisciplinary research institute now called the Institute for Physical Science and Technology (IPST). He has spent his entire career at the University of Maryland (College Park). From 1985 through 2001 he was the director of IPST. Currently he is a Distinguished University Professor in the Departments of Mathematics and Physics, and in IPST at the University of Maryland. He has earned an international reputation for his pioneering work in several different areas of mathematics and its applications. The famous papers entitled "On the existence of invariant measures for piecewise monotonic transformations" by A. Lasota and J.A. Yorke, which appeared in 1973 in volume 186 of the *Transactions of the American Mathematical Society*, and "Period three implies chaos" by T.-Y. Li and J.A. Yorke, which appeared in 1975 in volume 82 of the *American Mathematical Monthly* are just two examples. In particular, he has been uniquely innovative in his significant scientific contributions to applied topology, bifurcation theory, numerical methods, biology (including epidemiological modeling), theoretical nonlinear dynamics, and controlling chaos. He is coauthor of over 250 scientific papers and three books.

Yorke is a mathematician who likes to think of himself as a philosopher, but he is also the kind of mathematician who feels compelled to put his ideas of reality to some use, as the following examples demonstrate. In joint work with H.W. Hethcote and A. Nold, Yorke studied the transmission dynamics and control of gonorrhea and produced a report that persuaded the federal government to alter its national strategies for controlling the disease. During the gasoline crisis of the 1970s, Yorke gave testimony to the State of Maryland arguing correctly (but unpersuasively) that the even–odd system of limiting gasoline sales would only make lines longer. In the era of antiwar demonstrations, the government released a spy-plane photograph purporting to show sparse crowds around the Washington Monument at the height of the rally. However, Yorke proved that the photograph had actually been taken later when the rally was breaking up (about a half-hour later), by analyzing the monument's shadow in that photograph. Now he is working on a diverse collection of projects including understanding transmission patterns of HIV/AIDS, how to improve weather prediction, how better to determine genomes (the DNA sequences of plants and animals), and diverse mathematics projects. A recent example of his lively lecturing style was demonstrated during a lecture in which he jumped on a table and stamped a grapefruit flat just to illustrate the idea of a map from three-dimensional to two-dimensional phase space. Those in the audience will remember that lecture.

At the institute, Yorke enjoyed unusual freedom to work on a variety of problems and he had frequent contact with experts in a wide range of disciplines. One of these experts, a fluid dynamicist, encountered Lorenz's 1963 paper "Deterministic Nonperiodic Flow" in 1972 and gave a copy to Yorke. Lorenz's paper was the kind of scientific work that Yorke had been looking for without even knowing it. It was a vivid physical model, a picture of a fluid in motion, and Yorke realized that it had ideas he wanted physicists to see and understand. To a physicist, a "legitimate" example was a differential equation that could be written down in a simple form. When Yorke saw Lorenz's paper, he knew that it was an example that physicists would embrace. Yorke felt that physicists had learned not to see chaos. In daily life, the Lorenzian quality of sensitive dependence on initial conditions lurks everywhere. Small perturbations in one's daily trajectory can have large consequences. When trying to explain this idea, he may pose the question, "How did your parents meet?" Our very existence is the result of improbable events. Yorke understood that there is some disorder virtually everywhere. Physicists and mathematicians want to discover regularities, but one has to know about disorder if one is going to deal with it. Yorke believed that scientists and nonscientists alike can easily mislead themselves about complexity if they are not properly attuned to it and that many scientists must have seen chaotic behavior in innumerable circumstances for many years. In the past, when they were running a physical experiment, and the experiment behaved in an erratic manner, they either tried to fix it or gave up. They explained the erratic behavior by saying that it was due to the noise, or that the experiment was poorly designed. Yorke says that scientists were the last to find out about chaos. He wonders whether there are other phenomena that we are trained to ignore. His goal is to tell scientists what to look for. Some mathematicians prefer to prove results that clarify issues scientists have been discussing, but scientists, Yorke says, generally don't appreciate such rigorous work. They cannot read it and are likely to say "we already knew that." He believes that it is better to try to lead an elephant than to follow.

More than twenty-five years ago, Tien-Yien Li and James A. Yorke published their famous paper "Period three implies chaos." In that paper, "chaos" means (1) infinitely many periods, and (2) sensitivity to initial data. Most of the lines of proof in the paper were devoted to establishing the latter property. The evolution to the final version of the paper provides some insight into the working habits of Yorke. At that time, Li was a graduate student working with Yorke and had finished a draft of the paper. It remained untouched on Yorke's desk for a few months. Li worried that perhaps Yorke did not have a high opinion of the draft and that he was no longer interested, but Yorke was convinced that the results were important, since they illuminated and clarified the message in the works of Lorenz and Smale that physicists were not hearing. However, he did not think that there was need to rush; substantial revision was needed to make the paper more readable for the *Monthly*. Robert May received a copy of the draft when he gave a series of lectures at the University of Maryland in early 1974. May told Yorke about period doubling, and Yorke told May about chaos. (Later in 1974, May published a paper entitled "Biological populations with nonoverlapping generations, stable points, stable cycles and chaos" in volume 186 of the prestigious journal Science mentioning the Li–Yorke "period three implies chaos" theorem, referring to their paper as "in press." He credited Li and Yorke with coining the term "chaos.") An hour after May departed from the University of Maryland, Li remembers Yorke calling to say that they had to rush to finish the revisions so they could submit it the next Monday. This paper was important on its merits, though part of it turned out to be implied by an earlier result of Sarkovskii. Perhaps more than any previous work in the area, it grabbed the attention of nonmathematicians. In the end its most influential feature was its mysterious and mischievous title: "Period three implies chaos." Some colleagues advised them to choose something more sober, but Li and Yorke stuck with a word that came to stand for the whole growing business of deterministic disorder. Soon work of many authors from Maxwell, Poincaré, and Birkhoff to Arnol'd, Sinai, and Smale could be viewed as fitting under the umbrella term "chaos theory."

Jim Yorke always wears red socks. When asked why he says, "Red socks are a trivial reminder to try to think differently."

A Bird's-Eye View on the Variety of Yorke's Important Scientific Contributions

James A. Yorke's principal scientific contributions have been in the area of chaotic dynamical systems. Virtually all his work is with collaborators, and the results described here are equally due to them, but for brevity they are not mentioned. Although he has worked extensively in differential delay equations and in mathematical epidemiology, his interests turned mainly to the area of chaos. Chaos in dynamical systems refers to the fact that even simple deterministic systems can have time evolution that shares many of the properties of randomness. It has become increasingly clear that chaotic dynamics has very important implications for a broad variety of applications (e.g., weather prediction, celestial mechanics, plasmas, mechanical systems, solid state physics). The development of the theory of chaotic dynamics and its evident wide applicability in science and technology is an extremely important achievement of modern mathematics. Indeed, it is the opinion of many scientists that chaotic dynamics is one of the most significant new developments in mathematics in the last few decades. Yorke has played a key role in initiating investigations in fundamental and important questions in chaotic dynamics, always searching for results that scientists need. He has done this consistently over a period of years, starting with the by now famous papers "On the existence of invariant measures for piecewise monotonic transformations" by A. Lasota and J.A. Yorke in 1973, and "Period three implies chaos" by T.-Y. Li and J.A. Yorke in 1975. It is characteristic how differently these two papers approach chaos: one from the view of measure and operators on a Banach space, the other as topology of the real line. This latter paper first used the word "chaos" as a mathematical term to describe this type of dynamics, and thereby named this field of research. This paper introduced the ideas of chaos and sensitivity to initial data to a generation of mathematicians, physicists, engineers, and scientists from a variety of other disciplines. Yorke has been one of the most significant and consistently innovative contributors to the field of chaotic dynamical systems, and he is the driving force behind several of the major innovations. In addition, the theoretical and mathematical work of Yorke and his collaborators led to many applications in a variety of sciences, including physics, astrophysics, biology, chemistry, and economics.

Yorke's pioneering work has helped to bridge the gap between mathematics and the phenomenological approach of physical scientists. One of his goals has been to bring mathematical ideas to scientists by reformulating established concepts into new patterns that they might find useful, for example, the OGY method (Ott–Grebogi–Yorke method) for controlling chaos. He helped to establish a common vocabulary for approaching such problems, as we now illustrate. At the end of the 1970s, Yorke was the first to point out and explain the phenomenon now called "transient chaos," whereby orbits can behave in a chaotic manner before suddenly settling into a more regular steady state, giving both a mathematical theory and numerical experiments to explain how transient chaos phenomena manifest themselves in practice. These studies on transient chaos have paved the way for a host of significant subsequent work by other researchers and, more generally, have demonstrated the practical importance of chaotic sets that are not attractors.

The fractal dimension of strange attractors is a basic attribute of these attractors. Yorke's early work on fractal dimension managed to integrate diverse approaches and expedite measurements of dimensions for chaotic attractors. The Kaplan–Yorke conjecture concerns the idea of determining the fractal dimension of attractors from Lyapunov numbers. These investigations played a major role in initiating the now extensive developments with respect to this important question, to the point that fractal dimensions are now routinely measured in experiments.

Yorke and his collaborators developed methods for analyzing discontinuous changes in the structure of chaotic attractors and the fractal basin boundaries defined by them; the resulting terminology of "crisis" and "metamorphosis" is in common use today. Yorke has also made important and innovative contributions to bifurcation theory, including both local bifurcations and global bifurcations. For example, with collaborators he proved why cascades of period-doubling bifurcations occur in two-dimensional dynamical systems and why there are concurrent creation and annihilation of periodic orbits in any neighborhood of homoclinic tangencies. He has been influential in developing the view that to analyze systems that exhibit irregular behavior, the concept of mathematical modeling of systems needs to be expanded to include algorithms that make direct use of measured time series from the system and that these must be given a firm mathematical foundation. These ideas, which include observation, analysis and control of chaotic systems, have been widely implemented by experimental scientists.

Yorke has done pathbreaking work on the reliability of computer simulations of dynamical systems. He and his collaborators have developed a variety of reliable numerical methods, resulting in the software package "Dynamics." This software is a useful tool for exploring dynamical systems numerically in a way that leads to new fundamental properties and mathematical results of the dynamical system under investigation. The phenomenon of Wada basins in dynamical systems demonstrates that interaction between numerical investigations and development of new mathematical results is of crucial importance. (A Wada basin is a basin of attraction such that every boundary point of that basin is a boundary point of at least three different basins of attraction simultaneously.) Concerning higher-dimensional dynamics, although many mathematical results are known, many more phenomena and results are awaiting to be discovered. Indeed, it is the opinion of several leading mathematicians in dynamical systems that computer explorations are essential to theoretical development and understanding dynamical systems. As of this writing, he is working with collaborators on understanding AIDS in Africa, the variability of chaos in weather prediction, how better to assemble a genome, and a number of more strictly mathematical topics.

A selection of James A. Yorke's publications

Books

H.W. Hethcote and J.A. Yorke,
Gonorrhea Transmission Dynamics and Control,
Lecture Notes in Biomathematics #56,
Springer-Verlag, Berlin, 1984

E. Ott, T. Sauer, and J.A. Yorke (editors)
Coping with Chaos,
Wiley Series in Nonlinear Science
John Wiley & Sons Inc., New York, 1994

K. Alligood, T. Sauer, and J.A. Yorke,
Chaos: An Introduction to Dynamical Systems,
Textbooks in Mathematical Sciences
Springer-Verlag, New York, 1997

H.E. Nusse and J.A. Yorke,
Dynamics: Numerical Explorations,
Second Edition, accompanied by software program Dynamics 2, an interactive program for IBM compatible PC's and Unix computers, (the Unix version of the program is by B.R. Hunt and E.J. Kostelich),
Applied Mathematical Sciences 101,
Springer-Verlag, New York, 1998

Journal Papers (chronologically ordered; a partial but representative list selected by Yorke)

J.A. Yorke, A continuous differential equation in Hilbert space without existence, Funkcialaj Ekvacioj **13** (1970), 19–21

F.W. Wilson, Jr. and J.A. Yorke, Lyapunov functions and isolating blocks, J. Differential Equations **13** (1973), 106–123

W. London, M.D. and J.A. Yorke, Recurrent outbreaks of measles, chicken pox, and mumps, I. Seasonal variation in contact rates, and II. Systematic differences in contact rates and stochastic effects, Amer. J. Epidemiology **98** (1973), 453–468 and 469–482

A. Lasota and J.A. Yorke, The generic property of existence of solutions of differential equations in Banach space, J. Differential Equations **13** (1973), 1–12

A. Lasota and J.A. Yorke, On the existence of invariant measures for piecewise monotonic transformations, Trans. Amer. Math. Soc. **186** (1973), 481–488

T.-Y. Li and J.A. Yorke, Period three implies chaos, Amer. Math. Monthly **82** (1975), 985–992

J.C. Alexander and J.A. Yorke, The implicit function theorem and the global methods of cohomology, J. Functional Analysis, **21** (1976), 330–339

A. Lajmanovich and J.A. Yorke, A deterministic model for gonorrhea in a nonhomogeneous population, Math. Biosci. **28** (1976), 221–236

R.B. Kellogg, T.-Y. Li, and J.A. Yorke, A constructive proof of the Brouwer fixed-point theorem and computational results, SIAM J. Numer. Anal. **13** (1976), 473–383

J.L. Kaplan and J.A. Yorke, On the nonlinear differential delay equation $dx/dt = -f(x(t), x(t-1))$, J. Differential Equations **23** (1977), 293–314

T.Y. Li and J.A. Yorke, Ergodic transformations from an interval into itself, Trans. Amer. Math. Soc. **235** (1978), 183–192

J.C. Alexander and J.A. Yorke, Global bifurcation of periodic orbits, Amer. J. Math. **100** (1978), 263–292

S.N. Chow, J. Mallet-Paret, and J.A. Yorke, Finding zeroes of maps: homotopy methods that are constructive with probability one, Math. of Comp. **32** (1978), 887–899

J.A. Yorke, H.W. Hethcote, and A. Nold, Dynamics and control of the transmission of gonorrhea, Sexually Transmitted Diseases **5** (1978), 51–56

J.A. Yorke, N. Nathanson, G. Pianigiani, and J. Martin, Seasonality and the requirements for perpetuation and eradication of viruses in populations, Amer. J. Epidemiology **109** (1979), 103–123

G. Pianigiani and J.A. Yorke, Expanding maps on sets which are almost invariant: decay and chaos, Trans. Amer. Math. Soc. **252** (1979), 351–366

J.L. Kaplan and J.A. Yorke, Preturbulence: a regime observed in a fluid flow model of Lorenz, Comm. Math. Phys. **67** (1979), 93–108

J.A. Yorke and E.D. Yorke, Metastable chaos: the transition to sustained chaotic oscillations in the Lorenz model, J. Stat. Phys. **21** (1979), 263–277

J. Mallet-Paret and J.A. Yorke, Snakes: oriented families of periodic orbits, their sources, sinks, and continuation, J. Differential Equations **43** (1982), 419–450

C. Grebogi, E. Ott, and J.A. Yorke, Chaotic attractors in crisis, Phys. Rev. Lett. **48** (1982), 1507–1510

P. Frederickson, J.L. Kaplan, E.D. Yorke, and J.A. Yorke, The Lyapunov dimension of strange attractors, J. Differential Equations **49** (1983), 185–207

C. Grebogi, E. Ott, and J.A. Yorke, Crises, sudden changes in chaotic attractors, and transient chaos, Physica D **7** (1983), 181–200

J.D. Farmer, E. Ott, and J.A. Yorke, The dimension of chaotic attractors, Physica D **7** (1983), 153–180

J.A. Yorke and K.T. Alligood, Cascades of period doubling bifurcations: a prerequisite for horseshoes, Bull. Amer. Math. Soc. **9** (1983), 319–322

C. Grebogi, S.W. McDonald, E. Ott, and J.A. Yorke, Final state sensitivity: an obstruction to predictability, Phys. Lett. A **99** (1983), 415–418

J.L. Kaplan, J. Mallet-Paret, and J.A. Yorke, The Lyapunov dimension of a nowhere differentiable attracting torus, Ergodic Theory and Dyn. Sys. **4** (1984), 261–281

J.C. Alexander and J.A. Yorke, Fat baker's transformations, Ergodic Theory and Dyn. Sys. **4** (1984), 1–23

C. Grebogi, E. Ott, S. Pelikan, and J.A. Yorke, Strange attractors that are not chaotic, Physica D **13** (1984), 261–268

E. Ott, W.D. Withers, and J.A. Yorke, Is the dimension of chaotic attractors invariant under coordinate changes?, J. Stat. Phys. **36** (1984), 687–697

C. Grebogi, E. Ott, and J.A. Yorke, Attractors on an N-torus: quasiperiodicity versus chaos, Physica D **15** (1985), 354–373

S.W. McDonald, C. Grebogi, E. Ott, and J.A. Yorke, Fractal basin boundaries, Physica D **17** (1985), 125–153

S.W. McDonald, C. Grebogi, E. Ott, and J.A. Yorke, Structure and crises of fractal basin boundaries, Phys. Lett. A **107** (1985), 51–54

J.A. Yorke and K.T. Alligood, Period doubling cascades of attractors: a prerequisite for horseshoes, Comm. Math. Phys. **101** (1985), 305–321

C. Grebogi, S. W. McDonald, E. Ott, and J. A. Yorke, The exterior dimension of fat fractals, Phys. Lett. A **110** (1985), 1–4; **113** (1986), 495

C. Grebogi, E. Ott, and J.A. Yorke, Comment on "Sensitive dependence on parameters in nonlinear dynamics" and on "Fat fractals on the energy surface" Phys. Rev. Lett. **56** (1986), 266

C. Grebogi, E. Ott, and J.A. Yorke, Metamorphoses of basin boundaries in nonlinear dynamical systems, Phys. Rev. Lett. **56** (1986), 1011–1014

J.A. Yorke, E.D. Yorke, and J. Mallet-Paret, Lorenz-like chaos in a partial differential equation for a heated fluid loop, Physica D **24** (1987), 279–291

C. Grebogi, E. Ott, and J.A. Yorke, Basin boundary metamorphoses: changes in accessible boundary orbits, Physica D **24** (1987), 243–262

C. Grebogi, E. Ott, and J.A. Yorke, Chaos, strange attractors, and fractal basin boundaries in nonlinear dynamics, Science **238** (1987), 632–638

C. Grebogi, E. Ott, and J.A. Yorke, Unstable periodic orbits and the dimension of chaotic attractors, Phys. Rev. A **36** (1987), 3522–3524

S.M. Hammel, J.A. Yorke, and C. Grebogi, Do numerical orbits of chaotic dynamical processes represent true orbits?, J. of Complexity **3** (1987), 136–145

C. Grebogi, E. Ott, F. Romeiras, and J.A. Yorke, Critical exponents for crisis induced intermittency, Phys. Rev. A **36** (1987), 5365–5380

H.E. Nusse and J.A. Yorke, Is every approximate trajectory of some process near an exact trajectory of a nearby process?, Comm. Math. Phys. **114** (1988), 363–379

E.J. Kostelich and J.A. Yorke, Noise reduction in dynamical systems, Phys. Rev. A **38** (1988), 1649–1652

S.M. Hammel, J.A. Yorke, and C. Grebogi, Numerical orbits of chaotic processes represent true orbits, Bull. Amer. Math. Soc. **19** (1988), 465–469

C. Grebogi, E. Ott, and J.A. Yorke, Roundoff-induced periodicity and the correlation dimension of chaotic attractors, Phys. Rev. A **38** (1988), 3688–3692

T.-Y. Li, T. Sauer, and J.A. Yorke, Numerically determining solutions of systems of polynomial equations, Bull. Amer. Math. Soc. **18** (1988), 173–177

I. Kramer, E. D. Yorke, and J. A. Yorke, The AIDS epidemic's influence on the gay contact rate from analysis of gonorrhea incidence, Math. Comput. Modelling **12** (1989), 129–137

T.-Y. Li, T. Sauer, and J.A. Yorke, The cheater's homotopy: an efficient procedure for solving systems of polynomial equations, SIAM J. Numer. Anal. **26** (1989), 1241–1251

H.E. Nusse and J.A. Yorke, A procedure for finding numerical trajectories on chaotic saddles, Physica D **36** (1989), 137–156

W.L. Ditto, S. Rauseo, R. Cawley, C. Grebogi, G.-H. Hsu, E. Kostelich, E. Ott, H.T. Savage, R. Segnan, M. Spano, and J.A. Yorke, Experimental observation of crisis-induced intermittency and its critical exponent, Phys. Rev. Lett. **63** (1989), 923–926

E. Kostelich and J.A. Yorke, Noise reduction: finding the simplest dynamical system consistent with the data, Physica D **41** (1990), 183–196

C. Grebogi, S. M. Hammel, J.A. Yorke, and T. Sauer, Shadowing of physical trajectories in chaotic dynamics: containment and refinement, Phys. Rev. Lett. **65** (1990), 1527–1530

T. Shinbrot, E. Ott, C. Grebogi, and J.A. Yorke, Using chaos to direct trajectories to targets, Phys. Rev. Lett. **65** (1990), 3215–3218

E. Ott, C. Grebogi, and J.A. Yorke, Controlling chaos, Phys. Rev. Lett. **64** (1990), 1196–1199

M. Ding, C. Grebogi, E. Ott, and J.A. Yorke, Massive bifurcation of chaotic scattering, Phys. Lett. A **153** (1991), 21–26

J. Kennedy and J.A. Yorke, Basins of Wada, Physica D **51** (1991), 213–225

H.E. Nusse and J.A. Yorke, Analysis of a procedure for finding numerical trajectories close to chaotic saddle hyperbolic sets, Ergodic Theory and Dyn. Sys. **11** (1991), 189–208

B.R. Hunt and J.A. Yorke, Smooth dynamics on Weierstrass nowhere differentiable curves, Trans. Amer. Math. Soc. **325** (1991), 141–154

T. Sauer and J.A. Yorke, Rigorous verification of trajectories for the computer simulation of dynamical systems, Nonlinearity **4** (1991), 961–979

T. Sauer, J.A. Yorke, and M. Casdagli, Embedology, J. Stat. Phys. **65** (1991), 579–616

Z.-P. You, E.J. Kostelich, and J.A. Yorke, Calculating stable and unstable manifolds, Int. J. Bifurcation and Chaos **1** (1991), 605–623

K.T. Alligood, L. Tedeschini, and J.A. Yorke, Metamorphoses: sudden jumps in basin boundaries, Comm. Math. Phys. **141** (1991), 1–8

H.E. Nusse and J.A. Yorke, A numerical procedure for finding accessible trajectories on basin boundaries, Nonlinearity **4** (1991), 1183–1212

I. Kan, H. Koçak, and J.A. Yorke, Antimonotonicity: concurrent creation and annihilation of periodic orbits, Annals of Mathematics **136** (1992), 219–252

H.E. Nusse and J.A. Yorke, Border collision bifurcations including period two to period three bifurcation for piecewise smooth systems, Physica D **57** (1992), 39–57

T. Shinbrot, W. Ditto, C. Grebogi, E. Ott, M. Spano, and J.A. Yorke, Using the sensitive dependence of chaos (the Butterfly Effect) to direct orbits to targets in an experimental chaotic system, Phys. Rev. Lett. **68** (1992), 2863–2866

K.T. Alligood and J.A. Yorke, Accessible saddles on fractal basin boundaries, Ergodic Theory and Dyn. Sys. **12** (1992), 377–400

J.C. Alexander, J.A. Yorke, Z-P. You, and I. Kan, Riddled basins, Int. J. Bifurcation & Chaos **2** (1992), 795–813

B.R. Hunt, T. Sauer, and J.A. Yorke, Prevalence: a translation-invariant "almost every" on infinite dimensional spaces, Bull. Amer. Math. Soc. **27** (1992), 217–238

T. Shinbrot, C. Grebogi, E. Ott, and J.A. Yorke, Using small perturbations to control chaos, Nature **363** (1993), 411–417

M. Ding, C. Grebogi, E. Ott, T. Sauer, and J.A. Yorke, Plateau onset for correlation dimension: when does it occur?, Phys. Rev. Lett. **70** (1993), 3872–3873

B.R. Hunt and J.A. Yorke, Maxwell on chaos, Nonlinear Science Today **3** (1993), 2–4

J. Kennedy and J. A. Yorke, Pseudocircles in dynamical systems, Trans. Amer. Math. Soc. **343** (1994), 349–366

S.P. Dawson, C. Grebogi, T. Sauer, and J.A. Yorke, Obstructions to shadowing when a Lyapunov exponent fluctuates about zero, Phys. Rev. Lett. **73** (1994), 1927–1930

H.E. Nusse and J.A. Yorke, Border-collision bifurcations for piecewise smooth one-dimensional maps, Int. J. Bifurcation and Chaos **5** (1995), 189–207

J. Kennedy and J.A. Yorke, Bizarre Topology is Natural in Dynamical Systems, Bull. Amer. Math. Soc. **32** (1995), 309–316

H.E. Nusse, E. Ott, and J.A. Yorke, Saddle-node bifurcations on fractal basin boundaries, Phys. Rev. Lett. **75** (1995), 2482–2485

H.E. Nusse and J.A. Yorke, Wada basin boundaries and basin cells, Physica D **90** (1996), 242–261

H.E. Nusse and J.A. Yorke, Basins of attraction, Science **271** (1996), 1376–1380

U. Feudel, C. Grebogi, B. Hunt, and J.A. Yorke, A map with more than 100 coexisting low-period, periodic attractors, Phys. Rev. E **54** (1996), 71–81

E.J. Kostelich, J.A. Yorke, and Z. You, Plotting stable manifolds: error estimates and noninvertible maps, Physica D **93** (1996), 210–222

J. Kennedy and J.A. Yorke, Pseudocircles, diffeomorphisms, and perturbable dynamical systems, Ergodic Theory and Dyn. Sys. **16** (1996), 1031–1057

M. Sanjuan, J. Kennedy, C. Grebogi, and J.A. Yorke, Indecomposable continua in dynamical systems with noise: fluid flow past an array of cylinders, Int. J. Bifurcation & Chaos **7** (1997), 125–138

H.E. Nusse and J.A. Yorke, The structure of basins of attraction and their trapping regions, Ergodic Theory and Dyn. Sys. **17** (1997),463–482

T. Sauer, C. Grebogi,, and J.A. Yorke, How long do numerical chaotic solutions remain valid?, Phys. Rev. Lett. **79** (1997), 59–62

J. Kennedy and J.A. Yorke, The topology of stirred fluids, Topology and Its Applications **80** (1997), 201–238

T. Sauer and J.A. Yorke, Are the dimensions of a set and its image equal under typical smooth functions?, Ergodic Theory and Dyn. Sys. **17** (1997), 941–956

E. Kostelich, I. Kan, C. Grebogi, E. Ott, and J. A. Yorke, Unstable dimension variability: a source of nonhyperbolicity in chaotic systems, Physica D **109** (1997), 81–90

U. Feudel, C. Grebogi, L. Poon, and J.A. Yorke, Dynamical properties of a simple mechanical system with a large number of coexisting periodic attractors, Chaos, Solitons and Fractals **9** (1998), 171–180

C. Robert, K.T. Alligood, E. Ott, and J.A. Yorke, Outer tangency bifurcations of chaotic sets, Phys. Rev. Lett. **80** (1998), 4867–4870

G.-H. Yuan, S. Banerjee, E. Ott, and J.A. Yorke, Border-collision bifurcations in the Buck Converter, IEEE Trans. Circuits and Systems I: Fund. Theory and Appl. **45** (1998), 707–716

T. Sauer, J. Tempkin, and J.A. Yorke, Spurious Lyapunov exponents in attractor reconstruction, Phys. Rev. Lett. **81** (1998), 4341–4344

J. Kennedy, M.A.F. Sanjuan, J.A. Yorke, and C. Grebogi, The Topology of Fluid Flow Past a Sequence of Cylinders, Topology and Its Applications **94** (1999), 207–242

G.-C. Yuan and J.A. Yorke, An open set of maps for which every point is absolutely nonshadowable, Proc. Amer. Math. Soc. **128** (1999), 909–918

J. Miller and J.A. Yorke, Finding all periodic orbits of maps using Newton methods: sizes of basins, Physica D **135** (2000), 195–211

G.-C. Yuan and J.A. Yorke, Collapsing of chaos in one dimensional maps, Physica D **136** (2000), 18–30

C. Robert, K.T. Alligood, E. Ott, and J.A. Yorke, Explosions of Chaotic Sets, Physica D **144** (2000), 44–61

G.-C. Yuan, J.A. Yorke, T.L. Caroll, E. Ott, L.M. Pecora, Testing whether two chaotic one dimensional processes are dynamically identical, Phys. Rev. Lett. **85** (2000), 4265–4268

J. Kennedy and J.A. Yorke, Topological horseshoes, Trans. Amer. Math. Soc. **353** (2001), 2513–2530

D. Sweet, H.E. Nusse, and J.A. Yorke, Stagger and step method: detecting and computing chaotic saddles in higher dimensions, Phys. Rev. Lett. **86** (2001), 2261–2264

D.J. Patil, B.R. Hunt, E. Kalnay, J.A. Yorke, and E. Ott, Local low dimensionality of atmospheric dynamics Phys. Rev. Lett. **86** (2001), 5878–5881

J. Kennedy, S. Koçak, and J.A. Yorke, A chaos lemma, Am. Math. Monthly **108** (2001), 411–423

K. Alligood, E. Sander, and J.A. Yorke, Explosions: global bifurcations at heteroclinic tangencies, Ergodic Theory and Dyn. Sys. **22** (2002), 953–972

Original Contributions in Symposium Proceedings

J.A. Yorke, Spaces of solutions, and Invariance of contingent equations, both in Mathematical Systems Theory and Economics II, Springer-Verlag Lecture Notes in Operations Res. and Math. Econ. #12, 383–403 and 379–381: The Proceedings of International Conference for Mathematical Systems Theory and Economics in Varenna, Italy, June 1967

J.L. Kaplan and J.A. Yorke, Toward a unification of ordinary differential equations with nonlinear semi-group theory, International Conference on Ordinary Differential Equations, H. Antosiewicz, ed., Academic Press (1975), 424–433: The proceedings of a conference in Los Angeles, September 1974

J. Curry and J.A. Yorke, A transition from Hopf bifurcation to chaos: computer experiments with maps in R^2, in The Structure of Attractors in Dynamical Systems, Springer Lecture Notes in Mathematics #668, 48–66: The proceedings of the NSF regional conference in Fargo, ND, June 1977

J.L. Kaplan and J.A. Yorke, Chaotic behavior of multidimensional difference equations, in Functional Differential Equations and Approximation of Fixed Points, H. O. Peitgen and H. O. Walther, eds., Springer Lecture Notes in Mathematics #730 (1979), 204–227

S. N. Chow, J. Mallet-Paret, and J.A. Yorke, A homotopy method for locating all zeroes of a system of polynomials, ibid, 77–78

C. Grebogi, H.E. Nusse, E. Ott, and J.A. Yorke, Basic Sets: Sets that determine the dimension of basin boundaries, in Dynamical Systems, Proc. of Special Year at the University of Maryland, ed. J. Alexander, Springer Lecture Notes in Mathematics #1342 (1988), 220–250

G.-H. Yuan, B.R. Hunt, C. Grebogi, E. Kostelich, E. Ott, and J.A. Yorke, Design and control of shipboard cranes, in Proc. of the 16th ASME Biennial Conference on Mechanical Vibration and Noise, September 1997, Sacramento, CA

Part II: Chaos and SLYRB Measures: The Development of the Theory of Chaotic Attractors

This book is a collection of articles devoted to the theoretical development of chaotic attractors with an emphasis on measures on attractors. The term "SLYRB" in the title of this section is pronounced as the single word "slurb." This acronym stands for "Sinai, Lasota, Yorke, Ruelle, Bowen" and will be explained below in the text. (An asterisk following a citation means that the article appears in this book.) We relate the topics of "chaotic attractors" and the corresponding "natural invariant measures." Limitation of space has forced us to exclude many outstanding papers. Our goal is to have a coherent collection of articles that reflect the theoretical development of chaotic attractors, are interesting to read, and are appropriate for a graduate student seminar. All too often, graduate students preparing to write a thesis have no knowledge of how a field has developed and little knowledge of how individuals have built the field. The courses they take and books they read present the subject as a fully constructed entity. However, we assume that the reader is familiar with a few basic notions from nonlinear dynamics such as "Lyapunov exponents" and "invariant measures." If this is not the case, the reader is advised to consult any good book in nonlinear dynamics covering these notions. The order of the articles is based on the date of publication.

In the 1970s, Lorenz's work became known. Lorenz [L]* discovered sensitivity to initial data in a simple system of three differential equations that served as a simple model for weather prediction. He described the structure of a set that is now called the *Lorenz attractor*. To explain Lorenz's result, Li and Yorke [LiY1]* proved a theorem on sensitivity to initial data in one-dimensional maps, which is now known as the "period three implies chaos" theorem. (This result has considerable overlap with the Sarkovskii theorem [Sa] on periodic orbits, which appeared in 1964 but was virtually unknown in the West until it was featured in a 1977 paper by Stefan [St].) From a different perspective, Rössler [Ro] showed that the invariant set for the Lorenz equations contains horseshoes, so the dynamics display chaos. In 1976, May [Ma]* published a review article on simple models with complicated dynamics, and this article explains the period-doubling phenomena and the Li–Yorke chaos that occur in these models.

In the 1980s, considerable attention was given to attractors in dynamical systems, and conferences devoted to this single topic were held. The notion of "strange attractor" was introduced by Ruelle and Takens [RT], referring to an attractor that is not a fixed-point attractor, a periodic orbit, or a smooth manifold without boundary such as a circle or torus. Nowadays the terms "chaotic attractor" and "strange attractor" are quite common. For some early review articles on this topic, see the review articles by Ott [O1]*, Eckmann [E], and Eckmann and Ruelle [ER]*. For the purpose of this book, for any continuous map from a Euclidean space into itself, we call a compact set A a *Milnor attractor* if (1) A is mapped onto itself, and (2) every open neighborhood of A

contains a set of positive Lebesgue measure such that every initial condition in that set has the property that the distance between the nth iterate of that initial condition and A converges to zero as n goes to infinity; see [Mil]*. The second condition says that there is a positive probability that the trajectory of a randomly chosen point will be attracted to A. (In the literature there exist several (inequivalent) definitions of the notion "attractor.")

We now define "chaotic trajectory" and "chaotic attractor". The trajectory of a point x is called a *chaotic trajectory* if (1) the trajectory of x is bounded and x is not asymptotically periodic, and (2) the trajectory of x has a positive Lyapunov exponent. Therefore, a chaotic trajectory must have infinitely many distinct points. We call a compact set A a *chaotic attractor* if (a) A is a Milnor attractor, and (b) A contains a chaotic trajectory that is dense in A. If the trajectory of some initial point is a chaotic trajectory, may we conclude that there is a chaotic attractor? To illustrate the problem, consider the example of the Hénon map depending on parameters a and b. For specified parameter values (namely, $a = 1.4$ and $b = 0.3$), Hénon [H]* gave the coordinates of the four corner points of a quadrilateral having the property that it is mapped into itself when the map (the Hénon map) is applied once. Hence, the quadrilateral contains an attractor, referred to as the *Hénon attractor*. The Hénon attractor (for these parameter values) is believed to be a chaotic attractor because (according to detailed numerical calculations) it appears to contain a chaotic trajectory that is dense in the Hénon attractor and has a positive Lyapunov exponent. However, there exists no proof of such a fact. One has to prove that there exists a chaotic trajectory that is dense in the closure of the unstable manifold of the fixed point both of whose coordinates are positive. Numerical simulations give the strong impression that this is true. Later in this introduction we discuss the situation for other parameter values.

Generally, a chaotic attractor is a Milnor attractor on which the dynamics are chaotic; that is, the system has at least one positive Lyapunov exponent. Frequently, the two notions of chaotic attractor and "strange attractor" are used as synonyms. Indeed, a chaotic attractor is often a strange attractor in the sense of Ruelle and Takens. However, an example of a strange attractor that is not a chaotic attractor is the so-called *Feigenbaum attractor* (that is, a Milnor attractor that is a Cantor set and for which the Lyapunov exponent of Lebesgue almost every trajectory is zero), which occurs in the one-parameter family of quadratic maps. Also, one encounters in the literature the following. A *strange attractor* is a Milnor attractor that has a Cantor-like structure. There is a large literature on "strange nonchaotic attractors" in systems that have terms that are quasi-periodic of time; see, for example, Ott [O2]. A common way to characterize strange attractors is through their fractal (that is, not necessarily integer-valued) dimension. The articles by Farmer, Ott and Yorke [FOY]*, Grassberger and Procaccia [GP]*, and Grebogi, Ott and Yorke [GOY2]* are recommended. In an earlier paper, Kaplan and Yorke [KY] conjectured a connection between the fractal dimension of an attractor and its

Lyapunov exponents. Their formula is now often used to compute the "Lyapunov dimension" of experimental time series.

At the end of the 1970s, Rössler [Ro] gave an example and Yorke and Yorke [YY] investigated and explained the phenomenon now called "transient chaos", in which trajectories behave chaotically for a finite period of time before settling down to more regular motion. Based on numerical studies of the Lorenz equations, they gave a theory that has found application in many systems, and illustrated the importance of nonattracting chaotic sets. For work on sudden changes in chaotic attractors and the appearance of transient chaos, see [GOY1].

In many applications it is useful to consider not only the set that constitutes a chaotic attractor but also (if it exists) the asymptotic distribution of a typical trajectory converging to the attractor. Indeed, in the physics literature such a distribution is often assumed to exist. When it exists, it is called a "natural invariant measure" [FOY]*. We will discuss some aspects of natural invariant measures. The goal of the paper [HKLN]* is to relate the "Lasota–Yorke measure" for chaotic attractors in one-dimensional maps and the "Sinai–Ruelle–Bowen measure" for chaotic attractors in higher-dimensional dynamical systems, and to introduce the notion of "SLYRB measure," which is a natural invariant measure. Both approaches deal with the problem that attractors are proper subsets of the space and that their ergodic theory should reflect the average behavior of initial points, where "average" is taken with respect to Lebesgue measure. We now discuss briefly the two approaches to the invariant measures derived from Lebesgue measure. Let M be a piecewise smooth map from the space X into itself. One approach follows ideas of Sinai, who since the beginning of the 1970s systematically developed a Markov partition method for dynamical systems. In a 1972 paper [Si], he proved the existence of invariant measures for certain systems having a Markov partition. In fact, he proved this result for Anosov diffeomorphisms. Bowen and Ruelle [BR]* extended these techniques to Axiom A diffeomorphisms and flows in several papers that were published during the period 1973–1976. The resulting measure is often called an *SRB measure*. To describe such an invariant measure, let U be a set of positive Lebesgue measure in X that is mapped into itself and let A be an attractor that is contained in U such that for every initial condition x in U, the distance between the nth iterate of x and A converges to zero as n goes to infinity. They consider the average value along a trajectory of any continuous bounded function. They are interested in cases in which Lebesgue almost all initial conditions x in U yield the same measure. Hence, the significance of such an SRB measure is that it describes the asymptotic distribution of the trajectory of initial points in a set of positive Lebesgue measure in the phase space. The second approach (based on the Perron–Frobenius techniques) was used, for example, by Lasota and Yorke [LaY]*. The Perron–Frobenius operator for M is very useful for studying the evolution of densities under iteration of M. The construction of Lasota and Yorke is an example of a method that is now commonly used for constructing natural invariant measures. The main result of Lasota and Yorke [LaY]* says that piecewise-smooth expanding one-dimensional maps admit an absolutely continuous invariant measure. Lasota and Yorke prove this theorem by establishing convergence for functions of bounded variation and using the fact that these functions are dense in the space of integrable functions. Their original idea of considering the space of functions of bounded variation and the properties of the Perron–Frobenius operator on this space has been used many times since. We refer to the measure in their theorem as a *Lasota–Yorke measure*. This measure is an absolutely continuous invariant measure. It is not difficult to show that each chaotic attractor (with a dense trajectory) for one-dimensional maps has a unique absolutely continuous invariant measure and each Lasota–Yorke measure is a linear combination of these [LiY2]. Yorke reports privately that the goal of [LaY]* initially was only to handle tent maps, but their techniques allowed them to prove the result for more general scalar maps. The theorem of Krzyzewski and Szlenk [KS]* had considered only maps for which there is in fact a smooth invariant density. That is not the case for most tent maps.

The way the two approaches describe the measures appear quite different. However, if the system has one attractor, then the two measures resulting from the two approaches are equal. The theoretical construction of a global SRB measure consists in averaging the iterates of a local measure on an arbitrary unstable manifold. This construction resembles the construction by Lasota and Yorke of pushing forward Lebesgue measure. Hofbauer and Keller [HK]* provide a general setting in which to define the Perron–Frobenius operator of a map and use it to obtain results about invariant measures and equilibrium states. Gora and Boyarsky [GB]* generalized Lasota and Yorke's result to n-dimensional space, and Tsujii [T2]* proved that for every piecewise real-analytic map of a bounded region in the plane there exists a probabilistic absolutely continuous invariant measure. Dellnitz and Junge [DJ]* presented efficient techniques for the numerical approximation of SRB measures based on appropriate discretization of the Perron–Frobenius operator. If m denotes the SRB measure, then it means that a Lebesgue typical initial condition chosen near the attractor A generates a trajectory that is asymptotically distributed according to m; that is, m is a natural invariant measure. Invariant measures with this property have been called by a variety of names in the literature, perhaps putting different emphases on the contributions of Sinai, Ruelle, and Bowen with minor variations in the properties of these measures. They were called Bowen–Ruelle measures by Young [Y1]* and a Bowen–Ruelle–Sinai (or BRS) measures by Tsujii [T1]. On the other hand, Benedicks and Young [BY]* define a Sinai–Bowen–Ruelle (or SBR) measure as an invariant measure with a positive Lyapunov exponent and absolutely continuous conditional measures along unstable manifolds. In a recent paper, Young [Y2] limits the definition of SRB measures to the context of no zero Lyapunov exponents and states a definition of Sinai–Ruelle–Bowen measure. She notes that this definition is a generalization of the ideas in Sinai [Si] and that it first appeared in a paper by Ledrappier [Le]. See also Pesin's book [P] for another variation. The above discussion on the variety of terms and notions of invariant measures demonstrates that the term "SRB measure" is not yet stable.

Notice that all of these concepts require that the measure in question be unique. Generally, this hypothesis is quite difficult to verify. Below, we argue that the term SLYRB, introduced in [HKLN]* and described briefly below, is perhaps better.

We now discuss briefly natural invariant measures for one-dimensional quadratic maps. The family of quadratic maps is not uniformly expanding for any of its members. In fact, near the critical point the quadratic map transforms any bounded density in one step to a density with inverse square-root singularity. A very important result is the famous theorem by Jakobson [J] that says the following. The set of parameters for which the corresponding quadratic map admits an absolutely continuous invariant measure, has positive Lebesgue measure. For an alternative proof, see, for example [Ry1]*. The argument used in the proof of Jakobson's result is that the orbit of the critical point does not approach the critical point too closely too quickly, and the collection of those parameters for which this approach is sufficiently controlled has positive Lebesgue measure. In describing the distribution of the orbit of x, Tsujii [T1] uses a sequence of probability measures. If this sequence converges to a probability measure as n goes to infinity, he calls that measure the *asymptotic distribution* for the orbit of x. Tsujii calls a probability measure m on the interval the *Bowen–Ruelle–Sinai measure* if the asymptotic distribution of the orbit exists and equals m for almost every point in the interval with respect to Lebesgue measure. He studies how the BRS measure depends on the parameter in the family of quadratic maps and considers two cases. One of these cases is that the map admits an absolutely continuous invariant probability measure. In this case the measure is unique and it is the BRS measure for the quadratic map.

We now turn to natural invariant measures for Lozi maps and Hénon maps. For the Hénon attractor $H*$ (which is the attractor for the Hénon map with parameter values $a = 1.4$ and $b = 0.3$ chosen by Hénon [H]*), one may ask the following question: Does there exist a natural invariant Borel probability measure m such that (1) the support of the measure m is $H*$, and (2) the orbit of Lebesgue almost every initial condition in the region defined by Hénon has a well-defined distribution and that this distribution is independent of the choice of the initial condition? Unfortunately, the answer is still unknown. In 1978 Lozi [Lo] introduced a piecewise-linear map as an analogue of the Hénon map. The dynamics of the Lozi map may be easier to understand than those of the Hénon map. It appears that the Lozi map is an intermediate stage between the Axiom A systems and more complicated systems like the Hénon map. Misiurewicz [Mis] showed that for certain parameter values, the Lozi map has a chaotic attractor. The attractor is the closure of the unstable "manifold" of a fixed point P, that is, the collection of points whose orbits converge to P as the map is iterated backwards. The maximal smooth component of the unstable "manifold" of P containing P is called the *local unstable manifold* of P. Collet and Levy [CL]* constructed an invariant measure on the attractor as follows. They started off with Lebesgue measure on the local unstable manifold of P. Then they considered the iterates of this measure under the Lozi map. Since the attractor is compact, one can extract a subsequence that converges to an invariant Borel probability measure m. They showed that m is unique, that it is absolutely continuous with respect to the Lebesgue measure in the unstable direction, and that it is in fact a Bowen–Ruelle measure (that is, a measure generated by the orbits of almost all points). This approach resembles Lasota and Yorke's construction of pushing forward Lebesgue measure.

We continue with an example combining the LY and SRB approaches. Bowen and Ruelle [BR]* assumed that the maps were smooth in the expanding directions. The measure they described, however, was typically not smooth, because the map had contracting directions. In contrast, Lasota and Yorke [LaY]* studied maps that were expanding. Their measure was typically not smooth because the maps were only piecewise smooth, and the successive iterates of the nonsmooth points (points at which the map is not smooth) could be dense in a collection of intervals. The first papers to deal with maps having both properties were those by Rychlik [Ry2], [Ry3]*, Young [Y1]* and Collet and Levy [CL]*. They dealt with maps that had the properties of having a contracting direction and being nonsmooth in the expanding direction. Young [Y1]* introduces and investigates a class of piecewise smooth maps on the unit square R in the plane that are continuous, one-to-one, and map the region R into itself. These maps are hyperbolic piecewise smooth maps and she calls them "generalized Lozi maps." Recall that the above result of Lasota and Yorke states that piecewise expanding maps of the interval have absolutely continuous invariant measures (Lasota–Yorke measure). Young proves an analogous statement for her generalized Lozi maps. A part of the proof follows a combination of the methods used by Sinai, and by Lasota and Yorke. Young's result is that the generalized Lozi maps admit certain invariant measures that imply that the maps have what she calls *Bowen–Ruelle measures*, that is, measures generated by the orbits of almost all initial points.

The existence of invariant measures for the Lozi maps was shown independently by Rychlik [Ry2], [Ry3]* by a method independent of Young [Y1]* and Collet and Levy [CL]*. The results were reported in [Ry2]. The paper [Ry2] contains only sketches of the arguments, and the full proofs can be found in Rychlik's dissertation [Ry3]*. The paper [Ry2] and the dissertation [Ry3]* present a hybrid approach to the "Lasota–Yorke" measures and the "Sinai–Ruelle–Bowen" measures.

We now return to the Hénon map (with parameter values $a = 1.4$ and $b = 0.3$ chosen by Hénon) [H]*. Numerical simulations suggest that for almost every initial condition in a certain region, the orbit has a well-defined distribution as n goes to infinity, and this distribution is independent of the choice of initial condition. However, in this case, as well as for other parameter values for which there seems to be a strange attractor, the situation is very delicate. The question is whether there exists an invariant measure that reflects the statistical behavior of Lebesgue almost every point in a much larger set. But why should such an invariant measure exist? On the other hand, if (one can show that) such a measure exists, it is of great significance. In the proofs of the results by Sinai [Si] and by Bowen and Ruelle [BR]*, Markov partitions were used to connect the dynamics of F to certain one-dimensional lattice systems in statistical mechanics, and m was realized as a Gibbs state or equilibrium state. However, Markov partitions are not available for systems more general than Axiom A, such as the Hénon map. We now explain why it is so difficult to prove the existence of a natural invariant measure for the Hénon map. Let P be the fixed point of the Hénon map that has two positive coordinates. The unstable manifold of P is the collection of points whose orbits converge to P as the map is iterated backwards. In analogy to the construction of an invariant measure by Collet and Levy [CL]* for the Lozi map, suppose we start with the local unstable manifold of the fixed point P and push forward Lebesgue measure on the local unstable manifold. The definition of a local unstable manifold says that the backward nth iterates of points on the local unstable manifold converge to P as n goes to infinity, but generally, one has no idea what happens to points on the local unstable manifold under forward iteration of the map. For example, it is easy to imagine that parts of the unstable manifold belong to basins of attraction of periodic orbits, so the orbits of certain points of the local unstable manifold converge to a periodic orbit attractor when the Hénon map is applied repeatedly. Hence, the question of existence of natural invariant (SRB) measures is very delicate. It is in some sense analogous to the problem of existence of absolutely continuous invariant measures for nonuniformly expanding maps (Jakobson theorem), except that it is more complicated because one has to consider not only rates of expansion, but their directions as well. Despite these difficulties, there have been some major breakthroughs. Benedicks and Carleson [BC] developed an elaborate machinery for analyzing the dynamics of the Hénon map for a positive measure set of parameters. However, this set of parameters does not include the parameter values $a = 1.4$ and $b = 0.3$ chosen by Hénon [H]*. Subsequently, Benedicks and Young [BY]* constructed SRB measures for these attractors. More recently, Young [Y2] and Alves, Bonatti, and Viana [ABV]* investigated diffeomorphisms that are not hyperbolic, but partially hyperbolic. These papers contain breakthrough results. Under quite general assumptions, [Y2] proves that correlations for hyperbolic systems with singularities decay exponentially. If Lyapunov exponents are bounded away from zero, [ABV]* prove that there are only finitely many natural invariant (Sinai–Ruelle–Bowen) measures for such maps and that Lebesgue almost every point is in one of the corresponding basins.

We recall that Lasota–Yorke measures are absolutely continuous invariant measures, while Bowen–Ruelle–Sinai measures are not. Hence, from the point of view of measure theory applied to dynamics, those invariant measures are considered to be different. However, from the point of view of dynamics using measure theory, the existence of a chaotic attractor implies in certain systems the existence of the special invariant measures as discussed above. After comparing the construction of SRB measures and their properties with the construction and properties of Lasota–Yorke measure, in [HKLN]* it is suggested that the natural invariant measure supported on any chaotic attractor might be called a "SLYRB measure." We quote from [HKLN]*: "The SRB concept of measure can be motivated by asking how a trajectory from a typical initial point is distributed asymptotically. Similarly, the SLYRB concept of measure can be motivated by asking what the average distribution is for trajectories of a large collection of initial points in some region not necessarily restricted to a single basin. The latter is analogous to asking where all the raindrops from a rainstorm go, and the former asks about where a single raindrop goes, perhaps winding up distributed throughout a particular lake." In [HKLN]* the following is proposed. A measure m for a dynamical system is called a *SLYRB measure* if m is a weighted sum (with positive weights) of invariant measures such that each of these invariant measures is the asymptotic distribution of a trajectory for a set of initial points of positive measure. Consider a system that has exactly one attractor A and for which almost all initial points chosen near the attractor A generate trajectories that are asymptotically distributed according to a single measure m. Then m would be both an SRB measure (by some standards) and a SLYRB measure. An invariant measure with this latter property for certain two-dimensional systems has been called a Bowen–Ruelle measure by Young [Y1]* and for certain one-dimensional systems a Bowen–Ruelle–Sinai (or BRS) measure by Tsujii [T1], so these measures are special cases of SLYRB measures. Furthermore, the SBR measure defined by Benedicks and Young [BY]*, being an invariant measure with a positive Lyapunov exponent and absolutely continuous conditional measures along unstable manifolds, is also an example of a SLYRB measure. Obviously, the Lasota–Yorke measure is an example of a SLYRB measure. Finally, [HKLN]* proposes that a dynamical system displays *SLYRB chaos* whenever there exists a chaotic attractor that supports a SLYRB measure. Enjoy!

References

[ABV] J.F. Alves, C. Bonatti and M. Viana, SRB measures for partially hyperbolic systems whose central direction is mostly expanding, Invent. Math. **140** (2000), 351–398

[BC] M. Benedicks and L. Carleson, The dynamics of the Hénon map, Annals of Math. **133** (1991), 73–169

[BY] M. Benedicks and L.-S. Young, Sinai-Bowen-Ruelle measures for certain Hénon maps, Invent. Math. **112** (1993), 541–576

[BR] R. Bowen and D. Ruelle, The ergodic theory of Axiom A flows, Invent. Math. **29** (1975), 181–202

[CL] P. Collet and Y. Levy, Ergodic properties of the Lozi mappings, Commun. Math. Phys. **93** (1984), 461–481

[DJ] M. Dellnitz and O. Junge, On the approximation of complicated dynamical behavior, SIAM J. Numer. Anal. **36** (1999), 491–515

[E] J.-P. Eckmann, Roads to turbulence in dissipative dynamical systems, Rev. Modern Phys. **53** (1981), 643–654

[ER] J.-P. Eckmann and D. Ruelle, Ergodic theory of chaos and strange attractors, Rev. Modern Phys. **57** (1985), 617–656

[FOY] J.D. Farmer, E. Ott and J.A. Yorke, The dimension of chaotic attractors, Physica D **7** (1983), 153–180

[GB] P. Gora and A. Boyarsky, Absolutely continuous invariant measures for piecewise expanding C^2 transformation in R^N, Israel J. Math. **67** (1989), 272–286

[GP] P. Grassberger and I. Procaccia, Measuring the strangeness of strange attractors, Physica D **9** (1983), 189–208

[GOY1] C. Grebogi, E. Ott and J.A. Yorke, Crisis, sudden changes in chaotic attractors, and transient chaos, Physica D **7** (1983), 181–200

[GOY2] C. Grebogi, E. Ott, and J.A. Yorke, Unstable periodic orbits and the dimensions of multifractal chaotic attractors, Phys. Rev. A **37** (1988), 1711–1724

[H] M. Hénon, A two-dimensional mapping with a strange attractor, Commun. Math. Phys. **50** (1976), 69–77

[HK] F. Hofbauer and G. Keller, Ergodic properties of invariant measures for piecewise monotonic transformations, Math. Z. **180** (1982), 119–140

[HKLN] B.R. Hunt, J.A. Kennedy, T.-Y. Li and H.E. Nusse, SLYRB measures: natural invariant measures for chaotic systems, Physica D **170**(2002), 50–71

[J] M. Jakobson, Absolutely continuous invariant measures for one-parameter families of one-dimensional maps, Commun. Math. Phys. **81** (1981), 39–88

[KY] J.L. Kaplan and J.A. Yorke, Chaotic behavior of multidimensional difference equations, in Functional Differential Equations and Approximation of Fixed Points, H.O. Peitgen and H.O. Walther, eds., Springer Lecture Notes in Mathematics #730 (1979), 204–227

[KS] K. Krzyzewski and W. Szlenk, On invariant measures for expanding differentiable mappings, Studia Math. **33** (1969), 83–92

[LaY] A. Lasota and J.A. Yorke, On the existence of invariant measures for piecewise monotonic transformations, Trans. Amer. Math. Soc. **186** (1973), 481–488

[Le] F. Ledrappier, Proprietes ergodiques des mesures de Sinai, I.H.E.S. Publ. Math. **59** (1984), 163–188

[LiY1] T.-Y. Li and J.A. Yorke, Period three implies chaos, Amer. Math. Monthly **82** (1975), 985–992

[LiY2] T.-Y. Li and J.A. Yorke, Ergodic transformations from an interval into itself, Trans. Amer. Math. Soc. **235** (1978), 183–192

[L] E.N. Lorenz, Deterministic nonperiodic flow, J. Atm. Sc. **20** (1963), 130–

141
[Lo] R. Lozi, Un attracteur etrange (?) du type attracteur de Hénon, J. Physique (Paris) **39** (Coll. C5) (1978), 9–10
[Ma] R.M. May, Simple mathematical models with very complicated dynamics, Nature **261** (1976), 459–467
[Mil] J. Milnor, On the concept of attractor, Commun. Math. Phys. **99** (1985), 177–195; Comments "On the concept of attractor": corrections and remarks, Commun. Math. Phys. **102** (1985), 517–519
[Mis] M. Misiurewicz, Strange attractors for the Lozi mappings, in Nonlinear Dynamics, R.H.G. Helleman, ed., Ann. New York Acad. Sci. #357 (1980), 348–358
[O1] E. Ott, Strange attractors and chaotic motions of dynamical systems, Rev. Modern Phys. **53** (1981), 655–671
[O2] E. Ott, Chaos in Dynamical Systems, Cambridge University Press, 1993
[P] Ya.B. Pesin, Dimension Theory in Dynamical Systems, Contemporary Views and Applications, University of Chicago Press, 1997
[Ro] O. Rössler, Horseshoe map chaos in the Lorenz equation, Phys. Lett. A **60** (1977), 392–394
[RT] D. Ruelle and F. Takens, On the nature of turbulence, Commun. Math. Phys. **20** (1971), 167–192
[Ry1] M.R. Rychlik, Another proof of Jakobson's theorem and related results, Ergodic Theory Dynamical Systems 8 (1988), 93–109
[Ry2] M. Rychlik, Mesures invariantes et principe variationel pour les applications de Lozi, C.R. Acad. Sc. Paris, t. 296 (1983), 19–22
[Ry3] M. Rychlik, Invariant Measures and Variational Principle for Lozi Applications, Ph.D. dissertation, University of California at Berkeley (1983),
[Sa] A.N. Sarkovskii, Coexistence of cycles of a continuous map of the line into itself, Ukr. Mat. Z. **16** (1964), 61–71
[Si] Ya.G. Sinai, Gibbs measures in ergodic theory, Russian Mathematical Surveys **27** (1972), 21–69
[St] P. Stefan, A theorem of Sarkovskii on the existence of periodic orbits of continuous endomorphisms of the real line, Comm. Math. Phys. **54** (1977), 237–248
[T1] M. Tsujii, On continuity of Bowen-Ruelle-Sinai measures in families of one dimensional maps, Comm. Math. Phys. **177** (1996), 1–11
[T2] M. Tsujii, Absolutely continuous invariant measures for piecewise real-analytic expanding maps on the plane, Comm. Math. Phys. **208** (2000), 605–622
[YY] J.A. Yorke and E.D. Yorke, Metastable chaos: the transition to sustained chaotic behavior in the Lorenz model, J. Stat. Phys. **21** (1979), 263–277
[Y1] L.-S. Young, Bowen-Ruelle measures for certain piecewise hyperbolic maps, Trans. Amer. Math. Soc. **287** (1985), 41–48
[Y2] L.-S. Young, Statistical properties of dynamical systems with some hyperbolicity, Ann. of Math. (2) **147** (1998), 585–650

Reprinted from JOURNAL OF THE ATMOSPHERIC SCIENCES, Vol. 20, No. 2, March, 1963, pp. 130–141
Printed in U. S. A.

Deterministic Nonperiodic Flow[1]

EDWARD N. LORENZ

Massachusetts Institute of Technology

(Manuscript received 18 November 1962, in revised form 7 January 1963)

ABSTRACT

Finite systems of deterministic ordinary nonlinear differential equations may be designed to represent forced dissipative hydrodynamic flow. Solutions of these equations can be identified with trajectories in phase space. For those systems with bounded solutions, it is found that nonperiodic solutions are ordinarily unstable with respect to small modifications, so that slightly differing initial states can evolve into considerably different states. Systems with bounded solutions are shown to possess bounded numerical solutions.

A simple system representing cellular convection is solved numerically. All of the solutions are found to be unstable, and almost all of them are nonperiodic.

The feasibility of very-long-range weather prediction is examined in the light of these results.

1. Introduction

Certain hydrodynamical systems exhibit steady-state flow patterns, while others oscillate in a regular periodic fashion. Still others vary in an irregular, seemingly haphazard manner, and, even when observed for long periods of time, do not appear to repeat their previous history.

These modes of behavior may all be observed in the familiar rotating-basin experiments, described by Fultz, *et al.* (1959) and Hide (1958). In these experiments, a cylindrical vessel containing water is rotated about its axis, and is heated near its rim and cooled near its center in a steady symmetrical fashion. Under certain conditions the resulting flow is as symmetric and steady as the heating which gives rise to it. Under different conditions a system of regularly spaced waves develops, and progresses at a uniform speed without changing its shape. Under still different conditions an irregular flow pattern forms, and moves and changes its shape in an irregular nonperiodic manner.

Lack of periodicity is very common in natural systems, and is one of the distinguishing features of turbulent flow. Because instantaneous turbulent flow patterns are so irregular, attention is often confined to the statistics of turbulence, which, in contrast to the details of turbulence, often behave in a regular well-organized manner. The short-range weather forecaster, however, is forced willy-nilly to predict the details of the large-scale turbulent eddies—the cyclones and anticyclones—which continually arrange themselves into new patterns. Thus there are occasions when more than the statistics of irregular flow are of very real concern.

In this study we shall work with systems of deterministic equations which are idealizations of hydrodynamical systems. We shall be interested principally in nonperiodic solutions, i.e., solutions which never repeat their past history exactly, and where all approximate repetitions are of finite duration. Thus we shall be involved with the ultimate behavior of the solutions, as opposed to the transient behavior associated with arbitrary initial conditions.

A closed hydrodynamical system of finite mass may ostensibly be treated mathematically as a finite collection of molecules—usually a very large finite collection—in which case the governing laws are expressible as a finite set of ordinary differential equations. These equations are generally highly intractable, and the set of molecules is usually approximated by a continuous distribution of mass. The governing laws are then expressed as a set of partial differential equations, containing such quantities as velocity, density, and pressure as dependent variables.

It is sometimes possible to obtain particular solutions of these equations analytically, especially when the solutions are periodic or invariant with time, and, indeed, much work has been devoted to obtaining such solutions by one scheme or another. Ordinarily, however, nonperiodic solutions cannot readily be determined except by numerical procedures. Such procedures involve replacing the continuous variables by a new finite set of functions of time, which may perhaps be the values of the continuous variables at a chosen grid of points, or the coefficients in the expansions of these variables in series of orthogonal functions. The governing laws then become a finite set of ordinary differential

[1] The research reported in this work has been sponsored by the Geophysics Research Directorate of the Air Force Cambridge Research Center, under Contract No. AF 19(604)-4969.

equations again, although a far simpler set than the one which governs individual molecular motions.

In any real hydrodynamical system, viscous dissipation is always occurring, unless the system is moving as a solid, and thermal dissipation is always occurring, unless the system is at constant temperature. For certain purposes many systems may be treated as conservative systems, in which the total energy, or some other quantity, does not vary with time. In seeking the ultimate behavior of a system, the use of conservative equations is unsatisfactory, since the ultimate value of any conservative quantity would then have to equal the arbitrarily chosen initial value. This difficulty may be obviated by including the dissipative processes, thereby making the equations nonconservative, and also including external mechanical or thermal forcing, thus preventing the system from ultimately reaching a state of rest. If the system is to be deterministic, the forcing functions, if not constant with time, must themselves vary according to some deterministic rule.

In this work, then, we shall deal specifically with finite systems of deterministic ordinary differential equations, designed to represent forced dissipative hydrodynamical systems. We shall study the properties of nonperiodic solutions of these equations.

It is not obvious that such solutions can exist at all. Indeed, in dissipative systems governed by finite sets of *linear* equations, a constant forcing leads ultimately to a constant response, while a periodic forcing leads to a periodic response. Hence, nonperiodic flow has sometimes been regarded as the result of nonperiodic or random forcing.

The reasoning leading to these concludions is not applicable when the governing equations are nonlinear. If the equations contain terms representing advection—the transport of some property of a fluid by the motion of the fluid itself—a constant forcing can lead to a variable response. In the rotating-basin experiments already mentioned, both periodic and nonperiodic flow result from thermal forcing which, within the limits of experimental control, is constant. Exact periodic solutions of simplified systems of equations, representing dissipative flow with constant thermal forcing, have been obtained analytically by the writer (1962a). The writer (1962b) has also found nonperiodic solutions of similar systems of equations by numerical means.

2. Phase space

Consider a system whose state may be described by M variables $X_1, \cdots, X_M$. Let the system be governed by the set of equations

$$dX_i/dt = F_i(X_1, \cdots X_M), \quad i=1, \cdots, M, \tag{1}$$

where time t is the single independent variable, and the functions F_i possess continuous first partial derivatives. Such a system may be studied by means of *phase space*—an M-dimensional Euclidean space Γ whose coordinates are $X_1, \cdots, X_M$. Each *point* in phase space represents a possible instantaneous state of the system. A state which is varying in accordance with (1) is represented by a moving *particle* in phase space, traveling along a *trajectory* in phase space. For completeness, the position of a stationary particle, representing a steady state, is included as a trajectory.

Phase space has been a useful concept in treating finite systems, and has been used by such mathematicians as Gibbs (1902) in his development of statistical mechanics, Poincaré (1881) in his treatment of the solutions of differential equations, and Birkhoff (1927) in his treatise on dynamical systems.

From the theory of differential equations (e,g., Ford 1933, ch. 6), it follows, since the partial derivatives $\partial F_i/\partial X_j$ are continuous, that if t_0 is any time, and if $X_{10}, \cdots X_{M0}$ is any point in Γ, equations (1) possess a unique solution

$$X_i = f_i(X_{10}, \cdots, X_{M0}, t), \quad i=1, \cdots, M, \tag{2}$$

valid throughout some time interval containing t_0, and satisfying the condition

$$f_i(X_{10}, \cdots, X_{M0}, t_0) = X_{i0}, \quad i=1, \cdots, M. \tag{3}$$

The functions f_i are continuous in $X_{10}, \cdots, X_{M0}$ and t. Hence there is a unique trajectory through each point of Γ. Two or more trajectories may, however, approach the same point or the same curve asymptotically as $t \to \infty$ or as $t \to -\infty$. Moreover, since the functions f_i are continuous, the passage of time defines a continuous deformation of any region of Γ into another region.

In the familiar case of a conservative system, where some positive definite quantity Q, which may represent some form of energy, is invariant with time, each trajectory is confined to one or another of the surfaces of constant Q. These surfaces may take the form of closed concentric shells.

If, on the other hand, there is dissipation and forcing, and if, whenever Q equals or exceeds some fixed value Q_1, the dissipation acts to diminish Q more rapidly then the forcing can increase Q, then $(-dQ/dt)$ has a positive lower bound where $Q \geqq Q_1$, and each trajectory must ultimately become trapped in the region where $Q < Q_1$. Trajectories representing forced dissipative flow may therefore differ considerably from those representing conservative flow.

Forced dissipative systems of this sort are typified by the system

$$dX_i/dt = \sum_{j,k} a_{ijk} X_j X_k - \sum_j b_{ij} X_j + c_i, \tag{4}$$

where $\sum a_{ijk} X_i X_j X_k$ vanishes identically, $\sum b_{ij} X_i X_j$ is positive definite, and $c_1, \cdots, c_M$ are constants. If

$$Q = \tfrac{1}{2} \sum_i X_i^2, \tag{5}$$

and if $e_1, \cdots, e_M$ are the roots of the equations

$$\sum_j (b_{ij}+b_{ji})e_j = c_i, \tag{6}$$

it follows from (4) that

$$dQ/dt = \sum_{i,j} b_{ij}e_ie_j - \sum_{i,j} b_{ij}(X_i-e_i)(X_j-e_j). \tag{7}$$

The right side of (7) vanishes only on the surface of an ellipsoid E, and is positive only in the interior of E. The surfaces of constant Q are concentric spheres. If S denotes a particular one of these spheres whose interior R contains the ellipsoid E, it is evident that each trajectory eventually becomes trapped within R.

3. The instability of nonperiodic flow

In this section we shall establish one of the most important properties of deterministic nonperiodic flow, namely, its instability with respect to modifications of small amplitude. We shall find it convenient to do this by identifying the solutions of the governing equations with trajectories in phase space. We shall use such symbols as $P(t)$ (variable argument) to denote trajectories, and such symbols as P or $P(t_0)$ (no argument or constant argument) to denote points, the latter symbol denoting the specific point through which $P(t)$ passes at time t_0.

We shall deal with a phase space Γ in which a unique trajectory passes through each point, and where the passage of time defines a continuous deformation of any region of Γ into another region, so that if the points $P_1(t_0)$, $P_2(t_0)$, $\cdots$ approach $P_0(t_0)$ as a limit, the points $P_1(t_0+\tau)$, $P_2(t_0+\tau)$, $\cdots$ must approach $P_0(t_0+\tau)$ as a limit. We shall furthermore require that the trajectories be uniformly bounded as $t\rightarrow\infty$; that is, there must be a bounded region R, such that every trajectory ultimately remains with R. Our procedure is influenced by the work of Birkhoff (1927) on dynamical systems, but differs in that Birkhoff was concerned mainly with conservative systems. A rather detailed treatment of dynamical systems has been given by Nemytskii and Stepanov (1960), and rigorous proofs of some of the theorems which we shall present are to be found in that source.

We shall first classify the trajectories in three different manners, namely, according to the absence or presence of transient properties, according to the stability or instability of the trajectories with respect to small modifications, and according to the presence or absence of periodic behavior.

Since any trajectory $P(t)$ is bounded, it must possess at least one *limit point* P_0, a point which it approaches arbitrarily closely arbitrarily often. More precisely, P_0 is a limit point of $P(t)$ if for any $\epsilon>0$ and any time t_1 there exists a time $t_2(\epsilon,t_1)>t_1$ such that $|P(t_2)-P_0|<\epsilon$. Here absolute-value signs denote distance in phase space. Because Γ is continuously deformed as t varies, every point on the trajectory through P_0 is also a limit point of $P(t)$, and the set of limit points of $P(t)$ forms a trajectory, or a set of trajectories, called the *limiting trajectories* of $P(t)$. A limiting trajectory is obviously contained within R in its entirety.

If a trajectory is contained among its own limiting trajectories, it will be called *central*; otherwise it will be called *noncentral*. A central trajectory passes arbitrarily closely arbitrarily often to any point through which it has previously passed, and, in this sense at least, separate sufficiently long segments of a central trajectory are statistically similar. A noncentral trajectory remains a certain distance away from any point through which it has previously passed. It must approach its entire set of limit points asymptotically, although it need not approach any particular limiting trajectory asymptotically. Its instantaneous distance from its closest limit point is therefore a transient quantity, which becomes arbitrarily small as $t\rightarrow\infty$.

A trajectory $P(t)$ will be called *stable at a point* $P(t_1)$ if any other trajectory passing sufficiently close to $P(t_1)$ at time t_1 remains close to $P(t)$ as $t\rightarrow\infty$; i.e., $P(t)$ is stable at $P(t_1)$ if for any $\epsilon>0$ there exists a $\delta(\epsilon,t_1)>0$ such that if $|P_1(t_1)-P(t_1)|<\delta$ and $t_2>t_1$, $|P_1(t_2)-P(t_2)|<\epsilon$. Otherwise $P(t)$ will be called *unstable* at $P(t_1)$. Because Γ is continuously deformed as t varies, a trajectory which is stable at one point is stable at every point, and will be called a *stable* trajectory. A trajectory unstable at one point is unstable at every point, and will be called an *unstable* trajectory. In the special case that $P(t)$ is confined to one point, this definition of stability coincides with the familiar concept of stability of steady flow.

A stable trajectory $P(t)$ will be called uniformly stable if the distance within which a neighboring trajectory must approach a point $P(t_1)$, in order to be certain of remaining close to $P(t)$ as $t\rightarrow\infty$, itself possesses a positive lower bound as $t_1\rightarrow\infty$; i.e., $P(t)$ is uniformly stable if for any $\epsilon>0$ there exists a $\delta(\epsilon)>0$ and a time $t_0(\epsilon)$ such that if $t_1>t_0$ and $|P_1(t_1)-P(t_1)|<\delta$ and $t_2>t_1$, $|P_1(t_2)-P(t_2)|<\epsilon$. A limiting trajectory $P_0(t)$ of a uniformly stable trajectory $P(t)$ must be uniformly stable itself, since all trajectories passing sufficiently close to $P_0(t)$ must pass arbitrarily close to some point of $P(t)$ and so must remain close to $P(t)$, and hence to $P_0(t)$, as $t\rightarrow\infty$.

Since each point lies on a unique trajectory, any trajectory passing through a point through which it has previously passed must continue to repeat its past behavior, and so must be *periodic*. A trajectory $P(t)$ will be called *quasi-periodic* if for some arbitrarily large time interval τ, $P(t+\tau)$ ultimately remains arbitrarily close to $P(t)$, i.e., $P(t)$ is quasi-periodic if for any $\epsilon>0$ and for any time interval τ_0, there exists a $\tau(\epsilon,\tau_0)>\tau_0$ and a time $t_1(\epsilon,\tau_0)$ such that if $t_2>t_1$, $|P(t_2+\tau)-P(t_2)|$

$<\epsilon$. Periodic trajectories are special cases of quasi-periodic trajectories.

A trajectory which is not quasi-periodic will be called *nonperiodic*. If $P(t)$ is nonperiodic, $P(t_1+\tau)$ may be arbitrarily close to $P(t_1)$ for some time t_1 and some arbitrarily large time interval τ, but, if this is so, $P(t+\tau)$ cannot remain arbitrarily close to $P(t)$ as $t\rightarrow\infty$. Nonperiodic trajectories are of course representations of deterministic nonperiodic flow, and form the principal subject of this paper.

Periodic trajectories are obviously central. Quasi-periodic central trajectories include multiple periodic trajectories with incommensurable periods, while quasi-periodic noncentral trajectories include those which approach periodic trajectories asymptotically. Nonperiodic trajectories may be central or noncentral.

We can now establish the theorem that a trajectory with a stable limiting trajectory is quasi-periodic. For if $P_0(t)$ is a limiting trajectory of $P(t)$, two distinct points $P(t_1)$ and $P(t_1+\tau)$, with τ arbitrarily large, may be found arbitrary close to any point $P_0(t_0)$. Since $P_0(t)$ is stable, $P(t)$ and $P(t+\tau)$ must remain arbitrarily close to $P_0(t+t_0-t_1)$, and hence to each other, as $t\rightarrow\infty$, and $P(t)$ is quasi-periodic.

It follows immediately that a stable central trajectory is quasi-periodic, or, equivalently, that a nonperiodic central trajectory is unstable.

The result has far-reaching consequences when the system being considered is an observable nonperiodic system whose future state we may desire to predict. It implies that two states differing by imperceptible amounts may eventually evolve into two considerably different states. If, then, there is any error whatever in observing the present state—and in any real system such errors seem inevitable—an acceptable prediction of an instantaneous state in the distant future may well be impossible.

As for noncentral trajectories, it follows that a uniformly stable noncentral trajectory is quasi-periodic, or, equivalently, a nonperiodic noncentral trajectory is not uniformly stable. The possibility of a nonperiodic noncentral trajectory which is stable but not uniformly stable still exists. To the writer, at least, such trajectories, although possible on paper, do not seem characteristic of real hydrodynamical phenomena. Any claim that atmospheric flow, for example, is represented by a trajectory of this sort would lead to the improbable conclusion that we ought to master long-range forecasting as soon as possible, because, the longer we wait, the more difficult our task will become.

In summary, we have shown that, subject to the conditions of uniqueness, continuity, and boundedness prescribed at the beginning of this section, a central trajectory, which in a certain sense is free of transient properties, is unstable if it is nonperiodic. A noncentral trajectory, which is characterized by transient properties, is not uniformly stable if it is nonperiodic, and, if it is stable at all, its very stability is one of its transient properties, which tends to die out as time progresses. In view of the impossibility of measuring initial conditions precisely, and thereby distinguishing between a central trajectory and a nearby noncentral trajectory, all nonperiodic trajectories are effectively unstable from the point of view of practical prediction.

4. Numerical integration of nonconservative systems

The theorems of the last section can be of importance only if nonperiodic solutions of equations of the type considered actually exist. Since statistically stationary nonperiodic functions of time are not easily described analytically, particular nonperiodic solutions can probably be found most readily by numerical procedures. In this section we shall examine a numerical-integration procedure which is especially applicable to systems of equations of the form (4). In a later section we shall use this procedure to determine a nonperiodic solution of a simple set of equations.

To solve (1) numerically we may choose an initial time t_0 and a time increment Δt, and let

$$X_{i,n}=X_i(t_0+n\Delta t). \tag{8}$$

We then introduce the auxiliary approximations

$$X_{i(n+1)}=X_{i,n}+F_i(P_n)\Delta t, \tag{9}$$

$$X_{i((n+2))}=X_{i(n+1)}+F_i(P_{(n+1)})\Delta t, \tag{10}$$

where P_n and $P_{(n+1)}$ are the points whose coordinates are

$$(X_{1,n},\cdots,X_{M,n}) \quad\text{and}\quad (X_{1(n+1)},\cdots,X_{M(n+1)}).$$

The simplest numerical procedure for obtaining approximate solutions of (1) is the forward-difference procedure,

$$X_{i,n+1}=X_{i(n+1)}. \tag{11}$$

In many instances better approximations to the solutions of (1) may be obtained by a centered-difference procedure

$$X_{i,n+1}=X_{i,n-1}+2F_i(P_n)\Delta t. \tag{12}$$

This procedure is unsuitable, however, when the deterministic nature of (1) is a matter of concern, since the values of $X_{1,n},\cdots,X_{M,n}$ do not uniquely determine the values of $X_{1,n+1},\cdots,X_{M,n+1}$.

A procedure which largely overcomes the disadvantages of both the forward-difference and centered-difference procedures is the double-approximation procedure, defined by the relation

$$X_{i,n+1}=X_{i,n}+\tfrac{1}{2}[F_i(P_n)+F_i(P_{(n+1)})]\Delta t. \tag{13}$$

Here the coefficient of Δt is an approximation to the time derivative of X_i at time $t_0+(n+\frac{1}{2})\Delta t$. From (9) and (10), it follows that (13) may be rewritten

$$X_{i,n+1}=\tfrac{1}{2}(X_{i,n}+X_{i((n+2))}). \tag{14}$$

A convenient scheme for automatic computation is the successive evaluation of $X_{i(n+1)}$, $X_{i((n+2))}$, and $X_{i,n+1}$ according to (9), (10) and (14). We have used this procedure in all the computations described in this study.

In phase space a numerical solution of (1) must be represented by a jumping particle rather than a continuously moving particle. Moreover, if a digital computer is instructed to represent each number in its memory by a preassigned fixed number of bits, only certain discrete points in phase space will ever be occupied. If the numerical solution is bounded, repetitions must eventually occur, so that, strictly speaking, every numerical solution is periodic. In practice this consideration may be disregarded, if the number of different possible states is far greater than the number of iterations ever likely to be performed. The necessity for repetition could be avoided altogether by the somewhat uneconomical procedure of letting the precision of computation increase as n increases.

Consider now numerical solutions of equations (4), obtained by the forward-difference procedure (11). For such solutions,

$$Q_{n+1}=Q_n+(dQ/dt)_n\Delta t+\tfrac{1}{2}\sum_i F_i^2(P_n)\Delta t^2. \qquad (15)$$

Let S' be any surface of constant Q whose interior R' contains the ellipsoid E where dQ/dt vanishes, and let S be any surface of constant Q whose interior R contains S'.

Since $\sum F_i^2$ and dQ/dt both possess upper bounds in R', we may choose Δt so small that P_{n+1} lies in R if P_n lies in R'. Likewise, since $\sum F_i^2$ possesses an upper bound and dQ/dt possesses a *negative* upper bound in $R-R'$, we may choose Δt so small that $Q_{n+1}<Q_n$ if P_n lies in $R-R'$. Hence Δt may be chosen so small that any jumping particle which has entered R remains trapped within R, and the numerical solution does not blow up. A blow-up may still occur, however, if initially the particle is exterior to R.

Consider now the double-approximation procedure (14). The previous arguments imply not only that $P_{(n+1)}$ lies within R if P_n lies within R, but also that $P_{((n+2))}$ lies within R if $P_{(n+1)}$ lies within R. Since the region R is convex, it follows that P_{n+1}, as given by (14), lies within R if P_n lies within R. Hence if Δt is chosen so small that the forward-difference procedure does not blow up, the double-approximation procedure also does not blow up.

We note in passing that if we apply the forward-difference procedure to a conservative system where $dQ/dt=0$ everywhere,

$$Q_{n+1}=Q_n+\tfrac{1}{2}\sum_i F_i^2(P_n)\Delta t^2. \qquad (16)$$

In this case, for any fixed choice of Δt the numerical solution ultimately goes to infinity, unless it is asymptotically approaching a steady state. A similar result holds when the double-approximation procedure (14) is applied to a conservative system.

5. The convection equations of Saltzman

In this section we shall introduce a system of three ordinary differential equations whose solutions afford the simplest example of deterministic nonperiodic flow of which the writer is aware. The system is a simplification of one derived by Saltzman (1962) to study finite-amplitude convection. Although our present interest is in the nonperiodic nature of its solutions, rather than in its contributions to the convection problem, we shall describe its physical background briefly.

Rayleigh (1916) studied the flow occurring in a layer of fluid of uniform depth H, when the temperature difference between the upper and lower surfaces is maintained at a constant value ΔT. Such a system possesses a steady-state solution in which there is no motion, and the temperature varies linearly with depth, If this solution is unstable, convection should develop.

In the case where all motions are parallel to the $x-z$-plane, and no variations in the direction of the y-axis occur, the governing equations may be written (see Saltzman, 1962)

$$\frac{\partial}{\partial t}\nabla^2\psi=-\frac{\partial(\psi,\nabla^2\psi)}{\partial(x,z)}+\nu\nabla^4\psi+g\alpha\frac{\partial\theta}{\partial x}, \qquad (17)$$

$$\frac{\partial}{\partial t}\theta=-\frac{\partial(\psi,\theta)}{\partial(x,z)}+\frac{\Delta T}{H}\frac{\partial\psi}{\partial x}+\kappa\nabla^2\theta. \qquad (18)$$

Here ψ is a stream function for the two-dimensional motion, θ is the departure of temperature from that occurring in the state of no convection, and the constants g, α, ν, and κ denote, respectively, the acceleration of gravity, the coefficient of thermal expansion, the kinematic viscosity, and the thermal conductivity. The problem is most tractable when both the upper and lower boundaries are taken to be free, in which case ψ and $\nabla^2\psi$ vanish at both boundaries.

Rayleigh found that fields of motion of the form

$$\psi=\psi_0\sin(\pi aH^{-1}x)\sin(\pi H^{-1}z), \qquad (19)$$

$$\theta=\theta_0\cos(\pi aH^{-1}x)\sin(\pi H^{-1}z), \qquad (20)$$

would develop if the quantity

$$R_a=g\alpha H^3\Delta T\nu^{-1}\kappa^{-1}, \qquad (21)$$

now called the *Rayleigh number*, exceeded a critical value

$$R_c=\pi^4a^{-2}(1+a^2)^3. \qquad (22)$$

The minimum value of R_c, namely $27\pi^4/4$, occurs when $a^2=\tfrac{1}{2}$.

Saltzman (1962) derived a set of ordinary differential equations by expanding ψ and θ in double Fourier series in x and z, with functions of t alone for coefficients, and substituting these series into (17) and (18). He arranged the right-hand sides of the resulting equations in double-Fourier-series form, by replacing products of trigonometric functions of x (or z) by sums of trigonometric functions, and then equated coefficients of similar functions of x and z. He then reduced the resulting infinite system to a finite system by omitting reference to all but a specified finite set of functions of t, in the manner proposed by the writer (1960).

He then obtained time-dependent solutions by numerical integration. In certain cases all except three of the dependent variables eventually tended to zero, and these three variables underwent irregular, apparently nonperiodic fluctuations.

These same solutions would have been obtained if the series had at the start been truncated to include a total of three terms. Accordingly, in this study we shall let

$$a(1+a^2)^{-1}\kappa^{-1}\psi = X\sqrt{2}\sin(\pi a H^{-1}x)\sin(\pi H^{-1}z), \tag{23}$$

$$\pi R_c^{-1}R_a\Delta T^{-1}\theta = Y\sqrt{2}\cos(\pi a H^{-1}x)\sin(\pi H^{-1}z) - Z\sin(2\pi H^{-1}z), \tag{24}$$

where X, Y, and Z are functions of time alone. When expressions (23) and (24) are substituted into (17) and (18), and trigonometric terms other than those occurring in (23) and (24) are omitted, we obtain the equations

$$X^\cdot = -\sigma X + \sigma Y, \tag{25}$$

$$Y^\cdot = -XZ + rX - Y, \tag{26}$$

$$Z^\cdot = XY - bZ. \tag{27}$$

Here a dot denotes a derivative with respect to the dimensionless time $\tau = \pi^2 H^{-2}(1+a^2)\kappa t$, while $\sigma = \kappa^{-1}\nu$ is the *Prandtl number*, $r = R_c^{-1}R_a$, and $b = 4(1+a^2)^{-1}$. Except for multiplicative constants, our variables X, Y, and Z are the same as Saltzman's variables A, D, and G. Equations (25), (26), and (27) are the convection equations whose solutions we shall study.

In these equations X is proportional to the intensity of the convective motion, while Y is proportional to the temperature difference between the ascending and descending currents, similar signs of X and Y denoting that warm fluid is rising and cold fluid is descending. The variable Z is proportional to the distortion of the vertical temperature profile from linearity, a positive value indicating that the strongest gradients occur near the boundaries.

Equations (25)–(27) may give realistic results when the Rayleigh number is slightly supercritical, but their solutions cannot be expected to resemble those of (17) and (18) when strong convection occurs, in view of the extreme truncation.

6. Applications of linear theory

Although equations (25)–(27), as they stand, do not have the form of (4), a number of linear transformations will convert them to this form. One of the simplest of these is the transformation

$$X' = X, \quad Y' = Y, \quad Z' = Z - r - \sigma. \tag{28}$$

Solutions of (25)–(27) therefore remain bounded within a region R as $\tau \to \infty$, and the general results of Sections 2, 3 and 4 apply to these equations.

The stability of a solution $X(\tau)$, $Y(\tau)$, $Z(\tau)$ may be formally investigated by considering the behavior of small superposed perturbations $x_0(\tau)$, $y_0(\tau)$, $z_0(\tau)$. Such perturbations are temporarily governed by the linearized equations

$$\begin{bmatrix} x_0 \\ y_0 \\ z_0 \end{bmatrix}^\cdot = \begin{bmatrix} -\sigma & \sigma & 0 \\ (r-Z) & -1 & -X \\ Y & X & -b \end{bmatrix} \begin{bmatrix} x_0 \\ y_0 \\ z_0 \end{bmatrix}. \tag{29}$$

Since the coefficients in (29) vary with time, unless the basic state X, Y, Z is a steady-state solution of (25)–(27), a general solution of (29) is not feasible. However, the variation of the volume V_0 of a small region in phase space, as each point in the region is displaced in accordance with (25)–(27), is determined by the diagonal sum of the matrix of coefficients; specifically

$$V_0^\cdot = -(\sigma+b+1)V_0. \tag{30}$$

This is perhaps most readily seen by visualizing the motion in phase space as the flow of a fluid, whose divergence is

$$\frac{\partial X^\cdot}{\partial X} + \frac{\partial Y^\cdot}{\partial Y} + \frac{\partial Z^\cdot}{\partial Z} = -(\sigma+b+1). \tag{31}$$

Hence each small volume shrinks to zero as $\tau \to \infty$, at a rate independent of X, Y, and Z. This does not imply that each small volume shrinks to a point; it may simply become flattened into a surface. It follows that the volume of the region initially enclosed by the surface S shrinks to zero at this same rate, so that all trajectories ultimately become confined to a specific subspace having zero volume. This subspace contains all those trajectories which lie entirely within R, and so contains all central trajectories.

Equations (25)–(27) possess the steady-state solution $X = Y = Z = 0$, representing the state of no convection. With this basic solution, the characteristic equation of the matrix in (29) is

$$[\lambda + b][\lambda^2 + (\sigma+1)\lambda + \sigma(1-r)] = 0. \tag{32}$$

This equation has three real roots when $r > 0$; all are negative when $r < 1$, but one is positive when $r > 1$. The criterion for the onset of convection is therefore $r = 1$, or $R_a = R_c$, in agreement with Rayleigh's result.

When $r > 1$, equations (25)–(27) possess two additional steady-state solutions $X = Y = \pm\sqrt{b(r-1)}$, $Z = r - 1$.

For either of these solutions, the characteristic equation of the matrix in (29) is

$$\lambda^3+(\sigma+b+1)\lambda^2+(r+\sigma)b\lambda+2\sigma b(r-1)=0. \quad (33)$$

This equation possesses one real negative root and two complex conjugate roots when $r>1$; the complex conjugate roots are pure imaginary if the product of the coefficients of λ^2 and λ equals the constant term, or

$$r=\sigma(\sigma+b+3)(\sigma-b-1)^{-1}. \quad (34)$$

This is the critical value of r for the instability of steady convection. Thus if $\sigma<b+1$, no positive value of r satisfies (34), and steady convection is always stable, but if $\sigma>b+1$, steady convection is unstable for sufficiently high Rayleigh numbers. This result of course applies only to idealized convection governed by (25)–(27), and not to the solutions of the partial differential equations (17) and (18).

The presence of complex roots of (34) shows that if unstable steady convection is disturbed, the motion will oscillate in intensity. What happens when the disturbances become large is not revealed by linear theory. To investigate finite-amplitude convection, and to study the subspace to which trajectories are ultimately confined, we turn to numerical integration.

TABLE 1. Numerical solution of the convection equations. Values of X, Y, Z are given at every fifth iteration N, for the first 160 iterations.

N	X	Y	Z
0000	0000	0010	0000
0005	0004	0012	0000
0010	0009	0020	0000
0015	0016	0036	0002
0020	0030	0066	0007
0025	0054	0115	0024
0030	0093	0192	0074
0035	0150	0268	0201
0040	0195	0234	0397
0045	0174	0055	0483
0050	0097	−0067	0415
0055	0025	−0093	0340
0060	−0020	−0089	0298
0065	−0046	−0084	0275
0070	−0061	−0083	0262
0075	−0070	−0086	0256
0080	−0077	−0091	0255
0085	−0084	−0095	0258
0090	−0089	−0098	0266
0095	−0093	−0098	0275
0100	−0094	−0093	0283
0105	−0092	−0086	0297
0110	−0088	−0079	0286
0115	−0083	−0073	0281
0120	−0078	−0070	0273
0125	−0075	−0071	0264
0130	−0074	−0075	0257
0135	−0076	−0080	0252
0140	−0079	−0087	0251
0145	−0083	−0093	0254
0150	−0088	−0098	0262
0155	−0092	−0099	0271
0160	−0094	−0096	0281

7. Numerical integration of the convection equations

To obtain numerical solutions of the convection equations, we must choose numerical values for the constants. Following Saltzman (1962), we shall let $\sigma=10$ and $a^2=\frac{1}{2}$, so that $b=8/3$. The critical Rayleigh number for instability of steady convection then occurs when $r=470/19=24.74$.

We shall choose the slightly supercritical value $r=28$. The states of steady convection are then represented by the points $(6\sqrt{2}, 6\sqrt{2}, 27)$ and $(-6\sqrt{2}, -6\sqrt{2}, 27)$ in phase space, while the state of no convection corresponds to the origin (0,0,0).

We have used the double-approximation procedure for numerical integration, defined by (9), (10), and (14). The value $\Delta\tau=0.01$ has been chosen for the dimensionless time increment. The computations have been performed on a Royal McBee LGP-30 electronic com-

TABLE 2. Numerical solution of the convection equations Values of X, Y, Z are given at every iteration N for which Z possesses a relative maximum, for the first 6000 iterations.

N	X	Y	Z	N	X	Y	Z
0045	0174	0055	0483	3029	0117	0075	0352
0107	−0091	−0083	0287	3098	0123	0076	0365
0168	−0092	−0084	0288	3171	0134	0082	0383
0230	−0092	−0084	0289	3268	0155	0069	0435
0292	−0092	−0083	0290	3333	−0114	−0079	0342
0354	−0093	−0083	0292	3400	−0117	−0077	0350
0416	−0093	−0083	0293	3468	−0125	−0083	0361
0478	−0094	−0082	0295	3541	−0129	−0073	0378
0540	−0094	−0082	0296	3625	−0146	−0074	0413
0602	−0095	−0082	0298	3695	0127	0079	0370
0664	−0096	−0083	0300	3772	0136	0072	0394
0726	−0097	−0083	0302	3853	−0144	−0077	0407
0789	−0097	−0081	0304	3926	0129	0072	0380
0851	−0099	−0083	0307	4014	0148	0068	0421
0914	−0100	−0081	0309	4082	−0120	−0074	0359
0977	−0100	−0080	0312	4153	−0129	−0078	0375
1040	−0102	−0080	0315	4233	−0144	−0082	0404
1103	−0104	−0081	0319	4307	0135	0081	0385
1167	−0105	−0079	0323	4417	−0162	−0069	0450
1231	−0107	−0079	0328	4480	0106	0081	0324
1295	−0111	−0082	0333	4544	0109	0082	0329
1361	−0111	−0077	0339	4609	0110	0080	0334
1427	−0116	−0079	0347	4675	0112	0076	0341
1495	−0120	−0077	0357	4741	0118	0081	0349
1566	−0125	−0072	0371	4810	0120	0074	0360
1643	−0139	−0077	0396	4881	0130	0081	0376
1722	0140	0075	0401	4963	0141	0068	0406
1798	−0135	−0072	0391	5035	−0133	−0081	0381
1882	0146	0074	0413	5124	−0151	−0076	0422
1952	−0127	−0078	0370	5192	0119	0075	0358
2029	−0135	−0070	0393	5262	0129	0083	0372
2110	0146	0083	0408	5340	0140	0079	0397
2183	−0128	−0070	0379	5419	−0137	−0067	0399
2268	−0144	−0066	0415	5495	0140	0081	0394
2337	0126	0079	0368	5576	−0141	−0072	0405
2412	0137	0081	0389	5649	0135	0082	0384
2501	−0153	−0080	0423	5752	0160	0074	0443
2569	0119	0076	0357	5816	−0110	−0081	0332
2639	0129	0082	0371	5881	−0113	−0082	0339
2717	0136	0070	0395	5948	−0114	−0075	0346
2796	−0143	−0079	0402				
2871	0134	0076	0388				
2962	−0152	−0072	0426				

puting machine. Approximately one second per iteration, aside from output time, is required.

For initial conditions we have chosen a slight departure from the state of no convection, namely (0,1,0). Table 1 has been prepared by the computer. It gives the values of N (the number of iterations), X, Y, and Z at every fifth iteration for the first 160 iterations. In the printed output (but not in the computations) the values of X, Y, and Z are multiplied by ten, and then only those figures to the left of the decimal point are printed. Thus the states of steady convection would appear as 0084, 0084, 0270 and −0084, −0084, 0270, while the state of no convection would appear as 0000, 0000, 0000.

The initial instability of the state of rest is evident. All three variables grow rapidly, as the sinking cold fluid is replaced by even colder fluid from above, and the rising warm fluid by warmer fluid from below, so that by step 35 the strength of the convection far exceeds that of steady convection. Then Y diminishes as the warm fluid is carried over the top of the convective cells, so that by step 50, when X and Y have opposite signs, warm fluid is descending and cold fluid is ascending. The motion thereupon ceases and reverses its direction, as indicated by the negative values of X following step 60. By step 85 the system has reached a state not far from that of steady convection. Between steps 85 and 150 it executes a complete oscillation in its intensity, the slight amplification being almost indetectable.

The subsequent behavior of the system is illustrated in Fig. 1, which shows the behavior of Y for the first 3000 iterations. After reaching its early peak near step 35 and then approaching equilibrium near step 85, it undergoes systematic amplified oscillations until near step 1650. At this point a critical state is reached, and thereafter Y changes sign at seemingly irregular intervals, reaching sometimes one, sometimes two, and sometimes three or more extremes of one sign before changing sign again.

Fig. 2 shows the projections on the X-Y- and Y-Z-planes in phase space of the portion of the trajectory corresponding to iterations 1400–1900. The states of steady convection are denoted by C and C'. The first portion of the trajectory spirals outward from the vicinity of C', as the oscillations about the state of steady convection, which have been occurring since step 85, continue to grow. Eventually, near step 1650, it crosses the X-Z-plane, and is then deflected toward the neighborhood of C. It temporarily spirals about C, but crosses the X-Z-plane after one circuit, and returns to the neighborhood of C', where it soon joins the spiral over which it has previously traveled. Thereafter it crosses from one spiral to the other at irregular intervals.

Fig. 3, in which the coordinates are Y and Z, is based upon the printed values of X, Y, and Z at every fifth iteration for the first 6000 iterations. These values determine X as a smooth single-valued function of Y and Z over much of the range of Y and Z; they determine X

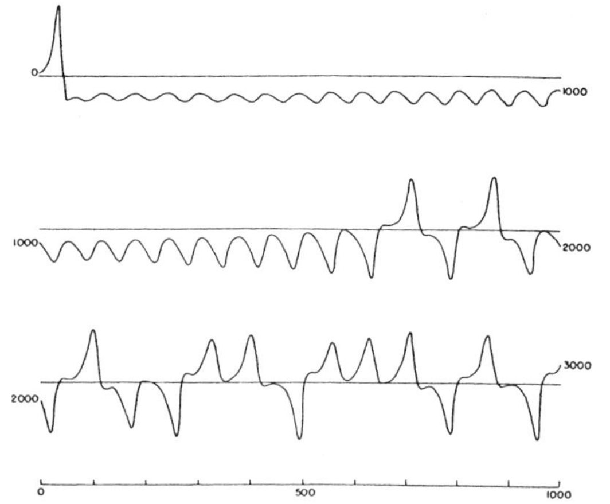

FIG. 1. Numerical solution of the convection equations. Graph of Y as a function of time for the first 1000 iterations (upper curve), second 1000 iterations (middle curve), and third 1000 iterations (lower curve).

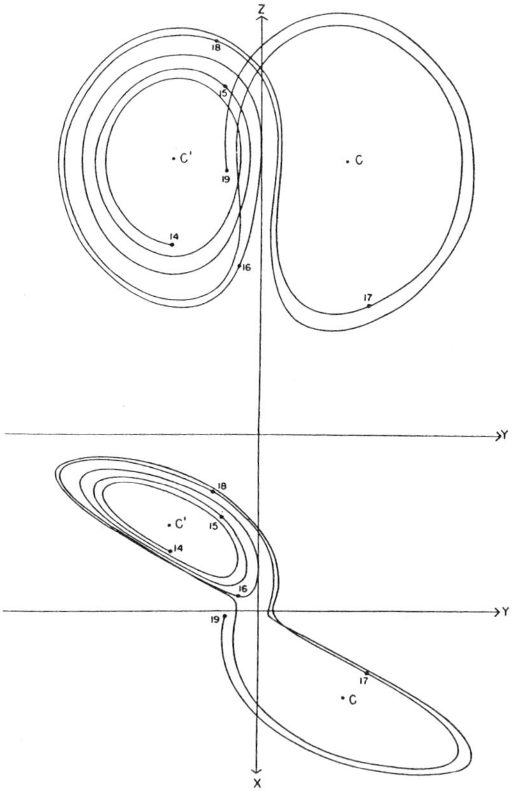

FIG. 2. Numerical solution of the convection equations. Projections on the X-Y-plane and the Y-Z-plane in phase space of the segment of the trajectory extending from iteration 1400 to iteration 1900. Numerals "14," "15," etc., denote positions at iterations 1400, 1500, etc. States of steady convection are denoted by C and C'.

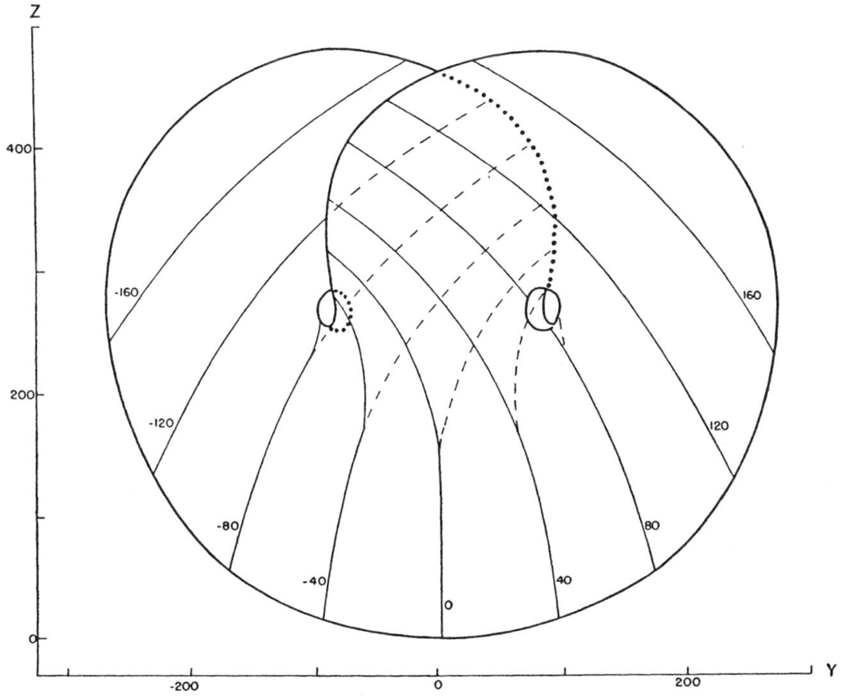

FIG. 3. Isopleths of X as a function of Y and Z (thin solid curves), and isopleths of the lower of two values of X, where two values occur (dashed curves), for approximate surfaces formed by all points on limiting trajectories. Heavy solid curve, and extensions as dotted curves, indicate natural boundaries of surfaces.

as one of two smooth single-valued functions over the remainder of the range. In Fig. 3 the thin solid lines are isopleths of X, and where two values of X exist, the dashed lines are isopleths of the lower value. Thus, within the limits of accuracy of the printed values, the trajectory is confined to a pair of surfaces which appear to merge in the lower portion of Fig. 3. The spiral about C lies in the upper surface, while the spiral about C' lies in the lower surface. Thus it is possible for the trajectory to pass back and forth from one spiral to the other without intersecting itself.

Additional numerical solutions indicate that other trajectories, originating at points well removed from these surfaces, soon meet these surfaces. The surfaces therefore appear to be composed of all points lying on limiting trajectories.

Because the origin represents a steady state, no trajectory can pass through it. However, two trajectories emanate from it, i.e., approach it asymptotically as $\tau \to -\infty$. The heavy solid curve in Fig. 3, and its extensions as dotted curves, are formed by these two trajectories. Trajectories passing close to the origin will tend to follow the heavy curve, but will not cross it, so that the heavy curve forms a natural boundary to the region which a trajectory can ultimately occupy. The holes near C and C' also represent regions which cannot be occupied after they have once been abandoned.

Returning to Fig. 2, we find that the trajectory apparently leaves one spiral only after exceeding some critical distance from the center. Moreover, the extent to which this distance is exceeded appears to determine the point at which the next spiral is entered; this in turn seems to determine the number of circuits to be executed before changing spirals again.

It therefore seems that some single feature of a given circuit should predict the same feature of the following circuit. A suitable feature of this sort is the maximum value of Z, which occurs when a circuit is nearly completed. Table 2 has again been prepared by the computer, and shows the values of X, Y, and Z at only those iterations N for which Z has a relative maximum. The succession of circuits about C and C' is indicated by the succession of positive and negative values of X and Y. Evidently X and Y change signs following a maximum which exceeds some critical value printed as about 385.

Fig. 4 has been prepared from Table 2. The abscissa is M_n, the value of the nth maximum of Z, while the ordinate is M_{n+1}, the value of the following maximum. Each point represents a pair of successive values of Z taken from Table 2. Within the limits of the round-off in tabulating Z, there is a precise two-to-one relation between M_n and M_{n+1}. The initial maximum $M_1=483$ is shown as if it had followed a maximum $M_0=385$, since maxima near 385 are followed by close approaches to the origin, and then by exceptionally large maxima.

It follows that an investigator, unaware of the nature of the governing equations, could formulate an empirical prediction scheme from the "data" pictured in Figs. 2 and 4. From the value of the most recent maximum of Z, values at future maxima may be obtained by repeated applications of Fig. 4. Values of X, Y, and Z between maxima of Z may be found from Fig. 2, by interpolating between neighboring curves. Of course, the accuracy of predictions made by this method is limited by the exactness of Figs. 2 and 4, and, as we shall see, by the accuracy with which the initial values of X, Y, and Z are observed.

Some of the implications of Fig. 4 are revealed by considering an idealized two-to-one correspondence between successive members of sequences M_0, M_1, $\cdots$, consisting of numbers between zero and one. These sequences satisfy the relations

$$\begin{aligned} M_{n+1}&=2M_n && \text{if } M_n<\tfrac{1}{2} \\ M_{n+1}& \text{ is undefined} && \text{if } M_n=\tfrac{1}{2} \\ M_{n+1}&=2-2M_n && \text{if } M_n>\tfrac{1}{2}. \end{aligned} \tag{35}$$

The correspondence defined by (35) is shown in Fig. 5, which is an idealization of Fig. 4. It follows from repeated applications of (35) that in any particular sequence,

$$M_n=m_n\pm 2^n M_0, \tag{36}$$

where m_n is an even integer.

Consider first a sequence where $M_0=u/2^p$, where u is odd. In this case $M_{p-1}=\frac{1}{2}$, and the sequence terminates. These sequences form a denumerable set, and correspond to the trajectories which score direct hits upon the state of no convection.

Next consider a sequence where $M_0=u/2^p v$, where u and v are relatively prime odd numbers. Then if $k>0$, $M_{p+1+k}=u_k/v$, where u_k and v are relatively prime and u_k is even. Since for any v the number of proper fractions u_k/v is finite, repetitions must occur, and the sequence is periodic. These sequences also form a denumerable set, and correspond to periodic trajectories.

The periodic sequences having a given number of distinct values, or phases, are readily tabulated. In particular there are a single one-phase, a single two-phase, and two three-phase sequences, namely,

$$\begin{aligned} &2/3, \cdots, \\ &2/5, 4/5, \cdots, \\ &2/7, 4/7, 6/7, \cdots, \\ &2/9, 4/9, 8/9, \cdots. \end{aligned}$$

The two three-phase sequences differ qualitatively in that the former possesses two numbers, and the latter only one number, exceeding $\frac{1}{2}$. Thus the trajectory corresponding to the former makes two circuits about C, followed by one about C' (or vice versa). The trajectory corresponding to the latter makes three circuits about C, followed by three about C', so that actually only Z varies in three phases, while X and Y vary in six.

Now consider a sequence where M_0 is not a rational fraction. In this case (36) shows that M_{n+k} cannot equal

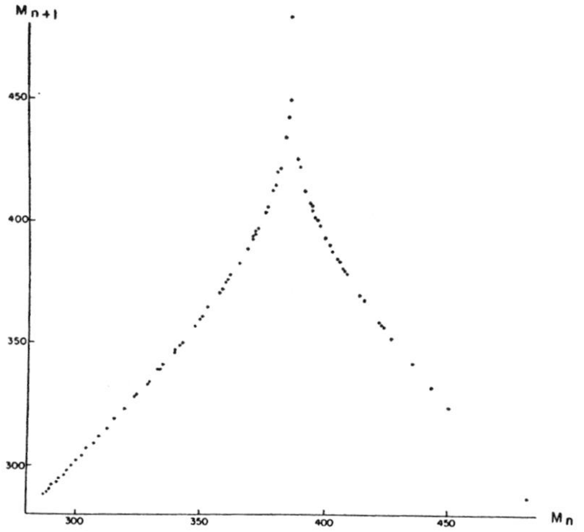

FIG. 4. Corresponding values of relative maximum of Z (abscissa) and subsequent relative maximum of Z (ordinate) occurring during the first 6000 iterations.

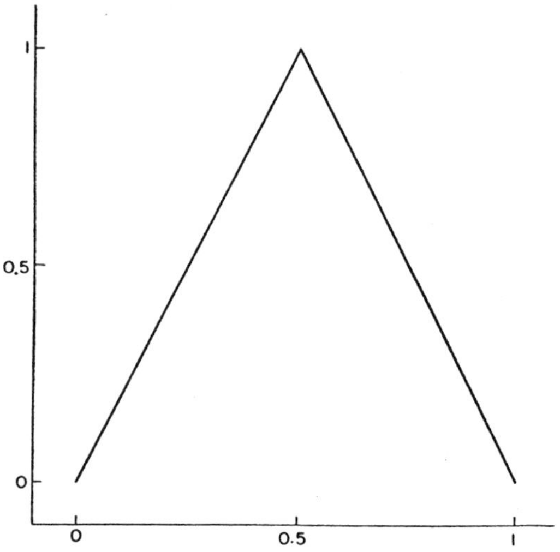

FIG. 5. The function $M_{n+1}=2M_n$ if $M_n<\frac{1}{2}$, $M_{n+1}=2-2M_n$ if $M_n>\frac{1}{2}$, serving as an idealization of the locus of points in Fig. 4.

M_n if $k>0$, so that no repetitions occur. These sequences, which form a nondenumerable set, may conceivably approach periodic sequences asymptotically and be quasi-periodic, or they may be nonperiodic.

Finally, consider two sequences M_0, M_1, $\cdots$ and M_0', M_1', $\cdots$, where $M_0'=M_0+\epsilon$. Then for a given k, if ϵ is sufficiently small, $M_k'=M_k\pm 2^k\epsilon$. All sequences are therefore unstable with respect to small modifications. In particular, all periodic sequences are unstable, and no other sequences can approach them asymptotically. All sequences except a set of measure zero are therefore nonperiodic, and correspond to nonperiodic trajectories.

Returning to Fig. 4, we see that periodic sequences analogous to those tabulated above can be found. They are given approximately by

$$
\begin{aligned}
&398, \cdots, \\
&377, 410, \cdots, \\
&369, 391, 414, \cdots, \\
&362, 380, 419, \cdots.
\end{aligned}
$$

The trajectories possessing these or other periodic sequences of maxima are presumably periodic or quasi-periodic themselves.

The above sequences are temporarily approached in the numerical solution by sequences beginning at iterations 5340, 4881, 3625, and 3926. Since the numerical solution eventually departs from each of these sequences, each is presumably unstable.

More generally, if $M_n'=M_n+\epsilon$, and if ϵ is sufficiently small, $M_{n+k}'=M_{n+k}+\Lambda\epsilon$, where Λ is the product of the slopes of the curve in Fig. 4 at the points whose abscissas are M_n, $\cdots$, M_{n+k-1}. Since the curve apparently has a slope whose magnitude exceeds unity everywhere, all sequences of maxima, and hence all trajectories, are unstable. In particular, the periodic trajectories, whose sequences of maxima form a denumerable set, are unstable, and only exceptional trajectories, having the same sequences of maxima, can approach them asymptotically. The remaining trajectories, whose sequences of maxima form a nondenumerable set, therefore represent deterministic nonperiodic flow.

These conclusions have been based upon a finite segment of a numerically determined solution. They cannot be regarded as mathematically proven, even though the evidence for them is strong. One apparent contradiction requires further examination.

It is difficult to reconcile the merging of two surfaces, one containing each spiral, with the inability of two trajectories to merge. It is not difficult, however, to explain the *apparent* merging of the surfaces. At two times τ_0 and τ_1, the volumes occupied by a specified set of particles satisfy the relation

$$V_0(\tau_1)=e^{-(\sigma+b+1)(\tau_1-\tau_0)}V_0(\tau_0), \tag{37}$$

according to (30). A typical circuit about C or C' requires about 70 iterations, so that, for such a circuit, $\tau_2=\tau_1+0.7$, and, since $\sigma+b+1=41/3$,

$$V_0(\tau_1)=0.00007\,V_0(\tau_0). \tag{38}$$

Two particles separated from each other in a suitable direction can therefore come together very rapidly, and appear to merge.

It would seem, then, that the two surfaces merely appear to merge, and remain distinct surfaces. Following these surfaces along a path parallel to a trajectory, and circling C or C', we see that each surface is really a pair of surfaces, so that, where they appear to merge, there are really four surfaces. Continuing this process for another circuit, we see that there are really eight surfaces, etc., and we finally conclude that there is an infinite complex of surfaces, each extremely close to one or the other of two merging surfaces.

The infinite set of values at which a line parallel to the X-axis intersects these surfaces may be likened to the set of all numbers between zero and one whose decimal expansions (or some other expansions besides binary) contain only zeros and ones. This set is plainly nondenumerable, in view of its correspondence to the set of all numbers between zero and one, expressed in binary. Nevertheless it forms a set of measure zero. The sequence of ones and zeros corresponding to a particular surface contains a history of the trajectories lying in that surface, a one or zero immediately to the right of the decimal point indicating that the last circuit was about C or C', respectively, a one or zero in second place giving the same information about the next to the last circuit, etc. Repeating decimal expansions represent periodic or quasi-periodic trajectories, and, since they define rational fractions, they form a denumerable set.

If one first visualizes this infinite complex of surfaces, it should not be difficult to picture nonperiodic deterministic trajectories embedded in these surfaces.

8. Conclusion

Certain mechanically or thermally forced nonconservative hydrodynamical systems may exhibit either periodic or irregular behavior when there is no obviously related periodicity or irregularity in the forcing process. Both periodic and nonperiodic flow are observed in some experimental models when the forcing process is held constant, within the limits of experimental control. Some finite systems of ordinary differential equations designed to represent these hydrodynamical systems possess periodic analytic solutions when the forcing is strictly constant. Other such systems have yielded nonperiodic numerical solutions.

A finite system of ordinary differential equations representing forced dissipative flow often has the property that all of its solutions are ultimately confined within the same bounds. We have studied in detail the properties of solutions of systems of this sort. Our principal results concern the instability of nonperiodic solutions. A nonperiodic solution with no transient component must be unstable, in the sense that solutions temporarily approximating it do not continue to do so. A nonperiodic solution with a transient component is sometimes stable, but in this case its stability is one of its transient properties, which tends to die out.

To verify the existence of deterministic nonperiodic flow, we have obtained numerical solutions of a system of three ordinary differential equations designed to represent a convective process. These equations possess three steady-state solutions and a denumerably infinite set of periodic solutions. All solutions, and in particular the periodic solutions, are found to be unstable. The remaining solutions therefore cannot in general approach the periodic solutions asymptotically, and so are nonperiodic.

When our results concerning the instability of nonperiodic flow are applied to the atmosphere, which is ostensibly nonperiodic, they indicate that prediction of the sufficiently distant future is impossible by any method, unless the present conditions are known exactly. In view of the inevitable inaccuracy and incompleteness of weather observations, precise very-long-range forecasting would seem to be non-existent.

There remains the question as to whether our results really apply to the atmosphere. One does not usually regard the atmosphere as either deterministic or finite, and the lack of periodicity is not a mathematical certainty, since the atmosphere has not been observed forever.

The foundation of our principal result is the eventual necessity for any bounded system of finite dimensionality to come arbitrarily close to acquiring a state which it has previously assumed. If the system is stable, its future development will then remain arbitrarily close to its past history, and it will be quasi-periodic.

In the case of the atmosphere, the crucial point is then whether analogues must have occurred since the state of the atmosphere was first observed. By analogues, we mean specifically two or more states of the atmosphere, together with its environment, which resemble each other so closely that the differences may be ascribed to errors in observation. Thus, to be analogues, two states must be closely alike in regions where observations are accurate and plentiful, while they need not be at all alike in regions where there are no observations at all, whether these be regions of the atmosphere or the environment. If, however, some unobserved features are implicit in a succession of observed states, two successions of states must be nearly alike in order to be analogues.

If it is true that two analogues have occurred since atmospheric observation first began, it follows, since the atmosphere has not been observed to be periodic, that the successions of states following these analogues must eventually have differed, and no forecasting scheme could have given correct results both times. If, instead, analogues have not occurred during this period, some accurate very-long-range prediction scheme, using observations at present available, may exist. But, if it does exist, the atmosphere will acquire a quasi-periodic behavior, never to be lost, once an analogue occurs. This quasi-periodic behavior need not be established, though, even if very-long-range forecasting is feasible, if the variety of possible atmospheric states is so immense that analogues need never occur. It should be noted that these conclusions do not depend upon whether or not the atmosphere is deterministic.

There remains the very important question as to how long is "very-long-range." Our results do not give the answer for the atmosphere; conceivably it could be a few days or a few centuries. In an idealized system, whether it be the simple convective model described here, or a complicated system designed to resemble the atmosphere as closely as possible, the answer may be obtained by comparing pairs of numerical solutions having nearly identical initial conditions. In the case of the real atmosphere, if all other methods fail, we can wait for an analogue.

Acknowledgments. The writer is indebted to Dr. Barry Saltzman for bringing to his attention the existence of nonperiodic solutions of the convection equations. Special thanks are due to Miss Ellen Fetter for handling the many numerical computations and preparing the graphical presentations of the numerical material.

REFERENCES

Birkhoff, G. O., 1927: *Dynamical systems.* New York, Amer. Math. Soc., Colloq. Publ., 295 pp.

Ford, L. R., 1933: *Differential equations.* New York, McGraw-Hill, 264 pp.

Fultz, D., R. R. Long, G. V. Owens, W. Bohan, R. Kaylor and J. Weil, 1959: Studies of thermal convection in a rotating cylinder with some implications for large-scale atmospheric motions. *Meteor. Monog,* **4**(21), Amer. Meteor. Soc., 104 pp.

Gibbs, J. W., 1902: *Elementary principles in statistical mechanics.* New York, Scribner, 207 pp.

Hide, R., 1958: An experimental study of thermal convection in a rotating liquid. *Phil. Trans. Roy. Soc. London,* (A), **250**, 441–478.

Lorenz, E. N., 1960: Maximum simplification of the dynamic equations. *Tellus,* **12**, 243–254.

——, 1962a: Simplified dynamic equations applied to the rotating-basin experiments. *J. atmos. Sci.,* **19**, 39–51.

——, 1962b: The statistical prediction of solutions of dynamic equations. *Proc. Internat. Symposium Numerical Weather Prediction,* Tokyo, 629–635.

Nemytskii, V. V., and V. V. Stepanov, 1960: *Qualitative theory of differential equations.* Princeton, Princeton Univ. Press, 523 pp.

Poincaré, H., 1881: Mémoire sur les courbes définies par une équation différentielle. *J. de Math.,* **7**, 375–442.

Rayleigh, Lord, 1916: On convective currents in a horizontal layer of fluid when the higher temperature is on the under side. *Phil. Mag.,* **32**, 529–546.

Saltzman, B., 1962: Finite amplitude free convection as an initial value problem—I. *J. atmos. Sci.,* **19**, 329–341.

STUDIA MATHEMATICA, T. XXXIII. (1969)

On invariant measures for expanding differentiable mappings

by

K. KRZYŻEWSKI and W. SZLENK (Warszawa)

This note concerns expanding differentiable mappings first studied by M. Shub, see [5] and [6]. These mappings are closely connected with Anosov diffeomorphisms. But while it is not known whether there always exists a finite Lebesgue measure invariant with respect to an Anosov diffeomorphism (see [1] and [6]), it turns out that such a measure always exists for any expanding differentiable mapping. The purpose of this note is to prove this fact. It seems that this may be of some interest and that is why we publish the proof although the arguments used in it have some points of similarity with the proof of Theorem 1 in [3], p. 483.

The authors are very much indebted to Professor J. G. Sinai for his valuable remarks concerning this paper.

In the sequel M will always denote a compact, connected differentiable manifold of class C^∞ unless stated otherwise. If φ is a map of class C^1 of M into itself, then $d\varphi$ will denote the derivative of φ which is the map of the tangent bundle $T(M)$ into itself. We shall say that φ *is expanding if* there exist a Riemannian metric $\|\cdot\|$ on M, a positive real number a and a real number c greater than 1 and such that

$$\|(d\varphi^n)(\mathfrak{a})\| \geqslant ac^n \|\mathfrak{a}\| \tag{1}$$

for each $\mathfrak{a} \in T(M)$ and $n = 1, 2, \ldots$

EXAMPLE. Let φ be a differentiable mapping of the 2-dimensional torus into itself given by the formula

$$\varphi(x, y) = \big(mx + ny + \varepsilon \cdot f(x, y),\, px + qy + \varepsilon g(x, y)\big) (\mathrm{mod}\, 1),$$
$$0 \leqslant x < 1,\ 0 \leqslant y < 1,$$

where

(i) m, n, p, q are integers;

(ii) the eigenvalues of the matrix $\begin{pmatrix} m, & n \\ p, & q \end{pmatrix}$ are real and their moduli are greater than 1;

(iii) f and g are the real functions of class C^1 on R^2, periodic with period 1 with respect to each variable;

(iv) ε is a real positive number.

It is easy to see that if ε is sufficiently small, then φ is expanding. For more general examples of expanding mappings see [5] and [6].

Let μ be a Borel measure on M. We shall say that μ is the *Lebesgue measure* if μ is equivalent to the Riemannian measure on M induced by a certain Riemannian metric on M. Now we may state the following

THEOREM. *If φ is an expanding map of class C^2 of M into itself, then*

(a) *there exists a normalised Lebesgue measure μ on M invariant with respect to φ;*

(b) *the dynamical system (μ, φ) is exact* (1) *and therefore ergodic;*

(c) *if $\bar{\mu}$ is any normalised Borel measure on M absolutely continuous with respect to a certain Riemannian measure and invariant with respect to φ, then $\bar{\mu} = \mu$.*

For the proof of the theorem the following lemmas will be needed.

LEMMA 1. *If φ is an expanding map of M into itself, there exist a Riemannian metric $\|\cdot\|_1$ of class C^∞ on M and a real number c_1 greater than 1 such that*

$$\|(d\varphi)(\mathfrak{a})\|_1 \geqslant c_1 \|\mathfrak{a}\|_1 \tag{2}$$

for each $\mathfrak{a} \in T(M)$.

Proof. First we shall prove that there exsits a Riemannian metric $\|\cdot\|_1$, not necessarily of class C^∞, such that (2) is satisfied. For this purpose let k be an integer such that $ac^k > 1$ and $k > 1$, where a and c are from (1). Then we may define the Riemannian metric $\|\cdot\|_1$ on M as follows:

$$\|\mathfrak{a}\|_1^2 = \|\mathfrak{a}\|^2 + \ldots + \|(d\varphi^{k-1})(\mathfrak{a})\|^2, \tag{3}$$

where $\mathfrak{a} \in T(M)$ and where $\|\cdot\|$ is from (1). Since M is compact, there exists a finite real number A such that

$$\|(d\varphi)(\mathfrak{a})\|^2 + \ldots + \|(d\varphi^{k-1})(\mathfrak{a})\|^2 \leqslant A\, \|\mathfrak{a}\|^2 \tag{4}$$

for $\mathfrak{a} \in T(M)$. Now let c_1 be any real number such that $1 < c_1 < ac^k$ and $a^2 c^{2k} - c_1^2 \geqslant A(c_1^2 - 1)$. Then it is easy to see that (2) is satisfied. Now we shall show that $\|\cdot\|_1$ may be chosen to be of class C^∞. For this purpose let us assume that (2) is satisfied for the Riemannian metric $\|\cdot\|_1$. If we prove that for any positive number ε there exists a Riemannian metric $\|\cdot\|_2$ of class C^∞ on M such that

$$\big|\|\mathfrak{a}\|_2^2 - \|\mathfrak{a}\|_1^2\big| \leqslant \varepsilon \|\mathfrak{a}\|_1^2 \tag{5}$$

for each $\mathfrak{a} \in T(M)$, the proof of the lemma will be completed. In fact, if we choose ε such that $1+\varepsilon < c_1^2(1-\varepsilon)$, then $\|\cdot\|_2$ will satisfy (2) with the constant $c_1\sqrt{\dfrac{1-\varepsilon}{1+\varepsilon}}$.

(1) For the definition of exact dynamical systems see [4].

If we apply the standard method based on the C^∞-partition of unity, the proof of (5) may be reduced to the proving of the following:

(6) for any $\varepsilon > 0$ and any point $p \in M$ there exists a Riemannian metric $\|\cdot\|_{2,U}$ of class C^∞ on some neighborhood U of p such that

$$\big|\|\mathfrak{a}\|_1^2 - \|\mathfrak{a}\|_{2,U}^2\big| \leqslant \varepsilon \|\mathfrak{a}\|_1^2 \quad \text{for } \mathfrak{a} \in T(M)/U.$$

Condition (6) may easily be proved by a uniform approximation of the coordinates of $\|\cdot\|_1$ in some coordinate system on M containing p by real functions of class C^∞. Therefore the proof of (6) will be omitted. Thus the proof of the lemma is completed.

From now on, φ will denote, unless stated otherwise, an expanding map of class C^2, $\|\cdot\|_1$ will denote the Riemannian metric on M given by Lemma 1, and $d_1(\cdot, \cdot)$, μ_1 will denote the natural metric and the Riemannian measure induced by $\|\cdot\|_1$ on M respectively.

The following two lemmas may be proved in the standard way and therefore we shall omit the proofs:

LEMMA 2. *Each expanding map is an N-fold covering, where $1 < N < \infty$.*

LEMMA 3. *If $f(\cdot)$ is a real function of class C^1 on M, then there exists a finite real number L such that*

$$|f(x) - f(y)| \leqslant L d_1(x, y) \quad \textit{for } x, y \in M.$$

Let φ be a map of class C^1 of a Riemannian manifold $(X, \|\cdot\|)$ of class C^∞ into a Riemannian manifold $(X_*, \|\cdot\|_*)$ of class C^∞, where $\dim X = \dim X_*$. Then we may define on X the function $D\varphi$ as

$$(D\varphi)(x) = \mu_*^{\varphi(x)}\big((d\varphi)(A_x)\big)$$

for $x \in X$, where A_x is any Borel set in $T_x(X)$ of measure μ^x equal to 1 and μ^x, μ_*^y are the natural measures induced by the Riemannian metrics $\|\cdot\|$ and $\|\cdot\|_*$ on $T_x(X)$ and $T_y(X_*)$, respectively. This function, as is easy to see, is well-defined and will be termed the *scalar derivative* of φ. If φ is of class C^2, then $D\varphi$ is of class C^1. Further, if φ is a diffeomorphism X onto X_*, then

$$\mu_*\big(\varphi(A)\big) = \int_A (D\varphi)(x)\, d\mu(x)$$

for each Borel set $A \subset X$, where μ, μ_* are the Riemannian measures induced by $\|\cdot\|$, $\|\cdot\|_*$ on X, X_*, respectively.

Now we may prove the following

LEMMA 4. *There exists a real finite number a such that*

$$\mu_1\big(\varphi^{-n}(A)\big) \leqslant a\mu_1(A). \tag{7}$$

for each Borel set $A \subset M$ and $n = 1, 2, \ldots$

Proof. In view of Lemma 2 and the well-known property of the Riemannian metric of class C^∞, there exists an open cover $\{U_i\}_{1\leqslant i\leqslant p}$ of M such that

(8) φ is a diffeomorphism of each component of $\varphi^{-1}(U_i)$ onto U_i for $i = 1, 2, \ldots, p$;

(9) for each pair of points x, y belonging to U_i, there exists a regular curve joining x and y, contained in U_i and such that $d_1(x, y)$ is equal to its length, $i = 1, \ldots, p$.

Let δ be the Lebesgue number of the cover $\{U_i\}_{1\leqslant i\leqslant p}$; then there exist a positive integer k and open sets $A_{i_0,\ldots,i_n}(1\leqslant i_0\leqslant k, 1\leqslant i_1\leqslant N, \ldots, 1\leqslant i_n\leqslant N)$ for $n = 0, 1, \ldots$, where N is from Lemma 2, such that

(10) $\mu_1\big(M - \bigcup_{i_0=1}^{k} A_{i_0}\big) = 0$ and $A_{i_0} \neq \emptyset$ for $i_0 = 1, \ldots, k$

(11) $A_{i_0,\ldots,i_n} \cap A_{i'_0,\ldots,i'_n} = \emptyset$ for each pair $(i_0, \ldots, i_n)$, $(i'_0, \ldots, i'_n)$ of different admissible $(n+1)$-tuples of indices for $n = 0, 1, \ldots$;

(12) $\varphi^{-1}(A_{i_0,\ldots,i_n}) = \bigcup_{i_{n+1}=1}^{N} A_{i_0,\ldots,i_n,i_{n+1}}$;

(13) φ is a diffeomorphism of $A_{i_0,\ldots,i_{n+1}}$ onto $A_{i_0,\ldots,i_n}$;

(14) $\operatorname{diam}(A_{i_0,\ldots,i_n}) < \delta/c_1^n$;

(15) $d_1(\varphi(x), \varphi(y)) \geqslant c_1 d_1(x, y)$ for each pair of points x, y belonging to $A_{i_0,\ldots,i_{n+1}}$ where $n = 0, 1, \ldots$

The above sets will be defined by induction. First we shall define A_{i_0} for $i_0 = 1, 2, \ldots, k$. For this purpose let us remark that there exists a cover $\{B_i\}_{1\leqslant i\leqslant k}$ of M such that B_i are open balls of radii not greater than $\delta/2$ and such that $\mu_1(\mathrm{Fr}(B_i)) = 0$ for $i = 1, 2, \ldots, k$. Then the open sets defined as

$$A_1 = B_1 \text{ and } A_s = B_s - \bigcup_{j=1}^{s-1} \bar{B}_j \quad \text{for } s = 2, 3, \ldots, k$$

have the required properties if one rejects empty sets. Let us now assume that the sets $A_{i_0,\ldots,i_j}$ have been defined for $1\leqslant i_0\leqslant k, 1\leqslant i_1\leqslant N, \ldots$ $\ldots, 1\leqslant i_j\leqslant N$, where $j = 0, 1, \ldots, n$. If $(i_0, \ldots, i_n)$ is any admissible $(n+1)$-tuple of indices, then in view of (14) there exists an $r_{i_0,\ldots,i_n}$ $(1\leqslant r_{i_0,\ldots,i_n}\leqslant p)$ such that $A_{i_0,\ldots,i_n} \subset U_{r_{i_0,\ldots,i_n}}$. On account of (8) $\varphi^{-1}(A_{i_0,\ldots,i_n})$ is equal to $\bigcup_{i_{n+1}=1}^{N} A_{i_0,\ldots,i_n,i_{n+1}}$ where $A_{i_0,\ldots,i_n,i_{n+1}}$ $(i_{n+1} = 1, \ldots$ $\ldots, N)$ are the intersections of $\varphi^{-1}(A_{i_0,\ldots,i_n})$ with the components of $\varphi^{-1}(U_{r_{i_0,\ldots,i_n}})$. It is easy to see that (11) and (13) are satisfied if one replaces n by $n+1$, and in view of (9) one obtains (15) and therefore (14) for $A_{i_0,\ldots,i_n,i_{n+1}}(i_{n+1} = 1, \ldots, N)$. Now we shall prove that there exists a finite real number β such that

$$\frac{(D\varphi^n)(y)}{(D\varphi^n)(x)} \leqslant \beta \tag{16}$$

for $x, y \in A_{i_0,\ldots,i_n}$, $n = 1, 2, \ldots$ For this purpose let us assume that $x, y \in A_{i_0,\ldots,i_n}$, where n is any positive integer. Then, in view of the chain rule for scalar derivatives, one obtains

$$\begin{aligned} \frac{(D\varphi^n)(y)}{(D\varphi^n)(x)} &= \prod_{i=0}^{n-1} \frac{(D\varphi)\big(\varphi^i(y)\big)}{(D\varphi)\big(\varphi^i(x)\big)} \\ &\leqslant \prod_{i=0}^{n-1} \left(1 + \frac{\big|(D\varphi)\big(\varphi^i(x)\big) - (D\varphi)\big(\varphi^i(y)\big)\big|}{(D\varphi)\big(\varphi^i(x)\big)}\right) \\ &\leqslant \exp\left\{\sum_{i=0}^{n-1} \frac{\big|(D\varphi)\big(\varphi^i(x)\big) - (D\varphi)\big(\varphi^i(y)\big)\big|}{(D\varphi)\big(\varphi^i(x)\big)}\right\}. \end{aligned} \tag{17}$$

On account of Lemma 3 inequality (17) implies that

$$\frac{(D\varphi^n)(y)}{(D\varphi^n)(x)} \leqslant \exp\left\{\frac{L}{\gamma} \sum_{i=0}^{n-1} d_1\big(\varphi^i(x), \varphi^i(y)\big)\right\},$$

where $\gamma = \inf_{x \in M} (D\varphi)(x)$, and L is the constant given by Lemma 3. From (15) it follows that

$$d_1\big(\varphi^{i+1}(x), \varphi^{i+1}(y)\big) \geqslant c_1 d_1\big(\varphi^i(x), \varphi^i(y)\big)$$

for $i = 0, 1, \ldots, n-1$. This implies that

$$d_1\big(\varphi^{n-1}(x), \varphi^{n-1}(y)\big) \geqslant c_1^{n-1-i} d_1\big(\varphi^i(x), \varphi^i(y)\big) \tag{18}$$

for $i = 0, 1, \ldots, n-1$. From (18) it follows that (16) is satisfied with the constant β equal to

$$\exp\left\{\frac{L}{\gamma} \operatorname{diam}(M) \frac{c_1}{c_1 - 1}\right\}.$$

Now we may prove (7). For this purpose let us remark that in view of (11), (12) and (13) it follows that

$$\mu_1\big(\varphi^{-n}(A)\big) = \sum_{\substack{1 \leqslant i_1 \leqslant N \\ \cdots\cdots\cdots \\ 1 \leqslant i_n \leqslant N}} \int_A (D\varphi^{-n}_{i_0,\ldots,i_n})(y)\, d\mu_1(y), \tag{19}$$

where A is any Borel set contained in A_{i_0} $(1 \leqslant i_0 \leqslant k)$, and $D\varphi^{-n}_{i_0,\ldots,i_n}$ denotes the scalar derivative of the inverse map $\varphi^{-n}_{i_0,\ldots,i_n}$ to $\varphi^n / A_{i_0,\ldots,i_n}$.

But (16) implies

$$\frac{\sup_{x\in A_{i_0}} (D\varphi_{i_0,\ldots,i_n}^{-n})(x)}{\inf_{x\in A_{i_0}} (D\varphi_{i_0,\ldots,i_n}^{-n})(y)} \leqslant \beta. \tag{20}$$

From (19) and (20) one obtains

$$\mu_1(\varphi^{-n}(A)) \leqslant \beta\cdot\mu_1(A) \sum_{\substack{1\leqslant i_1\leqslant N\\ \cdots\cdots\cdots\\ 1\leqslant i_n\leqslant N}} \inf_{x\in A_{i_0}} (D\varphi_{i_0,\ldots,i_n}^{-n})(x).$$

Since, in view of (24),

$$\mu_1(\varphi^{-n}(A_{i_0})) \geqslant \mu_1(A_{i_0}) \sum_{\substack{1\leqslant i_1\leqslant N\\ \cdots\cdots\cdots\\ 1\leqslant i_n\leqslant N}} \inf_{x\in A_{i_0}} (D\varphi_{i_0,\ldots,i_n}^{-n})(x),$$

taking into account (10) we find that (8) is satisfied with a constant a equal to $\beta\mu_1(M)/\min_{1\leqslant i_0\leqslant K}\mu_1(A_{i_0})$. Thus the lemma is completely proved.

LEMMA 5. *For each open non-empty set $U \subset M$ there exists a positive integer n_0 such that $\varphi^{n_0}(U) = M$.*

Proof. Let (M_*, π) denote the universal covering of M. Since π is a regular map of class C^∞, the Riemannian metric $\|\cdot\|_*$, defined on M_* as $\|\mathfrak{a}\|_* = \|(d\pi)(\mathfrak{a})\|_1$ for $\mathfrak{a}\in T(M_*)$ is of class C^∞. Let Γ denote the group of cover transformations of the covering (M_*, π). It is well known that Γ is the group of isometries of the Riemannian manifold $(M_*, \|\cdot\|_*)$. Now, since M is compact, there exsists an open set $Z \subset M_*$ such that

$$\pi(Z) = M \quad\text{and}\quad \operatorname{diam}(Z) = \delta_* < \infty \tag{21}$$

(the diameter of Z is in the metric $d_*(\cdot,\cdot)$ induced on M_* by $\|\cdot\|_*$). Now let us remark that

$$\pi(K_*(x_0, \delta_*)) = M, \tag{22}$$

where $K_*(x_0, \delta_*)$ is any closed ball of radius equal to δ_*. Indeed, in view of (21) there exists an $\bar{x}_0\in Z$ such that $\pi(x_0) = \pi(\bar{x}_0)$. Further, since Γ is transitive on each fibre of the covering (M_*, π), there exists a $g\in\Gamma$ such that $x_0 = g(\bar{x}_0)$. This implies that $g(K_*(\bar{x}_0, \delta_*)) = K_*(x_0, \delta_*)$ and in view of (21) we infer that (22) is satisfied.

Since (M_*, π) is universal, there exists a continuous map $\varphi_*\colon M_* \to M_*$ such that

$$\varphi\pi = \pi\varphi_*. \tag{23}$$

It is easy to see that (M_*, φ_*) is the covering of M_* and φ_* is a regular map of class C^2. Since M_* is simply connected, it follows that φ_* is a homeomorphism, and therefore a diffeomorphism of class C^2 of M_* onto itself. In view of the definition of $\|\cdot\|_*$, we obtain

$$\|(d\varphi_*)(\mathfrak{a})\|_* \geqslant c_1 \|\mathfrak{a}\|_* \tag{24}$$

for $\mathfrak{a} \in T(M_*)$. Since φ_* is a diffeomorphism, (24) implies

$$d_*\big(\varphi_*(x), \varphi_*(y)\big) \geqslant c_1 d_*(x, y) \quad \text{for } x, y \in M_*. \tag{25}$$

Now let us assume that U is any open non-empty set, $U \subset M$. There exists a closed ball $K_*(x_0, r)$ in M_* such that $\pi\big(K_*(x_0, r)\big) \subset U$, $K_*(x_0, r) \subset Z$. Now it suffices to show that

$$\varphi^{n_0}\Big(\pi\big(K_*(x_0, r)\big)\Big) = M \tag{26}$$

for a certain positive integer n_0. In view of (23), (26) is equivalent to

$$\pi\Big(\varphi_*^{n_0}\big(K_*(x_0, r)\big)\Big) = M. \tag{27}$$

To prove (27), let n_0 be such a positive integer that

$$c_1^{n_0} r > \delta_*. \tag{28}$$

Then, on account of (22) and (27) it suffices to show that

$$K_*\big(\varphi_*^{n_0}(x_0), \delta_*\big) \subset \varphi_*^{n_0}\big(K_*(x_0, r)\big). \tag{29}$$

To prove (29) let us assume that, on the contrary, there exists a $y \in K_*\big(\varphi_*^{n_0}(x_0), \delta_*\big) - \varphi_*^{n_0}\big(K_*(x_0, r)\big)$. Then there exists a regular curve $k: \langle 0, 1\rangle \to M_*$ such that $k(0) = \varphi_*^{n_0}(x_0)$, $k(1) = y$ and $L_*(k|_0^1) < \delta_* + \varepsilon$, where $L_*(k|_0^1)$ denote the length of k and ε is such that $c_1^{n_0} r \geqslant \delta_* + \varepsilon$, $\varepsilon > 0$. Further, it is easy to see that there exists a t_0, $0 < t_0 < 1$, such that $k(t_0) \in \mathrm{Fr}\Big(\varphi_*^{n_0}\big(K_*(x_0, r)\big)\Big)$. This implies that $L_*(k|_0^{t_0}) < \delta_* + \varepsilon$ and therefore $d_*\big(\varphi_*^{n_0}(x_0), k(t_0)\big) < \delta_* + \varepsilon$, but in view of formula (25) we have $d_*\big(\varphi_*^{n_0}(x_0), k(t_0)\big) \geqslant \delta_* + \varepsilon$. Thus the proof of the lemma is completed.

The following lemma is, in fact, the theorem on p. 525 in [4], suitably modified for our purposes. Therefore its proof will be omitted.

LEMMA 6. *Let φ be an endomorphism of a Lebesque space (M, μ) with the measure vanishing on points. If*

(i) *there exists a family $\mathscr{A}$ of measurable sets of positive measure such that finite sums of disjoint sets belonging to $\mathscr{A}$ are dense in the space of all measurable sets;*

(ii) *there exist real finite numbers L_1, L_2 and for each $A \in \mathscr{A}$ there exist positive integers n_1, n_2, $n_1 \leqslant n_2$, such that*

(ii$_a$) $\mu\big(\varphi^{n_1}(Z)\big) \leqslant L_1 \dfrac{\mu(Z)}{\mu(A)}$ *for each measurable set $Z \subset A$,*

(ii$_b$) $\mu\big(\varphi^{n_2}(A)\big) = 1$,

(ii$_c$) $\mu(\varphi^{n_2-n_1}(Z)) \leqslant L_2\mu(Z)$ *for each measurable set* $Z \subset \varphi^{n_1}(A)$; *then* φ *is exact.*

Proof of the theorem. We may assume that the measure μ_1 is normalised. From Lemma 4 it follows that

$$\frac{\mu_1(A)+\mu_1(\varphi^{-1}(A))+\ldots+\mu_1(\varphi^{-n+1}(A))}{n} \leqslant a\cdot\mu_1(A) \tag{30}$$

for each Borel set A and $n = 1, 2, \ldots$ From (30), in view of Theorem 9 in [2] on p. 667, it easily follows that the sequence

$$\frac{1}{n}\sum_{i=1}^{n}\mu_1(\varphi^{-i+1}(A))$$

is convergent (see the proof of Theorem 1 on p. 483 in [3]). Let us put

$$\mu(A) = \lim_{n\to\infty}\frac{\mu_1(A)+\ldots+\mu_1(\varphi^{-n+1}(A))}{n}.$$

From Corollary 4 in [2] on p. 160 it follows that μ is a normalised Borel measure on M. It is evident that μ is absolutely continuous with respect to μ_1. It remains to show that μ_1 is absolutely continuous with respect to μ. For this purpose we shall prove that

(31) if $\mu_1(\varphi^{-n}(A_{i_0})) \nrightarrow 0$ for some A_{i_0}, then the condition $\mu(A) = 0$ implies $\mu_1(A) = 0$, where A is a Borel set contained in A_{i_0}.

For this purpose let us assume that $\mu_1(\varphi^{-n}(A_{i_0})) \nrightarrow 0$. Then, in view of Lemma 4, it follows that

$$\inf_n \mu_1(\varphi^{-n}(A_{i_0})) = a_{i_0} > 0. \tag{32}$$

Further, keeping in mind the notation from the proof of Lemma 4, one obtains

$$\begin{aligned}\mu_1(\varphi^{-n}(A)) &\geqslant \mu_1(A)\sum_{\substack{1\leqslant i_1\leqslant N\\ \cdots\cdots\\ 1\leqslant i_n\leqslant N}}\inf_{x\in A_{i_0}}(D\varphi^{-n}_{i_0,\ldots,i_n})(x)\\ &\geqslant \frac{1}{\beta}\mu_1(A)\sum_{\substack{1\leqslant i_1\leqslant N\\ \cdots\cdots\\ 1\leqslant i_n\leqslant N}}\sup_{x\in A_{i_0}}(D\varphi^{-n}_{i_0,\ldots,i_n})(x).\end{aligned} \tag{33}$$

From (32) it follows that

$$\sum_{\substack{1\leqslant i_1\leqslant N\\ \cdots\cdots\\ 1\leqslant i_n\leqslant N}}\sup_{x\in A_{i_0}}(D\varphi^{-n}_{i_0,\ldots,i_n})(x) \geqslant \frac{a_{i_0}}{\mu_1(A_{i_0})}\mu_1(A). \tag{34}$$

From (33) and (34) we find that

$$\mu_1(\varphi^{-n}(A)) \geqslant \frac{1}{\beta}\frac{a_{i_0}}{\mu_1(A_{i_0})}\mu_1(A)$$

for $n = 1, 2, \ldots$ It is easy to see that this completes the proof of (31).

Part (a) of the theorem will be completely proved if we show that (31) is satisfied for each i_0, $1 \leqslant i_0 \leqslant k$. Suppose, on the contrary, that there exists an i_0', $1 \leqslant i_0' \leqslant k$, such that (31) is not satisfied for $A_{i_0'}$. Then, since

$$\mu_1(M) = \sum_{i_0=1}^{k} \mu_1(\varphi^{-n}(A_{i_0})),$$

there exists an i_0'' such that (31) is satisfied for $A_{i_0''}$. Then in view of Lemma 5, there exists a positive integer n_0 such that $\varphi^{n_0}(A_{i_0''}) \supset A_{i_0''}$. Let us put $B = \varphi^{-n_0}(A_{i_0'}) \cap A_{i_0''}$. The set B is open and non-empty; therefore in view of (31)

$$\mu_1(\varphi^{-n}(B)) \nrightarrow 0.$$

But $\varphi^{-n}(B) \subset \varphi^{-n-n_0}(A_{i_0'})$ for $n = 1, 2, \ldots$ This implies that $\mu_1(\varphi^{-n}(B)) \to 0$; thus we obtain a contradiction. This completes the proof of part (a) of the theorem.

Now we shall proceed to part (b). For this purpose it suffices to show that our dynamical system satisfies the hypothesis of Lemma 6 ([2]). To do so let $\mathscr{A}$ be the family of all sets $A_{i_0,\ldots,i_n} (n \geqslant 1)$, from the proof of Lemma 4. Then, to prove that (i) is satisfied, it suffices to show that for each $\varepsilon > 0$ and each open set G there exists a set G_ε, $G_\varepsilon \subset G$, equal to the finite sum of disjoint set belonging to $\mathscr{A}$ such that

$$\mu(G - G_\varepsilon) < \varepsilon. \tag{35}$$

For this purpose let us put $G_k = \{x\colon x \in G,\ d_1(\operatorname{Fr} G, x) > 1/k\}$, $k = 1, 2, \ldots$ Then $G_k \subset G_{k+1}$ for $k = 1, 2, \ldots$ and $\bigcup_{k=1}^{\infty} G_k = G$. Therefore there exists a k_0 such that $\mu(G - G_{k_0}) < \varepsilon$. In view of (14) there exists an n_0 such that

$$d_1(A_{i_0,\ldots,i_n}) < \frac{1}{k_0} \tag{36}$$

for each admissible (n_0+1)-tuple of indices. Now let us put $G_\varepsilon = \bigcup A_{i_0,\ldots,i_n}$, where the sum is over such admissible (n_0+1)-tuples of indices $(i_0, \ldots, i_n)$ that $G_{k_0} \cap A_{i_0,\ldots,i_{n_0}} \neq \emptyset$. Since

$$\mu(M - \bigcup_{(i_0,\ldots,i_{n_0})} A_{i_0,\ldots,i_{n_0}}) = 1$$

([2]) In fact, to apply Lemma 6 one has to complete the measure.

and in view of (36), it is easy to see that (35) is satisfied. Now we shall prove that (ii) is also satisfied. To do so, let us remark that from the proof of part (a) it follows that there exist two real finite numbers a, $\bar{a}$ such that $\mu(A) \leqslant a\mu_1(A)$ and $\mu_1(A) \leqslant \bar{a}\mu(A)$ for each Borel set A. Therefore it suffices to show that (ii) is satisfied when we replace the measure μ by μ_1. For this purpose let us put $L_1 = \beta$, where β is from (16). Further, in view of Lemma 5, it follows that there exists a positive integer s_0 such that $\varphi^{s_0}(A_{i_0}) = M$ for $i_0 = 1, 2, \ldots, k$, where A_{i_0} is from the proof of Lemma 4. Since φ^{s_0} is a local diffeomorphism and in view of the remarks preceding Lemma 4, it easily follows that there exists a finite real number L_2 such that $\mu_1(\varphi^{s_0}(Z)) \leqslant L_2\mu_1(Z)$ for each Borel set Z. Now let us put for each set $A_{i_0,\ldots,i_n}(n \geqslant 1)$ $n_1 = n$ and $n_2 = n + s_0$. Then ($\mathrm{ii_b}$) and ($\mathrm{ii_c}$) are satisfied; to prove ($\mathrm{ii_a}$) it suffices to apply (21) and act as in the proofs of Lemma 4 and (31) (see the proof of the theorem on p. 525 in [4]). Thus the proof of part (b) is completed.

It remains to prove (c). In fact, in view of (a) and (b), the part (c) follows from the well-known theorem which states that any invariant normalized measure absolutely continuous with respect to a normalized invariant ergodic measure is equal to this measure.

Added in proof. After the paper was submitted for publication, the paper [7] came to our attention. As we understand, there is announced the following result: for each expanding mapping of the compact manifold M into itself there exists an invariant regular Borel measure μ positive on each open set in M.

References

[1] Д. В. Аносов, *Геодезические потоки на замкнутых римановых многообразиях отрицательной кривизны*, Труды математ. инст. им. В. А. Стеклова 90 (1967).

[2] N. Dunford and T. Schwartz, *Linear operators (I)*, 1958.

[3] A. Rényi, *Representations for real numbers and their ergodic properties*, Acta Math. Acad. Scient. Hung. 8 (1957), p. 477-493.

[4] В. А. Рохлин, *Точные эндоморфизмы пространств Лебега*, Изв. Акад. Наук СССР, Серия мат., 25 (1961), 499-530.

[5] M. Shub, Thesis, University of California, Berkeley 1967.

[6] S. Smale, *Differentiable dynamical systems*, Bull. Amer. Math. Soc. 73 (1967), p. 747-817.

[7] A. Avez, *Propriétés ergodiques des endomorphismes dilatants des variétés compactes*, C. R. Acad. Sc. Paris 266 (1968), sér. A, p. 610-612.

Reçu par la Rédaction le 28. 5. 1968

TRANSACTIONS OF THE
AMERICAN MATHEMATICAL SOCIETY
Volume 186, December 1973

ON THE EXISTENCE OF INVARIANT MEASURES FOR PIECEWISE MONOTONIC TRANSFORMATIONS(1)

BY

A. LASOTA AND JAMES A. YORKE

ABSTRACT. A class of piecewise continuous, piecewise C^1 transformations on the interval $[0, 1]$ is shown to have absolutely continuous invariant measures.

1. **Introduction.** The purpose of this note is to prove the existence of absolutely continuous invariant measures for a class of point-transformations of the unit interval [0, 1] into itself. Our main result is Theorem 1 which generalizes some previous results of A. Rényi [5], A. O. Gel′fond [2], W. Parry [4] and A. Lasota [3]. It gives, also, a positive answer to a conjecture of S. Ulam [7, p. 74]. Theorem 1 is stated for a piecewise monotonic function with a finite number of discontinuities but it can be easily extended to some piecewise monotonic functions with infinite number of discontinuities.

Our method is different from the methods of the above mentioned authors. Firstly we explore the fact that the Frobenius-Perron operator corresponding to the point-transformation under consideration has the property of sometimes shrinking the variation of the function. Secondly to prove the existence of invariant measures we use the abstract ergodic theorem which enables us to make our proofs constructive. The advantage of this method is that we do not require that our mappings be local homeomorphisms nor that they generate an exact endomorphism in the sense of Rohlin [6], a property that has been the typical requirement for previous work. §4 describes some extensions, including an extension to higher dimensions.

2. **Existence theorem.** Denote by $(L_1, \| \|)$ the space of all integrable functions defined on the interval [0, 1]. Lebesgue measure on [0, 1] will be denoted by m. Let τ: $[0, 1] \to [0, 1]$ be a measurable nonsingular transformation.

Received by the editors November 9, 1972 and, in revised form, February 13, 1973.

AMS (MOS) subject classifications (1970). Primary 28A65.

Key words and phrases. Frobenius-Perron operator, invariant measures.

(1) The research of both authors was partially supported by the National Science Foundation under grant GP-31386x.

"Nonsingularity" means that $m(\tau^{-1}(A)) = 0$ whenever $m(A) = 0$ for a measurable set A. Given τ we define the Frobenius-Perron operator $P_\tau\colon L_1 \to L_1$ by the formula

$$P_\tau f(x) = \frac{d}{dx} \int_{\tau^{-1}([0,x])} f(s)\, ds.$$

It is well known that the operator P_τ is linear and continuous and satisfies the following conditions:

(a) P_τ is positive: $f \geq 0 \Rightarrow P_\tau f \geq 0$;

(b) P_τ preserves integrals

$$\int_0^1 P_\tau f\, dm = \int_0^1 f\, dm, \qquad f \in L_1;$$

(c) $P_{\tau^n} = P_\tau^n$ (τ^n denotes the nth iterate of τ);

(d) $P_\tau f = f$ if and only if the measure $d\mu = f\,dm$ is invariant under τ, that is $\mu(\tau^{-1}(A)) = \mu(A)$ for each measurable A.

A transformation $\tau\colon [0, 1] \to R$ will be called *piecewise* C^2, if there exists a partition $0 = a_0 < a_1 < \cdots < a_p = 1$ of the unit interval such that for each integer i ($i = 1, \ldots, p$) the restriction τ_i of τ to the open interval (a_{i-1}, a_i) is a C^2 function which can be extended to the closed interval $[a_{i-1}, a_i]$ as a C^2 function. τ need not be continuous at the points a_i.

Theorem 1. *Let $\tau\colon [0, 1] \to [0, 1]$ be a piecewise C^2 function such that* $\inf |\tau'| > 1$. *Then for any $f \in L_1$ the sequence*

$$\frac{1}{n} \sum_{k=0}^{n-1} P_\tau^k f$$

is convergent in norm to a function $f^* \in L_1$. The limit function has the following properties:

(1) $f \geq 0 \Rightarrow f^* \geq 0$.

(2) $\int_0^1 f^* dm = \int_0^1 f\, dm$.

(3) $P_\tau f^* = f^*$ *and consequently the measure $d\mu^* = f^* dm$ is invariant under τ.*

(4) *The function f^* is of bounded variation; moreover, there exists a constant c independent of the choice of initial f such that the variation of the limiting f^* satisfies the inequality*(2)

$$\bigvee_0^1 f^* \leq c\|f\|.$$

(2) Here and in what follows the symbol $\bigvee_a^b f$ as well as $\bigvee_{[a,b]} f$ denote the varition of f over the closed interval $[a, b]$.

We point out in Theorem 3 that it is sufficient to assume just that some iterate of τ satisfy the derivative condition.

Proof. Write $s = \inf|\tau'|$ and choose a number N such that $s^N > 2$. It is easy to see that the function $\phi = \tau^N$ is piecewise C^2. Denote by $b_0, \cdots, b_q$ the corresponding partition for ϕ. Writing ϕ_i for the corresponding C^2 functions we have

$$(5) \qquad |\phi_i'(x)| \geq s^N, \quad x \in [b_{i-1}, b_i],\ i = 1, \cdots, q.$$

Computing the Frobenius-Perron operator for ϕ we obtain

$$(6) \qquad P_\phi f(x) = \sum_{i=1}^{q} f(\psi_i(x))\sigma_i(x)\chi_i(x)$$

where $\psi_i = \phi_i^{-1}$, $\sigma_i(x) = |\psi_i'(x)|$ and χ_i is the characteristic function of the interval $J_i = \phi_i([b_{i-1}, b_i])$. From (5) it follows that

$$(7) \qquad |\sigma_i(x)| \leq s^{-N}, \quad x \in J_i,\ i = 1, \cdots, q.$$

By its very definition the operator P_ϕ is defined as a mapping from L_1 into L_1 but the formula (6) enables us to consider P_ϕ as a map from the space of functions defined on $[0, 1]$ into itself.

Let f be a given function of bounded variation over $[0, 1]$. From (6) and (7) it follows that

$$(8) \qquad \bigvee_0^1 P_\phi f \leq \sum_{i=1}^{q} \bigvee_{J_i} (f \circ \psi_i)\sigma_i + s^{-N} \sum_{i=1}^{q} (|f(b_{i-1})| + |f(b_i)|).$$

In order to evaluate the first sum we write

$$\bigvee_{J_i} (f \circ \psi_i)\sigma_i = \int_{J_i} |d((f \circ \psi_i)\sigma_i)|$$

$$\leq \int_{J_i} |f \circ \psi_i||\sigma_i'|\, dm + \int_{J_i} \sigma_i |d(f \circ \psi_i)|$$

$$\leq K \int_{J_i} |f \circ \psi_i|\sigma_i\, dm + s^{-N} \int_{J_i} |d(f \circ \psi_i)|$$

where $K = \max|\sigma_i'|/\min(\sigma_i)$. Changing the variables we obtain

$$(9) \qquad \bigvee_{J_i} (f \circ \psi_i)\sigma_i \leq K \int_{b_{i-1}}^{b_i} |f|\, dm + s^{-N} \int_{b_{i-1}}^{b_i} |df|.$$

In order to evaluate the second term in (8) we write

$$(10) \qquad |f(b_{i-1})| + |f(b_i)| \leq \bigvee_{b_{i-1}}^{b_i} f + 2d_i$$

where $d_i = \inf\{|f(x)| : x \in [b_{i-1}, b_i]\}$. On the other hand we have an obvious inequality

$$d_i \leq b^{-1} \int_{b_{i-1}}^{b_i} |f|\, dm \tag{11}$$

where $b = \min_i (b_i - b_{i-1})$. From (10), (11) it follows that

$$\sum_{i=1}^{q} (|f(b_{i-1})| + |f(b_i)|) \leq \bigvee_0^1 f + 2b^{-1}\|f\|. \tag{12}$$

Applying (12) and (9) to (8) we obtain $\bigvee_0^1 P_\phi f \leq \alpha\|f\| + \beta \bigvee_0^1 f$ where $\alpha = (K+2b^{-1})$ and $\beta = 2s^{-N} < 1$.

Now, for the same function f, let us write $f_k = P_\tau^k f$. Since $P_\tau^N = P_\phi$ we have

$$\bigvee_0^1 f_{Nk} \leq \alpha\|f_{N(k-1)}\| + \beta \bigvee_0^1 f_{N(k-1)} \leq \alpha\|f\| + \beta \bigvee_0^1 f_{N(k-1)}$$

and consequently

$$\limsup_{k\to\infty} \bigvee_0^1 f_{Nk} \leq \alpha(1-\beta)^{-1}\|f\|. \tag{13}$$

The last inequality and the condition $\|f_k\| \leq \|f\|$ (which follows from (a) and (b)) prove that the set $C = \{f_{Nk}\}_{k=0}^{\infty}$ is relatively compact in L_1. Since $\{f_k\}_{k=0}^{\infty} \subset \bigcup_{k=0}^{N-1} P_\tau^k C$, the whole sequence $\{f_k\}_{k=0}^{\infty}$ is relatively compact, too. By Mazur's theorem the same is true for the sequence

$$\left\{\frac{1}{n} \sum_{k=0}^{n-1} P_\tau^k f\right\}. \tag{14}$$

The set of functions of bounded variation is dense in L_1. We have proved that for any such function f the sequence (14) is relatively compact. Therefore, we are in a position to use the Kakutani-Yosida Theorem (see [1, VIII.5.3]) which says that for any $f \in L_1$ the sequence (14) converges strongly to a function f^* which is invariant under P_τ. From (a) and (b) it follows that f^* satisfies (1) and (2). Therefore it remains only to prove (4). Since the operator P_τ is given by a formula analogous to (6) it is easy to derive the inequality $\bigvee_0^1 P_\tau f \leq c_1 \bigvee_0^1 f + c_2\|f\|$ with some constants c_1 and c_2. This and (13) imply the inequality

$$\limsup_{k\to\infty} \bigvee_0^1 P_\tau^k f \leq c\|f\|$$

(with a positive constant c) which is valid for any f with bounded variation. Consequently for any such f we have also

$$\limsup_{k\to\infty} \bigvee_0^1 \left(\frac{1}{n}\sum_{k=1}^{n-1} P_\tau^k f\right) \le c\|f\|.$$

Writing $Q = \lim_n (1/n)\sum_{k=1}^{n-1} P_\tau^k$ and using Helly's theorem we have $\bigvee_0^1 Qf \le c\|f\|$, for f of bounded variation. The operator Q is linear and contractive. We may therefore apply Helly's theorem once more to extend this inequality for the closure of the set of functions of bounded variation, that is to all of L_1. This finishes the proof.

3. **A counterexample.** Now we shall show that our assumption $\inf|\tau'| > 1$ is essential. Consider the transformation

$$\gamma(x) = \begin{cases} x/(1-x) & \text{for } 0 \le x < \tfrac{1}{2}, \\ 2x - 1 & \text{for } \tfrac{1}{2} \le x \le 1 \end{cases}$$

for which the assumption $|\gamma'(x)| > 1$ is violated only at $x = 0$. We are going to prove that for any $f \in L^1$ the sequence $P_\gamma^n f$ converges in measure to zero. Therefore the equation $P_\gamma f = f$ has only the trivial solution and there is no absolutely continuous nontrivial measure invariant under γ.

The proof will be given in a few steps. First we prove that for $f_0 \equiv 1$ the sequence $g_n(x) = xf_n(x)$, where $f_n = P_\gamma^n f_0$, converges to a constant c_0. Then using the condition $\|f_n\| = 1$ we derive easily that $c_0 = 0$, and consequently $f_n \to 0$. Finally by an approximation argument we may extend this result to an arbitrary sequence $P_\gamma^n f$ with $f \in L_1$.

The Frobenius-Perron operator P_γ may be written in the form

$$P_\gamma f(x) = \frac{1}{(1+x)^2} f\left(\frac{x}{1+x}\right) + \frac{1}{2} f\left(\frac{1}{2} + \frac{x}{2}\right).$$

Thus for g_n we have the following recursive formula:

$$(15) \qquad g_{n+1}(x) = \frac{1}{1+x} g_n\left(\frac{x}{1+x}\right) + \frac{x}{1+x} g_n\left(\frac{1}{2} + \frac{x}{2}\right), \qquad g_0(x) \equiv x.$$

By an induction argument it is easy to check that $g_n' \ge 0$ for each n. Therefore all the functions g_n are positive and increasing. According to (15) we have

$$g_{n+1}(1) = \tfrac{1}{2} g_n(\tfrac{1}{2}) + \tfrac{1}{2} g_n(1) \le g_n(1).$$

This proves the existence of a limit $\lim_n g_n(1) \overset{\mathrm{df}}{=} c_0$. Write $z_0 = 1$ and $z_{k+1} = z_k/(1+z_k)$. According to (15) we obtain

$$g_{n+1}(z_k) = \frac{1}{1+z_k} g_n(z_{k+1}) + \frac{z_k}{1+z_k} g_n\left(\frac{1}{2} + \frac{z_k}{2}\right).$$

Fix k and suppose that $\lim_n g_n(x) = C_0$ for $z_k \le x \le 1$. (This is certainly true for $k = 0$.) Since $z_k \le \tfrac{1}{2} + \tfrac{1}{2} z_k$, we obtain at the limit as $n \to \infty$

$$C_0 = \frac{1}{1+z_k} \lim_n g_n(z_{k+1}) + \frac{z_k}{1+z_k} C_0.$$

Thus $\lim_n g_n(z_{k+1}) = C_0$. Since g_n are increasing, this proves that $\lim_n g_n(x) = C_0$ uniformly for all $x \in [z_{k+1}, 1]$. Therefore by an induction argument it follows that $\lim_n g_n(x) = C_0$ in any interval $[z_k, 1]$ and consequently, since $\lim_k z_k = 0$, we have $\lim_n g_n(x) = C_0$ for all $0 < x \leq 1$. Hence, $\lim_n f_n(x) = c_0/x$. We claim that $c_0 = 0$. If not there would exist $\epsilon > 0$ such that $\int_\epsilon^1 c_0/x\, dx > 1$ and

$$\lim_n \int_\epsilon^1 f_n(x)\, dx = \int_\epsilon^1 \frac{c_0}{x}\, dx > 1$$

which is impossible since $\|f_n\| = 1$ for each n. It can be easily proved by induction that each of the functions f_n is decreasing. Thus the convergence of f_n to zero is uniform on any interval $[\epsilon, 1]$ with $\epsilon > 0$.

Now let f be an arbitrary function. We may write $f = f^+ - f^-$ where $f^+ = \max(0, f)$ and $f^- = \max(0, -f)$. Given $\epsilon > 0$ consider a constant r such that

$$\int_0^1 (f^- - r)^+ dm + \int_0^1 (f^+ - r)^+ dm \leq \epsilon.$$

We have

$$\int_\epsilon^1 |P_\gamma^n f|\, dm = \int_\epsilon^1 P_\gamma^n f^+ dm + \int_\epsilon^1 P_\gamma^n f^- dm$$

$$\leq 2\int_\epsilon^1 P_\gamma^n r\, dm + \int_\epsilon^1 P^n(f^+ - r)\, dm + \int_\epsilon^1 P^n(f^- - r)\, dm$$

$$\leq 2r \int_\epsilon^1 P_\gamma^n 1\, dm + \epsilon.$$

Since $P_\gamma^n 1$ converges on $[\epsilon, 1]$ uniformly to zero we have

$$\lim_n \int_\epsilon^1 |P_\gamma^n f|\, dm = 0 \qquad \text{for } \epsilon > 0$$

which proves that the sequence $P_\gamma^n f$ converges in measure to zero.

4. Final remarks. Now we want to discuss some extensions of our method to other transformations. First of all we may prove an analogue of Theorem 1 for piecewise C^2 transformations with a countable number of pieces.

Let $\tau_i\colon \Delta_i \to [0, 1]$ be a countable sequence of C^2 functions where Δ_i is a sequence of closed intervals such that $\Sigma_i\, m(\Delta_i) = 1$, $m([0, 1] - \bigcup_i \Delta_i) = 0$. The function τ defined by the condition

$$\tau(x) = \tau_i(x), \qquad x \in \text{interior of } \Delta_i,$$

will be called countably piecewise C^2. Note that the values of τ on the set $[0, 1] \setminus \bigcup_i \operatorname{int} \Delta_i$ are arbitrary.

Theorem 2. *Let τ be a countably piecewise C^2 function such that*

(16) $$\inf|\tau'(x)| > 2, \qquad \sup|\tau''(x)| < \infty,$$

(17) $$\tau_i(\Delta_i) = [0, 1] \quad \textit{except for a finite number of intervals.}$$

Then for each $f \in L_1$ the sequence $(1/n)\sum_{k=0}^{n-1} P_\tau^k f$ is convergent in norm to a function f^* which satisfies conditions (1), (2), (3) and (4).

The proof of Theorem 2 is basically the same as the proof of Theorem 1. Thus it can be omitted. Let us only note that the condition (17) is essential. In fact it is easy to construct a countably piecewise linear function with the slope $\tau' > 3$ such that $\inf_{\epsilon \le x \le 1-\epsilon} \tau(x) - x$ is a positive number for each ϵ in $(0, 1/2)$. (The graph of τ lies over the diagonal.) It can be proved by elementary calculation that for any such function τ and $f \in L_1$, $P_\tau^n f \longrightarrow 0$ in measure as $n \longrightarrow \infty$.

A close look at the proof of Theorem 1 shows that we have used only the fact that $\sup|(\tau^N)'| > 2$. Therefore, in fact, we have proved the following result.

Theorem 3. *Let τ: $[0, 1] \longrightarrow [0, 1]$ be a piecewise C^2 function such that $\inf|(\tau^{n_0})'| > 1$ for a positive integer n_0. Then for any $f \in L_1$ the sequence $(1/n)\sum_{k=0}^{n-1} P_\tau^k f$ is convergent in norm to a function f^* which satisfies conditions* (1), (2) *and* (3). *If, in addition,* $\inf|\tau'| > 0$ *then condition* (4) *is also satisfied.*

Observe that in our counterexample the function γ has the property that $(\gamma^n)'_{x=0} = 0$ for each n. This is because the point $(0, \gamma(0))$ lies on the diagonal.

Our techniques can be easily used to obtain new proofs of known results in higher dimensions. See [8], [9], [10] for such results. In this case τ: $M \longrightarrow M$ is assumed C^1 on a compact manifold M and the variation of a C^1 function f is defined as $\int_M |\operatorname{grad} f(m)|\, dm$. Hence in this case we do not allow discontinuities in τ, or more generally if f is C^1 on $M \setminus \partial M$, we must make assumptions on τ guaranteeing $P_\tau(f)$ is C^1 on $M \setminus \partial M$. The techniques in [8], [9], [10] are quite different from the "bounded variation" approach of this paper.

The study of the functions τ described arose while investigating the design of more durable high speed oil well drilling bits. The invariant measure $f(x)\,dx$ describes the distribution of impacts on the surface of the bit. The durability and efficiency of the tool depends strongly on f. The first author is part of a team that has obtained patents in Poland for superior bits by slightly altering the bit shape to one with a better impact distribution f.

REFERENCES

1. N. Dunford and J. T. Schwartz, *Linear operators*. I. *General theory*, Pure and Appl. Math., vol. 7, Interscience, New York, 1958. MR 22 #8302.

2. A. O. Gel′fond, *A common property of number systems*, Izv. Akad. Nauk Ser. Mat. SSSR 23 (1959), 809–814. (Russian) MR 22 #702.

3. A. Lasota, *Invariant measures and functional equations*, Aequationes Math. (in press).

4. W. Parry, *On the β-expansion of real numbers*, Acta Math. Acad. Sci. Hungar. 11 (1960), 401–416. MR 26 #288.

5. A. Rényi, *Representation for real numbers and their ergodic properties*, Acta Math. Acad. Sci. Hungar. 8 (1957), 477–493. MR 20 #3843.

6. V. A. Rohlin, *Exact endomorphisms of Lebesgue spaces*, Izv. Akad. Nauk Ser. Mat. SSSR 25 (1961), 499–530; English transl., Amer. Math. Soc. Transl. (2) 39 (1964), 1–36. MR 26 #1423.

7. S. M. Ulam, *A collection of mathematical problems*, Interscience Tracts in Pure and Appl. Math., no. 8, Interscience, New York, 1960. MR 22 #10884.

8. M. S. Waterman, *Some ergodic properties of multidimensional F-expansions*, Z. Wahrscheinlichkeitstheorie und Verw. Gebiete 16 (1970), 77–103. MR 44 #173.

9. A. Avez, *Propriétés ergodiques des endomorphismes dilatants des variétés compactes*, C. R. Acad. Sci. Paris Sér. A–B 266 (1968), A610–A612. MR 37 #6944.

10. K. Krzyzewski and W. Szlenk, *On invariant measures for expanding differentiable mappings*, Studia Math. 33 (1969), 83–92. MR 39 #7067.

DEPARTMENT OF MATHEMATICS, JAGELLONIAN UNIVERSITY, CRACOW, POLAND

INSTITUTE FOR FLUID DYNAMICS AND APPLIED MATHEMATICS, UNIVERSITY OF MARYLAND, COLLEGE PARK, MARYLAND 20742

Inventiones math. 29, 181–202 (1975)

The Ergodic Theory of Axiom A Flows

Rufus Bowen* (Berkeley) and David Ruelle** (Bures-sur-Yvette)

1. Introduction

Let M be a compact (Riemann) manifold and (f^t): $M \to M$ a differentiable flow. A closed (f^t)-invariant set $\Lambda \subset M$ containing no fixed points is hyperbolic if the tangent bundle restricted to Λ can be written as the Whitney sum of three (Tf^t)-invariant continuous subbundles

$$T_\Lambda M = E + E^s + E^u$$

where E is the one-dimensional bundle tangent to the flow, and there are constants $c, \lambda > 0$ so that

(a) $\|Tf^t(v)\| \leqq c\, e^{-\lambda t} \|v\|$ for $v \in E^s$, $t \geqq 0$ and

(b) $\|Tf^{-t}(v)\| \leqq c\, e^{-\lambda t} \|v\|$ for $v \in E^u$, $t \geqq 0$.

We can choose $t_0 > 0$ and change λ so that the above conditions hold with $c = 1$ when $t \geqq t_0$. We can also assume that, for such t, Tf^t (resp. Tf^{-t}) expands E at a smaller rate than it expands any element of E^u (resp. E^s). It is then said that the metric is *adapted* (see [14]) to f^{t_0}. We will always assume that $t_0 \leqq 1$ – this can be achieved by a rescaling of t $(t \to t' = t/t_0)$ which does not affect our main results.

A closed invariant set Λ is a *basic hyperbolic* set if

(a) Λ contains no fixed points and is hyperbolic;

(b) the periodic orbits of $f^t|\Lambda$ are dense in Λ;

(c) $f^t|\Lambda$ is a topologically transitive flow; and

(d) there is an open set $U \supset \Lambda$ with $\Lambda = \bigcap_{t \in \mathbb{R}} f^t U$.

These sets are the building blocks of the Axiom A flows of Smale [27]. We will especially be interested in *attractors*, basic hyperbolic sets Λ for which the U in (d) can be found satisfying $f^t U \subset U$ for all $t \geqq T_0$ (T_0 fixed) and hence $\Lambda = \bigcap_{t \geqq 0} f^t U$. This paper will study the average asymptotic behavior of orbits of points in the neighborhood U of a C^2-attractor.

Precisely we will find an ergodic probability measure μ_φ on a C^2 attractor Λ so that for almost all $x \in U$ w.r.t. Lebesgue measure and all continuous g: $U \to \mathbb{R}$ one has

$$\lim_{T \to \infty} \frac{1}{T} \int_0^T g(f^t x)\, dt = \int g\, d\mu_\varphi \tag{1}$$

* Math. Dept. U.C. Berkeley. Partially supported by NSF GP 14519.
** Institut des Hautes Etudes Scientifiques, Bures-sur-Yvette (France).

(see Theorem 5.1). The measure μ_φ will be described as the unique equilibrium state for a certain function $\varphi=\varphi^{(u)}$ (defined by (2), Section IV) on Λ, i.e. the unique f^t-invariant probability measure μ on Λ which maximizes the expression

$$h_\mu(f^1)+\int \varphi \, d\mu$$

where $h_\mu(f^1)$ is measure theoretic entropy. This variational principle (which is formally identical with one in statistical mechanics [21]) is useful because it gives a description of μ_φ which persists when one lifts μ_φ to a symbol space for closer study.

This paper carries over to flows results previously obtained for diffeomorphisms with regard to equilibrium states [6, 7, 24] and attractors [24]. For Anosov flows ($\Lambda=M$) the measure μ_φ has been studied in [9, 16, 17, 20, 25, 26] and the theory of Gibbs states (a slightly different formalism from equilibrium states which yields the same measures for basic hyperbolic sets) has been developed in [26]. Some results obtained here for flows are new even for diffeomorphisms; this is the case of Theorem 5.6. Results for diffeomorphisms can be obtained from those for flows via suspension (or directly by simplification of the proofs).

The determination of the asymptotic behavior of orbits is a significant problem in the study of differentiable dynamical systems. In particular the asymptotic behavior of solutions of a differential equation is of central interest in physical applications. Here we consider only the case of Axiom A flows. In that case it is known that $f^t x$ often depends in a very sensitive or "unstable" manner on the initial condition x, and (1)—which describes the time-average of an "observable" g—is probably the best way of expressing the asymptotic behavior of $f^t x$. It is a natural problem to extend (1) to non Axiom A situations.

We shall show that $\mu_{\varphi^{(u)}}$ depends continuously on the flow (f^t) (Proposition 5.4). In the same direction, Sinai [26] has proved the stability of μ_φ under small stochastic perturbations for Anosov flows[1]. (1) holds almost everywhere for x in the basin of an Axiom A attractor; one can prove that, for a C^2 Axiom A flow, these basins (and those of point attractors) cover M up to a set of Lebesgue measure zero. Equivalently: if a basic set is not an attractor, its stable manifold has measure zero (Theorem 5.6).

It can be seen that, unless Λ is a periodic orbit, the entropy of μ_φ does not vanish; this indicates "strong ergodic properties" of the system (μ_φ, f^t). In fact, if (f^t) restricted to Λ is C-dense, (μ_φ, f^t) is a Bernoulli flow (see Remark 3.5). The correlation functions

$$\rho_{gg'}(t)=\int (g\circ f^t)\cdot g'\, d\mu_\varphi-\int g\, d\mu_\varphi\cdot\int g'\, d\mu_\varphi$$

are interesting to consider in physical applications. In the C-dense case we have $\lim_{t\to\infty}\rho_{gg'}(t)=0$ if $g, g'\in L^2(\mu_\varphi)$ (Remark 3.5). Assuming that g, g' are C^1, does $\rho_{gg'}(t)$ tend to zero exponentially when $t\to\infty$? The methods of the present paper do not seem capable of answering this question. A positive answer has been obtained for diffeomorphisms ([24, 26]).

---

[1] The corresponding problem for attractors for Axiom A diffeomorphisms has been treated by Kifer (Sinai, private communication).

Terminology

The manifold M and the Riemann metric on M are C^∞. The flow (f^t) is called C^r $(r \geqq 1)$ if it corresponds to a C^r vector field on M; a basic hyperbolic set Λ for (f^t) is then called a C^r basic hyperbolic set. The flow (f^t) restricted to Λ is topologically transitive if it has a dense orbit.

For easy reference, we collect here the definitions of stable manifolds

$$W_x^s = \{y \in M : \lim_{t \to \infty} d(f^t x, f^t y) = 0\}$$

$$W_x^{cs} = \bigcup_{t \in \mathbb{R}} W_{f^t x}^s .$$

A distance on M is defined by

$$\delta_T(x, y) = \sup_{0 \leqq t \leqq T} d(f^t x, f^t y)$$

when $0 \leqq T < \infty$; $B_x(\varepsilon, T)$ is the closed ε-neighbourhood of x for that distance; also

$$W_x^s(\varepsilon) = W_x^s \cap B_x(\varepsilon, \infty).$$

Replacing t by $-t$ and s by u we obtain the definition of unstable manifolds. We also write

$$W_\Lambda^s(\varepsilon) = \bigcup_{x \in \Lambda} W_x^s(\varepsilon), \qquad \text{etc.}$$

The basic hyperbolic set Λ is C-dense if $W_x^s \cap \Lambda$ is dense in Λ for some (hence for all) $x \in \Lambda$.

In general we write $f^* \mu$ the image of a measure μ by a continuous map f.

2. Symbolic Dynamics

Let us recall the symbolic dynamics of a basic hyperbolic set Λ [4]. For $A = [A_{ij}]$ an $n \times n$ matrix of 0's and 1's we define

$$\Sigma_A = \{\mathbf{x} = (x_i)_{i=-\infty}^{+\infty} \in \{1, \ldots, n\}^{\mathbb{Z}} : A_{x_i x_{i+1}} = 1 \, \forall i \in \mathbb{Z}\}$$

and $\sigma_A : \Sigma_A \to \Sigma_A$ by $\sigma_A(\mathbf{x}) = (x_i')_{i=-\infty}^{\infty}$ where $x_i' = x_{i+1}$. If we give $\{1, \ldots, n\}$ the discrete topology and $\{1, \ldots, n\}^{\mathbb{Z}}$ the product topology, then Σ_A becomes a compact metrizable space and σ_A a homeomorphism. σ_A (or Σ_A) is called a *subshift of finite type* if $\sigma_A : \Sigma_A \to \Sigma_A$ is topologically transitive (i.e. for U, V non-empty open sets there is an $n > 0$ with $f^n U \cap V \neq \emptyset$).

For $\psi : \Sigma_A \to \mathbb{R}$ a positive continuous function one can define a special (or suspension) flow as follows. Let

$$Y = \{(\mathbf{x}, s) : s \in [0, \psi(\mathbf{x})], \mathbf{x} \in \Sigma_A\} \subset \Sigma_A \times \mathbb{R} .$$

Identify the points $(\mathbf{x}, \psi(\mathbf{x}))$ and $(\sigma_A(\mathbf{x}), 0)$ for all $\mathbf{x} \in \Sigma_A$ to get a new space $\Lambda(A, \psi)$. Then $\Lambda(A, \psi)$ is a compact metric space (see [8] for a metric) and one can define a flow g^t on $\Lambda(A, \psi)$ by

$$g^t(\mathbf{x}, s) = (\mathbf{x}, s + t) \qquad \text{for } s + t \in [0, \psi(\mathbf{x})]$$

and remembering identifications. More precisely, if $z = q(\mathbf{x}, s)$ where $q: Y \to \Lambda(A, \psi)$ is the quotient map, then $g^t(z) = q(\sigma_A^k \mathbf{x}, v)$ where k is chosen so that

$$v = t + s - \sum_{j=0}^{k-1} \psi(\sigma_A^j \mathbf{x}) \in [0, \psi(\sigma_A^k \mathbf{x})].$$

The flow g^t on $\Lambda(A, \psi)$ will be important to us with ψ satisfying an additional condition. For $\psi: \Sigma_A \to \mathbb{R}$ let

$$\operatorname{var}_n \psi = \sup \{ |\psi(\mathbf{x}) - \psi(\mathbf{y})| : \mathbf{x}, \mathbf{y} \in \Sigma_A, \; x_i = y_i \forall |i| \leqq n \}.$$

Let

$$\mathscr{F}_A = \{ \psi \in C(\Sigma_A) : \exists b > 0, \; \alpha \in (0, 1) \text{ so that } \operatorname{var}_n \psi \leqq b \alpha^n \text{ for all } n \geqq 0 \}.$$

2.1. **Lemma.** *Let Λ be a basic hyperbolic set. Then there is a topologically mixing*[2] *subshift of finite type $\sigma_A: \Sigma_A \to \Sigma_A$, a positive $\psi \in \mathscr{F}_A$ and a continuous surjection $\rho: \Lambda(A, \psi) \to \Lambda$ so that*

$$\begin{array}{ccc} \Lambda(A, \psi) & \xrightarrow{\;g^t\;} & \Lambda(A, \psi) \\ \downarrow \rho & & \downarrow \rho \\ \Lambda & \xrightarrow{\;f^t\;} & \Lambda \end{array}$$

commutes.

This is from [4], Section 2, except for the mixing condition on σ_A. If $\sigma_A: \Sigma_A \to \Sigma_A$ is *not* mixing, then for some $m > 0$ $\Sigma_A = X_1 \cup \cdots \cup X_m$, a disjoint union of closed sets with $\sigma_A(X_i) = X_{i+1}$ and $\sigma_A^m | X_i : X_i \to X_i$ conjugate to a mixing subshift of finite type (see e.g. 2.7 of [2]).

Identifying $\sigma_A^m: X_1 \to X_1$ with some $\sigma_B: \Sigma_B \to \Sigma_B$ and defining $\psi': \Sigma_B \to \mathbb{R}$ by

$$\psi'(x) = \psi(x) + \psi(\sigma_A x) + \cdots + \psi(\sigma_A^{m-1} x)$$

one can see that $\Lambda(B, \psi')$ is homeomorphic to $\Lambda(A, \psi)$ in a natural way and $\psi' \in \mathscr{F}_B$.

There are other properties of the map ρ which we shall recall as we need them. Throughout the remainder of the paper ψ will always denote a positive function in $\mathscr{F}_A$ and σ_A a mixing subshift of finite type.

For any homeomorphism f the set of f-invariant Borel probability measures will be denoted $M(f)$. If $F = (f^t)_{t \in \mathbb{R}}$ is a continuous flow we will write $M(F) = \bigcap_{t \in \mathbb{R}} M(f^t)$.

3. Equilibrium States

Let us review the definition of topological pressure for a homeomorphism $f: X \to X$ of a compact metric space and a continuous function $\varphi: X \to \mathbb{R}$ [23, 28]. For given $\varepsilon > 0$ and $n > 0$, a subset $E \subset X$ is called *(ε, n)-separated* if

$$x, y \in E, \quad x \neq y \Rightarrow d(f^k x, f^k y) > \varepsilon \quad \text{for some } k \in [0, n].$$

[2] A homeomorphism $F: X \to X$ is topologically mixing if, for U, V open nonempty in X, $U \cap F^n V \neq \emptyset$ for all sufficiently large n.

One defines

$$Z_n(f, \varphi, \varepsilon) = \sup\left\{\sum_{x\in E} \exp \sum_{k=0}^{n-1} \varphi(T^k x) \colon E \text{ is } (\varepsilon, n)\text{-separated}\right\}$$

$$P(f, \varphi, \varepsilon) = \limsup_{n\to\infty} \frac{1}{n} \log Z_n(f, \varphi, \varepsilon)$$

and

$$P(f, \varphi) = \lim_{\varepsilon\to 0} P(f, \varphi, \varepsilon).$$

For $\varphi = 0$ the number $P(f, \varphi)$ is just the topological entropy $h(f)$ of f; the theory of topological pressure generalizes that of topological entropy. The main general result is that

$$P(f, \varphi) = \sup_{\mu\in M(f)} \left(h_\mu(f) + \int \varphi \, d\mu\right).$$

This was proved by Walters [28]; for f expansive there is a $\mu \in M(f)$ with $h_\mu(f) + \int \varphi \, d\mu = P(f, \varphi)$.

An *equilibrium state* for $\varphi\colon X \to \mathbb{R}$ with respect to $f\colon X \to X$ is a $\mu \in M(f)$ with $h_\mu(f) + \int \varphi \, d\mu = P(f, \varphi)$, i.e. a $\mu \in M(f)$ maximizing the quantity $h_\mu(f) + \int \varphi \, d\mu$

Now we will consider the case of a flow $F = (f^t\colon X \to X)$ and $\varphi\colon X \to \mathbb{R}$. A set $E \subset X$ is (ε, T)-*separated* if

$$x, y \in E, \quad x \neq y \;\Rightarrow\; d(f^t x, f^t y) > \varepsilon \quad \text{for some } t \in [0, T].$$

Then we define

$$Z_T(F, \varphi, \varepsilon) = \sup\left\{\sum_{x\in E} \exp \int_0^T \varphi(f^t x)\, dt \colon E \text{ is } (\varepsilon, T)\text{-separated}\right\}$$

$$P(F, \varphi, \varepsilon) = \limsup_{T\to\infty} \frac{1}{T} \log Z_T(F, \varphi, \varepsilon)$$

and

$$P(F, \varphi) = \lim_{\varepsilon\to 0} P(F, \varphi, \varepsilon).$$

The definition of $P(F, \varphi)$ is independent of the choice of metric on M. It is a straightforward exercise to check that if one lets $\varphi^1(x) = \int_0^1 \varphi(f^t x)\, dt$, then $P(F, \varphi) = P(f^1, \varphi^1)$; also, for $\mu \in M(F)$ one has $\int \varphi \, d\mu = \int \varphi^1 \, d\mu$. As $M(F) \subset M(f^1)$ one has

$$h_\mu(f^1) + \int \varphi \, d\mu = h_\mu(f^1) + \int \varphi^1 \, d\mu \leqq P(f^1, \varphi^1) = P(F, \varphi)$$

for $\mu \in M(F)$. There is an argument ([28], see also [10], p. 359–360) to show that for any $\mu' \in M(f^1)$ one can find a $\mu \in M(F)$ with

$$h_\mu(f^1) + \int \varphi^1 \, d\mu \geqq h_{\mu'}(f^1) + \int \varphi^1 \, d\mu'.$$

Hence it follows that

$$P(F, \varphi) = \sup_{\mu\in M(F)} \left(h_\mu(f^1) + \int \varphi \, d\mu\right).$$

By an *equilibrium state* for φ (with respect to F) we mean a $\mu \in M(F)$ with

$$h_\mu(f^1) + \int \varphi \, d\mu = P(F, \varphi).$$

For $G=\{g^t\}$ the special flow on $\Lambda(A,\psi)$ there is a well-known bijection between $M(G)$ and $M(\sigma_A)$. For $\nu\in M(\sigma_A)$ and m Lebesgue measure, $\nu\times m$ gives measure 0 to the identifications on $Y\to\Lambda(A,\psi)$ and so $\mu_\nu=(\nu\times m(Y))^{-1}\,\nu\times m|Y$ gives a probability measure on $\Lambda(A,\psi)$. One can check that $\nu\in M(\sigma_A)$ implies $\mu_\nu\in M(G)$ and that $\nu\to\mu_\nu$ defines a bijection $M(\sigma_A)\to M(G)$.

It is known that any function $\gamma\in\mathscr{F}_A$ has a unique equilibrium state ν w.r.t. σ_A [6, 11, 22, 24] and that ν depends continuously on γ (weak topology for ν, uniform topology for γ). We will now state the corresponding condition on φ: $\Lambda(A,\psi)\to\mathbb{R}$ which guarantees a unique equilibrium state.

3.1 **Proposition**[3]. *Let* φ: $\Lambda(P,\psi)\to\mathbb{R}$ *be continuous,* $\Phi(\mathbf{x})=\int_0^{\psi(\mathbf{x})}\varphi(\mathbf{x},t)\,dt$ *and* $c=P(G,\varphi)$. *Assume that* $\Phi\in\mathscr{F}_A$. *Then there is a measure* $\mu_\varphi\in M(G)$ *so that*

(a) μ_φ *is the unique equilibrium state for* φ *with respect to* G.

(b) $\mu_\varphi=\mu_{\nu_0}$ *where* ν_0 *is the unique equilibrium state for* $\Phi-c\psi$ *on* Σ_A.

(c) μ_φ *is ergodic and positive on non-empty open sets and*

(d) *for* $\varepsilon>0$ *there is a* $C_\varepsilon>0$ *so that*

$$\mu_\varphi(B_{x,G}(\varepsilon,T))\geqq C_\varepsilon\exp\left(-cT+\int_0^T\varphi(g^t x)\,dt\right)$$

for all $x\in\Lambda(A,\psi)$, $T\geqq 0$ *where*

$$B_{x,G}(\varepsilon,T)=\{y\in\Lambda(A,\psi)\colon\ d(g^t y,g^t x)\leqq\varepsilon \text{ for all } t\in[0,T]\}.$$

Let $\gamma=\Phi-c\psi$. As $\Phi,\psi\in\mathscr{F}_A$ we have $\gamma\in\mathscr{F}_A$. This guarantees that γ has a unique equilibrium state ν_0. By Fubini's theorem, for any $\nu\in M(\sigma_A)$, $(\nu\times m)(Y)=\int\psi\,d\nu$ and $\int\varphi\,d\mu_\nu=\dfrac{\int\Phi\,d\nu}{\int\psi\,d\nu}$. A theorem of Abramov [1] states that

$$h_{\mu_\nu}(g^1)=\frac{h_\nu(\sigma_A)}{\int\psi\,d\nu}.$$

Hence

$$\begin{aligned} c=P(G,\varphi)&=\sup_{\mu\in M(G)}\left(h_\mu(g^1)+\int\varphi\,d\mu\right)\\ &=\sup_{\nu\in M(\sigma_A)}\frac{h_\nu(\sigma_A)+\int\Phi\,d\nu}{\int\psi\,d\nu}.\end{aligned}$$

Thus $P(\sigma_A,\gamma)=\sup_\nu\left(h_\nu(\sigma_A)+\int(\Phi-c\psi)\,d\nu\right)=0$ with ν attaining the supremum (i.e. $\nu=\nu_0$) precisely if μ_ν is the unique equilibrium state for φ. This shows that $\mu_\varphi=\mu_{\nu_0}$ satisfies (a) and (b); (c) is true because ν_0 has these same properties ([6] or [24], Appendix B).

We now verify (d). Let $x\in\Lambda(A,\psi)$ be represented by $x=(\mathbf{x},t_1)$, $t_1\in[0,\psi(\mathbf{x}))$ and $g^T x=(\sigma_A^n\mathbf{x},t_2)$, where $n=n(x)$ is such that

$$t_2=T+t_1-\sum_{k=0}^{n-1}\psi(\sigma_A^k\mathbf{x})\in[0,\psi(\sigma_A^n\mathbf{x})).$$

[3] Another proof of (a), (b), (d) has now been obtained by E. Franco-Sanchez, Berkeley thesis, 1974.

Given $\varepsilon>0$ one can find $\delta_\varepsilon>0$ and $s_\varepsilon>0$ (not depending on x or T) so that[4]

$$B_{\mathbf{x},G}(\varepsilon,T)\supset\{(\mathbf{y},t)\colon |t-t_1|\leqq s_\varepsilon \text{ and } d(\sigma_A^k\mathbf{y},\sigma_A^k\mathbf{x})\leqq\delta_\varepsilon \text{ for all } k\in[0,n-1]\}.$$

We do not go through the details but do point out that, for any $\alpha>0$, $\psi\in\mathscr{F}_A$ implies that for δ small enough

$$d(\sigma_A^k\mathbf{x},\sigma_A^k\mathbf{y})\leqq\delta\,\forall k\in[0,n-1] \Rightarrow \sum_{k=0}^{n-1}|\psi(\sigma_A^k\mathbf{x})-\psi(\sigma_A^n\mathbf{y})|<\alpha.$$

Then

$$\mu_\varphi B_{\mathbf{x},G}(\varepsilon,T)\geqq\frac{s_\varepsilon}{\int\psi\,d\nu_0}\nu_0\{\mathbf{y}\colon d(\sigma_A^k\mathbf{y},\sigma_A^k\mathbf{x})\leqq\delta_\varepsilon\forall k\in[0,n-1]\}.$$

By [6] Lemma 5, this right side is at least

$$a_\varepsilon\exp\left(\sum_{k=0}^{n-1}\gamma(\sigma_A^k\mathbf{x})-nP(\sigma_A,\gamma)\right)=a_\varepsilon\exp\sum_{k=0}^{n-1}\gamma(\sigma_A^k\mathbf{x})$$

for some $a_\varepsilon>0$. Since

$$\sum_{k=0}^{n-1}\psi(\sigma_A^k\mathbf{x})+t_2=T+t_1$$

and

$$\int_0^{t_1}\varphi(\mathbf{x},t)\,dt+\int_0^T\varphi(g^t x)\,dt-\int_0^{t_2}\varphi(\sigma_A^n\mathbf{x},t)\,dt=\sum_{k=0}^{n-1}\Phi(\sigma_A^k\mathbf{x}),$$

one sees that $\sum_{k=0}^{n}\gamma(\sigma_A^k\mathbf{x})$ differs from $-cT+\int_0^T\varphi(g^t x)\,dt$ by at most $2\|\psi\|(|c|+\|\varphi\|)$.

This proves (d).

3.2. *Remark.* Special flows are simple enough that parts (a) and (b) above could have been derived without appealing so much to general (and harder) results on topological pressure.

3.3 **Theorem.** *Assume that Λ is a basic hyperbolic set for F and that $\varphi\colon\Lambda\to\mathbb{R}$ satisfies a Hölder condition of positive exponent. Then φ has a unique equilibrium state μ_φ. Furthermore, μ_φ is ergodic and positive on non-empty open sets of Λ, and for any $\varepsilon>0$ there is a $C_\varepsilon>0$ so that*

$$\mu_\varphi(B_{x,F|\Lambda}(\varepsilon,T))\geqq C_\varepsilon\exp\left(-P(F|\Lambda,\varphi)\,T+\int_0^T\varphi(f^t x)\,dt\right)$$

for all $x\in\Lambda$, $T\geqq0$.

We apply the preceding proposition to the function $\varphi^*=\varphi\circ\rho$ on $\Lambda(A,\psi)$. There are $b_1>0$ and $\tau\in(0,1)$ so that

$$d(\rho(\mathbf{x},0),\rho(\mathbf{y},0))\leqq b_1\tau^N$$

if $x_i=y_i$ for all $|i|\leqq N$ (see [4], Lemma 2.2.(i)).

[4] In this formula $(\mathbf{y},t)$ has to be replaced by $(\sigma_A^{-1}\mathbf{y},\psi(\sigma_A^{-1}\mathbf{y})+t)$ resp. by $(\sigma_A\mathbf{y},t-\psi(\mathbf{y}))$ when $t<0$ resp. $t>\psi(\mathbf{y})$.

Since F is a differentiable flow, there is a constant b_2 so that

$$d(f^t x, f^t y) \leqq b_2\, d(x, y) \quad \text{provided } t \in [0, \|\psi\|].$$

The Hölder condition on φ states that

$$|\varphi(x) - \varphi(y)| \leqq b_3\, d(x, y)^\alpha \quad \text{with} \quad \alpha > 0.$$

Combining these estimates, when $x_i = y_i \forall i \in [-N, N]$ we have

$$\left| \int_0^{\psi(\mathbf{x})} \varphi^*(\mathbf{x}, t)\, dt - \int_0^{\psi(\mathbf{y})} \varphi^*(\mathbf{y}, t)\, dt \right|$$
$$\leqq \|\varphi\|\, |\psi(\mathbf{x}) - \psi(\mathbf{y})| + \int_0^{\psi(\mathbf{x})} |\varphi(f^t \rho(\mathbf{x}, 0)) - \varphi(f^t \rho(\mathbf{y}, 0))|\, dt$$
$$\leqq \|\varphi\|\, |\psi(\mathbf{x}) - \psi(\mathbf{y})| + \|\psi\|\, b_3 (b_2 b_1)^\alpha (\tau^\alpha)^N.$$

Since $\psi \in \mathscr{F}$, this gives $\Phi^* \in \mathscr{F}$ where $\Phi^*(\mathbf{x}) = \int_0^{\psi(\mathbf{x})} \varphi^*(\mathbf{x}, t)\, dt$. So φ^* has a unique equilibrium state μ_{φ^*} as in the preceding proposition.

We recall that there are closed subsets $A_s = \rho^{-1}(\Delta^s \mathscr{M})$ and $A_u = \rho^{-1}(\Delta^u \mathscr{M})$ of $\Lambda(A, \psi)$ so that (see [4])

(a) $A_s \neq \Lambda(P, \psi) \neq A_u$

(b) $g^t A_s \subset A_s$, $g^{-t} A_u \subset A_u \forall t \geqq 0$ and

(c) ρ is one-to-one off $\bigcup_{t \in \mathbb{R}} g^t (A_u \cup A_s) = \bigcup_{n \in \mathbb{Z}} g^n (A_u \cup A_s)$.

Because μ_{φ^*} is positive on non-empty open sets $\mu_{\varphi^*}(A_s) \neq 1 \neq \mu_{\varphi^*}(A_u)$; since μ_{φ^*} is ergodic and each of these sets is invariant under one direction of time, $\mu_{\varphi^*}(A_s) = 0 = \mu_{\varphi^*}(A_u)$. By (c) then ρ gives a conjugacy of the measurable flows (G, μ_{φ^*}) and $(F(\Lambda, \mu_\varphi))$ where $\mu_\varphi = \rho^* \mu_{\varphi^*}$. In particular $h_{\mu_\varphi}(f^1) = h_{\mu_{\varphi^*}}(g^1)$ and

$$h_{\mu_\varphi}(f^1) + \int \varphi\, d\mu_\varphi = h_{\mu_{\varphi^*}}(g^1) + \int \varphi^*\, d\mu_{\varphi^*} = P(G, \varphi^*).$$

As $F|\Lambda$ is the quotient of G and $\varphi^* = \varphi \circ \rho$, one has $P(F|\Lambda, \varphi) \leqq P(G, \varphi^*)$ (see Walters [28], Theorem 2.2); because $h_{\mu_\varphi}(f^1) + \int \varphi\, d\mu_\varphi = P(G, \varphi^*)$ one has $P(F|\Lambda, \varphi) = P(G, \varphi^*)$ and μ_φ is an equilibrium state for φ. If μ were another equilibrium state for φ, then $\mu = \rho^* \mu'$ for some $\mu' \in M(G)$ (by an easy application of the Hahn-Banach and Markov-Kakutani theorems) and

$$h_{\mu'}(g^1) + \int \varphi^*\, d\mu' \geqq h_\mu(f^1) + \int \varphi\, d\mu = P(F|\Lambda, \varphi) = P(G, \varphi^*).$$

So μ' is an equilibrium state for φ^*, $\mu' = \mu_{\varphi^*}$ and $\mu = \mu_\varphi$. Thus μ_φ is the unique equilibrium state. The remaining properties for μ_φ follow from the corresponding ones for μ_{φ^*} in Proposition 3.1.

3.4 *Remark.* For $\varphi = 0$ the uniqueness of equilibrium state just says that $F|\Lambda$ has a unique invariant measure maximizing entropy. This was proved earlier in [5].

3.5. *Remark.* It was proved in [2] and [24] that (σ_A, ν_0) is isomorphic to a Bernoulli shift where ν_0 is the equilibrium state of $\gamma \in \mathscr{F}_A$. The corresponding result for $(F|\Lambda, \mu_\varphi)$ follows from results proved elsewhere. If $F|\Lambda$ is C-dense (i.e. $W^u(x) \cap \Lambda$ is dense in Λ for every $x \in \Lambda$ where $W^u(x) = \{y\colon d(f^{-t} x, f^{-t} y) \to 0$ as $t \to +\infty\}$), then G is also C-dense and Sinai [26] p. 48–9, applied a theorem of

Gurevič [12] to show that $(G, \mu_{\varphi*})$ is a K-flow. We mention that, although Sinai uses the formalism of Gibbs states instead of equilibrium states, the measure $\mu_{\varphi*}$ is the same as the one he constructs. M. Ratner [20] (also Bunimovič [9]) has proved in this C-dense case that $(G, \mu_{\varphi*})$ is actually Bernoulli (i.e. $(g^t, \mu_{\varphi*})$ is isomorphic to a Bernoulli shift for each $t \neq 0$). Since $(F|\Lambda, \mu_\varphi) \approx (G, \mu_{\varphi*})$, $(F|\Lambda, \mu_\varphi)$ is Bernoulli when $F|\Lambda$ is C-dense and φ is Hölder continuous. In that case we have

$$\lim_{t\to\infty} \int (g \circ f^t) \cdot g' \, d\mu_\varphi = \int g \, d\mu_\varphi \cdot \int g' \, d\mu_\varphi$$

for all $g, g' \in L^2(\mu_\varphi)$. [This follows from the fact that (μ_φ, f^t) is equivalent to a Bernoulli shift for each t, and the continuity of the flow (f^t)].

4. Attractors

Now assume Λ is a C^2 basic hyperbolic set. For $x \in \Lambda$ let $\lambda_t(x)$ be the Jacobian of the linear map $D f^t \colon E^u_x \to E^u_{f^t x}$ using inner products induced by the Riemannian metric.

Define

$$\varphi^{(u)}(x) = -\frac{d \ln \lambda_t(x)}{dt}\bigg|_{t=0} = -\frac{d\lambda_t(x)}{dt}\bigg|_{t=0} \tag{2}$$

which exists and depends differentiably on E^u_x (hence continuously on x) as f^t is a C^2 flow. Since $\lambda_{T+t}(x) = \lambda_t(f^T x)\, \lambda_T(x)$ one has

$$-\ln \lambda_t(f^T x) = -\ln \lambda_{T+t}(x) + \ln \lambda_T(x)$$

and so

$$\varphi^{(u)}(f^T x) = -\frac{d \ln \lambda_s(x)}{ds}\bigg|_{s=T}.$$

This implies that

$$\int_0^T \varphi^{(u)}(f^t x)\, dt = -\ln \lambda_T(x).$$

This integral is the one appearing in Theorem 3.3 for $\varphi = \varphi^{(u)}$.

4.1. **Lemma.** *For Λ a C^2 basic hyperbolic set and $\varphi^{(u)} \colon \Lambda \to \mathbb{R}$ as above, $\varphi^{(u)}$ satisfies a Hölder condition of positive exponent.*

$x \to E^u_x$ is Hölder continuous (3.1 of [19]) and $E^u_x \to \varphi^{(u)}(x)$ is differentiable, so the composition $x \to \varphi^{(u)}(x)$ is Hölder.

4.2. **Lemma** (Volume lemma). *Let Λ be a C^2 basic hyperbolic set and define*

$$B_x(\varepsilon, T) = \{y \in M \colon d(f^t x, f^t y) \leqq \varepsilon \text{ for all } t \in [0, T]\}.$$

For small $\varepsilon > 0$ there is a constant $c_\varepsilon > 1$ so that

$$m(B_x(\varepsilon, T))\, \lambda_T(x) \in [c_\varepsilon^{-1}, c_\varepsilon]$$

for all $x \in \Lambda$ and $T \geqq 0$, where m is the measure on M derived from the Riemann metric.

4.3. **Lemma** (Second volume lemma). *For small* ε, $\delta>0$ *there is* $d=d(\varepsilon,\delta)>0$ *(*d *independent of* n*) so that*

$$m(B_y(\delta,n))\geqq d\cdot m(B_x(\varepsilon,n))$$

whenever $x\in\Lambda$ *and* $y\in B_x(\varepsilon,n)$.

These two lemmas are proved in the Appendix.

4.4. **Proposition.** (a) *Let* Λ *be a* C^2 *basic hyperbolic set and define*

$$B_\Lambda(\varepsilon,T)=\bigcup_{x\in\Lambda}B_x(\varepsilon,T).$$

Then (for sufficiently small ε*)*

$$P(F|\Lambda,\varphi^{(u)})=\limsup_{T\to\infty}\frac{1}{T}\log m(B_\Lambda(\varepsilon,T))\leqq 0. \tag{3}$$

(b) *Define*

$$W_x^s(\varepsilon)=\{y\in M:\ \lim_{t\to\infty}d(f^t y,f^t x)=0 \text{ and } d(f^t y,f^t x)\leqq\varepsilon\,\forall t\geqq 0\}$$

when $x\in\Lambda$, *and let* $W_\Lambda^s(\varepsilon)=\bigcup_{x\in\Lambda}W_x^s(\varepsilon)$. *If* $m(W_\Lambda^s(\varepsilon))>0$, *then* $P(F|\Lambda,\varphi^{(u)})=0$ *and*

$$h_{\mu_{\varphi^{(u)}}}(f^1)=-\int\varphi^{(u)}\,d\mu_{\varphi^{(u)}}. \tag{4}$$

This is true in particular if Λ *is an attractor.*

Let $0<\delta\leqq\varepsilon$. If E is a maximal (δ,T)-separated set for $F|\Lambda$, then

$$\bigcup_{x\in E}B_x(\delta/2,T)\subset B_\Lambda(\varepsilon,T)\subset\bigcup_{x\in E}B_x(\delta+\varepsilon,T)$$

where the $B_x(\delta/2,T)$ are disjoint, and the second inclusion follows from $\Lambda\subset\bigcup_{x\in E}B_x(\delta,T)$. Thus, assuming ε small enough and using the volume lemma,

$$c_{\delta/2}^{-1}\sum_{x\in E}\lambda_T(x)^{-1}\leqq m(B_\Lambda(\varepsilon,T))\leqq c_{\delta+\varepsilon}\sum_{x\in E}\lambda_T(x)^{-1}.$$

Therefore

$$c_{\delta/2}^{-1}Z_T(F|\Lambda,\varphi^{(u)},\delta)\leqq m(B_\Lambda(\varepsilon,T))\leqq c_{\delta+\varepsilon}Z_T(F|\Lambda,\varphi^{(u)},\delta)$$

and

$$P(F|\Lambda,\varphi^{(u)},\delta)=\limsup_{T\to\infty}\frac{1}{T}\log m(B_\Lambda(\varepsilon,T)).$$

The limit $\delta\to 0$ yields (3), proving (a).

By Theorem 3.3 and Lemma 4.1, $\varphi=\varphi^{(u)}$ has a unique equilibrium state $\mu_{\varphi^{(u)}}$. By the definition of equilibrium states, (4) is equivalent to $P(F|\Lambda,\varphi^{(u)})=0$. The latter statement follows from (3) with

$$m(B_\Lambda(\varepsilon,T))\geqq m(W_\Lambda^s(\varepsilon))>0.$$

There is a neighbourhood V of Λ so that

$$W_\Lambda^s(\varepsilon)\supset\{y\in M:\ f^t y\in V \text{ for all } t\geqq 0\}.$$

Indeed 5.1 of [13] gives this for diffeomorphisms and [13] indicates how to do the proof for flows. If Λ is an attractor, one can find a small neighbourhood U' of Λ so that $f^t y\in V$ for all $t\geqq 0$ whenever $y\in U'$. Then $U'\subset W_\Lambda^s(\varepsilon)$ and therefore $m(W_\Lambda^s(\varepsilon))>0$.

4.5. *Remark.* A rescaling of t: $t \to t' = t/t_0$ does not change the invariant measures, it replaces $h_\mu(f^1)$ by $h_\mu(f^{t_0}) = t_0\, h_\mu(f^1)$ (see [1]) and $\varphi^{(u)}$ by $\varphi'^{(u)} = t_0\, \varphi^{(u)}$. Therefore – as indicated in the Introduction – rescaling of t does not change $\mu_{\varphi^{(u)}}$ or the main results below. A change of Riemann metric on M changes $\varphi^{(u)}$ but not $\mu_{\varphi^{(u)}}$, or $P(F|\Lambda, \varphi^{(u)})$ as one readily sees.

4.6. **Corollary.** *Let Λ be a C^2 basic hyperbolic set. For sufficiently small $\varepsilon > 0$ there is a constant c'_ε such that*

$$m(B_x(2\varepsilon, T)) \leqq c'_\varepsilon\, \mu_{\varphi^{(u)}}(B_x(\varepsilon, T))$$

for all $x \in \Lambda$, $T \geqq 0$.

This follows from Theorem 3.3, Lemma 4.2, and Proposition 4.4(a), with $c'_\varepsilon = c_{2\varepsilon}/C_\varepsilon$.

5. Main Results

5.1. **Theorem.** *Let Λ be a C^2 hyperbolic attractor, W^s_Λ its basin. Then for m-almost all points $x \in W^s_\Lambda$ one has*

$$\lim_{T \to \infty} \frac{1}{T} \int_0^T g(f^t x)\, dt = \int g\, d\mu_{\varphi^{(u)}}$$

for all continuous g: $M \to \mathbb{R}$ (i.e. x is a generic point for $\mu_{\varphi^{(u)}}$).

We can replace W^s_Λ by a neighbourhood U of Λ such that $f^t U \subset U$ for all $t \geqq t_0$ and $\bigcap_{t \geqq 0} f^t U = \Lambda$.

Let us write

$$\bar{g}(T, x) = \frac{1}{T} \int_0^T g(f^t x)\, dt \quad \text{and} \quad \bar{g} = \int g\, d\mu_\varphi .$$

Let

$$E(g, \delta) = \{x \in U \colon \limsup_{T \to \infty} |\bar{g}(T, x) - \bar{g}| \geqq \delta\}.$$

Choose $\varepsilon > 0$ so small that $|g(f^t x) - g(f^t x')| < \delta/4$ whenever $d(x, x') \leqq \varepsilon$ and $0 \leqq t \leqq 1$. If we set $C_n(g, \delta') = \{x \in U \colon |\bar{g}(n, x) - \bar{g}| > \delta'\}$, then

$$E(g, \delta) \subset \bigcap_{N=0}^{\infty} \bigcup_{n=N}^{\infty} C_n\left(g, \frac{3\delta}{4}\right) \subset E\left(g, \frac{3\delta}{4}\right).$$

Now fix $N > 0$ and choose finite subsets $S_N, S_{N+1}, \ldots,$ of Λ successively as follows. Let $S_n (n \geqq N)$ be a maximal subset of $C_n(g, \delta/2) \cap \Lambda$ satisfying the conditions:

(a) $B_x(\varepsilon, n) \cap B_y(\varepsilon, k) = \emptyset$ for $x \in S_n$, $y \in S_k$, $N \leqq k < n$ and

(b) $B_x(\varepsilon, n) \cap B_{x'}(\varepsilon, n) = \emptyset$ for $x, x' \in S_n$, $x \neq x'$.

(Notice that each S_n is finite.) Choose $\alpha > 0$ so that

$$B_\Lambda(\alpha) \subset W^s_\Lambda(\varepsilon) = \bigcup_{z \in \Lambda} W^s_z(\varepsilon)$$

($B_\Lambda(\alpha)$ is the closed α-neighbourhood of Λ). If

$$y \in B_\Lambda(\alpha) \cap C_n\left(g, \frac{3\delta}{4}\right) \quad (n \geqq N)$$

and $y \in W_z^s(\varepsilon)$ with $z \in \Lambda$, then $z \in C_n(g, \delta/4)$ because of the way ε was chosen. By the maximality of S_n one has

$$B_z(\varepsilon, n) \cap B_x(\varepsilon, k) \neq \emptyset \quad \text{for some } x \in S_k,\ N \leqq k \leqq n$$

and then $B_x(2\varepsilon, k) \supset B_z(\varepsilon, n) \supset W_z^s(\varepsilon) \ni y$. Thus one has

$$B_\Lambda(\alpha) \cap \bigcup_{n=N}^{\infty} C_n\left(g, \frac{3\delta}{4}\right) \subset \bigcup_{k=N}^{\infty} \bigcup_{x \in S_k} B_x(2\varepsilon, k).$$

Using Lemma 4.2 one has

$$m\left(B_\Lambda(\alpha) \cap \bigcup_{n=N}^{\infty} C_n\left(g, \frac{3\delta}{4}\right)\right) \leqq A'_{2\varepsilon} \sum_{k=N}^{\infty} \sum_{x \in S_k} \exp \int_0^k \varphi(f^t x)\, dt. \tag{5}$$

The definition of S_n implies that $V_N = \bigcup_{k=N}^{\infty} \bigcup_{x \in S_k} B_x(\varepsilon, k)$ is a disjoint union. The choice of ε gives that $B_\varepsilon(x, k) \subset C_k(g, \delta/4)$ for $x \in S_k \subset C_k(g, \delta/2)$ and so $V_N \subset \bigcup_{k=N}^{\infty} C_k(g, \delta/4)$. Because $\mu_{\varphi^{(u)}}$ is ergodic,

$$0 = \mu_{\varphi^{(u)}}\left(E\left(g, \frac{\delta}{4}\right)\right) \geqq \mu_{\varphi^{(u)}}\left(\bigcap_{N=0}^{\infty} \bigcup_{k=N}^{\infty} C_k\left(g, \frac{\delta}{4}\right)\right) = \lim_{N \to \infty} \mu_{\varphi^{(u)}}\left(\bigcup_{k=N}^{\infty} C_k\left(g, \frac{\delta}{4}\right)\right)$$

and $\lim_{N \to \infty} \mu_{\varphi^{(u)}}(V_N) = 0$. By Theorem 3.3 (and $P(F|\Lambda, \varphi^{(u)}) = 0$, confer Proposition 4.4)

$$\mu_{\varphi^{(u)}}(V_N) \geqq C_\varepsilon \sum_{k=N}^{\infty} \sum_{x \in S_k} \exp \int_0^k \varphi(f^t x)\, dt.$$

Hence the sum on the right converges to 0 as $N \to \infty$ and using (5) above we get

$$\lim_{N \to \infty} m\left(B_\Lambda(\alpha) \cap \bigcup_{n=N}^{\infty} C_n\left(g, \frac{3\delta}{4}\right)\right) = 0.$$

This in turn gives $m(B_\Lambda(\alpha) \cap E(g, \delta)) = 0$.

Now $f^t E(g, \delta) \subset E(g, \delta)$ for all $t \geqq 0$ and $f^t(U) \subset B_\Lambda(\alpha)$ for some $t > 0$. As f^t is a diffeomorphism, $m(f^t E(g, \delta)) \leqq m(B_\Lambda(\alpha) \cap E(g, \delta)) = 0$ implies $m(E(g, \delta)) = 0$. Letting $\{g_k\}_{k=1}^{\infty}$ be a dense sequence of continuous functions $\bar{U} \to \mathbb{R}$, we get that for x outside the m-null set

$$\bigcup_{k, m \geqq 1} E\left(g_k, \frac{1}{m}\right)$$

one has $\lim_{T \to \infty} \bar{g}_k(T, x) = \bar{g}_k$; as the g_k are dense, it follows that $\lim_{T \to \infty} \bar{g}(T, x) = \bar{g}$ for all continuous $g: \bar{U} \to \mathbb{R}$.

5.2. *Remark.* For the special case of an Anosov flow ($\Lambda = M$) with an invariant (probability) measure μ' absolutely continuous w.r.t. m, this theorem implies the known fact that $\mu' = \mu_{\varphi^{(u)}}$.

5.3. **Theorem.** *Let Λ be a C^2 attractor, W_Λ^s its basin, and let ν be a probability measure absolutely continuous with respect to m and with support in W_Λ^s. If the flow F restricted to Λ is C-dense, then*

$$\lim_{t \to \infty} \int (g \circ f^t)\, d\nu = \int g\, d\mu_{\varphi^{(u)}}$$

for all continuous $g: M \to \mathbb{R}$.

We may choose U as in the proof of Theorem 5.1, and assume that supp $\nu \subset U$. Define $f^{*t}\nu$ by

$$(f^{*t}\nu)(g) = \nu(g \circ f^t)$$

and write $\mu_{\varphi^{(u)}} = \mu$. We have to prove that weak $\lim_{t \to \infty} f^{*t}\nu = \mu$. In showing this we may assume that $\nu = r \cdot m$ where $r \geqq 0$ is bounded (by density of bounded functions in L^1).

Given $\varepsilon > 0$ we find as in the proof of Proposition 4.4 that if U' is a sufficiently small neighbourhood of Λ, then $U' \subset W^s_\Lambda(\varepsilon)$. We can choose $t(\varepsilon) > 0$ so that $f^{t(\varepsilon)}\, U \subset U'$ and therefore

$$\operatorname{supp} f^{*t(\varepsilon)}\nu \subset W^s_\Lambda(\varepsilon).$$

Let $E \subset \Lambda$ be a maximal (T, ε)-separated set for $F|\Lambda$, we have thus

$$\operatorname{supp} f^{*t(\varepsilon)}\nu \subset \bigcup_{x \in E} B_x(2\varepsilon, T).$$

Let $(\psi_x)_{x \in E}$ be a non-negative measurable partition of unity on $\operatorname{supp} f^{*t(\varepsilon)}\nu$ subordinate to the covering by the $B_x(2\varepsilon, T)$ and let χ_x be the characteristic function of $B_x(\varepsilon, T)$. We write

$$\nu_{\varepsilon, T} = \sum_{x \in E} \left(\frac{\int \psi_x \, d(f^{*t(\varepsilon)}\nu)}{\int \chi_x \, d\mu} \right) \cdot \chi_x \, \mu.$$

The measure $\nu_{\varepsilon, T}$ is a probability measure absolutely continuous with respect to μ, with density bounded independently of T. This is because

$$\frac{\int \psi_x \, d(f^{*t(\varepsilon)}\nu)}{\int \chi_x \, d\mu} \leqq \|r_{t(\varepsilon)}\|_\infty \frac{m(B_x(2\varepsilon, T))}{\mu(B_x(\varepsilon, T))} \leqq C'_\varepsilon \|r_{t(\varepsilon)}\|_\infty$$

by Corollary 4.6 ($r_{t(\varepsilon)}$ denotes the density of $f^{*t(\varepsilon)}\nu$ with respect to m).

Notice that $\nu_{\varepsilon, T}$ is obtained by redistributing the mass of $f^{*t(\varepsilon)}\nu$ in such a manner that all that goes to $B_x(\varepsilon, T)$ comes from $B_x(2\varepsilon, T)$. Therefore also $f^{*t}\nu_{\varepsilon, T}$ is obtained by redistributing the mass of $f^{*(t+t(\varepsilon))}\nu$ in such a manner that all that goes to $f^t B_x(\varepsilon, T)$ comes from $f^t B_x(2\varepsilon, T)$. The diameter of $f^t B_x(2\varepsilon, T)$ is at most 4ε when $0 \leqq t \leqq T$; therefore if $\mathcal{N}$ is a closed weak neighbourhood of the origin in the space of real measures on M we have

$$f^{*(t+t(\varepsilon))}\nu - f^{*t}\nu_{\varepsilon, T} \in \mathcal{N}$$

when $0 \leqq t \leqq T$, provided ε has been chosen sufficiently small.

Remember now that $\nu_{\varepsilon, T} = s_{\varepsilon, T} \cdot \mu$ where $s_{\varepsilon, T} \in L^\infty(\mu)$ and $\|s_{\varepsilon, T}\|_\infty$ is bounded independently of T. We can thus choose $T_n \to \infty$ such that $s_{\varepsilon, T_n} \to s_\varepsilon$ in $L^\infty(\mu)$ with its topology of weak dual of $L^1(\mu)$. We have thus

$$f^{*(t+t(\varepsilon))}\nu - f^{*t}(s_\varepsilon \cdot \mu) \in \mathcal{N} \tag{6}$$

for all $t \geqq 0$. We use now Remark 3.5: (μ, f^t) is a Bernoulli flow and

$$\lim_{t \to \infty} \int s_\varepsilon \cdot (g \circ f^t) \, d\mu = \mu(g)$$

for all continuous $g: \Lambda \to \mathbb{R}$. There is thus $t_{\mathcal{N}}$ such that

$$f^{*t}(s_\varepsilon \cdot \mu) - \mu \in \mathcal{N} \tag{7}$$

for $t \geqq t_{\mathcal{N}}$. From (6) and (7) we obtain

$$f^{*t} \nu - \mu \in 2\mathcal{N}$$

when $t \geqq t(\varepsilon) + t_{\mathcal{N}}$. Therefore $f^{*t} \nu$ tends weakly to μ when $t \to \infty$.

5.4. **Proposition.** *Let Λ be a C^2 basic hyperbolic set. The measure $\mu_{\varphi^{(u)}}$ depends continuously on the C^2 flow F for the weak topology on measures and the C^1 topology on flows. Also the pressure of $\varphi^{(u)}$ and the entropy of $\mu_{\varphi^{(u)}}$ with respect to F depend continuously on F for the C^1 topology on flows.*

Let $\mu = \mu_{\varphi^{(u)}}$, and μ' be the corresponding measure for a flow F'. We have to show that $\mu' \to \mu$ weakly when $F' \to F$ in the C^1 sense; μ' is a measure carried by the F'-basic set Λ' close to Λ.

Going back to Lemma 2.1 and using [4] we have a special flow G' on $\Lambda'(A, \psi')$ and a continuous surjection ρ': $\Lambda'(A, \psi') \to \Lambda'$. By [4] and the Ω-stability theorem [5], we can construct $\Lambda'(A, \psi')$ from the same subshift $\sigma_A: \Sigma_A \to \Sigma_A$ which was used for $\Lambda(P, \psi)$. When $F' \to F$, we have $\psi' \to \psi$ uniformly, and $\rho'(\mathbf{x}, t) \to \rho(\mathbf{x}, t)$ (uniformly in $(\mathbf{x}, t)$ for $0 \leqq t \leqq \min\{\psi(\mathbf{x}), \psi'(\mathbf{x})\}$). Furthermore $E'^{u}_{\rho'(\mathbf{x}, t)} \to E^{u}_{\rho(\mathbf{x}, t)}$ where E'^{u} is the unstable subbundle for F' (by Theorem (6.1) of [15]).

We know that

$$\Phi(\mathbf{x}) = \int_0^{\psi(\mathbf{x})} \varphi^{(u)}(\rho(\mathbf{x}, t))\, dt = \int_0^{\psi(\mathbf{x})} \varphi^{(u)}(f^t \rho(\mathbf{x}, 0))\, dt$$
$$= -\ln \lambda_{\psi(\mathbf{x})}(\rho(\mathbf{x}, 0))$$

and correspondingly

$$\Phi'(\mathbf{x}) = -\ln \lambda'_{\psi'(\mathbf{x})}(\rho'(\mathbf{x}, 0)).$$

Therefore when $F' \to F$, we have $\Phi' \to \Phi$ uniformly.

According to Section 3 we have

$$P(F|\Lambda, \varphi^{(u)}) = P(G, \varphi^{(u)} \circ \rho) = \sup_{\nu \in M(\sigma_A)} \frac{h_\nu(\sigma_A) + \int \Phi\, d\nu}{\int \psi\, d\nu}.$$

Therefore $P(F|\Lambda, \varphi^{(u)})$ depends continuously on F. Using the notation of Proposition 3.1, we let ν_0 be the unique equilibrium state for $\Phi - P(F|\Lambda, \varphi^{(u)}) \cdot \psi$. By the continuity of the equilibrium state indicated just before Proposition 3.1, $\nu_0' \to \nu_0$ when $F' \to F$. Thus $\mu_{\nu_0'} \to \mu_{\nu_0}$ and $\mu' = \rho'^* \mu_{\nu_0'} \to \mu = \rho^* \mu_{\nu_0}$.

Finally, the entropy of $\mu_{\varphi^{(u)}}$ is

$$P(F|\Lambda, \varphi^{(u)}) - \int \varphi^{(u)}\, d\mu_{\varphi^{(u)}}$$

where

$$\int \varphi^{(u)}\, d\mu_{\varphi^{(u)}} = \int (\varphi^{(u)} \circ \rho)\, d\mu_{\nu_0} = \int \Phi\, d\nu_0 / \int \psi\, d\nu_0$$

depends continuously on F.

5.5. **Proposition.** *Let Λ be a C^1 basic hyperbolic set and let $\varepsilon>0$.*

(a) *If $W_x^u(\varepsilon)\subset\Lambda$ for some $x\in\Lambda$, then Λ is an attractor.*

(b) *If Λ is not an attractor, there exists $\gamma>0$ such that for all $x\in\Lambda$, there is $y\in W_x^u(\varepsilon)$ with $d(y,\Lambda)>\gamma$.*

If $W_x^u(\varepsilon)\subset\Lambda$, and $u>0$, the set

$$U_x=\bigcup\{W_y^s(\varepsilon):\ y\in\bigcup_{|t|\leqq u} f^t W_x^u(\varepsilon)\} \tag{8}$$

is a neighbourhood of x in M (see [13], Lemma 4.1). Choose a periodic point $p\in U_x\cap\Lambda$ and let t_0 be its period. For some $\beta\in(0,\varepsilon]$, we have $W_p^u(\beta)\subset U_x$, hence

$$W_p^u(\beta)\subset W_\Lambda^s(\varepsilon)\cap W_\Lambda^u(\varepsilon)=\Lambda$$

(by [18], Theorem 3.2). Now

$$W_p^u=\bigcup_{n=1}^{\infty} f^{-nt_0}\, W_p^u(\beta)\subset\Lambda$$

and

$$W_p^{cu}=\bigcup_{0\leqq t\leqq t_0} W_{f^t p}^u$$

is dense in Λ (see for instance [3], p. 11–13).

For each $x\in W_p^{cu}$, the set U_x defined by (8) is a neighbourhood of x in M. As $W_x^s(\varepsilon)$, $W_x^u(\varepsilon)$ depend continuously on $x\in\Lambda$, one can find $\delta>0$ independent of x such that $U_x\supset B_x(2\delta)$ for all $x\in W_p^{cu}$ (see [13], Lemma 4.1). In view of this, and the density of W_p^{cu} in Λ,

$$B_\Lambda(\delta)\subset\bigcup\{U_x:\ x\in W_p^{cu}\}.$$

Therefore, if $z\in B_\Lambda(\delta)$ there exist $x\in W_p^{cu}$ and $y\in\bigcup_{|t|\leqq u} f^t W_x^u(\varepsilon)\subset W_p^{cu}\subset\Lambda$ such that $z\in W_y^s(\varepsilon)$. When $t\to\infty$ then $d(f^t z, f^t y)\to 0$ uniformly in z. Therefore

$$\bigcap_{t\geqq 0} f^t B_\Lambda(\delta)=\Lambda,$$

which shows that Λ is an attractor and proves (a).

To prove (b), notice that the set

$$V_\gamma=\{x\in\Lambda:\ d(y,\Lambda)>\gamma \text{ for some } y\in W_x^u(\varepsilon)\}$$

is open in Λ since $W_x^u(\varepsilon)$ varies continuously with x. Also V_γ increases as γ decreases and, by part (a) of the present proposition, $\bigcup_{\gamma>0} V_\gamma=\Lambda$. Therefore, by compactness, $V_\gamma=\Lambda$ for some $\gamma>0$.

5.6. **Theorem**[5]. *Let Λ be a C^2 basic hyperbolic set. The following conditions are equivalent:*

(a) *Λ is an attractor;*

(b) $m(W_\Lambda^s)>0$;

(c) $P(F|\Lambda,\varphi^{(u)})=0$.

[5] Some of the ideas in the proof of this theorem were earlier discovered by J. Franks and R.F. Williams.

Since $W_\Lambda^s = \bigcup_{n=0}^{\infty} f^{-n} W_\Lambda^s(\varepsilon)$, (b) can be replaced by $m(W_\Lambda^s(\varepsilon))>0$ for any small $\varepsilon>0$. In Proposition 4.4 (b) we have seen that (a) $\Rightarrow$ (b) $\Rightarrow$ (c). To complete the proof we assume that Λ is not an attractor, and show that $P(F|\Lambda, \varphi^{(u)})<0$.

Given a small $\varepsilon>0$, choose γ as in Proposition 5.5(b). There is $t>0$ such that if $x\in\Lambda$, $f^t W_x^u(\gamma/4) \supset W_{f^t x}^u(\varepsilon)$. Let $E\subset\Lambda$ be (γ, T)-separated. For any $x\in E$ and $T>0$, there is $y(x, T)\in B_x(\gamma/4, T)$ such that $d(f^{t+T}(y(x, T), \Lambda)>\gamma$ [because $f^T B_x(\gamma/4, T)\supset W_x^u(\gamma/4)$, hence $f^{T+t} B_x(\gamma/4, T)\supset W_{f^t x}^u(\varepsilon)$]. Choose $\delta\in(0, \gamma/4]$ such that $d(f^t z, f^t y)<\gamma/2$ whenever $d(z, y)\leqq\delta$. Then

$$B_{y(x,T)}(\delta, T)\subset B_x(\gamma/2, T)$$

$$f^{T+t} B_{y(x,T)}(\delta, T)\cap B_\Lambda(\gamma/2)=\emptyset,$$

hence

$$B_{y(x,T)}(\delta, T)\cap B_\Lambda(\gamma/2, T+t)=\emptyset.$$

Using the second volume lemma we have thus

$$\begin{aligned} m(B_\Lambda(\gamma/2, T))-m(B_\Lambda(\gamma/2, T+t)) &\geqq \sum_{x\in E} m(B_{y(x,T)}(\delta, T)) \\ &\geqq d(3\gamma/2, \delta)\sum_{x\in E} m(B_x(3\gamma/2, T))\geqq d(3\gamma/2, \delta)\, m(B_\Lambda(\gamma/2, T)) \end{aligned}$$

and therefore

$$m(B_\Lambda(\gamma/2, T+t))\leqq(1-d(3\gamma/2, \delta))\, m(B_\Lambda(\gamma/2, T))$$

so that by Proposition 4.4(a)

$$P(F|\Lambda, \varphi^{(u)})\leqq\frac{1}{t}\log(1-d(3\gamma/2, \delta))<0.$$

5.7. **Corollary.** *Let F be a C^2 Axiom A flow on the compact manifold M.*

(a) *The closures of the basins of the attractors cover M.*

(b) *If Λ is a basic hyperbolic set and $m(\Lambda)>0$, then Λ is a connected component of M and $F|\Lambda$ is an Anosov flow.*

Since F satisfies Axiom A, $M=\bigcup\{W_\Lambda^s\colon \Lambda$ is a basic hyperbolic set$\}$.

The complement of the basins of the attractors is $\bigcup\{W_\Lambda^s\colon \Lambda$ is not an attractor$\}$, it has measure 0, and therefore contains no open set. This proves (a).

If Λ is a basic set and $m(\Lambda)>0$, then $m(W_\Lambda^s)>0$ and $m(W_\Lambda^u)>0$ so that Λ is an attractor for both F and the opposite flow F^{-1}. Since Λ is an attractor for F, then $W_\Lambda^u=\Lambda$. Since Λ is an attractor for F^{-1}, then W_Λ^u is open. Therefore Λ is open and closed. Also, Λ is connected since W_p^{cu} is dense in Λ for periodic p (see [3], p. 11–13). This proves (b).

Appendix

Throughout what follows Λ will be a hyperbolic set for the C^r flow (f^t) on the manifold M $(r\geqq 1)$.

We recall that, by assumption, the Riemann metric on M is adapted to f^1 (see Introduction). Notice also that there is $K>0$ such that $\|Tf^n|E\|\leqq K$ for all $n\geqq 0$.

Denote by indices 0, 1, 2 the components in E_x, E_x^s, E_x^u of a vector in $T_x M$. We shall use in this Appendix a new scalar product in $T_x M$ defined by

$$\langle u, v\rangle_x = \sum_0^2 (u_i, v_i). \tag{A.1}$$

We write $\|u\| = \|u\|_x = (\langle u, u\rangle_x)^{1/2}$. If E' is a subspace of $T_x M$, $E'(\varepsilon)$ will denote the closed ε-ball centered at the origin of E' for this metric.

A.1. **Convenient Charts.** *For sufficiently small $\varepsilon > 0$ and each $x \in \Lambda$, let us define a C^r chart φ_x: $T_x M(\varepsilon) \to M$ such that*[6]

$$\varphi_x(E_x + E_x^s)(\varepsilon) \subset W_x^{cs}, \qquad \varphi_x(E_x + E_x^u)(\varepsilon) \subset W_x^{cu}$$

and the map $F = \varphi_{fx}^{-1} \circ f^1 \circ \varphi_x$ is tangent to $T_x f^1$ at the origin of $T_x M$.

If $\exp_x^{-1} W_x^{cs}$ is (in a neighbourhood of the origin of $T_x M$) the graph of a function ψ': $E_x \times E_x^s \to E_x^u$ we set

$$\varphi'(u) = u_0 + u_1 + \big(u_2 + \psi'(u_0, u_1)\big).$$

If $\varphi'^{-1} \exp_x^{-1} W_x^{cu}$ is the graph of ψ'': $E_x \times E_x^u \to E_x^s$ we set

$$\varphi''(u) = u_0 + \big(u_1 + \psi''(u_0, u_2)\big) + u_2.$$

Then $\varphi_x = \exp_x \circ \varphi' \circ \varphi''$ has the desired properties.

If $u \in T_x M(\varepsilon)$, ε sufficiently small, Taylor's formula yields

$$\|F_2(u)\| = \|F_2(u_0 + u_1 + u_2) - F_2(u_0 + u_1)\| \geqq \gamma^{-1} \|u_2\| \tag{A.2}$$

for some $\gamma \in (0, 1)$ independent of x. Similarly if $0 < \omega \leqq 1$, ε can be chosen so small that

$$\|F_0(u) + F_1(u) - F_0(v) - F_1(v)\| \leqq \omega \|F_2(u) - F_2(v)\| \tag{A.3}$$

whenever

$$\|u_0 + u_1 - v_0 - v_1\| \leqq \omega \|u_2 - v_2\| \quad \text{and} \quad u, v \in T_x M(\varepsilon).$$

We define

$$D_x(\varepsilon, n) = \{u \in T_x M: \|F^k u\|_{f^k x} \leqq \varepsilon \text{ for } k = 0, 1, \dots, n\}. \tag{A.4}$$

Let $u \in D_x(\varepsilon, n)$, then (A.2) yields

$$\|(F^k)_2(u)\| \leqq \gamma^{n-k} \|(F^n)_2(u)\| \leqq \varepsilon \gamma^{n-k}$$

for $k = 0, 1, \dots, n$. Let $v = u_0 + u_1$. We may assume that $v \in D_x\big((K+1)\varepsilon, n\big)$ and, for ε suitably small, apply (A.3) with $\omega = 1$.

We obtain

$$\|(F^k)_0(u) + (F^k)_1(u) - (F^k)_0(v) - (F^k)_1(v)\| \leqq \|(F^k)_2(u)\| \leqq \varepsilon \gamma^{n-k}$$

and therefore

$$\|F^k(u) - F^k(v)\| \leqq 2\varepsilon \gamma^{n-k} \tag{A.5}$$

for $k = 0, 1, \dots, n$. (In particular $v \in D_x(3\varepsilon, n)$.)

[6] $W_x^{cs} = \bigcup \{W_y^s\colon y \in \text{orbit of } x\}$, $W_x^{cu} = \bigcup \{W_y^u\colon y \in \text{orbit of } x\}$.

A.2. **Lemma.** *Let $r=1$ and $\psi\colon M\to\mathbb{R}$ be C^1. Given $\theta>0$ there is $\delta>0$ such that: if $x\in\Lambda$, $y\in M$, $n>0$ and $d(f^k y, f^k x)\leqq\delta$ for $k=0,1,\ldots,n$, then*

$$\left|\int_0^n \psi(f^t y)\,dt-\int_0^n \psi(f^t x)\,dt\right|<\theta.$$

Because of the C^1 assumptions there exists $C>0$ such that

$$\left|\int_0^1 \psi(f^t p)\,dt-\int_0^1 \psi(f^t q)\,dt\right|\leqq C\,d(p,q).$$

Given ε, one can choose δ so small that $d(x,y)<\delta$ implies $u=\varphi_x^{-1}y\in T_x M(\varepsilon/2)$ for $x\in\Lambda$ and $y\in M$. Define $v=u_0+u_1$ and $z=\varphi_x v\in W_x^{cs}$. We may then assume that $z\in W_{f^s x}^s(\varepsilon)$ with $|s|<C_0\,\varepsilon$ (C_0 independent of ε). Then

$$d(f^k z, f^{k+s} x)<C_1\,\varepsilon\,\gamma'^k$$

for some $\gamma'\in(0,1)$. If

$$d(f^k x, f^k y)\leqq\delta \qquad \text{for } k=0,1,\ldots,n,$$

(A.5) yields

$$d(f^k y, f^k z)<C_2\,\varepsilon\,\gamma^{n-k}$$

with $C_2>0$. Thus

$$\begin{aligned}\left|\int_0^n \psi(f^t y)-\int_0^n \psi(f^t x)\right| &\leqq \sum_{k=0}^{n-1}\left|\int_0^1 \psi(f^t f^k y)-\int_0^1 \psi(f^t f^k z)\right|\\ &\quad+\sum_{k=0}^{n-1}\left|\int_0^1 \psi(f^t f^k z)-\int_0^1 \psi(f^t f^{k+s} x)\right|\\ &\quad+\left|\int_0^n \psi(f^{t+s} x)-\int_0^n \psi(f^t x)\right|\\ &\leqq C\left[\frac{C_2\,\varepsilon}{1-\gamma}+\frac{C_1\,\varepsilon}{1-\gamma'}\right]+2C_0\,\varepsilon\,\|\psi\|.\end{aligned}$$

A.3. **Lemma.** *Let $\pi\colon G^q M\to M$ be the Grassmannian bundle of q-dimensional subspaces in TM and $G^q f^t\colon G^q M\to G^q M$ be the diffeomorphism induced by Tf^t. We assume that $r=2$ so that $G^q f$ is a C^1 flow on $G^q M$. If $q=\dim E_x^u$ and $\Lambda^*=\{E_x^u\colon x\in\Lambda\}$, then Λ^* is a hyperbolic set for the flow $G^q f$.*

For $x\in\Lambda$, define the manifolds

$$V_x^{*s}=\{E\in\pi^{-1}(W_x^s(\varepsilon))\colon d(E-E_x^u)<\varepsilon\};\qquad V_x^{*u}=TW_x^u(\varepsilon)$$

in $G^q M$. The manifold $\pi^{-1}(x)$ contains E_x^u, and $G^q f^t[\pi^{-1}(x)]=\pi^{-1}(f^t x)$. It is known (and easily seen) that $G^q f^1$ contracts a neighbourhood of E_x^u in $\pi^{-1}(x)$ (see B.1 of [19]). Since f^1 contracts $W_x^s(\varepsilon)$, it follows that, when ε is sufficiently small, Gf^1 contracts V_x^{*s}. Thus TGf^1 is a contraction of $E^{*s}=TV_x^{*s}$. Clearly $G^q f^{-1}$ contracts V^{*u} and therefore $TG^q f^{-1}$ is a contraction of $E^{*u}=TV_x^{*u}$. We have

$$\begin{aligned}\dim E^{*s}+\dim E^{*u}&=\dim W_x^s(\varepsilon)+\dim\pi^{-1}(x)+\dim W_x^u(\varepsilon)\\ &=\dim M-1+\dim\pi^{-1}(x)=\dim G^q M-1\end{aligned}$$

which concludes the proof.

A.4. **Lemma.** *Let $r=2$. Given $\theta>0$, there is $\varepsilon>0$ so that the following holds.*

If $x\in\Lambda$ and $v\in D_x(\varepsilon,n)$ (defined by (A.4)), let $E_v^*\in G^qM$ be the tangent at φ_x^v to the manifold $\varphi_x(v+E_x^u(\varepsilon))$. Then

$$e^{-\theta}\leqq\frac{\operatorname{Jac}Tf^n|E_v^*}{\operatorname{Jac}Tf^n|E_x^u}\leqq e^{\theta} \tag{A.6}$$

and

$$e^{-\theta}\leqq\frac{\operatorname{Jac}D_v(F^n|v+E_x^u)}{\operatorname{Jac}D_0(F^n|E_x^u)}\leqq e^{\theta}. \tag{A.7}$$

(D_0, D_v are derivatives in charts.)

The Jacobian in (A.7) is computed with respect to the scalar product (A.1). Clearly the estimate (A.7) differs from (A.6) only by bounded factors and it suffices to prove (A.6). To do this we apply Lemma A.2 to the hyperbolic set Λ^* for the C^1 flow G^qf (cf. Lemma A.3), with the replacement $x\to E_x^u$, $y\to E_v^*$. We have to check that (for sufficiently small ε),

$$d(Tf^kE_x^u, Tf^kE_v^*)\leqq\delta$$

for $k=0,1,\ldots,n$. Using the charts φ_x, this results from (A.4) and (A.3). To conclude the proof it suffices to define

$$\psi(E)=\frac{d}{dt}\ln(\operatorname{Jac}Tf^t|E)\bigg|_{t=0}$$

and remark that

$$\int_0^n\psi(Tf^tE_v)\,dt=\ln\operatorname{Jac}Tf^n|E_v.$$

A.5. **Proof of the Volume Lemma.** We shall show that for sufficiently small $\varepsilon>0$ there exist $b_\varepsilon, b'_\varepsilon>0$ so that

$$b_\varepsilon\leqq m_x(D_x(\varepsilon,n))\cdot\operatorname{Jac}D_0(F^n|E_x^u)\leqq b'_\varepsilon \tag{A.8}$$

for all $x\in\Lambda$, $n>0$. Here m_x denotes the measure on T_xM associated with the scalar product (A.1). This will prove the volume lemma because the use of the charts φ_x multiplies all distances, measures and Jacobians (see Lemma A.4) by positive factors bounded away from 0 and ∞.

If $v\in(E_x+E_x^s)(\varepsilon)$ define

$$N_v(\varepsilon,n)=\{u\in T_xM:\ u_0+u_1=v \text{ and } (F^k)_2(u)\in E_{f^kx}^u(\varepsilon) \text{ for } k=0,1,\ldots,n\}.$$

If ε is sufficiently small, $\|F^kv\|\leqq(K+1)\varepsilon$ for all $k\geqq0$.

Also, using (A.3) with $\omega\leqq1$ and induction on n, we find that $F^nN_v(\varepsilon,n)$ is the graph of a C^1 function $g\colon E_{f^nx}^u(\varepsilon)\to(E_{f^nx}+E_{f^nx}^s)((K+2)\varepsilon)$ such that $\|Dg\|\leqq\omega$[7]. In particular we obtain the second inclusion of:

$$D_x(\varepsilon,n)\subset\bigcup_v N_v(\varepsilon,n)\subset D_x((K+3)\varepsilon,n).$$

---

[7] This is an easy adaptation of the first part of the proof of Theorem 2.3 of [14].

To prove (A.8) it suffices thus to show that

$$c_\varepsilon \leqq m_x\Big(\bigcup_v N_v(\varepsilon, n)\Big) \cdot \operatorname{Jac} D_0(F^n | E_x^u) \leqq c'_\varepsilon \tag{A.9}$$

for some $c_\varepsilon, c'_\varepsilon > 0$.

Since $\|Dg\| \leqq \omega$, the measure of $F^n N_v(\varepsilon, n)$ (induced on the manifold by the metric (A.1) on $T_{f^n x} M$) is contained between bounds $d_\varepsilon, d'_\varepsilon > 0$. In view of (A.7), the measure of $N_v(\varepsilon, n)$ multiplied by $\operatorname{Jac} D_0(F^n | E_x^u)$ is contained between $d_\varepsilon e^{-\theta}$ and $d'_\varepsilon e^\theta$.

Finally, using Fubini's theorem to integrate over $v \in (E_x + E_x^s)(\varepsilon)$ yields (A.9).

A.6. **Proof of the Second Volume Lemma.** Let $w \in D_x(\varepsilon, n)$, and define

$$D_{xw}(\delta, n) = \{u \in T_x M \colon \|F^k u - F^k w\|_{f^k x} \leqq \delta \text{ for } 0, 1, \ldots, n\}.$$

It will suffice to show that there is $b_\delta > 0$ so that

$$m_x(D_{xw}(\delta, n)) \cdot \operatorname{Jac} D_0(F^n | E_x^u) \geqq b_\delta \tag{A.10}$$

for all $x \in \Lambda$, $w \in D_x(\varepsilon, n)$, $n > 0$. Furthermore it suffices to prove (A.10) for $\delta < \varepsilon$.

Let

$$\Delta = \{u \in T_{f^n x} M \colon u_0 = (F^n w)_0,\ u_2 = (F^n w)_2,\ u_1 \in E_{f^n x}^s(3\varepsilon)\}.$$

For each $v \in \Delta$, let $\Gamma_v = \{f^t \cdot \varphi_{f^n x} v \colon |t| < \alpha\}$ so that $\varphi_{f^n x}^{-1} \Gamma_v$ is a "piece of trajectory" through v. Then, for small ε and suitable α,

$$W = \bigcup_{v \in \Delta} (\varphi_{f^n x}^{-1} \Gamma_v)$$

is a C^2 manifold in $T_{f^n x} M$, which is the graph of a function ψ defined on a subset of $E_{f^n x} + E_{f^n x}^s$ with values in $E_{f^n x}^u$ and such that $\|D\psi\| \leqq \omega$ with $\omega \in (0, 1)$.

We may assume that the domain of ψ contains $E_{f^n x}(2\varepsilon) + E_{f^n x}^s(2\varepsilon)$ and let W' be the graph of the restriction of ψ to $E_{f^n x}(2\varepsilon) + E_{f^n x}^s(2\varepsilon)$. We write

$$W' = \bigcup_{\tau \colon |\tau| < 2\varepsilon} W'_\tau$$

where $W'_\tau \subset \{u \colon u_0 = \tau\}$. Let

$$W''_\tau = \{u \in T_x M \colon F^n u \in W'_\tau \text{ and } (F^k)_1(u) \in E_{f^k x}^s(2\varepsilon) \text{ for } k = 0, 1, \ldots, n\}$$

$$W'' = \bigcup_{\tau \colon |\tau| < 2\varepsilon} W''_\tau.$$

Applying (A.3) and the argument in A.5 to $F^n W''_\tau$, F^{-1} instead of $N_v(\varepsilon, n)$, F we find that W''_τ is the graph of a function $g_\tau \colon E_x^s(2\varepsilon) \to (E_x + E_x^u)(2(K+1)\varepsilon)$ such that $\|Dg_\tau\| \leqq \omega$. On the other hand W'' is a union of "pieces of trajectories" which are graphs of maps $E_x \to E_x^s + E_x^u$ with derivative $\leqq \omega$ (for sufficiently small ε). The C^2 manifold W'' is thus the graph of a function ψ'' defined on a subset of $E_x + E_x^s$ with values in E^u and such that $\|D\psi''\| \leqq 1$ (for small ε, hence small ω). The manifold W'' imitates a piece of center-stable manifold through w.

Notice that F^k contracts or expands a "piece of trajectory" in W'' by a factor bounded away from 0 and ∞ (contained between $(K+1)^{-1}$ and $(K+1)$, say). From this and the above properties of W'' it follows that the domain of ψ'' contains a ball B of radius β around $w_0 + w_1$, in $E_x + E_x^s$ for sufficiently small β

and (taking $\beta<\delta/4(K+1)$) we have

$$d(F^k\psi''v, F^k w)\leqq\frac{\delta}{2}$$

for $k=0,1,\ldots,n$ whenever $v\in B$.

For each $v\in B$, define

$$N_v^*\left(\frac{\delta}{2},n\right)' = \{u\in T_xM\colon u_0+u_1=v \text{ and } \|(F^k)_2(u)-(F^k)_2(\psi''v)\|\leqq\frac{\delta}{2} \text{ for } k=0,1,\ldots,n\}.$$

We have

$$\bigcup_{v\in B} N_v^*\left(\frac{\delta}{2},n\right)'\subset D_{xw}(\delta,n)$$

and proceeding as in A.5 we find that

$$m_x\left(\bigcup_v N_v^*\left(\frac{\delta}{2},m\right)\right)\cdot \operatorname{Jac} D_0(F^n|E_x^u)\geqq c_\delta$$

for some $c_\delta>0$. This proves (A.10) and therefore the second volume lemma.

References

1. Abramov, L.M.: On the entropy of a flow. A.M.S. Translations **49**, 167–170 (1966)
2. Bowen, R.: Periodic points and measures for Axiom A diffeomorphisms. Trans. A.M.S. **154**, 377–397 (1971)
3. Bowen, R.: Periodic orbits for hyperbolic flows. Amer. J. Math. **94**, 1–30 (1972)
4. Bowen, R.: Symbolic dynamics for hyperbolic flows. Amer. J. Math. **95**, 429–459 (1973)
5. Bowen, R.: Maximizing entropy for a hyperbolic flow. Math. Systems Theory. To appear
6. Bowen, R.: Some systems with unique equilibrium states. Math. Systems Theory. To appear
7. Bowen, R.: Bernoulli equilibrium states for Axiom A diffeomorphisms. Math. Systems Theory. To appear
8. Bowen, R., Walters, P.: Expansive one-parameter flows. J. Diff. Equs. **12**, 180–193 (1972)
9. Bunimovič, L.A.: Imbedding of Bernoulli shifts in certain special flows (in Russian). Uspehi mat. Nauk **28**, 171–172 (1973)
10. Dinaburg, E.I.: On the relations among various entropy characteristics of dynamical systems. Math. USSR Izvestia **5**, 337–378 (1971)
11. Dobrušin, R.L.: Analyticity of correlation functions in one-dimensional classical systems with slowly decreasing potentials. Commun. math. Phys. **32**, 269–289 (1973)
12. Gurevič, B.M.: Some existence conditions for K-decompositions for special flows. Trans. Moscow Math. Soc. **17**, 99–128 (1967)
13. Hirsch, M., Palis, J., Pugh, C., Shub, M.: Neighborhoods of hyperbolic sets. Inventiones Math. **9**, 121–134 (1970)
14. Hirsch, M., Pugh, C.: Stable manifolds and hyperbolic sets. Proc. Symp. in Pure Math. **14**, 133–163 (1970)
15. Hirsch, M., Pugh, C., Shub, M.: Invariant manifolds. To appear
16. Livšic, A.N., Sinai, Ya.G.: On invariant measures compatible with the smooth structure for transitive U-systems. Soviet Math. Dokl. **13**, 1656–1659 (1972)
17. Margulis, G.A.: Certain measures associated with U-flows on compact manifolds. Func. Anal. and its Appl. **4**, 55–67 (1970)
18. Pugh, C., Shub, M.: The Ω-stability theorem for flows. Inventiones math. **11**, 150–158 (1970)
19. Pugh, C.: Ergodicity of Anosov Actions. Inventiones math. **15**, 1–23 (1972)

20. Ratner, M.: Anosov flows with Gibbs measures are also Bernoullian. Israel J. Math. To appear
21. Ruelle, D.: Statistical Mechanics. New York: Benjamin 1969
22. Ruelle, D.: Statistical mechanics of a one-dimensional lattice gas. Commun. Math. Phys. **9**, 267–278 (1968)
23. Ruelle, D.: Statistical mechanics on a compact set with Z^ν action satisfying expansiveness and specification. Trans. A.M.S. To appear
24. Ruelle, D.: A measure associated with Axiom *A* attractors. Amer. J. Math. To appear
25. Sinai, Ya. G.: Markov partitions and *Y*-diffeomorphisms. Func. Anal. and its Appl. **2**, 64–69 (1968)
26. Sinai, Ya. G.: Gibbs measures in ergodic theory. Russ. Math. Surveys **166**, 21–69 (1972)
27. Smale, S.: Differentiable dynamical systems. Bull. A.M.S. **73**, 747–817 (1967)
28. Walters, P.: A variational principle for the pressure of continuous transformations. Amer. J. Math. To appear

R. Bowen
University of California
Department of Mathematics
Berkeley, Calif. 94720, USA

David Ruelle
Institut des Hautes Etudes Scientifiques
35, Route de Chartres
F-91449 Bures-sur-Yvette/France

(Received May 18, 1974)

Reprinted from the AMERICAN MATHEMATICAL MONTHLY
Vol. 82, No. 10, December 1975
pp. 985–992

PERIOD THREE IMPLIES CHAOS

TIEN-YIEN LI AND JAMES A. YORKE

1. Introduction. The way phenomena or processes evolve or change in time is often described by differential equations or difference equations. One of the simplest mathematical situations occurs when the phenomenon can be described by a single number as, for example, when the number of children susceptible to some disease at the beginning of a school year can be estimated purely as a function of the number for the previous year. That is, when the number x_{n+1} at the beginning of the $n+1$st year (or time period) can be written

$$x_{n+1} = F(x_n), \tag{1.1}$$

where F maps an interval J into itself. Of course such a model for the year by year progress of the disease would be very simplistic and would contain only a shadow of the more complicated phenomena. For other phenomena this model might be more accurate. This equation has been used successfully to model the distribution of points of impact on a spinning bit for oil well drilling, as mentioned in **[8, 11]**, knowing this distribution is helpful in predicting uneven wear of the bit. For another example, if a population of insects has discrete generations, the size of the $n+1$st generation will be a function of the nth. A reasonable model would then be a generalized logistic equation

$$x_{n+1} = rx_n[1 - x_n/K]. \tag{1.2}$$

A related model for insect populations was discussed by Utida in **[10]**. See also Oster *et al* **[14, 15]**.

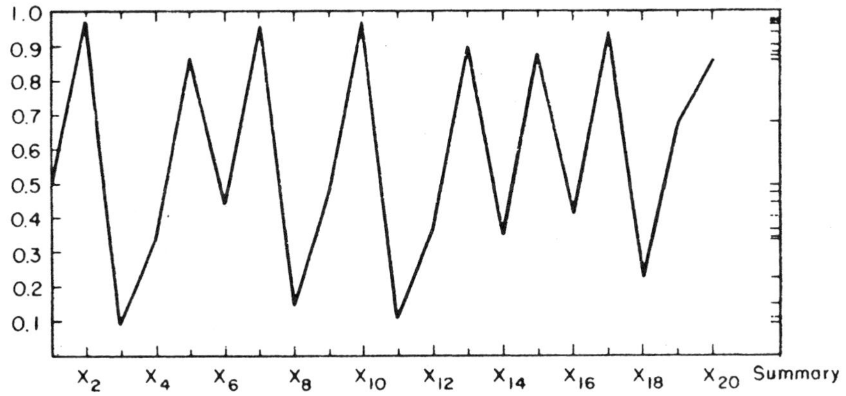

FIG. 1. For $K = 1$, $r = 3.9$, with $x_1 = .5$, the above graph is obtained by iterating Eq. (1.2) 19 times. At right the 20 values are repeated in summary. No value occurs twice. While $x_2 = .975$ and $x_{10} = .973$ are close together, the behavior is not periodic with period 8 since $x_{18} = .222$.

These models are highly simplified, yet even this apparently simple equation (1.2) may have surprisingly complicated dynamic behavior. See Figure 1. We approach these equations with the viewpoint that irregularities and chaotic oscillations of complicated phenomena may sometimes be understood in terms of the simple model, even if that model is not sufficiently sophisticated to allow accurate numerical predictions. Lorenz **[1–4]** took this point of view in studying turbulent behavior in a fascinating series of papers. He showed that a certain complicated fluid flow could be modelled by such a sequence $x, F(x), F^2(x), \cdots$, which retained some of the chaotic aspects of the original flow. See Figure 2. In this paper we analyze a situation in which the sequence $\{F^n(x)\}$ is non-periodic and might be called "chaotic." Theorem 1 shows that chaotic behavior for (1.1) will result in any situation in which a "population" of size x can grow for two or more successive generations and then having reached an unsustainable height, a population bust follows to the level x or below.

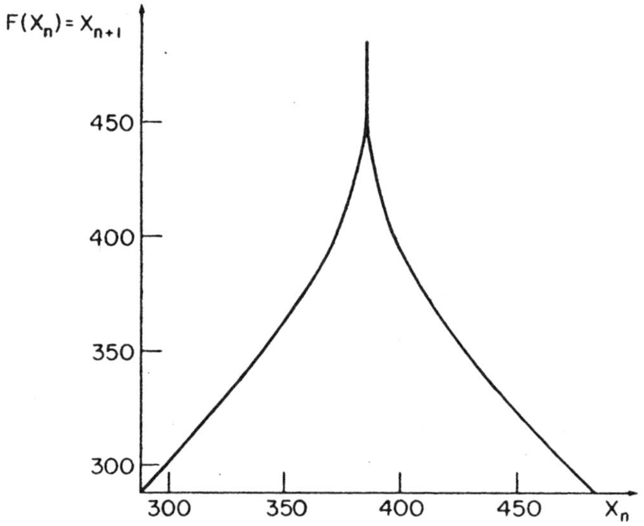

FIG. 2. Lorenz [1] studied the equations for a rotating water-filled vessel which is circularly symmetric about its vertical axis. The vessel is heated near the rim and cooled near its center. When the vessel is annular in shape and the rotation rate high, waves develop and alter their shape irregularly. From a simplified set of equations solved numerically, Lorenz let X_n be in essence the maximum kinetic energy of successive waves. Plotting X_{n+1} against X_n, and connecting the points, the above graph is obtained.

In section 3 we give a well-known simple condition which guarantees that a periodic point is stable and then in section 4 we quote a result applicable when F is like the one in Figure 2. It implies that there is an interval $J_\infty \subset J$ such that for almost every $x \in J$, the set of limit points of the sequence $\{F^n(x)\}$ is J_∞.

A number of questions remain unanswered. For example, is the closure of the periodic points an interval or at least a finite union of intervals? Other questions are mentioned later.

Added in proof. May has recently discovered other strong properties of these maps in his independent study of how the behavior changes as a parameter is varied [17].

2. The main theorem. Let $F: J \to J$. For $x \in J$, $F^0(x)$ denotes x and $F^{n+1}(x)$ denotes $F(F^n(x))$ for $n = 0, 1, \cdots$. We will say p is a **periodic point with period** n if $p \in J$ and $p = F^n(p)$ and $p \neq F^k(p)$ for $1 \leqq k < n$. We say p is **periodic** or is a **periodic point** if p is periodic for some $n \geqq 1$. We say q is **eventually periodic** if for some positive integer m, $p = F^m(q)$ is periodic. Since F need not be one-to-one, there may be points which are eventually periodic but are not periodic. Our objective is to understand the situations in which iterates of a point are very irregular. A special case of our main result says that if there is a periodic point with period 3, then for each integer $n = 1, 2, 3, \cdots$, there is a periodic point with period n. Furthermore, there is an uncountable subset of points x in J which are not even "asymptotically periodic."

THEOREM 1. *Let J be an interval and let $F: J \to J$ be continuous. Assume there is a point $a \in J$ for which the points $b = F(a)$, $c = F^2(a)$ and $d = F^3(a)$, satisfy*

$$d \leqq a < b < c \text{ (or } d \geqq a > b > c).$$

Then

T1: *for every $k = 1, 2, \cdots$ there is a periodic point in J having period k.*

Furthermore,

T2: *there is an uncountable set $S \subset J$ (containing no periodic points), which satisfies the following conditions:*

(A) *For every $p, q \in S$ with $p \neq q$,*

$$\limsup_{n \to \infty} |F^n(p) - F^n(q)| > 0 \tag{2.1}$$

and

$$\liminf_{n \to \infty} |F^n(p) - F^n(q)| = 0. \tag{2.2}$$

(B) *For every $p \in S$ and periodic point $q \in J$,*

$$\limsup_{n \to \infty} |F^n(p) - F^n(q)| > 0.$$

REMARKS. Notice that if there is a periodic point with period 3, then the hypothesis of the theorem will be satisfied.

An example of a function satisfying the hypotheses of the theorem is $F(x) = rx[1 - x/K]$ as in (1.2) for $r \in (3.84, 4]$ with $J = [0, K]$ and for $r > 4$, $F(x) = \max\{0, rx[1 - x/K]\}$ with $J = [0, K]$. See **[2]** for a detailed description of iterates of this function for $r \in [0, 4)$. The case $r = 4$ is discussed in **[6, 7, 12]**.

While the existence of a point of period 3 implies the existence of one of period 5, the converse is false. (See Appendix 1).

We say $x \in J$ is **asymptotically periodic** if there is a periodic point p for which

$$F^n(x) - F^n(p) \to 0 \quad \text{as} \quad n \to \infty. \tag{2.3}$$

It follows from (B) that the set S contains no asymptotically periodic points. We remark that it is unknown what the infimum of r is for which the equation (1.2) has points which are not asymptotically periodic.

Proof of Theorem 1. The proof of T1 introduces the main ideas for both T1 and T2. We now give the proof of T1 with necessary lemmas and relegate the tedious proof of T2 to Appendix 2.

LEMMA 0. *Let $G: I \to R$ be continuous, where I is an interval. For any compact interval $I_1 \subset G(I)$ there is a compact interval $Q \subset I$ such that $G(Q) = I_1$.*

Proof. Let $I_1 = [G(p), G(q)]$, where $p, q \in I$. If $p < q$, let r be the last point of $[p, q]$ where $G(r) = G(p)$ and let s be the first point after r where $G(s) = G(q)$. Then $G([r, s]) = I_1$. Similar reasoning applies when $p > q$.

LEMMA 1. *Let $F: J \to J$ be continuous and let $\{I_n\}_{n=0}^{\infty}$ be a sequence of compact intervals with $I_n \subset J$ and $I_{n+1} \subset F(I_n)$ for all n. Then there is a sequence of compact intervals Q_n such that $Q_{n+1} \subset Q_n \subset I_0$ and $F^n(Q_n) = I_n$ for $n \geqq 0$. For any $x \in Q = \cap Q_n$ we have $F^n(x) \in I_n$ for all n.*

Proof. Define $Q_0 = I_0$. Then $F^0(Q_0) = I_0$. If Q_{n-1} has been defined so that $F^{n-1}(Q_{n-1}) = I_{n-1}$, then $I_n \subset F(I_{n-1}) = F^n(Q_{n-1})$. By Lemma 0 applied to $G = F^n$ on Q_{n-1} there is a compact interval $Q_n \subset Q_{n-1}$ such that $F^n(Q_n) = I_n$. This completes the induction.

The technique of studying how certain sequences of sets are mapped into or onto each other is often used in studying dynamical systems. For instance, Smale uses this method in his famous "horseshoe example" in which he shows how a homeomorphism on the plane can have infinitely many periodic points [13].

LEMMA 2. *Let $G: J \to R$ be continuous. Let $I \subset J$ be a compact interval. Assume $I \subset G(I)$. Then there is a point $p \in I$ such that $G(p) = p$.*

Proof. Let $I = [\beta_0, \beta_1]$. Choose $\alpha_i (i = 0, 1)$ in I such that $G(\alpha_i) = \beta_i$. It follows $\alpha_0 - G(\alpha_0) \geqq 0$ and $\alpha_1 - G(\alpha_1) \leqq 0$ and so continuity implies $G(\beta) - \beta$ must be 0 for some β in I.

Assume $d \leqq a < b < c$ as in the theorem. The proof for the case $d \geqq a > b > c$ is similar and so is omitted. Write $K = [a, b]$ and $L = [b, c]$.

Proof of T1: Let k be a positive integer. For $k > 1$ let $\{I_n\}$ be the sequence of intervals $I_n = L$ for $n = 0, \cdots, k-2$ and $I_{k-1} = K$, and define I_n to be periodic inductively, $I_{n+k} = I_n$ for $n = 0, 1, 2, \cdots$. If $k = 1$, let $I_n = L$ for all n.

Let Q_n be the sets in the proof of Lemma 1. Then notice that $Q_k \subset Q_0$ and $F^k(Q_k) = Q_0$ and so by Lemma 2, $G = F^k$ has a fixed point p_k in Q_k. It is clear that p_k cannot have period less than k for F; otherwise we would need to have $F^{k-1}(p_k) = b$, contrary to $F^{k+1}(p_k) \in L$. The point p_k is a periodic point of period k for F.

3. Behavior near a periodic point. For some functions F, the asymptotic behavior of iterates of a point can be understood simply by studying the periodic points. For

$$F(x) = ax(1-x) \tag{3.1}$$

a detailed discussion of the points of period 1 and 2 may be found in [1] for $a \in [0, 4]$ and we now summarize some of those results. For $a \in [0, 4]$, F: $[0, 1] \to [0, 1]$.

For $a \in [0, 1]$, $x = 0$ is the only point of period 1; in fact, for $x \in [0, 1]$, the sequence $F^n(x) \to 0$ as $n \to \infty$.

For $a \in (1, 3]$, there are two points of period 1, namely 0 and $1 - a^{-1}$, and for $x \in (0, 1)$, $F^n(x) \to 1 - a^{-1}$ as $n \to \infty$.

For $a > 3$ there are also two points of period 2 which we may call p and q and of course $F(p) = q$ and $F(q) = p$. For $a \in (3, 1 + \sqrt{6} \approx 3.449)$ and $x \in (0, 1)$, $F^{2n}(x)$ converges to either p or q while $F^{2n+1}(x)$ converges to the other, except for those x for which there is an n for which $F^n(x)$ equals the point $1 - a^{-1}$ of period 1. There are only a countable number of such points so that the behavior of $\{F^n(x)\}$ can be understood by studying the periodic points.

For $a > 1 + \sqrt{6}$, there are 4 points of period 4 and for a slightly greater than $1 + \sqrt{6}$, $F^{4n}(x)$ tends to one of these 4 unless for some n, $F^n(x)$ equals one of the points of period 1 or 2. Therefore we may summarize this situation by saying that each point in $[0, 1]$ is asymptotically periodic.

For those values of a for which each point is asymptotically periodic, it is sufficient to study only the periodic points and their "stability properties." For any function F a point $y \in J$ with period k is said to be **asymptotically stable** if for some interval $I = (y - \delta, y + \delta)$ we have

$$|F^k(x) - y| < |x - y| \quad \text{for all} \quad x \in I.$$

If F is differentiable at the points $y, F(y), \cdots, F^{k-1}(y)$, there is a simple condition that will guarantee this behavior, namely

$$\left| \frac{d}{dx} F^k(x) \right| < 1.$$

By the chain rule

$$\frac{d}{dx}F^k(y)=\frac{d}{dx}F(F^{k-1}(y))\cdot\frac{d}{dx}F^{k-1}(y)$$

(3.2)
$$=\frac{d}{dx}F(F^{k-1}(y))\times\frac{d}{dx}F(F^{k-2}(y))\times\cdots\times\frac{d}{dx}F(y)$$

$$=\prod_{n=0}^{k-1}\frac{d}{dx}F(y_n),$$

where y_n is the nth iterate, $F^n(y)$. Therefore y is asymptotically stable if

$$\left|\prod_{i=0}^{k-1}\frac{d}{dx}F(y_i)\right|<1, \qquad \text{where} \qquad y_i=F^i(y).$$

This condition of course guarantees nothing about the limiting behavior of points which do not start "near" the periodic point or one of its iterates. The function in Figure 2 which was studied by Lorenz has the opposite behavior, namely, where the derivative exists we have

$$\left|\frac{d}{dx}F(x)\right|>1.$$

For such a function every periodic point is "unstable" since for x near a periodic point y of period k, the kth iterate $F^k(x)$ is further from y than x is. To see this, approximate $F^k(x)$ by

$$F^k(y)+\frac{d}{dx}F^k(y)[y-x]=y+\frac{d}{dx}F^k(y)[y-x].$$

Thus for x near y, $|F^k(x)-y|$ is approximately $|x-y|\,|(d/dx)F^k(y)|$. From (3.2) $|(d/dx)F^k(y)|$ is greater than 1. Therefore $F^n(x)$ is further from y than x is.

We do not know when values of a begin to occur for which F in (3.1) has points which are not asymptotically periodic. For $a=3.627$, F has a periodic point (which is asymptotically stable) of period 6 (approx. $x=.498$). This x is therefore a point of period 3 for F^2 and so Theorem 1 may be applied to F^2. Since F^2 has points which are not asymptotically periodic, the same is true of F.

In order to contrast the situations in this section with other possible situations discussed in the next section, we define the limit set of a point x. The point y is a **limit point** of a sequence $\{x_n\}\subset J$ if there is a subsequence x_{n_i} converging to y. The **limit set** $L(x)$ is defined to be the set of limit points of $\{F^n(x)\}$. If x is asymptotically periodic, then $L(x)$ is the set $\{y, F(y),\cdots,F^{k-1}(y)\}$ for some periodic point y of period k.

4. Statistical properties of $\{F^n(x)\}$. Theorem 1 establishes the irregularity of the behavior of iterates of points. What is also needed is a description of the regular behavior of the sequence $\{F^n(x)\}$ when F is piecewise continuously differentiable (as is Lorenz's function in Figure 2) and

(4.1)
$$\inf_{x\in J_1}\left|\frac{dF}{dx}\right|>1 \qquad \text{where} \qquad J_1=\left\{x:\frac{dF}{dx}\text{ exists}\right\}.$$

One approach to describing the asymptotic behavior for such functions is to describe $L(x)$, if possible. A second approach, which turns out to be related, is to examine the average behavior of $\{F^n(x)\}$. The fraction of the iterates $\{x,\cdots,F^{N-1}(x)\}$ of x that are in $[a_1,a_2]$ will be denoted by $\phi(x,N,[a_1,a_2])$. The limiting fraction will be denoted

$$\phi(x,[a_1,a_2])=\lim_{N\to\infty}\phi(x,N,[a_1,a_2])$$

when the limit exists. The subject of ergodic theory, which studies transformation on general spaces, motivates the following definition. We say g is the **density of** x (for F) if the limiting fraction satisfies

$$\phi(x,[a_1,a_2]) = \int_{a_1}^{a_2} g(x)dx \quad \text{for all} \quad a_1, a_2 \in J; \quad a_1 < a_2.$$

The techniques for the study of densities use non-elementary techniques of measure theory and functional analysis, so that we shall only summarize the results. But their value lies in the fact that for certain F almost all $x \in J$ have the same density. Until recently the existence of such densities had not been proved, except for the simplest of functions F. The following result has recently been proved:

THEOREM 2. **[5]**. *Let F: $J \to J$ satisfy the following conditions:*
1) *F is continuous.*
2) *Except at one point $t \in J$, F is twice continuously differentiable.*
3) *F satisfies* (4.1).

Then there exists a function g: $J \to [0,\infty)$, such that for almost all $x \in J$, g is the density of x. Also for almost all $x \in J$, $L(x) = \{y: g(y) > 0\}$ which is an interval. Moreover, the set $J_\infty = \{y: g(y) > 0\}$ is an interval, and $L(x) = J_\infty$ for almost all x.

The proof makes use of results in **[8]**. The problem of computationally finding the density is solved in **[9]**.

A detailed discussion of (3.1) is given in **[16]**, describing how $L(x)$ varies as the parameter a in (3.1) varies between 3.0 and 4.0.

A major question left unsolved is whether (for some nice class of functions F) the existence of a stable periodic point implies that almost every point is asymptotically periodic.

Appendix 1: Period 5 does not imply period 3. In this Appendix we give an example which has a fixed point of period 5 but no fixed point of period 3.

Let F: $[1,5] \to [1,5]$, be defined such that $F(1) = 3$, $F(2) = 5$, $F(3) = 4$, $F(4) = 2$, $F(5) = 1$ and on each interval $[n, n+1]$, $1 \leqq n \leqq 4$, assume F is linear. Then

$$F^3([1,2]) = F^2([3,5]) = F([1,4]) = [2,5].$$

Hence, F^3 has no fixed points in $[1,2]$. Similarly, $F^3([2,3]) = [3,5]$ and $F^3([4,5]) = [1,4]$, so neither of these intervals contains a fixed point of F^3. On the other hand,

$$F^3([3,4]) = F^2([2,4]) = F([2,5]) = [1,5] \supset [3,4].$$

Hence, F^3 must have a fixed point in $[3,4]$. We shall now demonstrate that the fixed point of F^3 is unique and is also a fixed point of F.

Let $p \in [3,4]$ be a fixed point of F^3. Then $F(p) \in [2,4]$. If $F(p) \in [2,3]$, then $F^3(p)$ would be in $[1,2]$ which is impossible since then p could not be a fixed point. Hence $F(p) \in [3,4]$ and $F^2(p) \in [2,4]$. If $F^2(p) \in [2,3]$ we would have $F^3(p) \in [4,5]$, an impossibility. Hence p, $F(p)$, $F^2(p)$ are all in $[3,4]$. On the interval $[3,4]$, F is defined linearly and so $F(x) = 10 - 2x$. It has a fixed point $10/3$ and it is easy to see that F^3 has a unique fixed point, which must be $10/3$. Hence there is no point of period 3.

Appendix 2. Proof of T2 of Theorem 1. Let $\mathcal{M}$ be the set of sequences $M = \{M_n\}_{n=1}^{\infty}$ of intervals with

(A.1) $$M_n = K \quad \text{or} \quad M_n \subset L, \quad \text{and} \quad F(M_n) \supset M_{n+1}$$

$$\text{if} \quad M_n = K \quad \text{then}$$

(A.2) $$n \text{ is the square of an integer and } M_{n+1}, M_{n+2} \subset L,$$

where $K = [a, b]$ and $L = [b, c]$. Of course if n is the square of an integer, then $n+1$ and $n+2$ are not, so the last requirement in (A.2) is redundant. For $M \in \mathcal{M}$, let $P(M, n)$ denote the number of i's in $\{1, \cdots, n\}$ for which $M_i = K$. For each $r \in (3/4, 1)$ choose $M^r = \{M_n^r\}_{n=1}^{\infty}$ to be a sequence in $\mathcal{M}$ such that

(A.3) $$\lim_{n\to\infty} P(M^r, n^2)/n = r.$$

Let $M_0 = \{M^r : r \in (3/4, 1)\} \subset M$. Then $\mathcal{M}_0$ is uncountable since $M^{r_1} \neq M^{r_2}$ for $r_1 \neq r_2$. For each $M^r \in \mathcal{M}_0$, by Lemma 1, there exists a point x_r with $F^n(x_r) \in M_n^r$ for all n. Let $S = \{x_r : r \in (3/4, 1)\}$. Then S is also uncountable. For $x \in S$, let $P(x, n)$ denote the number of i's in $\{1, \cdots, n\}$ for which $F^i(x) \in K$. We can never have $F^k(x_r) = b$, because then x_r would eventually have period 3, contrary to (A.2). Consequently $P(x_r, n) = P(M^r, n)$ for all n, and so

$$\rho(x_r) = \lim_{n\to\infty} P(X_r, n^2) = r$$

for all r. We claim that

(A.4) for $p, q \in S$, with $p \neq q$, there exist infinitely many n's such that $F^n(p) \in K$ and $F^n(q) \in L$ or vice versa.

We may assume $\rho(p) > \rho(q)$. Then $P(p, n) - P(q, n) \to \infty$, and so there must be infintely many n's such that $F^n(p) \in K$ and $F^n(q) \in L$.

Since $F^2(b) = d \leqq a$ and F^2 is continuous, there exists $\delta > 0$ such that $F^2(x) < (b+d)/2$ for all $x \in [b-\delta, b] \subset K$. If $p \in S$ and $F^n(p) \in K$, then (A.2) implies $F^{n+1}(p) \in L$ and $F^{n+2}(p) \in L$. Therefore $F^n(p) < b - \delta$. If $F^n(q) \in L$, then $F^n(q) \geqq b$ so

$$|F^n(p) - F^n(q)| > \delta.$$

By claim (A.4), for any $p, q \in S$, $p \neq q$, it follows

$$\limsup_{n\to\infty} |F^n(p) - F^n(q)| \geqq \delta > 0.$$

Hence (2.1) is proved. This technique may be similarly used to prove (B) is satisfied.

Proof of 2.2. Since $F(b) = c$, $F(c) = d \leqq a$, we may choose intervals $[b^n, c^n]$, $n = 0, 1, 2, \cdots$, such that

(a) $[b, c] = [b^0, c^0] \supset [b^1, c^1] \supset \cdots \supset [b^n, c^n] \supset \cdots$,
(b) $F(x) \in (b^n, c^n)$ for all $x \in (b^{n+1}, c^{n+1})$,
(c) $F(b^{n+1}) = c^n$, $F(c^{n+1}) = b^n$.

Let $A = \bigcap_{n=0}^{\infty} [b^n, c^n]$, $b^* = \inf A$ and $c^* = \sup A$, then $F(b^*) = c^*$ and $F(c^*) = b^*$, because of (c).

In order to prove (2.2) we must be more specific in our choice of the sequences M^r. In addition to our previous requirements on $M \in M$, we will assume that if $M_k = K$ for both $k = n^2$ and $(n+1)^2$ then $M_k = [b^{2n-(2j-1)}, b^*]$ for $k = n^2 + (2j-1)$, $M_k = [c^*, c^{2n-2j}]$ for $k = n^2 + 2j$ where $j = 1, \cdots, n$. For the remaining k's which are not squares of integers, we assume $M_k = L$.

It is easy to check that these requirements are consistent with (A.1) and (A.2), and that we can still choose M^r so as to satisfy (A.3). From the fact that $\rho(x)$ may be thought of as the limit of the fraction of n's for which $F^{n^2}(x) \in K$, it follows that for any $r^*, r \in (3/4, 1)$ there exist infinitely many n such that $M_k^r = M_k^{r^*} = K$ for both $k = n^2$ and $(n+1)^2$. To show (2.2), let $x_r \in S$ and $x_{r^*} \in S$. Since $b^n \to b^*$, $c^n \to c^*$ as $n \to \infty$, for any $\varepsilon > 0$ there exists N with $|b^n - b^*| < \varepsilon/2$, $|c^n - c^*| < \varepsilon/2$ for all $n > N$. Then, for any n with $n > N$ and $M_k^r = M_k^{r^*} = K$ for both $k = n^2$ and $(n+1)^2$, we have

$$F^{n^2+1}(x_r) \in M_k^r = [b^{2n-1}, b^*]$$

with $k = n^2+1$ and $F^{n^2+1}(x_r)$ and $F^{n^2+1}(x_{r^*})$ both belong to $[b^{2n-1}, b^*]$. Therefore, $|F^{n^2+1}(x_r) - F^{n^2+1}(x_{r^*})| < \varepsilon$. Since there are infinitely many n with this property, $\liminf_{n\to\infty} |F^n(x_r) - F^n(x_{r^*})| = 0$. □

REMARK. The theorem can be generalized by assuming that $F: J \to R$ without assuming that $F(J) \subset J$ and we leave this proof to the reader. Of course $F(J) \cap J$ would be nonempty since it would contain the points a, b, and c, assuming that b, c, and d, are defined.

Research partially supported by National Science Foundation grant GP-31386X.

References

1. E. N. Lorenz, The problem of deducing the climate from the governing equations, Tellus, 16 (1964) 1–11.
2. ———, Deterministic nonperiodic flows, J. Atmospheric Sci., 20 (1963) 130–141.
3. ———, The mechanics of vacillation, J. Atmospheric Sci., 20 (1963) 448–464.
4. ———, The predictability of hydrodynamic flow, Trans. N.Y. Acad. Sci., Ser. II, 25 (1963) 409–432.
5. T. Y. Li and J. A. Yorke, Ergodic transformations from an interval into itself, (submitted for publication).
6. P. R. Stein and S. M. Ulam, Nonlinear transformation studies on electronic computers, Los Alamos Sci. Lab., Los Alamos, New Mexico, 1963.
7. S. M. Ulam, A Collection of Mathematical Problems, Interscience, New York, 1960, p. 150.
8. A. Lasota and J. A. Yorke, On the existence of invariant measures for piecewise monotonic transformations, Trans. Amer. Math. Soc., 186 (1973) 481–488.
9. T. Y. Li, Finite approximation for the Frobenius-Perron operator—A Solution to Ulam's conjecture, (submitted for publication).
10. Syunro Utida, Population fluctuation, an experimental and theoretical approach, Cold Spring Harbor Symposia on Quantitative Biology, 22 (1957) 139–151.
11. A. Lasota and P. Rusek, Problems of the stability of the motion in the process of rotary drilling with cogged bits, Archium Gornictua, 15 (1970) 205–216 (Polish with Russian and German Summary).
12. N. Metropolis, M. L. Stein and P. R. Stein, On infinite limit sets for transformations on the unit interval, J. Combinatorial Theory Ser. A 15, (1973) 25–44.
13. S. Smale, Differentiable dynamical systems, Bull. A.M.S., 73 (1967) 747–817 (see §1.5).
14. G. Oster and Y. Takahashi, Models for age specific interactions in a periodic environment, Ecology, in press.
15. D. Auslander, G. Oster and C. Huffaker, Dynamics of interacting populations, a preprint.
16. T. Y. Li and James A. Yorke, The "simplest" dynamics system, (to appear).
17. R. M. May, Biological populations obeying difference equations, stable cycles, and chaos, J. Theor. Biol. (to appear).

DEPARTMENT OF MATHEMATICS, UNIVERSITY OF UTAH, SALT LAKE CITY, UT 84112.
INSTITUTE FOR FLUID DYNAMICS AND APPLIED MATHEMATICS, UNIVERSITY OF MARYLAND, COLLEGE PARK, MD 20742.

(Reprinted from Nature, Vol. 261, No. 5560, pp. 459–467, June 10, 1976)

Simple mathematical models with very complicated dynamics

Robert M. May*

First-order difference equations arise in many contexts in the biological, economic and social sciences. Such equations, even though simple and deterministic, can exhibit a surprising array of dynamical behaviour, from stable points, to a bifurcating hierarchy of stable cycles, to apparently random fluctuations. There are consequently many fascinating problems, some concerned with delicate mathematical aspects of the fine structure of the trajectories, and some concerned with the practical implications and applications. This is an interpretive review of them.

THERE are many situations, in many disciplines, which can be described, at least to a crude first approximation, by a simple first-order difference equation. Studies of the dynamical properties of such models usually consist of finding constant equilibrium solutions, and then conducting a linearised analysis to determine their stability with respect to small disturbances: explicitly nonlinear dynamical features are usually not considered.

Recent studies have, however, shown that the very simplest nonlinear difference equations can possess an extraordinarily rich spectrum of dynamical behaviour, from stable points, through cascades of stable cycles, to a regime in which the behaviour (although fully deterministic) is in many respects "chaotic", or indistinguishable from the sample function of a random process.

This review article has several aims.

First, although the main features of these nonlinear phenomena have been discovered and independently rediscovered by several people, I know of no source where all the main results are collected together. I have therefore tried to give such a synoptic account. This is done in a brief and descriptive way, and includes some new material: the detailed mathematical proofs are to be found in the technical literature, to which signposts are given.

Second, I indicate some of the interesting mathematical questions which do not seem to be fully resolved. Some of these problems are of a practical kind, to do with providing a probabilistic description for trajectories which seem random, even though their underlying structure is deterministic. Other problems are of intrinsic mathematical interest, and treat such things as the pathology of the bifurcation structure, or the truly random behaviour, that can arise when the nonlinear function $F(X)$ of equation (1) is not analytical. One aim here is to stimulate research on these questions, particularly on the empirical questions which relate to processing data.

Third, consideration is given to some fields where these notions may find practical application. Such applications range from the abstractly metaphorical (where, for example, the transition from a stable point to "chaos" serves as a metaphor for the onset of turbulence in a fluid), to models for the dynamic behaviour of biological populations (where one can seek to use field or laboratory data to estimate the values of the parameters in the difference equation).

*King's College Research Centre, Cambridge CB2 1ST; on leave from Biology Department, Princeton University, Princeton 08540.

Fourth, there is a very brief review of the literature pertaining to the way this spectrum of behaviour—stable points, stable cycles, chaos—can arise in second or higher order difference equations (that is, two or more dimensions; two or more interacting species), where the onset of chaos usually requires less severe nonlinearities. Differential equations are also surveyed in this light; it seems that a three-dimensional system of first-order ordinary differential equations is required for the manifestation of chaotic behaviour.

The review ends with an evangelical plea for the introduction of these difference equations into elementary mathematics courses, so that students' intuition may be enriched by seeing the wild things that simple nonlinear equations can do.

First-order difference equations

One of the simplest systems an ecologist can study is a seasonally breeding population in which generations do not overlap[1–4]. Many natural populations, particularly among temperate zone insects (including many economically important crop and orchard pests), are of this kind. In this situation, the observational data will usually consist of information about the maximum, or the average, or the total population in each generation. The theoretician seeks to understand how the magnitude of the population in generation $t+1$, X_{t+1}, is related to the magnitude of the population in the preceding generation t, X_t: such a relationship may be expressed in the general form

$$X_{t+1} = F(X_t) \tag{1}$$

The function $F(X)$ will usually be what a biologist calls "density dependent", and a mathematician calls nonlinear; equation (1) is then a first-order, nonlinear difference equation.

Although I shall henceforth adopt the habit of referring to the variable X as "the population", there are countless situations outside population biology where the basic equation (1) applies. There are other examples in biology, as, for example in genetics[5,6] (where the equation describes the change in gene frequency in time) or in epidemiology[7] (with X the fraction of the population infected at time t). Examples in economics include models for the relationship between commodity quantity and price[8], for the theory of business cycles[9], and for the temporal sequences generated by various other economic quantities[10]. The general equation (1) also is germane to the social sciences[11], where it arises, for example, in theories of learning (where X may be the number of bits of information that can be remembered after an interval t), or in the propagation of rumours in variously structured societies (where X is the number of people to have heard the rumour after time t). The imaginative reader will be able to invent other contexts for equation (1).

In many of these contexts, and for biological populations in particular, there is a tendency for the variable X to increase from one generation to the next when it is small, and for it to decrease when it is large. That is, the nonlinear function $F(X)$ often has the following properties: $F(0)=0$; $F(X)$ increases monotonically as X increases through the range $0<X<A$ (with $F(X)$ attaining its maximum value at $X=A$); and $F(X)$ decreases monotonically as X increases beyond $X=A$. Moreover, $F(X)$ will usually contain one or more parameters which "tune" the severity of this nonlinear behaviour; parameters which tune the steepness of the hump in the $F(X)$ curve. These parameters will typically have some biological or economic or sociological significance.

A specific example is afforded by the equation[1,4,12–23]

$$N_{t+1} = N_t(a - bN_t) \tag{2}$$

This is sometimes called the "logistic" difference equation. In the limit $b=0$, it describes a population growing purely exponentially (for $a>1$); for $b\neq 0$, the quadratic nonlinearity produces a growth curve with a hump, the steepness of which is tuned by the parameter a. By writing $X=bN/a$, the equation may be brought into canonical form[1,4,12–23]

$$X_{t+1} = aX_t(1 - X_t) \tag{3}$$

In this form, which is illustrated in Fig. 1, it is arguably the simplest nonlinear difference equation. I shall use equation (3) for most of the numerical examples and illustrations in this article. Although attractive to mathematicians by virtue of its extreme simplicity, in practical applications equation (3) has the disadvantage that it requires X to remain on the interval $0<X<1$; if X ever exceeds unity, subsequent iterations diverge towards $-\infty$ (which means the population becomes extinct). Furthermore, $F(X)$ in equation (3) attains a maximum value of $a/4$ (at $X=\frac{1}{2}$); the equation therefore possesses non-trivial dynamical behaviour only if $a<4$. On the other hand, all trajectories are attracted to $X=0$ if $a<1$. Thus for non-trivial dynamical behaviour we require $1<a<4$; failing this, the population becomes extinct.

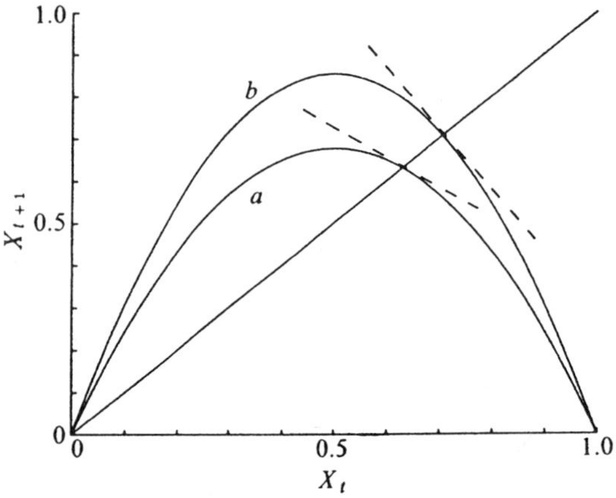

Fig. 1 A typical form for the relationship between X_{t+1} and X_t described by equation (1). The curves are for equation (3), with $a = 2.707$ (*a*); and $a = 3.414$ (*b*). The dashed lines indicate the slope at the "fixed points" where $F(X)$ intersects the 45° line: for the case *a* this slope is less steep than $-45°$ and the fixed point is stable; for *b* the slope is steeper than $-45°$, and the point is unstable.

Another example, with a more secure provenance in the biological literature[1,23–27], is the equation

$$X_{t+1} = X_t \exp[r(1 - X_t)] \tag{4}$$

This again describes a population with a propensity to simple exponential growth at low densities, and a tendency to decrease at high densities. The steepness of this nonlinear behaviour is tuned by the parameter r. The model is plausible for a single species population which is regulated by an epidemic disease at high density[28]. The function $F(X)$ of equation (4) is slightly more complicated than that of equation (3), but has the compensating advantage that local stability implies global stability[1] for all $X>0$.

The forms (3) and (4) by no means exhaust the list of single-humped functions $F(X)$ for equation (1) which can be culled from the ecological literature. A fairly full such catalogue is given, complete with references, by May and Oster[1]. Other similar mathematical functions are given by Metropolis *et al.*[16]. Yet other forms for $F(X)$ are discussed under the heading of "mathematical curiosities" below.

Dynamic properties of equation (1)

Possible constant, equilibrium values (or "fixed points") of X in equation (1) may be found algebraically by putting $X_{t+1}=X_t=X^*$, and solving the resulting equation

$$X^* = F(X^*) \tag{5}$$

An equivalent graphical method is to find the points where the curve $F(X)$ that maps X_t into X_{t+1} intersects the 45° line, $X_{t+1}=X_t$, which corresponds to the ideal nirvana of zero population growth; see Fig. 1. For the single-hump curves discussed above, and exemplified by equations (3) and (4), there are two such points: the trivial solution $X=0$, and a non-trivial solution X^* (which for equation (3) is $X^*=1-[1/a]$).

The next question concerns the stability of the equilibrium point X^*. This can be seen[24,25,19–21,1,4] to depend on the slope of the $F(X)$ curve at X^*. This slope, which is illustrated by the dashed lines in Fig. 1, can be designated

$$\lambda^{(1)}(X^*) = [dF/dX]_{X=X^*} \tag{6}$$

So long as this slope lies between 45° and $-45°$ (that is, $\lambda^{(1)}$ between $+1$ and -1), making an acute angle with the 45° ZPG line, the equilibrium point X^* will be at least locally stable, attracting all trajectories in its neighbourhood. In equation (3), for example, this slope is $\lambda^{(1)}=2-a$: the equilibrium point is therefore stable, and attracts all trajectories originating in the interval $0<X<1$, if and only if $1<a<3$.

As the relevant parameters are tuned so that the curve $F(X)$ becomes more and more steeply humped, this stability-determining slope at X^* may eventually steepen beyond $-45°$ (that is, $\lambda^{(1)}<-1$), whereupon the equilibrium point X^* is no longer stable.

What happens next? What happens, for example, for $a>3$ in equation (3)?

To answer this question, it is helpful to look at the map which relates the populations at successive intervals 2 generations apart; that is, to look at the function which relates X_{t+2} to X_t. This second iterate of equation (1) can be written

$$X_{t+2} = F[F(X_t)] \tag{7}$$

or, introducing an obvious piece of notation,

$$X_{t+2} = F^{(2)}(X_t) \tag{8}$$

The map so derived from equation (3) is illustrated in Figs 2 and 3.

Population values which recur every second generation (that

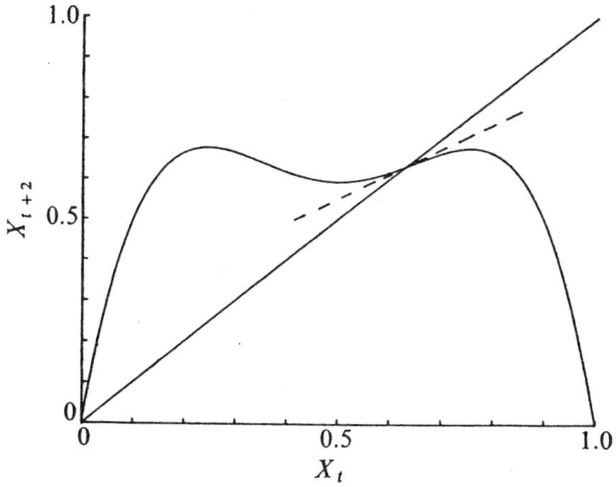

Fig. 2 The map relating X_{t+2} to X_t, obtained by two iterations of equation (3). This figure is for the case (*a*) of Fig. 1, $a = 2.707$: the basic fixed point is stable, and it is the only point at which $F^{(2)}(X)$ intersects the 45° line (where its slope, shown by the dashed line, is less steep than 45°).

is, fixed points with period 2) may now be written as X^*_2, and found either algebraically from

$$X^*_2 = F^{(2)}(X^*_2) \tag{9}$$

or graphically from the intersection between the map $F^{(2)}(X)$ and the 45° line, as shown in Figs 2 and 3. Clearly the equilibrium point X^* of equation (5) is a solution of equation (9); the basic fixed point of period 1 is a degenerate case of a period 2 solution. We now make a simple, but crucial, observation[1]: the slope of the curve $F^{(2)}(X)$ at the point X^*, defined as $\lambda^{(2)}(X^*)$ and illustrated by the dashed lines in Figs 2 and 3, is the square of the corresponding slope of $F(X)$

$$\lambda^{(2)}(X^*) = [\lambda^{(1)}(X^*)]^2 \tag{10}$$

This fact can now be used to make plain what happens when the fixed point X^* becomes unstable. If the slope of $F(X)$ is less than $-45°$ (that is, $|\lambda^{(1)}|<1$), as illustrated by curve *a* in Fig. 1, then X^* is stable. Also, from equation (10), this implies $0<\lambda^{(2)}<1$ corresponding to the slope of $F^{(2)}$ at X^* lying between 0° and 45°, as shown in Fig. 2. As long as the fixed point X^* is stable, it provides the only non-trivial solution to equation (9). On the other hand, when $\lambda^{(1)}$ steepens beyond $-45°$ (that is, $|\lambda^{(1)}|>1$), as illustrated by curve *b* in Fig 1, X^* becomes unstable. At the same time, from equation (10) this implies $\lambda^{(2)}>1$, corresponding to the slope of $F^{(2)}$ at X^* steepening beyond 45°, as shown in Fig. 3. As this happens, the curve $F^{(2)}(X)$ must develop a "loop", and two new fixed points of period 2 appear, as illustrated in Fig. 3.

In short, as the nonlinear function $F(X)$ in equation (1) becomes more steeply humped, the basic fixed point X^* may become unstable. At exactly the stage when this occurs, there are born two new and initially stable fixed points of period 2, between which the system alternates in a stable cycle of period 2. The sort of graphical analysis indicated by Figs 1, 2 and 3, along with the equation (10), is all that is needed to establish this generic result[1,4].

As before, the stability of this period 2 cycle depends on the slope of the curve $F^{(2)}(X)$ at the 2 points. (This slope is easily shown to be the same at both points[1,20], and more generally to be the same at all *k* points on a period *k* cycle.) Furthermore, as is clear by imagining the intermediate stages between Figs 2 and 3, this stability-determining slope has the value $\lambda=+1$ at the birth of the 2-point cycle, and then decreases through zero towards $\lambda=-1$ as the hump in $F(X)$ continues to steepen. Beyond this point the period 2 points will in turn become unstable, and bifurcate to give an initially stable cycle of period 4. This in turn gives way to a cycle of period 8, and thence to a hierarchy of bifurcating stable cycles of periods 16, 32, 64, . . ., 2^n. In each case, the way in which a stable cycle of period *k* becomes unstable, simultaneously bifurcating to produce a new and initially stable cycle of period $2k$, is basically similar to the process just adumbrated for $k=1$. A more full and rigorous account of the material covered so far is in ref. 1.

This "very beautiful bifurcation phenomenon"[22] is depicted in Fig. 4, for the example equation (3). It cannot be too strongly emphasised that the process is generic to most functions $F(X)$ with a hump of tunable steepness. Metropolis *et al.*[16] refer to this hierarchy of cycles of periods 2^n as the harmonics of the fixed point X^*.

Although this process produces an infinite sequence of cycles with periods 2^n ($n\to\infty$), the "window" of parameter values wherein any one cycle is stable progressively diminishes, so that the entire process is a convergent one, being bounded above by some critical parameter value. (This is true for most, but not all, functions $F(X)$: see equation (17) below.) This critical parameter value is a point of accumulation of period 2^n cycles. For equation (3) it is denoted a_c: $a_c=3.5700\ldots$

Beyond this point of accumulation (for example, for $a>a_c$ in equation (3)) there are an infinite number of fixed points with different periodicities, and an infinite number of different periodic cycles. There are also an uncountable number of initial points X_0 which give totally aperiodic (although bounded) trajectories; no matter how long the time series generated by $F(X)$ is run out, the pattern never repeats. These facts may be established by a variety of methods[1,4,20,22,29]. Such a situation, where an infinite number of different orbits can occur, has been christened "chaotic" by Li and Yorke[20].

As the parameter increases beyond the critical value, at first all these cycles have even periods, with X_t alternating up and down between values above, and values below, the fixed point X^*. Although these cycles may in fact be very complicated (having a non-degenerate period of, say, 5,726 points before repeating), they will seem to the casual observer to be rather like a somewhat "noisy" cycle of period 2. As the parameter value continues to increase, there comes a stage (at $a=3.6786\ldots$ for equation (3)) at which the first odd period cycle appears. At first these odd cycles have very long periods, but as the parameter value continues to increase cycles with smaller and smaller odd periods are picked up, until at last the three-point

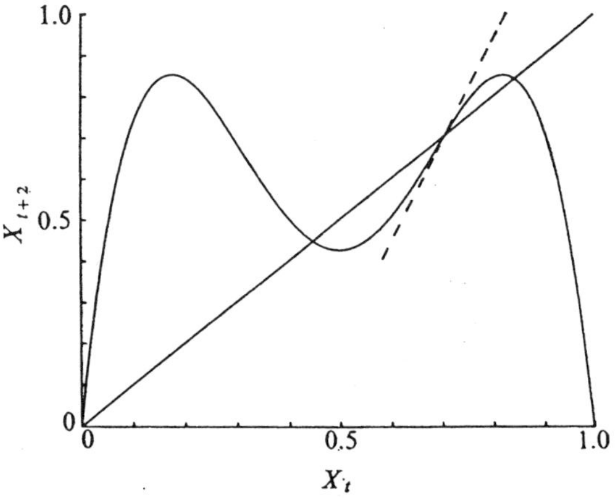

Fig. 3 As for Fig. 2, except that here $a = 3.414$, as in Fig. 1*b*. The basic fixed point is now unstable: the slope of $F^{(2)}(X)$ at this point steepens beyond 45°, leading to the appearance of two new solutions of period 2.

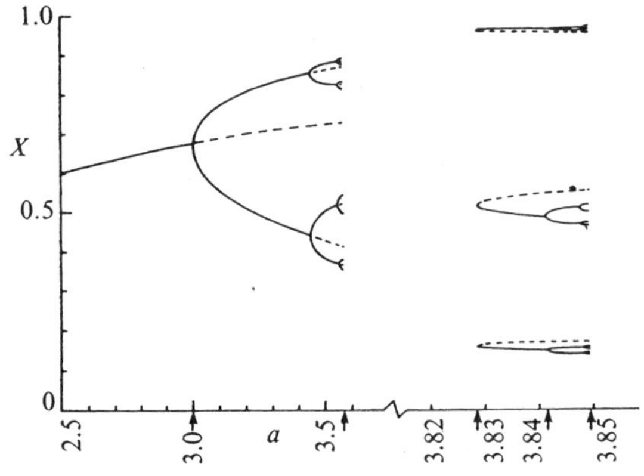

Fig. 4 This figure illustrates some of the stable (———) and unstable (— — — —) fixed points of various periods that can arise by bifurcation processes in equation (1) in general, and equation (3) in particular. To the left, the basic stable fixed point becomes unstable and gives rise by a succession of pitchfork bifurcations to stable harmonics of period 2^n; none of these cycles is stable beyond $a = 3.5700$. To the right, the two period 3 cycles appear by tangent bifurcation: one is initially unstable; the other is initially stable, but becomes unstable and gives way to stable harmonics of period 3×2^n, which have a point of accumulation at $a = 3.8495$. Note the change in scale on the a axis, needed to put both examples on the same figure. There are infinitely many other such windows, based on cycles of higher periods.

cycle appears (at $a=3.8284$ . . for equation (3)). Beyond this point, there are cycles with every integer period, as well as an uncountable number of asymptotically aperiodic trajectories: Li and Yorke[20] entitle their original proof of this result "Period Three Implies Chaos".

The term "chaos" evokes an image of dynamical trajectories which are indistinguishable from some stochastic process. Numerical simulations[12,15,21,23,25] of the dynamics of equation (3), (4) and other similar equations tend to confirm this impression. But, for smooth and "sensible" functions $F(X)$ such as in equations (3) and (4), the underlying mathematical fact is that for any specified parameter value there is one unique cycle that is stable, and that attracts essentially all initial points[22,29] (see ref. 4, appendix A, for a simple and lucid exposition). That is, there is one cycle that "owns" almost all initial points; the remaining infinite number of other cycles, along with the asymptotically aperiodic trajectories, own a set of points which, although uncountable, have measure zero.

As is made clear by Tables 3 and 4 below, any one particular stable cycle is likely to occupy an extraordinarily narrow window of parameter values. This fact, coupled with the long time it is likely to take for transients associated with the initial conditions to damp out, means that in practice the unique cycle is unlikely to be unmasked, and that a stochastic description of the dynamics is likely to be appropriate, in spite of the underlying deterministic structure. This point is pursued further under the heading "practical applications", below.

The main messages of this section are summarised in Table 1, which sets out the various domains of dynamical behaviour of the equations (3) and (4) as functions of the parameters, a and r respectively, that determine the severity of the nonlinear response. These properties can be understood qualitatively in a graphical way, and are generic to any well behaved $F(X)$ in equation (1).

We now proceed to a more detailed discussion of the mathematical structure of the chaotic regime for analytical functions, and then to the practical problems alluded to above and to a consideration of the behavioural peculiarities exhibited by non-analytical functions (such as those in the two right hand columns of Table 1).

Fine structure of the chaotic regime

We have seen how the original fixed point X^* bifurcates to give harmonics of period 2^n. But how do new cycles of period k arise?

The general process is illustrated in Fig. 5, which shows how period 3 cycles originate. By an obvious extension of the notation introduced in equation (8), populations three generations apart are related by

$$X_{t+3} = F^{(3)}(X_t) \tag{11}$$

If the hump in $F(X)$ is sufficiently steep, the threefold iteration will produce a function $F^{(3)}(X)$ with 4 humps, as shown in Fig. 5 for the $F(X)$ of equation (3). At first (for $a < 3.8284$ . . in equation 3) the 45° line intersects this curve only at the single point X^* (and at $X=0$), as shown by the solid curve in Fig. 5. As the hump in $F(X)$ steepens, the hills and valleys in $F^{(3)}(X)$ become more pronounced, until simultaneously the first two valleys sink and the final hill rises to touch the 45° line, and then to intercept it at 6 new points, as shown by the dashed curve in Fig. 5. These 6 points divide into two distinct three-point cycles. As can be made plausible by imagining the intermediate stages in Fig. 5, it can be shown that the stability-determining slope of $F^{(3)}(X)$ at three of these points has a common value, which is $\lambda^{(3)}=+1$ at their birth, and thereafter steepens beyond $+1$: this period 3 cycle is never stable. The slope of $F^{(3)}(X)$ at the other three points begins at $\lambda^{(3)}=+1$, and then decreases towards zero, resulting in a stable cycle of period 3. As $F(X)$ continues to steepen, the slope $\lambda^{(3)}$ for this intially stable three-point cycle decreases beyond -1; the cycle becomes unstable, and gives rise by the bifurcation process discussed in the previous section to stable cycles of period 6, 12, 24, . . ., 3×2^n. This birth of a stable and unstable pair of period 3 cycles, and the subsequent harmonics which arise as the initially stable cycle becomes unstable, are illustrated to the right of Fig. 4.

Table 1 Summary of the way various "single-hump" functions $F(X)$, from equation (1), behave in the chaotic region, distinguishing the dynamical properties which are generic from those which are not

The function $F(X)$ of equation (1)	$aX(1-X)$	$X\exp[r(1-X)]$	aX; if $X<\frac{1}{2}$ $a(1-X)$; if $X>\frac{1}{2}$	λX; if $X<1$ λX^{1-b}; if $X>1$
Tunable parameter	a	r	a	b
Fixed point becomes unstable	3.0000	2.0000	1.0000*	2.0000
"Chaotic" region begins [point of accumulation of cycles of period 2^n]	3.5700	2.6924	1.0000	2.0000
First odd-period cycle appears	3.6786	2.8332	1.4142	2.6180
Cycle with period 3 appears [and therefore every integer period present]	3.8284	3.1024	1.6180	3.0000
"Chaotic" region ends	4.0000†	∞‡	2.000†	∞‡
Are there stable cycles in the chaotic region?	Yes	Yes	No	No

* Below this a value, $X = 0$ is stable.
† All solutions are attracted to $-\infty$ for a values beyond this.
‡ In practice, as r or b becomes large enough, X will eventually be carried so low as to be effectively zero, thus producing extinction in models of biological populations.

Table 2 Catalogue of the number of periodic points, and of the various cycles (with periods $k = 1$ up to 12), arising from equation (1) with a single-humped function $F(X)$

k	1	2	3	4	5	6	7	8	9	10	11	12
Possible total number of points with period k	2	4	8	16	32	64	128	256	512	1,024	2,048	4,096
Possible total number of points with non-degenerate period k	2	2	6	12	30	54	126	240	504	990	2,046	4,020
Total number of cycles of period k, including those which are degenerate and/or harmonics and/or never locally stable	2	3	4	6	8	14	20	36	60	108	188	352
Total number of non-degenerate cycles (including harmonics and unstable cycles)	2	1	2	3	6	9	18	30	56	99	186	335
Total number of non-degenerate, stable cycles (including harmonics)	1	1	1	2	3	5	9	16	28	51	93	170
Total number of non-degenerate, stable cycles whose basic period is k (that is, excluding harmonics)	1	–	1	1	3	4	9	14	28	48	93	165

There are, therefore, two basic kinds of bifurcation processes[1,4] for first order difference equations. Truly new cycles of period k arise in pairs (one stable, one unstable) as the hills and valleys of higher iterates of $F(X)$ move, respectively, up and down to intercept the 45° line, as typified by Fig. 5. Such cycles are born at the moment when the hills and valleys become tangent to the 45° line, and the initial slope of the curve $F^{(k)}$ at the points is thus $\lambda^{(k)}=+1$: this type of bifurcation may be called[1,4] a tangent bifurcation or a $\lambda=+1$ bifurcation. Conversely, an originally stable cycle of period k may become unstable as $F(X)$ steepens. This happens when the slope of $F^{(k)}$ at these period k points steepens beyond $\lambda^{(k)}=-1$, whereupon a new and initially stable cycle of period $2k$ is born in the way typified by Figs 2 and 3. This type of bifurcation may be called a pitchfork bifurcation (borrowing an image from the left hand side of Fig. 4) or a $\lambda=-1$ bifurcation[1,4].

Putting all this together, we conclude that as the parameters in $F(X)$ are varied the fundamental, stable dynamical units are cycles of basic period k, which arise by tangent bifurcation, along with their associated cascade of harmonics of periods $k2^n$, which arise by pitchfork bifurcation. On this basis, the constant equilibrium solution X^* and the subsequent hierarchy of stable cycles of periods 2^n is merely a special case, albeit a conspicuously important one (namely $k=1$), of a general phenomenon. In addition, remember[1,4,22,29] that for sensible, analytical functions (such as, for example, those in equations (3) and (4)) there is a unique stable cycle for each value of the parameter in $F(X)$. The entire range of parameter values ($1<a<4$ in equation (3), $0<r$ in equation (4)) may thus be regarded as made up of infinitely many windows of parameter values—some large, some unimaginably small—each corresponding to a single one of these basic dynamical units. Tables 3 and 4, below, illustrate this notion. These windows are divided from each other by points (the points of accumulation of the harmonics of period $k2^n$) at which the system is truly chaotic, with no attractive cycle: although there are infinitely many such special parameter values, they have measure zero on the interval of all values.

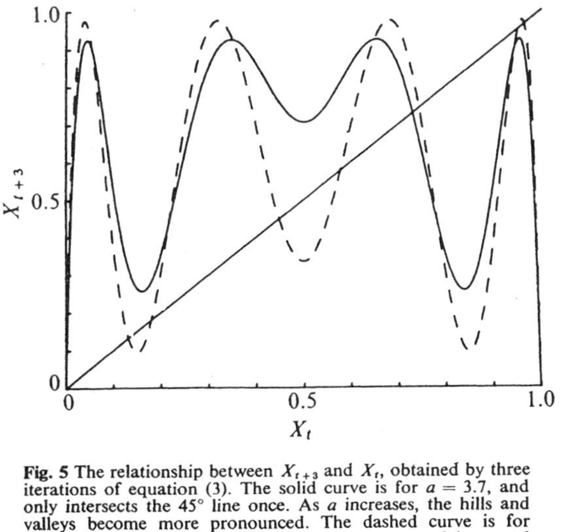

Fig. 5 The relationship between X_{t+3} and X_t, obtained by three iterations of equation (3). The solid curve is for $a = 3.7$, and only intersects the 45° line once. As a increases, the hills and valleys become more pronounced. The dashed curve is for $a = 3.9$, and six new period 3 points have appeared (arranged as two cycles, each of period 3).

How are these various cycles arranged along the interval of relevant parameter values? This question has to my knowledge been answered independently by at least 6 groups of people, who have seen the problem in the context of combinatorial theory[16,30], numerical analysis[13,14], population biology[1], and dynamical systems theory[22,31] (broadly defined).

A simple-minded approach (which has the advantage of requiring little technical apparatus, and the disadvantage of being rather clumsy) consists of first answering the question, how many period k points can there be? That is, how many distinct solutions can there be to the equation

$$X^*_k = F^{(k)}(X^*_k)? \qquad (12)$$

If the function $F(X)$ is sufficiently steeply humped, as it will be once the parameter values are sufficiently large, each successive iteration doubles the number of humps, so that $F^{(k)}(X)$ has 2^{k-1} humps. For large enough parameter values, all these hills and valleys will intersect the 45° line, producing 2^k fixed points of period k. These are listed for $k \leqslant 12$ in the top row of Table 2. Such a list includes degenerate points of period k, whose period is a submultiple of k; in particular, the two period 1 points ($X=0$ and X^*) are degenerate solutions of equation (12) for all k. By working from left to right across Table 2, these degenerate points can be subtracted out, to leave the total number of non-degenerate points of basic period k, as listed in the second row of Table 2. More sophisticated ways of arriving at this result are given elsewhere[13,14,16,22,30,31].

For example, there eventually are $2^6=64$ points with period 6. These include the two points of period 1, the period 2 "harmonic" cycle, and the stable and unstable pair of triplets of points with period 3, for a total of 10 points whose basic period is a submultiple of 6; this leaves 54 points whose basic period is 6.

The 2^k period k points are arranged into various cycles of period k, or submultiples thereof, which appear in succession by either tangent or pitchfork bifurcation as the parameters in $F(X)$ are varied. The third row in Table 2 catalogues the total number of distinct cycles of period k which so appear. In the fourth row[14], the degenerate cycles are subtracted out, to give the total number of non-degenerate cycles of period k: these numbers must equal those of the second row divided by k. This fourth row includes the (stable) harmonics which arise by pitchfork bifurcation, and the pairs of stable–unstable cycles arising by tangent bifurcation. By subtracting out the cycles which are unstable from birth, the total number of possible stable cycles is given in row five; these figures can also be obtained by less pedestrian methods[13,16,30]. Finally we may subtract out the stable cycles which arise by pitchfork bifurcation, as harmonics of some simpler cycle, to arrive at the final

Table 3 A catalogue of the stable cycles (with basic periods up to 6) for the equation $X_{t+1} = aX_t(1-X_t)$

Period of basic cycle	*a* value at which: Basic cycle first appears	Basic cycle becomes unstable	Subsequent cascade of "harmonics" with period $k2^n$ all become unstable	Width of the range of *a* values over which the basic cycle, or one of its harmonics, is attractive
1	1.0000	3.0000	3.5700	2.5700
3	3.8284	3.8415	3.8495	0.0211
4	3.9601	3.9608	3.9612	0.0011
5(*a*)	3.7382	3.7411	3.7430	0.0048
5(*b*)	3.9056	3.9061	3.9065	0.0009
5(*c*)	3.99026	3.99030	3.99032	0.00006
6(*a*)	3.6265	3.6304	3.6327	0.0062
6(*b*)	3.937516	3.937596	3.937649	0.000133
6(*c*)	3.977760	3.977784	3.977800	0.000040
6(*d*)	3.997583	3.997585	3.997586	0.000003

row in Table 2, which lists the number of stable cycles whose basic period is k.

Returning to the example of period 6, we have already noted the five degenerate cycles whose periods are submultiples of 6. The remaining 54 points are parcelled out into one cycle of period 6 which arises as the harmonic of the only stable three-point cycle, and four distinct pairs of period 6 cycles (that is, four initially stable ones and four unstable ones) which arise by successive tangent bifurcations. Thus, reading from the foot of the column for period 6 in Table 2, we get the numbers 4, 5, 9, 14.

Using various labelling tricks, or techniques from combinatorial theory, it is also possible to give a generic list of the order in which the various cycles appear[1,13,16,22]. For example, the basic stable cycles of periods 3, 5, 6 (of which there are respectively 1, 3, 4) must appear in the order 6, 5, 3, 5, 6, 6, 5, 6: compare Tables 3 and 4. Metropolis *et al.*[16] give the explicit such generic list for all cycles of period $k \leq 11$.

As a corollary it follows that, given the most recent cycle to appear, it is possible (at least in principle) to catalogue all the cycles which have appeared up to this point. An especially elegant way of doing this is given by Smale and Williams[22], who show, for example, that when the stable cycle of period 3 first originates, the total number of other points with periods k, N_k, which have appeared by this stage satisfy the Fibonacci series, $N_k=2$, 4, 5, 8, 12, 19, 30, 48, 77, 124, 200, 323 for $k=1, 2, \ldots, 12$: this is to be contrasted with the total number of points of period k which will eventually appear (the top row of Table 2) as $F(X)$ continues to steepen.

Such catalogues of the total number of fixed points, and of their order of appearance, are relatively easy to construct. For any particular function $F(X)$, the numerical task of finding the windows of parameter values wherein any one cycle or its harmonics is stable is, in contrast, relatively tedious and inelegant. Before giving such results, two critical parameter values of special significance should be mentioned.

Hoppensteadt and Hyman[21] have given a simple graphical method for locating the parameter value in the chaotic regime at which the first odd period cycle appears. Their analytic recipe is as follows. Let α be the parameter which tunes the steepness of $F(X)$ (for example, $\alpha=a$ for equation (3), $\alpha=r$ for equation (4)), $X^*(\alpha)$ be the fixed point of period 1 (the non-trivial solution of equation (5)), and $X_{max}(\alpha)$ the maximum value attainable from iterations of equation (1) (that is, the value of $F(X)$ at its hump or stationary point). The first odd period cycle appears for that value of α which satisfies[21,31]

$$X^*(\alpha) = F^{(2)}(X_{max}(\alpha)) \tag{13}$$

As mentioned above, another critical value is that where the period 3 cycle first appears. This parameter value may be found numerically from the solutions of the third iterate of equation (1): for equation (3) it is[14] $a=1+\sqrt{8}$.

Myrberg[13] (for all $k \leq 10$) and Metropolis *et al.*[16]. (for all $k \leq 7$) have given numerical information about the stable cycles in equation (3). They do not give the windows of parameter values, but only the single value at which a given cycle is maximally stable; that is, the value of a for which the stability-determining slope of $F^{(k)}(X)$ is zero, $\lambda^{(k)}=0$. Since the slope of the k-times iterated map $F^{(k)}$ at any point on a period k cycle is simply equal to the product of the slopes of $F(X)$ at each of the points X^*_k on this cycle[1,8,20], the requirement $\lambda^{(k)}=0$ implies that $X=A$ (the stationary point of $F(X)$, where $\lambda^{(1)}=0$) is one of the periodic points in question, which considerably simplifies the numerical calculations.

For each basic cycle of period k (as catalogued in the last row of Table 2), it is more interesting to know the parameter values at which: (1) the cycle first appears (by tangent bifurcation); (2) the basic cycle becomes unstable (giving rise by successive pitchfork bifurcations to a cascade of harmonics of periods $k2^n$); (3) all the harmonics become unstable (the point of accumulation of the period $k2^n$ cycles). Tables 3 and 4 extend the work of May and Oster[1], to give this numerical information for equations (3) and (4), respectively. (The points of accumulation are not ground out mindlessly, but are calculated by a rapidly convergent iterative procedure, see ref. 1, appendix A.) Some of these results have also been obtained by Gumowski and Mira[32].

Practical problems

Referring to the paradigmatic example of equation (3), we can now see that the parameter interval $1<a<4$ is made up of a one-dimensional mosaic of infinitely many windows of a-values, in each of which a unique cycle of period k, or one of its harmonics, attracts essentially all initial points. Of these windows, that for $1<a<3.5700$ . . corresponding to $k=1$ and its harmonics is by far the widest and most conspicuous. Beyond the first point of accumulation, it can be seen from Table 3 that these windows are narrow, even for cycles of quite low periods, and the windows rapidly become very tiny as k increases.

As a result, there develops a dichotomy between the underlying mathematical behaviour (which is exactly determinable) and the "commonsense" conclusions that one would draw from numerical simulations. If the parameter a is held constant at one value in the chaotic region, and equation (3) iterated for an arbitrarily large number of generations, a density plot of the observed values of X_t on the interval 0 to 1 will settle into k equal spikes (more precisely, delta functions) corresponding to the k points on the stable cycle appropriate to this a-value. But for most a-values this cycle will have a fairly large period, and moreover it will typically take many thousands of generations before the transients associated with the initial conditions are damped out: thus the density plot produced by numerical simulations usually looks like a sample of points taken from some continuous distribution.

An especially interesting set of numerical computations are due to Hoppensteadt (personal communication) who has combined many iterations to produce a density plot of X_t for each one of a sequence of a-values, gradually increasing from 3.5700 . . to 4. These results are displayed as a movie. As can be expected from Table 3, some of the more conspicuous cycles do show up as sets of delta functions: the 3-cycle and its first few harmonics; the first 5-cycle; the first 6-cycle. But for most values of a the density plot looks like the sample function of a random process. This is particularly true in the neighbourhood of the a-value where the first odd cycle appears ($a=3.6786$ . .), and again in the neighbourhood of $a=4$: this is not surprising, because each of these locations is a point of accumulation of points of accumulation. Despite the underlying discontinuous changes in the periodicities of the stable cycles, the observed density pattern tends to vary smoothly. For example, as a increases toward the value at which the 3-cycle appears, the density plot tends to concentrate around three points, and it smoothly diffuses away from these three points after the 3-cycle and all its harmonics become unstable.

I think the most interesting mathematical problem lies in designing a way to construct some approximate and "effectively continuous" density spectrum, despite the fact that the exact density function is determinable and is always a set of delta functions. Perhaps such techniques have already been developed in ergodic theory[33] (which lies at the foundations of statistical mechanics), as for example in the use of "coarse-grained observers". I do not know.

Such an effectively stochastic description of the dynamical properties of equation (4) for large r has been provided[28], albeit by tactical tricks peculiar to that equation rather than by any general method. As r increases beyond about 3, the trajectories generated by this equation are, to an increasingly good approximation, almost periodic with period $(1/r)\exp(r-1)$.

The opinion I am airing in this section is that although the exquisite fine structure of the chaotic regime is mathematically fascinating, it is irrelevant for most practical purposes. What seems called for is some effectively stochastic description of the deterministic dynamics. Whereas the various statements about the different cycles and their order of appearance can be made in generic fashion, such stochastic description of the actual dynamics will be quite different for different $F(X)$: witness the difference between the behaviour of equation (4), which for large r is almost periodic "outbreaks" spaced many generations apart, versus the behaviour of equation (3), which for $a \to 4$ is not very different from a series of Bernoulli coin flips.

Mathematical curiosities

As discussed above, the essential reason for the existence of a succession of stable cycles throughout the "chaotic" regime is that as each new pair of cycles is born by tangent bifurcation (see Fig. 5), one of them is at first stable, by virtue of the way the smoothly rounded hills and valleys intercept the 45° line. For analytical functions $F(X)$, the only parameter values for which the density plot or "invariant measure" is continuous and truly ergodic are at the points of accumulation of harmonics, which divide one stable cycle from the next. Such exceptional parameter values have found applications, for example, in the use of equation (3) with $a=4$ as a random number generator[34,35]: it has a continuous density function proportional to $[X(1-X)]^{-\frac{1}{2}}$ in the interval $0<X<1$.

Non-analytical functions $F(X)$ in which the hump is in fact a spike provide an interesting special case. Here we may imagine spikey hills and valleys moving to intercept the 45° line in Fig. 5, and it may be that both the cycles born by tangent bifurcation are unstable from the outset (one having $\lambda^{(k)}>1$, the other $\lambda^{(k)}<-1$), for all $k>1$. There are then no stable cycles in the chaotic regime, which is therefore literally chaotic with a continuous and truly ergodic density distribution function.

One simple example is provided by

$$X_{t+1} = aX_t; \text{ if } X_t<\tfrac{1}{2} \qquad (14)$$
$$X_{t+1} = a(1-X_t); \text{ if } X_t>\tfrac{1}{2}$$

defined on the interval $0<X<1$. For $0<a<1$, all trajectories are attracted to $X=0$; for $1<a<2$, there are infinitely many periodic orbits, along with an uncountable number of aperiodic trajectories, none of which are locally stable. The first odd period cycle appears at $a=\sqrt{2}$, and all integer periods are represented beyond $a=(1+\sqrt{5})/2$. Kac[36] has given a careful discussion of the case $a=2$. Another example, this time with an extensive biological pedigree[1-3], is the equation

$$X_{t+1} = \lambda X_t; \text{ if } X_t<1 \qquad (15)$$
$$X_{t+1} = \lambda X_t^{1-b}; \text{ if } X_t>1$$

If $\lambda>1$ this possesses a globally stable equilibrium point for $b<2$. For $b>2$ there is again true chaos, with no stable cycles: the first odd cycle appears at $b=(3+\sqrt{5})/2$, and all integer periods are present beyond $b=3$. The dynamical properties of equations (14) and (15) are summarised to the right of Table 2.

The absence of analyticity is a necessary, but not a sufficient, condition for truly random behaviour[31]. Consider, for example,

$$X_{t+1} = (a/2)X_t; \text{ if } X_t<\tfrac{1}{2}$$
$$X_{t+1} = aX_t(1-X_t)\ ; \text{ if } X_t>\tfrac{1}{2} \qquad (16)$$

This is the parabola of equation (3) and Fig. 1, but with the left hand half of $F(X)$ flattened into a straight line. This equation does possess windows of a values, each with its own stable cycle, as described generically above. The stability-determining slopes $\lambda^{(k)}$ vary, however, discontinuously with the parameter a, and the widths of the simpler stable regions are narrower than for equation (3): the fixed point becomes unstable at $a=3$; the point of accumulation of the subsequent harmonics is at $a=3.27$ . .; the first odd cycle appears at $a=3.44$ . .; the 3-point cycle at $a=3.67$. . (compare the first column in Table 1).

These eccentricities of behaviour manifested by non-analytical functions may be of interest for exploring formal questions in ergodic theory. I think, however, that they have no relevance to models in the biological and social sciences, where functions such as $F(X)$ should be analytical. This view is elaborated elsewhere[37].

As a final curiosity, consider the equation

$$X_{t+1} = \lambda X_t[1+X_t]^{-\beta} \qquad (17)$$

Table 4 Catalogue of the stable cycles (with basic periods up to 6) for the equation $X_{t+1} = X_t \exp[r(1-X_t)]$

Period of basic cycle	r value at which: Basic cycle first appears	Basic cycle becomes unstable	Subsequent cascade of "harmonics" with period $k2^n$ *all* become unstable	Width of the range of r values over with the basic cycle, or one of its harmonics, is attractive
1	0.0000	2.0000	2.6924	2.6924
3	3.1024	3.1596	3.1957	0.0933
4	3.5855	3.6043	3.6153	0.0298
5(*a*)	2.9161	2.9222	2.9256	0.0095
5(*b*)	3.3632	3.3664	3.3682	0.0050
5(*c*)	3.9206	3.9295	3.9347	0.0141
6(*a*)	2.7714	2.7761	2.7789	0.0075
6(*b*)	3.4558	3.4563	3.4567	0.0009
6(*c*)	3.7736	3.7745	3.7750	0.0014
6(*d*)	4.1797	4.1848	4.1880	0.0083

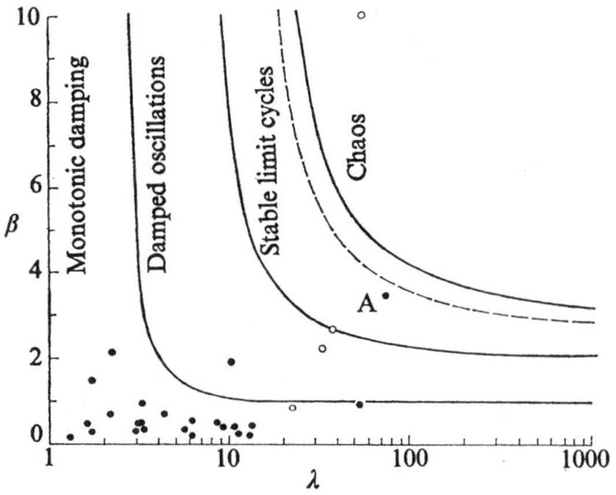

Fig. 6 The solid lines demarcate the stability domains for the density dependence parameter, β, and the population growth rate, λ, in equation (17); the dashed line shows where 2-point cycles give way to higher cycles of period 2^n. The solid circles come from analyses of life table data on field populations, and the open circles from laboratory populations (from ref. 3, after ref. 39).

This has been used to fit a considerable amount of data on insect populations[38,39]. Its stability behaviour, as a function of the two parameters λ and β, is illustrated in Fig. 6. Notice that for $\lambda < 7.39\ldots$ there is a globally stable equilibrium point for all β; for $7.39\ldots < \lambda < 12.50\ldots$ this fixed point becomes unstable for sufficiently large β, bifurcating to a stable 2-point cycle which is the solution for all larger β; as λ increases through the range $12.50\ldots < \lambda < 14.77\ldots$ various other harmonics of period 2^n appear in turn. The hierarchy of bifurcating cycles of period 2^n is thus truncated, and the point of accumulation and subsequent regime of chaos is not achieved (even for arbitrarily large β) until $\lambda > 14.77\ldots$

Applications

The fact that the simple and deterministic equation (1) can possess dynamical trajectories which look like some sort of random noise has disturbing practical implications. It means, for example, that apparently erratic fluctuations in the census data for an animal population need not necessarily betoken either the vagaries of an unpredictable environment or sampling errors: they may simply derive from a rigidly deterministic population growth relationship such as equation (1). This point is discussed more fully and carefully elsewhere[1].

Alternatively, it may be observed that in the chaotic regime arbitrarily close initial conditions can lead to trajectories which, after a sufficiently long time, diverge widely. This means that, even if we have a simple model in which all the parameters are determined exactly, long term prediction is nevertheless impossible. In a meteorological context, Lorenz[15] has called this general phenomenon the "butterfly effect": even if the atmosphere could be described by a deterministic model in which all parameters were known, the fluttering of a butterfly's wings could alter the initial conditions, and thus (in the chaotic regime) alter the long term prediction.

Fluid turbulence provides a classic example where, as a parameter (the Reynolds number) is tuned in a set of deterministic equations (the Navier–Stokes equations), the motion can undergo an abrupt transition from some stable configuration (for example, laminar flow) into an apparently stochastic, chaotic regime. Various models, based on the Navier–Stokes differential equations, have been proposed as mathematical metaphors for this process[15,40,41]. In a recent review of the theory of turbulence, Martin[42] has observed that the one-dimensional difference equation (1) may be useful in this context. Compared with the earlier models[15,40,41], it has the disadvantage of being even more abstractly metaphorical, and the advantage of having a spectrum of dynamical behaviour which is more richly complicated yet more amenable to analytical investigation.

A more down-to-earth application is possible in the use of equation (1) to fit data[1,2,3,38,39,43] on biological populations with discrete, non-overlapping generations, as is the case for many temperate zone arthropods. Figure 6 shows the parameter values λ and β that are estimated[39] for 24 natural populations and 4 laboratory populations when equation (17) is fitted to the available data. The figure also shows the theoretical stability domains: a stable point; its stable harmonics (stable cycles of period 2^n); chaos. The natural populations tend to have stable equilibrium point behaviour. The laboratory populations tend to show oscillatory or chaotic behaviour; their behaviour may be exaggeratedly nonlinear because of the absence, in a laboratory setting, of many natural mortality factors. It is perhaps suggestive that the most oscillatory natural population (labelled A in Fig. 6) is the Colorado potato beetle, whose present relationship with its host plant lacks an evolutionary pedigree. These remarks are only tentative, and must be treated with caution for several reasons. Two of the main caveats are that there are technical difficulties in selecting and reducing the data, and that there are no single species populations in the natural world: to obtain a one-dimensional difference equation by replacing a population's interactions with its biological and physical environment by passive parameters (such as λ and β) may do great violence to the reality.

Some of the many other areas where these ideas have found applications were alluded to in the second section, above[5–11]. One aim of this review article is to provoke applications in yet other fields.

Related phenomena in higher dimensions

Pairs of coupled, first-order difference equations (equivalent to a single second-order equation) have been investigated in several contexts[4,44–46], particularly in the study of temperate zone arthropod prey–predator systems[2–4,23,47]. In these two-dimensional systems, the complications in the dynamical behaviour are further compounded by such facts as: (1) even for analytical functions, there can be truly chaotic behaviour (as for equations (14) and (15)), corresponding to so-called "strange attractors"; and (2) two or more different stable states (for example, a stable point and a stable cycle of period 3) can occur together for the same parameter values[4]. In addition, the manifestation of these phenomena usually requires less severe nonlinearities (less steeply humped $F(X)$) than for the one-dimensional case.

Similar systems of first-order ordinary differential equations, or two coupled first-order differential equations, have much simpler dynamical behaviour, made up of stable and unstable points and limit cycles[48]. This is basically because in continuous two-dimensional systems the inside and outside of closed curves can be distinguished; dynamic trajectories cannot cross each other. The situation becomes qualitatively more complicated, and in many ways analogous to first-order difference equations, when one moves to systems of three or more coupled, first-order ordinary differential equations (that is, three-dimensional systems of ordinary differential equations). Scanlon (personal communication) has argued that chaotic behaviour and "strange attractors", that is solutions which are neither points nor periodic orbits[48], are typical of such systems. Some well studied examples arise in models for reaction–diffusion systems in chemistry and biology[49], and in the models of Lorenz[15] (three dimensions) and Ruelle and Takens[40] (four dimensions) referred to above. The analysis of these systems is, by virtue of their higher dimensionality, much less transparent than for equation (1).

An explicit and rather surprising example of a system which

has recently been studied from this viewpoint is the ordinary differential equations used in ecology to describe competing species. For one or two species these systems are very tame: dynamic trajectories will converge on some stable equilibrium point (which may represent coexistence, or one or both species becoming extinct). As Smale[50] has recently shown, however, for 3 or more species these general equations can, in a certain reasonable and well-defined sense, be compatible with any dynamical behaviour. Smale's[50] discussion is generic and abstract: a specific study of the very peculiar dynamics which can be exhibited by the familiar Lotka-Volterra equations once there are 3 competitors is given by May and Leonard[51].

Conclusion

In spite of the practical problems which remain to be solved, the ideas developed in this review have obvious applications in many areas.

The most important applications, however, may be pedagogical.

The elegant body of mathematical theory pertaining to linear systems (Fourier analysis, orthogonal functions, and so on), and its successful application to many fundamentally linear problems in the physical sciences, tends to dominate even moderately advanced University courses in mathematics and theoretical physics. The mathematical intuition so developed ill equips the student to confront the bizarre behaviour exhibited by the simplest of discrete nonlinear systems, such as equation (3). Yet such nonlinear systems are surely the rule, not the exception, outside the physical sciences.

I would therefore urge that people be introduced to, say, equation (3) early in their mathematical education. This equation can be studied phenomenologically by iterating it on a calculator, or even by hand. Its study does not involve as much conceptual sophistication as does elementary calculus. Such study would greatly enrich the student's intuition about nonlinear systems.

Not only in research, but also in the everyday world of politics and economics, we would all be better off if more people realised that simple nonlinear systems do not necessarily possess simple dynamical properties.

I have received much help from F. C. Hoppensteadt, H. E. Huppert, A. I. Mees, C. J. Preston, S. Smale, J. A. Yorke, and particularly from G. F. Oster. This work was supported in part by the NSF.

1 May, R. M., and Oster, G. F., *Am. Nat.*, **110** (in the press).
2 Varley, G. C., Gradwell, G. R., and Hassell, M. P., *Insect Population Ecology* (Blackwell, Oxford, 1973).
3 May, R. M. (ed.), *Theoretical Ecology: Principles and Applications* (Blackwell, Oxford, 1976).
4 Guckenheimer, J., Oster, G. F., and Ipaktchi, A., *Theor. Pop. Biol.* (in the press).
5 Oster, G. F., Ipaktchi, A., and Rocklin, I., *Theor. Pop. Biol.* (in the press).
6 Asmussen, M. A., and Feldman, M. W., *J. theor. Biol.* (in the press).
7 Hoppensteadt, F. C., *Mathematical Theories of Populations: Demographics, Genetics and Epidemics* (SIAM, Philadelphia, 1975).
8 Samuelson, P. A., *Foundations of Economic Analysis* (Harvard University Press, Cambridge, Massachusetts, 1947).
9 Goodwin, R. E., *Econometrica*, **19**, 1–17 (1951).
10 Baumol, W. J., *Economic Dynamics*, 3rd ed. (Macmillan, New York, 1970).
11 See, for example, Kemeny, J., and Snell, J. L., *Mathematical Models in the Social Sciences* (MIT Press, Cambridge, Massachusetts, 1972).
12 Chaundy, T. W., and Phillips, E., *Q. Jl Math. Oxford*, **7**, 74–80 (1936).
13 Myrberg, P. J., *Ann. Akad. Sc. Fennicae*, *A*, I, No. 336/3 (1963).
14 Myrberg, P. J., *Ann. Akad. Sc. Fennicae*, *A*, I, No. 259 (1958).
15 Lorenz, E. N., *J. Atmos. Sci.*, **20**, 130–141 (1963); *Tellus*, **16**, 1–11 (1964).
16 Metropolis, N., Stein, M. L., and Stein, P. R., *J. Combinatorial Theory*, 15(A), 25–44 (1973).
17 Maynard Smith, J., *Mathematical Ideas in Biology* (Cambridge University Press, Cambridge, 1968).
18 Krebs, C. J., *Ecology* (Harper and Row, New York, 1972).
19 May, R. M., *Am. Nat.*, **107**, 46–57 (1972).
20 Li, T-Y., and Yorke, J. A., *Am. Math. Monthly*, **82**, 985–992 (1975).
21 Hoppensteadt, F. C., and Hyman, J. M. (Courant Institute, New York University: preprint, 1975).
22 Smale, S., and Williams, R. (Department of Mathematics, Berkeley: preprint, 1976).
23 May, R. M., *Science*, **186**, 645–647 (1974).
24 Moran, P. A. P., *Biometrics*, **6**, 250–258 (1950).
25 Ricker, W. E., *J. Fish. Res. Bd. Can.*, **11**, 559–623 (1954).
26 Cook, L. M., *Nature*, **207**, 316 (1965).
27 Macfadyen, A., *Animal Ecology: Aims and Methods* (Pitman, London, 1963).
28 May, R. M., *J. theor. Biol.*, **51**, 511–524 (1975).
29 Guckenheimer, J., *Proc. AMS Symposia in Pure Math.*, *XIV*, 95–124 (1970).
30 Gilbert, E. N., and Riordan, J., *Illinois J. Math.*, **5**, 657–667 (1961).
31 Preston, C. J. (King's College, Cambridge: preprint, 1976).
32 Gumowski, I., and Mira, C., *C. r. hebd. Séanc. Acad. Sci., Paris*, **281a**, 45–48 (1975); **282a**, 219–222 (1976).
33 Layzer, D., *Sci. Am.*, **233**(6), 56–69 (1975).
34 Ulam, S. M., *Proc. Int. Congr. Math.1950, Cambridge, Mass.; Vol. II*, pp. 264–273 (AMS, Providence R.I., 1950).
35 Ulam, S. M., and von Neumann, J., *Bull. Am. math. Soc.* (abstr.), **53**, 1120 (1947).
36 Kac, M., *Ann. Math.*, **47**, 33–49 (1946).
37 May, R. M., *Science*, **181**, 1074 (1973).
38 Hassell, M. P., *J. Anim. Ecol.*, **44**, 283–296 (1974).
39 Hassell, M. P., Lawton, J. H., and May, R. M., *J. Anim. Ecol.* (in the press).
40 Ruelle, D., and Takens, F., *Comm. math. Phys.*, **20**, 167–192 (1971).
41 Landau, L. D., and Lifshitz, E. M., *Fluid Mechanics* (Pergamon, London, 1959).
42 Martin, P. C., *Proc. Int. Conf. on Statistical Physics, 1975, Budapest* (Hungarian Acad. Sci., Budapest, in the press).
43 Southwood, T. R. E., in *Insects, Science and Society* (edit. by Pimentel, D.), 151–199 (Academic, New York, 1975).
44 Metropolis, N., Stein, M. L., and Stein, P. R., *Numer. Math.*, **10**, 1–19 (1967).
45 Gumowski, I., and Mira, C., *Automatica*, **5**, 303–317 (1969).
46 Stein, P. R., and Ulam, S. M., *Rosprawy Mat.*, **39**, 1–66 (1964).
47 Beddington, J. R., Free, C. A., and Lawton, J. H., *Nature*, **255**, 58–60 (1975).
48 Hirsch, M. W., and Smale, S., *Differential Equations, Dynamical Systems and Linear Algebra* (Academic, New York, 1974).
49 Kolata, G. B., *Science*, **189**, 984–985 (1975).
50 Smale. S. (Department of Mathematics, Berkeley: preprint, 1976).
51 May, R. M., and Leonard, W. J., *SIAM J. Appl. Math.*, **29**, 243–253 (1975).

Commun. math. Phys. 50, 69—77 (1976)

Communications in
Mathematical
Physics

A Two-dimensional Mapping with a Strange Attractor

M. Hénon
Observatoire de Nice, F-06300 Nice, France

Abstract. Lorenz (1963) has investigated a system of three first-order differential equations, whose solutions tend toward a "strange attractor". We show that the same properties can be observed in a simple mapping of the plane defined by: $x_{i+1}=y_i+1-ax_i^2$, $y_{i+1}=bx_i$. Numerical experiments are carried out for $a=1.4$, $b=0.3$. Depending on the initial point (x_0, y_0), the sequence of points obtained by iteration of the mapping either diverges to infinity or tends to a strange attractor, which appears to be the product of a one-dimensional manifold by a Cantor set.

1. Introduction

Lorenz (1963) proposed and studied a remarkable system of three coupled first-order differential equations, representing a flow in three-dimensional space. The divergence of the flow has a constant negative value, so that any volume shrinks exponentially with time. Moreover, there exists a bounded region R into which every trajectory becomes eventually trapped. Therefore, all trajectories tend to a set of measure zero, called *attractor*. In some cases the attractor is simply a point (which is then a stable equilibrium point) or a closed curve (known as a limit cycle). But in other cases the attractor has a much more complex structure; it appears to be locally the product of a two-dimensional manifold by a Cantor set. This is known as a *strange attractor*. Inside the attractor, trajectories wander in an apparently erratic manner. Moreover, they are highly sensitive to initial conditions. These phenomena are of interest for weather prediction (Lorenz, 1963) and more generally for turbulence theory (Ruelle and Takens, 1971; Ruelle, 1975). Further numerical explorations of the Lorenz system have been made by Lanford (1975) and Pomeau (1976).

We present her a "reductionist" approach in which we try to find a model problem which is as simple as possible, yet exhibits the same essential properties as the Lorenz system. Our aim is (i) to make the numerical exploration faster and more accurate, so that solutions can be followed for a longer time, more

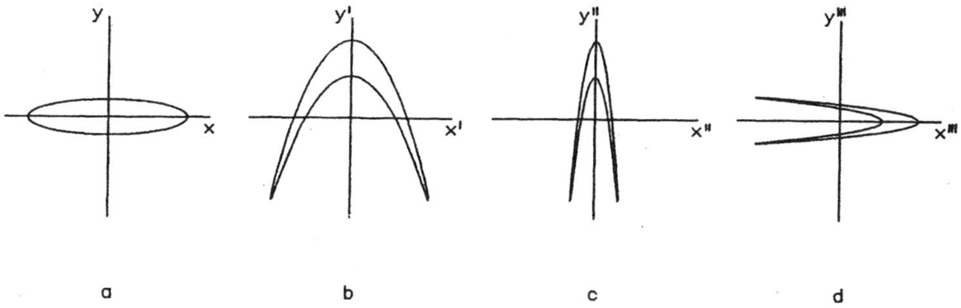

Fig. 1. The initial area a is mapped by T' into b, then by T'' into c, and finally by T''' into d

detailed explorations can be conducted, etc.; (ii) to provide a model which might lend itself more easily to mathematical analysis.

2. The Model

Our first step is classical (Birkhoff, 1917) and consists in considering not the whole trajectories in the three-dimensional space, but only their successive intersections with a two-dimensional *surface of section S*. We define a mapping T of S into itself as follows: given a point A of S, we follow the trajectory which originates from A until it intersects S again; this new point is $T(A)$. This mapping is sometimes called a *Poincaré map*. A trajectory is thus replaced by an infinite set of points in S, obtained by repeated application of the mapping T. The essential properties of the trajectory are reflected into corresponding properties of the set of points. We have thus formally reduced the problem to the study of a two-dimensional mapping.

At this point, however, the only advantage really gained is in clarity of presentation of the results; the actual computation of the mapping still requires the numerical integration of the differential equations. Now comes the second and decisive step: we forget about the differential system, and we define a mapping T by explicit equations, giving directly $T(A)$ when A is known. This of course simplifies the computation drastically. The new mapping T does not any more correspond to the Lorenz system; however, by choosing it carefully we may hope to retain the essential properties which we wish to study. Past experience in the measure-preserving case (see Hénon, 1969, and references therein) has shown indeed that the same features are found in dynamical systems defined by differential equations and in mappings defined as such.

The third step consists in specifying T. Here we have been inspired by the numerical results of Pomeau (1976) on the Lorenz system, which show clearly how a volume is stretched in one direction, and at the same time folded over itself, in the course of one revolution. This folding effect has been also described by Ruelle (1975, Fig. 5 and 6). We simulate it by the following chain of three mappings of the (x, y) plane onto itself. Consider a region elongated along the x axis (Fig. 1a). We begin the folding by

$$T': x' = x, \quad y' = y + 1 - ax^2, \tag{1}$$

which produces Figure 1b; a is an adjustable parameter. We complete the folding by a contraction along the x axis:

$$T'' : x'' = bx' , \quad y'' = y' , \tag{2}$$

which produces Figure 1c; b is another parameter, which should be less than 1 in absolute value. Finally we come back to the orientation along the x axis by

$$T'' : x''' = y'' , \quad y''' = x'' , \tag{3}$$

which results in Figure 1d.

Our mapping will be defined as the product $T = T'''T''T'$. We write now (x_i, y_i) for (x, y) and (x_{i+1}, y_{i+1}) for (x''', y''') (as a reminder that the mapping will be iterated) and we have

$$T : x_{i+1} = y_i + 1 - ax_i^2 , \quad y_{i+1} = bx_i . \tag{4}$$

This mapping has some interesting properties. Its Jacobian is a constant:

$$\frac{\partial(x_{i+1}, y_{i+1})}{\partial(x_i, y_i)} = -b . \tag{5}$$

The geometrical interpretation is quite simple: T' preserves areas; T''' also preserves areas but reverses the sign; and T'' contracts areas, multiplying them by the constant factor b. The property (5) is welcome because it is the natural counterpart of the constant negative divergence in the Lorenz system.

A polynomial mapping satisfying (5) is known as an *entire Cremona transformation*, and the inverse mapping is also given by polynomials (Engel, 1955, 1958). Indeed we have here

$$T^{-1} : x_i = b^{-1} y_{i+1} , \quad y_i = x_{i+1} - 1 + ab^{-2} y_{i+1}^2 . \tag{6}$$

Thus T is a one-to-one mapping of the plane onto itself. This is also a welcome property, because it is the natural counterpart of the fact that in the Lorenz system there is a unique trajectory through any given point.

The selection of T could have been approached in a different way, by looking for the "simplest" non-trivial mapping. It is natural then to consider polynomial mappings of progressively increasing order. Linear mappings are trivial, so the polynomials must be at least of degree 2. The most general quadratic mapping is

$$\begin{aligned} x_{i+1} &= f + ax_i + by_i + cx_i^2 + dx_iy_i + ey_i^2 , \\ y_{i+1} &= f' + a'x_i + b'y_i + c'x_i^2 + d'x_iy_i + e'y_i^2 \end{aligned} \tag{7}$$

and depends on 12 parameters. But if we impose the condition that the Jacobian is a constant, some relations must be satisfied by these parameters. We can further reduce the number of parameters by an appropriate linear change of coordinates in the plane. In this way, by a slight extension of the results of Engel (1958), it can be shown that the general form (7) is reducible to a "canonical form" depending on two parameters only. This is a generalization of our earlier result (Hénon, 1969) that a quadratic *area-preserving* mapping can be brought into a form depending on one parameter only. The canonical form can be written in several different ways; and one of them turns out to be identical with (4), which is thus reached by an entirely different road! The mapping (4), which was initially constructed in empirical fashion, is in fact the most general quadratic mapping with constant Jacobian.

One difference with the Lorenz problem is that the successive points obtained by repeated application of T do not always converge towards an attractor; sometimes they "escape" to infinity. This is because the quadratic term in (4) dominates when the distance from the origin becomes large. However, for particular values of a and b it is still possible to prove the existence of a bounded "trapping region" R, from which the points can never escape once they have entered it (see below Section 5).

T has two invariant points, given by

$$x=(2a)^{-1}[-(1-b)\pm\sqrt{(1-b)^2+4a}], \qquad y=bx. \tag{8}$$

These points are real for

$$a>a_0=(1-b)^2/4. \tag{9}$$

When this is the case, one of the points is always linearly unstable, while the other is unstable for

$$a>a_1=3(1-b)^2/4. \tag{10}$$

3. Choice of Parameters

We select now particular values of a and b for a numerical study. b should be small enough for the folding described by Figure 1 to occur really, yet not too small if one wishes to observe the fine structure of the attractor. The value $b=0.3$ was found to be adequate. A good value of a was found only after some experimenting. For $a<a_0$ or $a>a_3$, where a_0 is given by (9) and a_3 is of the order of 1.55 for $b=0.3$, the points always escape to infinity: apparently there exists no attractor in these cases. For $a_0<a<a_3$, depending on the initial values (x_0, y_0), either the points escape to infinity or they converge towards an attractor, which appears to be unique for a given value of a. We concentrate now on this attractor. For $a_0<a<a_1$, where a_1 is given by (10), the attractor is the stable invariant point. When a is increased over a_1, at first the attractor is still simple and consists of a periodic set of p points. (An equivalent attractor in the Lorenz problem would be a limit cycle intersecting the surface of section p times). The value of p increases through successive "bifurcations" as a increases, and appears to tend to infinity as a approaches a critical value a_2, of the order of 1.06 for $b=0.3$. For $a_2<a<a_3$, the attractor is no more simple, and the behaviour of the points becomes erratic. This is the case in which we are interested. We adopt the following values:

$$a=1.4, \qquad b=0.3. \tag{11}$$

4. Numerical Results

Figure 2 shows the result of plotting 10000 successive points, obtained by iteration of T, starting from the arbitrarily chosen initial point $x_0=0$, $y_0=0$; the vertical scale is enlarged to give a better picture. Figure 3 shows the result of 10000

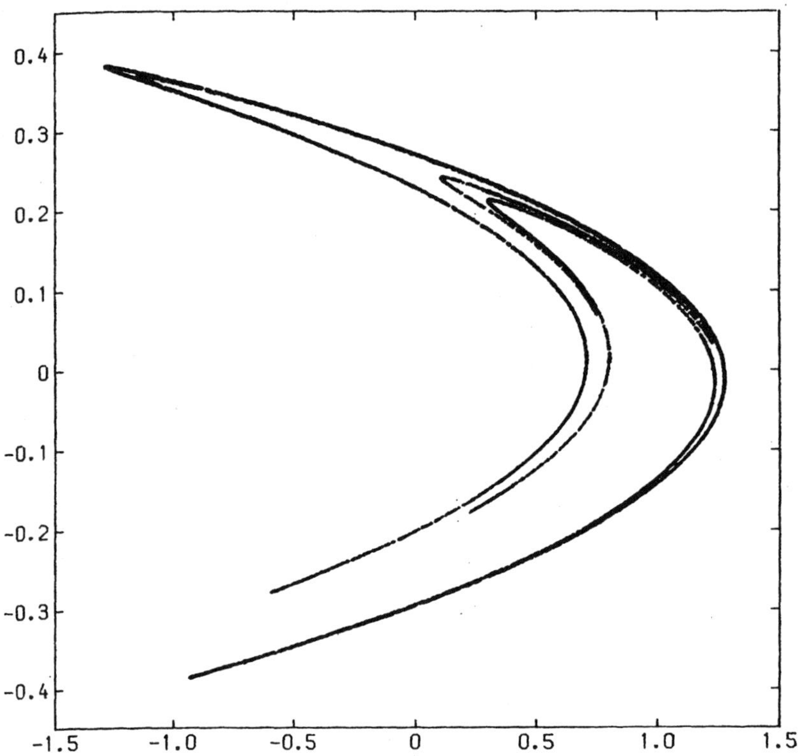

Fig. 2. 10000 successive points obtained by iteration of the mapping T starting from $x_0=0$, $y_0=0$

iterations of T again, starting from a different point: $x_0=0.63135448$, $y_0=0.18940634$ (this choice will be explained below). The two figures are seen to be almost identical. This suggests strongly that what we see in both figures is essentially the attractor itself: the successive points quickly approach the attractor and soon become undistinguishable from it at the scale of the figure. This is confirmed if one looks at the first few points on Figure 2. The initial point at $x_0=0$, $y_0=0$ and the first iterate at $x_1=1$, $y_1=0$ are clearly visible; the second iterate is still visible at $x_2=-0.4$, $y_2=0.3$; the third iterate can barely be distinguished at $x_3=1.076$, $y_3=-0.12$; and the fourth iterate at $x_4=-0.7408864$, $y_4=0.3228$ is already lost inside the attractor at the resolution of Figure 2. The following points then wander over the attractor in an apparently erratic manner.

One of the two unstable invariant points has the coordinates, given by (8):

$$x=0.63135448\ldots, \quad y=0.18940634\ldots . \tag{12}$$

This point appears to belong to the attractor. The two eigenvalues λ_1, λ_2 and the slopes p_1, p_2 of the corresponding eigenvectors are

$$\lambda_1=0.15594632\ldots, \quad p_1=1.92373886\ldots,$$
$$\lambda_2=-1.92373886\ldots, \; p_2=-0.15594632\ldots . \tag{13}$$

The instability is due to λ_2. The corresponding slope p_2 appears to be tangent to the "curves" in Figure 2.

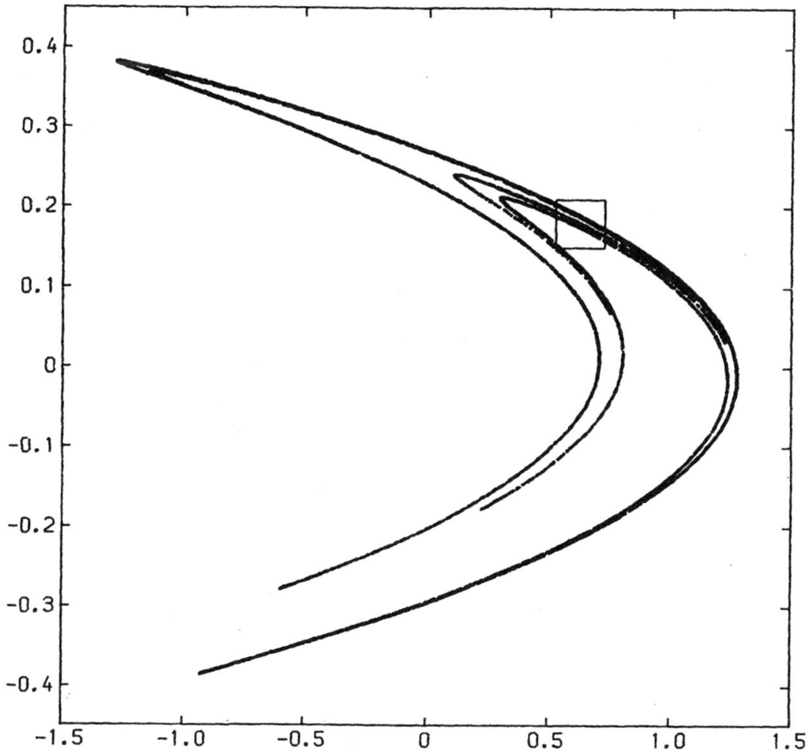

Fig. 3. Same as Figure 2, but starting from $x_0=0.63135448$, $y_0=0.18940634$

These properties allow us to eliminate the "transient regime" in which the points approach the attractor, and which is not of much interest: we simply start from the close vicinity of the unstable point (12), by rounding off its coordinates to 8 digits. This is done in Figure 3 and in the following figures. The points quickly move away along the line of slope p_2 since $|\lambda_2|$ is appreciably larger than 1.

The attractor appears to consist of a number of more or less parallel "curves"; the points tend to distribute themselves densely over these curves. The few gaps that can still be seen on Figures 2 and 3 have probably no particular significance. Their locations are not the same on the two figures. They are simply due to statistical fluctuations in the quasi-random distribution of points, and they would disappear if more moints were plotted. Thus, the *longitudinal structure* of the attractor (along the curves) appears to be simple, each curve being essentially a one-dimensional manifold.

The *transversal structure* (across the curves) appears to be entirely different, and much more complex. Already on Figures 2 and 3 a number of curves can be seen, and the visible thickness of some of them suggests that they have in fact an underlying structure. Figure 4 is a magnified view of the small square of Figure 3: some of the previous "curves" are indeed resolved now into two or more components. The number n of iterations has been increased to 10^5, in order to have a sufficient number of points in the small region examined. The small square in Figure 4 is again magnified to produce Figure 5, with n increased to 10^6: again the

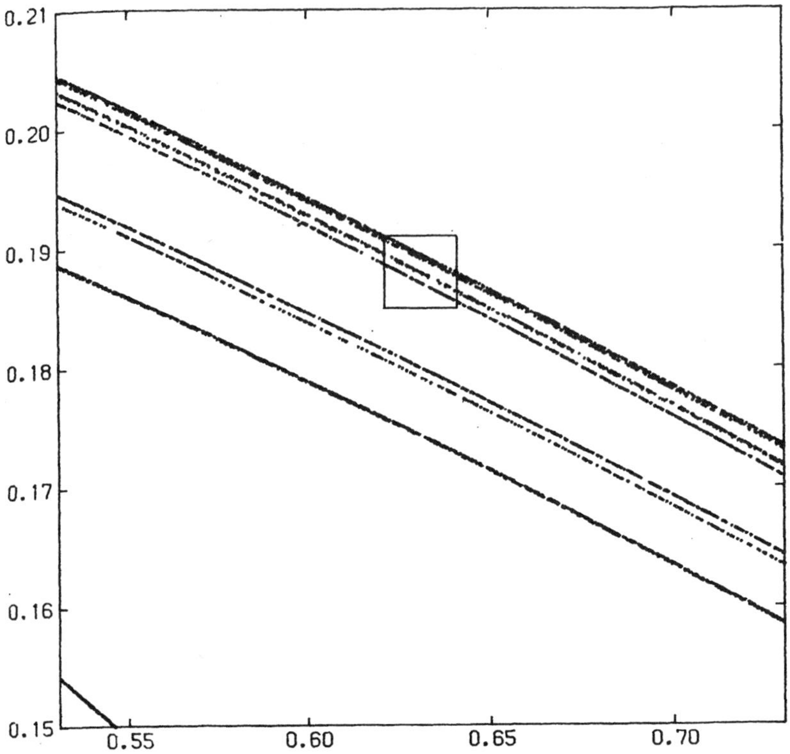

Fig. 4. Enlargement of the squared region of Figure 3. The number of computed points is increased to $n = 10^5$

number of visible "curves" increases. One more enlargement results in Fig. 6, with $n = 5 \times 10^6$: the points become sparse but new curves can still easily be traced.

These figures strongly suggest that the process of multiplication of "curves" will continue indefinitely, and that each apparent "curve" is in fact made of an infinity of quasi-parallel curves. Moreover, Figures 4 to 6 indicate the existence of a hierarchical sequence of "levels", the structure being practically identical at each level save for a scale factor. This is exactly the structure of a Cantor set.

The frames of Figures 4 to 6 have been chosen so as to contain the invariant point (12). This point appears to lie on the upper boundary of the attractor. Surprisingly, its presence is completely invisible on the figures; this contrasts with the area-preserving case, were stable and unstable invariant points play a very conspicuous role (see for instance Hénon, 1969). On the other hand, the presence of the invariant point explains, locally at least, the hierarchy of similar structures: at each application of the mapping, the scale of the transversal structure is multiplied by λ_1 given by (13). At the same time, the points spread out along the curves, as dictated by the value of λ_2.

5. A Trapping Region

The fact that even after 5×10^6 iterations the points have not diverged to infinity suggests that there is a region of the plane from which the points cannot escape.

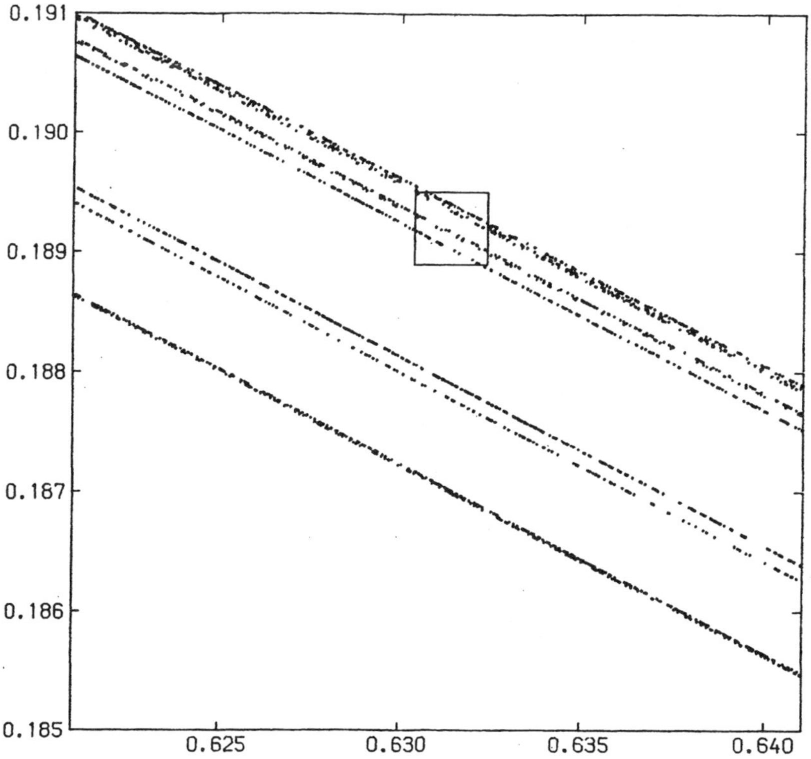

Fig. 5. Enlargement of the squared region of Figure 4; $n=10^6$

This can be actually proved by finding a region R which is mapped inside itself. An example of such a region is the quadrilateral $ABCD$ defined by

$$x_A=-1.33, \quad y_A=0.42, \qquad x_B=1.32, \qquad y_B=0.133,$$

$$x_C=1.245, \quad y_C=-0.14, \quad x_D=-1.06, \quad y_D=-0.5. \tag{14}$$

The image of $ABCD$ is a region bounded by four arcs of parabola, and it can be shown by elementary algebra that this image lies inside $ABCD$. Plotting the quadrilateral on Figure 2 or 3, one can verify that it encloses the observed attractor.

6. Conclusions

The simple mapping (4) appears to have the same basic properties as the Lorenz system. Its numerical exploration is much simpler: in fact most of the exploratory work for the present paper was carried out with a programmable pocket computer (HP-65). For the more extensive computations of Figures 2 to 6, we used a IBM 7040 computer, with 16-digit accuracy. The solutions can be followed over a much longer time than in the case of a system of differential equations. The accuracy is also increased since there are no integration errors.

Lorenz (1963) inferred the Cantor-set structure of the attractor from reasoning, but could not observe it directly because the contracting ratio after one "circuit"

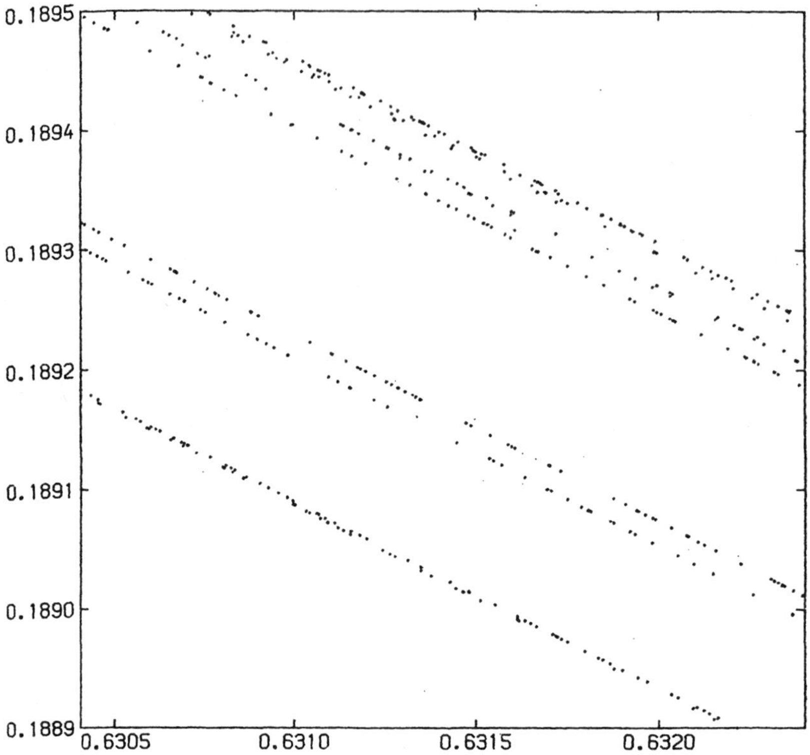

Fig. 6. Enlargement of the squared region of Figure 5; $n = 5 \times 10^6$

was too small: 7×10^{-5}. A similar experience was reported by Pomeau (1976). In the present mapping, the contracting ratio after one iteration is 0.3, and one can easily observe a number of successive levels in the hierarchy. This is also facilitated by the larger number of points.

Finally, for mathematical studies the mapping (4) might also be easier to handle than a system of differential equations.

References

Birkhoff, G. D.: Trans. Amer. Math. Soc. **18**, 199 (1917)
Engel, W.: Math. Annalen **130**, 11 (1955)
Engel, W.: Math. Annalen **136**, 319 (1958)
Hénon, M.: Quart. Appl. Math. **27**, 291 (1969)
Lanford, O.: Work cited by Ruelle, 1975
Lorenz, E. N.: J. atmos. Sci. **20**, 130 (1963)
Pomeau, Y.: to appear (1976)
Ruelle, D., Takens, F.: Comm. math. Phys. **20**, 167; **23**, 343 (1971)
Ruelle, D.: Report at the Conference on "Quantum Dynamics Models and Mathematics" in Bielefeld, September 1975

Communicated by K. Hepp

Received March 25, 1976

Strange attractors and chaotic motions of dynamical systems

Edward Ott

Department of Physics and Astronomy, University of Maryland, College Park, Maryland 20742
and Department of Electrical Engineering, University of Maryland, College Park, Maryland 20742

A review is presented of recent work related to strange attractors and chaotic motions of dynamical systems. First, simple systems capable of displaying chaotic behavior are discussed. In order of increasing dimensionality of the system, they are one-dimensional noninvertible maps, two-dimensional invertible maps, and autonomous systems of three coupled ordinary differential equations. The concept of fractional dimension of the strange attractor is stressed. Several physical examples well be reviewed, along with the possible relevance to turbulence in systems, such as fluids or plasmas, that are described by partial differential equations.

CONTENTS

I. INTRODUCTION

In this report, a review will be presented of recent developments related to strange attractors and chaotic motions of dynamical systems. Emphasis will be placed on those aspects that may prove useful for physical scientists.

At this point it might be appropriate to discuss some of the terms used in the title. A *dynamical system* may be thought of as any set of equations giving the time evolution of the state of a system from a knowledge of its previous history. Examples are Maxwell's equations, the Navier-Stokes equations, and Newton's equations of motion for a particle with suitably specified forces. The adjective *chaotic* is used here to describe a type of time evolution resulting from a dynamical system. In particular, it describes motions which are commonly thought of as "turbulent," i.e., motions whose time evolution appears, on detailed examination, to be very complex. For such motions, one often has the feeling that a statistical description may be of more use than actual knowledge of the true evolution. (A more precise definition of the term *chaotic* appears in Sec. II.) At this point I shall not attempt to define a *strange attractor*. Rather, we only note that its presence can lead to chaotic motion. Thus, the topic under discussion in this review is related to the occurrence of turbulent-type motions in physical systems. More specifically, we shall be concerned with nonconservative systems. [Chaotic motion in Hamiltonian (conservative) systems is not within the scope of the present review.]

It is becoming increasingly clear that the topic of strange attractors is one that will find abundant applications in a wide variety of physical situations. The list of such applications is already large, including problems in the onset of turbulence in fluids (Lorenz, 1963; McLaughlin and Martin, 1975; Ruelle and Takens, 1971), chemically reacting systems (Tomita and Kai, 1979), buckling beams (Holmes, 1980), nonlinear wave interactions in plasmas (Adam, Bussac, and Laval, 1980; Wersinger, Finn, and Ott, 1980a, 1980b, 1980c; Vyshkind, 1978; Vyshkind and Rabinovich, 1976; Russell and Ott, 1980; Wang, 1980; Masui and Wang, 1980), solid-state physics (Huberman and Crutchfield, 1979), lasers (Haken, 1975), self-generation of the earth's magnetic field (Robbins, 1977), magnetohydrodynamic flow (Maschke and Saramito, 1980; Treve and Manley, 1980), etc. (The above is only a partial listing and many other relevant references exist.)

As can be surmised from the dates of the references just mentioned, virtually all the activity in this field (at least when restricted to problems in the physical sciences) has occurred since 1975. The notable exceptions to this statement are the papers of Lorenz (1963) and of Ruelle and Takens (1971). These two papers, independently and from quite different points of view, originally suggested the relevance of strange attractors to the onset of turbulence in fluid flows. Lorenz was interested in explaining the presence of chaotic behavior in numerical solutions of a model system of three coupled, first-order, nonlinear, ordinary differential equations which modeled the nonlinear evolution of the Benard instability, i.e., the instability which results when a fluid layer subjected to gravity is heated sufficiently strongly from below. By a combination of careful analysis of the computer generated solutions and analytical reasoning, Lorenz was able to deduce that the solution of his equations was eventually trapped in a region of the phase space of the system which had a very intricate (strange) geometric structure. The now-recognized general implications of Lorenz's paper were not widely appreciated until many years after its publication. In 1971, Ruelle and Takens, making use of then recent developments in mathematics, offered a possible mechanism by which turbulent solutions to the Navier-Stokes equations could appear as a parameter is varied (e.g., as the Reynolds number is increased). In particular, they showed on the basis of quite general arguments that a strange attractor could appear. It is to be anticipated that the application of the type of considerations initiated by these two papers will lead to new insights in a variety of fields of physics. The purpose of the present paper is to facilitate this process by providing an elementary introduction designed for researchers and students in the physical sciences.

Some previous related reviews from various points of view are those of Treve (1978), Swinney and Gollub (1978), Rabinovich (1978), Ruelle (1977), Sinai (1977), Holmes (1977), Helleman (1980), Shaw (1981), and Yorke and Yorke (1981).

We shall proceed by discussing dynamical systems of progressively higher dimension (all of which can display chaotic behavior) [Ruelle (1977)]:

(1) One-dimensional noninvertible maps,

(2) Two-dimensional invertible maps,

(3) First-order systems of autonomous ordinary differential equations,

$$\frac{dx_i(t)}{dt}=f_i[x_1(t),x_2(t),\ldots,x_n(t)]\quad i=1,2,\ldots,n,\ \text{with } n>3.$$

(4) Partial differential equations.

One-dimensional maps will be discussed in the next section (Sec. II). These are relations of the form $x_{n+1}=F(x_n)$; thus, for some initial x_0, a sequence, $x_0, x_1, x_2, \ldots$, is generated. Surprisingly, very simple one-dimensional maps will turn out to yield rather good qualitative models for behavior in two-dimensional maps (Sec. III), ordinary differential equations (Sec. IV), and partial differential equations (Sec. V).

II. ONE-DIMENSIONAL NONINVERTIBLE MAPS

To begin, we consider one-dimensional maps

$$x_{n+1}=F(x_n), \tag{2.1}$$

where $F(x)$ is a scalar function. [Some relevant references on one-dimensional maps are Li and Yorke (1975); May (1976); Guckenheimer *et al.* (1977); Feigenbaum (1978); and Guckenheimer (1979). In addition, a monograph by Coullet and Eckmann (1980) has been published on the subject.] We only consider the case in which the sequence $x_0, x_1, x_2, \ldots$, generated by F is bounded, $P<x_n<Q$, for all n. We will often say that such a sequence is "chaotic" or "turbulent," by which we mean that it has the following properties: (1) sensitive dependence on initial conditions (if two initial points x_0^a and x_0^b are chosen very close to each other, the distance between their successive images under F initially diverges exponentially); (2) the average correlation function for a given sequence satisfies $C(m)\to 0$ as $m\to\infty$, where

$$C(m)=\lim_{N\to\infty}\frac{1}{N}\sum_{m=1}^{N}(x_n-\langle x\rangle)(x_{n+m}-\langle x\rangle),$$

$$\langle x\rangle=\lim_{N\to\infty}\frac{1}{N}\sum_{n=1}^{N}x_n;$$

(3) The sequence is nonperiodic.

We now introduce the following two one-dimensional maps, each of which typifies a broad class,

$$F_1(x)\equiv a(1-2|x-\tfrac{1}{2}|),\quad 0<a\leq 1 \tag{2.2}$$

$$F_2(x)=4bx(1-x),\quad 0<b\leq 1 \tag{2.3}$$

where a and b are constants. The map F_1 has a sharp peak, while the map F_2 has a rounded smooth maximum. For $0<(a,b)<1$, both F_1 and F_2 map the interval 0 to 1 into itself, and we shall only consider this range of x (i.e., initial conditions are always assumed to satisfy $1>x_0>0$). Both F_1 and F_2 are noninvertible, since, given x_{n+1}, one cannot solve $x_{n+1}=F(x_n)$ for x_n. (From Fig. 1 it is seen that for each value of x_{n+1} there are *two* possible values of x_n.) Thus one may say that it is possible to go forward in time but not backward in time. This represents a basic difference with ordinary differential equations, $\dot{\mathbf{x}}=\mathbf{f}(\mathbf{x})$, which may, in principal, be integrated either forward or backward in time.

First consider F_1. There are essentially two cases of interest, $0<a<\frac{1}{2}$ and $\frac{1}{2}<a\leq 1$, both illustrated in Fig. 1(a). In Fig. 1, the dashed line represents $x_n=x_{n+1}$. For $0<a<\frac{1}{2}$, Fig. 1(a) shows that $x_{n+1}<x_n$, since, over

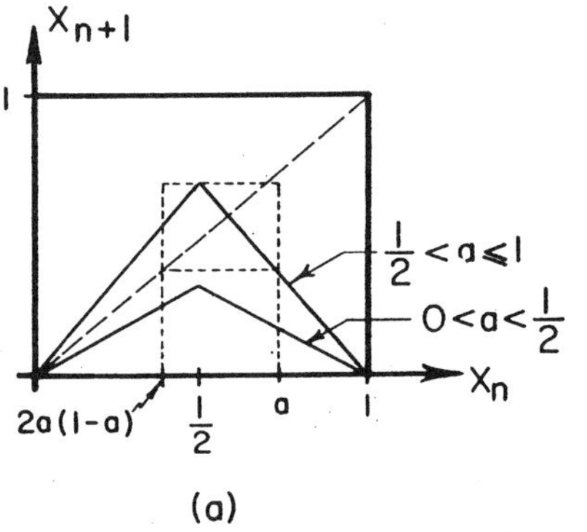

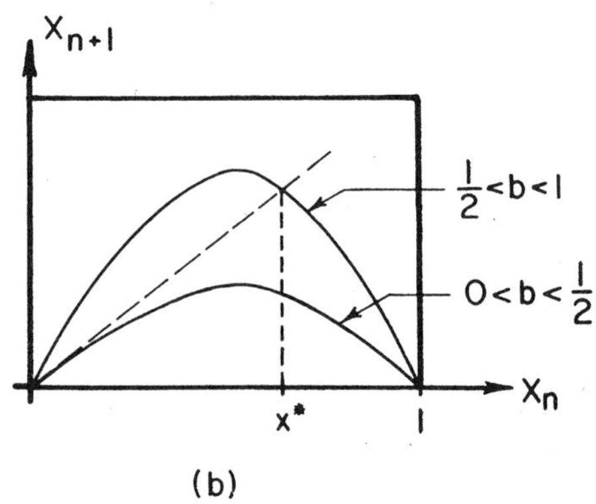

FIG. 1. (a) Map F_1 [Eq. (2.2)], and (b) the map F_2 [Eq. (2.3)].

the entire range, $F_1(x_n)$ lies below the line $x_n = x_{n+1}$. In this case it is clear that x_n converges to zero as n increases. Now, consider $\frac{1}{2} < a \leq 1$. For any initial x_0 between 0 and 1, the sequence eventually becomes trapped in the interval $2a(1-a) < x < a$ through which it will typically wander chaotically. To illustrate this, consider the particularly simple case $a = 1$ for which the chaotic interval becomes $0 < x < 1$. In this case, we may consider the map to represent two steps: (1) a uniform stretching of the interval 0 to 1 to twice its original length, and (2) a folding in half of the stretched interval so that it now has its original length. These steps are illustrated in Fig. 2. The stretching property leads to exponential separation of nearby points and hence, sensitive dependence on initial conditions. The folding property keeps the generated sequence bounded, but also causes the map to be noninvertible, since it causes two different x_n points to be mapped into one x_{n+1} point.

Conversely, for a general one-dimensional map, in order to have the distance between nearby points separate exponentially, it is necessary for the map to be, on the average, stretching. On the other hand, to have the sequence remain bounded (confined between 0 and 1 in the case of F_1), folding must take place. Thus we conclude that in order for a one-dimensional map to exhibit chaotic behavior, it must be noninvertible. Figure 3 illustrates the stretching and folding properties of F_1 for a value of a less than one ($\frac{1}{2} < a < 1$). From Fig. 3(a) we see that after one application of F_1, there are no points in $a < x < 1$. From Fig. 3(b), we see that the interval 0 to $2a(1-a)$ is stretched but that no points are folded back onto it. Thus any point in $0 < x < 2a(1-a)$ will eventually leave that interval and never return. Thus the generated sequence is eventually trapped in $2a(1-a) < x < a$. An alternative way of seeing the fact that F_1 has sensitive dependence on initial conditions is illustrated in Fig. 4 for the case $a = 1$. Figure 4(a) shows x_{n+2} versus x_n obtained from two applications of F_1; that is, $x_{n+2} = F_1[F_1(x_n)]$. We will use the notation $x_{n+2} = F_1^{(2)}(x_n)$ to denote $F_1[F_1(x_n)]$. For x_{n+m} versus x_n, Fig. 4(a) generalizes to Fig. 4(b), $x_{n+m} = F_1^{(m)}(x_n)$. From Fig. 4(b), one can see that if the initial condition has an uncertainty $\pm\varepsilon$, then after $m \sim \ln_2(1/\varepsilon)$ iterations of F_1, we will have essentially no clue as to where x lies in the interval 0 to 1 (example: for $\varepsilon \cong 10^{-12}$, $m \cong 40$).

We now turn to a consideration of the map F_2 [Eq.

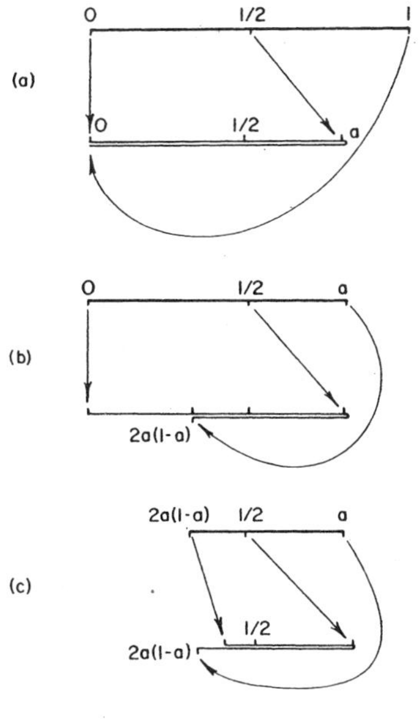

FIG. 3. Map F_1 for $a = 0.8$; (a) mapping of the interval 0 to 1, (b) 0 to a, and (c) $2a(1-a)$ to a.

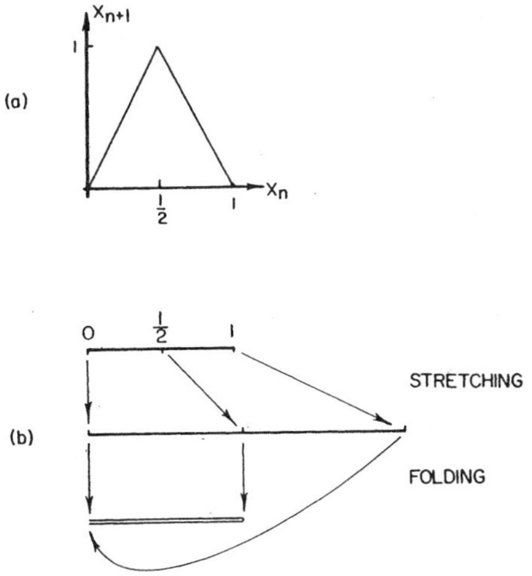

FIG. 2. (a) F_1 map at $a = 1$. (b) Illustration of the stretching and folding properties of F_1 for $a = 1$.

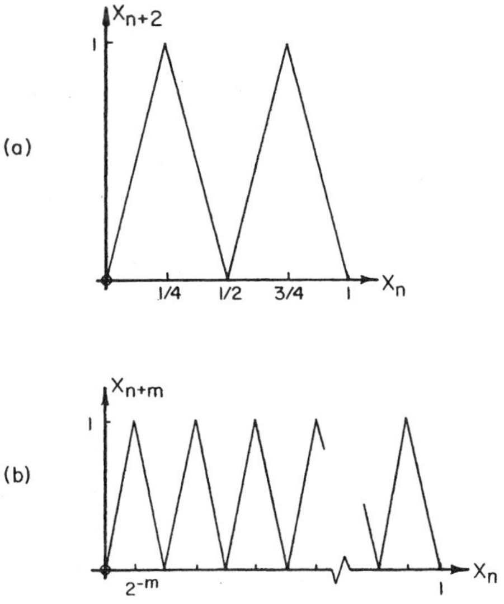

FIG. 4. (a) x_{n+1} vs x_n for F_1 with $a = 1$. (b) x_{n+m} vs x_n for F_1 with $a = 1$ (the length along the x_n axis is expanded).

(2.3) and Fig. 1(b)], which has a smooth rounded maximum at $x_n=\frac{1}{2}$ as opposed to the sharp peak in F_1 [Fig. 1(a)]. From Fig. 1(b) we note that the line $x_n=x_{n+1}$ intersects the map at $x_n=0$ for $0<b<\frac{1}{2}$ and at two points, $x_n=0$ and $x_n=x^*$, for $\frac{1}{2}<b<1$, where, from (2.3), $x^*=1-(4b)^{-1}$. These points of intersection are fixed points of the map; that is, if the initial point is chosen to be a fixed point, then successive applications of the map leave it unmoved. It is important to discover whether the fixed points are stable to small perturbations. Let $\bar{x}$ be a fixed point, $\bar{x}=F(\bar{x})$, and consider a perturbation from it, $x_n=\bar{x}+\delta_n$. From (2.1), $\bar{x}+\delta_{n+1}=F(\bar{x}+\delta_n)$. For δ_n small, we can Taylor-series expand $F(\bar{x}+\delta_n)$ as $F(\bar{x})+F'(\bar{x})\delta_n=\bar{x}+F'(\bar{x})\delta_n$, from which we obtain

$$\delta_{n+1}/\delta_n=F'(\bar{x}), \tag{2.4}$$

where $F'\equiv dF/dx$. Thus, if $|F'(\bar{x})|>1$, images under F of points near $\bar{x}$ successively move farther away from it, and $\bar{x}$ is unstable. For $|F'(\bar{x})|<1$, points near $\bar{x}$ converge to it, and $\bar{x}$ is stable. (For example, for F_1 given by Eq. (2.2), $|F_1'(x)|=2a$ for $x\neq\frac{1}{2}$, and thus $\bar{x}=0$ is stable for $0<a<\frac{1}{2}$, while zero and the second intersection [cf. Fig. 1(a)] are both unstable for $\frac{1}{2}<a\leq 1$.) For F_2, Eq. (2.3) shows that $F'(0)=4b$, and $F'(x^*)=2(1-2b)$. Thus, zero is stable for $b<\frac{1}{4}$. The fixed point $x=x^*$ first appears at $b=\frac{1}{4}$, and, simultaneously with its appearance, the zero fixed point loses its stability. The fixed point $x^*=1-(4b)^{-1}$ is stable in $\frac{3}{4}>b>\frac{1}{4}$, since $|F'(x^*)|<1$ in this range. Corresponding to these results, it can be shown that the sequence generated by F_2 converges to zero for $0\leq b<\frac{1}{4}$ and to x^* for $\frac{1}{4}<b<\frac{3}{4}$. $F'(x^*)=1$ at $b=\frac{1}{4}$ and decreases as b increases, becoming zero at $b=\frac{1}{2}$, minus one at $b=\frac{3}{4}$, and less than minus one (unstable) for $b>\frac{3}{4}$. The question which then arises is what happens in the range $\frac{3}{4}<b<1$ for which both 0 and x^* are unstable. To begin answering this question it is instructive to examine the map $x_{n+2}=F_2^{(2)}(x_n)$ shown in Fig. 5 for b slightly below $\frac{3}{4}$ and for b slightly above $\frac{3}{4}$. Values of x which recur every second iteration, i.e., in a series $e,f,e,f,e,f,\ldots$ are fixed points of $F^{(2)}$, $e=F^{(2)}(e)$, and $f=F^{(2)}(f)$, with $e=F(f)$ and $f=F(e)$. For the special case $e=f$, e will also be a fixed point of F itself. Now consider the stability of the sequence $e,f,e,f,\ldots$. Taking $x_n=e+\delta_n$, $x_{n+2}=e+\delta_{n+2}$, we have $e+\delta_{n+2}=F^{(2)}(e+\delta_n)=F[F(e+\delta_n)]$, which, when Taylor-series expanded, yields

$$\delta_{n+2}/\delta_n=F'(e)F'(f)=F^{(2)\prime}(e)=F^{(2)\prime}(f). \tag{2.5}$$

Applying (2.5) to a fixed point of F, we have $F^{(2)\prime}(x^*)=[F'(x^*)]^2$. Thus, when x^* loses stability, the slope of $F^{(2)}$ at x^* becomes greater than one. Referring, now, to Fig. 5 for F_2, it is seen that, when this occurs, two new intersections of $F_2^{(2)}$ with $x_{n+2}=x_n$, e and f, simultaneously appear. Furthermore, when these points appear, they initially have $F_2^{(2)\prime}(e)=F_2^{(2)\prime}(f)=1$, and the slopes decrease as b is raised. Thus, the new fixed points of $F^{(2)}$ are initially stable, and it is found that, when b slightly exceeds $\frac{3}{4}$, the generated sequence converges to an alternating one, $e,f,e,f,\ldots$. As b increases further, however, $F_2^{(2)\prime}(e)=F_2^{(2)\prime}(f)$ decreases; eventually, past $b=0.862\ldots$, $F_2^{(2)\prime}(f)$ becomes less than minus one, and the e,f cycle becomes unstable.

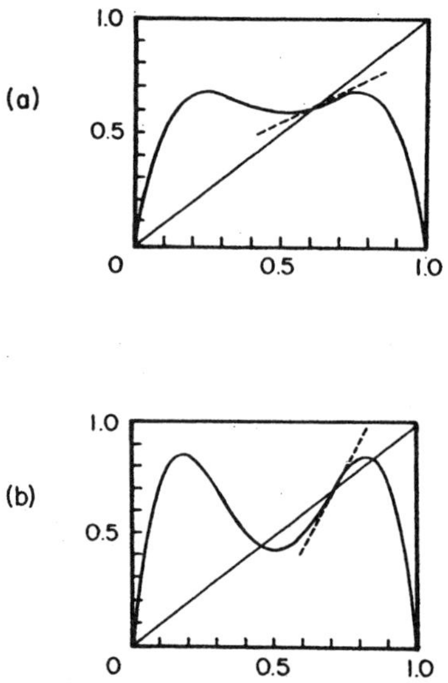

FIG. 5. x_{n+2} vs x_n for F_2 with (a) $b=0.678$ and (b) $b=0.854$. The dashed lines are the slope at $x=x^*$. The intersections of $x_{n+2}=x_n$ with $x_{n+2}-F_2^{(2)}(x_n)$ are $e>x^*>f$.

What happens next can be deduced, in an analogous way, from the map $x_{n+4}=F_2^{(2)}[F_2^{(2)}(x_n)]\equiv F_2^{(4)}(x_n)$. When the e,f cycle loses stability, a stable four-point periodic cycle simultaneously appears: $g,h,i,j,g,h,i,j,g,h,\ldots$, which then gives way ("bifurcates") to an eight-point cycle, which then gives way to a 16-point cycle, etc. Furthermore, the band of b values over which a given 2^k-point cycle is stable decreases geometrically with k, so that

$$\frac{b_k-b_{k-1}}{b_{k+1}-b_k}\rightarrow 4.669\,201\ldots \tag{2.6}$$

for $k\rightarrow\infty$, where b_k is the value of b at the point where the 2^k-point cycle bifurcates to a 2^{k+1}-point cycle. Also $b_k\rightarrow 0.892\ldots$ for $k\rightarrow\infty$; that is, there is an accumulation point of an infinite number of bifurcations at $b_\infty=0.892\ldots$. Fiegenbaum (1978) has derived Eq. (2.6) using arguments based on scale invariance near b_∞, and has also obtained other properties of the generated sequences for b near b_∞. These properties apply independently of the detailed functional form of the map as functional form of the map as long as it has a quadratic maximum as does F_2. Thus Eq. (2.6) applies to a wide class of maps.

Just past b the orbit generated by F_2 looks like a noisy cycle of periodicity 2^p with $p\rightarrow\infty$ as b approaches b_∞ from above. By a "noisy cycle of periodicity 2^p" we mean that the orbit is confined to 2^p disjoint intervals in $1>x>0$ which it visits in a sequential order. Thus the orbit always comes back to the same interval after 2^p iterations. On the other hand, if one looks at the points generated by $F_2^{(2p)}$ with an initial condition in one of these intervals, then the orbit looks completely chaotic in this interval. As b increases, these intervals merge in pairs so that a noisy 2^p cycle goes into a noisy 2^{p-1} cycle as b increases past a critical value $\bar{b}_p$. Furthermore, $\bar{b}_p$ obeys the same scaling relation[1] as in Eq. (2.6) (Coullet and Tresser, 1980). As b increases past $\bar{b}_1$, chaotic motion over a single connected band emerges.

In addition to the noisy 2^p cycles, narrow windows in b also exist within $b_\infty < b < 1.0$ for which the generated sequence is exactly periodic. Generally, these periodic sequences first appear with some period N and then go through a sequence of period doubling bifurcations, creating periods $2^k N$, with an accumulation point at $k \to \infty$ ending the particular periodic window. The widest such window is for 3×2^k periodic cycles, $0.9571 < b < 0.9624$. Other periodic windows are exceedingly narrow in b, and most of the range $b_\infty < b < 1$ appears to be chaotic. As an example of how these cycles first appear, consider the onset of the $N = 3$ period cycle. Figure 6 shows the map x_{n+3} versus x_n [obtained from $F_2^{(3)}(x_n)$] for values of b just below and just above that for which the $N = 3$ cycle first appears. As b increases past its critical value, the minimum of $F_2^{(3)}$ lowers until two new intersections of $F_2^{(3)}$ with $x_{n+3} = x_n$ are created near this minimum, one with slope greater than one (unstable) and one with slope less than one (initially stable). This type of phenomenon, whereby a periodic orbit appears after a region of chaotic motion, is called a *tangent bifurcation*. For values of b where numerically generated sequences appear to be chaotic, it is not, at present, known whether they are truly chaotic, or whether, in fact, they are really periodic, but with exceedingly large periods and very long transients required to settle down. Recent numerical results do, however, strongly suggest that the sequences are truly chaotic (Lorenz, 1979). Figure 7 summarizes some of the previously described results for F_2.

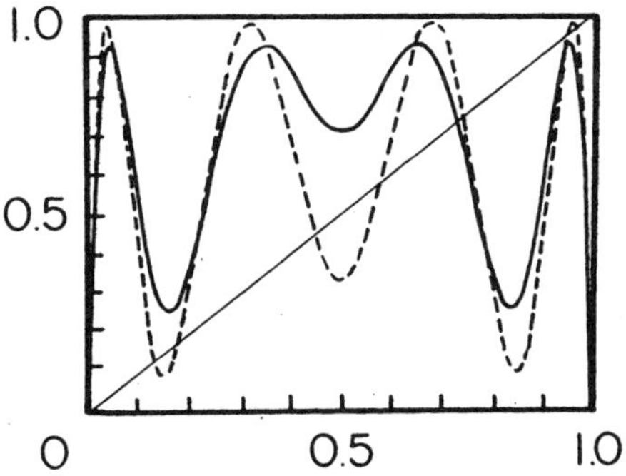

FIG. 6. x_{n+3} vs x_n for $b = 0.975$ (dashed curve) and for $b = 0.925$ (solid curve).

[1]The subject of the universal scaling properties of maps with a quadratic maximum is currently a very active research topic. Recent work includes the study of scaling of the Lyapunov number past b_∞ (Chang and Wright, 1981; Huberman and Rudnick, 1980), the noise power spectrum past b_∞ (Huberman and Zisook, 1981; Wolf and Swift, 1981), and scaling behavior with the addition of random noise (Crutchfield *et al.*, 1981; Shraiman *et al.*, 1981). Furthermore, universal scaling properties of conservative systems with period doubling have also recently been examined (for example, Green *et al.*, 1981).

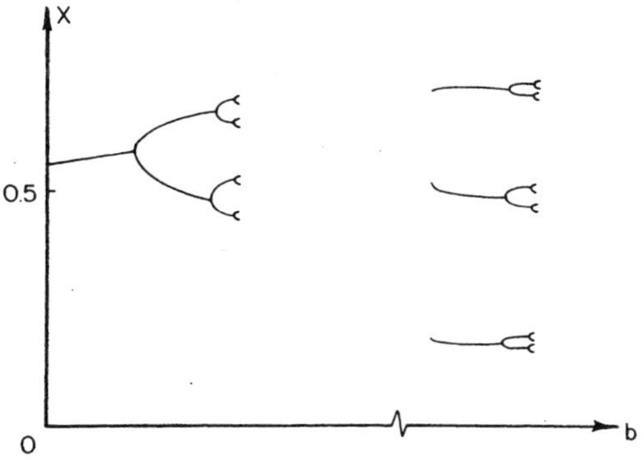

FIG. 7. Summary diagram for some of the bifurcations of $F^{(2)}$. The scale about the period 3×2^k cycles has been greatly expanded.

III. TWO-DIMENSIONAL INVERTIBLE MAPS

A general two-dimensional map can be written as

$$x_{n+1} = f_1(x_n, y_n), \quad y_{n+1} = f_2(x_n, y_n). \tag{3.1}$$

The map is invertible if (3.1) can be solved uniquely for x_n and y_n as functions of x_{n+1} and y_{n+1}, $x_n = g_1(x_{n+1}, y_{n+1})$ and $y_n = g_2(x_{n+1}, y_{n+1})$. That is, it is possible to go either backwards or forwards in time. A two-dimensional invertible map can easily be constructed from a one-dimensional noninvertible map as follows:

$$x_{n+1} = F(x_n) + y_n, \tag{3.2a}$$

$$y_{n+1} = \beta x_n, \tag{3.2b}$$

where $F(x)$ is noninvertible. For $\beta = 0$, $x_{n+1} = F(x_n)$, and the noninvertible one-dimensional map is recovered. However, as long as $\beta \neq 0$, no matter how small it is, the map (3.2) is invertible: $x_n = y_{n+1}/\beta$ and $y_n = x_{n+1} - F(y_{n+1}/\beta)$. On the other hand, if β is sufficiently small, the variation of x is well described by the one-dimensional map, $x_{n+1} = F(x_n)$. Furthermore, for small β, the range of variation of y_n will be small compared to that for x_n [cf. Eq. (3.2b)], and thus, if the points generated by (3.2) are plotted on the xy plane, the generated sequence will appear to lie on a line (the x axis) with some small spread about the line. For very small β, the spread may, in practical terms, be unmeasurable, so that (3.2) becomes indistinguishable from a one-dimensional noninvertible map. (This will, indeed, turn out to be similar to what happens in certain differential equation examples to be discussed in the next section.)

It is of interest to compute the Jacobian of the map (3.2),

$$J \equiv \begin{vmatrix} \dfrac{\partial x_{n+1}}{\partial x_n} & \dfrac{\partial x_{n+1}}{\partial y_n} \\ \dfrac{\partial y_{n+1}}{\partial x_n} & \dfrac{\partial y_{n+1}}{\partial y_n} \end{vmatrix} = -\beta .$$

Thus, for $|\beta| < 1$, we see that areas will contract by the factor $|\beta|$ on each application of the mapping (3.2). Thus, if the generated sequence of pairs (x_n, y_n) remains in a bounded region of the xy plane, then the sequence must asymptotically approach a subset of the original bounded xy region which has zero area. This subset is called an *attractor*. For example, if the sequence becomes attracted to an N-point periodic cycle, then the attractor would be the N points plotted in the xy plane, clearly a subset of zero area. Another possible subset of zero area that a sequence might asymptote to is a curve. At first sight, these two possibilities, a zero-dimensional subset (points) and a one-dimensional subset (a curve), might appear to exhaust all possibilities for zero-area attractors. This is, however, not the case. There can be attractors which have noninteger dimension (at least according to the definition of dimension that we will use). Such attractors would be termed strange. The relevant definition of dimension is that due to Hausdorf[2] [see, for example, Mandelbrot (1977)]:

$$d=\lim_{\varepsilon\to 0} \ln N(\varepsilon)/\ln\left(\frac{1}{\varepsilon}\right), \tag{3.3}$$

where, if the set in question is a subset of a p-dimensional ordinary space, then $N(\varepsilon)$ is the number of p-dimensional cubes of side ε needed to cover the set. Alternatively, (3.3) implies that for small ε, $N(\varepsilon) \cong K\varepsilon^{-d}$. Thus, if one is content to know where the set lies to within an accuracy ε, then, to specify the location of the set, we need only specify the positions of the $N(\varepsilon)$ cubes covering the set. Hence, the dimension may be viewed as telling us how much information is necessary to specify the location of the set to within a given accuracy. If the set has complicated fine-scale structure, then, as a practical matter, it may be advantageous to introduce some coarse-graining into the description of the set, and then ε may be thought of as specifying the degree of coarse-graining. As an example of the application of Eq. (3.3), if the set in question is a point, then $N(\varepsilon)=1$, and, according to (3.3), the Hausdorf dimension is zero; if the set in question is the section of the xy plane given by $0<x<1$ and $0<y<1$, then $N(\varepsilon)=\varepsilon^{-2}$, and the Hausdorf dimension is two; if the set is a straight line joining (0, 0) and (1.0), then $N(\varepsilon)=\varepsilon^{-1}$, and the Hausdorf dimension is one. These examples all yield the obvious results, so that (3.3) conforms to our intuition in these cases. As an example of a set with a noninteger dimension, consider the following construction of a Cantor set (illustrated in Fig. 8): take a line of unit length, $0\leq x\leq 1$, and remove the middle third $\frac{1}{3}<x<\frac{2}{3}$; then take the two remaining intervals between 0 and $\frac{1}{3}$ and between $\frac{2}{3}$ and 1, divide them in thirds, and remove the two middle thirds; in the limit as this process is repeated an infinite number of times, what is left is a set that has zero net length and an uncountable number of elements. To apply (3.3) to this set, we note the following which is evident from Fig. 8,

$$\varepsilon=\tfrac{1}{3},\quad N=2$$

[2]Actually, the Hausdorf dimension has a somewhat more involved definition than that given by Eq. (3.3). More precisely, d given by (3.3) defines the "capacity" of the set. For cases of interest to us here, however, Eq. (3.3) probably gives the same result as would the actual Hausdorf dimension definition.

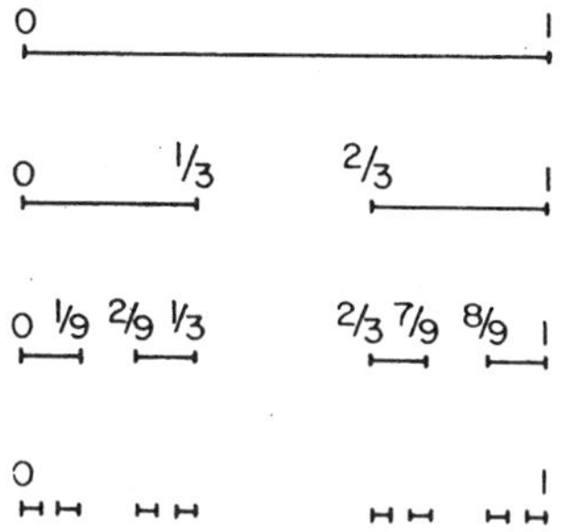

FIG. 8. First few steps in the construction of an example of a Cantor set.

$$\varepsilon=\tfrac{1}{9},\quad N=4$$

$$\varepsilon=\tfrac{1}{3}^{p},\quad N=2^{p}.$$

Thus from (3.3)

$$d=(\ln 2)/(\ln 3)=0.630.$$

Note that the Cantor set just constructed has the property of scale invariance. That is, by the nature of the construction, the set between 0 and 1 will look precisely the same as that part of it between 0 and $\frac{1}{3}$, if the latter is examined under a magnifying glass which magnifies by a factor of three.

As a concrete example, we now consider a mapping first studied by Henon (1976):

$$x_{n+1}=1-cx_n^2+y_n,\quad y_{n+1}=\beta x_n. \tag{3.4}$$

This mapping is essentially equivalent to Eq. (3.2) with F given by Eq. (2.3). [A study of the map obtained from (3.2) and (2.2) has been presented by Lozi (1978), but will not be discussed here.] Figure 9 shows the results of plotting 10^4 successive points obtained by iterating the map (3.4) with $c=1.4$ and $\beta=0.3$ from an initial point $x_0=0.631$, $y_0=0.189$. Similarly obtained

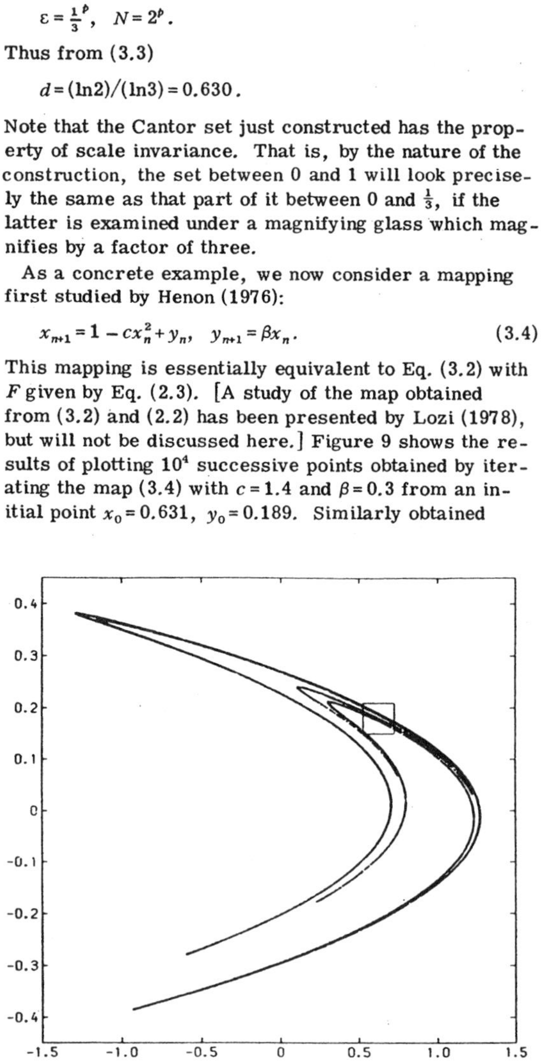

FIG. 9. Iterated points of the map, Eq. (3.4), 10^4 iterations.

plots starting with other initial values are almost identical (except for an initial transient), suggesting that Fig. 9 is, in fact, essentially a picture of the attractor. That is, as the map is iterated, points come closer and closer to the attractor and eventually become indistinguishable from it on the scale of the figure. Figure 10(a) shows a blow-up of the squared region in Fig. 9; Figs. 10(b) and 10(c) are successive blow-ups of the squared regions in the preceding figure. Scale invariant, Cantor-set-like structure transverse to the linear structure is evident. Thus, the attractor is probably strange with dimension between one and two. In fact, a recent study (Russell *et al.*, 1980) gives $d \cong 1.26$. Further studies of the map (3.4) have been carried out by Feit (1978), Curry (1979), and Simó (1979), who have explicitly verified exponential divergence of initially close points. In addition, Feit (1978) and Simó (1979) have studied how the character of the generated sequences changes as c is varied with β fixed. It is found that, dispersed among intervals of c where the motion is chaotic, there are many small subintervals where the motion is periodic. On each such subinterval, there appear attractors of period $k, 2k, 4k, \ldots, 2^n k, \ldots$, similar to the phenomenon observed for the one-dimensional map, Eq. (3.4) with $\beta = 0$ [which is equivalent to Eq. (2.3)].

Bridges and Rowlands (1977) have given a procedure for investigating maps of the type (3.2) essentially by a power series in β. For example, to lowest order in β (3.2a) gives $x_{n+1} = F(x_n)$ (as previously noted), which, when substituted into (3.2b), yields the result that the attractor lies in the vicinity of the curve

$$x = F(y/\beta). \tag{3.5}$$

Higher-order approximations yield increasing detail to the curve. For Henon's map, Eq. (3.4), with $c = 1.4$ and $\beta = 0.3$, Eq. (3.5) yields a surprisingly good first-order approximation of the attractor.

The reader may be aware that two-dimensional maps have been extensively utilized as models of Hamiltonian systems, and, in particular, to model ergodic behavior in Hamiltonian systems. Since Hamilton's equations conserve phase-space volume (by Liouville's theorem), two-dimensional maps modeling Hamiltonian systems do not cause uniform contraction of areas, and chaotic motions (i.e., ergodicity) generated by such maps generally fill up a two-dimensional area. Thus, no strange (fractional dimensional) set is involved. If dissipation (e.g., friction) is added to a Hamiltonian system, phase-space contraction may be expected to result. Thus it is of interest to consider the relation of ergodicity in Hamiltonian systems to the possible

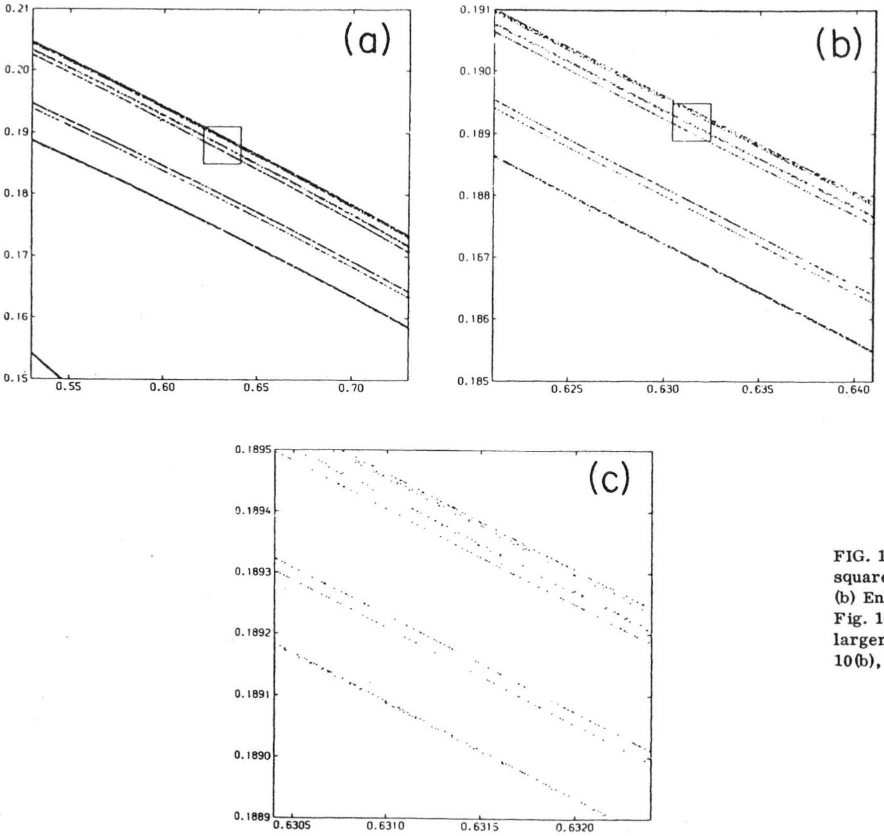

FIG. 10. (a) Enlargement of the square in Fig. 9, 10^5 iterations; (b) Enlargement of the square in Fig. 10(a), 10^6 iterations; (c) Enlargement of the square in Fig. 10(b), 5×10^6 iterations.

appearance of strange attractors when dissipation is added to these systems (McLaughlin, 1979; Zaslavskii, 1978). Recent studies addressed to this question (Zaslavskii, 1978; Zaslavskii and Rachko, 1979) consider the two-dimensional map

$$y_{n+1} = e^{-\Gamma}(y_n + \varepsilon \cos 2\pi x_n), \tag{3.6a}$$

$$x_{n+1} = \left\langle x_n + \frac{\Omega}{2\pi} + \frac{\alpha\Omega}{2\pi\Gamma}(1-e^{-\Gamma})y_n + \frac{K}{\Gamma}(1-e^{-\Gamma})\cos 2\pi x_n \right\rangle, \tag{3.6b}$$

where $\langle \cdots \rangle$ denotes the fractional part of the argument and $K = \alpha\varepsilon\Omega/2\pi$. The Jacobian of the map (3.6) is $\exp(-\Gamma)$. Thus, the map contracts areas for $\Gamma > 0$ and is area preserving for $\Gamma = 0$. In the absence of dissipation, $\Gamma = 0$, and (3.6) reduces to

$$y_{n+1} = y_n + \varepsilon \cos 2\pi x_n, \tag{3.7a}$$

$$x_{n+1} = \left\langle x_n + \frac{\Omega}{2\pi} + \frac{\alpha\Omega}{2\pi} y_n + K\cos 2\pi x_n \right\rangle, \tag{3.7b}$$

which has been extensively studied as a basic model of stochasticity in Hamiltonian systems (Rosenbluth *et al.*, 1966; Stix, 1973; Rechester and Stix, 1976; Chirikov, 1979; Greene, 1979) (and also as a model for stochastic magnetic field line topologies in magnetic confinement controlled thermonuclear fusion devices). The onset of ergodic behavior in the map (3.7) can be estimated from Chirikov's island overlap condition which, for small ε, yields ergodicity for $K \simeq 1$. By qualitative arguments based upon the exponential divergence of nearby chaotic orbits Zaslavskii is able to estimate the condition for the appearance of chaotic orbits and a strange attractor for the map (3.6), $K\mu \gtrsim 1$, where $\mu = (1-e^{-\Gamma})\Gamma^{-1}$. Numerical results confirm this rough estimate.

In connection with chaotic maps a useful notion is that of the Lyapunov numbers. An illustration of the Lyapunov numbers is given in Fig. 11. For a two-dimensional map, the Lyapunov numbers, λ_1 and λ_2, are the average principal stretching factors for a very small circular area; more formally

$$(\lambda_1, \lambda_2) = \lim_{n\to\infty}[\text{magnitude of the eigenvalues of } \underline{J}(x_n, y_n)\underline{J}(x_{n-1}, y_{n-1})\cdots\underline{J}(x_1, y_1)]^{1/n}, \tag{3.8}$$

where $\underline{J}(x,y)$ is the Jacobian matrix of the map:

$$\underline{J}(x,y) = \begin{bmatrix} \frac{\partial f_1(x,y)}{\partial x} & \frac{\partial f_1(x,y)}{\partial y} \\ \frac{\partial f_2(x,y)}{\partial x} & \frac{\partial f_2(x,y)}{\partial y} \end{bmatrix}$$

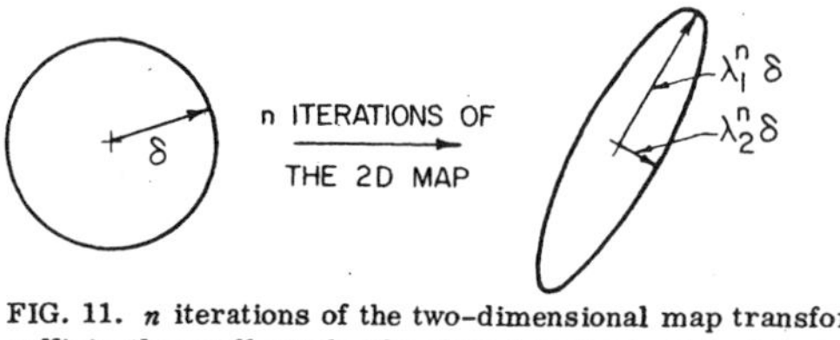

FIG. 11. n iterations of the two-dimensional map transform a sufficiently small circle of radius δ approximately into an ellipse with major and minor radii $\lambda_1^n\delta$ and $\lambda_2^n\delta$, where λ_1 and λ_2 are Lyapunov numbers, for $n\to\infty$.

[f_1 and f_2 are defined by Eq. (2.2)], and (x_1, y_1), (x_2, y_2), $\ldots$, (x_n, y_n) is a sequence generated by the map. Thus, the Lyapunov numbers specify the average stretching rate of nearby points. Say $\lambda_1 > \lambda_2$. If the map is chaotic, then λ_1 must exceed unity, so that the distance between almost all nearby points increases on successive mappings. If the map is area contracting, $\lambda_1\lambda_2 < 1$; if it is area preserving, $\lambda_1\lambda_2 = 1$. The calculation of λ_1 and λ_2 from Eq. (3.8) is fairly easy using a digital computer and can then be used as a criterion for chaos, $\lambda_1 > 1$. An interesting conjecture concerning Lyapunov numbers has been put forward by Frederickson *et al.* (1980) [see also Kaplan and Yorke (1979a)], namely, that the Hausdorff dimension of a strange attractor of a map with the eigenvalues of $\underline{J}$ independent of x and y is related to its Lyapunov numbers. For a two-dimensional map with Lyapunov numbers $\lambda_1 > 1 > \lambda_2$, $\lambda_1\lambda_2 < 1$, their conjecture states that the dimension is given by

$$d = 1 + (\ln\lambda_1)/(\ln\lambda_2^{-1}). \tag{3.9}$$

For example, for the Henon map, Eq. (3.4), with $c = 1.4$ and $b = 0.3$, $\lambda_1 \cong 0.2$, and Eq. (3.9) gives $d \cong 1.26$. Recently, numerical experiments have been performed which tend to confirm (3.9) (Russell *et al.*, 1980). Furthermore, we note that these numerical experiments also tested maps for which the eigenvalues $\underline{J}$ are not independent of x and y and still obtained excellent agreement with Eq. (3.9). While it is known by counterexample that Eq. (3.9) fails in general if the eigenvalues of $\underline{J}$ depend on x and y, the results of these computer experiments indicate that (3.9) still yields a surprisingly good approximation to the dimension in some cases.

A possible motivation for (3.9) is depicted in Fig. 12, which considers a map of the unit square into itself. The mapping consists of two steps. The first step is a stretching along y by $\lambda_1 > 1$ and a contraction along x by $\lambda_2 < 1$. For the second step, we assume that λ_1 is an integer ($\lambda_1 = 3$ for Fig. 12), and move the stretched and contracted area back into the unit square, as shown. This may be represented analytically by the map

$$y_{n+1} = \lambda_1 y_n \bmod 1,$$

$$x_{n+1} = \lambda_2 x_n + y_n - \lambda_1^{-1}(\lambda_1 y_n \bmod 1).$$

From the construction in Fig. 12, it is clear that λ_1 and λ_2 are the Lyapunov numbers of this map. If the proc-

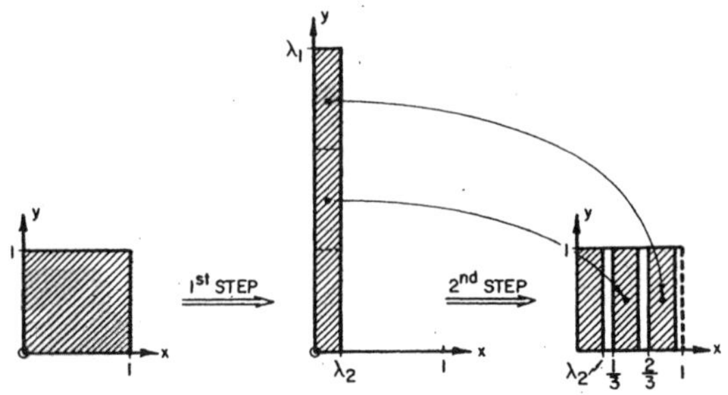

FIG. 12. Map motivating Eq. (3.9). The illustrated map may be represented analytically as $y_{n+1} = 3y_n \bmod 1$, $x_{n+1} = \lambda_2 x_n + y_n - y_{n+1}/3$.

ess in Fig. 12 is repeated p times, λ_1^p vertical strips of width λ_2^p will be created. To apply Eq. (3.3) for the dimension, let $\varepsilon = \lambda_2^p$. Then, the number of squares of side ε needed to cover the strips is $N \cong \lambda_1^p \lambda_2^{-p}$, and Eq. (3.3) yields (3.9). In addition, the conjecture can be used to predict the dimension of strange attractors for systems of ordinary differential equations (Russell *et al.*, 1980). The map of Fig. 12 is also interesting, since it demonstrably yields a strange attractor, while for maps such as (3.4) and (3.6) one can only say that numerical results suggest the presence of a strange attractor.

IV. THREE-DIMENSIONAL SYSTEMS OF ORDINARY DIFFERENTIAL EQUATIONS

A. Background

Here we consider systems of three coupled, nonlinear, autonomous, differential equations,

$$dx_i(t)/dt = f_i[x_1(t), x_2(t), x_3(t)], \quad i = 1, 2, 3. \tag{4.1}$$

The system is "autonomous" because f_i does not depend explicitly on t but only on x_i. Alternatively, a nonautonomous system of two coupled equations, $dx_1/dt = g_1(x_1, x_2, t)$ and $dx_2/dt = g_2(x_1, x_2, t)$, can be written in the form (4.1) by defining $f_1 = g_1(x_1, x_2, x_3)$; $f_2 = g_2(x_1, x_2, x_3)$; $f_3 = 1$.

We now define the Poincaré map of a system like (4.1). The Poincaré map represents a reduction of a system like (4.1) to a two-dimensional map, such as those studied in Sec. III. Figure 13 illustrates the construction of a Poincaré map. Consider a particular solution of (4.1) to generate an orbit in x_1, x_2, x_3 phase space. We now assume that some appropriate surface (the "surface of section") in this space has been chosen, and we study intersections of the orbit with the chosen surface. In Fig. 13 the chosen surface is the plane $x_2 = K$. Every time the orbit crosses the chosen surface in a particular direction ($dx_2/dt < 0$ for Fig. 13) we record the crossing point, e.g., points A and B in Fig. 13. For Fig. 13, it is clear that point A uniquely determines point B, since the solution of (4.1) is unique. Likewise, point B determines point A by time reversal of (4.1). Thus the Poincaré map in this illustration represents an invertible transformation of a point in the plane $x_2 = K$ into another point, i.e., it is an invertible two-dimensional map.

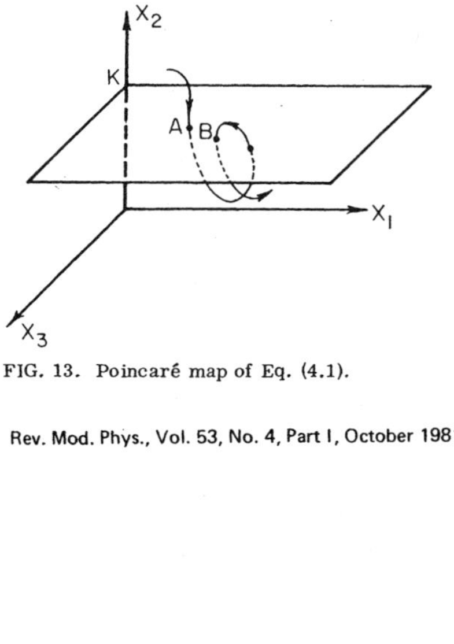

FIG. 13. Poincaré map of Eq. (4.1).

As an example, we consider the following system of two nonautonomous ordinary differential equations:

$$\frac{dp}{dt} = f(q) \sum_{n=-\infty}^{+\infty} \delta(\omega t - 2n\pi) - \nu(p - p_0),$$

$$\frac{dq}{dt} = p,$$

where $\delta(\theta)$ denotes the delta function of θ. This can be written as three autonomous equations

$$\frac{dp}{dt} = f(q) \sum_{n=-\infty}^{+\infty} \delta(\theta - 2n\pi) - \nu(p - p_0), \tag{4.2a}$$

$$\frac{dq}{dt} = p, \tag{4.2b}$$

$$\frac{d\theta}{dt} = \omega. \tag{4.2c}$$

We take the surface of section to be $\theta = 2m\pi - \varepsilon$, where $\varepsilon \to 0^+$. Defining $(p_m, q_m) = \lim_{\varepsilon \to 0^+} [p(t_m - \varepsilon), q(t_m - \varepsilon)]$, and $t_m = 2m\pi/\omega$, Eqs. (4.2) then yield by simple integration, the following two-dimensional map:

$$y_{m+1} = e^{-\Gamma}[y_m + f(q_m)] \tag{4.3a}$$

$$q_{m+1} = q_m + (1 - e^{-\Gamma})\nu^{-1}[y_m + f(q_m)] + 2\pi p_0/\omega, \tag{4.3b}$$

where $\Gamma = 2\pi\nu/\omega$, and $y_m = p_m - p_0$. By a suitable choice of $f(q)$ this two-dimensional map may be reduced to that studied by Zaslavskii (1978) and discussed in the preceding section.

Now, we turn to a discussion of the evolution of phase-space volumes as governed by the system (4.1). That is, we consider the volume (V) enclosed by some closed surface S in the x_1, x_2, x_3 phase space, and let the surface evolve by having each point on the surface follow an orbit generated by (4.1). By the divergence theorem

$$\frac{dV}{dt} = \int_V \left(\sum_{i=1}^{3} \frac{\partial f_i}{\partial x_i} \right) dx_1 dx_2 dx_3. \tag{4.4}$$

In the special case where divergence of the phase-space flow, $\sum_{i=1}^{3} \partial f_i/\partial x_i$, is a negative constant, $\sum_{i=1}^{3} \partial f_i/\partial x_i = -k$, Eq. (4.4) yields $dV/dt = -kV$, so that

$$V(t) = V(0)\exp(-kt). \tag{4.5}$$

Thus phase-space volumes shrink exponentially in time. Many of the physical examples yielding a system of the form (4.1) that have been investigated for strange attractors also happen to have a constant negative divergence. We shall restrict our discussion in the remainder of this section to this case. [In fact, Eqs. (4.2) fall in this class with the flow divergence being $-\nu$.] In Sec. V, more general cases, without negative divergence, will be discussed.

The special case of three ordinary autonomous differential equations with negative phase-space flow divergence presents a very clear case for the necessity of introducing the concept of a strange attractor. Since phase-space volume contracts to zero in the limit of large time, it follows that any attractor must have zero volume. A natural assumption might then be that the attractor would have to be a surface (two dimensional), a curve (one dimensional), or a point (zero dimensional). However, none of these allows chaotic motion. In particular, not even the highest dimension of the above three possibilities (two) allows chaos. For example, for orbits within a finite section of a plane, the Poincaré-Bendixson theorem shows that the only possible attractor for the orbit must be either a point, a simple closed curve, or a self-intersecting closed curve[3] (e.g., a figure eight (e.g., Hirsch and Smale, 1974). Thus if one observes chaotic motion in the system (4.1), and if (4.1) has negative phase-space flow divergence, then one is faced with something of a paradox. One way out is to realize that attractors with zero volume need not only have dimension zero, one, or two, but can, in fact, have noninteger dimension. In particular, chaotic motion is possible if Eqs. (4.1) have an attractor of dimension greater than two but less than three (the latter so that the volume of the attractor is zero), i.e., a strange attractor. We will now outline some work on physically interesting systems exhibiting strange attractors. The three examples which we will discuss are (1) the so-called Lorentz system, which represents a simple model of the convective motions that result when a temperature difference is maintained across a fluid layer which is subjected to gravity, (2) a simple model for the saturation of an unstable mode by coupling energy through quadratic nonlinearities from the unstable mode to damped modes, and (3) a model similar to (2) but with cubic nonlinearities (resulting, for example, from a nonlinear Schrödinger equation).

B. Examples

The first two examples, (1) and (2), are for parameters such that the phase-space contraction rate is large. This has the consequence that the attractor in the surface of section appears to be one dimensional. Actually, closer examination under magnification would reveal thickness containing structure within the attractor. Furthermore, the dimension in the surface of section must be $1<d<2$, as for any invertible two-dimensional map with a strange attractor. Here, however, d is only slightly larger than one. This leads to the result that the system dynamics can be well approximated by a one-dimensional noninvertible map of the types discussed in Sec. II. [Recall that, as discussed in Sec. III, this also is true in the analogous case of the two-dimensional map, Eq. (3.2) with β small.] The example in (1) yields a one-dimensional map that is like that given by Eq. (2.2); it has a sharp (nondifferentiable) maximum, and, in the notation of Sec. II, $|F'(x)|>1$ for all points on the chaotic orbit generated by $x_{n+1}=F(x_n)$. The example in (2) yields a one-dimensional map that is like that given in Eq. (2.3). As a consequence, example (2) leads to a pattern of period doubling bifurcations, tangent bifurcations, and chaotic orbits that is essentially the same as that for the quadratic map, Eq. (2.3) (cf. Fig. 7). Example (3) will be discussed for a range of parameters for which the contraction rate is not large. Thus, for example (3), non-one-dimensional structure will be readily evident in the surface of section. Furthermore, it will be shown that this structure appears to have approximate scale-invariant properties upon magnification, in analogy to the Cantor set example of Sec. III (cf. Fig. 8) and to the Henon map (cf. Figs. 9 and 10). Thus these three examples serve to illustrate the relevance of essential features of one- and two-dimensional chaotic maps to systems of ordinary differential equations modeling physical systems.

[3]For motion described by more than three autonomous ordinary differential equations with phase-space contraction, $\sum_i \partial f_i/\partial x_i<0$, or by three autonomous equations where $\sum \partial f_i/\partial x_i$ is not everywhere negative, an attractor which is a toroidal surface is possible. Motion on the toroidal surface is doubly periodic; that is, the solution of (4.1) can be represented as $x_i(t)=l_i(t,t)$, where l_i is periodic in both variables, $l_i(t+T_1,t+T_2)=l_i(t,t)$, and T_1/T_2 is not a rational number. Alternatively, if the orbit is doubly periodic, then the only frequency components of the Fourier spectrum of $x_i(t)$ are $(m/T_1)+(n/T_2)$, where m and n are integers. Doubly periodic motion in three dimensions with phase-space contraction is not possible, because the volume V enclosed by the toroidal attractor would be time independent, contrary to Eq. (4.5). At any rate, doubly periodic orbits are not chaotic, either.

1. Lorenz's treatment of the Benard instability

Consider two rigid plane parallel walls at $z=0,L$ with a fluid occupying the space in between. Gravity is in the negative z direction, and the plate at $z=0$ is maintained at a higher temperature than the plate at $z=L$, $T_0>T_L$. A possible equilibrium of this system is one in which the fluid is at rest and heat is transported from $z=0$ to $z=L$ via thermal conduction. Lord Rayleigh studied the linear stability of this equilibrium and found that if (T_0-T_L) exceeds a critical value, then the system becomes unstable to perturbations in the form of circulating fluid flow. The linear analysis cannot, however, be used to specify the ultimate nonlinear state of the fluid once instability sets in. Considering variations only in two dimensions, Saltzman (1962) derived a set of nonlinear ordinary differential equations by expanding the stream function and the temperature perturbation in double spatial Fourier series, with coefficients functions of t alone. By substituting the series into the original governing set of partial differential equations and truncating the infinite sum to a finite number of terms, he obtained a set of ordinary differential equations. Lorenz (1963) further examined this problem and added much insight. The paper by Lorenz has greatly added to the understanding of this type of problem. Lorenz considered a truncation to only three Fourier modes, for which the describing equations become

$$dX/dt=-\sigma X+\sigma Y\,, \tag{4.6a}$$

$$dY/dt=-XZ+rX-Y\,, \tag{4.6b}$$

$$dZ/dt=XY-bZ\,, \tag{4.6c}$$

where σ, r, and b are dimensionless parameters of the system. X is proportional to the circulatory fluid flow velocity, Y characterizes the temperature difference between ascending and descending fluid elements, and Z is proportional to the distortion of the vertical temperature profile from its equilibrium (which is linear with height).

Setting $dX/dt=dY/dt=dZ/dt=0$, we find that (4.6) possesses steady-state solutions, $X=Y=Z=0$, and if $r>1$,

$X=Y=\pm[b(r-1)]^{1/2}$, $Z=r-1$. The equilibrium $X=Y=Z=0$ represents the case of no fluid flow, while the two possible equilibria for $r>1$ represent steady circulating convection. Linearization of Eqs. (4.6) about these equilibria reveals that the $X=Y=Z=0$ equilibrium looses stability when $r>1$, while the steadily convecting equilibria become unstable if

$$r>\sigma(\sigma+b+3)(\sigma-b-1)\equiv r_c \tag{4.7}$$

and $\sigma>b+1$.

Lorenz numerically considered the solution of (4.6) for a case in which (4.7) is satisfied: $\sigma=10$, $b=\frac{8}{3}$, $r=28$. [Note that from (4.7) $r_c=24.74$ in this case.] For these values of the parameters he obtained a chaotic solution and examined its properties in detail. It is instructive also briefly to consider the character of the solutions of (4.6) for other parameters. In particular, say $\sigma=10$, $b=\frac{8}{3}$, and increase r. For $r<1$ all initial conditions eventually decay to the convection-free equilibrium. For $1<r\lesssim 24.06$ all initial conditions eventually settle into one of the two stable convective equilibria. For $24.07\lesssim r\lesssim 24.74$, depending on initial conditions, the solution settles into either a chaotic motion or into one of the two stable convective equilibria. This phenomenon of dependence upon initial conditions is called hysteresis. For $r>24.74$ the eventual orbit is chaotic for all initial conditions. [For r still larger, say $r\gtrsim 50$, the behavior can again change (Shimzu and Morioda, 1978).] Even in the range $r<24.06$ interesting behavior arises, in particular, the phenomenon of "chaotic transients" first noted by Yorke and Yorke (1979) and by Kaplan and Yorke (1979b). These authors find that the evolution from a given initial condition initially looks very much like the time dependence of the solution in the chaotic regime, after which the solution quickly settles into one of the two stable equilibria. Furthermore, they find that the time to settle into equilibrium is sensitive to the initial condition, and can be fairly long with its duration increasing with r. Similar phenomena are to be expected in other cases (e.g., Shimzu and Morioda, 1978) and in other systems.

We now return to a description of Lorenz's results for $r=28$. Figure 14 shows a projection of the phase-space orbit of the system onto the YZ plane. The points labeled C and C' are the steady convection equilibria points. Evidently, the orbit spirals outward from one of the points C or C' until it exceeds some critical distance from the origin, at which point it starts spiraling about the other point. If one were to make a sequential list of the number of circuits the solution makes around one point before it switches to the other point, the sequence would appear to be chaotic. By examination of the solution, Lorenz has deduced that the orbit appears to be confined to a surface. Actually this apparent "surface" must have some small thickness, inside of which is embedded the more complicated structure of the strange attractor. In fact, if one were to pass a line through this surface normal to it, one would find that the intersection of the line with the surface is a set of dimension $0<d<1$, i.e., like the Cantor set of Fig. 8. However, since the thickness of the strange attractor is small, the presence of structure in this intersection would only be visible upon magnification, and, unmagnified, it would appear to be a point.

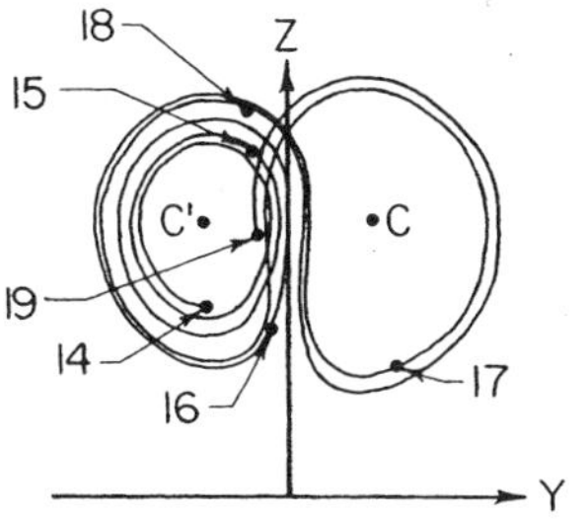

FIG. 14. Projections of an orbit for $r=28$ onto the YZ plane. Numerals 14, 15, etc., denote positions at iterations 1400, 1500, etc.

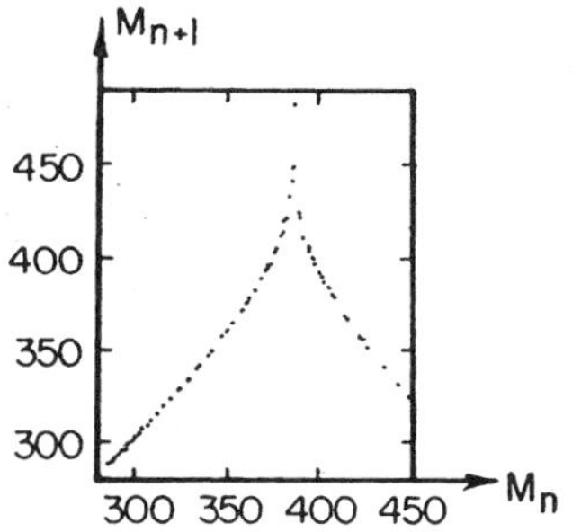

FIG. 15. Maxima vs subsequent maxima of Z occurring during 6000 iterations.

Figure 15 shows a plot, obtained by Lorenz, of M_n, the nth maximum of Z, versus M_{n+1}, the value of the following maximum. It is clear that an (approximate) one-dimensional map is generated. Furthermore, $|dM_{n+1}/dM_n|>1$, which is similar to the result for Eq. (2.2) with $a>\frac{1}{2}$. Thus, as for Eq. (2.2), we expect this one-dimensional map to generate a chaotic sequence. For later comparison with the results of example (2), note that the maxima of Z may be regarded to lie in a surface of section, $bZ=XY$ [put $dZ/dt=0$ in (4.6c)]. For further discussion of the Lorenz attractor see Lanford (1976), Afraymovich *et al.* (1977), and Bunimovich and Sinai (1977).

2. Instability saturation by quadratically nonlinear mode coupling

An important problem in plasma physics (and in other fields as well) is that of determining the nonlinear state resulting from a linearly unstable wave. An elementary process by which saturation can occur is that of resonant three-wave mode coupling of energy in the linearly unstable wave to two other waves which are linearly damped. The following normalized system of equations describes this process:

$$dC_1/dt=C_1+C_2C_3\exp(i\delta t)\,,$$

$$dC_{2,3}/dt=-\gamma_{2,3}C_{2,3}-C_1C^*_{3,2}\exp(i\delta t)\,,$$

where C_i $(i=1,2,3)$ are time-dependent complex wave amplitudes, C_i^* is the complex conjugate of C_i, times have been normalized to the growth rate of wave 1 (the higher frequency wave), δ represents the effect of a mismatch in the frequency resonance, and amplitudes have been normalized so that the coefficients of the nonlinear terms are one. Introducing $a_1 \exp(i\phi_1) = C_1$, $a_{2,3} \exp(i\phi_{2,3}) = C_{2,3} \exp(i\delta t/2)$, and $\phi = \phi_1 - \phi_2 - \phi_3$, where a_i and ϕ_i are real, the previous system gives four real equations for $a_{1,2,3}$ and ϕ. These equations readily yield

$$d(a_2^2 - a_3^2)/dt = -2(\gamma_2 a_2^2 - \gamma_3 a_3^2).$$

Thus in the special case $\gamma_2 = \gamma_3$, $a_2^2 - a_3^2$ decreases exponentially. Restricting consideration to $\gamma_2 = \gamma_3 \equiv \gamma$ and $a_2 = a_3$, the basic equations then become (Vyshkind and Rabinovich, 1976)

$$da_1/dt = a_1 + a_2^2 \cos\phi \,,$$

$$da_2/dt = -a_2(\gamma + a_1 \cos\phi)\,,$$

$$d\phi/dt = -\delta + a_1^{-1}(2a_1^2 - a_2^2)\sin\phi \,.$$

For definiteness in examining this system, Wersinger *et al.* (1980a, 1980b, 1980c) chose $\delta = 2$ and studied the properties of numerical solutions[4] for a range of γ from $\gamma = 1$ to $\gamma = 25$. For $\gamma \lesssim 3$ the damping was not strong enough to arrest the instability, and the system evolution was apparently unbounded. For $3 \leq \gamma \leq 8.5$, all initial conditions led to a simple periodic limit cycle. Using $\phi(t) = \pi/2$ as the surface of section, this limit cycle was manifested as a single point in the surface of section. For γ somewhat larger than 8.5 a limit cycle is still observed, but the single fixed point in the surface of section that previously manifested the limit cycle splits into two points that are visited alternately. Correspondingly, the single-peak-per-period function for $3 \lesssim \gamma \lesssim 8.5$ becomes a function with two alternating maxima (cf. Figs. 16 and 17). As γ is increased, the two-point cycle splits into a four-point cycle (for $\gamma \gtrsim 11.9$), which splits into an eight-point cycle (for $\gamma \cong 13.15$), ..., followed by a region of γ $(\gamma \lesssim 13.4)$ for which the generated sequence appears chaotic[5] (cf. Fig. 18). For still larger γ a sequence from a three-point cycle to a six-point cycle, etc., briefly appears. To summarize, the phenomenology is precisely the same as that depicted in Fig. 7 for the one-dimensional quadratic map,[6] Eq. (2.3)! Similar results for entirely different physical situations have also been obtained [see, for example, Tomita and Kai (1979)]. To make the correspondence with Eq. (2.3) more concrete, first consider results from the numerically generated surface of section shown in Fig. 19 for a value of γ in the chaotic regime, $\gamma = 15$. It is seen that the points generated in the surface of section *appear* to lie on an arc. Since this arc has no visible thickness, it is natural to attempt an approximate reduction to a one-dimensional map. This can be done, for example, by plotting $x_{n+1} \equiv a_2(t_{n+1})$ versus $x_n \equiv a_2(t_n)$, where t_n is the nth time at which the system orbit pierces the surface of section. The points so generated (cf. Fig. 20) lie along a curve $x_{n+1} = F(x_n)$. By a change of variables (e.g., $\bar{x}_n \equiv (\text{const}) - x_n$) the one-dimensional map in Fig. 20 can be turned upside down so that it has a smooth rounded maximum (rather than

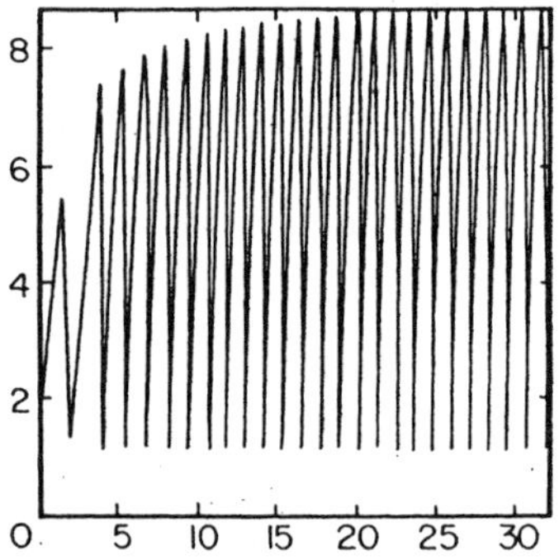

FIG. 16. a_1 vs t for $\gamma = 3$. After an initial transient, the solution settles into a limit cycle.

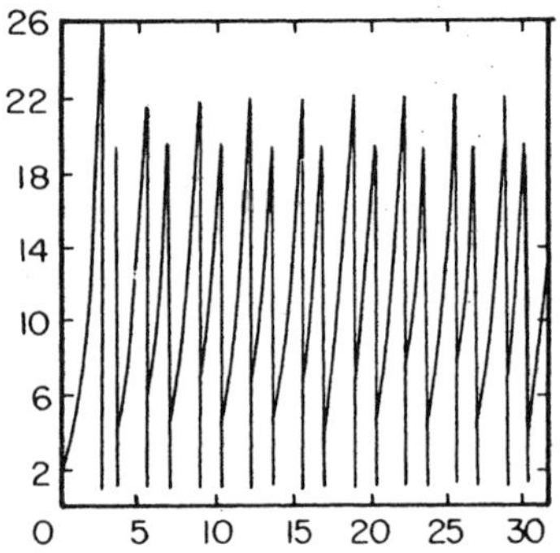

FIG. 17. a_1 vs t for $\gamma = 9$.

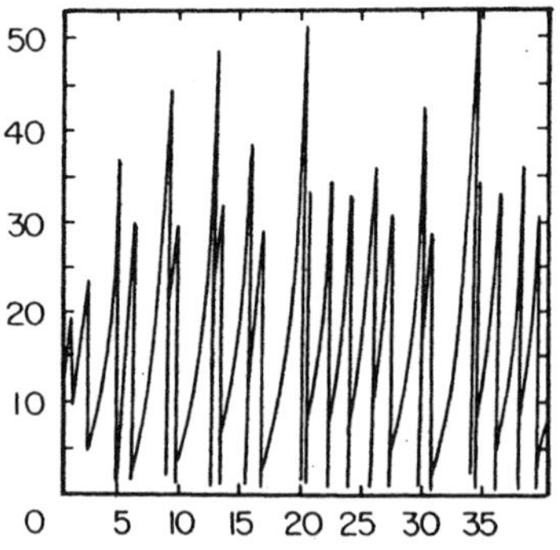

FIG. 18. a_1 vs t for $\gamma = 15$.

[4]For larger values of δ time-independent equilibrium solutions of the equations exist.

[5]Feigenbaum (1979) has shown that the frequency power spectrum at the accumulation point of the period doubling bifurcations of a differential equation system has universal properties.

[6]This includes the phenomenon of noisy 2^p cycles studied by Crutchfield *et al.* (1980) by computations on another set of differential equations.

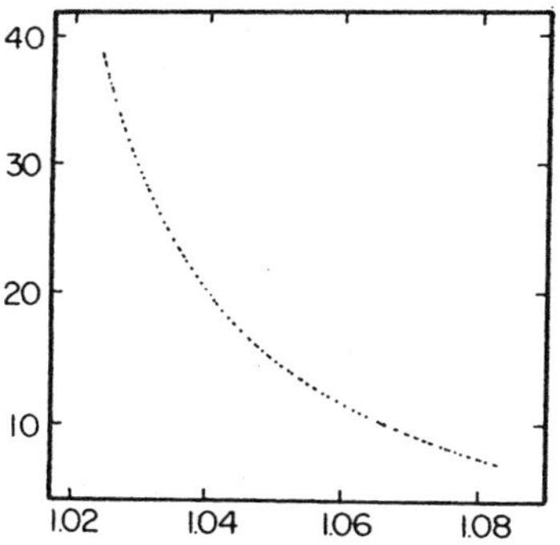

FIG. 19. Surface of section for $\gamma = 15$.

minimum). Thus we see that the one-dimensional map generated by the particular system discussed is of the same general character as the simple quadratic map discussed in Sec. II, Eq. (2.3), i.e., it is concave down with a smooth rounded maximum.

3. Instability saturation by cubicly nonlinear mode coupling

Here we very briefly describe another study of nonlinear mode coupling saturation of an unstable plasma wave. The physical situation, however, is somewhat different from that considered in example (2) in that a resonant three-wave process is precluded by the wave dispersion relation. The physical system is assumed to be well modeled by the following normalized, one-dimensional, nonlinear Schröedinger equation with linear wave growth and damping included,

$$i\left(\frac{\partial E}{\partial t} + \hat{\gamma}E\right) + \frac{\partial^2 E}{\partial x^2} + [\,|E|^2 - |E|_0^2]E = 0\,,$$

where E is the complex amplitude coefficient of the x-directed electric field, $|E|_0^2$ denotes the spatial average of $|E|^2$, and $\hat{\gamma}$ is a linear growth-damping operator defined so that the Fourier transform of $\hat{\gamma}E(x,t)$ is $\gamma(k)E_k(t)$ with $E_k(t)$ the Fourier coefficient of E and $\gamma(k)$ the linear damping rate of a wave of wave number k [$\gamma(k) < 0$ for growth]. This equation may be used to model a situation where an electron beam with a thermal spread is injected into a plasma [cf. Russell and Ott (1980) and references therein]. Under suitable conditions on $\gamma(k)$ it is possible to consider $E(x,t)$ to consist of just three wave number components, k_0, k_1, k_2, with $2k_0 = k_1 + k_2$. There then results a system of ordinary differential equations for the three complex wave amplitudes corresponding to the three wave numbers,

$$\frac{dE_0}{dt} = -\gamma(k_0)E_0 + i[\,|E_1|^2E_0 + |E_2|^2E_0 + 2E_0^*E_1E_2\exp(2i\delta t)]\,,$$

$$\frac{dE_1}{dt} = -\gamma(k_1)E_1 + i[\,|E_0|^2E_1 + |E_2|^2E_1 + E_0^2E_2^*\exp(-2i\delta t)]\,,$$

$$\frac{dE_2}{dt} = -\gamma(k_2)E_2 + i[\,|E_0|^2E_2 + |E_1|^2E_2 + E_0^2E_1^*\exp(-2i\delta t)]\,,$$

where we assume $\gamma(k_0) < 0$ and $\gamma(k_{1,2}) > 0$. As in example (2), if $\gamma(k_1) = \gamma(k_2)$, these three complex equations can be reduced to a system of three real equations:

$$\frac{da_0}{dt} = a_0 + 2a_0a_1^2\sin\phi\,,$$

$$\frac{da_1}{dt} = -\gamma a_1 - a_0^2a_1\sin\phi\,,$$

$$\frac{d\phi}{dt} = -2\delta + 2(a_1^2 - a_0^2) + 2(2a_1^2 - a_0^2)\cos\phi\,.$$

The above system has been examined analytically and numerically as a function of the dimensionless parameters of the system (δ and γ). It was found that the model exhibits a wealth of characteristic dynamical behavior, including stationary equilibria, bifurcations from stationary equilibria to periodic orbits, period doubling bifurcations, chaotic solutions on a strange attractor, tangent bifurcations from chaotic to periodic solutions, transient chaos, and hysteresis. It is not our purpose here to detail the behavior of this system. Rather, we wish to use it as an illustration of scale invariant nature in a strange attractor. For a range of the parameters δ and γ, this system exhibits chaotic time dependence with only a moderate phase-space volume contraction rate. Thus structure of the associated strange attractor is evident in the surface of section (Fig. 21). Furthermore, upon magnification, it is evi-

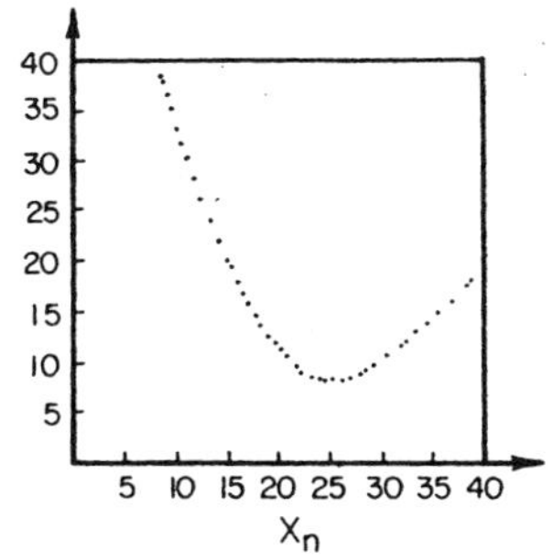

FIG. 20. Points x_{n+1} vs x_n lie approximately on a curve defining a one-dimensional map.

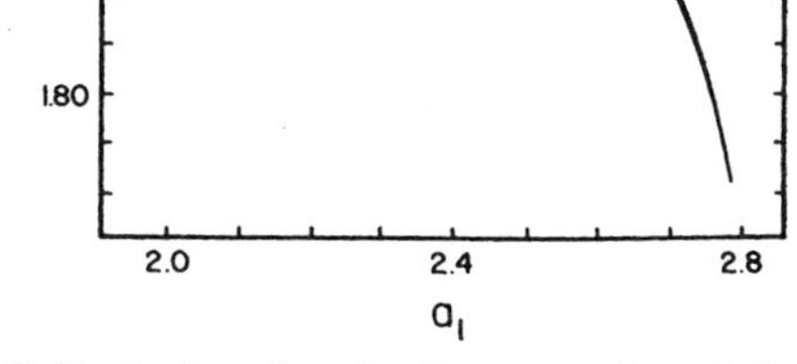

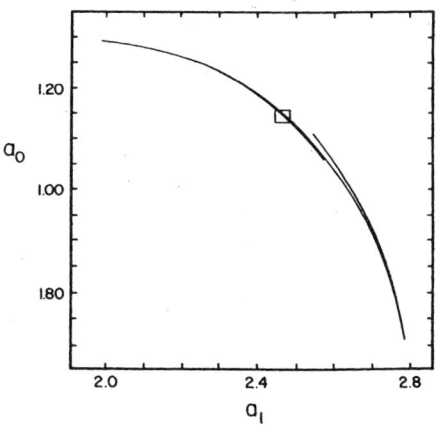

FIG. 21. Surface of section for a particular set of parameters for example (3).

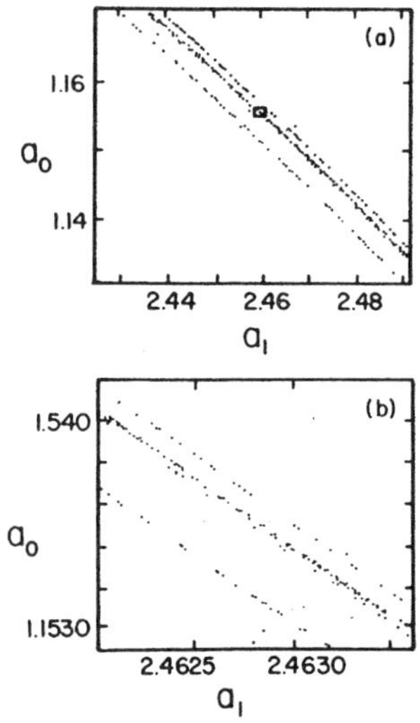

FIG. 22. (a) Magnification of the region in Fig. 21 shown in the rectangle, and (b) magnification of the rectangular region indicated in Fig. 22(a). Scale invariance is evident.

dent that the structure appears to be approximately scale invariant. This is illustrated in Fig. 22 which is to be considered analogous to Fig. 10 for the Henon map.

V. PARTIAL DIFFERENTIAL EQUATIONS, ORDINARY DIFFERENTIAL EQUATIONS, AND TURBULENCE

It is natural to ask whether the phenomena revealed by the examples of the preceding section carry over to more complicated systems. In particular, what happens if the restriction of phase-space volume contraction is lifted and if the dimension of the system is larger than three? Furthermore, the examples in Sec. IV were meant as approximate models for phenomena which are more exactly described only by partial differential equations. Partial differential equations may often be thought of as infinite systems of ordinary differential equations; e.g., it is frequently possible to expand the dependent variables of a partial differential equation in an infinite discrete set of Fourier spatial modes and to derive an infinite set of coupled nonlinear ordinary differential equations for the time dependence of the Fourier coefficients.

It seems clear that such more general systems might display the same characteristic phenomena as those discussed in Sec. IV, but could also reveal additional phenomena ruled out by the specific constraints adopted in Sec. IV. For example, we have found in Sec. IV cases where chaotic solutions occur for a strange attractor of dimension between two and three. For higher dimensional systems, strange attractors are possible with higher dimensionality—e.g., one might have a strange attractor of dimension between five and six if a system of n differential equations with $n \geq 6$ were investigated. (Such higher dimensional strange attractors might be quite difficult to diagnose in actual situations.) At any rate, it seems reasonable to suppose that, as a parameter which characterizes the strength of destabilizing forces in a system described by partial differential equations is cranked up (e.g., the Reynolds number), the general (although not uniform) tendency would be toward motion on attractors of increasing dimension. For example, a stable attracting stationary equilibrium point (zero-dimensional attractor) might bifurcate to a periodic orbit (one-dimensional attractor), which then proceeds via an infinite number of period doubling bifurcations to a strange attractor of dimension between two and three, which then becomes a strange attractor with dimension between three and four, etc. Another possible sequence might be a stationary point (zero dimensions) that bifurcates to a periodic orbit (one dimension), which bifurcates to a doubly periodic orbit (a two-dimensional attractor formed by the surface of a torus), which then bifurcates to a strange attractor (dimension> 2). The former route to a strange attractor [cf. Fig. 23(a)] has been demonstrated in Sec. IV.B.2. The route to a strange attractor via transition from a doubly periodic orbit (motion on a two-dimensional toroidal surface) is illustrated in Fig. 23(b) and was first discussed in a pioneering paper by Ruelle and Takens (1971) [cf. also Newhouse *et al.*, 1978). (Note that this route to a strange attractor [Fig. 23(b)] is specifically ruled out for the examples in Sec. IV, since, as previously mentioned, doubly periodic orbits are not possible in a system of three ordinary differential equations with phase-space contraction.)

Ruelle and Takens (1971) conjectured that small nonlinearities would destroy triply periodic motions. They therefore reasoned that, as the Reynolds number of a fluid flow is increased, the sequence in Fig. 23(b) ought to occur. In particular, they concluded that the last step, (doubly periodic motion) → (strange attractor), is quite likely, since doubly periodic motions cannot bifurcate to triply periodic motions if the latter are unstable. Although the reasoning leading to their conclusion that the onset of turbulence may be associated with a strange attractor is indirect, this paper and that of Lorenz are the first to point out the possible relevance of strange attractors in a physical context. Furthermore, their picture of turbulence onset is a fundamental departure from that advocated by Landau (1941) (cf. also Landau and Lifshitz, 1959). Landau argued that turbulence in fluid flow may be viewed as a hierarchy of instabilities. As the Reynolds number, R, is increased from zero, the basic state becomes unstable to a mode of frequency $\tilde{\omega}_1$ which saturates in a nonlinear periodic state for which dependent variables can be written in the form $\sum_l a_l \exp(-il\omega_1 t)$; as R is further increased, another instability appears at $\tilde{\omega}_2$ and the saturated state becomes doubly periodic, $\sum_{l,m} a_{lm} \times \exp(-il\omega_1 t - im\omega_2 t)$; further increase of R leads to the successive appearance of more and more discrete frequencies so that doubly periodic flow (ω_1, ω_2) transforms to triply periodic, then to quadruply periodic, etc. Thus, as R is increased, more and more frequencies are present, and the flow pattern becomes more and more complicated. Thus in this model the

(A)

{STATIONARY POINT} → {BIFURCATION TO A PERIODIC ORBIT} → {PERIOD DOUBLING BIFURCATIONS} → {ACCUMULATION POINT OF AN INFINITE NUMBER OF PERIOD DOUBLING BIFURCATIONS} → {STRANGE ATTRACTOR}

(B)

{STATIONARY POINT} → {BIFURCATION TO A PERIODIC ORBIT} → {BIFURCATION TO A DOUBLY PERIODIC ORBIT} → {STRANGE ATTRACTOR}

FIG. 23. Two possible routes to a strange attractor.

Fourier spectrum is always discrete and approximates a continuum only in the case where a large number of discrete frequencies are present. Furthermore, the flow is never truly chaotic, since the time correlation functions of multiply periodic functions do not tend to zero for large argument. On the other hand, the appearance of a strange attractor (as, for example, by the sequence of events in Fig. 23) can lead to turbulent motions in a direct way. Recent experiments on the onset of turbulence induced by various instabilities lend support to the idea that turbulence onset may be due to the appearance of a strange attractor (Swinney and Gollub, 1978; Ahlers and Behringer, 1978; Fenstermacher *et al.*, 1979; Walden and Donnelly, 1979; Lashinsky, 1980; Donnelly *et al.*, 1980; Moon and Holmes, 1980).

We now discuss some evidence that the sequence to a strange attractor via a route something like that depicted in Fig. 23(b) can occur. To begin we note the recent study by Curry (1978). In this paper, the author considers the same physical problem as that considered by Lorenz, but includes a greater number of Fourier spatial modes before truncating the series representation of the dependent variables. The result is a system of 14 coupled ordinary differential equations, rather than the three studied by Lorenz. The numerical results of Curry are somewhat different than those of Lorenz. In particular, he finds that chaotic time dependence is preceded by doubly periodic motion on a two-dimensional toroidal surface that is embedded in the full 14-dimensional phase space. The existence of the torus is found by using the surface of section technique: a surface of section $x_1 = \text{const}$ was chosen (where x_1 is one of the 14 variables), and another two of the 14 variables, which we denote by x_2 and x_3, were singled out. Every time the orbit crossed the surface of section the x_2, x_3 coordinates of the crossing point were plotted. Figure 24 shows a typical result for the case where a doubly periodic orbit exists. As can be seen from Fig. 24, the points lie on a closed curve. This is the result to be expected for doubly periodic behavior, since the intersection of a torus with a plane (the surface of section) is a closed curve. Related numerical studies have been performed by Yahata (1978) and Sherman and McLaughlin (1978).

Another relevant study is that of Curry and Yorke (1978), who consider a simple two-dimensional map that models the onset of chaos preceded by doubly periodic motion. The mapping that they study is one which may conveniently be represented as the result of two successive transformations, the first in polar coordinates

$$(\rho, \theta) \rightarrow [\varepsilon \ln(1+\rho), \theta + \theta_0] \tag{5.1a}$$

and the second in rectangular coordinates

$$(x, y) \rightarrow (x, y + x^2), \tag{5.1b}$$

where $\varepsilon \geq 0$ and $\theta_0 \geq 0$ are parameters to be chosen. Note that for $\varepsilon > 1$, the mapping $\rho \rightarrow \varepsilon \ln(1+\rho)$ has two fixed points, one of which is $\rho = 0$ and the other of which (denoted by ρ_ε) is the positive root of $\rho_\varepsilon = \varepsilon \ln(1+\rho_\varepsilon)$. For $1 \geq \varepsilon > 0$, the only fixed point is $\rho = 0$. Consider first the iteration of the polar map, $(\rho, \theta) \rightarrow [\varepsilon \ln(1+\rho), \theta + \theta_0]$, alone. For $\varepsilon > 1$, all points tend asymptotically to the circle $\rho = \rho_\varepsilon > 0$. Furthermore, if the rotation angle θ_0 is not 2π times a rational number, then the orbit fills up the entire circle $\rho = \rho_\varepsilon$. As ε approaches one from above, $\rho_\varepsilon \rightarrow 0$, so that is may be considered that this map undergoes a bifurcation from a stable fixed point to an attracting orbit on the circle $\rho = \rho_\varepsilon$.

Thus the map $(\rho, \theta) \rightarrow [\varepsilon \ln(1+\rho), \theta + \theta_0]$ can be viewed as modeling a surface of section for some differential equation that undergoes a bifurcation of a stable periodic orbit to a doubly periodic orbit on a torus (where $\rho = \rho_\varepsilon$ represents the intersection of this torus with the surface of section and $\rho = 0$ represents the intersection of the periodic orbit with the surface of section).

For ε only slightly larger than one, $\rho_\varepsilon \ll 1$, and the map $(x, y) \rightarrow (x, y + x^2)$ is almost the identity map, since

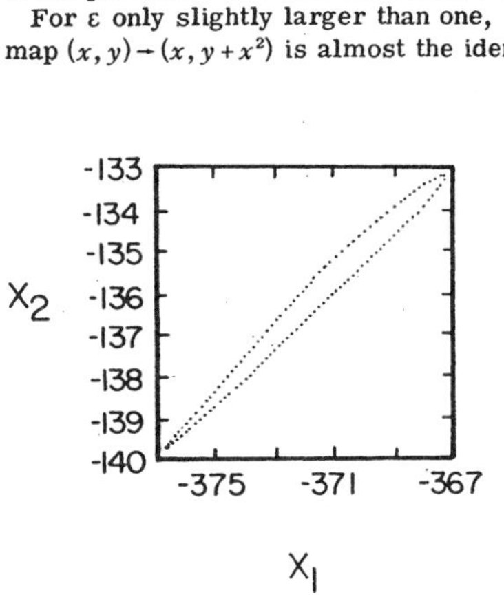

FIG. 24. Orbit intersections with the surface of section, obtained by Curry (1978).

the quadratic term, x^2, is comparatively small. Thus if one considers the map resulting from the successive application of the two above-mentioned transformations, then for $\varepsilon > 1$ the map $(x,y) \to (x, y+x^2)$ introduces a nonlinearity that becomes stronger as ε is increased. Curry and Yorke have described a sequence of numerical studies of the attracting set for the composite map, Eqs. (5.1), with $\theta_0 = 2$. At $\varepsilon = 1.01$ a nearly circular loop encircles the origin. As ε increases, the loop distorts and increases in size until $\varepsilon = 1.28$. For $1.39 \gtrsim \varepsilon > 1.28$ an attracting period three orbit is found. Immediately after $\varepsilon \cong 1.3953$, an apparently connected loop returns. This apparent loop, however, actually has a complicated structure and is, in fact, a strange attractor.

As ε increases somewhat, the strange aspect of the attractor becomes more apparent, as evidenced by Fig. 25 for $\varepsilon = 1.45$. The bump containing the point P_1 in Fig. 25 is mapped to P_2, to P_3, to P_4, ... to P_{11}. The cusplike structure at P_4 shows signs of being flattened against the rest of the set. Each successive cusp is more flattened and more elongated. Soon the cusps are so flattened that they are no longer discernable. This sequence, P_i, is, however, infinite. Note, too, that the successive elongation of the cusps represents a stretching apart of two nearby points. As discussed in Sec. III, such stretching should be expected to lead to chaotic behavior, and this is in fact the case here. Similar behavior has also been observed in a map studied by Coullet *et al.* (1980).

The character of the strange attractor for the example of Curry and Yorke appears to be that of a very complicated curve folded over on itself an infinite number of times and with an infinite length. In this connection, it may be of some interest to give a simple example of an infinite length curve with Hausdorf dimension between one and two (Mandelbrot, 1977). Consider the sequence of operations shown in Fig. 26. We start with an equilateral triangle, divide each of the sides in thirds, and erect smaller equilateral triangles on the middle thirds on each side. The process is repeated twice in the figure. On each application of the process the length of the bounding curve increases by $\frac{4}{3}$. In the limit that the process is repeated an infinite number of times, the length of the bounding curve approaches infinity, although the curve remains in a bounded region of the plane. Furthermore, it may be verified that the Hausdorf dimension of this curve is $\ln 4/\ln 3 \cong 1.26$.

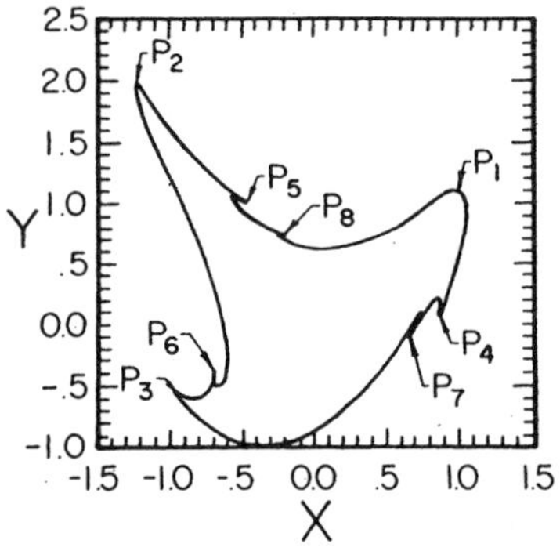

FIG. 25. The attractor for the map of Curry and Yorke with $\theta_0 = 2$ and $\varepsilon = 1.45$.

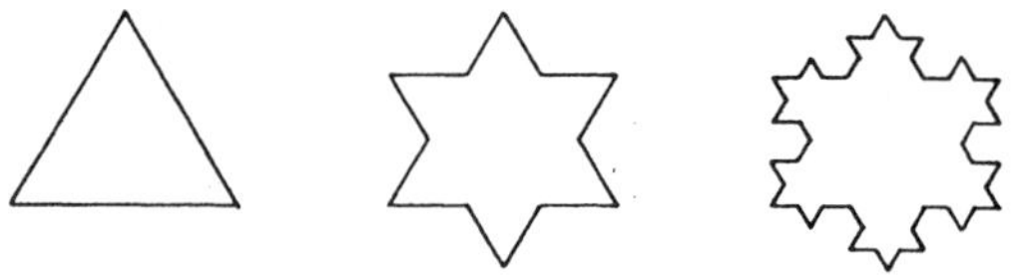

FIG. 26. Illustration of the construction of a curve of dimension ln4/ln3.

VI. CONCLUSIONS

This subject matter is certain to receive much future attention from researchers in the physical sciences and in mathematics. It seems clear that there is a great need for further work in order that the subject develop to the point that theory can provide answers to many of the most practical questions. For example, given a system of equations, can one predict the occurrence of a strange attractor? To what extent can properties of the strange attractor such as its dimension, associated distribution function, power spectra, and correlation functions be predicted? What is the distribution function on a strange attractor and what is the most efficient way to find it and characterize it? Further work will also certainly be done identifying physical systems which exhibit chaotic motions associated with strange attractors. The list of such systems (partially enumerated in the Introduction) is already impressive, and, as it grows, it is to be expected that interest on the part of physical scientists will also grow.

ACKNOWLEDGMENTS

The author wishes to thank Professor J. A. Yorke for many informative discussions. This work was supported by the U. S. Department of Energy.

REFERENCES

Adam, J. C., M. N. Bussac, and G. Laval, 1980, in *Intrinsic Stochasticity in Plasmas*, edited by G. Laval and D. Gresillon (Les Editions de Physique Courtaboeuf, Orsay, France), p. 415.

Afraymovich, V. S., V. V. Bykov, and L. P. Shilnikov, 1977, Dokl. Adak. Nauk, SSSR **234**, 336.

Ahlers, G., and R. P. Behringer, 1978, Phys. Rev. Lett. **40**, 712.

Bridges, R., and G. Rowlands, 1977, Phys. Lett. A **63**, 189.

Bunimovich, L. A., and Ya. G. Sinai, 1977, in *Proceedings of the Spring School on Nonlinear Waves, Gorki, U.S.S.R.*

Chang, S.-J., and J. Wright, 1981, Phys. Rev. **23**, 1419.

Chirikov, B. V., 1979, Phys. Rep. **52**, 463.

Coullet, P., and J.-P. Eckmann, 1980, *Iterated Maps on the Interval as Dynamical Systems* (Birkhauser, Boston).

Coullet, P., and C. Tresser, 1980, J. Phys. (Paris), Lett. **41**, L255.

Coullet, P., C. Tesser, and A. Arneodo, 1980, Phys. Lett. A **77**, 327.

Crutchfield, J., D. Farmer, N. Packard, R. Shaw, G. Jones, and R. J. Donnelly, 1980, Phys. Lett. A **76**, 1.

Crutchfield, J., M. Nauenberg, and J. Rudnick, 1981, Phys.

Rev. Lett. **46**, 933.
Curry, J. H., 1979, Commun. Math. Phys. **68**, 129.
Curry, J. H., and J. A. Yorke, 1978, in *The Structure of Attractors in Dynamical Systems*, edited by N. G. Markley, J. C. Martin, and W. Perrizo, Lecture Notes in Mathematics (Springer, Berlin), Vol. 668, p. 48.
Donnelly, R. J., K. Park, R. Shaw, and R. W. Walden, 1980, Phys. Rev. Lett. **44**, 987.
Feigenbaum, M. J., 1978, J. Stat. Phys. **19**, 25.
Feigenbaum, M. J., 1979, Phys. Lett. A **74**, 375.
Feit, S. D., 1978, Commun. Math. Phys. **61**, 249.
Fenstermacher, P. R., H. L. Swinney, and J. P. Gollub, 1979, J. Fluid Mech. **94**, 103.
Frederickson, P., J. L. Kaplan, and J. A. Yorke, 1980, preprint.
Greene, J. M., R. S. MacKay, F. Vivaldi, and M. J. Feigenbaum, 1981, preprint.
Greene, J. M., 1979, J. Math. Phys. **20**, 1183.
Guckenheimer, J., 1979, Commun. Math. Phys. **70**, 133.
Guckenheimer, J., G. Oster, and Ipaktchi, 1977, J. Math. Biol. **4**, 101.
Haken, H., 1975, Phys. Lett. A **53**, 77.
Helleman, R. H. G., 1980, in *Fundamental Problems in Statistical Mechanics 5*, edited by E. G. D. Cohen (North-Holland, Amsterdam), p. 165.
Henon, M., 1976, Commun. Math. Phys. **50**, 69.
Hirsch, M. W., and S. Smale, 1974, *Differential Equations, Dynamical Systems and Linear Algebra* (Academic, New York).
Holmes, P. J., 1977, Appl. Math. Modeling **1**, 362.
Holmes, P. J., 1980, preprint.
Huberman, B. A., and J. P. Cruchfield, 1979, Phys. Rev. Lett. **43**, 1743.
Huberman, B. A., and J. Rudnick, 1980, Phys. Rev. Lett. **45**, 154.
Huberman, B. A., and A. B. Zisook, 1981, Phys. Rev. Lett. **46**, 626.
Kaplan, J. L., and J. A. Yorke, 1979a, in *Functional differential Equations and Approximation of Fixed Points*, edited by H.-O. Peitzen and H.-O. Walther, Lecture Notes in Mathematics (Springer, Berlin), Vol. 730, p. 228.
Kaplan, J. L., and J. A. Yorke, 1979b, Commun. Math. Phys. **67**, 93.
Landau, L. D., 1941, C. R. (Dokl.) Acad. Sci. URSS **44**, 311.
Landau, L. D., and E. M. Lifshitz, 1959, *Fluid Mechanics* (Pergamon, London), Chap. 3.
Lanford, O. E., III, 1976, in *Statistical Mechanics and Dynamical Systems* (Mathematics Department, Duke University, Durham, North Carolina), Chap. 4.
Lashinsky, J., in *Intrinsic Stochasticity in Plasmas*, 1980, edited by G. Laval and D. Gresillon (Les Editions de Physique Courtaboeuf, Orsay, France), p. 425.
Li, T.-Y., and J. A. Yorke, Am. Math. Mon. **82**, 985 (1975).
Lorentz, E. N., 1979, presentation at the International Conference on Nonlinear Dynamics, New York Academy of Sciences (unpublished).
Lorenz, E. N., 1963, J. Atmos. Sci. **20**, 130.
Lozi, R., 1978, J. Phys. (Paris) Suppl. 8, tome 35, p. C5–9.
Mandelbrot, B. B., 1977, *Fractals* (Freeman, San Francisco).
Maschke, E. K., and B. Saramito, 1980, in *Intrinsic Stochasticity in Plasmas*, edited by G. Laval and D. Gresillon (Les Editions de Physique Courtaboeuf, Orsay, France), p. 383.
Masui, K., and P. K. C. Wang, 1980 (unpublished).
May, R. M., 1976, Nature **261**, 459.
McLaughlin, J., 1979, Phys. Rev. A **20**, 2114.
McLaughlin, J. B., and P. C. Martin, 1975, Phys. Rev. A **12**, 186.
Moon, F. C., and P. J. Holmes, 1979, J. Sound Vib. **65**, 275.
Newhouse, S., D. Ruelle, and F. Takens, 1978, Commun. Math. Phys. **64**, 35.
Rabinovitch, M. I., 1978, Sov. Phys.—Usp. **21**, 443.
Rechester, A. B., and T. H. Stix, 1976, Phys. Rev. Lett. **35**, 597.
Robbins, K. A., 1977, Math. Proc. Cambridge Philos. Soc. **82**, 209.
Rosenbluth, M. N., R. Z. Sagdeev, J. B. Taylor, and G. M. Zaslavskii, 1966, Nucl. Fusion **6**, 297.
Ruelle, D., 1977, in *Mathematical Problems in Theoretical Physics*, edited by G. Dell'Antonio, S. Doplicher, and G. Jona-Lasinio, Lecture Notes in Physics (Springer, Berlin), Vol. 80, p. 341.
Ruelle, D., and F. Takens, 1971, Commun. Math. Phys. **20**, 167.
Russell, D. A., J. D. Hanson, and E. Ott, 1980, Phys. Rev. Lett. **45**, 1175.
Russell, D. A., and E. Ott, 1980, preprint.
Saltzman, B., 1962, J. Atmos. Sci. **19**, 329.
Shaw, R., 1981, Z. Naturforsch. A **36**, 80.
Sherman, J., and J. B. McLaughlin, 1978, Commun. Math. Phys. **58**, 9.
Shimzu, T., and N. Morioda, 1978, Phys. Lett. A **69**, 148.
Shraiman, B., C. E. Wayne, and P. C. Martin, 1981, Phys. Rev. Lett. **46**, 935.
Simó, 1979, J. Stat. Phys. **21**, 465.
Sinai, Ya. G., 1977, in *Proceedings of the Spring School on Nonlinear Waves, Gorki, U.S.S.R.*
Stix, T. H., 1973, Phys. Rev. Lett. **30**, 833.
Swinney, H. L., and J. P. Gollub, 1978, Phys. Today **31** (8), 41.
Tomita, K., and T. Kai, 1979, J. Stat. Phys. **21**, 65.
Treve, Y., 1978, in *Topics in Nonlinear Dynamics*, edited by S. Jorna, AIP Conference Proceedings (American Institute of Physics, New York), no. 46, p. 147.
Treve, Y., and O. P. Manley, 1980, in *Intrinsic Stochasticity in Plasmas*, edited by G. Laval and D. Gresillon (Les Editions de Physique Courtaboeuf, Orsay, France), p. 393.
Vyshkind, S. Ya., 1978, Radiofizika **21**, 850 [Sov. Radiophys. **21**, 600].
Vyshkind, S. Ya., and M. I. Rabinovich, 1976, Zh. Eksp. Teor. Fiz. **71**, 557 [Sov. Phys.—JETP **44**, 292 (1976)].
Walden, R. W., and R. J. Donnelly, 1979, Phys. Rev. Lett. **42**, 301.
Wang, P. K. C., 1980, J. Math. Phys. **21**, 398.
Wersinger, J. M., J. M. Finn, and E. Ott, 1980a, in *Intrinsic Stochasticity in Plasmas*, edited by G. Laval and D. Gresillon (Les Editions de Physique Courtaboeuf, Orsay, France), p. 403.
Wersinger, J. M., J. M. Finn, and E. Ott, 1980b, Phys. Rev. Lett. **44**, 453.
Wersinger, J. M., J. M. Finn, and E. Ott, 1980c, Phys. Fluids **23**, 1142.
Wolf, A., and J. Swift, 1981, preprint.
Yahata, H., 1978, Prog. Theor. Phys. Suppl. **64**, 165.
Yorke, J. A., and E. D. Yorke, 1979, J. Stat. Phys. **21**, 263.
Yorke, J. A., and E. D. Yorke, 1981, in *Hydrodynamic Instabilities and the Transition to Turbulence*, edited by H. L. Swinney and J. P. Gollub (Springer, Berlin), p. 77.
Zaslavskii, G. M., 1978, Phys. Lett. A **69**, 145.
Zaslavskii, G. M., and Kh.-R. Ya Rachko, 1979, Zh. Eksp. Teor. Fiz. **76**, 2052 [Sov. Phys.—JETP **49**, 1039 (1979)].

Math. Z. 180, 119–140 (1982)

Mathematische Zeitschrift

Ergodic Properties of Invariant Measures for Piecewise Monotonic Transformations

Franz Hofbauer[1,*] and Gerhard Keller[2,**]

[1] Institut für Mathematik, Universität Wien, Strudlhofgasse 4, A-1090 Wien, Austria
[2] Institut für Mathematische Statistik der Universität, Lotzestraße 13, D-3400 Göttingen, Federal Republic of Germany

Introduction

During the last decade a lot of research has been done on onedimensional dynamics. Different kinds of invariant measures for certain classes of piecewise monotonic transformations have been considered. In most of these cases the Perron-Frobenius-operator plays an important role. In this paper we try to unify these different examples, which are discussed in detail below. First we give a discription of the results proved in this paper.

The general setting we propose is the following.

X is a totally ordered, order-complete set and $T: X \to X$ a transformation which is piecewise monotonic and order-continuous. P is a Perron-Frobenius-operator on the space of all bounded measurable functions on X given by

$$Pf(x) = \sum_{y \in T^{-1}x} g(y) f(y),$$

where g is a function of bounded variation on X with $0 < g(x) \leqq d < 1$. Furthermore we assume the existence of a Borel probability measure m on X, which is invariant for P, i.e. $m(Pf) = m(f)$ for all bounded measurable $f: X \to \mathbb{R}$.

Under these assumptions we give a kind of spectral representation for P (Theorem 1) which is fundamental for the subsequent results:

(1) There is a "greatest" T-invariant m-absolutely continuous probability μ (also denoted by $\mu \ll m$), $\mu = hm$, where h is of bounded variation. "Greatest" means that all other T-invariant, m-absolutely continuous probabilities are absolutely continuous with respect to μ (Theorem 1).

(2) The σ-algebra of T-invariant sets is μ-a.s. finite (Theorem 2). This gives the possibility to split up μ into its finitely many ergodic components. Furthermore, there is a k such that the finitely many ergodic components $W_i \subset X$ of

---

* Research for this paper was done, when the first author visited the Institut für mathematische Statistik, Göttingen

** The second author was supported by the Deutsche Forschungsgemeinschaft

(X, μ, T^k) are already weak mixing. We refer to $(W_i, \mu | W_i, T^k | W_i)$ as a weak mixing component of μ (Remark 2).

(3) The asymptotic σ-algebra $\mathfrak{B}_\infty = \bigcap_{j=0}^{\infty} T^{-j}\mathfrak{B}$ of T, where $\mathfrak{B}$ denotes the Borel sets of X, is μ-a.s. finite. It is trivial for a weak mixing component of μ and hence such a component is exact (Theorem 3).

(4) Each finite partition into intervals is a weak Bernoulli partition for every weak mixing component W_i of μ. The corresponding mixing-coefficient decreases exponentially (Theorem 4 and its Corollary). For an $f: W_i \to \mathbb{R}$ of bounded p-variation the stationary stochastic process $\xi_n = f \circ T^{kn}$ $(n \geqq 0)$ satisfies a central limit theorem and an a.s. invariance principle with respect to μ (Theorem 5).

The examples of invariant measures, which fit in our situation, are the following:

(i) Lasota and Yorke [11] consider piecewise monotonic and C^2 transformations T on $[0, 1]$ such that $\inf |T'| > 1$. If one sets $g = |1/T'|$, then the operator P satisfies $\lambda(Pf) = \lambda(f)$ for the Lebesguemeasure λ. In [11] the existence of a function h with $Ph = h$ is proved, which implies the existence of an invariant $\mu \ll \lambda$. In [13] μ is split up into its finitely many ergodic components. Bowen [3] shows that μ is weak Bernoulli, if it is weak mixing. Wong [23] generalizes the results of [11] and [13] to the class of T's, which are piecewise monotonic and C^1, such that T' is of bounded variation and satisfies $\inf |T'| > 1$. Wagner [20] shows that the asymptotic σ-algebra $\mathfrak{B}_\infty$ is μ-a.s. finite. For the class of T considered in [11] the spectral decomposition of P can be found in [10] (cf. also [14]), from which a central limit theorem for stationary processes $f \circ T^n$ on $([0, 1], \mu)$ is deduced, if μ is weak mixing (a special case of this is shown in [24]).

As $g = |1/T'|$, where T is as in [23], satisfies our requirements, all these results are contained in this paper. Furthermore, we show for a weak mixing component of μ, that the mixing coefficients in the weak Bernoulli property decrease at an exponential rate and that the process $f \circ T$ satisfies for certain f a central limit theorem and an almost sure invariance principle.

(ii) Next we consider any piecewise monotonic transformation T on $[0, 1]$ with $h_{\text{top}}(T) > 0$. In [6] and [7] it is shown that every maximal measure can be split up into finitely many ergodic components, which are isomorphic to Markov measures on a finite type subshift with finite or countable alphabet.

If one sets $g(x) = \exp(-h_{\text{top}}(T)) < 1$ for all $x \in [0, 1]$, it is shown in §5 of Chap. III that there is an m with $m(Pf) = m(f)$ and that μ is then a measure with maximal entropy for T. Hence the results of this paper give a decomposition of the maximal measure into its finitely many weak mixing components, each of which is weak Bernoulli and the process $f \circ T^n$ on such a component satisfies for certain f a central limit theorem and an almost sure invariance principle.

(iii) Now let T be the transformation $x \to \beta x \pmod 1$ on $[0, 1]$. Using a limit theorem for the operator P, Walters [22] shows, that there is a unique equilibrium state μ for every Lipschitz continuous function $\varphi: [0, 1] \to \mathbb{R}$, and that μ is weak Bernoulli.

In §4 of Chap. III we consider $\varphi\colon [0,1] \to \mathbb{R}$ of bounded variation, which satisfy $\sum_{i=1}^{\infty} \operatorname{var}_i \varphi < \infty$, where $\operatorname{var}_i \varphi = \sup\{|\varphi(x) - \varphi(y)|\colon x$ and y are contained in the same inerval, on which T^i is monotone$\}$. Using the same technique as Walters [22], we show the existence of a probability m, a $\lambda > 0$ and an $n \in \mathbb{N}$ such that $m(Pf) = m(f)$ and $\|g\|_\infty < 1$ is satisfied for $g(x) = \exp\left(\sum_{i=0}^{n-1} \varphi(T^i x)\right) \Big/ \lambda^n$. Furthermore we get that μ is then an equilibrium state for φ.

This means that we generalize the results of Walters to a larger class of functions φ and show besides the weak Bernoulli property also a central limit theorem and an almost sure invariance principle for certain processes $f \circ T^n$.

(iv) Also subshifts X of finite type with finite alphabet are piecewise monotonic, if one introduces the lexicographic ordering in X.

Using a limit theorem for the operator P, Bowen [2] has shown the weak Bernoulli property with exponentially decreasing coefficient and a central limit theorem for equilibrium states of Hölder continuous functions $\varphi\colon X \to \mathbb{R}$, i.e. $\operatorname{var}_i \varphi \leqq C q^i$ for some $C > 0$ and $0 < q < 1$.

In §3 of Chap. III we show for every $\varphi\colon X \to \mathbb{R}$ of bounded variation, which satisfies $\sum_{i=1}^{\infty} \operatorname{var}_i \varphi < \infty$, the existence of a probability measure m, a $\lambda > 0$ and an $n \in \mathbb{N}$ such that $m(Pf) = m(f)$ and $\|g\|_\infty < 1$ is satisfied for $g(x) = \exp\left(\sum_{i=0}^{n=1} \varphi(T^i x)\right) \Big/ \lambda^n$. Furthermore μ is then an equilibrium state for φ. Hence for subshifts of finite type we show Bowen's results for a slightly different class of φ's.

Finally we should like to thank M. Denker for encouraging discussions during the preparation of the paper.

I. Spectral Decomposition of the Operator P and the Measure μ

Let X be a totally ordered, complete set. Then the order topology, this is that one generated by open intervals, makes X into a compact space [see e.g. Birkhoff, Lattice Theory, A.M.S. Colloquium Publications 25, Chap. X, §7, Theorem 12]. We call a transformation T on X *piecewise monotonic*, if $X = \bigcup_{i=1}^{N} I_i$, the I_i are disjoint intervals and $T|I_i$ is continuous and preserves or reverses the ordering such that $T(I_i)$ is again an interval. We consider $g\colon X \to (0, d]$ with

(a) $$d < 1 \quad \text{and} \quad V_X g < \infty$$

where the variation $V_X g$ of g over X is defined by

$$\sup\left\{\sum_{i=1}^{k-1} |g(a_i) - g(a_{i+1})|\colon k \geqq 2,\ a_1 < \ldots < a_k,\ a_i \in X\right\}.$$

The *Perron-Frobenius-operator* P on the set $\mathfrak{F}$ of all bounded complex-valued functions on X, which are measurable with respect to the Borel-σ-algebra $\mathfrak{B}$, is defined by

$$Pf(x) = \sum_{y \in T^{-1}x} g(y) f(y) = \sum_{i=1}^{N} g(T_i^{-1}x) f(T_i^{-1}x) 1_{T(I_i)}(x)$$

where $T_i = T \mid I_i$. Finally we assume that there is a Borel-probability-measure m on X which satisfies

(b) $$m(Pf) = m(f) \quad \text{for all } f \in \mathfrak{F}.$$

§ 1. We begin with the investigation of the support of the measure m.

Lemma 1. *For $B \in \mathfrak{B}$, $m(B)=0$ implies that $m(TB)=0$.*

Proof. We can assume that $B \subset I_i$ for some i. Suppose that $m(TB)>0$. We set $A_n = \left\{ x \in T(I_i) : g(T_i^{-1}x) \geqq \frac{1}{n} \right\}$. As $g(x)>0$ for all $x \in X$, it follows that $\bigcup_{n=1}^{\infty} A_n = T(I_i)$. This implies that there is a k with $m(A_k \cap TB)>0$. Then we have

$$m(B) = m(P 1_B) = m\left(\sum_{j=1}^{N} g(T_j^{-1}x) 1_B(T_j^{-1}x) 1_{TI_j}(x) \right)$$
$$= m(g(T_i^{-1}x) 1_{TB}(x)) \geqq \frac{1}{k} m(TB \cap A_k) > 0.$$

This contradicts $m(B)=0$. Hence $m(TB)=0$.

Let $\mathfrak{P} = \{I_1, \ldots, I_N\}$ denote the partition of X into intervals of continuity and monotonicity of T. Then $\mathfrak{P}_0^k = \bigvee_{i=0}^{k-1} T^{-i}\mathfrak{P}$ is the finite partition of X into intervals, on which T^k is monotone and continuous.

Lemma 2. $m(A) \leqq d^k$ *for all* $A \in \mathfrak{P}_0^k$. *In particular,* $m(\{x\})=0$ *for all* $x \in X$.

Proof. This is proved by induction. Let $A \in \mathfrak{P}_0^n$ with $n \geqq 1$ be such that $A \subset I_i$. Then

$$m(A) = m(P 1_A) = m(g(T_i^{-1}(x)) 1_{TA}(x)) \leqq d\, m(TA).$$

One sees easily that TA is contained in an element of $\mathfrak{P}_0^{n-1}$. Hence $m(TA) \leqq d^{n-1}$ by induction hypothesis and $m(A) \leqq d^n$.

Lemma 3. *Let $Y = \bigcap \{F : F$ is a closed subset of X with $m(F)=1\}$ be the support of m. T can be changed at a finite number of points in such a way that then $T(Y) \subseteq Y$ and $T^{-1}\{x\} \subseteq Y$ for all $x \in Y$ except finitely many ones.*

Proof. Let J be a subinterval of X, which is contained in I_i for some i. We show that $m(TJ)>0$ if and only if $m(J)>0$. It follows from Lemma 1 that $m(TJ)>0$ implies $m(J)>0$. The other direction follows from

$$m(T^{-1}B) = m(1_B \circ T) = m(P(1_B \circ T)) = m(1_B \cdot P1) \leqq N d\, m(B)$$

because $P1(x) \leqq Nd$ for all $x \in X$ ($g(x) \leqq d$ and the set $T^{-1}\{x\}$ contains at most N elements).

If x is now in the interior of some I_i then one takes for J open subintervals of I_i containing x and it follows that $x \in Y$ if and only if $Tx \in Y$. In particular, $T^{-1}\{x\} \subseteq Y$ for all $x \in Y$, which are not endpoints of some TI_i. As we shall change T only at endpoints of I_i's, this gives the second assertion. By Lemma 2, Y contains no isolated points. Hence, for an endpoint x of some I_i, we can redefine $T(x)$ as a one-sided limit of $T(t)$ for $t \in Y$. Then $T(Y) \subseteq Y$, because Y is closed and $T(t) \in Y$ for all $t \in Y$, which are in the interior of some I_i. This proves the lemma.

If T is changed on a finite set, $Pf(x)$ changes only for finitely many x. By Lemma 2, we have then $m(Pf) = m(f)$ also for the modified T of Lemma 3. We consider now the modified T, restrict it to the closed set $Y \subseteq X$, which is T-invariant by Lemma 3, and denote it again by T. As the sets $T^{-1}\{x\}$ for the original and the restricted T differ only for finitely many $x \in Y$ by Lemma 3, it follows from Lemma 2 that $m(Pf) = m(f)$ holds also for the restricted T. Y is again a totally ordered set, which is compact and has the advantage, that $m(J) > 0$ for every nontrivial interval $J \subseteq Y$ (important for Lemmas 5 and 7 below). T on Y is again piecewise monotonic. Hence we consider from now on only the dynamical system (Y, T).

For $f\colon Y \to \mathbb{C}$ we define the *variation over a subset J of Y* by

$$V_J f = \sup\left\{\sum_{i=1}^{k-1} |f(a_i) - f(a_{i+1})| \colon k \geqq 2,\ a_1 < \ldots < a_k,\ a_i \in J\right\}.$$

We shall need a variation $v(f)$ for $f \in L^1_m$, the set of all equivalence classes of complex-valued, m-integrable functions on Y. We define

$$v(f) = \inf\{V_Y \tilde{f} \colon \tilde{f} \text{ is a version of } f\}.$$

Remark 1. If $\tilde{f}$ is defined on $Y \setminus N$, if $Y \setminus N$ is dense in Y – this is satisfied, if N has measure zero, because every interval has positive m-measure – and if $V_{Y \setminus N} \tilde{f} < \infty$, then one shows easily that for all $x \in Y$ at least one of the limits $\tilde{f}(x^+) = \lim_{z \downarrow x} \tilde{f}(z)$ and $\tilde{f}(x^-) = \lim_{z \uparrow x} \tilde{f}(z)$ exists, since each $x \in Y$ can be approximated from above or below (Y contains no isolated points). Extend $\tilde{f}$ to the whole of Y such that for each $x \in N$ $\tilde{f}(x) = \tilde{f}(x^+)$ or $\tilde{f}(x) = \tilde{f}(x^-)$. Then $V_Y \tilde{f} = V_{Y \setminus N} \tilde{f}$ as is easily seen.

Now set $BV = \{f \in L^1_m \colon v(f) < \infty\}$, which is a linear subspace of L^1_m, but it is not closed with respect to $\| \ \|_1$. We define for $f \in BV$

$$\|f\|_v = \int |f|\, dm + v(f) = \|f\|_1 + v(f).$$

It is easy to see that $\| \ \|_v$ is a norm on BV. The following lemma is needed for Lemma 5.

Lemma 4. *Let $f \in L^1_m$ satisfy $v(f) \leqq c < \infty$. Then $\int |f - E_m(f \mid \mathfrak{A}_n)|\, dm \leqq c d^n$, where $E_m(f \mid \mathfrak{A}_n)$ denotes the conditional expectation of f with respect to the measure m and the σ-algebra $\mathfrak{A}_n$ generated by $\mathfrak{P}_0^n$.*

Proof.

$$\int |f-E_m(f\,|\,\mathfrak{A}_n)|\,dm \leqq \sum_{A\in\mathfrak{P}_0} m(A)(\operatorname*{ess\,sup}_{x\in A} f(x)-\operatorname*{ess\,inf}_{x\in A} f(x)) \leqq d^n\cdot v(f) \leqq c d^n.$$

For the second inequality we have used Lemma 2.

The following lemma and Lemmas 6 and 7 will be requirements for the theorem of Ionescu-Tulcea and Marinescu (Theorem 1 below).

Lemma 5. (i) *For every $c>0$ the set $E=\{f\in L^1_m\colon \|f\|_v\leqq c\}$ is compact in $(L^1_m, \|\ \|_1)$.*

(ii) *$(BV, \|\ \|_v)$ is a Banach space*

(iii) *BV is dense in $(L^1_m, \|\ \|_1)$.*

Proof. (i) Let f_n be in E $(n\geqq 1)$. From Lemma 4 and Theorem IV.8.18 in [4] it follows that there is a subsequence (g_n) of (f_n) and a $f\in L^1_m$ with $\lim_{n\to\infty}\|g_n-f\|_1=0$. We must show that $f\in E$: Let $\varepsilon>0$ be arbitrary. There are versions $\tilde g_n$ of g_n such that $V_Y(\tilde g_n)\leqq v(g_n)+\varepsilon$ for all n. As g_n converges to f in L^1_m, we can find a subsequence $(\tilde h_n)$ of $(\tilde g_n)$ and a set $N\subseteq Y$ of measure zero such that $\lim_{n\to\infty}\tilde h_n(x)=\tilde f(x)$ for all $x\in Y\setminus N$, where $\tilde f$ is a version of f defined on $Y\setminus N$. The corresponding subsequence of (g_n) will be called (h_n). We thus have

$$\begin{aligned} V_{Y\setminus N}(\tilde f) &\leqq \varepsilon+\sum_{i=1}^{k}|\tilde f(a_i)-\tilde f(a_{i-1})| \qquad \text{for suitable } k\in\mathbb{N} \text{ and } a_i\in Y\setminus N,\\ &=\varepsilon+\lim_{n\to\infty}\sum_{i=1}^{k}|\tilde h_n(a_i)-\tilde h_n(a_{i-1})|\\ &\leqq \varepsilon+\limsup_{n\to\infty} V_Y(\tilde h_n)\\ &\leqq 2\varepsilon+\limsup_{n\to\infty} v(h_n). \end{aligned}$$

Extending $\tilde f$ to the whole of Y as in Remark 1 we obtain

$$v(f)\leqq V_Y(\tilde f)=V_{Y\setminus N}(\tilde f)\leqq \limsup_{n\to\infty} v(h_n),$$

as $\varepsilon>0$ was arbitrary and $h_n\in E$. Together with $\lim_{n\to\infty}\|h_n-f\|_1=0$ this implies $\|f\|_v\leqq \limsup_{n\to\infty}\|h_n\|_v\leqq c$, i.e. $f\in E$. This proves (i).

(ii) follows immediately from (i) and I.1.6 in [18].

(iii) follows from the fact that the functions constant on the elements of some $\mathfrak{P}^n_0$ are dense in $(L^1_m, \|.\|_1)$, which is a standard result of measure theory, as $\bigcup_{n=1}^{\infty}\mathfrak{P}^n_0$ generates the Borel-σ-algebra m-a.e.

§2. Now we turn to the investigation of the operator P. The following result is an easy consequence of (b) and the formula $\varphi\cdot Pf=P(f\cdot(\varphi\circ T))$, which is easiliy verified.

Lemma 6. (i) *P is an operator on L^1_m satisfying $m(Pf)=m(f)$ for $f\in L^1_m$.*

(ii) *$\|Pf\|_1\leqq\|f\|_1$ for all $f\in L^1_m$.*

Proof. We show (ii) first for $f \in \mathfrak{F}$ using (b). We have $|Pf(x)| = Pf(x) \cdot \varphi(x)$, where $\varphi(x) = Pf(x)/|Pf(x)|$. Since $|\varphi(x)| = 1$ for all $x \in Y$, we have

$$\begin{aligned}\|Pf\|_1 &= \int (Pf)\, \varphi\, dm = \int P(f \cdot (\varphi \circ T))\, dm \\ &= \int f(\varphi \circ T)\, dm \leqq \int |f| \cdot |\varphi \circ T|\, dm = \|f\|_1.\end{aligned}$$

If one changes f on a set B of measure zero, then Pf is changed on TB, which has measure zero by Lemma 1. Hence P can be considered as an operator on the set of all bounded $f \in L^1_m$. Because of $\|Pf\|_1 \leqq \|f\|_1$ for $f \in \mathfrak{F}$ and as $\mathfrak{F}$ is dense in L^1_m, P can be extended continuously to all $f \in L^1_m$. This proves the lemma.

Now we choose an $M \in \mathbb{N}$ such that $\|g_M\|_\infty = \sup\{g_M(x) : x \in Y\} < \frac{1}{3}$, where g_n for $n \geqq 1$ is given by $g_n(x) = g(x)\, g(Tx) \dots g(T^{n-1}x)$. Such an M exists, because $\|g\|_\infty \leqq d < 1$ (cf. (a)). The following result is a modification of a result in [23] (cf. also [11]).

Lemma 7. *There is an $\alpha \in (0, 1)$ and a $\beta \in (0, \infty)$ with*

$$v(P^M f) \leqq \alpha v(f) + \beta \|f\|_1 \qquad \textit{for all } f \in L^1_m.$$

Proof. Writing Q for P^M and U for T^M, we have $Qf(x) = \sum_{y \in U^{-1}x} g_M(y) f(y)$. Let $(J_1, \dots, J_{\tilde{N}})$ be a partition of Y into intervals such that U/J_i is monotone and continuous and that $V_{J_i} g_M \leqq \|g_M\|_\infty$. As $T(I_i)$ is an interval in Y, also $U(J_i)$ is an interval in Y. For $\varepsilon > 0$, let $\tilde{f}$ be a version of $f \in L^1_m$ such that $V_Y \tilde{f} \leqq v(f) + \varepsilon$. As $m(J_i) > 0$ for $1 \leqq i \leqq \tilde{N}$ and as $U(J_i)$ is an interval in Y, the same computation as in Theorem 1 of [23] works for Q and $\tilde{f}$. Hence we get $V_Y Q\tilde{f} \leqq \alpha V_Y \tilde{f} + \beta \|\tilde{f}\|_1$, where $\alpha = 3\|g_M\|_\infty < 1$ and $\beta = 3\|g_M\|_\infty / \min\{m(J_i) : 1 \leqq i \leqq \tilde{N}\}$. Since $Q\tilde{f}$ is a version of Qf (cf. the proof of Lemma 6) we have

$$v(Qf) \leqq V_Y Q\tilde{f} \leqq \alpha V_Y \tilde{f} + \beta \|\tilde{f}\|_1 \leqq \alpha \cdot v(f) + \beta \|f\|_1 + \alpha\varepsilon.$$

Because ε was arbitrary, we get the desired result.

The properties of the space BV and of the operator P, which are listed in Lemma 5, 6 and 7 allow to apply an ergodic theorem of Ionescu-Tulcea and Marinescu [9], stating in an abstract Banach space setting, the following results (they are formulated here for the special situation under consideration).

Theorem 1. *Under the above assumptions on P, T, X and m we have*

(i) *$P: L^1_m \to L^1_m$ only has a finite number of eigenvalues $\lambda_1, \dots, \lambda_r$ of modulus* 1.

(ii) *Set $E_i = \{f \in L^1_m : Pf = \lambda_i f\}$ for $1 \leqq i \leqq r$. Then $E_i \subset BV$ and* $\dim(E_i) < \infty$.

(iii) *The operator P can be represented as*

$$P = \sum_{i=1}^{r} \lambda_i \phi_i + \psi$$

where the ϕ_i are projections onto the eigenspaces E_i, $\|\phi_i\|_1 \leqq 1$, and ψ is a linear operator on L^1_m with $\sup_{n \geqq 1} \|\psi^n\|_1 < \infty$. Furthermore $\phi_i \phi_j = 0 (i \neq j)$ and $\phi_i \psi = 0$ (all i).

(iv) *$\psi(BV) \subset BV$ and, considered as a linear operator on $(BV, \| \ \|_v)$, ψ satisfies $\|\psi^n\|_v \leqq H q^n$ $(n \geqq 1)$ for some constants H and q with $H > 0$ and $0 < q < 1$.*

(v) *1 is an eigenvalue of P. We assume $\lambda_1 = 1$.*

(vi) *Writing $h := \phi_1(1)$ and $\mu = hm$ one has $h \geqq 0$, $\int h\, dm = 1$ and $Ph = h$. Thus μ is a T-invariant probability on Y and hence on X. If μ' is T-invariant and $\mu' \ll m$, then $\mu' \ll \mu$.*

Proof. (i) to (iv) are immediate consequences of the theorem of Ionescu-Tulcea and Marinescu. It remains to prove (v) and (vi).

Set $h_n = \frac{1}{n} \sum_{k=0}^{n-1} P^k(1)$. Then $h_n \geqq 0$, because P is a positive operator and $\int h_n\, dm = 1$ because of $m(Pf) = m(f)$. Using (iii) and evaluating geometric series it follows that h_n converges in L^1_m. The limit function is invariant under P, nonnegative and has integral 1. Thus 1 is an eigenvalue of P, say $\lambda_1 = 1$, and it follows from (iii) as above that h_n converges to $\phi_1(1) = h$. That μ is T-invariant follows from the following equalities, because $Ph = h$:

$$\mu(f \circ T) = m(h \cdot (f \circ T)) = m(P(h \cdot (f \circ T))) = m(f \cdot Ph)$$
$$\mu(f) = m(fh).$$

If $\mu' = h'm$ is T-invariant, it follows from these equalities that $Ph' = h'$. Thus $h' \in BV$ by (ii) and h' is bounded by some constant c. This implies $h' = \frac{1}{n} \sum_{k=0}^{n-1} P^k(h') \leqq c h_n$, hence $h' \leqq ch$ and $\mu' \ll \mu$.

§3. Once having obtained the invariant measure μ for T, we focus attention to the dynamical system (T, μ). Sometimes we shall adopt the Hilbert-space-viewpoint, interpreting T as a linear operator on the complex Hilbert space L^2_μ and writing $\langle f, g \rangle$ for $\int f \bar{g}\, d\mu$, such that $\langle Tf, Tg \rangle = \langle f, g \rangle$.

The next lemma will provide a description of the adjoint T^* of T in terms of P.

Lemma 8. (i) *$T^* f = \dfrac{P(f \cdot h)}{h}$ for all $f \in L^2_\mu$.*

(ii) *$T^* f = \lambda f \Leftrightarrow P(fh) = \lambda f h$ for all $f \in L^2_\mu$ and $\lambda \in \mathbb{C}$ (all identities are L^2_μ-identities).*

Proof. (i) for $f, g \in L^2_\mu$ one has

$$\langle T^* f, g \rangle = \langle f, Tg \rangle = \int f \cdot (\bar{g} \circ T)\, h\, dm = \int \frac{P(fh)}{h}\, \bar{g}\, d\mu = \left\langle \frac{P(fh)}{h}, g \right\rangle.$$

(ii) $T^* f = \lambda f \Leftrightarrow P(fh) = \lambda f h$ because of (i).

Lemma 9. (i) *$(T^*)^n \circ T^n = Id$ for each $n \in \mathbb{N}$.*

(ii) *$T^n \circ (T^*)^n = Pr_n$ for each n, where Pr_n denotes the orthogonal projection onto the subspace $T^n(L^2_\mu)$.*

Proof. (i) It suffices to show that $T^* T = Id$. But for $f, g \in L^2_\mu$ one has $\langle T^* Tf, g \rangle = \langle Tf, Tg \rangle = \langle f, g \rangle$.

(ii) Since $(T^n \circ (T^*)^n)^* = T^n \circ (T^*)^n$ and $(T^n \circ (T^*)^n)^2 = T^n \circ (T^*)^n$ (because of (i)), $T^n \circ (T^*)^n$ is an orthogonal projection. On the other hand,

$$(T^n \circ (T^*)^n)(L^2_\mu) \subset T^n(L^2_\mu) \quad \text{and} \quad (T^n \circ (T^*)^n)(T^n f) = T^n f$$

for each $f \in L^2_\mu$. Thus $T^n \circ (T^*)^n = Pr_n$.

Now we can easily clarify the ergodic structure of (T, μ) up to the level of weak mixing (for weak mixing cf. [21]).

Theorem 2. *Let $G \subseteq \{z \in \mathbb{C} : |z| = 1\}$ be the set of eigenvalues of T. Then*

(i) $G = \{\lambda_1, \ldots, \lambda_r\}$

(ii) $Tf = \lambda f \Leftrightarrow T^*\bar{f} = \lambda \bar{f}$ *and* $|\lambda| = 1$ *for all* $f \in L^2_\mu$ *and* $\lambda \in \mathbb{C}$.

(iii) $\dim(E_{(T,\mu)}(\lambda)) < \infty$ *for each* $\lambda \in G$, *where* $E_{(T,\mu)}(\lambda) = \{f \in L^2_\mu : Tf = \lambda f\}$.

Proof. Let us first assume that $Tf = \lambda f$ for some $f \in L^2_\mu$, $f \neq 0$. Then $|\lambda| = 1$, as μ is T-invariant, and $T^*\bar{f} = \lambda T^*(T\bar{f}) = \lambda \bar{f}$. On the other hand, if $T^* f = \lambda f$ for some $f \in L^2_\mu$, $f \neq 0$ and $\lambda \in \mathbb{C}$, $|\lambda| = 1$, we have

$$\langle f, f \rangle = \langle T^* f, \lambda f \rangle = \langle f, \lambda T f \rangle \quad \text{and} \quad \langle \lambda Tf, \lambda Tf \rangle = |\lambda|^2 \langle Tf, Tf \rangle = \langle f, f \rangle.$$

This implies $\langle \lambda Tf - f, f \rangle = 0$ and $\langle \lambda Tf - f, \lambda Tf \rangle = 0$, hence $\lambda Tf = f$ or, equivalently, $T\bar{f} = \lambda \bar{f}$. This proves (ii).

If now $Tf = \lambda f$, we get from (ii) and Lemma 8 that $|\lambda| = 1$ and $P(\bar{f}h) = \lambda(\bar{f}h)$. Hence it follows from Theorem 1 that $G \subseteq \{\lambda_1, \ldots, \lambda_r\}$. If $Pg = \lambda_i g$ for some i, then $g \in BV$ by Theorem 1 and $|g| = |P^n g| \leqq P^n |g| \leqq \|g\|_\infty \cdot P^n(1)$. Hence

$$|g| = \frac{1}{n} \sum_{i=0}^{n-1} |P^i g| \leqq \|g\|_\infty \frac{1}{n} \sum_{i=0}^{n-1} P^i(1) \to \|g\|_\infty h \qquad \text{for } n \to \infty.$$

This gives $\frac{|g|}{h} \leqq \|g\|_\infty$ such that $f := \frac{g}{h} \in L^2_\mu$. Furthermore it follows from Lemma 8 and (ii) that $T\bar{f} = \lambda_i \bar{f}$. Hence $G \supseteq \{\lambda_1, \ldots, \lambda_r\}$ proving (i). The same arguments show also that $\dim(E_{(T,\mu)}(\lambda_i)) = \dim(E_i) < \infty$. This is (iii).

The next theorem describes the structure of (T, μ) up to the level of exactness (for exactness see [17]).

Theorem 3. *Let $\mathfrak{B}_\infty(T) := \bigcap_{n=1}^{\infty} T^{-n}\mathfrak{B}$, where $\mathfrak{B}$ denotes the Borel-σ-algebra on Y. Then*

(i) $\{f \in L^2_\mu : f$ *is* $\mathfrak{B}_\infty(T)$-*measurable*$\} = \bigcap_{n=1}^{\infty} T^n(L^2_\mu) = \sum_{i=1}^{r} E_{(T,\mu)}(\lambda_i)$. *In particular,* $\mathfrak{B}_\infty(T)$ *is finite* μ-a.s.

(ii) *If (T, μ) is weakly mixing, then $\mathfrak{B}_\infty(T)$ is trivial, i.e. (T, μ) is exact.*

Proof. (i) The first of the above equalities is generally true. Hence we only have to show the second one. First observe that for $f = \sum_{i=1}^{r} f_i$, $f_i \in E_{(T,\mu)}(\lambda_i)$, we have $f = T^n\left(\sum_{i=1}^{r} \bar{\lambda}_i^n f_i\right)$ such that $\sum_{i=1}^{r} E_{(T,\mu)}(\lambda_i) \subseteq \bigcap_{n=1}^{\infty} T^n(L^2_\mu)$.

The other direction is a bit harder to prove. Let $g \in L^2_\mu \cap BV$. Then $gh \in L^1_m \cap BV$ and since $P(\phi_i(gh)) = \lambda_i \phi_i(gh)$, we have $T^*(\phi_i(gh)/h) = \lambda_i \phi_i(gh)/h$ by Lemma 8. Hence $T(\phi_i(gh)/h) = \bar{\lambda}_i \phi_i(gh)/h$ by Theorem 2.

As $G = \{\lambda_1, \ldots, \lambda_r\}$ generates a finite subgroup of the circle, there is a $p \in \mathbb{N}$ such that $\lambda_i^p = 1$ for $1 \leqq i \leqq r$. This gives for any $n \in \mathbb{N}$

$$\begin{aligned} Pr_{np}(g) &= T^{np}(T)^{*np}(g) = T^{np}\left(\sum_{i=1}^{r} \phi_i(gh)/h + \psi^{np}(gh)/h\right) \\ &= \sum_{i=1}^{r} \phi_i(gh)/h + T^{np}(\psi^{np}(gh)/h). \end{aligned}$$

We have used Lemma 9 and 8 and (iii) of Theorem 1. Now one has for any $f \in \bigcap_{n=1}^{\infty} T^n(L^2_\mu)$, any $g \in L^2_\mu \cap BV$ and any $n \in \mathbb{N}$ that

$$\begin{aligned} \int \left| f - \sum_{i=1}^{r} \phi_i(fh)/h \right| d\mu &\leqq \int |Pr_{np}(f) - Pr_{np}(g)|\, d\mu \\ &\quad + \sum_{i=1}^{r} \int |\phi_i(fh)/h - \phi_i(gh)/h|\, d\mu + \int |\psi^{np}(gh)/h| \circ T^{np}\, d\mu \\ &\leqq \|Pr_{np}(f-g)\|_{2,\mu} + r \int |f-g|\, h\, dm + \int |\psi^{np}(gh)|\, dm \\ &\leqq (r+1)\, \|f-g\|_{2,\mu} + H q^{np}\, \|gh\|_v. \end{aligned}$$

Since BV is dense in L^2_μ this shows that $f = \sum_{i=1}^{r} \phi_i(fh)/h$ μ-a.e., i.e. $f \in \sum_{i=1}^{r} E_{(T,\mu)}(\lambda_i)$.

(ii) If (T, μ) is weakly mixing, then 1 is the only eigenvalue, and the space of fix-elements of T only contains the constant functions. Hence $\mathfrak{B}_\infty(T)$ is trivial by (i).

Remark 2. Let $W \subset Y$ satisfy $T^{-1} W = W$, i.e. $1_W \circ T = 1_W$. Define

$$m'(A) = m(A \cap W)/m(W).$$

Then

$$\begin{aligned} m'(Pf) &= m(1_W Pf)/m(W) = m(P(f \cdot (1_W \circ T)))/m(W) \\ &= m(f \cdot (1_W \circ T))/m(W) = m(f \cdot 1_W)/m(W) = m'(f). \end{aligned}$$

Similarly one can show that $P(h \cdot 1_W) = h 1_W$. Hence every ergodic component of μ is again of the form $h \cdot m$ for some m and h satisfying $m(Pf) = m(f)$ and $Ph = h$.

If one considers T^k instead of T, where k is such that $\lambda_i^k = 1$ for all i, then an ergodic component of μ is already weak mixing for T^k (cf. Theorem 2). Hence every weak mixing component of μ is again of the form hm for some m and h with $m(Pf) = m(f)$ and $Ph = h$. Here one has to take T^k instead of T and g_k instead of g in the definition of P. Therefore we can assume that μ is weak mixing. We shall do this in the next chapter.

Remark 3. The spectral representation of Theorem 1 together with Theorem 2 allows an easy proof of a central limit theorem for stationary processes of the kind $\xi_n = f \circ T^n$ with $f \in BV$. The proof is completely analogous to the one in [10], where it is shown how to reduce this problem to a central limit theorem of Gordin [5]. As in the next chapter we shall even derive almost sure invariance principles, an exact statement of this theorem is omitted.

II. Weak Bernoullicity and Invariance Principles

In this chapter we shall show that, if (T, μ) is weakly mixing (cf. Remark 2), then the natural partition $\mathfrak{P} = (I_1, \ldots, I_N)$ of Y into intervals of continuity and monotonicity of T (and even more each finite partition of Y into intervals) is a weak-Bernoulli-partition for T. This means that the associated "label-process" (ξ_n), where $\xi_n(x) = i$, if $T^n x \in I_i$, is absolutely regular. The corresponding mixing-coefficients will be shown to decrease at an exponential rate.

§1. We start with the investigation of the weak Bernoulli property. Technically we can follow very closely Ratner's article [16]. Remember that, for all $k \geqq 1$, $\mathfrak{P}_0^k = \bigvee_{i=0}^{k-1} T^{-i}\mathfrak{P}$ is a finite partition of Y into intervals of m-measure $\leqq d^k$ (cf. Lemma 2) such that m-measure plays the same role here as Lebesgue-measure (respectively "length") does in the article of Ratner [16]. In that paper the invariant measure also is called μ, but its density is called p and the function $\varphi(x) = 1/|f'(x)|$ plays the role of our $g(x)$.

The first lemma is proved in Bowen's paper [3] and is easily seen to be true also in this situation.

Lemma 10. *If $A \in \mathfrak{P}_0^k$, and if $\bar{A}$ contains no endpoint of an interval $I_i \in \mathfrak{P}$, then $TA \in \mathfrak{P}_0^{k-1}$.*

The next lemma is the same as Lemma 3.2 in [16].

Lemma 11. *There is a constant $L > 0$ such that, given $\varepsilon > 0$, we can find for all $n > L \cdot \log \frac{1}{\varepsilon}$ a collection $\alpha_n \subseteq \mathfrak{P}_0^n$ with $\mu(\bigcup \alpha_n) > 1 - \varepsilon$ such that for any $x, y \in A \in \alpha_n$*

$$\frac{h(x)}{h(y)} \in [e^{-\varepsilon}, e^{\varepsilon}] \quad \textit{and} \quad \frac{g(x)}{g(y)} \in [e^{-\varepsilon}, e^{\varepsilon}].$$

Proof. The proof is quite analogous to the one in [16] except for the estimate of the measure of the set $S_l = \{x \in Y \colon g(x) < l\}$ $(l > 0)$, which must be replaced by

$$\begin{aligned} \mu(S_l) &\leqq \|h\|_\infty \cdot m(S_l) = \|h\|_\infty \int P(1_{S_l})\, dm \\ &\leqq \|h\|_\infty \sum_{i=1}^{N} \|(g \circ T_i^{-1}) \cdot 1_{T_i(S_l)}\|_\infty, \qquad \text{where } T_i = T/I_i \\ &\leqq \|h\|_\infty \cdot N \cdot l, \qquad \text{by definition of } S_l. \end{aligned}$$

The next lemma, which is basic for our further considerations, can be proved just as the corresponding one (3.3) in [16], too.

Lemma 12. *There is a constant $D>0$ such that for all sufficiently large $M\in\mathbb{N}$ and all $m\geqq 0$ one can find a collection of atoms $\beta=\beta_{m,M}\subseteq\mathfrak{P}_0^{m+M}$ with*

(i) $T^m B\in\mathfrak{P}_0^M$ *for* $B\in\beta$

(ii) $\mu(\bigcup\beta)>1-e^{-M/D}$

(iii) $\left|\frac{\mu(T^m\tilde{B})}{\mu(T^m B)}-\frac{\mu(\tilde{B})}{\mu(B)}\right|<e^{-M/D}\frac{\mu(\tilde{B})}{\mu(B)}$ *for any measurable* $\tilde{B}\subseteq B\in\beta$ *with* $\mu(B)>0$.

Proof. The proof is again as in [16] using Lemmas 10 and 11 instead of the corresponding ones in that paper. One only has to observe that for any measurable $\tilde{B}\subset B\in\mathfrak{P}_0^n$ (n arbitrary) the following holds. Put $q(x):=h(T^n x)/(h(x)\,g_n(x))$, where $g_n(x)=g(x)\,g(Tx)\dots g(T^{n-1}x)$. Then

$$\begin{aligned}\textstyle\int_{\tilde{B}} q\,d\mu &= \textstyle\int P^n(1_{\tilde{B}}\cdot(h\circ T^n)/g_n)\,dm\\ &= \int \sum_{y\in T^{-n}x} g_n(y)\,1_{\tilde{B}}(y)(h(T^n y)/g_n(y))\,dm(x)\\ &= \int\Big(\sum_{y\in T^{-n}x} 1_{\tilde{B}}(y)\Big)\,h(x)\,dm(x)\\ &= \textstyle\int 1_{T^n\tilde{B}}(x)\,h(x)\,dm(x) \qquad \text{since } \tilde{B}\subseteq B\in\mathfrak{P}_0^n\\ &= \mu(T^n\tilde{B})\end{aligned}$$

and

$$\frac{q(y)}{q(z)}=\frac{h(T^n y)}{h(T^n z)}\cdot\frac{h(z)}{h(y)}\cdot\prod_{k=0}^{n-1}\frac{g(T^k z)}{g(T^k y)}.$$

The next lemma "translates" the information about P, which is contained in the spectral representation, into the "language" of partitions. We introduce some new notation before. Let $\mathfrak{R}$ and $\mathfrak{S}$ be two finite collections of pairwise disjoint measurable subsets of Y. Define

$$D(\mathfrak{R},\mathfrak{S}):=\sum_{R\in\mathfrak{R}}\sum_{S\in\mathfrak{S}}|\mu(R\cap S)-\mu(R)\,\mu(S)|.$$

Lemma 13. *Under the assumption that (T,μ) is weakly mixing there is a constant $C>0$ such that for all M, n, $l\geqq 0$*

$$D(\mathfrak{P}_0^M,\mathfrak{P}_{M+l}^{M+l+n})\leqq C\cdot N^M q^l$$

where q is the constant of (iv) *of Theorem* 1 *and N the number of elements of $\mathfrak{P}$.*

Proof. Set $\mathfrak{R}=\mathfrak{P}_0^M$ and $\mathfrak{S}=\mathfrak{P}_{M+l}^{M+l+n}=T^{-l}(\mathfrak{P}_M^{M+n})$. Then

$$D(\mathfrak{R},\mathfrak{S})=\sum_{R\in\mathfrak{R}}\{|\mu(R\cap S_R^+)-\mu(R)\,\mu(S_R^+)|+|\mu(R\cap S_R^-)-\mu(R)\,\mu(S_R^-)|\},$$

where

$$S_R^+=\cup\{S\in\mathfrak{S}:\mu(R\cap S)>\mu(R)\,\mu(S)\}$$

and

$$S_R^-=\cup\{S\in\mathfrak{S}:\mu(R\cap S)\leqq\mu(R)\,\mu(S)\}=Y\setminus S_R^+.$$

Because of $\mathfrak{S}=T^{-l}(\mathfrak{P}_M^{M+n})$, there are measurable sets $\tilde{S}_R^+$ and $\tilde{S}_R^-$ with $S_R^+=T^{-l}(\tilde{S}_R^+)$ and $S_R^-=T^{-l}(\tilde{S}_R^-)$. Hence

$$D(\mathfrak{R}, \mathfrak{S}) = \sum_{R \in \mathfrak{R}} \{|\int 1_{S_R^+}(1_R - \mu(R))\, h\, dm| + |\int 1_{S_R^-}(1_R - \mu(R))\, h\, dm|\}$$
$$= \sum_{R \in \mathfrak{R}} \{|\int 1_{\bar{S}_R^+} P^l((1_R - \mu(R))\, h)\, dm| + |\int 1_{\bar{S}_R^-} P^l((1_R - \mu(R))\, h)\, dm|\}$$
$$\leqq 2 \sum_{R \in \mathfrak{R}} \|P^l((1_R - \mu(R))\, h)\|_v.$$

Since (T, μ) is weakly mixing, it follows from Theorems 2 and 1 that $P^l = \phi_1 + \psi^l$. Furthermore this implies $\phi_1((1_R - \mu(R))\, h) = (\int (1_R - \mu(R))\, h\, dm) \cdot h \equiv 0$. Thus

$$D(\mathfrak{R}, \mathfrak{S}) \leqq 2 \sum_{R \in \mathfrak{R}} \|\psi^l((1_R - \mu(R))\, h)\|_v$$
$$\leqq 2 \sum_{R \in \mathfrak{R}} H q^l \|(1_R - \mu(R))\, h\|_v$$
$$\leqq 2 |\mathfrak{R}|\, H q^l 3 \|h\|_v$$
$$\leqq 6H \|h\|_v N^M q^l = C N^M q^l.$$

The third inequality follows because the function $1_R - \mu(R)$ has the value $1 - \mu(R)$ on the interval R and the value $-\mu(R)$ on $Y \setminus R$. Hence $v((1_R - \mu(R))\, h) \leqq 3 v(h)$. The last inequality follows from $|\mathfrak{R}| \leqq N^M$, which is true, because $\mathfrak{R} = \mathfrak{P}_0^M$.

Now we can derive an estimate of $D(\mathfrak{P}_0^n, \mathfrak{P}_{n+l}^{2n+l})$ independent of n.

Theorem 4. *If (T, μ) is weakly mixing, then there are constants $K > 0$ and p with $0 < p < 1$ such that*

$$D(\mathfrak{P}_0^n, \mathfrak{P}_{n+l}^{n+l+k}) \leqq K p^l$$

for arbitrary n and k.

Proof. Fix $\delta > 0$ such that $\tilde{q} = q N^\delta < 1$ (q as in Lemma 13). It suffices to show Theorem 4 for $n \geqq [\delta l] + 1$. The case $n \leqq [\delta l]$ reduces immediately to $n = [\delta l] + 1$ using the triangle inequality.

Set $M = [\delta l] + 1$ and $m = n - M \geqq 0$. Then

$$D(\mathfrak{P}_0^n, \mathfrak{P}_{n+l}^{n+l+k}) \leqq D(\beta_{m,M}, \mathfrak{P}_{n+l}^{n+l+k}) + 2e^{-M/D}$$

by Lemma 12. We set $\mathfrak{S} = \mathfrak{P}_{n+l}^{n+l+k}$ and $\hat{\mathfrak{S}} = \mathfrak{P}_{M+l}^{M+l+k}$. Because $\mu(T^m S) = \mu(S)$ for every $S \in \mathfrak{S}$, we get $(\beta = \beta_{m,M})$

$$D(\beta, \mathfrak{S}) = \sum_{B \in \beta} \mu(B) \sum_{S \in \mathfrak{S}} \left\{ \left| \frac{\mu(B \cap S)}{\mu(B)} - \frac{\mu(T^m(B \cap S))}{\mu(T^m B)} \right| + \left| \frac{\mu(T^m(B \cap S))}{\mu(T^m B)} - \mu(T^m S) \right| \right\}$$
$$\leqq \sum_{B \in \beta} \sum_{S \in \mathfrak{S}} e^{-M/D} \mu(B \cap S) + \sum_{\hat{B} \in \mathfrak{P}_0^M} \left\{ \sum_{\substack{B \in \beta \\ T^m B = \hat{B}}} \mu(B) \sum_{\hat{S} \in \hat{\mathfrak{S}}} \left| \frac{\mu(\hat{B} \cap \hat{S})}{\mu(\hat{B})} - \mu(\hat{S}) \right| \right\}$$

by Lemma 12

$$\leqq e^{-M/D} + \sum_{\hat{B} \in \mathfrak{P}_0^M} \sum_{S \in \mathfrak{S}} |\mu(\hat{B} \cap \hat{S}) - \mu(\hat{B})\, \mu(\hat{S})|$$
$$\leqq e^{-M/D} + D(\mathfrak{P}_0^M, \mathfrak{P}_{M+l}^{M+l+k})$$
$$\leqq e^{-M/D} + C N^M q^l \quad \text{by Lemma 13.}$$

Putting together the two estimates and observing that $M=[\delta l]+1$ we get

$$D(\mathfrak{P}_0^n, \mathfrak{P}_{n+l}^{n+l+k}) \leqq 3e^{-l\delta/D} + CN\tilde{q}^l.$$

This gives the desired result if one sets $p=\max\{\tilde{q}, e^{-\delta/D}\}$ and $K=3+CN$.

Corollary. *If (T,μ) is weakly mixing and $\mathfrak{Q}$ is an arbitrary finite partition of Y into intervals, then there are constants $\tilde{K}$ and $\tilde{p}$, $0<\tilde{p}<1$ such that $D(\mathfrak{Q}_0^n, \mathfrak{Q}_{n+l}^{n+l+k}) \leqq \tilde{K}\tilde{p}^l$ for any k and n.*

Proof. The refinement $\mathfrak{P} \vee \mathfrak{Q}$ of $\mathfrak{P}$ and $\mathfrak{Q}$ is again a finite partition of Y into intervals, on which T is monotone and continuous. Hence Theorem 4 is also true for $\mathfrak{P} \vee \mathfrak{Q}$ instead of $\mathfrak{P}$ (with different constants $\tilde{K}$ and $\tilde{p}$). But then it is also true for each coarser partition, in particular for $\mathfrak{Q}$.

§**2.** In order to show central limit theorems, a.s. invariance principles and other limit theorems for stationary processes governed by the dynamical system (T,μ) one only has to observe the following: For $\mathfrak{P}=\{I_1,\dots,I_N\}$ let $(\xi_n)_{n\geqq 0}$ be the associated label-process, i.e. $\xi_n(x)=i$, if $T^n x \in I_i$. If $\mathfrak{F}_a^b$ denotes the σ-algebra generated by $\{\xi_n: a\leqq n<b\}$, then $\mathfrak{F}_a^b$ is the σ-algebra generated by $\mathfrak{P}_a^b$. Define

$$\beta(\mathrm{n}) := \sup_{t\geqq 1} E_\mu[\sup_{A\in\mathfrak{F}_{t+n}^\infty} |\mu(A\,|\,\mathfrak{F}_0^t)-\mu(A)|].$$

If $\beta(n)\to 0$, (ξ_n) is called *absolutely regular*. One easily checks that (cf. [19])

$$2\beta(n) = \sup_{t\geqq 1}\sup_{k\geqq 1} D(\mathfrak{P}_0^t, \mathfrak{P}_{t+n}^{t+n+k}).$$

Hence $\beta(n)\leqq Kp^n$ by Theorem 4.

This enables us to apply some well-known limit theorems to the process (ξ_n) and to functionals of it. First define that $f: Y\to\mathbb{R}$ is of *bounded p-variation* on an interval $A\subseteq Y$ if

$$V_A^p f := \sup \sum_{i=1}^n |f(a_i)-f(a_{i-1})|^p < \infty,$$

where the supremum is taken over all finite subsets $a_0<a_1<\dots<a_n$ of A.

Theorem 5. *Assume that (T,μ) is weakly mixing. Let $f: Y\to\mathbb{R}$ be a function of bounded p-variation, $p\geqq 1$, and $\int f\,d\mu=0$. Define $S(t) := \sum_{i=0}^{t-1} f\circ T^i$, which is a stochastic process on (Y,μ). Then the series*

$$\sigma^2 := \int f^2\,d\mu + 2\sum_{k=1}^\infty \int f(f\circ T^k)\,d\mu$$

converges absolutely, $\int S(t)^2\,d\mu = t\sigma^2+O(1)$, and if $\sigma^2\neq 0$ the following holds

(i) $\sup_{z\in\mathbb{R}} \left|\mu\left\{\frac{1}{\sigma\sqrt{t}}S(t)\leqq z\right\} - \frac{1}{\sqrt{2\pi}}\int_{-\infty}^z e^{-x^2/2}\,dx\right| = O(t^{-\nu})$ *for some* $\nu>0$.

(ii) *Without changing its distribution one can redefine the process $(S(t))_{t\geqq 0}$ on a richer probability space together with standard Brownian motion $(B(t))_{t\geqq 0}$ such that*

$$|\sigma^{-1} S(t) - B(t)| = O(t^{1/2-\lambda}) \qquad \mu\text{-a.s.}$$

for some $\lambda > 0$.

Proof. We show that theorems in [15] and [1] are applicable here. First observe that the Borel-σ-algebra on Y is generated by the family $(\mathfrak{P}_0^k : k \in \mathbb{N})$, because each interval in Y has positive m-measure and $m(A) \leqq d^k$ for each $A \in \mathfrak{P}_0^k$ (see Lemma 2). Thus f can be considered to be μ-a.e. a functional of the label-process (ξ_n). Furthermore

(i) as a function of bounded p-variation, f is bounded.

(ii) For each $r \geqq p$

$$\begin{aligned}\int |f - E_\mu(f \mid \mathfrak{F}_0^k)|^r \, d\mu &\leqq \sum_{A \in \mathfrak{P}_0^k} \mu(A) \cdot \sup_{x, y \in A} |f(x) - f(y)|^r \\ &\leqq d^k \, \|h\|_\infty \sum_{A \in \mathfrak{P}_0^k} V_A^p f \cdot (2 \|f\|_\infty)^{r-p} \qquad \text{by Lemma 2} \\ &\leqq d^k \, \|h\|_\infty \, V_Y^p f \cdot (2 \|f\|_\infty)^{r-p}.\end{aligned}$$

(iii) Set $R(k) := \int f(f \circ T^k) \, d\mu$. Then

$$\begin{aligned}|R(k)| &\leqq \int |f - E_\mu(f \mid \mathfrak{F}_0^{[k/2]})| \cdot |f \circ T^k| \, d\mu + \int |E_\mu(f \mid \mathfrak{F}_0^{[k/2]}) \cdot (f \circ T^k)| \, d\mu \\ &\leqq (d^{[k/2]} \, \|h\|_\infty \, V_Y^p f)^{1/p} \, \|f\|_\infty + 4 \, \|f\|_\infty^2 \cdot \beta([k/2])\end{aligned}$$

by Theorem 17.2.1 of [8]. Hence $|R(k)|$ decreases exponentially with $k \to \infty$ such that $\sigma^2 = R(0) + 2 \sum_{k=1}^{\infty} R(k)$ converges absolutely, and

$$\begin{aligned}\int S(t)^2 \, d\mu &= t \cdot R(0) + 2 \sum_{k=1}^{t-1} (t-k) R(k) \\ &= t\sigma^2 - 2t \sum_{k=t}^{\infty} R(k) - 2 \sum_{k=1}^{t-1} k \cdot R(k) \\ &= t\sigma^2 + O(1)\end{aligned}$$

since $|R(k)|$ decreases exponentially.

Thus Theorem 7.1 in [15] can be applied to obtain (ii), while (i) follows from [1].

Remark 4. (i) The order of approximation in (ii) of Theorem 5 is good enough to imply integral tests, log-log-laws, and weak invariance principles for the process $(f \circ T^i)_{i \geqq 0}$ (see e.g. [15]).

(ii) Each Hölder-continuous function on [0, 1] with exponent $\frac{1}{p}$ is of bounded p-variation.

III. Equilibrium States

In this chapter we show first that μ is an equilibrium state for the function $\log g$, i.e. $h(\mu) + \mu(\log g) = \sup\{h(\nu) + \nu(\log g) : \nu$ is a T-invariant probability on

$Y\}$, where $h(\mu)$ denotes the entropy of μ. Then we give a method, how one can show the existence of a measure m and a function g such that (a) and (b) of Chap. I are satisfied. This method is then applied in three different situations.

§1. In order to show that μ is an equilibrium state for $\log g$, we need the following lemmas. For a function $f\colon Y\to\mathbb{R}$ let f^+ be defined by $f^+(x) = \max\{f(x), 0\}$.

Lemma 14. *If $f\colon Y\to\mathbb{R}$ is measurable and ν is a T-invariant probability on Y, then $(f-f\circ T)^+\in L^1_\nu$ implies $f-f\circ T\in L^1_\nu$ and $\int (f-f\circ T)\,d\nu=0$.*

Proof. Setting $F:=(f-f\circ T)^+$, we have $0\leqq F+f\circ T-f$, $F\geqq 0$ and $F\in L^1_\nu$. Define $f_n\colon Y\to\mathbb{R}$ by $f_n(x)=\min\{f(x), n\}$. Then we have $0\leqq F+f_n\circ T-f_n$ and

$$\lim_{n\to\infty}(f_n\circ T-f_n)=f\circ T-f.$$

Hence by Fatou's lemma

$$\begin{aligned}\int (F+f\circ T-f)\,d\nu&=\int \liminf(F+f_n\circ T-f_n)\,d\nu\\&\leqq \liminf\int (F+f_n\circ T-f_n)\,d\nu=\int F\,d\nu<\infty.\end{aligned}$$

This says that $F+f\circ T-f\in L^1_\nu$ and hence $f\circ T-f\in L^1_\nu$. As $|f_n\circ T-f_n|\leqq|f\circ T-f|$, it follows from Lebesgue's theorem of dominated convergence that

$$\int (f\circ T-f)\,d\nu=\lim_{n\to\infty}\int (f_n\circ T-f_n)\,d\nu=0.$$

Lemma 15. *Set $\bar g:=gh/(h\circ T)$. If ν is a T-invariant probability on Y, then*

$$\int \log\bar g\,d\nu=\int \log g\,d\nu.$$

Proof. It follows from $g(x)\leqq d<1$ (cf. (a)) that $\log g(x)<0$ for all $x\in Y$ and from $\bar g(x)\geqq 0$ and $\sum_{y\in T^{-1}x}\bar g(y)=1$, which is implied by $Ph=h$, that $\log\bar g(x)\leqq 0$ for all $x\in Y$.

If $\int\log g\,d\nu=\int\log\bar g\,d\nu=-\infty$, we have nothing to show. If $\int\log g\,d\nu>-\infty$, we have $\log g\in L^1_\nu$. Because of $\log h-\log h\circ T=\log\bar g-\log g$, it follows from $\log\bar g\leqq 0$ that $(\log h-\log h\circ T)^+\leqq|\log g|$ and hence from Lemma 14 that $\int(\log h-\log h\circ T)\,d\nu=0$. This implies $\log\bar g\in L^1_\nu$ and $\int\log g\,d\nu=\int\log\bar g\,d\nu$. If $\int\log\bar g\,d\nu>-\infty$, we apply the same arguments to $\log h^{-1}-\log h^{-1}\circ T=\log g-\log\bar g$ and get again $\int\log g\,d\nu=\int\log\bar g\,d\nu$.

Theorem 6. *μ is an equilibrium state for* $\log g$ *on Y.*

Proof. Define the operator $\bar P$ by $\bar Pf(x)=\sum_{y\in T^{-1}x}\bar g(y)f(y)$. It follows from $m(Pf)=m(f)$ and $Ph=h$ that $\mu(\bar Pf)=\mu(f)$ and $\bar P1=1$. Now we can use a result of Ledrappier (Theorem 1 of [12]), which says that μ is then an equilibrium state for the function $\log\bar g$, i.e. $h(\mu)+\mu(\log\bar g)=\sup_\nu\{h(\nu)+\nu(\log\bar g)\}$. But by Lemma 15, we have $\nu(\log\bar g)=\nu(\log g)$ for every T-invariant probability ν on Y and hence $h(\mu)+\mu(\log g)=\sup_\nu\{h(\nu)+\nu(\log g)\}$, which says that μ is an equilibrium state for $\log g$.

§2. Let T be a piecewise monotonic transformation on a totally ordered set X, such that the order topology is compact and separable. Let $\varphi \in C(X)$ satisfy $\sum_{i=1}^{\infty} \operatorname{var}_i \varphi < \infty$, where $\operatorname{var}_i \varphi = \sup \{|\varphi(x) - \varphi(y)|: x, y \in A, A \in \mathfrak{P}_0^i\}$. A method, to show that an equilibrium state for φ fits in the situation described at the beginning of Chap. I, is the following one. It is adapted from [22].

We have to modify X a little bit. Set $\tilde{W} = \bigcup_{j=0}^{\infty} T^{-j}\{\lim_{t\uparrow x} T^k(t), \lim_{t\downarrow x} T^k(t): k \geqq 1$ and x is an endpoint of some $I_i \in \mathfrak{P}, 1 \leqq i \leqq N\}$.

Let W be the set of those elements of $\tilde{W}$, which are not an endpoint of an interval, which is open and closed. Now we consider $X' = X \cup W'$, where W' is a copy of W disjoint from X. We extend the order relation to X'. If $y < x < z$ in X, $x \in W$ and x' is the element of W' corresponding to x, then we define $y < x < x' < z$ or $y < x' < x < z$, such that we can extend T continuously to X'. T is then continuous on X' and if one defines $\tilde{W}$ for X', all elements of $\tilde{W}$ are endpoints of intervals, which are open and closed. The order topology of X' is again compact and separable. We extend also φ continuously to X' and define $\tilde{P}$ by $\tilde{P}f(x) = \sum_{y \in T^{-1}x} e^{\varphi(y)} f(y)$ for $x \in X'$.

Because the intervals $T(I_i)$ for $1 \leqq i \leqq N$ are open and closed in X', $\tilde{P}$ maps $C(X')$ into $C(X')$. We can apply the Schauder-Tychonoff-theorem (see Dunford and Schwartz, Linear operators I, p. 456) to the continuous map $m \mapsto \tilde{P}^* m / \tilde{P}^* m(1)$ on the set of all Borel-probability measures, which is a compact, convex subset of the dual of $C(X')$ with w^*-topology. $\tilde{P}^*$ denotes the dual operator of $\tilde{P}$. We get a fixed point m for this map, i.e. $m(\tilde{P}f) = \lambda m(f)$ for all $f \in C(X')$, where $\lambda = \tilde{P}^* m(1)$.

Now set $\tilde{g}(x) = e^{\varphi(x)}/\lambda$. The problem is to show that $\|\tilde{g}_n\|_\infty < 1$ for some n ($\tilde{g}_n(x) = \tilde{g}(x)\,\tilde{g}(Tx) \ldots \tilde{g}(T^{n-1}x)$) and that $m(W') = 0$. To this end we prove two lemmas.

Lemma 16. *Let A be a Borel subset of X' such that T^n/A is monotone for all n. Suppose that for every $j \geqq 1$ there is a B_j, which is a Borel subset of some element of $\mathfrak{P}_0^{k_j}$ and which satisfies $T^{k_j}(B_j) = A$ for some k_j. Furthermore suppose that $B_i \cap B_j = \varnothing$ for $i \neq j$ and that there is an $l \in \mathbb{N}$ such that B_j and A are contained in the same element of $\mathfrak{P}_0^{k_j - l}$ for all $j \geqq 1$. Then $m(A) = 0$.*

Proof. We have $m(B_j) = \lambda^{-k_j} m(\tilde{P}^{k_j} 1_{B_j}) \geqq \inf\{\tilde{g}_{k_j}(y): y \in B_j\} \cdot m(A)$ and similarly $m(A) \leqq \sup\{\tilde{g}_{k_j}(y): y \in A\} \cdot m(T^{k_j} A)$. This implies

$$m(B_j)\, m(T^{k_j} A) \geqq (\inf\{\tilde{g}_{k_j}(y): y \in B_j\} / \sup\{\tilde{g}_{k_j}(y): y \in A\})\, m(A)^2$$
$$\geqq \exp\left(-\sum_{i=0}^{\infty} \operatorname{var}_i \varphi - 2l\, \|\varphi\|_\infty\right) \cdot m(A)^2$$

because A and B_j are contained in the same element of $\mathfrak{P}_0^{k_j - l}$. As the B_j are disjoint, we have $m(B_j) \to 0$ as $j \to \infty$. Of course, $m(T^{k_j} A) \leqq 1$, hence $m(A) = 0$.

Lemma 17. *Let J be a closed interval, which is a subset of some element of $\mathfrak{P}_0^n$. Then*

$$\sup_J \tilde{g}_n \leqq K m(J)/m(T^n J), \qquad \textit{where} \quad K = \exp\left(\sum_{i=1}^{\infty} \operatorname{var}_i \varphi\right).$$

Proof.

$$\begin{aligned} m(J) &= \lambda^{-n} m(\tilde{P}^{n_1} J) \geqq (\inf_J \tilde{g}_n)\, m(T^n J) \\ &\geqq (\sup_J \tilde{g}_n) \exp\left(-\sum_{i=1}^{n} \operatorname{var}_i \varphi\right) \cdot m(T^n J) \\ &\geqq K^{-1} (\sup_J \tilde{g}_n)\, m(T^n J). \end{aligned}$$

This implies the desired inequality.

We shall apply these lemmas in three special cases in §§ 3–5.

§3. Let $X = \{x_0 x_1 \ldots \in \{1, 2, \ldots, N\}^{\mathbb{N}} : M(x_i x_{i+1}) = 1\}$ be a finite type subshift, which is topologically transitive, i.e. the transition matrix M is irreducible. Furthermore we assume that $h_{\mathrm{top}}(X) > 0$, i.e. X consists not only of a periodic orbit. If we introduce the lexicographic ordering, the shift transformation σ becomes a piecewise monotinic transformation on X. We have here $W = \varnothing$ and $X' = X$. We apply Lemma 16 to $A = \{x\}$ for $x \in X$. Because X is not only a periodic orbit, we can find an inverse image $y_0 y_1 \ldots y_{m-1} x_0 x_1 \ldots$ of $x = x_0 x_1 \ldots$, which is not periodic. As M is irreducible, there is a k_0 such that for any x_j, $x_0 \in \{1, \ldots, N\}$ there are $i_0 = x_j$, $i_1, \ldots, i_k = y_0$ with $k \leqq k_0$ and $M(i_r, i_{r+1}) = 1$ for $0 \leqq r \leqq k = 1$. Then among the points

$$x_0 x_1 \ldots x_j i_1 \ldots i_{k-1} y_0 y_1 \ldots y_{m-1} x_0 x_1 x_2 \ldots \in X \qquad \text{for } j = 1, 2, 3, \ldots$$

there are infinitely many different ones. They can be used in Lemma 16 as B_j $(l = m + k_0)$ to get $m(\{x\}) = 0$.

Set $q_k = \max\{m(A) : A \in \mathfrak{P}_0^k\}$. In this case $\mathfrak{P}_0^k$ is the set of all cylinders $[x_0 \ldots x_{k-1}] = \{y_0 y_1 \ldots \in X : y_i = x_i \text{ for } 0 \leqq i \leqq k-1\}$. We have $q_k \to 0$ $(k \to \infty)$, because otherwise one finds an $x \in X$ and an $\varepsilon > 0$, such that each neighbourhood of x contains an $A \in \mathfrak{P}_0^k$ for some k with $m(A) \geqq \varepsilon$. This implies $m(\{x\}) \geqq \varepsilon$, a contradiction.

Next we show that the support Y of m is all of X. There is at least one i with $m([i]) > 0$, because otherwise $m(X) = 0$. Choose some cylinder $[j_0 \ldots j_{n-1}] \subset X$. As above one finds $i_0 = j_{n-1}$, $i_1, \ldots, i_k = i$ with $M(i_r, i_{r+1}) = 1$. Then

$$m([j_0 \ldots j_{n-1} i_1 \ldots i_k]) \geqq (\inf \tilde{g})^{k+n-1} m([i_k] > 0$$

and hence $m([j_0 \ldots j_{n-1}]) > 0$. This gives $Y = X$. In particular $m([i]) > 0$ for $1 \leqq i \leqq N$.

Now set $J := [x_0 \ldots x_{n-1}]$. Then $T^n(J) \supset [i]$ for some i, hence

$$m(T^n J) \geqq \min\{m([i]) : 1 \leqq i \leqq N\} = : c > 0.$$

Lemma 17 implies then that $\sup_J \tilde{g}_n \leqq \frac{K}{c} q_n$. As J was an arbitrary element of $\mathfrak{P}_0^n$, this shows that there is an n with $\|\tilde{g}_n\|_\infty < 1$. Setting $g = \tilde{g}_n$, we have $P = \tilde{P}^n / \lambda^n$ and (a) and (b) are satisfied, if we assume additionally that φ and hence also $\tilde{g}$ and g are of bounded variation. As $Y=X$ and as $h(\mu, T^n) = nh(\mu, T)$, it follows from Theorem 6 that μ is then an equilibrium state for φ. (cf. the results of Chap. 1 of [2]).

§4. We consider the transformation $T: x \mapsto \beta x \pmod 1$ on $[0, 1)$. Let $\chi: [0, 1) \to \{1, \dots, N\}^{\mathbb{N}}$ (N such that $\beta \leqq N < \beta + 1$) be the β-expansion $x \mapsto i_0 i_1, \dots, i_j$ such that $i_j - 1 \leqq \beta T^j(x) < i_j$. Set $e_0 e_1 \dots = \lim_{t \uparrow 1} \chi(t)$ and

$$X = \{x_0 x_1 \dots \in \{1, \dots, N\}^{\mathbb{N}}: x_k x_{k+1} \dots \leqq e_0 e_1 \dots \text{ for } k \geqq 0\},$$

where $\leqq$ denotes the lexicographic ordering. It follows from Theorem 1 of [6] that $\overline{\chi([0, 1))} = X$. Furthermore $\chi \circ T = \sigma \circ \chi$ and $X \setminus \chi([0, 1))$ is countable. Hence we can consider (X, σ) instead of $([0, 1), T)$, because m will have no atoms. It follows from the definitions, that $e_k e_{k+1} \dots \neq 111 \dots$ for all $k \geqq 0$.

We proceed as in §3. Let $x \in X'$ and $y = y_0 \dots y_{m-1} x_0 x_1 \dots \in X'$ be an inverse image of x, which is not periodic. We choose a sequence (k_j) and integers z_j with $1 \leqq z_j \leqq x_{k_j}$ as follows. If $\sigma^i(y) \neq e_0 e_1 \dots$ for all i then choose k_1 such that $x_i = e_i$ for $0 \leqq i \leqq k_1 - 1$ and $x_{k_1} < e_{k_1}$. Having chosen $k_1 < \dots < k_{n-1}$ choose k_n such that $x_{k_{n-1}+i} = e_i$ for $0 \leqq i \leqq k_n - k_{n-1} - 1$ and $x_{k_n} < e_{k_n - k_{n-1}}$. Using the definition of X, one checks easily that $x_0 \dots x_{k_i}$ does not end with an initial segment of $e_0 e_1 \dots$. Set $z_j = x_{k_j}$. If $\sigma^i(y) = e_0 e_1 \dots$ for some i choose the $k_j \geqq \max\{i, m\}$ such that $e_{k_j - i} = x_{k_j} > 1$. This is possible, because $e_j e_{j+1} \dots \neq 11 \dots$ for all j. Set $z_j = x_{k_j} - 1$. One has again that $x_0 \dots x_{k_j - 1} z_j$ does not end with an initial segment of $e_0 e_1 \dots$. We set

$$B_j = \{x_0 \dots x_{k_j - 1} z_j y_0 \dots y_{m-1} x_0 x_1 \dots\} \subset X'.$$

Then the requirements of Lemma 16 are satisfied for these B_j, $A = \{x\}$ and $l = m+1$. Hence $m(\{x\}) = 0$. In particular $m(W') = 0$ and we can consider again X instead of X'.

As in §3 we get that $q_k = \sup\{m(A): A \in \mathfrak{P}_0^k\}$ tends to zero as $k \to \infty$. We set $J = [x_0 \dots x_{n-1}]$. If $x_0 \dots x_{n-1}$ does not end with an initial segment of $e_0 e_1 \dots$, then $\sigma^n(J) = X$. Hence by Lemma 17 we get

$$\sup_J \tilde{g}_n \leqq K q_n.$$

If $x_0 \dots x_{n-1}$ ends with an initial segment of $e_0 e_1 \dots$, choose k minimal such that $x_{n-k} \dots x_{n-1} = e_0 \dots e_{k-1}$. Then $J' = [x_0 \dots x_{n-k-1} (x_{n-k} - 1) x_{n-k+1} \dots x_{n-1}]$ satisfies $\sigma^n(J') = X$. Hence

$$\sup_J \tilde{g}_n \leqq (\sup_{J'} \tilde{g}_n) \cdot \exp\left(\sum_{i=1}^{\infty} \operatorname{var}_i \varphi + 2\|\varphi\|_\infty\right) \leqq K q_n \cdot \tilde{K}.$$

Therefore we find again an n such that $\|\tilde{g}_n\|_\infty < 1$. We set $g = \tilde{g}_n$. As in §3, one can show that the support Y of m is all of X. Hence by Theorem 6, μ is an equilibrium state for φ.

This means that we have generalized the results of [22] from the class of all Lipschitz continuous φ to the class of φ satisfying $\sum \operatorname{var}_i \varphi < \infty$ and $V_X \varphi < \infty$.

§5. Now let X be the interval $[0, 1]$ and T any piecewise monotonic transformation on X with positive topological entropy. For φ we take the constant function zero. Then $\tilde{g}$ is the constant function λ^{-1}. We have to show that $\lambda > 1$. Then we can set $g = \tilde{g}$ and (a) and (b) are satisfied.

To this end we need some results of [6] (cf. also [7]). Let D be the following set of subintervals of $I_i \subset X'$ $(1 \leqq i \leqq N)$, on which T is monotone. An interval K is called a successor of an interval $J \subset I_i$, if K is one of the nonempty intervals among $T(J) \cap I_j$ for $1 \leqq j \leqq N$. We define D in the following way: D contains I_i for $1 \leqq i \leqq N$ and with every interval also its successors. Because all images of I_i's are open and closed subsets of X', all elements of D are open and closed. In [6] the elements of D are subsets of a shift space isomorphic to $([0, 1], T)$, but this makes no difference. Furthermore we define a $D \times D$-matrix M with entries 0 and 1 by $M(J, K) = 1$ if and only if K is a successor of J, for $J, K \in D$. It happens that elements of D are successors of more than one element of D.

It is shown in [6] that every irreducible submatrix L of M corresponds to a closed T-invariant subset Ω of $[0, 1]$. If $F \subset D$ is the index set of L, i.e. $L = M/F$, then $I = \cup\{J : J \in \tilde{F}\}$, where $\tilde{F} := \{J \in D$: there are $J_0, J_1, \ldots, J_n = J$ with $J_0 \in F$ and $M(J_i, J_{i+1}) = 1\}$, and $I' = \cup\{J : J \in \tilde{F} \setminus F\}$ are T-invariant sets, which are finite unions of intervals and $\Omega = \bigcap_{i=0}^{\infty} T^{-i}\overline{(I \setminus I')}$. We choose an Ω, such that $r(L) = r(M)$, where r denotes the spectral radius of the l^1-operators L and M. Because I is T-invariant, T/I is again a piecewise monotonic transformation. Hence we can take I as X and assume that m is concentrated on I.

Lemma 18. (i) *If* $M(J_i J_{i+1}) = 1$ *for* $0 \leqq i \leqq k-1$ *then* $T^k(J_0 \cap \ldots \cap T^{-k} J_k) = J_k$.

(ii) *For every* $J \in F$, *we have* $m(J) > 0$.

Proof. (i) is proved by induction. The induction step is as follows

$$\begin{aligned} T^k(J_0 \cap T^{-1} J_1 \cap \ldots \cap T^{-k} J_k) &= T^{k-1}(T J_0 \cap J_1 \cap \ldots \cap T^{-k+1} J_k) \\ &= T^{k-1}(J_1 \cap \ldots \cap T^{-k+1} J_k) \end{aligned}$$

because $T J_0 \supset J_1$, since J_1 is a successor of J_0.

In order to show (ii) note that there is a $K \in \tilde{F}$ with $m(K) > 0$, since $m(I) = 1$. We can find $J_0 = J, J_1, \ldots, J_k = K$ with $M(J_i, J_{i+1}) = 1$ and $J \in F$. As L is irreducible, we can do this for all $J \in F$. Set $Z = J_0 \cap T^{-1} J_1 \cap \ldots \cap T^{-k} J_k \subset J$. Then $m(Z) = \lambda^{-k} m(\tilde{P}^k 1_Z) = \lambda^{-k} m(T^k Z) = \lambda^{-k} m(K) > 0$, using (i). Hence $m(J) > 0$ proving (ii).

It follows from §3 of [6] that $r(M) = \exp h_{\text{top}}(X)$. Thus $r(L) = r(M) > 1$, and L consists not only of a cycle. Therefore we can find $J_0, J_1, \ldots, J_n = J_0 \in F$ and $K_0 = J_0$, $K_1, \ldots, K_k = J_r \in F$ for some $n, k \geqq 0$ and $0 \leqq r \leqq n-1$ such that $M(J_i, J_{i+1}) = M(K_i, K_{i+1}) = 1$. Now set

$$A_k = \bigcap_{i=0}^{k-1} T^{-ni}\left(\bigcap_{j=0}^{n-1} T^{-j} J_j\right), \qquad A = \bigcap_{k=0}^{\infty} A_k,$$

$$B_j = \bigcap_{i=0}^{jn} T^{-i} J_{i(\text{mod } n)} \cap \bigcap_{i=1}^{k} T^{-jn-i} K_i \cap \bigcap_{i=1}^{\infty} T^{-jn-k-i} J_{r+i(\text{mod } n)}.$$

The B_j are disjoint, because $T^{jn} B_j \subset K_1$, $T^{jn} B_{j+m} \subset J_1$ and K_1 and J_1 are disjoint as different successors of J_0. Furthermore $T^{jn+k+n-r} B_j = A$ and A and B_j are contained in the same element of $\mathfrak{P}_0^{jn}$. Hence it follows from Lemma 16 that $m(A)=0$. Now we apply Lemma 17 and get

$$\sup_{A_k} \tilde{g}_{kn} \leqq K m(A_k)/m(T^{kn} A_k).$$

Because of $T^{kn} A_k = T(J_{n-1}) \supset J_0$ ((i) of Lemma 18) and $m(J_0) > 0$ ((ii) of Lemma 18), the right hand side of this inequality tends to zero. Hence there is a k with $\sup_{A_k} \tilde{g}_{kn} < 1$. Because of $\tilde{g}_{kn} \equiv \lambda^{-kn}$, we get $\lambda > 1$, the desired result. It follows now from Lemma 2 that $m(\{x\})=0$ for all $x \in X'$. Hence $m(W')=0$.

A similar proof as that of (ii) of Lemma 18 shows that the support Y of m contains Ω. As $h_{\text{top}}(\Omega) = \log r(L) = \log r(M) = h_{\text{top}}(X)$, it follows from Theorem 6, that μ is a measure with maximal entropy.

Acknowledgement. We thank the referee for some comments which simplified the proof of Lemma 5.

References

1. Babbel, B.: Diplomarbeit, Institut für Mathematische Statistik und Wirtschaftsmathematik der Universität Göttingen 1980
2. Bowen, R.: Equilibrium states and the ergodic theory of Anosov diffeomorphisms. Lecture Notes in Mathematics **470**. Berlin-Heidelberg-New York: Springer 1975
3. Bowen, R.: Bernoulli maps of an interval. Israel J. Math. **28**, 298–314 (1978)
4. Dunford, N., Schwartz, J.T.: Linear Operators, Part I. New York: Interscience 1957
5. Gordin, M.I.: The central limit theorem for stationary processes. Soviet Math. Dokl. **10**, 1174–1176 (1969)
6. Hofbauer, F.: On intrinsic ergodicity of piecewise monotonic transformations with positive entropy. Israel J. Math. **34**, 213–237 (1979)
7. Hofbauer, F.: On intrinsic ergodicity of piecewise monotonic transformations with positive entropy II. Israel J. Math. (To appear)
8. Ibragimov, I.A., Linnik, Y.V.: Independent and stationary sequences of random variables. Groningen: Wolters-Noordhoff 1971
9. Ionescu-Tulcea, C., Marinescu, G.: Théorie ergodique pour des classes d'opérations non complètement continues. Ann. of Math. (2) **52**, 140–147 (1950)
10. Keller, G.: Un théorème de la limite centrale pour une classe de transformations monotones per morceaux. C.R. Acad. Sci. Paris Sér. A **291**, 155–158 (1980)
11. Lasota, A., Yorke, J.: On the existence of invariant measures for piecewise monotonic transformations. Trans. Amer. Math. Soc. **186**, 481–488 (1973)
12. Ledrappier, F.: Principe variationnel et systemes dynamiques symboliques. Z. Wahrscheinlichkeitstheorie verw. Gebiete **30**, 185–202 (1974)
13. Li, T., Yorke, J.: Ergodic transformations from an interval into itself. Trans. Amer. Math. Soc. **235**, 183–192 (1978)
14. Li, T., Yorke, J.: Iterating piecewise expanding maps: Asymptotic dynamics of probability densities. Preprint
15. Philipp, W., Stout, W.: Almost sure invariance principles for partial sums of weakly dependent random variables. Mem. Amer. Math. Soc. **161** (1975)
16. Ratner, M.: Bernoulli flows over maps of the interval. Israel J. Math. **31**, 298–314 (1978)
17. Rohlin, V.A.: Exact endomorphisms of Lebesgue spaces. Amer. Math. Soc. Transl. (2) **39**, 1–36 (1964)

18. Schäfer, H.H.: Topological Vector Spaces. New York: MacMillan 1966
19. Volkonski, V.A., Rozanov, Y.A.: Some limit theorems for random functions II. Theor. Probability Appl. **6**, 186–198 (1961)
20. Wagner, G.: The ergodic behavior of piecewise monotonic transformations. Z. Wahrscheinlichkeitstheorie und verw. Gebiete **46**, 317–324 (1979)
21. Walters, P.: Ergodic theory. Lecture Notes in Mathematics **458**. Berlin-Heidelberg-New York: Springer 1975
22. Walters, P.: Equilibrium states for β-transformations and related transformations. Math. Z. **159**, 65–88 (1978)
23. Wong, S.: Some metric properties of piecewise monotonic mappings of the unit interval. Trans. Amer. Math. Soc. **246**, 493–500 (1978)
24. Wong, S.: A central limit theorem for piecewise monotonic mappings of the unit interval. Ann. Probability **7**, 500–514 (1979)

Received March 25, 1981; received in final form December 7, 1981

Physica 7D (1983) 153–180
North-Holland Publishing Company

THE DIMENSION OF CHAOTIC ATTRACTORS

J. Doyne FARMER
Center for Nonlinear Studies and Theoretical Division, MS B258, Los Alamos National Laboratory, Los Alamos, New Mexico 87545, USA

Edward OTT
Laboratory of Plasma and Fusion Energy Studies, University of Maryland, College Park, Maryland, USA

and

James A. YORKE
Institute for Physical Science and Technology and Department of Mathematics, University of Maryland, College Park, Maryland, USA

Dimension is perhaps the most basic property of an attractor. In this paper we discuss a variety of different definitions of dimension, compute their values for a typical example, and review previous work on the dimension of chaotic attractors. The relevant definitions of dimension are of two general types, those that depend only on metric properties, and those that depend on the frequency with which a typical trajectory visits different regions of the attractor. Both our example and the previous work that we review support the conclusion that all of the frequency dependent dimensions take on the same value, which we call the "dimension of the natural measure", and all of the metric dimensions take on a common value, which we call the "fractal dimension". Furthermore, the dimension of the natural measure is typically equal to the Lyapunov dimension, which is defined in terms of Lyapunov numbers, and thus is usually far easier to calculate than any other definition. Because it is computable and more physically relevant, we feel that the dimension of the natural measure is more important than the fractal dimension.

Table of contents

1. Introduction

It is the purpose of this paper to discuss and review questions relating to the dimension of chaotic attractors. Before doing so, however, we should first say what we mean by the work "attractor".

1.1. *Attractors*

In this paper we consider dynamical systems such as maps (discrete time, n)

$$x_{n+1} = F(x_n),$$

or ordinary differential equations (continuous

time, t)

$$\frac{dx(t)}{dt} = G(x(t)),$$

where in both cases x is a vector. Thus, given an initial value of x (at $n = 0$ for the map or $t = 0$ for the differential equations) an orbit is generated $((x_1, x_2, \ldots, x_n, \ldots)$ for the map and $x(t)$ for the differential equations). We shall be interested in attractors for such systems. Loosely speaking, an attractor is something that "attracts" initial conditions from a region around it once transients have died out. More precisely, an *attractor* is a compact set, A, with the property that there is a neighborhood of A such that for almost every* initial condition the limit set of the orbit as n or $t \to +\infty$ is A. Thus, almost every trajectory in this neighborhood of A passes arbitrarily close to every point of A. The *basin of attraction* of A is the closure of the set of initial conditions that approach A.

We are primarily interested in *chaotic* attractors. We give a definition of chaos in section 3, but the reader may also wish to see the reviews given in references 1–4.

1.2. *Why study dimension?*

The dimension of an attractor is clearly the first level of knowledge necessary to characterize its properties. Generally speaking, we may think of the dimension as giving, in some way, the amount of information necessary to specify the position of a point on the attractor to within a given accuracy (cf. section 2). The dimension is also a lower bound on the number of essential variables needed to model the dynamics. For an extensive discussion of dimension in many contexts, see Mandelbrot [5, 6, 46].

For simple attractors, defining and determining the dimension is easy. For example, using any reasonable definition of dimension, a stationary time independent equilibrium (fixed point) has dimension zero, a stable periodic oscillation (limit cycle) has dimension one, and a doubly periodic attractor (2-torus) has dimension two. It is because their structure is very regular that the dimension these simple attractors takes on integer values.

Chaotic (strange) attractors, however, often have a structure that is not simple; they are often not manifolds, and frequently have a highly fractured character. For chaotic attractors, intuition based on properties of regular, smooth examples does not apply. The most useful notions of dimension take on values that are typically not integers.

To fully understand the properties of a chaotic attractor, one must take into account not only the attractor itself, but also the "distribution" or "density" of points on the attractor. This is more precisely discussed in terms of what we shall call the *natural measure* associated with a given attractor. The natural measure provides a notion of the relative frequency with which an orbit visits different regions of the attractor. Just as chaotic attractors can have very complicated properties, the natural measures of chaotic attractors often have complicated properties that make the relevant assignment of a dimension a nontrivial problem.

Precise definitions of such terms as "natural measure" follow, but we would first like to give an example in order to motivate the central questions we are addressing in this paper.

Consider the following two dimensional map†:

$$\begin{aligned} x_{n+1} &= x_n + y_n + \delta \cos 2\pi y_n \quad \bmod 1, \\ y_{n+1} &= x_n + 2y_n \qquad\qquad\qquad \bmod 1. \end{aligned} \tag{1}$$

For small values of δ, Sinai [7] has shown that the attractor of this map is the entire square, and is thus of dimension 2. Therefore almost every initial condition generates a trajectory that eventually comes arbitrarily close to every point on the square. However, consider the typical trajectory

*The phrase "almost every" here signifies that the set of initial conditions in this neighborhood for which the corresponding limit set is not A can be covered by a set of cubes of arbitrarily small volume (i.e. has Lebesgue measure zero).

†Mod 1 means that the values of x and y are truncated to be less than or equal to one and their integer part are discarded, so that the map is defined on the unit square.

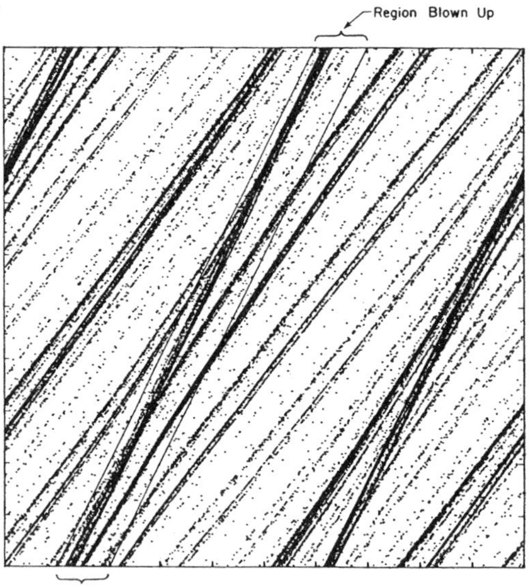

Fig. 1. Successive iterates of the initial point $x_0 = 0.5$, $y_0 = 0.5$ using eq. (1) with $\delta = 0.1$. 80,000 points are shown. Almost any initial condition gives a qualitatively similar plot; the location of the individual points of course changes, but the location of the dark bands does not. The density of these points is described by the *natural measure* of this attractor. (For example, the outlined parallelogram (which is blown up in fig. 2) contains approximately 27% of the points of a typical trajectory, and thus can be said to have a natural measure of approximately 0.27.)

Fig. 2. A blow-up of the strip marked in fig. 1. This strip was chosen in order to follow one of the dark bands; the blow-up was made by expanding the strip in a direction perpendicular to its long sides (roughly horizontally), and the top and bottom were trimmed to make the result square. What appears to be a single band in fig. 1 is now seen as a collection of bands.

shown in fig. 1. Certain regions are visited far more often than others. The natural measure of a given region is proportional to the frequency with which it is visited (see section 2.2.2), in this case the natural measure is highly concentrated in diagonal bands whose density of points is much greater than the average*. Furthermore, as shown in fig. 2, if a small piece of the attractor is magnified, the same sort of structure is still seen.

For this map we do not know if the value of δ chosen to construct fig. 1 is small enough to insure that the dimension of the attractor is two. For practical purposes, though, this may be irrelevant. Even if a trajectory eventually comes arbitrarily close to any given point, the amount of time required for this to happen may be enormous. In order to assign a relevant dimension that will characterize the trajectories on the attractor, the natural measure must be taken into account. For this example the dimension that characterizes properties of the natural measure is between one and two.

These considerations are not as esoteric as they might seem. One may not be as interested in whether the dimension of a given attractor is 3.1 or 3.2 as in whether it is on the order of three or on the order of thirty. As we shall see, a proper understanding of *probabilistic* notions of dimension leads to an efficient method of computing the dimension of chaotic attractors, that provides the best known method of answering such questions.

The main points of this paper can be summarized as follows:

1) Although there are a variety of different definitions of dimension, the relevant definitions

* In fact, for small values of δ, Sinai [7] has shown that for any $\epsilon > 0$, there exists a collection of tiny squares whose total area is less than ϵ, and such that almost every trajectory spends $1 - \epsilon$ of the time inside this collection of squares. These squares cover what is called the *core* of the attractor. (See section 7).

Table I
Current evidence indicates that typically the first two dimensions take on the same value, called the fractal dimension, while the next five dimensions take on another typically smaller value, called the dimension of the measure.

Name of dimension	Symbol	Generic name	Symbol
Capacity	d_C	fractal	d_F
Hausdorff dimension	d_H	dimension	
Information dimension	d_I		
ϑ-capacity	$d_C(\vartheta)$		
ϑ-Hausdorff dimension	$d_H(\vartheta)$	dimension of the	d_μ
Pointwise dimension	d_p	natural measure	
Hausforff dimension of the core	d_H(core)		
Lyapunov dimension	d_L		

are of two types, those which only depend on metric properties, and those which depend on metric and probabilistic properties (i.e., they involve the natural measure of the attractor).

2) Current evidence supports the conclusion that all of the metric dimensions typically take on the same value, and all of the frequency dependent dimensions take on another, typically smaller, common value.

3) Current evidence supports a conjectured relationship whereby the dimension of the natural measure can be found from a knowledge of the stability properties of an orbit on the attractor (i.e., knowledge of the Lyapunov numbers).

4) For typical chaotic attractors we conjecture that the distribution of frequencies with which an orbit visits different regions of the attractor is, in a certain sense, log-normal (section 5).

Points 1–3 are summarized in table I. The first two entries in the table are metric dimensions, while the next five are frequency dependent dimensions. Under the hypothesis that all the metric dimensions yield the same value (point 2), we call this value the *fractal dimension* and denote it d_F. Similarly, if all the probabilistic dimensions yield the same value, we call this value the *dimension of the natural measure*, and denote it d_μ. Although in special cases d_F equals d_μ, typically $d_F > d_\mu$. Finally, the last entry in table I, the Lyapunov dimension, is by definition the predicted value of d_μ obtained from the Lyapunov numbers (cf. Point 3). The Lyapunov dimension is in a different category than the other dimensions listed, since it is defined in terms of dynamical properties of an attractor, rather than metric and natural measure properties.

1.3. *Outline*

This paper is organized as follows: In section 2 we give several definitions of dimension. Section 3 reviews conjectures relating Lyapunov numbers to dimension. These conjectures are particularly important because the Lyapunov numbers provide the only known efficient method to compute dimension. In sections 4, 5, 6, and 7, we compute all the dimensions discussed here for an explicitly soluble example, the generalized baker's transformation. In addition, based on this example, in section 5 we propose a new conjecture concerning the frequency with which different values of the probability occur. Section 7 gives a discussion of the "core" of attractors, and section 8 gives another example supporting the connection between Lyapunov numbers and dimension (an attractor which is topologically a torus but is nowhere differentiable). Section 9 reviews relevant results from numerical computations of the dimension of chaotic attractors. Concluding remarks are given in section 10.

In general terms, this paper has two functions. One is to present a review of the current status of research on the dimension of chaotic attractors. The other purpose is to present new results (sections 4–6).

2. Definitions of dimension

In this section we define and discuss six different concepts of dimension. The first two of these, the capacity and the Hausdorff dimension, require only a metric (i.e., a concept of distance) for their definition, and consequently we refer to them as "metric dimensions". The other dimensions we will discuss in this section are the information dimension, the ϑ-capacity, the ϑ-Hausdorff dimension, and the pointwise dimension. These dimensions require both a metric and a probability measure for their definition, and hence we will refer to them as "probabilistic dimensions".

In this paper we compute the values of these dimensions for an example that we believe is general enough to be "typical" of chaotic attractors, at least regarding the question of dimension. We find that the metric dimensions take on a common value. Whenever this is the case, we will refer to this common value d_F as the *fractal dimension**. For our example we also find that the probabilistic dimensions take on a common value d_μ, which we will refer to as the *dimension of the natural measure*. As we summarize in conjecture 1, we feel that this equality is a general property, true for typical cases.

Conjecture 1. For a typical chaotic attractor the capacity and Hausdorff dimensions have a common value d_F, and the information dimension, ϑ-capacity, ϑ-Hausdorff dimension, and pointwise dimensions have a common value d_μ, i.e., in the notation of table I,

$$d_C = d_H \equiv d_F$$

and

$$d_I = d_C(\vartheta) = d_H(\vartheta) = d_P \equiv d_\mu.$$

Note: For the case of diffeomorphisms† in two dimensions, L.S. Young has rigorously proven that information dimension, pointwise dimension, and the Hausdorff dimension of the core (see section 7) all take on the same value [12].

In addition to the dimensions defined in this section, we will also discuss three others‡. The Lyapunov dimension, the capacity of the core, and the Hausdorff dimension of the core. Lyapunov dimension is discussed in section 3, and the latter two dimensions are discussed in section 7. For our example the Lyapunov dimension and Hausdorff dimension of the core are equal to d_μ, while the capacity of the core is equal to d_F.

2.1. *Metric dimensions*

We begin by discussing two concepts of dimension which apply to sets in spaces on which a concept of distance, i.e., a metric is defined. In particular we begin by discussing the capacity and the Hausforff dimension.

2.1.1. *Capacity*

The capacity of a set was originally defined by Kolmogorov [14]. It is given by

$$d_C = \lim_{\epsilon \to 0} \frac{\log N(\epsilon)}{\log(1/\epsilon)}, \tag{2}$$

* The term *fractal* was originally coined by Mandelbrot [5]. However, he uses "fractal dimension" as a synonym for Hausdorff dimension. We should also mention that in some of our previous papers on this subject [8–11], we used the term "fractal dimension" as a synonym for capacity, rather than our current usage as described in the text.

† A diffeomorphism is a differentiable invertible mapping whose Jacobian has non-zero determinant everywhere.

‡ Note that in this paper we will not discuss the concept of *topological dimension*, since its application to chaotic dynamics is not clear. Its value is an integer and it is generally equal to neither d_F nor d_μ. For discussions of topological dimension, we refer the reader to Hurwicz and Wallman [13].

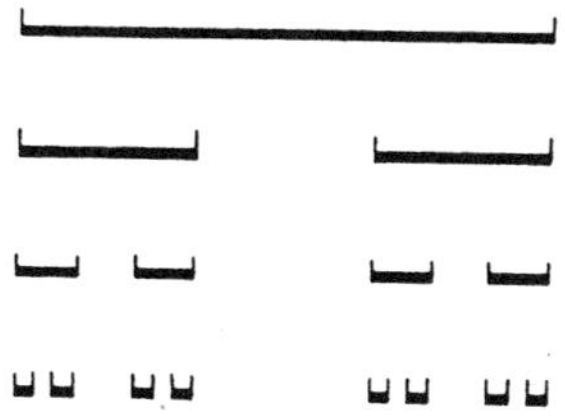

Fig. 3. The first few steps in the construction of the classic example of a Cantor set.

where, if the set in question is a bounded subset of a p-dimensional Euclidean space $\mathbb{R}^p$, then $N(\epsilon)$ is the minimum number of p-dimensional cubes of side ϵ needed to cover the set. For a point, a line, and an area, $N(\epsilon)=1$, $N(\epsilon)\sim\epsilon^{-1}$, and $N(\epsilon)\sim\epsilon^{-2}$, and eq. (2) yields $d_C=0$, 1, and 2, as expected. However, for more general sets (dubbed *fractals* by Mandelbrot), d_C can be noninteger*. For example, consider the Cantor set obtained by the limiting process of deleting middle thirds, as, illustrated in fig. 3. If we choose $\epsilon=(1/3)^m$, then $N=2^m$, and eq. (2) yields

$$d_C=\frac{\log 2}{\log 3}=0.630\ldots.$$

If one is content to know where the set lies to within an accuracy ϵ, then to specify the location of the set, we need only specify the position of the $N(\epsilon)$ cubes covering the set. Eq. (2) implies that for small ϵ, $\log N(\epsilon)\approx d_C\log(1/\epsilon)$. Hence, the dimension tells us how much information is necessary to specify the location of the set to within a given accuracy. If the set has a very fine-scaled structure (typical of chaotic attractors), then it may be advantageous to introduce some coarse-graining into the description of the set. In this case, ϵ may be thought of as specifying the degree of coarse-graining.

* Sets can be constructed for which the limit of eq. (2) does not exist. We would then say that the capacity is not defined.

† For example, for the set of numbers 1, 1/2, 1/3, 1/4, ... , the Hausdorff dimension is zero while (2) yields $d_C=\frac{1}{2}$.

‡ To show that $d_C\geqslant d_H$, consider a covering consisting of cubes of equal side $\epsilon_i=\epsilon$. Then due to the infimum in the definition of $l_d(\epsilon)$, we see that $\bar{l}_d(\epsilon)=\Sigma\epsilon^d=N(\epsilon)\epsilon^d$ satisfies $\bar{l}_d(\epsilon)\geqslant l_d(\epsilon)$. Thus taking the limit $\epsilon\to 0$ and making use of eq. (2) we see that $d_H\leqslant d_C$.

2.1.2. *Hausdorff dimension*

The capacity may be viewed as a simplified version of the Hausdorff dimension, originally introduced by Hausdorff in 1919 [15]. (We have reversed historical order and defined capacity before Hausdorff dimension because the definition of Hausdorff dimension is more involved.) We believe that for attractors these two dimensions are generally equal. While it is possible to construct simple examples of sets where the Hausdorff dimension and the capacity are unequal†, these do not seem to apply to attractors. (Although they may apply to the core of attractors. See section 7.)

To define the Hausdorff dimension of a set lying in a p-dimensional Euclidean space, consider a covering of it with p-dimensional cubes of variable edge length ϵ_i. Define the quantity $l_d(\epsilon)$ by

$$l_d(\epsilon)=\inf\sum_i\epsilon_i^d,$$

where the infimum (i.e. minimum) extends over all possible coverings subject to the constraint that $\epsilon_i\leqslant\epsilon$. Now let

$$l_d=\lim_{\epsilon\to 0}l_d(\epsilon).$$

Hausdorff showed that there exists a critical value of d above which $l_d=0$ and below which $l_d=\infty$. This critical value, $d=d_H$, is the Hausdorff dimension. (Precisely at $d=d_H$, l_d may be either 0, ∞, or a positive finite number.) This concept of dimension will be used in sections 4, 6, and 7. It is easy to see that $d_C\geqslant d_H$‡.

2.2. *Dimensions for the natural measure*

2.2.1. *The natural measure on an attractor*

Note that, in computing d_C from eq. (2), all cubes used in covering the attractor are equally important even though the frequencies with which an orbit on the attractor visits these cubes may be very different. In order to take the frequency with which each cube is visited into account, we need to consider not only the attractor itself, but the relative frequency with which a typical orbit visits different regions of the attractor as well. We can say that some regions of the attractor are more probable than others, or alternatively we may speak of a measure on the attractor*. We define the natural measure of an attractor as follows: For each cube C and initial condition x in the basin of attraction, define $\mu(x, C)$ as the fraction of time that the trajectory originating from x spends in C†. If almost every such x gives the same value of $\mu(x, C)$, we denote this value $\mu(C)$ and call μ the *natural measure* of the attractor [16]. The natural measure gives the relative probability of different regions of the attractor as obtained from time averages, and therefore is the "natural" measure to consider. We will assume throughout that any attractor we consider has a natural measure, at least whenever C is one of the cubes we are using to cover the attractor.

The four definitions discussed in the remainder of this section are defined for attractors with a metric and a natural measure defined on them.

2.2.2. *Information dimension*

The *information dimension*, d_I, is a generalization of the capacity that takes into account the relative probability of the cubes used to cover the set. This dimension was originally introduced by Balatoni and Renyi [17].

The information dimension is given by

* Although there are many measures possible for a given attractor, we are only interested in one of them, the natural measure.

† $\mu(x, C) = \lim_{\tau\to\infty}\mu_\tau(x, C)$, where $\mu_\tau(X, C)$ is the fraction of time spent in C up to some finite time τ.

$$d_I = \lim_{\epsilon\to 0}\frac{I(\epsilon)}{\log(1/\epsilon)}, \tag{3}$$

where

$$I(\epsilon) = \sum_{i=1}^{N(\epsilon)} P_i \log\frac{1}{P_i}$$

and P_i is the probability contained within the ith cube. Letting the ith cube of side ϵ be C_i, $P_i = \mu(C_i)$. Note that if all cubes have equal probability then $I(\epsilon) = \log N(\epsilon)$, and hence $d_C = d_I$. However, for unequal probabilities $I(\epsilon) < \log N(\epsilon)$. Thus, in general, $d_C \geqslant d_I$.

In information theory the quantity $I(\epsilon)$ defined in eq. (3) has a specific meaning [18]. Namely, it is the amount of information necessary to specify the state of the system to within an accuracy ϵ, or equivalently, it is the information obtained in making a measurement that is uncertain by an amount ϵ. Since for small ϵ, $I(\epsilon) \approx d_I \log(1/\epsilon)$, we may view d_I as telling how fast the information necessary to specify a point on the attractor increases as ϵ decreases. (For a more extensive discussion of the physical meaning of the information dimension, see refs. 9 and 10.)

2.2.3. *ϑ-Capacity*

Another definition of dimension which we shall be interested in is what we will call the ϑ-capacity, $d_C(\vartheta)$. Essentially, this quantity is the capacity of that part of the attractor of highest probability,

$$d_C(\vartheta) = \lim_{\epsilon\to 0}\frac{\log N(\epsilon;\vartheta)}{\log(1/\epsilon)}, \tag{4}$$

where $N(\epsilon;\vartheta)$ is the minimum number of cubes of side ϵ needed to cover at least a fraction ϑ of the natural measure of the attractor. In other words, the cubes must be chosen so that their combined natural measure is at least ϑ. Thus $d_C(1) = d_C$. For the examples we study here, we find that for any value of $\vartheta < 1$, the ϑ-capacity is independent of ϑ, but that $d_C(\vartheta)$ for $\vartheta < 1$ may differ from its value at $\vartheta = 1$. In particular $d_C(\vartheta) = d_\mu$ for $\vartheta < 1$ and

$d_C(\vartheta) = d_C$ for $\vartheta = 1$. ϑ-capacity was originally defined by Frederickson et al. [8]. Similar quantities have also been defined by Ledrappier [18], and Mandelbrot [6, 45].

2.2.4. *ϑ-Hausdorff dimension*

In analogy with the relationship between capacity (a metric dimension) and ϑ-capacity (a probability dimension), we introduce here a probability dimension based on the Hausdorff dimension. We call this new dimension the ϑ-Hausdorff dimension and denote it $d_H(\vartheta)$. To define the ϑ-Hausdorff dimension, modify the definition of Hausdorff dimension as follows: Define $l_d(\epsilon, \vartheta)$ by

$$l_d(\epsilon, \vartheta) = \inf \sum_i \epsilon_i^d,$$

where now the infimum extends over all possible $\epsilon_i < \epsilon$ which cover a fraction ϑ of the total probability of the set. We define $d_H(\vartheta)$ as that value of d below which $l_d(\vartheta) = \infty$ and above which $l_d(\vartheta) = 0$, where $l_d(\vartheta) = \lim_{\epsilon \to 0} l_d(\epsilon, \vartheta)$. This concept of dimension will be used in section 6.

2.2.5. *Pointwise dimension*

Roughly speaking, the pointwise dimension d_p is the exponent with which the total probability contained in a ball decreases as the radius of the ball decreases. To make this notion more precise, let μ denote the natural probability measure on the attractor, and let $B_\epsilon(x)$ denote a ball of radius ϵ centered about a point x on the attractor. Roughly speaking, $\mu(B_\epsilon(x)) \sim \epsilon^{d_p}$. More precisely, define this dimension as

$$d_p(x) = \lim_{\epsilon \to 0} \frac{\log \mu(B_\epsilon(x))}{\log \epsilon}. \tag{5}$$

If $d_p(x)$ is independent of x for almost all x with respect to the measure μ*, we call $d_p(x) = d_p$ the *pointwise* dimension. Similar definitions of dimension have also been given by Takens [20], Billingsley [31], Young [11], and Janssen and Tjon [21].

* By "almost all x with respect to the measure μ" we mean that the set of x which does not satisfy this is a set of μ measure zero.

2.3. *Using a grid of cubes to compute dimension*

Some of the definitions we have used, such as the capacity, allow any location or orientation of the cubes used to cover the attractor. In a numerical experiment, however, it is much more convenient to select the cubes used to cover the attractor out of a fixed grid, as shown in fig. 4. For these dimensions (d_C, d_I, and $d_C(\vartheta)$) it can be shown that selecting from a fixed grid of cubes gives the same value of the dimension as an optimal collection of cubes. For example, for the case of an attractor in a two-dimensional space, using a fixed grid to compute $N(\epsilon)$ in eq. (2) results in an increase of at most a factor of four in $N(\epsilon)$, which has no effect on the value of the dimension. Note that this is *not* true for the Hausdorff dimension, which requires a more general cover.

In principle, the definitions of dimension given in this section and the use of a fixed grid provide specific prescriptions for obtaining capacity, information dimension, and ϑ-capacity. To find approximate values for these dimensions, one can generate an orbit on the attractor using a computer, and then divide the space containing the orbit into cubes of side ϵ in order to estimate the numbers $N(\epsilon)$, $I(\epsilon)$, or $N(\epsilon; \vartheta)$. By examining how

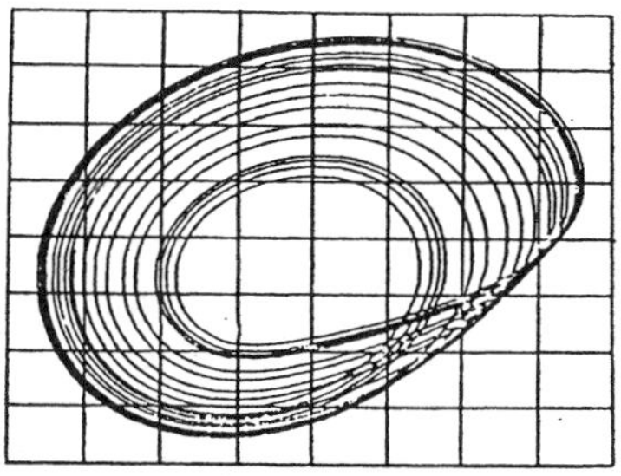

Fig. 4. The region of phase space containing an attractor can be divided with a fixed grid of cubes (in this case squares), which can be used to compute capacity, information dimension, or ϑ-capacity.

$N(\epsilon)$, $I(\epsilon)$, and $N(\epsilon; \vartheta)$ vary as ϵ is decreased the value of these dimensions can be estimated.

As discussed in section 9, however, in practice the agenda described above for computing dimension may be difficult, costly, or impossible. Thus it is of interest to consider other means of obtaining the dimension of chaotic attractors. The next section deals with this question. In particular, we discuss a conjecture that the dimension of chaotic attractors can be determined directly from the dynamics in terms of Lyapunov numbers.

3. Lyapunov numbers and Lyapunov dimension

The Lyapunov numbers quantify the average stability properties of an orbit on an attractor. For a fixed point attractor of a mapping, the Lyapunov numbers are simply the absolute values of the eigenvalues of the Jacobian matrix evaluated at the fixed point. The Lyapunov numbers generalize this notion for more complicated attractors. As we shall see, for a typical attractor there is a connection between average stability properties and dimension. The possibility of such a connection was first pointed out by Kaplan and Yorke [22] and later by Mori [23].

3.1. *Definition of Lyapunov numbers*

For expository purposes, for most of this paper we shall consider p-dimensional maps,

$$x_{n+1} = F(x_n),$$

where x is a p-dimensional vector. We emphasize, however, that similar considerations to those below apply to flows (e.g., systems of differential equations), including infinite-dimensional systems such as partial differential equations. To define the Lyapunov numbers, let $J_n = [J(x_n)J(x_{n-1}) \ldots J(x_1)]$ where $J(x)$ is the Jacobian matrix of the map, $J(x) = (\partial F/\partial x)$, and let $j_1(n) \geq j_2(n) \geq \cdots \geq j_p(n)$ be the magnitudes of the eigenvalues of J_n. The Lyapunov numbers are

$$\lambda_i = \lim_{n\to\infty} [j_i(n)]^{1/n}, \quad i = 1, 2, \ldots, p, \tag{6}$$

where the positive real nth root is taken. The Lyapunov numbers generally depend on the choice of the initial condition x_1. The Lyapunov numbers were originally defined by Oseledec [24]. We have the convention

$$\lambda_1 \geq \lambda_2 \geq \cdots \geq \lambda_p.$$

For a two-dimensional map, for example, λ_1 and λ_2 are the average principal stretching factors of an infinitesimal circular are (cf. fig. 5). For a chaotic attractor on the average nearby points initially diverge at an exponential rate, and hence at least one of the Lyapunov numbers is greater than one. This makes quantitative the notion of "sensitive dependence onititial conditions". We will take $\lambda_1 > 1$ as our definition of *chaos*. (Note that many authors refer to *Lyapunov exponents* rather than Lyapunov numbers. The Lyapunov exponents are simply the logarithms of the Lyapunov numbers.)

In this paper we assume that *almost every* initial condition in the basin of any attractor that we consider has the same Lyapunov numbers. Thus, the spectrum of Lyapunov numbers may be considered to be a property of an attractor. This assumption is supported by numerical experiments [25]. Exceptional trajectories, such as unstable fixed points on the attractor, typically do not sample the whole attractor and thus typically have Lyapunov numbers that are different from those of the attractor. Those points in the basin of attraction that have different Lyapunov numbers or for

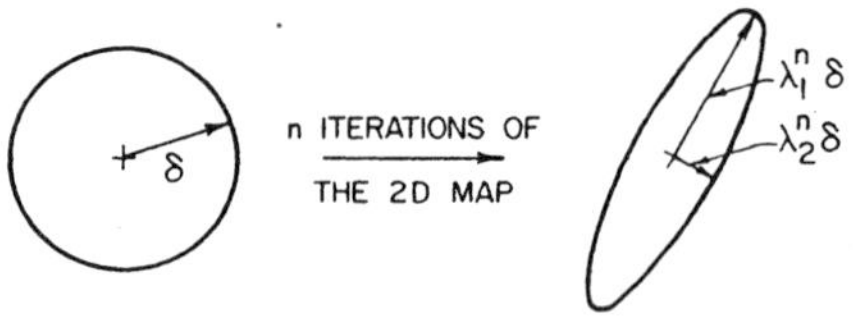

Fig. 5. n iterations of a two-dimensional map transform a sufficiently small circle of radius δ approximately into an ellipse with major and minor radii $(\lambda_1)^n\delta$ and $(\lambda_2^n)\delta$, where λ_1 and λ_2 are the Lyapunov numbers.

which Lyapunov numbers do not exist are here assumed to be of measure zero. (In other words, they may be covered by a collection of cubes of varying size having arbitrarily small total volume).

3.2. *Definition of Lyapunov dimension*

The following discussion contains a heuristic argument that motivates a connection between Lyapunov numbers and dimension. Consider a two-dimensional map. Suppose we wish to compute the capacity of a chaotic attractor, for which $\lambda_1 > 1 > \lambda_2$. Cover the attractor with $N(\epsilon)$ squares of side ϵ. Now, iterate the map q times. For q fixed and ϵ small enough, the action of the mapping is roughly linear over the square, and each square will be stretched into a long thin parallelogram. From the definition of the Lyapunov numbers, the average length of these parallelograms is $(\lambda_1)^q\epsilon$, and the average width is $(\lambda_2)^q\epsilon$. Now, suppose we had used a finer cover of squares of side $(\lambda_2)^q\epsilon$. (See fig. 6.) To cover each parallelogram takes about $(\lambda_1/\lambda_2)^q$ smaller squares. Thus, *if it is supposed* that all squares on the attractor behave in this typical way, then one is lead to the estimate

$$N(\lambda_2^q\epsilon) \approx \left(\frac{\lambda_1}{\lambda_2}\right)^q N(\epsilon). \qquad (7)$$

Motivated by eq. (2), assume $N(\epsilon) \approx k(1/\epsilon)^{d_C}$, and

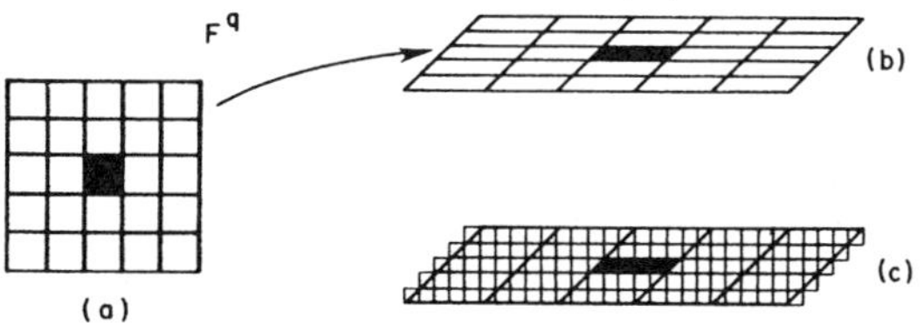

Fig. 6. A schematic illustration of the heuristic argument for the Lyapunov dimension. The image of each small square in (a) is approximately a parallelogram which has been stretched horizontally be a factor of λ_1^q and contracted vertically by a factor λ_2^q. The images in (b) thus have a smaller cover of squares as shown in (c).

substitute into both sides of eq. (7). This gives

$$k\left(\frac{1}{\lambda_2^q\epsilon}\right)^{d_C} \approx k\left(\frac{\lambda_1}{\lambda_2}\right)^q\left(\frac{1}{\epsilon}\right)^{d_C}.$$

Collecting terms, taking logarithms, and solving for d_C gives

$$d_C = 1 + \frac{\log \lambda_1}{\log(1/\lambda_2)}.$$

We will see that this expression is often meaningful even when this heuristic derivation is invalid, so we will call it the *Lyapunov dimension* d_L.

$$d_L = 1 + \frac{\log \lambda_1}{\log(1/\lambda_2)}. \qquad (8)$$

Generalization of the above heuristic argument to p-dimensional maps gives (cf. ref 7)

$$d_L = k + \frac{\log(\lambda_1\lambda_2\ldots\lambda_k)}{\log(1/\lambda_{k+1})}, \qquad (9)$$

where k is the largest value for which $\lambda_1\lambda_2\ldots\lambda_k \geqslant 1$. If $\lambda_1 < 1$, define $d_L = 0$; if $\lambda_1\lambda_2\ldots\lambda_p \geqslant 1$, define $d_L = p$. We shall refer to d_L as the *Lyapunov dimension*. This quantity was originally defined by Kaplan and Yorke [22], who originally gave it as a lower bound on the fractal dimension.

From the above argument one might be tempted to guess that $d_C = d_L$. The Lyapunov numbers are *average* quantities, however, and to compute an average, each cube must be weighted according to its probability. The capacity does not distinguish between probable and improbable cubes. To understand how some cubes might have vastly different probabilities than others, consider an atypical square of a two-dimensional map. If the area of the images of this square decreases half as fast as the average for k iterations, then its kth image will be 2^k times larger than the image of a typical square, and the number of squares needed to cover it will be 2^k times greater than the typical value. In fact, as will be evident from considerations of explicit examples (cf. section 5), it is commonly the case that the vast majority of cubes needed to cover the attractor are atypical, and do not represent the properties of time averages. By this we mean that all the atypical cubes taken together contain an extremely small fraction of the total probability on the attractor yet account for almost all of $N(\epsilon)$. Furthermore, this tendency increases as ϵ decreases. The behavior of the atypical cubes under iteration is in general not described by the Lyapunov numbers. It is clear, then, that in order for this estimate to be valid, we must consider only the more probable cubes, i.e., the estimate should be in terms of the dimension of the natural measure rather than the capacity. Assuming the equality of probabilistic dimensions (conjecture 1), we are led to the following conjecture:

Conjecture 2. For a typical* attractor $d_\mu = d_L$.

In the following six sections we present evidence supporting this conjecture. Also, L.S. Young has proved some rigorous results along these lines, which are reviewed in the next subsection.

In the special case that *every* initial condition on the attractor generates the same Lyapunov numbers, we will say that the attractor has *absolute* Lyapunov numbers. In this case it is not necessary to distinguish probable from improbable cubes, and the above conjecture can be made in terms of the fractal dimension rather than the dimension of the natural measure. We call this conjecture 3,

Conjecture 3. If *every* (not just almost every) initial condition generates the same set of p Lyapunov numbers $\lambda_1, \lambda_2, \ldots \lambda_p$, and if $\lambda_1 > 1$, then for a typical attractor of this type $d_F = d_L = d_\mu$.

The requirement of conjecture 3 that every initial condition on the attractor generate the same Lyapunov numbers is very restrictive and only holds for special cases. For example, it holds if the Jacobian matrix of the map is independent of x. In more general cases, the requirement of conjecture 3 would be expected to fail because of the existence of unstable fixed and periodic points on the attractor. For example, if x_1 is chosen to be precisely on an unstable fixed point, the Lyapunov numbers generated will simply be the eigenvalues of $J(x_1)$. These will typically be different from those generated by a chaotic orbit on the attractor. Examples for which conjecture 3 is valid will be special cases of the more general example presented in the following section. In addition, an example for which conjecture 3 can be proven to hold is given in section 8.

3.3. *Review of rigorous results concerning Lyapunov dimension*

In addition to the analytic and numerical evidence we will give for conjectures 1–3 in the remainder of this paper, there are several rigorous results supporting these statements which are reviewed in this section. For example, Ledrappier [19] has proven an inequality that is somewhat similar to conjecture 2. In particular, he defines a dimension that we will call d_{Led}, which is the ϑ-capacity in the limit as ϑ goes to one, i.e.

$$d_{Led} = \lim_{\vartheta \to 1} d_C(\vartheta).$$

For C^2 diffeomorphisms he has shown that

$$d_L \geqslant d_{Led}.$$

* The reason for the word "*typical*" is that there exist examples of maps that do not satisfy $d_\mu = d_L$. These maps are exceptional, however, in that arbitrarily small perturbations of them restore the conjectured equality of d_μ and d_L. An example of such an atypical case is where a point x_0 is attracting and yet has $\lambda_1 = 1$ (i.e., the Jacobian matrix $\partial F/\partial x$ has an eigenvalue $+1$ at x_0). The attraction here is due to higher order terms. The attractor is a point and so has dimension zero, yet $d_L \geqslant 1$. Small perturbations, however, will destroy this delicate balance. For example, the one-dimensional map $x_{i+1} = F(x_i) \equiv \alpha x_i - x_i^3$ has a fixed point at $x = 0$ with $\lambda_1 = 1$ for $\alpha = 1$ yet $x = 0$ is attracting. This situation is changed, however, as soon as $\alpha \neq 1$. When $|\alpha| < 1$, $d_L = 0$, and when $|\alpha| > 1$, $x = 0$ is no longer attracting.

The proof is a rigorous version of the heuristic argument that we have given (fig. 6). Also, Douady and Oesterle [26] have proven that an upper bound for the fractal dimension can be obtained yielding an expression like eq. (8), where the numbers they use are basically upper bounds for the Lyapunov numbers.

L.S. Young [12] has proven several results that strongly support conjectures 1 and 2. Particularly relevant are the following two theorems*.

1. If d_p exists then

$$d_p = d_I = d_H(\text{core}) = d_{\text{Led}}. \qquad (10)$$

2. For two-dimensional C^2 diffeomorphisms with $\lambda_1 > 1 > \lambda_2$, d_p exists, and

$$d_p = \frac{h_\mu}{\log \lambda_1}\left(1 + \frac{\log \lambda_1}{\log(1/\lambda_2)}\right). \qquad (11)$$

(See section 7 for a definition of d_H(core).) h_μ denotes the Kolmogorov entropy† of the attractor taken with respect to the measure μ, and λ_1 and λ_2 are the Lyapunov numbers with respect to μ. (More precisely, almost every initial condition x with respect to μ give λ_1 and λ_2 as the Lyapunov numbers.)

For Axiom-A attractors Bowen and Ruelle [16] have shown that there is a natural measure such that h_μ with respect to this measure is the sum of the logarithms of the Lyapunov numbers that are greater than one. For attractors with only one Lyapunov number greater than one, this implies that $h_\mu = \log \lambda_1$. Thus, for axiom-A attractors of two-dimensional maps, eqs. (9)–(11) yield $d_\mu = d_L$. Therefore Young has shown that conjecture 2 holds for this case. (It has been conjectured that the relationship between h_μ and the positive λ_i holds for non-axiom-A attractors that have a natural measure.) This result for the case of axiom-A attractors of two-dimensional maps has also been obtained independently by Pelikan [30].

* For these results Young does not require the existence of a natural measure, but rather assumes simply the existence of some invariant measure μ. In this case the Lyapunov numbers are those obtained when starting at almost every initial point with respect to μ.

† The Kolmogorov entropy, originally defined by Shannon [18] and applied to dynamical systems by Kolmogorov [27] and Sinai [28], puts a quantitative value on the average amount of new information obtained from a sequence of measurements. See [10] or [29] for physically motivated reviews. Note that this is also called metric entropy. The name *metric* entropy derives from the invariance properties of this quantity; in fact, the definition of metric entropy does not require a metric (but does require a measure).

‡ Except for the ϑ-Hausdorff dimension, for which we only obtain an upper bound.

4. Generalized baker's transformation: scaling

4.1. *Definition of generalized baker's transformation*

In this section we define the example which we will study in detail in this and the following four sections. Although we feel that this example is general enough to be typical of low-dimensional chaotic attractors (at least concerning its dimensional properties), it is also simple enough that all of the dimensions discussed in this paper can be analytically calculated‡. Thus, for this example, we shall be able to verify conjectures 1–3 in a case where generally $d_F \neq d_\mu$. As we shall show in section 5, another nice property of this map is that it allows us to investigate certain properties of the natural probability distribution in detail.

The map to be considered is

$$x_{n+1} = \begin{cases} \lambda_a x_n, & \text{if } y_n < \alpha, \\ \frac{1}{2} + \lambda_b x_n, & \text{if } y_n > \alpha; \end{cases} \qquad (12a)$$

$$y_{n+1} = \begin{cases} \frac{1}{\alpha} y_n, & \text{if } y_n < \alpha, \\ \frac{1}{1-\alpha}(y_n - \alpha), & \text{if } y_n > \alpha; \end{cases} \qquad (12b)$$

where we shall assume $0 \leqslant x_n \leqslant 1$ and $0 \leqslant y_n \leqslant 1$. If this condition is satisfied initially it is also satisfied

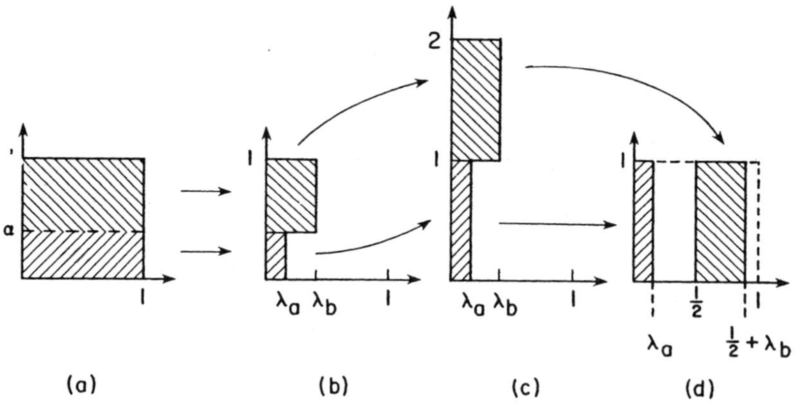

Fig. 7. The generalized baker's transformation. One iteration of the map takes us from (a) to (d). Steps (b) and (c) are conceptual intermediate stages.

at all subsequent iterates. Fig. 7 illustrates the action of this map on the unit square. As shown in fig. 7, we take α, λ_a, $\lambda_b \leqslant \frac{1}{2}$, and $\lambda_b \geqslant \lambda_a$. Fig. 8 shows the result of applying the map two times to the unit square. From fig. 8 it is seen that, if the x interval $[0, \lambda_\alpha]$ is magnified by a factor $1/\lambda_a$, it becomes a precise replica of fig. 7d. Similarly, if the x interval $[\frac{1}{2}, \frac{1}{2} + \lambda_b]$ is magnified by $1/\lambda_b$, a replica of fig. 7d again results. This self similarity property of eq. (12) will subsequently be used to obtain d_C, d_I, d_H, and d_p.

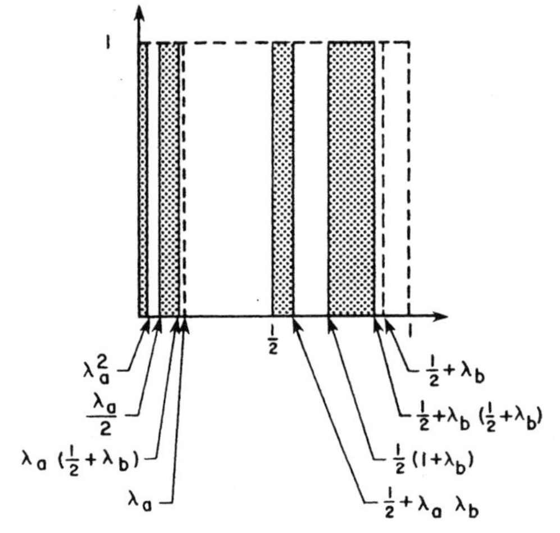

Fig. 8.

4.2. *Lyapunov numbers of generalized baker's transformation*

Now we consider the Lyapunov numbers. Eq. (12b) involves y alone and consists of a linear stretching on each of the y intervals $[0, \alpha]$ and $[\alpha, 1]$. Thus almost every y initial condition in $[0, 1]$ will generate an ergodic orbit in y with uniform density in $[0, 1]$. The Jacobian of eq. (12) is diagonal and depends only on y.

$$J = \begin{pmatrix} L_2(y) & 0 \\ 0 & L_1(y) \end{pmatrix},$$

where

$$L_2(y) = \begin{cases} \lambda_a, & \text{if } y < \alpha, \\ \lambda_b, & \text{if } y > \alpha, \end{cases}$$

and

$$L_1(y) = \begin{cases} \dfrac{1}{\alpha}, & \text{if } y < \alpha, \\[2ex] \dfrac{1}{1-\alpha}, & \text{if } y > \alpha. \end{cases}$$

Thus applying eq. (6) we have

$$\lambda_1 = \lim_{n\to\infty} [L_1(y_n) \ldots L_1(y_1)]^{1/n},$$

or

$$\log \lambda_1 = \lim_{n\to\infty} \left\{ \frac{n_\alpha}{n} \log \frac{1}{\alpha} + \frac{n_\beta}{n} \log \frac{1}{\beta} \right\},$$

where $\beta = 1 - \alpha$. n_α is the number of times the orbit has been in the set $y < \alpha$, and n_β is the number of times the orbit has been in the set $y > \alpha$. Since for *almost* any y_1, the orbit in y is ergodic with uniform density in [0, 1], $\lim_{n\to\infty} n_\alpha/n = \alpha$, and similarly $\lim_{n\to\infty} n_\alpha/n = \beta$. Thus

$$\log \lambda_1 = \alpha \log \frac{1}{\alpha} + \beta \log \frac{1}{\beta}. \qquad (13)$$

Similarly, we obtain for λ_2

$$\log \lambda_2 = \alpha \log \lambda_a + \beta \log \lambda_b. \qquad (14)$$

To simplify notation in this and subsequent expressions, let

$$H(\alpha) = \alpha \log \frac{1}{\alpha} + (1 - \alpha) \log \frac{1}{1 - \alpha}. \qquad (15)$$

$H(\alpha)$ is called the *binary entropy function* and is the amount of information contained in a coin-toss where heads has a probability α.

The Lyapunov dimension of the attractor of the generalized Baker's transformation (eq. (12)) is

$$d_L = 1 + \frac{H(\alpha)}{\alpha \log(1/\lambda_a) + \beta \log(1 + \lambda_b)}. \qquad (16)$$

In the following sections we compute the values of the dimensions defined in this paper, and show that all the probabilistic dimensions take on the value given in eq. (16).

For all but special values of λ_a, λ_b, and α, there exist unstable periodic orbits whose Lyapunov numbers are different from those given in eqs. (13) and (14)*. Thus, in general we expect that conjecture 2 rather than conjecture 3 applies and $d_F \neq d_\mu$.

* To see that for almost all parameter values the Lyapunov numbers of the generalized baker's transformation are not absolute, consider the special initial condition on the attractor with y-value $y_1 = y_a$ where $y_a = \alpha^2(1 - \alpha + \alpha^2)^{-1}$. This initial condition corresponds to one of the points on the unstable period 2 orbit, $(y_a, y_b, y_a, y_b, \ldots)$, where $y_b = \alpha^{-1} y_a$. Since $0 < y_a < \alpha < y_b < 1$, we have $n_\alpha = n_\beta = \frac{1}{2}$, and the Lyapunov numbers generated by this initial condition are $\lambda_1 = (\alpha\beta)^{-1/2}$ and $\lambda_2 = (\lambda_a \lambda_b)^{1/2}$, rather than those given by eq. (13) and (14).

4.3. *Capacity of generalized Baker's transformation*

To calculate d_C we first note that the attractor is a product of a Cantor set along x and the interval [0, 1] along y. Thus the capacity, or any of the other dimensions, are in the form $d_C \equiv 1 + \bar{d}_C$, where $\bar{d}_C$ is the dimension of the attractor in the x-direction. We will generally use a bar over a dimension to refer to the dimension along the x-direction.

We now obtain $\bar{d}_C$ by making use of the scaling property of the generalized Baker's transformation, discussed at the end of section 4.1. We write $N(\epsilon)$ as

$$N(\epsilon) = N_a(\epsilon) + N_b(\epsilon),$$

where $N_a(\epsilon)$ is the number of x-intervals of length ϵ needed to cover that part of the attractor which lies in the x-interval $[0, \lambda_a]$, and $N_b(\epsilon)$ is the analogous quantity for the x-interval $[\frac{1}{2}, \frac{1}{2} + \lambda_b]$. From the scaling property, $N_a(\epsilon) = N(\epsilon/\lambda_a)$, and similarly $N_b(\epsilon) = N(\epsilon/\lambda_b)$. Thus

$$N(\epsilon) = N(\epsilon/\lambda_a) + N(\epsilon/\lambda_b). \qquad (17)$$

Assuming heuristically that $N(\epsilon) \approx k\epsilon^{-\bar{d}_C}$ for small ϵ, substituting into eq. (17) gives

$$k\left(\frac{1}{\epsilon}\right)^{\bar{d}_C} = k\left(\frac{\lambda_a}{\epsilon}\right)^{\bar{d}_C} + k\left(\frac{\lambda_b}{\epsilon}\right)^{\bar{d}_C},$$

implying that

$$1 = \lambda_a^{\bar{d}_C} + \lambda_b^{\bar{d}_C}, \qquad (18)$$

which is a transcendental equation for $\bar{d}_C$. As

expected, eqs. (16) and (18) show that, in general, $1+\bar{d}_C = d_C \neq d_L$. However, for the special choice $\lambda_a = \lambda_b$, $\alpha = \frac{1}{2}$, corresponding to eq. (12) with $\lambda_a = \lambda_1 = 2$, the two agree. Note that for this case the Jacobian matrix is constant, the Lyapunov numbers are therefore absolute, and conjecture 3 applies.

In obtaining eq. (18), in order to keep the argument simple, we have made the strong assumption that $N(\epsilon) \approx k\epsilon^{-\bar{d}_C}$ for small ϵ, which implies the existence of the limit given in the definition of capacity, eq. (2). We can, however, show that the limit given in eq. (2) exists and $\bar{d}_C$ must satisfy eq. (18) in a rigorous manner, as follows:

Define $E_C(\epsilon)$ by

$$N(\epsilon) = E_C(\epsilon)\epsilon^{-\bar{d}},$$

where $\bar{d}$ is defined by $1 = \lambda_a^{\bar{d}} + \lambda_b^{\bar{d}}$. Substituting this into eq. (17) then yields

$$E_C(\epsilon) = \bar{\alpha} E_C\left(\frac{\epsilon}{\lambda_a}\right) + \bar{\beta} E_C\left(\frac{\epsilon}{\lambda_b}\right), \tag{19}$$

where $\bar{\alpha} = \lambda_a^{\bar{d}}$ and $\bar{\beta} = \lambda_b^{\bar{d}}$, and are independent of ϵ. Notice that by definition $\bar{\alpha} + \bar{\beta} = 1$, so the above expression says that $E_C(\epsilon)$ is a weighted average of its values at ϵ/λ_a and ϵ/λ_b. Choose ϵ_1 and ϵ_2 so that $\epsilon_1 > \epsilon_2 > 0$. Since $N(\epsilon)$ and hence $E_C(\epsilon)$ are finite and positive for any finite ϵ, there exist finite non-zero numbers $B_1 > B_2 > 0$ such that $B_2 < E_C(\epsilon) < B_1$ for $\epsilon_1 > \epsilon > \epsilon_2$. We can assume that ϵ_1 and ϵ_2 are chosen so that ϵ_1/ϵ_2 is large. Since $\bar{\alpha} + \bar{\beta} = 1$, eq. (19) implies that $B_2 < E_C(\epsilon) < B_1$ also applies to the wider interval $\epsilon_1 > \epsilon > \lambda_b\epsilon_2$. Repeating this argument increases the domain of validity of the bound to $\epsilon_1 > \epsilon > \lambda_b^2\epsilon_2$, and so on. Hence $E_C(\epsilon)$ is bounded uniformly from above and below for arbitrarily small ϵ. Thus the limit of eq. (2) exists and $\bar{d}_C = \bar{d}$. (In fact it can be shown that eq. (19) implies that $\lim_{\epsilon\to 0} E_C(\epsilon)$ is a constant if $\log\lambda_a/\log\lambda_b$ is an irrational number.) Note that in eq. (18), since both terms on the right-hand side are monotonically decreasing, d_C obtained from solving this equation is unique.

4.4. *Computation of Hausdorff dimension*

The Hausdorff dimension d_H can be calculated by an argument that is very similar to the one used above in computing the capacity. Let $\bar{d}_H \equiv d_H - 1$, the Hausdorff dimension along x. Applying the scaling property of the map to the quantity $l_d(\epsilon)$ (defined in section 2), we obtain

$$l_d(\epsilon) = (\lambda_a)^d l_d\left(\frac{\epsilon}{\lambda_a}\right) + (\lambda_b)^d l_d\left(\frac{\epsilon}{\lambda_b}\right).$$

Substituting $l_d(\epsilon) = E_H(\epsilon)\epsilon^{-(\bar{d}-d)}$ into the above equation, we again find that $E_H(\epsilon)$ satisfies eq. (19). Thus the limit $\epsilon \to 0$ yields $l_d = \infty$ or $l_d = 0$ for $d < \bar{d}_C$ or $d > \bar{d}_C$, respectively. Hence, as predicted in section 2, the Hausdorff dimension and capacity are equal, $d_H = d_C$.

4.5. *Calculation of information dimension*

The information dimension d_I can also be calculated by a scaling argument similar to that used above in computing the capacity. Once again, let $d_I = 1 + \bar{d}_I$ and express the summation for $I(\epsilon)$ in eq. (3) as the sum of contributions from the two strips in fig. 7d,

$$I(\epsilon) = I_a(\epsilon) + I_b(\epsilon). \tag{20}$$

The total probability contained in strip $[0, \lambda_a]$ is α, and that in strip $[\frac{1}{2}, \lambda_b + \frac{1}{2}]$ is β. Assuming that it takes $N(\epsilon)$ strips of width ϵ to cover the whole attractor, then from the scaling property of eq. (12), covering the strip $[0, \lambda_a]$ at resolution $\epsilon\lambda_a$ also requires $N(\epsilon)$ strips. Thus

$$I_a(\epsilon\lambda_a) = \sum_{i=1}^{N(\epsilon)} \alpha P_i \log\frac{1}{\alpha P_i}$$

$$= \alpha\left[\log\frac{1}{\alpha} + I(\epsilon)\right].$$

Hence, replacing $\epsilon\lambda_a$ by ϵ in the above,

$$I_a(\epsilon) = \alpha \log\frac{1}{\alpha} + \alpha I\left(\frac{\epsilon}{\lambda_a}\right),$$

$$I_b(\epsilon) = \beta \log\frac{1}{\beta} + \beta I\left(\frac{\epsilon}{\lambda_b}\right).$$

Thus

$$I(\epsilon) = \alpha I\left(\frac{\epsilon}{\lambda_a}\right) + \beta I\left(\frac{\epsilon}{\lambda_b}\right) + H(\alpha), \tag{21}$$

where $H(\alpha)$ is given by eq. (15). Motivated by eq. (3), if we assume that $I(\epsilon) = \bar{d}_I \log(1/\epsilon)$ for small ϵ, and substitute for $I(\epsilon)$, $I(\epsilon/\lambda_a)$, and $I(\epsilon/\lambda_b)$ in the above equation we obtain

$$\bar{d}_I = \frac{H(\alpha)}{\alpha \log(1/\lambda_a) + \beta \log(1/\lambda_b)},$$

which is in turn equal to $\bar{d}_L$. The assumption that $I(\epsilon) = \bar{d}_I \log(1/\epsilon)$ can be made rigorous in the limit as $\epsilon \to 0$ using an argument that is completely analogous to that used in deriving the capacity in the last part of subsection 4.3.

We should mention that Alexander and Yorke [11] have computed the Lyapunov and information dimensions of the generalized baker's transformation for the special case $\alpha = \frac{1}{2}$, $\lambda = \lambda_a = \lambda_b$, where $\lambda > \frac{1}{2}$. In this case $d_L = 2$. For uncountably many values of λ they find that also $d_I = 2$, although there are certain special values of λ for which $d_I < 2$.

In order to calculate the other probability dimensions listed in table I more information concerning the probability distribution is required. This is dealt with in section 5, and we therefore defer calculation of the remaining dimensions to the sections following section 5.

5. Distribution of probability

In this section we derive the form of the probability distribution $\{P_i(\epsilon)\}$ associated with the natural measure μ of the generalized baker's transformation. Here P_i denotes the probability of the ith cube C_i of edge ϵ, i.e., $P_i = \mu(C_i)$. The collection of numbers $\{P_i(\epsilon)\}$ may be also be thought of as the result of coarse graining the natural measure. This probability distribution is interesting both for its own sake, and because it is needed to compute some of the dimensions that we are interested in. In what follows we restrict ourselves to the case in which $\lambda_a = \lambda_b \equiv \lambda_2$, which keeps the width of all the strips the same. Thus a particularly convenient partition for computing $\{P_i\}$ is the set of 2^n nonempty strips obtained by iterating the unit square n times.

Starting with a uniform probability distribution, on one application of the map two strips are produced, one with total probability α and the other with total probability β. (See fig. 7d.) If the map is applied again (fig. 8), there results one strip of probability α^2, one of probability β^2, and two of probability $\alpha\beta$. In general, after n applications of the map, there result 2^n strips of width $(\lambda_2)^n$ and probabilities $\alpha^m\beta^{(n-m)}$, $m = 0, 1, 2, \ldots, n$. The number of strips with probability $\alpha^m\beta^{(n-m)}$ is

$$Z(n, m) = \frac{n!}{(n-m)!m!}, \tag{22}$$

i.e., the binomial coefficient. Since we take $\alpha < \frac{1}{2} < \beta$, lower m corresponds to more probable strips, i.e. strips of greater natural measure. The total probability contained in these $Z(n, m)$ strips is

$$W(n, m) \equiv \alpha^m\beta^{(n-m)}Z(n, m). \tag{23}$$

Note the similarity to a sequence of coin tosses. Using a coin with probability α of heads and β of tails, for a sequence of n flips the total number of sequences with m occurrences of heads is given by eq. (22), and the likelihood of all such sequences is given by eq. (23).

For large n (small ϵ) it is convenient to have smooth estimates for $Z(n, m)$ and $W(n, m)$. Using

Sterling's approximation, i.e.

$$\log n! = (n + \tfrac{1}{2}) \log(n + 1) - (n + 1) + \log(2\pi)^{1/2} + \mathcal{O}(n^{-1}),$$

we obtain from eq. (22)

$$\log Z \approx (n + \tfrac{1}{2}) \log(m + 1) - \log(2\pi)^{1/2} + 1.$$

Expanding this expression in a Taylor series about its maximum value, $m = n/2$, yields

$$Z(n, m) \approx \frac{2^n}{\sqrt{2\pi}} \sqrt{\frac{4}{n}} \exp\left\{ -\frac{1}{2}\left[\frac{4}{n}\left(m - \frac{n}{2}\right)^2\right]\right\}. \quad (24)$$

Similarly, from eq. (23), $W(n, m)$ is

$$W(n, m) \approx \frac{1}{\sqrt{2\pi n\alpha\beta}} \exp\left\{ -\frac{(m - n\alpha)^2}{2n\alpha\beta}\right\}. \quad (25)$$

Note that, since these expressions were obtained by Taylor series expansion, eq. (24) is only valid for $|m/n - \frac{1}{2}| \ll 1$, and eq. (25) is only valid for $|m/n - \alpha| \ll 1$. However, since the width of these Gaussians is $\mathcal{O}(1/n^{1/2})$, eq. (24) is valid for most of the strips, and eq. (25) is valid for most of the probability.

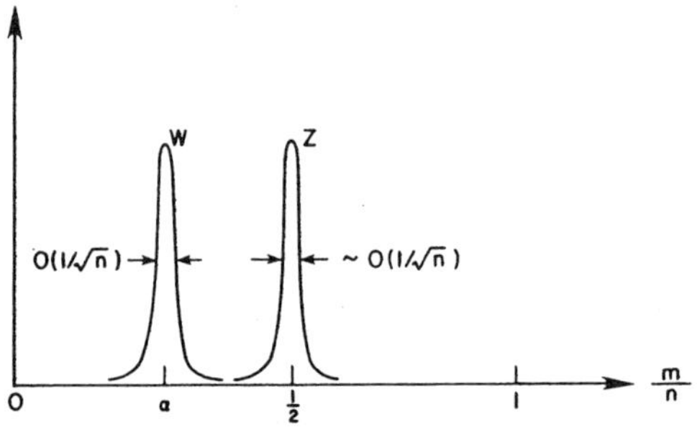

Fig. 9. A schematic representation of the distribution of probabilities on the attractor. $Z(n, m)$ is the number of cubes with probability $p = \alpha^m\beta^{(n-m)}$, and $W(n, m)$ is the sum of the probability contained in cubes of probability p. For large n and m/n close to its mean value, these are both approximately Gaussian distributions in m/n whose width is proportional to n. In the limit as $n \to \infty$, W and Z become delta functions, and no longer overlap.

Fig. 9 shows a schematic plot of Z and W. It is clear from this figure that, for large n, almost all of the probability is contained in a very small fraction of the total number of strips. Furthermore, the situation is accentuated as ϵ gets smaller (n gets larger), since the width of the Gaussians given in eqs. (24) and (25) decreases according to $n^{1/2}$. In the limit as $\epsilon \to 0$ these Gaussians approach delta functions, and they do not overlap. We feel that the above properties are typical features of chaotic attractors.

5.1. *Log-normal distribution of probabilities*

It is instructive to rewrite eq. (25) in another form. Let $p = \alpha^m\beta^{(n-m)}$ denote the probability of a strip, and reexpress eq. (25) in terms of $u = \log(1/p)$ rather than m. Noting that m is proportional to u, and letting $\epsilon = \lambda_2^n$ $W(n, m)$ becomes

$$F(u) = \frac{1}{\sqrt{2\pi}\sigma} e^{-(u-u_0)^2/2\sigma^2}, \quad (26)$$

where

$$\sigma^2 = \frac{[\alpha\beta(\log(\beta/\alpha))^2 \log(1/\epsilon)]}{\log(1/\lambda_2)}$$

and

$$u_0 = \bar{d}_L \log\frac{1}{\epsilon}, \quad (27)$$

with d_L given by eq. (16). Eq. (26) is only valid if

$$\frac{(u - u_0)^2}{\sigma^2} \ll \log\frac{1}{\epsilon}, \quad (28)$$

corresponding to $|m/n - \alpha| \ll 1$. $F(u)\mathrm{d}u$ is the total probability contained in strips whose values of $u = \log(1/p)$ fall between u and $u + \mathrm{d}u$. Thus we see that the values the logarithm of p asymptotically have a Gaussian distribution, or in other words, the values of p asymptotically have a log-normal distribution. We believe that this is typically true of chaotic attractors. In particular, we offer the following conjecture*:

Conjecture 4. Let A be a chaotic attractor of a p-dimensional *invertible* dynamical system, and assume that this attractor has a natural measure μ. Cover A with a fixed grid of p-dimensional cubes of side length ϵ. Assign each nonempty cube C_i probability $P_i = \mu(C_i)$, and let $U_i = \log(1/P_i)$. Let u_0 be the mean of the numbers U_i, and let σ^2 be the variance. For typical chaotic attractors, in the limit as $\epsilon \to 0$, values of U_i sufficiently close to the mean (in the sense of eq. (28)) approach a Gaussian distribution. In other words, the corresponding values of P_i approach a log-normal distribution.

Note that U_i is the information obtained in a measurement that finds the orbit inside of the ith cube [1, 9, 10]. Thus, conjecture 4 states that for chaotic attractors the information is approximately normally distributed for small ϵ.

The function $Z(n, m)$ given in eq. (24), can also be reexpressed in terms of p rather than m. When this is done, with similar restrictions to those of eq. (28), the result is also a Gaussian in terms of $\mu = \log(1/p)$. When recast in the more general setting of conjecture 4, this says that the number of cubes C_i whose values U_i lie between u and $u + du$ are given by a Gaussian distribution. (Similar restrictions to those given in conjecture 4 apply.)

6. Computation for the natural measure dimensions

In this section we verify conjectures 1 and 2 for the generalized baker's transformation by explicitly computing all of the probability dimensions defined in section 2. In order to simplify the computations, for all but the ϑ-Hausdorff dimension we restrict ourselves to the case in which $\lambda_a = \lambda_b \equiv \lambda_2$. For the ϑ-Hausdorff dimension we treat the most general case in which $\lambda_a \neq \lambda_b$, but are only able to obtain an upper bound for the dimension.

* The form of this conjecture was developed in collaboration with Erica Jen.

6.1. *Alternate derivation of information dimension*

Now that we know the probability distribution for the generalized baker's transformation for $\lambda_a = \lambda_b \equiv \lambda_2$, we can obtain the information dimension directly from its definition. From eq. (3) and eq. (26), $I(\epsilon)$ is the average value of $\log(1/P_i)$ or

$$I(\epsilon) = \int uF(u)\,du = u_0.$$

Since from eq. (27) $u_0 = d_L \log(1/\epsilon)$, eq. (3) yields $d_I = d_L$ (previously shown in section 4 for the more general case $\lambda_a \neq \lambda_b$). Thus the mean value of the log-normal distribution is simply the information contained in the probability distribution, and its scaling rate is the dimension of the nature measure, i.e., $I(\epsilon) \approx d_\mu \log(1/\epsilon)$.

6.2. *Determination of ϑ-capacity*

Here we calculate $d_C(\vartheta)$ for $\lambda_a = \lambda_b = \lambda_2$. We choose ϵ equal to the width of a strip, $\epsilon = \lambda_2^n$. As usual, for convenience we compute the ϑ-capacity of the attractor projected onto the x-axis, i.e. $\tilde{d}_C(\vartheta) = d_C(\vartheta) - 1$. The ϑ-capacity $\tilde{d}_C(\vartheta)$ is defined in terms of the minimum number of intervals $N(\epsilon; \vartheta)$ of width ϵ that have total natural measure at least ϑ,

$$N(\epsilon; \vartheta) = \sum_{m=0}^{m_\vartheta} Z(n, m), \tag{29}$$

where m_ϑ is the largest integer such that

$$\sum_{m=0}^{m_\vartheta - 1} W(n, m) \leqslant \vartheta, \tag{30}$$

To find m_ϑ we use eq. (25) and approximate the sum in eq. (30) by an integral,

$$\vartheta \approx \frac{1}{\sqrt{2\pi\alpha\beta n}} \int_{-\infty}^{m_\vartheta} e^{-(m-n\alpha)^2/2n\alpha\beta}\,dm.$$

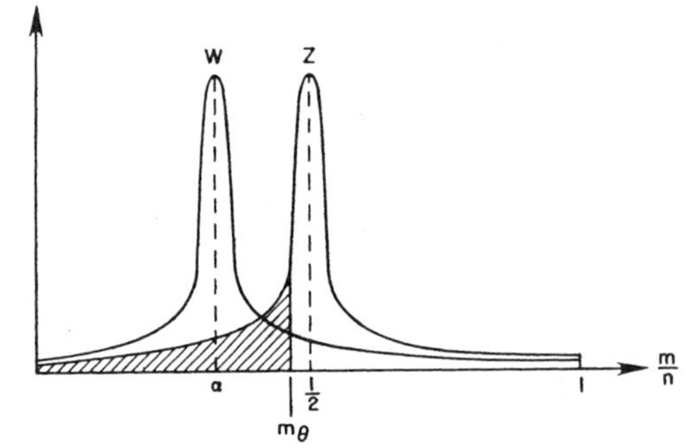

Fig. 10. The principal contribution to the sum needed to compute $N(\epsilon; \vartheta)$ (eq. (29)) comes from values of m near m_ϑ.

Thus for fixed ϑ we obtain

$$\frac{m_\vartheta}{n} \approx \alpha + \operatorname{erfc}^{-1}(\vartheta)\sqrt{\frac{\alpha\beta}{n}}, \tag{31}$$

where $\operatorname{erfc}(x) = (1/\sqrt{2\pi})\int_{-\infty}^{x} e^{-x^2/2}\,dx$. Now, consider eq. (29). The principal contribution to the sum will come from m values very close to m_ϑ, as depicted in fig. 10. Thus we use eqs. (23) and (25) to approximate $Z(n, m)$ as

$$Z(n, m) \approx \frac{\beta^{-n}(\beta/\alpha)^m}{\sqrt{2\pi n\alpha\beta}}\, e^{-(m-n\alpha)^2/2n\alpha\beta}. \tag{32}$$

The term $(\beta/\alpha)^m$ decreases as m decreases away from m_ϑ, and this decrease is very rapid compared to the variation of $e^{-(m-n\alpha)^2/2n\alpha\beta}$. Thus in performing the sum in eq. (29), we may approximate $e^{-(m-n\alpha)^2/2n\alpha\beta}$ as being constant and equal to its value at $m = m_\vartheta$. Hence the only m dependent term in the sum is $(\beta/\alpha)^m$. Since

$$\sum_{m=0}^{m_\vartheta}\left(\frac{\beta}{\alpha}\right)^m \approx \left(\frac{\beta}{\alpha}\right)^{m_\vartheta}\frac{\beta}{(\beta-\alpha)},$$

we find that

$$N(\epsilon; \vartheta) \sim \beta^{-(n-m_\vartheta)}\alpha^{-m_\vartheta} n^{-1/2}.$$

From eq. (4) and $n = \log(1/\epsilon)/\log(1/\lambda_2)$, the above estimate of $N(\epsilon; \vartheta)$ yields $d_C(\vartheta) = d_L$, in agreement with conjecture 2.

6.3. *Computation of ϑ-Hausdorff dimension*

In this section we obtain an upper bound on the ϑ-Hausdorff dimension of the generalized baker's transformation with $\lambda_a \neq \lambda_b$. (Recall that for our work in the previous section we took $\lambda_a = \lambda_b$.) We obtain an inequality for the ϑ-Hausdorff dimension by using a specific covering along x to compute the sum

$$l_d^*(\epsilon, \vartheta) = \sum_i \epsilon_i^d,$$

where the $\epsilon_i < \epsilon$ cover a fraction ϑ of the natural measure of the attractor. Our choice for the ϵ_i is specified below. Taking the limit as $\epsilon \to 0$, we find that there is a value of d at which $l_d^*(\epsilon, \vartheta)$ crosses over from ∞ to 0. For the partition we have chosen, we find that crossover occurs at $d = \bar{d}_L$. We believe that the value we obtain is in fact the true ϑ-Hausdorff dimension. However, we cannot be sure that the particular covering we have chosen gives the lowest possible value of d, and thus we can only say that we have obtained an upper limit.

After n iterations of the map, an initially uniform probability distribution in the unit square is transformed to 2^n strips with widths $\lambda_a^m\lambda_b^{(n-m)}$ and probabilities $\alpha^m\beta^{(n-m)}$, $m = 0, 1, 2, \ldots, n$. As shown in eq. (22), the number of such strips is $Z(n, m)$. We shall choose the ϵ_i to cover the most probable strips so that ϑ of the total probability is covered. ϵ for our covering is equal to the width of the widest strip, which is either $(\lambda_a)^n$ or $(\lambda_b)^n$, whichever is larger. Letting $U_d(n, m)$ be

$$U_d(n, m) = (\lambda_b^{m-n}\lambda_a^m)^d Z(n, m),$$

we have that

$$l_d^*(\epsilon, \vartheta) = \sum_i \epsilon_i^d = \sum_m U_d(n, m), \tag{33}$$

We still have yet to specify which m values are to be included in the sum. To do this, we expand $U_d(n,m)$ about its maximum value (as done for Z and W in section 5), and obtain

$$U_d(n,m) \approx \frac{[\lambda_a^d + \lambda_b^d]^n}{\sqrt{2\pi n \dfrac{\lambda_a^d \lambda_b^d}{(\lambda_a^d+\lambda_b^d)(\lambda_a^d+\lambda_b^d)}}} \times \exp -\frac{1}{2}\left\{\frac{\left[\dfrac{m}{n} - \dfrac{\lambda_a^d}{\lambda_a^d+\lambda_b^d}\right]^2}{\dfrac{\lambda_a^d\lambda_b^d}{n(\lambda_a^d+\lambda_b^d)(\lambda_a^d+\lambda_b^d)}}\right\} \qquad (34)$$

In order to compute $l_d^*(\epsilon, \vartheta)$, we must consider the natural measure as well as $U_d(n,m)$. Note that for the general case we are considering now with $\lambda_a \neq \lambda_b$, $W(n,m)$ obtained in eq. (25) continues to be the correct expression for the distribution of probabilities in each strip. Depending on the values of α, d, λ_a, and λ_b, W may peak at a value of m that is smaller, larger, or equal to the value of m at the peak of U_d. Comparing the location of the peaks of the Gaussians in eq. (34) (for U) and in eq. (25) (for W), we see that there are three cases:

Case 1: $\alpha < \dfrac{\lambda_a^d}{(\lambda_a^d+\lambda_b^d)}$,

Case 2: $\alpha > \dfrac{\lambda_a^d}{(\lambda_a^d+\lambda_b^d)}$,

Case 3: $\alpha = \dfrac{\lambda_a^d}{(\lambda_a^d+\lambda_b^d)}$.

Cases 1 and 2 may be shown to be equivalent as follows. From the case 2 inequality and the fact that $\alpha + \beta = 1$, we obtain $\beta < \lambda_b^d/(\lambda_a^d + \lambda_b^d)$. But, if we define m' by $m = n - m'$, and change the sums over m to sums over m', then the roles of (α, λ_a) and (β, λ_b) are interchanged, and case 2 is converted to case 1. We shall not consider case 3 here; suffice it to say that it does not alter the results obtained from consideration of cascs 1 and 2. Therefore it is sufficient to compute the ϑ-Hausdorff dimension for case 1.

For case 1, selecting the best covering of intervals that contain ϑ of the total probability is easy. Since W remains valid, we get a covering that includes ϑ of the total natural measure by including intervals whose value of m is less than m_ϑ, just as we did for the computation of ϑ-capacity. Furthermore, since U_d peaks at a larger value of m/n than W does, this selection gives the smallest value of l_d^*. The situation is analogous to the computation of ϑ-capacity, except that here the role of Z is played by U_d (cf. fig. 10). To evaluate

$$l_d^* = \sum_{m=0}^{m_\vartheta} U_d(n,m), \qquad (35)$$

we note that, as for the analogous evaluation for ϑ-capacity in the previous subsetion, the principal contribution to the sum comes from m-values close to m_ϑ. Thus we approximate $U_d(n,m)$ as

$$U_d(n,m) \approx \frac{\lambda_a^m \lambda_b^{n-m}}{\alpha^m \beta^{(n-m)}} W(n,m),$$

with W approximated by eq. (25). Proceeding as in section 6.2 we obtian an estimate for $l_d^*(\epsilon, \vartheta)$,

$$l_d^*(\epsilon,\vartheta) \sim n^{-1/2}\left(\frac{\beta}{\lambda_b^d}\right)^{m_\vartheta - n}\left(\frac{\alpha}{\lambda_a^d}\right)^{-m_\vartheta},$$

or

$$\log[l_d^*(\epsilon,\vartheta)] \approx -n[d - (\bar{d}_L)]\log\left(\frac{1}{\lambda_2}\right).$$

For $\epsilon \to 0$ (i.e., $n \to \infty$) we obtain $l_d^*(\vartheta) = 0$ for $d > \bar{d}_L$ and $l_d^*(\vartheta) = \infty$ for $d < \bar{d}_L$. Thus remembering that $d_H(\vartheta) = \bar{d}_H(\vartheta) + 1$,

$$d_H(\vartheta) \leqslant d_L. \qquad (36)$$

As already mentioned, we expect that the above inequality is really an equality. This expectation is reinforced by the fact that when $\vartheta = 1$ we recover the exact expression for the Hausdorff dimension computed in eq. (18). To see that this is true, replace m_9 in eq. (35) by n. From the form of U_d, this sum is simply the binomial expansion of $(\lambda_a^d + \lambda_b^d)^n$. As $n \to \infty$, this quantity is 0 or ∞ for $d > \bar{d}_H$ or $d < \bar{d}_H$, where $\bar{d}_H$ satisfies $\lambda_a^{\bar{d}_H} + \lambda_b^{\bar{d}_H} = 1$, which is the same as eq. (18). That is, for the specific choice of ϵ_i that we have used, we obtain the correct value of d_H. Since the same choice of the ϵ_i was used in obtaining $d_H(\vartheta)$, it seems plausible that the equality might apply in eq. (36).

6.4. *Computation of the pointwise dimension*

We now consider the pointwise dimension for the generalized baker's transformation with $\lambda_a = \lambda_b < \frac{1}{2}$, and we show that d_p exists and is equal to d_L.

As previously noted in section 5, application of the map n times to the unit square produces 2^n strips of widths $(\lambda_a)^n$. (Recall that we are assuming $\lambda_a = \lambda_b$.) In order to compute the pointwise dimension, we choose a point x at random with respect to the natural measure μ, compute the natural measure contained in an ϵ ball centered about x, (i.e. $\mu(B_\epsilon(x))$), and compute the ratio of $\log \mu(B_\epsilon(x))$ to $\log \epsilon$ in the limit as ϵ goes to zero (cf. eq. (5)). The simplest case for this computation occurs when $\lambda_a < \frac{1}{4}$, so that the gaps between strips are bigger than the strips themselves, as pictured in fig. 11a. Choosing a point x at random with respect to the natural measure μ, let S_n denote the nth order strip of width $(\lambda_a)^n$ that the point x lies in. Letting $\epsilon = (\lambda_a)^n$, the natural measure contained in a ball of

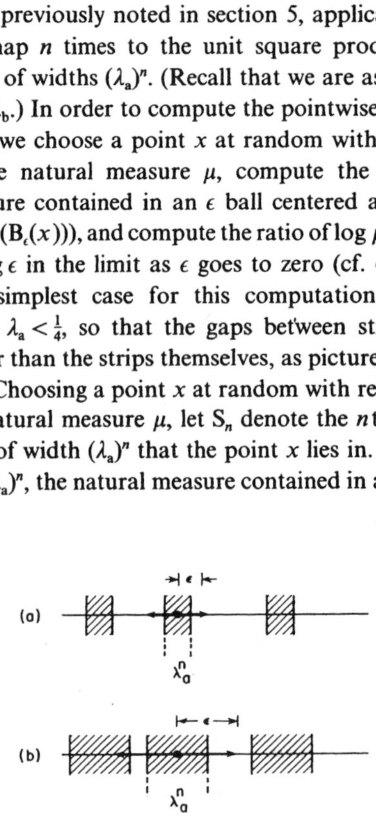

Fig. 11. (a) Computing the pointwise dimension for the case that $\lambda_a < \frac{1}{4}$. (b) The case $\lambda_a > \frac{1}{4}$, in which the computation is a little more complicated.

radius ϵ around x (i.e., the x-interval $[x - (\lambda_a)^n, x + (\lambda_a)^n]$) will be equal to the natural measure of the strip S_n, regardless of where in the strip x lies. (See fig. 11a.) The natural measure contained in a given strip is $\alpha^m \beta^{(n-m)}$, where $n \geqslant m \geqslant 0$, where m depends on the particular strip that x happens to lie in. (See section 5.) Thus, we have

$$\lim_{n\to\infty} \frac{\mu(B_\epsilon(x))}{\log \epsilon} = \lim_{n\to\infty} \frac{\log \mu(S_n)}{n \log \lambda_a}$$

$$= \lim_{n\to\infty} \frac{m \log \alpha + (n-m) \log \beta}{n \log \lambda_a}. \tag{37}$$

In the limit as n grows large, as shown in section 5 (see fig. 9), the total probability $W(n, m)$ contained in strips of a given m value is distributed as a Gaussian centered about $m/n = \alpha$. Thus, in the limit as $n \to \infty$ it becomes overwhelmingly likely that $m/n = \alpha$. Thus for almost every x with respect to the natural measure μ, $\lim_{n\to\infty} m/n = \alpha$. (This is just a statement of the law of large numbers.) Putting this into eq. (37) gives

$$d_p = \lim_{n\to\infty} \frac{\mu(B_\epsilon(x))}{\log \epsilon} = \frac{\alpha \log \alpha + \beta \log \beta}{\log \lambda_a}$$

$$= \frac{H(\alpha)}{\log(1/\lambda_a)} = \bar{d}_L. \tag{38}$$

(See eqs. (15) and (16).)

To extend this computation of the pointwise dimension to the case that $\frac{1}{2} > \lambda_a > \frac{1}{4}$, for any $\lambda_a < \frac{1}{2}$ choose a k such that $\lambda_a^{k+1} \leqslant \frac{1}{2} - \lambda_a$ (e.g., for $\lambda_a \leqslant \frac{1}{4}$ this relation is satisfied for any $k \geqslant 0$; for $\lambda_a \leqslant 0.365\ldots$, for any $k \geqslant 1$; etc.). Then we can show $\mu(B_\epsilon(x)) \leqslant \alpha^{-k} \mu(S_n)$, where without loss of generality, we have assumed $\alpha \leqslant \beta$. Since $B_\epsilon(x) \supset S_n$, we have also $\mu(B_\epsilon(x)) \geqslant \mu(S_n)$. Thus $\mu(S_n) \leqslant \mu(B_\epsilon(x)) \leqslant \alpha^{-k} \mu(S_n)$ which with eq. (37) yields $d_p = \bar{d}_L$. (Our evaluation of eq. (37) holds not for $\epsilon \to 0$ but rather holds for the restricted set of $\epsilon = \lambda_a^n$, $n = 1, 2, \ldots$, however, it is not hard to show that that in fact implies eq. (38) for every sequence of ϵ going to 0.)

Thus we have shown that for the generalized baker's transformation the pointwise dimension is equal to the dimension of the natural measure. (Although we have only shown this for $\lambda_a = \lambda_b$, it is not hard to extend this result to $\lambda_a \neq \lambda_b$.)

7. The core of attractors

As shown in section 5, for the generalized baker's transformation, typically almost all of the probability is contained in a very small fraction of the total number of cubes needed to cover the attractor. In the limit as ϵ goes to zero, this fraction goes to zero. Thus, the natural measure of the attractor is *concentrated* on a subset of the attractor. We will call this subset the *core* of the attractor.

To get a better feel for why this comes about, and to see how the properties of the core are related to those of the attractor and its natural measure, consider the special case of the generalized baker's transformation where $\lambda_a = \lambda_b = \frac{1}{2}$. As we have already seen, at the nth level of approximation the natural measure consists of 2^n vertical strips of probability $\alpha^m \beta^{n-m}$. For large n and $\beta > \alpha$, a small fraction of the strips whose m values are close to αn contain much more of the natural measure than all other strips. Fig. 12 shows a plot

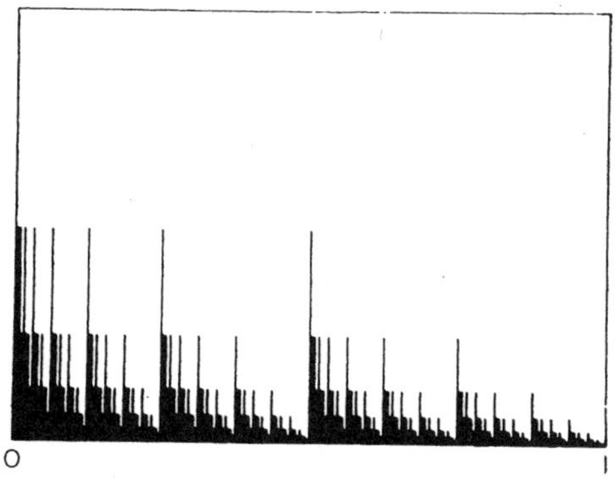

Fig. 12. The natural probability distribution of the generalized baker's transformation projected onto the x-axis, and coarse grained using intervals of width $\epsilon = 2^{-10}$. In this case $\alpha < \frac{1}{2}$, and $\lambda_a = \lambda_b = \frac{1}{2}$.

of the nth level approximation to the probability distribution as a function of x with $\lambda_a = \lambda_b = \frac{1}{2}$, and $\alpha < \frac{1}{2}$ and $n = 10$. The probability distribution looks as though it were made up of spikes, showing that already at $n = 10$ the natural measure has become quite concentrated in certain cubes (in this case intervals).

To understand the form of this probability distribution, it is instructive to represent the probability distribution of these strips in terms of x rather than m. To do this, approximate x using its first n binary digits, i.e. as a binary decimal truncated after n digits. Let m be the number of ones contained in the first n digits of the binary expansion of x. The natural measure of the strip $S_n(x)$ containing x is then $\mu(S_n(x)) = \alpha^m \beta^{(n-m)}$. (See the discussion at the beginning of section 5.) As we have already shown (see fig. 9), when written in terms of m, for large n the natural measure is approximately a Gaussian centered about αn, and in the limit where n is large almost all the measure is contained in strips with $m \approx \alpha n$. In other words, the natural measure of the generalized baker's transformation for $\lambda_a = \lambda_b = \frac{1}{2}$ is concentrated on those values of x that have 1's in their binary expansions in the fraction α, or equivalently, 0's in the fraction β. In the limit $n \to \infty$, *all* the natural measure is contained in this set, which we will call the core of this attractor.

For this case ($\lambda_a = \lambda_b = \frac{1}{2}$) the attractor is the entire unit square. The core of this attractor is dense on the attractor. In other words, any point of the attractor has points of the core arbitrarily close to it. Hence any covering of the core must also be a covering of the attractor, and vice versa. Thus the capacity of the core is the same as that of the attractor. The Hausdorff dimension, in contrast, is more subtle, and in fact, computing the Hausdorff dimension of the set of numbers whose binary expansions have a given fraction of ones is a classic problem in the study of Hausdorff dimension [31]. The Hausdorff dimension of this set is

$$d_H = \frac{H(\alpha)}{\log 2}.$$

(See eq. (15).) This result was conjectured by Good in 1941 [32] and proved by Eggleston in 1949 [33]. Also, the Hausdorff dimension of a very similar example (involving ternary rather than binary expansions) was proven by Besicovitch in 1931 [34].

Thus, for this example we see that the Hausdorff dimension of the core is equal to the dimension of the natural measure, and the capacity of the core is equal to the fractal dimension of the attractor (cf. eq. (16)). For the case of diffeomorphisms of the plane, the former result has been proven by Young [12]. We suspect that this is a property of typical attractors.

8. An attractor that is a nowhere differentiable torus

This section contains a review of the work of Kaplan, Mallet-Paret, and Yorke [35] on the dimension of a chaotic attractor in a setting that is quite different from that of the generalized baker's transformation. The attractor described below has the same topological form as a torus, and yet is nowhere differentiable, thus providing an interesting example of the nonanalytic forms that can be produced by chaotic dynamics.

Consider the following map:

$$\begin{aligned} x_{n+1} &= 2x_n + y_n \mod 1, \\ y_{n+1} &= x_n + y_n \mod 1, \\ z_{n+1} &= \lambda z_n + p(x_n, y_n). \end{aligned} \tag{39}$$

where x and y are taken mod 1, z can be any real number, and p is periodic in x and y with period 1 and is at least five times differentiable. (For example, $p(x, y) = \cos 2\pi x$.) In order to keep z bounded, λ is chosen between 0 and 1. Note that the eigenvalues and eigenvectors of the Jacobian matrix of eq. (39) are independent of x, y, and z. Thus *every* initial condition has the same Lyapunov numbers, i.e., the Lyapunov numbers are absolute, so that in this case conjecture 3 is relevant, and we expect that the fractal dimension and the dimension of the natural measure should be equal.

The equations for x and y are independent of z, and in fact are the classic Anasov or "cat" map [36],

$$\begin{pmatrix} x_{n+1} \\ y_{n+1} \end{pmatrix} = A \begin{pmatrix} x_n \\ y_n \end{pmatrix} \mod 1,$$

where

$$A = \begin{pmatrix} 2 & 1 \\ 1 & 1 \end{pmatrix}.$$

Thus, the x–y dynamics are chaotic, and are unaffected by the value of z.

To understand the shape of the attractor in the z-direction, put a sample initial condition into eq. (39). For example, consider $(x_0, y_0, 0)$. z_n takes on the form

$$z_n = \sum_{k=1}^{n} \lambda^{k-1} p(x_{n-k}, y_{n-k}).$$

Making use of the fact that $\binom{x_{n-k}}{y_{n-k}} = A^{-k}\binom{x_n}{y_n}$, and letting n go to infinity, it can be shown that the surface given by

$$z(x, y) = \sum_{k=1}^{\infty} \lambda^{k-1} p\left(A^{-k}\begin{pmatrix} x \\ y \end{pmatrix}\right),$$

is invariant and is the unique attractor of this dynamical system.

$z(x, y)$ has some very interesting properties. For $\lambda < 1/R$, where $R = (3 + \sqrt{5})/2$, $z(x, y)$ is smooth and has dimension 2. If $\lambda > 1/R$, however, for most choices of p, $z(x, y)$ is nowhere differentiable. A typical cross section of $z(x, y)$ is shown in fig. 13. To understand intuitively how the nondifferentiability of $z(x, y)$ comes about, notice that $z(x, y)$ is the sum of an infinite number of periodic functions whose arguments are the successive iterates of the cat map. Unless λ is small enough to diminish the effect of higher order iterates, the value of the sum can swing wildly as x or y change.

The Lyapunov numbers of the map given in eq.

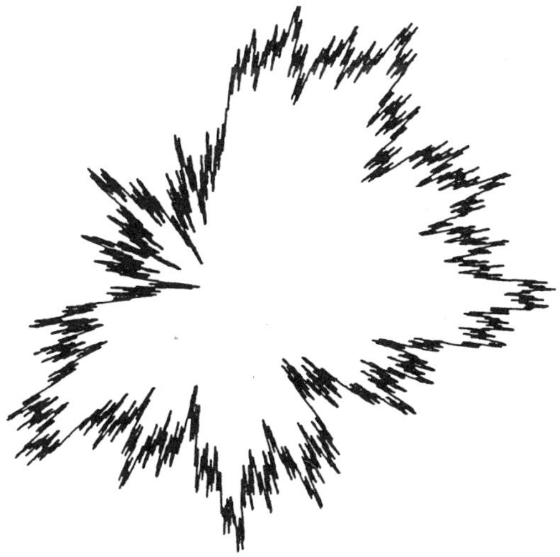

Fig. 13. A cross section of a nowhere differentiable torus, made using eq. (39) with $p(x, y) = \cos 2\pi x$ and λ chosen so that $d_C = 2\frac{1}{2}$.

(39) are $\lambda_1 = R$, $\lambda_2 = \lambda$, and $\lambda_3 = 1/R$, where $R = (3 + \sqrt{5})/2$, as given above. Kaplan, Mallet-Paret, and Yorke [35] have shown that there are two possibilities for the dimension of $z(x, y)$: Either

(i) $z(x, y)$ is nowhere differentiable and

$$d_C = d_L$$

or

(ii) $z(x, y)$ is differentiable and $d_C = 2$.

For given p, the nowhere differentiable case occurs for nearly every choice of λ. Thus we see that conjecture 3 is satisfied for this example.

9. Numerical computations

In this section we discuss some aspects of the numerical computation of dimension. First we will discuss the basic ideas behind numerical computations of dimension, secondly we will discuss some of the problems encountered in such computations, and finally we will review some previous numerical work.

The methods to compute dimension vary considerably depending on the dimension that one wishes to compute. Thus far, we are aware of numerical computations only of capacity [37–41], Lyapunov dimension [37–40], and Hausdorff dimension [42]. Of these, only the studies involving the capacity and the Lyapunov dimension were applied to attractors of dynamical systems. In each case, the computations follow from the definitions. As we shall see, the capacity is (in principle) straightforward to compute, but is in practice unfeasible to compute for all but very low dimensional attractors. The Lyapunov dimension, in contrast, is much more feasible to compute. We will begin the discussion with a description of the computation of Lyapunov dimension, and then go on to discuss the computation of capacity.

9.1. *Numerical computation of Lyapunov dimension*

The Lyapunov dimension is defined in terms of the Lyapunov numbers. (See section 3.) Thus, the work involved in computing Lyapunov dimension is in computing the Lyapunov numbers. Numerical methods for doing this have been discussed by Bennetin et al. [43], Shimada and Nagashima [44], and in infinite dimensions by Farmer [38]. With appropriate numerical caution, the largest k Lyapunov numbers can be computed by following the evolution of k nearby trajectories simultaneously and measuring their rate of separation. There are various numerical problems with this method, however, and a better method is to follow only one trajectory, but also follow k trajectories of the associated equations for the evolution of vectors in the tangent space. These methods have been successfully used in a variety of numerical studies.

For low-dimensional cases, such as two-dimensional maps or systems of three autonomous ordinary differential equations, with a modern computer and plenty of computer time, numerical computation of the dimensions we discuss here directly from their definitions is feasible, as discussed in the next subsection. Even in such low-dimensional cases, however, the computation of Lyapunov dimension is by far less costly in terms of computer time and memory than the computation of other dimensions. For higher dimensional attractors it appears that only the Lyapunov dimension is computationally feasible. The key reason that the Lyapunov dimension is feasible to compute numerically even for attractors of rather high dimension (e.g. $d_L \approx 10$) is that the difficulty of the computation scales linearly with the dimension of the attractor times the dimension of the space it lies in, rather than exponentially as it does for a computation of the fractal dimension, or any of the other dimensions discussed in this paper. The memory needed to compute the largest j Lyapunov numbers is equal to the memory needed to numerically integrate the equations under study, multiplied by $j+1$. (Memory requirements are usually a problem only in computations involving partial differential equations.) The computer time needed is the time needed to compute a time average to the desired accuracy (which depends, among other things, on the irregularity of the natural measure of the attractor), multiplied by $j+1$. Fortunately it is only necessary to compute the largest Lyapunov numbers, and the number of these needed depends on the dimension of the attractor rather than the dimension of the phase space. (See eq. (9).) This linear dependence on the dimension of the attractor has allowed computation of the Lyapunov dimension for attractors of dimension as large as twenty [38].

We should mention one disadvantage concerning Lyapunov dimension. Namely, it is not presently known how the Lyapunov dimension can be determined directly from a physical experiment. The difficulty comes about because, in some sense, in order to determine Lyapunov numbers it is necessary to be able to follow adjacent trajectories. To determine all the necessary Lyapunov numbers, it is necessary to follow some trajectories (at least one) which are not on the attractor. Thus it is not possible to compute the Lyapunov dimension by simply observing behavior on the attractor; one must perturb the system from the attractor, and do so in a very well defined way. This poses a very severe problem in the computation of dimension from experimental data, one that is not present in the computation of other dimensions.

9.2. *Computation of fractal dimension*

In principle, it is quite straightforward to use the definition of capacity, eq. (2), to compute the fractal dimension. The region of phase space surrounding the attractor is divided up into a grid of cubes of size ϵ, the equations are iterated, and the number of cubes $N(\epsilon)$ that contain part of the attractor are counted. ϵ is decreased and the process is repeated. If $\log N(\epsilon)$ is plotted against $\log \epsilon$, in the limit as ϵ goes to zero the slope is the fractal dimension.

The difficulty with this method is that one must use values of ϵ small enough to insure that the asymptotic scaling has been reached. (See Froehling et al. [40] and Greenside et al. [39].) The total number of cubes containing part of the attractor scales roughly as

$$N(\epsilon) \sim \epsilon^{-d_C}. \tag{40}$$

Thus, the number of cubes increases *exponentially* with the fractal dimension of the attractor. To get a feel for the seriousness of this problem, plug in some typical numbers: If $\epsilon = 0.01$ and $d_C = 3$, then $N \approx 10^6$, exceeding the core memory of all but the biggest current computers. Thus, computations of fractal dimension are currently not feasible for attractors of dimension significantly greater than three.

In addition, there is another potential problem involved in computing capacity. In counting cubes, how can one be sure that all the nonempty cubes have been counted? This problem is compounded by the highly nonuniform distribution of probability on an attractor. In particular, if our hypothesis that the probability is distributed log-normally is correct, in order to count the highly improbable cubes present in the wings of the distribution requires that a large number of points on the attractor must be generated. Furthermore, this number increases rapidly as ϵ decreases.

The conclusion is that a great deal of care must be taken in the computation of fractal dimension, and in particular, a sufficiently large number of points on the attractor must be generated to insure that low probability cubes are not left out in the determination of $N(\epsilon)$.

Although there are as yet no extensive results on direct computations of the dimension of the natural measure, it may be easier to reliably compute than the fractal dimension. The reason for this is that very improbable cubes are irrelevant for a computation of the dimension of the natural measure. Numerical experiments on this topic are currently in progress.

9.3. *Summary of past numerical experiments*

In this section we summarize previous numerical experiments on dimension computation. The two studies most relevant to the topic under discussion are those of Russel et al. [37] and Farmer [38]. Both of these were made in an attempt to test the Kaplan–Yorke conjecture [8, 22]. (See section 3.) In both of these studies, the capacity of chaotic attractors was computed directly from the definition. The Lyapunov dimension was also computed, and compared to the capacity.

In the study of Russel et al., five examples were examined. In each case, the computed capacity agree with the computed Lyapunov dimension to within experimental accuracy. These computations were done on the Crayl, a state of the art mainframe computer; at the smallest value of $\epsilon = 2^{-14}$, more than 10^5 cubes were counted.

The numerical experiments of Farmer were done using high-dimensional approximations to an infinite dimensional dynamical system. Because the equations under study were more time consuming to integrate, and because the capacity computations were done on a minicomputer, it was only possible to achieve about two significant figures of accuracy. The computed capacity and Lyapunov dimension agreed to this accuracy at the two parameter values tested.

In 1980, Mori [23] conjectured an alternate formula relating the fractal dimension to the spectrum of Lyapunov numbers. For attractors in a low-dimensional phase space, such as those studied by Russel et al. [37], Mori's formula and the Kaplan–Yorke formula (eq. (9)) predict the same value. For higher dimensional phase spaces, however, the two formulas no longer agree. Farmer's results support the Kaplan–Yorke formula.

One puzzling aspect of both of these numerical experiments is the striking agreement between the computed value of capacity and the Lyapunov dimension. The Kaplan–Yorke conjecture equates the Lyapunov dimension to the dimension of the natural measure, and therefore only gives a lower bound on the fractal dimension. Why, then, was such good agreement obtained between the computed capacity and the computed Lyapunov dimension? We do not yet understand the answer to this question, though further numerical experiments may resolve the question.

10. Conclusions

We have given several different definitions of dimension. These divide into two types, those that require a probability measure for their definition, and those that do not. (Refer back to table I.) For an example that we believe is typical of chaotic attractors, i.e., the generalized baker's transformation, our computations of dimension show that all of the probabilistic definitions take on one value, which we call the dimension of the natural measure, while the definitions that do not require a probability measure take on another value, which we call the fractal dimension of the attractor. We believe that this is true for typical attractors.

If the probability distribution on the attractor is "coarse grained" by covering the attractor with cubes, for the generalized baker's transformation we find that the probability contained in these cubes is distributed nearly log-normally when the cubes are sufficiently small. In other words, the total probability contained in cubes whose natural measure is between $u = \log p_i$ and $u + du$ has a distribution that is nearly Gaussian, and as the size of the cubes is decreased, it becomes more nearly Gaussian. Furthermore, the number of cubes in a given interval of u also has a Gaussian distribution, but with a different mean and variance. (See fig. 9.) As ϵ decreases, both of these distributions become narrower in a relative sense, in that the ratio of their variance to their mean decreases. In the limit as ϵ goes to zero, both distributions approach delta functions; since their means are different, in this limit the two distributions typically do not overlap. Thus, almost all of the natural measure is contained in almost none of the cubes, and the natural measure is concentrated on a core set. The capacity of the core is the fractal dimension of the attractor, while the Hausdorff dimension of the core is the dimension of the natural measure. Once again, although we have demonstrated the results mentioned in this paragraph only for the generalized baker's transformation, we feel that they are true for typical chaotic attractors.

Most of the dimensions that we have defined are difficult to compute numerically. The Lyapunov dimension, however, is much easier to compute numerically than any of the other dimensions. We compute the Lyapunov dimension for the generalized baker's transformation, and show that it is equal to the dimension of the natural measure obtained from any of the other probabilistic dimensions that we have investigated. This supports the conjecture of Kaplan and Yorke [22].

References

[1] R. Shaw, "Strange Attractors, Chaotic Behavior, and Information Flow", Z. Naturforsch. 36a (1981) 80.

[2] E. Ott, "Strange Attractors and Chaotic Motions of Dynamical Systems", Rev. Mod. Phys. 53 (1981) 655.

[3] R. Helleman, "Self-Generated Chaotic Behavior in Nonlinear Mechanics", Fundamental Problems in Stat. Mech. 5, E.G.D. Cohen, ed. (North-Holland, Amsterdam and New York, 1980) 165–233.

[4] J.A. Yorke and E.D. Yorke, "Chaotic Bahavior and Fluid Dynamics", Hydrodynamic Instabilities and the Transition to Turbulence, H.L. Swinney and J.P. Gollub eds., Topics in Applied Physics 45 (Springer, Berlin, 1981) pp. 77–95.

[5] B. Mandelbrot, Fractals: Form, Chance, and Dimension (Freeman, San Francisco, 1977).

[6] B. Mandelbrot, The Fractal Geometry of Nature (Freeman, San Francisco, 1982.)

[7] Ja. Sinai, "Gibbs Measure in Ergodic Theory", Russ. Math. Surveys 4 (1972) 21–64.

[8] P. Frederickson, J. Kaplan, E. Yorke, and J. Yorke, "The Lyapunov Dimension of Strange Attractors", J. Diff. Eqns. in press.

[9] J.D. Farmer, "Dimension, Fractal Measures, and Chaotic Dynamics", Evolution of Order and Chaos, H. Haken, ed., (Springer, Berlin, 1982), p. 228.

[10] J.D. Farmer, "Information Dimension and the Probabilistic Structure of Chaos", first chapter of UCSC doctoral disseration (1981) and Z. Naturforsch. 37a (1982) 1304–1325.

[11] J. Alexander and J. Yorke, "The Fat Baker's Transformation", U. of Maryland preprint (1982).

[12] L.S. Young, "Dimension, Entropy, and Lyapunov Exponents", to appear in Ergodic Theory and Dynamical Systems.

[13] W. Hurwicz and H. Wallman, Dimension Theory (Princeton Univ. Press, Princeton, 1948).

[14] A.N. Kolmogroov, "A New Invariant for Transitive Dynamical Systems", Dokl. Akad. Nauk SSSR 119 (1958) 861–864.

[15] Hausdorff, "Dimension und Außeres Maß", Math. Annalen. 79 (1918) 157.

[16] R. Bowen and D. Ruelle, "The Ergodic Theory of Axiom-A Flows", Inv. Math. 29 (1975) 181–202.

[17] J. Balatoni and A. Renyi, Publ. Math. Inst. of the Hungarian Acad. of Sci. 1 (1956) 9 (Hungarian). English translation, Selected Papers of A. Renyi, vol. 1 (Academiai Budapest, Budapest, 1976), p. 558. See also A. Renyi, Acta Mathematica (Hungary) 10 (1959) 193.

[18] C. Shannon, "A Mathematical Theory of Communication", Bell Tech. Jour. 27 (1948) 379–423, 623–656.

[19] F. Ledrappier, "Some Relations Between Dimension and Lyapunov Exponents" Comm. Math. Phys. 81 (1981) 229–238.

[20] F. Takens, "Invariants Related to Dimension and Entropy", to appear in Atas do 13 Colognio Brasiliero de Mathematica.

[21] T. Janssen and J. Tjon, "Bifurcations of Lattice Structure", Univ. of Utrecht preprint (1982).

[22] J. Kaplan and J. Yorke, Functional Differential Equations and the Approximation of Fixed Points, Proceedings, Bonn, July 1978, Lecture Notes in Math. 730, H.O. Peitgen and H.O. Walther, eds., (Springer, Berlin, 1978), p. 228.

[23] H. Mori, "Prog. Theor. Phys. 63 (1980) 3.

[24] V.I. Oseledec, "A Multiplicative Ergodic Theorem. Lyapunov Characteristic Numbers for Dynamical Systems", Trans. Moscow Math. Soc. 19 (1968) 197.
[25] C. Grebogi, E. Ott, and J. Yorke, "Chaotic Attractors in Crisis", in this volume.
[26] A. Douady and J. Oesterle, "Dimension de Hausdorff des Attracteurs, Comptes Rendus des Seances de L'academie des Sciences 24 (1980) 1135–38.
[27] A.N. Kolmogorov, Dolk. Akad. Nauk SSSR 124 (1959) 754. English summary in MR 21, 2035.
[28] Ya. G. Sinai, Dolk. Akad. Nauk SSSR 124 (1959) 768. English summary in MR 21, 2036.
[29] J. Crutchfield and N. Packard, "Symbolic Dynamics of One-Dimensional Maps: Entropies, Finite Precision, and Noise", Int'l. J. Theo. Phys. 21 (1982) 433.
[30] S. Pelikan, private communication.
[31] P. Billingsley, Ergodic Theory and Information, New York, 1965).
[32] I.J. Good, "The Fractional Dimensional Theory of Continued Fractions", Proc. Camb. Phil. Soc. 37 (1941) 199–228.
[33] H.G. Eggleston, "The Fractional Dimension of a Set Defined by Decimal Properties", Quart. J. Math. Oxford Ser. 20 (1949) 31–36.
[34] A. Besicovitch, "On the Sum of Digits of Real Numbers Represented in the Dyadic System", Math. Annalen. 110 (1934) 321.
[35] J.L. Kaplan, J. Mallet-Paret, and J.A. Yorke, "The Lyapunov Dimension of a Nowhere Differentiable Attracting Torus", Univ. of Maryland Preprint (1982).
[36] V.I. Arnold and Avez, Ergodic Theory in Classical Mechanics (New York, 1968).
[37] D. Russel, J. Hansen, and E. Ott, "Dimensionality and Lyapunov Numbers of Strange Attractors", Phys. Rev. Lett. 45 (1980) 1175.
[38] J.D. Farmer, "Chaotic Attractors of an Infinite Dimensional Dynamical System", Physica 4D (1982) 366–393.
[39] H. Greenside, A. Wolf, J. Swift, and T. Pignataro, "The Impracticality of a Box Counting Algorithm for Calculating the Dimensionality of Strange Attractors", Phys. Rev. A25 (1982) 3453.
[40] H. Froehling, J. Crutchfield, J.D. Farmer, N. Packard, and R. Shaw, "On Determining the Dimension of Chaotic Flows", Physica 3D (1981) 605.
[41] R. Kautz, private communication.
[42] A.J. Chorin, "The Evolution of a Turbulent Vortex", Comm. Math. Phys. 83 (1982) 517–535.
[43] G. Bennetin, L. Galgani and J. Strelcyn, Phys. Rev. A 14 (1976) 2338; also see G. Benettin, L. Galgani, A. Giorgilli and J. Strelcyn, Meccanica 15 (1980) 9.
[44] I. Shimada and T. Nagashima, Prog. Theor. Phys. 61 (1979) 228.
[45] B.B. Mandelbrot, "Intermittent turbulence in self-similar cascades: Divergence of high moments and dimension of the carrier". J. Fluid Mechanics 62 (1974) 331–358. See 351, 354 for a discussion of a frequency dependent dimension.
[46] B.B. Mandelbrot, "Fractals and turbulence: attractors and dispersion". Turbulence Seminar, Berkeley 1976/7. Lecture Notes in Mathematics 615, 83–93 (Springer, New York).

Physica 9D (1983) 189–208
North-Holland Publishing Company

MEASURING THE STRANGENESS OF STRANGE ATTRACTORS

Peter GRASSBERGER† and Itamar PROCACCIA
Department of Chemical Physics, Weizmann Institute of Science, Rehovot 76100, Israel

Received 16 November 1982
Revised 26 May 1983

We study the correlation exponent ν introduced recently as a characteristic measure of strange attractors which allows one to distinguish between deterministic chaos and random noise. The exponent ν is closely related to the fractal dimension and the information dimension, but its computation is considerably easier. Its usefulness in characterizing experimental data which stem from very high dimensional systems is stressed. Algorithms for extracting ν from the time series of a single variable are proposed. The relations between the various measures of strange attractors and between them and the Lyapunov exponents are discussed. It is shown that the conjecture of Kaplan and Yorke for the dimension gives an upper bound for ν. Various examples of finite and infinite dimensional systems are treated, both numerically and analytically.

1. Introduction

It is already an accepted notion that many nonlinear dissipative dynamical systems do not approach stationary or periodic states asymptotically. Instead, with appropriate values of their parameters, they tend towards strange attractors on which the motion is chaotic, i.e. not (multiply) periodic and unpredictable over long times, being extremely sensitive on the initial conditions [1–4].

A natural question is by which observables this situation is most efficiently characterized. Even more basically, when observing a seemingly strange behaviour, one would like to have clear-cut procedures which could exclude that the attractor is indeed multiply periodic, or that the irregularities are e.g. caused by external noise [5].

The first possibility can be ruled out by making a Fourier analysis, but for the second one has to turn to some other measures. These measures should be sensitive to the *local* structure, in order to distinguish the blurred tori of a noisy (multi-) periodic motion from the strictly deterministic motion on a fractal. Also, they should be able to distinguish between different strange attractors.

In this paper we shall propose such a measure. Before doing so we shall discuss however the existing approaches to the subject.

In a system with F degrees of freedom, an attractor is a subset of F-dimensional phase space towards which almost all sufficiently close trajectories get "attracted" asymptotically. Since volume is contracted in dissipative flows, the volume of an attractor is always zero, but this leaves still room for extremely complex structures.

Typically, a strange attractor arises when the flow does not contract a volume element in *all* directions, but stretches it in some. In order to remain confined to a bounded domain, the volume element gets folded at the same time, so that it has after some time a multisheeted structure. A closer study shows that it finally becomes (locally) Cantor-set like in some directions, and is accordingly a fractal in the sense of Mandelbrot [6].

Ever since the notion of strange attractors has been introduced, it has been clear that the Lyapunov exponents [7, 8] might be employed in studying them. Consider an infinitesimally small F-dimensional ball in phase space. During its

† Permanent address: Department of Physics, University of Wuppertal, W. Germany.

evolution it will become distorted, but being infinitesimal, it will remain an ellipsoid. Denote the principal axes of this ellipsoid by $\epsilon_i(t)$ $(i = 1, \ldots, F)$. The Lyapunov exponents λ_i are then determined by

$$\epsilon_i(t) \approx \epsilon_i(0)\, e^{\lambda_i t}. \tag{1.1}$$

The sum of the λ_i, describing the contraction of volume, has of course to be negative. But since a strange attractor results from a stretching and folding process, it requires at least one of the λ_i to be positive. Inversely, a positive Lyapunov exponent implies sensitive dependence on initial conditions and therefore chaotic behaviour.

One drawback of the λ_i's is that they are not easily measured in experimental situations. Another limitation is that while they describe the *stretching* needed to generate a strange attractor, they don't say much about the *folding*.

That these two are at least partially independent is best seen by looking at a horshoe-like map† embedded in 3-dimensional space (fig. 1). Assume that each step of the evolution consists of (i) stretching in the x-direction by a factor of 2, (ii) squeezing in the y- and z-direction by different factors $\mu_z < \mu_y < \frac{1}{2}$, and (iii) folding in the (x, y) plane (fig. 1a) or in the (x, z) plane (fig. 1b). From fig. 1 one realizes already that the attractor will in both cases be a Cantorian set of lines, being more "plane-filling" in the first case than in the second case. Indeed, using the results of Section 7, one finds easily that the fractal dimensions are $D_a = 1 + \ln 2/|\ln\mu_y|$ and $D_b = 1 + \ln 2/|\ln\mu_z|$, respectively.

It is this fractal (or Hausdorff–Besikovich) dimension which has until now attracted most attention [9–14] as a measure of the local structure of fractal attractors. In order to define it [5], one first covers the attractor by F-dimensional hypercubes of side length l and considers the limit $l \to 0$. If the minimal number of cubes needed for the covering grows like

$$M(l) \underset{l \to 0}{\simeq} l^{-D}, \tag{1.2}$$

the exponent D is called the Hansdorff dimension of the attractor [5].

Being a purely geometric measure, D is independent of the frequency with which a typical trajectory visits the various parts of the attractor.

† Notice that this is not a Smale's horseshoe. We also neglect in the following the bent parts of the horseshoe, in comparison to the parallel parts (i.e. we assume $L_x \gg L_y, L_z$; see fig. 1).

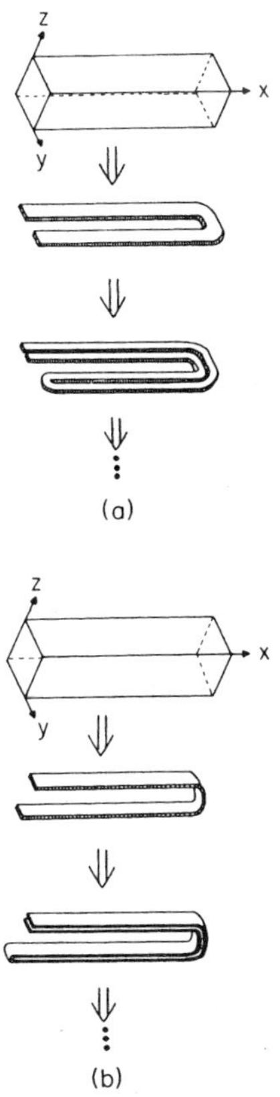

Fig. 1. Shape of an originally rectangular volume element after two iterations, each consisting of stretching, squeezing and folding. In fig. 1a (1b), the folding is in the $y(z)$-direction, which is the direction of lesser (stronger) squeezing.

Even if these frequencies are very inequal, developing maybe even singularities somewhere, all parts contribute to D equally. It has been documented [12, 14] that the calculation of D is exceedingly hard and in fact impractical for higher dimensional systems.

Another measure which has been considered and which is sensitive to the frequency of visiting, is the information entropy of the attractor. By "information entropy" here we understand the information gained by an observer who measures the actual state $X(t)$ of the system with accuracy l, and who knows all properties of the system but not the initial condition $X(0)$. This is very similar to the entropy in statistical mechanics if we relate $X(t)$ to the microstate ($F \approx 10^{23}$), and the "system" to the macrostate. It is *not* the Kolmogorov entropy which is essentially the sum of all positive Lyapunov exponents.

Using the above partition of phase space into cells with length l, the information entropy can be written as

$$S(l) = -\sum_{i=1}^{M(l)} p_i \ln p_i , \tag{1.3}$$

where p_i is the probability for $X(t)$ to fall into the ith cell. For all attractors studied so far, $S(l)$ increases logarithmically with $1/l$ as $l \to 0$, and we shall accordingly make the ansatz

$$S(l) \simeq S_0 - \sigma \ln l . \tag{1.4}$$

The constant σ will be called, following ref. 8, the information dimension. It is always a lower bound to the Hausdorff dimension, and in most cases they are almost the same within numerical errors.

The measure on which we shall concentrate mostly in this paper, has been recently introduced by the present authors [15]. It is obtained from the correlations between random points on the attractor. Consider the set $\{X_i, i = 1 \cdots N\}$ of points on the attractor, obtained e.g. from a time series, i.e. $X_i \equiv X(t + i\tau)$ with a fixed time increment τ between successive measurements. Due to the exponential divergence of trajectories, most pairs (X_i, X_j) with $i \neq j$ will be *dynamically* uncorrelated pairs of essentially random points. The points lie however on the attractor. Therefore they will be spatially correlated. We measure this spatial correlation with the correlation integral $C(l)$, defined according to

$$C(l) = \lim_{N\to\infty} \frac{1}{N^2} \times \{\text{number of pairs } (i,j) \text{ whose distance } |X_i - X_j| \text{ is less than } l\} . \tag{1.5}$$

The correlation integral is related to the standard correlation function

$$c(\mathbf{r}) = \lim_{N\to\infty} \frac{1}{N^2} \sum_{\substack{i,j=1 \\ i\neq j}}^{N} \delta^F(X_i - X_j - \mathbf{r}) \tag{1.6}$$

by

$$C(l) = \int_0^l \mathrm{d}^F r\, c(\mathbf{r}) . \tag{1.7}$$

One of the central aims of this paper is to establish that for small l's $C(l)$ grows like a power

$$C(l) \sim l^{\nu} , \tag{1.8}$$

and that this "correlation exponent" can be taken as a most useful measure of the local structure of a strange attractor. It seems that ν is more relevant, in this respect, than D. In any case, its calculation yields also an estimate of σ and D, since we shall argue that in general one has

$$\nu \leqslant \sigma \leqslant D . \tag{1.9}$$

We found that the inequalities are rather tight in most cases, but not in all. Given an experimental signal, if one finds eq. (1.8) with $\nu < F$, one knows that the signal stems from deterministic chaos rather than random noise, since random noise will always result in $C(l) \sim l^F$. Explicit algorithms will be proposed below.

One of the main advantages of ν is that it can easily be measured, at least more easily than either σ or D. This is particularly true for cases where the fractal dimension is large ($\gtrsim 3$) and a covering by small cells becomes virtually impossible. We thus expect that the measure ν will be used in experimental situations, where typically high dimensional systems exist.

In theoretical cases, when the evolution law is known analytically, the easiest quantities to evaluate are the Lyapunov exponents. General formulae expressing D in terms of the λ_i have been proposed by Mori [9] and by Kaplan and Yorke [10]. If they were correct, they would obviously be very useful. They have been verified in simple cases [11, 14]. But Mori's formula was shown to be wrong in one case by Farmer [8], and the above example shown in fig. 1 shows that also the Kaplan–Yorke formula

$$D = D_{KY} \equiv j + \frac{\lambda_1 + \lambda_2 + \cdots + \lambda_j}{|\lambda_{j+1}|} \tag{1.10}$$

does not hold even in all those cases where $\nu = \sigma = D$. Here, the exponents are ordered in descending order $\lambda_1 \geq \lambda_2 \geq \cdots \geq \lambda_F$, and j is the largest integer for which $\lambda_1 + \lambda_1 + \ldots + \lambda_j \geq 0$.

In section 7 we shall take up this question again. We shall show that the counterexample in fig. 1b is not generic. We shall however claim that eq. (1.10) cannot generally be expected to be correct, and that in fact D_{KY} is an upper bound, if $\nu = \sigma = D$.

In the next section, we shall present numerical results for several simple models, for which the fractal dimensions are known from the literature. This will serve to illustrate the scaling law (1.8), and to verify the inequality $\nu \leqslant D$. This inequality and its stronger version, eq. (1.9), will be derived in section 3. The case of one-dimensional maps at infinite bifurcation (Feigenbaum [16]) points is special in that there the information dimension σ and the exponent ν can be calculated exactly, with the result $\nu \neq \sigma \neq D$. It is treated in section 4. Section 5 is dedicated to an important modification which allows to extract ν from a time series of one single variable, instead of from the series $\{X_i\}$. This is of course most important for infinite-dimensional systems, but it is also very useful in low-dimensional cases where it diminishes systematic errors. Among others, we shall apply this method in section 6 to the Mackey–Glass [17] delay equation studied in great detail in ref. 8.

In section 7 we discuss the relation of ν to the Lyapunov exponents, and establish the result

$$\nu \leqslant D_{KY}\,. \tag{1.11}$$

A summary and a discussion of the actual method of treating experimental signals is offered in section 8.

2. Case studies of low-dimensional systems

In this section we shall establish that $C(l)$ can be very well represented by a power law l^ν, by exhibiting numerical results for a number of low dimensional systems. These results are summarized in table I. In section 5 we shall show that this is the case also in high (and infinite) dimensional systems. Details of the numerical algorithms are discussed in appendix A.

2.1. *One-dimensional maps*

The simplest cases of chaotic system are represented by maps of some interval into itself, as e.g. the logistic map [2]

$$x_{n+1} = ax_n(1 - x_n)\,. \tag{2.1}$$

We shall study this map both at the point of onset of chaos via period doubling bifurcations, i.e. when $a = a_\infty = 3.5699456\ldots$ and for the case $a = 4.0$. In fig. 2 we show the result for the first case. It is well known [2, 16] that for this map the attractor* is

* Note that the term "attractor" would not be universally accepted here due to the fact that in any neighbourhood there exist trajectories which do not tend towards it asymptotically.

Table I

	ν	No. of iterations, time increment τ	D	σ
Hénon map $a = 1.4$, $b = 0.3$	1.21 ± 0.01[d] 1.25 ± 0.02[e]	15000	1.26 (ref. 11)	-
Kaplan–Yorke map $\alpha = 0.2$	1.42 ± 0.02	15000	1.431(ref. 11)	-
Logistic eq., $b = 3.5699456\cdots$	0.500 ± 0.005 $0.4926 < \nu < 0.5024$[f]	25000	0.538(ref. 13)	0.5170976
Lorenz eq.[a]	2.05 ± 0.01	15000; $\tau = 0.25$	2.06 ± 0.01	-
Rabinovich–[b] Fabrikant eq.	2.19 ± 0.01	15000; $\tau = 0.25$	-	-
Zaslavskii map[c]	(≈ 1.5)	25000	1.39(ref. 11)	-

[a])Parameters as in refs. 7 and 11.
[b])Parameters as in section 3 of ref. 20.
[c])Parameters as in ref. 11.
[d])From eqs. (1.5) and (1.8).
[e])From single variable time series, with $f = 3$.
[f])Exact analytic bound.

Cantor-like with a fractal dimension satisfying the exact bound [13] $0.5376 < D < 0.5386$. In section 4 we shall prove exactly that $\sigma = 0.517097\ldots$, and that $0.4926 < \nu < 0.5024$ while from Fig. 2 we find $\nu = 0.500 \pm 0.005$. For very small distances, the data for $C(l)$ deviate from a power law, but that was to be expected: the behaviour at $a = a_\infty$ is not yet chaotic, and therefore the values x_n are strongly correlated. We verified that indeed the powerlaw holds down to smaller values of l if we increase N or use only values $x_i, x_{i+p}, x_{i+2p}, x_{i+2p}, \ldots$ with p being a large odd number.

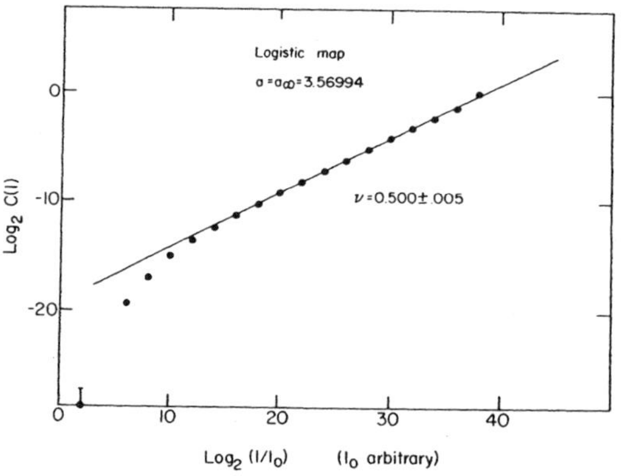

Fig. 2. Correlation integral for the logistic map (2.1) at the infinite bifurcation point $a = a_\infty = 3.699\ldots$ The starting point was $x_0 = \frac{1}{2}$, the number of points was $N = 30.000$.

The same map can be used also to introduce the important issue of corrections to scaling. These are found for the parameter value $a = 4$. It is well known that in this the attractor* consists of the interval [0, 1], and that the invariant probability density is equal to

$$p(x) \equiv \lim_{N\to\infty} \frac{1}{N} \sum_{i=1}^{N} \delta(x_i - x) \tag{2.2}$$

$$= \frac{1}{\pi} [x(1-x)]^{-1/2} . \tag{2.3}$$

From this, one finds easily

$$\nu = \sigma = D = 1 . \tag{2.4}$$

Notice, however, that while the scaling laws (1.2) and (1.4) are exact, the scaling law (1.8) for $C(l)$

* Again, the term is questionable, as no point outside the interval [0, 1] gets attracted towards it. We shall ignore this irrelevant point, which could be avoided by using $a = 4 - \epsilon$.

requires logarithmic corrections, due to the singular behaviour of $p(x)$:

$$C(l) = \int_0^1\!\!\int dx\, dy p(x) p(y) \theta(|x-y|-l)$$

$$\underset{l\to 0}{\simeq} \frac{4}{\pi^2} l \ln 1/l \,. \qquad (2.5)$$

Thus, a numerical calculation of ν is expected to converge very slowly. This problem and a remedy for it are discussed further in section 5.

2.2. *Maps of the plane*

Here we examined the Hénon [18] map

$$x_{n+1} = y_n + 1 - a x_n^2 \,, \quad y_{n+1} = b x_n \,, \qquad (2.6)$$

with $a = 1.4$ and $b = 0.3$, the Kaplan–Yorke [10] map

$$x_{n+1} = 2x_n \pmod 1 \,, \quad y_{n+1} = \alpha y_n + \cos 4\pi x_n \qquad (2.7)$$

with $\alpha = 0.2$, and the Zaslavskii [19] map

$$x_{n+1} = [x_n + \nu(1 + \mu y_n) + \epsilon\nu\mu \cos 2\pi x_n] \pmod 1),$$
$$y_{n+1} = e^{-\Gamma}(y_n + \epsilon \cos 2\pi x_n) \,, \qquad (2.8)$$

with the parameters

$$\mu = \frac{1 - e^{-\Gamma}}{\Gamma} \qquad (2.9)$$

and $\Gamma = 3.0$, $\nu = 400/3$, and $\epsilon = 0.3$ taken from ref. 11.

Figs. 3–5 exhibit the results for the correlation integrals. In the first two cases, we find excellent agreement with a power law; while for the Kaplan–Yorke map we find $\nu = 1.42 \pm .02$ in agreement with the published [11] value of D, a fit to the Hénon map yields $\nu_{\text{eff}} = 1.21$, smaller than the value [11] $D = 1.261 \pm 0.003$. We shall argue in sectin 5 that actually the value of ν for the Hénon map is underestimated here, and that instead $\nu = 1.25 \pm 0.02 \approx D$.

The case of the Zaslavskii map is exceptional as it was the only system for which we did not find

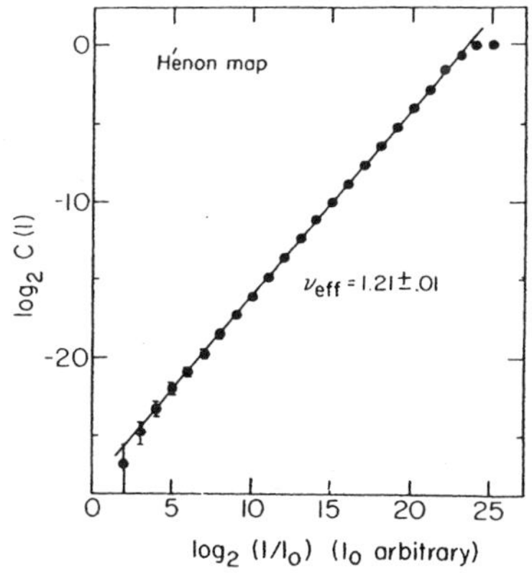

Fig. 3. Correlation integral for the Hénon map (2.6) with $a = 1.4$, $b = 0.03$ and $N = 15.000$.

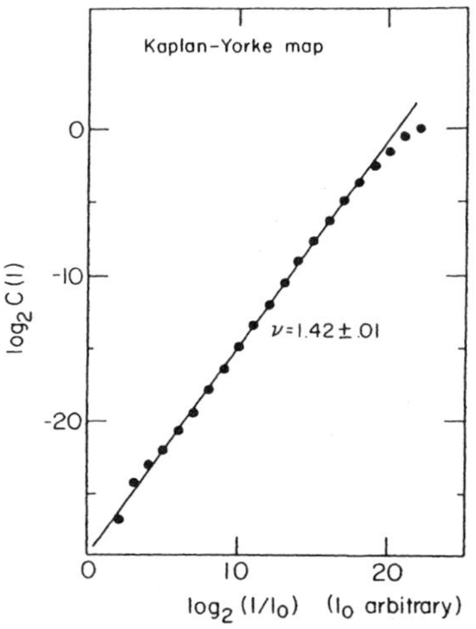

Fig. 4. Same as fig. 3, but for Kaplan–Yorke map (2.7) with $\alpha = 0.2$.

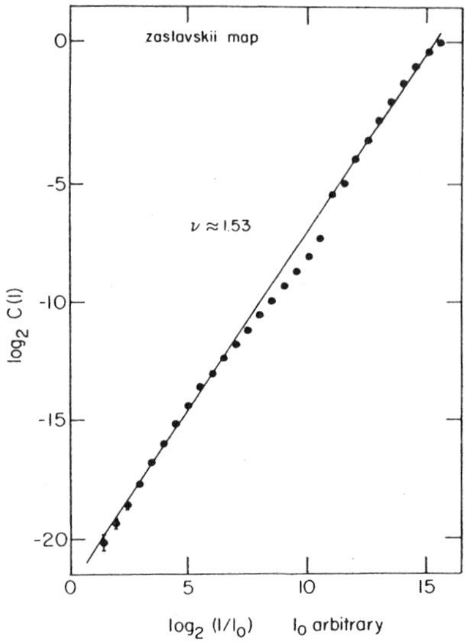

Fig. 5. Correlation integral for Zaslavskii map (eqs. (2.8), (2.9)); $N = 25.000$, parameters as in the text. For faster scaling, the y-coordinate was blown up by a factor of 25, rendering the attractor square-like at low resolution (see fig. 6; without this, the attractor would have looked effectively 1-dimensional for $l \gtrsim l_{max}/25$).

clear-cut power behaviour. Also, an (admittedly poor) fit would yield $\nu \approx 1.5$, in clear violation of the bound $\nu < D$. The reasons why our method has to fail for this map – with the parameters as quoted above – becomes clear when looking at fig. 6. Call l_0 the outer length scale. From fig. 6a one sees that the attractor looks 2-dimensional for $l \gtrsim l_0 \times 2^{-5}$ and $\approx$ 1-dimensional for $l_0 \times 2^{-5} \gtrsim l \gtrsim l_0 \times 2^{-9}$. From fig. 6b one sees that it looks $\approx$ 2-dimensional again down to $\approx l_0 \times 2^{-14}$, scaling behaviour setting in only at about that scale (which is beyond our resolution). It seems to us that the box-counting algorithm of ref. 11 in which D is evaluated, should confront the same problem†.

† Note added: Dr. Russel kindly provided us with the original data of $M(\epsilon)$ versus ϵ. From these, it seems that indeed a similar phenomenon occurs and that accordingly a value $D \approx 1.5$ cannot be excluded.

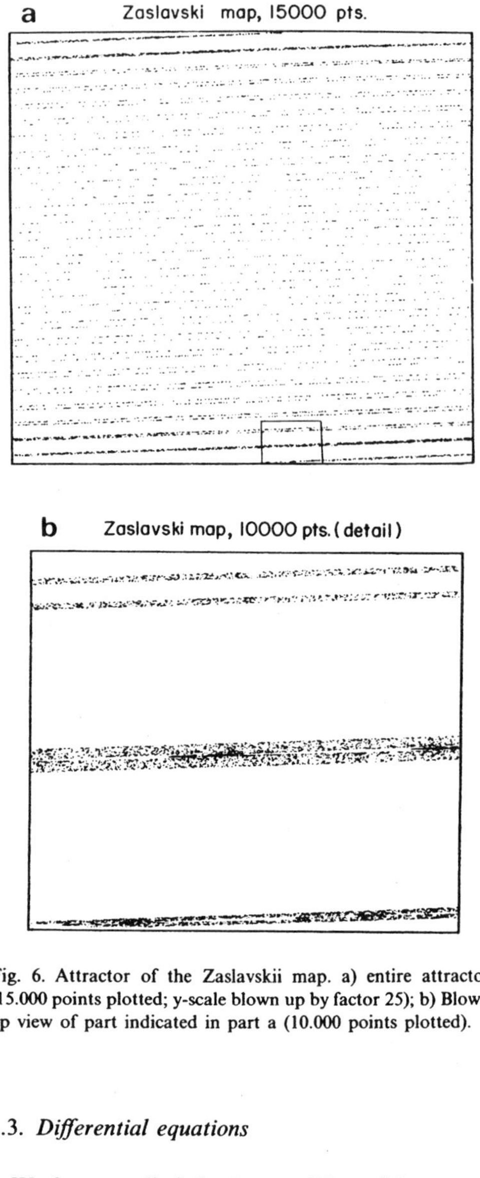

Fig. 6. Attractor of the Zaslavskii map. a) entire attractor (15.000 points plotted; y-scale blown up by factor 25); b) Blown up view of part indicated in part a (10.000 points plotted).

2.3. *Differential equations*

We have studied the Lorenz [1] model

$$\dot{x} = \sigma(y - x),$$
$$\dot{y} = -y - xz + Rx, \qquad (2.10)$$
$$\dot{z} = xy - bz,$$

with $R = 28$, $\sigma = 10$, and $b = 8/3$, and the Rabinovich–Fabrikant [20] equations

$$\begin{aligned} \dot{x} &= y(z - 1 + x^2) + \gamma x\,, \\ \dot{y} &= x(3z + 1 - x^2) + \gamma y\,, \\ \dot{z} &= -2z(\alpha + xy)\,, \end{aligned} \tag{2.11}$$

with $\gamma = 0.87$ and $\alpha = 1.1$.

As seen in fig. 7 we get adequate power laws for $C(l)$, and in the case of the Lorenz model, where D is known [11], we obtain $\nu \simeq D$.

Further examples will be studied in section 6, in the context of higher dimensional systems.

It should be stressed that the algorithm used to calculate ν converged quite rapidly. Although each entry in table I and figs. 2–7 were based on $\approx$15.000–25.000 points each, reasonable results (i.e. results for ν within $\pm 5\%$) were obtained in most cases already with only a few thousand points. This should be contrasted with the difficulties associated with estimating D in box-counting algorithms [11, 14].

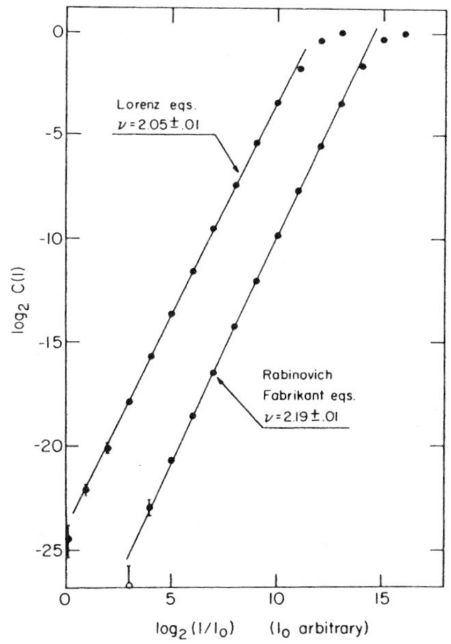

Fig. 7. Correlation integrals for the Lorenz equations (eq. (2.10); dots) and for the Rabinovich–Fabrikant equation (eq. (2.11); open circles). In both cases, $N = 15.000$ and $\tau = 0.25$.

Summarizing this section, we can say that except for the logistic map at $a = a_\infty$ ("Feigenbaum attractor") we found in all cases that $\nu \approx D$ within the limits of accuracy. We now turn to a theoretical analysis of the relations between ν, σ and D.

3. Relations between ν, σ and D

In this section we shall establish the inequalities (1.9). We shall do this in 3 steps.

a) The easiest inequality to prove is $\sigma \leqslant D$. Consider a covering of the attractor by hypercubes ("cells") of edge length l, and a time series $\{X_k;\, k = 1, \ldots, N\}$. The probabilities p_i for an arbitrary X_k to fall into cell i are simply

$$p_i = \lim_{N\to\infty} \frac{1}{N} \mu_i\,. \tag{3.1}$$

where μ_i is the number of points X_k which fall into cell i.

If the coverage of the attractor is uniform, one has,

$$p_i = \frac{1}{M(l)}\,, \tag{3.2}$$

where $M(l)$ is the number of cells needed to cover the attractor, and one finds from eqs. (1.3) and (1.2)

$$S(l) = S^{(0)}(l) = \ln M(l) = \text{const} - D \ln l\,. \tag{3.3}$$

In the general case, one uses the convexity of $x \ln x$ in the usual way to prove that $S(l) \leqslant S^{(0)}(l)$. Invoking the ansatz $S(l) = \text{const} - \sigma \ln l$, we find $\sigma \leqslant D$.

b) Instead of showing immediately $\nu \leqslant \sigma$, let us proceed slowly and show first that $\nu \leqslant D$.

From the definition of $C(l)$, we get up to a factor

of order unity

$$C(l) \simeq \lim_{N\to\infty} \frac{1}{N^2} \sum_{i=1}^{M(l)} \mu_i^2 = \sum_{i=1}^{M(l)} p_i^2 . \quad (3.4)$$

Here, we have replaced the number of pairs with distance $< l$ by the number of pairs which fall into the same cell of length l. The error committed should be independent on l, and thus should not affect the estimation of ν. Using the Schwartz inequality we get

$$C(l) = M(l)\langle p_i^2\rangle \geqslant M(l)\langle p_i\rangle^2 = \frac{1}{M(l)} \sim l^D . \quad (3.5)$$

In this equation square brackets denote average over all cells. Comparing eqs. (3.5) and (1.8) we find immediately $\nu \leqslant D$.

c) In order to derive $\nu \leqslant \sigma$, consider two nested coverings with cubes of lengths l and $2l$. The numbers of cubes that contain a piece of the attractor are then related by

$$M(l) = 2^D M(2l) . \quad (3.6)$$

Denote by p_i the probability to fall in cube i of the finer coverage, and by P_j the probability to fall in cube j of the coarser. Define $\omega_i (i = 1, \ldots M(l))$ by

$$p_i = \omega_i P_j \quad (i \in j) . \quad (3.7)$$

Evidently we have

$$P_j = \sum_{i\in j} p_i, \quad \sum_{i\in j} \omega_i = 1 . \quad (3.8)$$

We can then write the correlation integral as

$$C(l) \simeq \sum_{i=1}^{M(l)} p_i^2 = \sum_{j=1}^{M(2l)} P_j^2 \sum_{i\in j} \omega_i^2 . \quad (3.9)$$

Consider now the ratio

$$\frac{C(l)}{C(2l)} = \frac{\sum_j P_j^2 \sum_{i\in j} \omega_i^2}{\sum_j P_j^2} , \quad (3.10)$$

and compare it to the entropy difference

$$S(2l) - S(l) = \sum_{i=1}^{M(l)} p_i \ln p_i - \sum_{j=1}^{M(2l)} P_j \ln P_j$$
$$= \sum_{j=1}^{M(2l)} P_j \sum_{i\in j} \omega_i \ln \omega_i . \quad (3.11)$$

In order to estimate eq. (3.10) in terms of eq. (3.11), we have to introduce a new assumption. We assume that the ω_i's are distributed independently of the P_j. This means essentially that locally the attractor looks the same in regions where it is rather dense (P_j large) as in regions where P_j is small. Although we cannot further justify this assumption, it seems to us very natural. It leads immediately to

$$\frac{C(l)}{C(2l)} = \frac{\langle \omega^2\rangle}{\langle \omega\rangle} = 2^D \langle \omega^2\rangle , \quad (3.12)$$

and to

$$S(2l) - S(l) = 2^D \langle \omega \ln \omega\rangle . \quad (3.13)$$

Define now a normalized variable W by

$$W = \frac{\omega}{\langle \omega\rangle} = 2^D \omega . \quad (3.14)$$

Using the inequality [21]

$$\langle W^2\rangle > \exp\langle W \ln W\rangle , \quad (3.15)$$

we establish

$$\frac{C(l)}{C(2l)} \geqslant \exp[S(2l) - S(l)] \quad (3.16)$$

and thus

$$\nu \leqslant \sigma . \quad (3.17)$$

Remarks. From the proofs it is clear that if the attractor is uniformly covered, one has equalities

$$\nu = \sigma = D . \quad (3.18)$$

It is an interesting question how non-uniform the coverage must be in order to break them. With the exception of the Feigenbaum map (logistic map with $a = a_\infty$), which is however not generic, all examples of the last section were compatible with eq. (3.18).

In cases where $\nu \neq D$, we claim that indeed ν is the more relevant observable. In these cases, the neighbourhoods of certain points have higher "seniority" in the sense that they are visited more often than others. The fractal dimension is ignorant of seniority, being a purely geometric concept. But both the correlation integral and the entropy dimension weight regions according to their seniority.

Eqs. (1.9) and (3.18) have been used previously in the context of fully developed homogeneous turbulence [22]. The connection

$$c(l) \propto l^{D-F}, \quad l \in R^f$$

following from $\nu = D$ has been used previously also in percolation theory [23] and in a model for dendritic growth [24].

4. Information entropy and ν of the Feigenbaum attractor

In this section we shall compute exactly the information dimension and ν of one-dimensional maps

$$x_{n+1} = F(x_n) \tag{4.1}$$

at the onset of chaos. The method follows closely the one of ref. 13.

It is well known that such maps – provided they have a unique quadratic maximum – have universal scaling features, studied in most detail by Feigenbaum [16]. This behaviour is most easily described by observing that the iterations

$$F^{(2^n)}(x) = \underbrace{F(F(\ldots.F(x)\ldots))}_{2^n \text{ times}} \tag{4.2}$$

tend after a suitable rescaling towards a universal function

$$g(x) = \lim_{n\to\infty} \frac{1}{F^{(2^n)}(0)} F^{(2^n)}(xF^{(2^n)}(0)). \tag{4.3}$$

This "Feigenbaum function" $g(x)$ satisfies the exact scaling relation

$$g(g(x)) = -\frac{1}{\alpha} g(\alpha x), \tag{4.4}$$

with $\alpha = 2.50290\ldots$, and the normalization condition $g(0) = 1$. We have here assumed that the maximum of $F(x)$ is at $x = 0$, which can always be achieved by a change of variables. In order to obtain the information dimension of the logistic map at $a = a_\infty = 3.5699345\ldots$, it is thus sufficient to compute σ for the Feigenbaum map.

The "attractor" (see the reservations in section 2) of $g(x)$ consists of the sequence $\{\xi_n, n = 0, 1, 2, \ldots\}$ with

$$\xi_0 = 0 \tag{4.5}$$

and

$$\xi_{n+1} = g(\xi_n). \tag{4.6}$$

The first few ξ_k's are shown in fig. 8. There, it is also indicated how they build up the Cantorian structure of the attractor: the points $\xi_1, \xi_2, \xi_3, \ldots, \xi_{2^k+1}$ form the end-points of 2^k intervals, and the following ξ_k's fall all into these intervals. Furthermore, any sequence $\{\xi_n, \xi_{n+1} \ldots \xi_{n+2^k-1}\}$ of 2^k successive points visits each of these intervals exactly once. Thus, the a priori probabilities $p_i (i = 1, \ldots, 2^k)$ for an arbitrary x_n to fall into the ith interval are all equal to $p_i = 2^{-k}$.

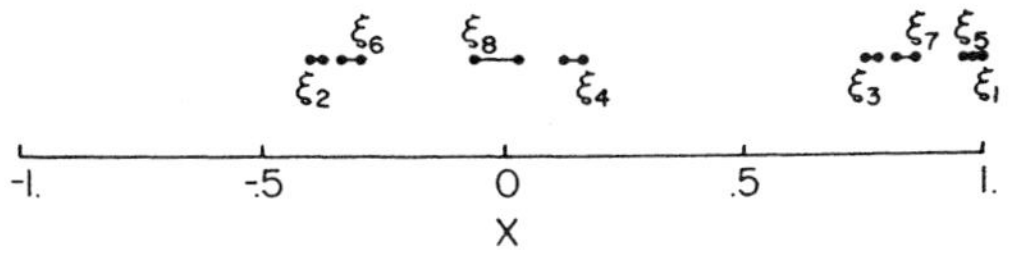

Fig. 8. First 16 points $\xi_1, \ldots, \xi_{16}$ of the attractor of the Feigenbaum equation describing the onset of chaos in 1-dimensional systems.

By the grouping axiom, we can first write the information entropy as

$$S(l) = \tfrac{1}{2}[S_{[2,4]}(l) + S_{[3,1]}(l)] + \ln 2\,, \tag{4.7}$$

where we denote by $S_{[i,j]}$ the information needed to specify the point on the interval $[\xi_i, \xi_j]$, and where we have used the fact that an arbitrary x_n has equal probability to be on $[\xi_2, \xi_4]$ or on $[\xi_3, \xi_1]$. From eq. (4.4) we find, however, that

$$\xi_{2n} = -\frac{1}{\alpha}\xi_n\,. \tag{4.8}$$

Thus, the interval $[\xi_2, \xi_4]$ is a down-scaled image of the whole attractor, and we have

$$S_{[2,4]}(l) = S(\alpha l) \approx S(l) - \sigma \ln \alpha\,, \tag{4.9}$$

where we have used the scaling ansatz (1.4).

In order to estimate $S_{[3,1]}(l)$, we decompose the interval [3, 1] into the 2^{k-1} subintervals discussed above, defined by the ξ_n with odd n's:

$$S_{[3,1]}(l) = (k-1)\ln 2 + 2^{-k+1}\sum_{i=1}^{2^{k-1}} S_i(l)\,.$$

Again, we have applied the grouping axiom, using that $p_i = 2^{-k}$. The $S_i(l)$ are the informations needed to pin down x_n provided one knows that it falls into the ith subinterval. Since each subinterval maps onto one on the left-hand piece $[\xi_2, \xi_4]$, each $S_i(l)$ is equal to the information $\tilde{S}_i(|g_i'|l)$ needed to pin x_{n+1} on the corresponding interval on the left-hand side. Here, g_i' is some average derivative of $g(x)$ in the ith subinterval. Using that $\tilde{S}_i(|g_i'|l) \simeq \tilde{S}_i(l) - \sigma \ln|g_i'|$, we obtain

$$\begin{aligned} S_{[3,1]}(l) &= (k-1)\ln 2 + 2^{-k+1}\sum_{i=1}^{2^{k-1}} \tilde{S}_i(l) \\ &\quad - \sigma \sum_{i=1}^{2^{k-1}} \ln|g_i'| \\ &= S_{[2,4]}(l) - \sigma \sum_{i=1}^{2^{k-1}} \ln|g_i'|\,. \end{aligned} \tag{4.10}$$

Inserting this and eq. (4.9) into eq. (4.7), we find after a few manipulations and after taking the limit $k \to \infty$

$$\sigma = \lim_{k\to\infty} \frac{\ln 2}{\ln\alpha + \dfrac{1}{2^{k+1}}\sum_{i=1}^{2^k} |g'(\xi_{2i-1})|}\,. \tag{4.11}$$

The limit converges very quickly, leading (for $k > 7$) to

$$\sigma = 0.5170976\,. \tag{4.12}$$

The calculation of the correlation exponent, or rather of the exponent of the Renyi entropy (see eq. (3.4))

$$R(l) = \sum_{i=1}^{M(l)} p_i^2 \tag{4.13}$$

follows even more closely the one in ref. 13.

As in that paper, we obtain a nested set of bounds. The first (and least stringent) is obtained by writing

$$R(l) = \tfrac{1}{4}\{R_{[2,4]}(l) + R_{[3,1]}(l)\} \tag{4.14}$$

and using $R_{[2,4]}(l) = R(\alpha . l)$ and $R_{[3,1]}(l) = R(\alpha g' l)$ with

$$|g'(\xi_3)| < g' < |g'(\xi_1)|\,. \tag{4.15}$$

Assuming $R(l) \sim l^\nu$, we obtain

$$1 + |g'(\xi_3)|^\nu < \frac{4}{\alpha^\nu} < 1 + |g'(\xi_1)|^\nu\,, \tag{4.16}$$

leading to $0.4857 < \nu < 0.5235$.

For the next more stringent bounds, we write further

$$R_{[3,1]}(l) = \tfrac{1}{4}\{R_{[3,7]}(l) + R_{[5,1]}(l)\}\,, \tag{4.17}$$

with

$$R_{[3,7]}(l) + R(\alpha^2 g^{(1)} . l), \quad |g'(\xi_3)| < g^{(1)} < |g'(\xi_7)| \tag{4.18}$$

and

$$R_{[5,1]}(l) + R_{[3,1]}(\alpha g^{(2)} l), \quad |g'(\xi_5)| < g^{(2)} < |g'(\xi_1)| . \tag{4.19}$$

Some algebra leads then to

$$|g'(\xi_5)|^\nu + \frac{\alpha^\nu}{4-\alpha^\nu} |g'(\xi_3)|^\nu < \frac{4}{\alpha^\nu}$$
$$< |g'(\xi_1)|^\nu + \frac{\alpha^\nu}{4-\alpha^\nu} |g'(\xi_7)|^\nu , \tag{4.20}$$

with the result

$$0.4926 < \nu < 0.5024 , \tag{4.21}$$

in agreement with the numerical value $\nu = 0.500 \pm 0.005$.

5. Using a single-variable time series

Very often one does not have access to a time series $\{X_n\}$ of F-dimensional vectors. Instead one follows only one or at most a few components of X_n. This is particularly relevant for real (as opposed to computer) experiments where the number of degrees of freedom often is very high if not infinite. Such systems nevertheless can have low-dimensional attractors. It would be very desirable to have a reliable method which allows a characterization of this attractor from a single-variable time series. $\{x_i,\ i = 1, \ldots, N;\ x_i \in R\}$.

The essential idea [25, 26] consists in constructing d-dimensional vectors

$$\xi_i = (x_i, x_{i+1}, \ldots, x_{i+d-1}) \tag{5.1}$$

and using ξ-space instead of X-space. The correlation integral would e.g. be

$$C(l) = \lim_{N\to\infty} \frac{1}{N^2} \sum_{i,j=1}^{N} \theta(l - |\xi_i - \xi_j|) . \tag{5.2}$$

More generally, one can use

$$\xi_i = (x(t_i), x(t_i + \tau) \ldots x(t_i + (d-1)\tau)) , \tag{5.3}$$

with τ some fixed interval. The magnitude of τ should not be chosen too small since otherwise $x_i \approx x_{i+\tau} \approx x_{i+2\tau} \approx \cdots$ so that the attractor in ξ-space would be stretched along the diagonal and thus difficult to disentangle. On the other hand, τ should not be chosen too large since distant values in the time series are not strongly correlated (due to the exponential divergence of trajectories and unavoidable small errors).

A similar compromise must be chosen for the dimension d. Clearly, d must be larger than the Hausdorff dimension D of the attractor (otherwise, $C(l) \sim l^d$). If the attractor is Cantorian in more than one dimension, this might however not be sufficient. Also, it might be that, when looked at in d dimension, the density

$$p(\xi) = \lim_{N\to\infty} \frac{1}{N} \sum_{i=1}^{N} \delta(\xi_i - \xi) \tag{5.4}$$

develops singularities which are absent in more than d dimensions (such singularities occur e.g. when one projects a sphere with constant density, $p(\xi) = p\delta(x^2 + y^2 + z^2 - R^2)$, onto the x–y plane: the new density $\tilde{p}(x, y)$ is infinite at $x^2 + y^2 = R^2$).

On the other hand, one cannot make d too large without getting lost in experimental errors and lack of statistics.

In the next section, we shall study an infinite-dimensional system from this point of view. In the remainder of the present section, we shall apply these considerations to the logistic map with $a = 4$, and to the Hénon map.

In the logistic map, we have seen that there are logarithmic corrections to the power law $C(l) \sim l^\nu$. They result precisely from singularities of $p(x)$, at $x = 0$ and $x = 1$. While embedding the attractor in a higher dimensional space does not completely remove these singularities, it substantially reduces their influence. The reason is that embedding in higher dimensional space always results in stretching the attractor. However, the portions which are most strongly stretched are those which are most densely populated at the lower dimension. For example in the logistic map with $a = 4$ the "attractor" is the interval [0, 1] in $1d$ but is the parabola in $2d$. The parabola has highest slopes at the end points, exhibiting the stronger stretching associated with regions of singular distributions at a lower dimension. A similar effect appears when going from $d = 2$ to $d = 3$. We thus expect that the importance of the singularities in the distribution would be reduced in higher dimensions.

In order to check this, we have calculated for the logistic map at $a = 4$ the original correlation integral and the modified integral obtained by embedding in a 2- and 3-dimensional space. The results are shown in fig. 9. We observe indeed the expected decrease of systematic error when increasing d, accompanied by an increase of the statistical error.

Analogous results for the Hénon map are shown in fig. 10. There, we used as time series the series $\{x_n, x_{n+2}, x_{n+4}, \ldots\}$. While the 2-dimensional correlation integral gives an effective ν in agreement with the result of section 2, the 3-dimensional embedding gives a larger values $\nu = 1.25 \pm 0.02$ which agrees with the value of D found in refs. 11 and 14.

No such effects were observed in the Lorenz model, where both the originally defined $C(l)$ and the modified correlation integral using only a single coordinate time series gave values of ν which agreed with D [15].

The conclusion drawn from these examples is that it is often useful to represent the attractor in a higher dimensional space than absolutely necessary, in order to reduce systematic errors. These errors result from a strongly non-uniform coverage of the attractor, provided this non-uniformity is not so strong as to make $\nu \neq D$.

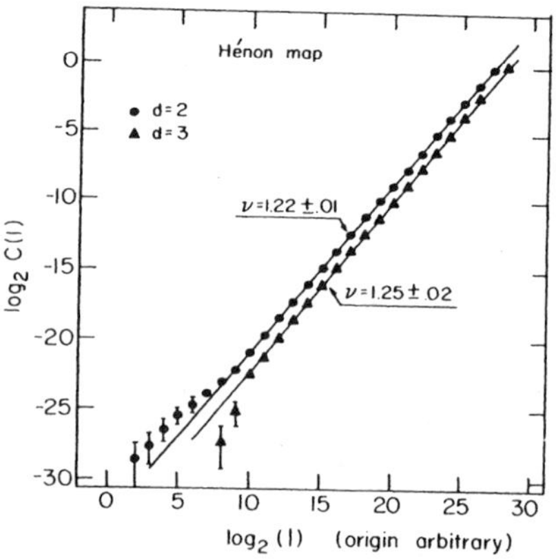

Fig. 10. Modified correlation integrals for the Hénon map (2.6). The time series consisted of coordinates x_n, x_{n+2}, $x_{n+4}, \ldots$, and $\xi = (x_n, x_{n+2}, \ldots, x_{n+2(d-1)})$ for each d. For $d = 2$, we took $N = 30.000$; for $d = 3$, we took $N = 20.000$.

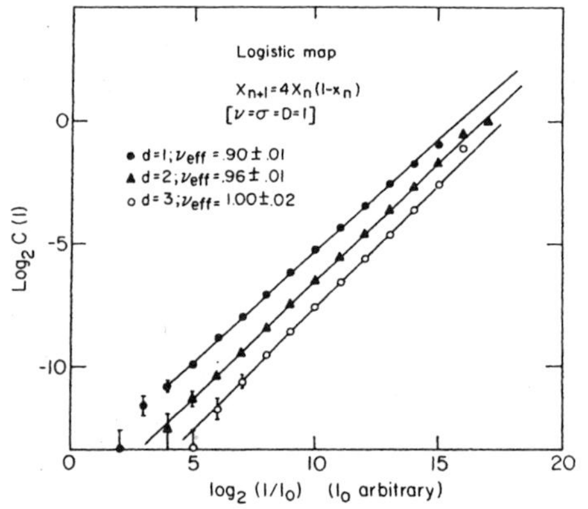

Fig. 9. Modified correlation integrals for the logistic map (2.1) with $a = 4$. The distance l between 2 points ξ_n and ξ_m on the attractor is defined as $l^2 = (\xi_n - \xi_m)^2 = (x_n - x_m)^2 + \ldots + (x_{n+d-1} - x_{m+d-1})^2$. For each value of d, we took $N = 15.000$.

6. Infinite-dimensional systems: an example

An extremely convenient way of generating very high dimensional systems is to consider delay differential equations of the type

$$\frac{dx(t)}{dt} = F(x(t), x(t-\tau)), \tag{6.1}$$

where τ is a given time delay. Such a delay equation is in fact infinite dimensional, as is most easily seen from the initial conditions necessary to solve eq. (6.1): they consist of the function $x(t)$ over a whole interval of length τ.

Following ref. 8, we shall study a particular example, introduced by Mackey and Glass [17] as a model for regeneration of blood cells in patients with leukemia. It is

$$\dot{x}(t) = \frac{ax(t-\tau)}{1+[x(t-\tau)]^{10}} - bx(t)\,. \tag{6.2}$$

As in ref. 8, we shall keep $a = 0.2$ and $b = 0.1$ fixed, and study the dependence on the delay time τ.

For the numerical investigation, eq. (6.2) is turned into an n-dimensional set of difference equations, with $n = 600$–1200. Details are described in the appendix. The time series was always chosen as $\{x(t), x(t+\tau), x(t+2\tau), \ldots\}$ except for some runs with $\tau = 100$, where we took points at times $t, t+\tau/2, t+2\tau/2, \ldots$.

The results for the correlation integral are shown in figs. 11–14. Estimated values of ν are given in table II, together with values of D obtained in ref. 8 by applying the defining eq. (1.2) to a Poincaré return map. Also shown in table II are the

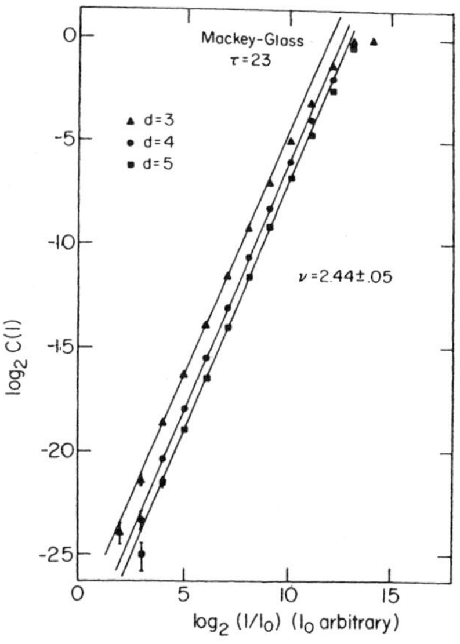

Fig. 12. Same as fig. 11, but for $\tau = 23$.

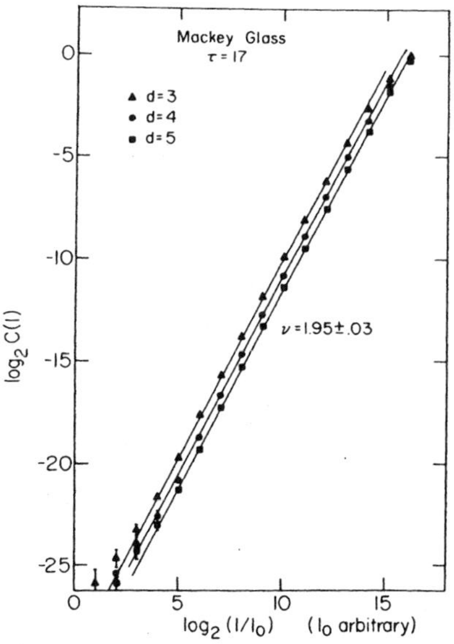

Fig. 11. Modified correlation integrals for the Mackey–Glass delay equation (6.2), with delay $\tau = 17$. The time series consisted of $\{X(t+i\tau);\ i = 1, \ldots, 25.000\}$.

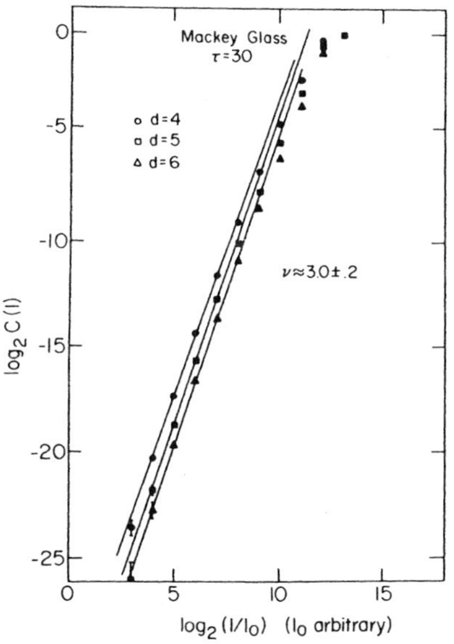

Fig. 13. Same as fig. 11, but for $\tau = 30$.

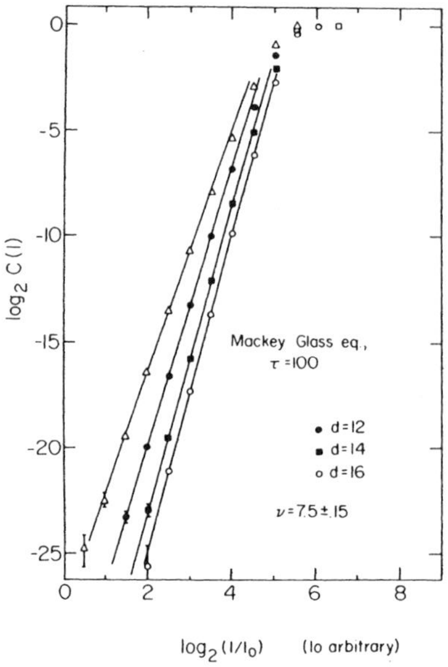

Fig. 14. Same as fig. 11, but for $\tau = 100$. For $d = 16$, the time series consisted of points $\{X(t + i\tau/2);\ i = 1, \ldots, 25.000\}$.

Kaplan–Yorke dimension D_{KY} (see eq. (1.10)) which will in the next section be shown to be an upper bound to ν, and the number of positive Lyapunov exponents, both taken from ref. 8. It is obvious that this latter number, called D_{LB}, is a lower bound to D. If the density of trajectories on the attractor is not too non-uniform, we expect that D_{LB} yields also a lower bound to ν.

From table II we see that indeed in all cases

$$D_{LB} \leqslant \nu \leqslant D \leqslant D_{KY}\,, \tag{6.3}$$

except for $\tau = 17$ where ν is slightly less than D_{LB}. However, for those small values of τ for which box-counting according to the definition of D had been feasible, our values of ν are considerably smaller than the values of D found in ref. 8, while the values of D were fairly close to D_{KY}.

In all cases, the linearity of the plot of log $C(l)$ versus log l improved substantially when increasing d above its minimal required value. For increasing values of d, the effective exponent at first also

Table II
Estimates of the correlation exponent ν for the Mackey–Glass equation (6.2) with $a = 0.2$, $b = 0.1$. Values for D_{LB}, D and D_{KY} are from ref. 8. For $\tau = 100$ the value of ν saturated at $d = 16$

τ	D_{LB}	ν	D	D_{KY}
17.0	2	1.95 ± 0.03 ($d = 3$)	2.13 ± 0.03	2.10 ± 0.02
		1.35 ± 0.03 ($d = 4$)		
		1.95 ± 0.03 ($d = 5$)		
23.0	2	2.38 ± 0.15 ($d = 3$)	2.76 ± 0.06	2.92 ± 0.03
		2.43 ± 0.05 ($d = 4$)		
		2.44 ± 0.05 ($d = 5$)		
		2.42 ± 0.1 ($d = 6$)		
30.0	3	2.87 ± 0.3 ($d = 4$)	> 2.94	3.58 ± 0.04
		3.0 ± 0.2 ($d = 5$)		
		3.0 ± 0.2 ($d = 6$)		
		2.8 ± 0.3 ($d = 7$)		
100.0	6	5.8 ± 0.3 ($d = 10$)		≈ 10.0
		6.6 ± 0.2 ($d = 12$)		
		7.2 ± 0.2 ($d = 14$)		
		7.5 ± 0.15 ($d = 16$)		

increases, but settles at a value which we assume to be the true value of ν. We must stress that we have no *proof* that the values of ν obtained with the highest chosen d represent the "true" exponent. We feel however that they surely represent reasonable estimates even for attractors with dimensions as high as ≈ 7.

In real experiments, where Lyapunov exponents are not available and thus D_{LB} and D_{KY} not easily obtained, our method seems the only one which could distinguish such an attractor from a system where the stochasticity is due to random noise. In that case, one would expect $C(l) \sim l^d$ as the trajectory is space-filling, in clear distinction from what we observe.

7. Relation to Lyapunov exponents and the Kaplan–Yorke conjecture

As we already mentioned in the introduction, the Lyapunov exponents are related to the evolution of the shape of an infinitesimal F-dimensional ball in phase space: being infinitesimal, it depends only on the linearized part of the flow, and thus becomes an ellipsoid with exponentially shrinking or growing axes. Denoting the principal axes by $\epsilon_i(t)$, the Lyapunov exponents are given by

$$\lambda_i = \lim_{t\to\infty} \lim_{\epsilon_i(0)\to 0} \frac{1}{t} \ln \frac{\epsilon_i(t)}{\epsilon_i(0)} . \tag{7.1}$$

Directions associated with positive Lyapunov exponents are called "unstable", those associated with negative exponents are called "stable".

Originally [10], Kaplan and Yorke had conjectured that D_{KY} is equal to D. In a recent preprint [27], they claim that D_{KY} is generically equal to a "probabilistic dimension", which seems to be the same as σ.

This latter claim has been partially supported in ref. 28, where essentially D_{KY} is proven to be an upper bound to the probabilistic dimension.

As shown by the counter example mentioned in the introduction, there are (possibly exceptional) cases where this bound is not saturated. In this section, we shall elucidate this question by giving a heuristic proof for the inequality $\nu \leqslant D$. From this, we see necessary conditions for the Kaplan–Yorke conjecture to hold, and which do not seem to be met generally.

Consider two infinitesimally close-by trajectories $X(t)$ and $X'(t) = X(t) + \Delta(t)$, where the latter could indeed be $X'(t) = X(t+T)$, which for sufficiently large T is essentially independent of $X(t)$. We assume that $\Delta_i(t)$ increase exponentially, without any fluctuations, as

$$\Delta_i(t) = \Delta_i(0)\, e^{\lambda_i t} , \tag{7.2}$$

where the components are along the principal axes discussed above. This is of course a strong assumption which would imply, in particular, that $\nu = \sigma = D$. Corrections to it will be treated in a forthcoming paper, but our main conclusion will remain unchanged. Conservation of the number of trajectories implies that the correlation function increases like

$$c(\Delta(t)) = \left|\frac{\partial(\Delta(0))}{\partial(\Delta(t))}\right| c(\Delta(0)) = e^{-t\sum_{i=1}^{F}\lambda_i} c(\Delta(0)) . \tag{7.3}$$

To proceed further, we need a scaling assumption which generalizes the scaling ansatz

$$c(|\Delta|) \sim |\Delta|^{\nu - F} . \tag{7.4}$$

Observing that the attractor is locally a topological product of an R^n with Cantor sets, and that the relevant axes are the principal axes, we associate with each axis an exponent ν_i, $0 < \nu_i < 1$, and make the ansatz

$$c(\Delta) \approx \prod_{i=1}^{F} c_i(\Delta_i) , \tag{7.5}$$

with

$$c_i(x) \propto \begin{cases} x^{\nu_i - 1}, & \text{if} \quad 0 < \nu_i \leqslant 1 , \\ \delta(x) , & \text{if} \quad \nu_i = 0 . \end{cases} \tag{7.6}$$

If $v_i = 0$, this means that the motion along this axis dies asymptotically (example: directions normal to a limit cycle). Directions with $v_i = 1$ are the unstable directions, with the continuous density. Directions with $0 < v_i < 1$, finally, are either Cantorian or, in exceptional cases, directions along which the distribution is continuous but singular at $\Delta_i = 0$. Notice that $v_i > 1$ is impossible.

Substituting eq. (7.5) into (7.3), we find

$$\prod_i (\Delta_i(0)^{v_i - 1} e^{t\lambda_i(v_i - 1)}) = e^{-t\Sigma_i \lambda_i} \prod_i \Delta_i(0)^{v_i - 1}, \quad (7.7)$$

or

$$\sum_{i=1}^{F} \lambda_i v_i = 0 . \quad (7.8)$$

In addition we have, from eqs. (7.5) and (7.4),

$$\sum_{i=1}^{F} v_i = v , \quad (7.9)$$

and

$$0 \leqslant v_i \leqslant 1 . \quad (7.10)$$

It is now easy to find the maximum of v subject to the constraints (7.8)–(7.10). It is obtained when

$$v_i = \begin{cases} 1, & \text{for } i \leqslant j , \\ 0, & \text{for } i \geqslant j+2 , \end{cases} \quad (7.11a)$$

and

$$v_{j+1} = \frac{1}{|\lambda_{j+1}|} \sum_{i<j} \lambda_i . \quad (7.11b)$$

Here, we have used that $\lambda_1 \geqslant \lambda_2 \geqslant \ldots$, and that $\Sigma_j \lambda_i < 0$. Expressed in words, the distribution (7.11) means that the attractor is the most extended along the most unstable directions. Inserting it into eq. (7.9), we obtain

$$v \leq j + \frac{\sum_{i \leq j} \lambda_i}{|\lambda_{j+1}|} \equiv D_{\mathrm{KY}} . \quad (7.12)$$

as we had claimed.

From the derivation it is clear that the Kaplan–Yorke conjectures $\sigma = D_{\mathrm{KY}}$ or $D = D_{\mathrm{KY}}$ cannot be expected to hold when *either the attractor is Cantorian in more than one dimension, or if the folding occurs in a direction which is not the minimally contracting one*. The latter was indeed the case for example b in fig. 1. But example b of fig. 1 is not generic, the generic case being the one where the folding is in a plane which encloses an arbitrary angle ϕ with the z-axis (see fig. 15). It is easy to convince oneself that $D = D_{\mathrm{KY}}$ whenever $\phi \neq 0$, i.e. nearly always. A still more general case is obtained if we fold in each $(2n)$th iteration in a plane characterized by ϕ_1, and each $(2n+1)$st iteration in a different plane. Again, it seems that $D = D_{\mathrm{KY}}$ is generic.

The examples might suggest that indeed $D = D_{\mathrm{KY}}$ in all those generic cases in which $v = D$, but we consider it as not very likely in high-dimensional cases. For invertible two-dimensional maps, the above conditions are of course satisfied, and thus $\sigma = D_{\mathrm{KY}}$ if $v = \sigma = D$ (see ref. 29).

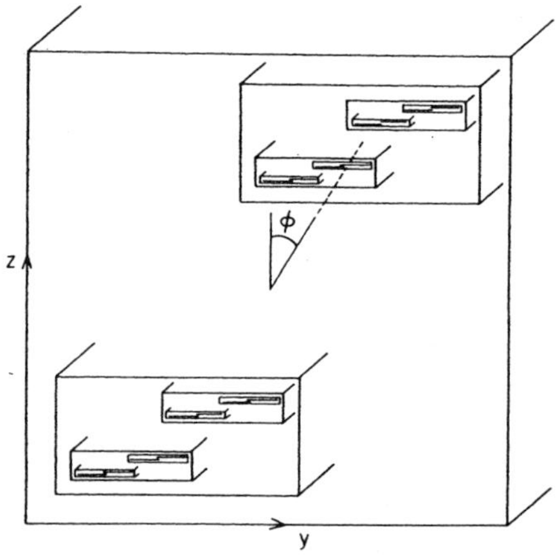

Fig. 15. Cross section through a rectangular volume element and its first 4 iterations under a map which stretches in x-direction, contracts in y- and z-directions (factors $\frac{1}{2}$ and $\frac{1}{4}$, respectively), and folds back under an angle ϕ with respect to the z-direction.

8. Conclusions

The theoretical arguments of section 3 and 6 of this paper have shown (though not with mathematical rigour) that the correlation exponent ν introduced in this paper is closely related to other quantities measuring the local structure of strange attractors.

The numerical results presented in section 2, 5 and 7 have yielded proof that ν can indeed be calculated with reasonable efforts. While all results presented in this paper were based on time series of 10.000–30.000 points, reasonable estimates of ν can already be obtained with series of a few thousand points, in most cases. Surely, for higher dimensional attractors one needs longer time series. However, rather than taking longer time series, we found it in general more important to embed the attractor in higher dimensional spaces, and to choose this embedding dimension judiciously. Compared to box-counting algorithms used previously by other authors, our method has two advantages: First, our storage requirements are drastically reduced. Secondly, in a box-counting algorithm one should iterate until *all* non-empty boxes of a given size l have been visited. This is clearly impractical, in particular if l is very small. Thus, one has systematic errors even if the number of iterations N is excessively large. In our method, there is no such problem. In particular, the finiteness of N induces no systematic errors beyond the corrections to the scaling law $C(l) \sim l^{\nu}$.

We found that in most cases ν was very close to the Hausdorff dimensions D and to the information dimension σ, with two notable exceptions. One was the Feigenbaum map, corresponding to the onset of chaos in 1 dimension. In that case, we were able to compute σ exactly in an analytic way, with the result $\sigma \neq D$, supporting the numerical evidence for $\nu < \sigma$.

The other exception was the Mackey–Glass delay equation, where we found numerically $\nu < D$. The information dimension has not been calculated directly in this case. Accepting the claim made in ref. 8 that the Kaplan–Yorke formula (1.10) predicts correctly σ, we would have $\nu < \sigma = D = D_{KY}$. This seems somewhat surprising, since we argued in section 7 that a rather direct connection (as an inequality $\nu \leqslant D_{KY}$) exists between ν and D_{KY}, while a connection between σ and D_{KY} seems less evident to us.

The main conclusion of this paper, as far as experiments are concerned, is that one can distinguish deterministic chaos from random noise. By analyzing the signal as explained in section 5, and embedding the attractor in an increasingly high dimensional space, one finds whether $C(l)$ scales like l^{ν} or l^{d}. With a random noise the slope of log $C(l)$ vs. log l will increase indefinitely as d is increased. For a signal that comes from a strange attractor the slope will reach a value of ν and will then become d independent.

An issue of experimental importance is the effect of random noise *on top* of the deterministic chaos. The treatment of this question is beyond the scope of this paper and is treated elsewhere [30]. Here we just remark that when there is an external noise of a given mean square magnitude, a plot of log $C(l)$ vs. log l has two regions. For length scales above those on which the random component blurrs the fractal structure, $C(l)$ continues to scale like l^{ν}. On length scales below those that are affected by the random jitter of the trajectory, $C(l)$ scales like l^{d}. The analysis of experimental signals along these lines can therefore yield simultaneously a characterization of the strange attractor *and* and estimate of the size of the random component. For more details see ref. 30.

It is thus our hope that the correlation exponent will indeed be measured in experiments whose dynamics is governed by strange attractors.

Acknowledgements

This work has been supported in part by the Israel Comission for Basic Research. P.G. thanks the Minerva Foundation for financial support. We thanks Drs. H.G.E. Hentschel and R.M. Mazo for a number of useful discussions.

Appendix A

All numerical calculations were performed in double precision arithmetic on an IBM 370/165 at the Weizmann Institute.

The integrations of the Lorenz and Rabinovich–Fabrikant equations were done using a standard Merson–Runge–Kutta subroutine of the NAG library.

In order to integrate the Mackey–Glass delay equation we approximated it by a N-dimensional set of difference equations by introducing a time step

$$\Delta t = \tau / n \,, \tag{A.1}$$

with n being some large integer, and writing

$$x(t + \Delta t) \approx x(t) + \frac{\Delta t}{2} (\dot{x}(t) + \dot{x}(t + \Delta t)) \,. \tag{A.2}$$

Notice that this, being the optimal second-order approximation, is a very efficient algorithm – provided we can compute $\dot{x}(t + \Delta t)$. In the present case we can, due to the special form

$$\dot{x}(t) = f(x(t - \tau)) - bx(t) \,. \tag{A.3}$$

Inserting this in eq. (A.2) and rearranging terms, we arrive at

$$x(t + \Delta t) = \frac{2 - b\Delta t}{2 + b\Delta t} x(t) + \frac{\Delta t}{2 + b\Delta t} \times \{f(x(t - \tau)) + f(x(t - \tau + \Delta t))\} \,. \tag{A.4}$$

In all runs shown in this paper, we used $n = 600$ (corresponding to $0.03 \lesssim \Delta t \lesssim 0.15$), except for the runs with $\tau = 100$, where we used $n = 1200$ and with $n = 600$, finding no appreciable differences.

We also performed control runs with a fourth-order approximation instead of eq. (A.2). The correlation integral was unchanged within statistical errors, and the stability of the solutions did not seem to improve much. This could result from the very large higher derivatives of x, resulting from the tenth power in eq. (6.2).

In order to ensure that all $\boldsymbol{x}_i$ are on the attractor, the first 100–200 iterations were discarded.

Generating the time series $\{\boldsymbol{X}_i\}_{i=1}^{N}$ was indeed the less time-consuming part of our computation, the more important part consisting of calculating the $N(N-1)/2 \gtrsim 10^8$ pairs of distances $r_{ij} = |\boldsymbol{X}_i - \boldsymbol{X}_j|$ and summing them up to get the correlation integral.

In particular, we found that an efficient algorithm for the latter was instrumental in applying the method advocated in this paper.

Such a fast algorithm was found using the fact that floating-point numbers are stored in a computer in the form

$$r = \pm \text{ mantissa} \cdot \text{base}^{+\exp} \,. \tag{A.5}$$

with base = 16 in our case 1/base < mantissa < 1, and exp being an integer. If one can extract the exponent, one can bin the r_{ij}'s in bins of widths increasing geometrically. By extracting the exponent of an arbitrary power r^p of r, one can furthermore choose the width of this binning arbitrarily. Access to the exponent is made very easy and fast by using the shifting and masking operations available e.g. in extended IBM and in CDC Fortran. After having computed the numbers N_K of pairs (i, j) in the interval $2^{k-1} < r_{ij} < 2^k$, the correlation integrals are obtained by

$$c(r = 2^k) = \frac{1}{N^2} \sum_{k' = -\infty}^{k} N_{k'} \,. \tag{A.6}$$

We found this method to be nearly an order of magnitude faster than computing e.g. the logarithmics of r_{ij} directly, and binning by taking their integer parts. A typical run with 20.000 points took – depending on the model studied – between 15 and 30 minutes CPU time.

References

[1] E.N. Lorenz, J. Atmos. Sci. 20 (1963) 130.
[2] R.M. May, Nature 261 (1976) 459.
[3] D. Ruelle and F. Takens, Commun. Math. Phys. 20 (1971) 167.

[4] E. Ott, Rev. Mod. Phys. 53 (1981) 655.
[5] J. Guckenheimer, Nature 298 (1982) 358.
[6] B. Mandelbrot, *Fractals – Form, Chance and Dimension* (Freeman, San Francisco, 1977).
[7] V.I. Oseledec, Trans. Moscow Math. Soc. 19 (1968) 197. D. Ruelle, Proc. N.Y. Acad. Sci. 357 (1980) 1 (R.H.G. Helleman, ed.).
[8] J.D. Farmer, Physica 4D (1982) 366.
[9] H. Mori, Progr. Theor. Phys. 63 (1980) 1044.
[10] J.L. Kaplan and J.A. Yorke, in: Functional Differential Equations and Approximations of Fixed Points, H.-O. Peitgen and H.-O. Walther, eds. Lecture Notes in Math. 730 (Springer, Berlin, 1979) p. 204.
[11] D.A. Russel, J.D. Hanson and E. Ott, Phys. Rev. Lett. 45 (1980) 1175.
[12] H. Froehling, J.P. Crutchfield, D. Farmer, N.H. Packard and R. Shaw, Physica 3D (1981) 605.
[13] P. Grassberger, J. Stat. Phys. 26 (1981) 173.
[14] H.S. Greenside, A. Wolf, J. Swift and T. Pignataro, Phys. Rev. A25 (1982) 3453.
[15] P. Grassberger and I. Procaccia, Phys. Rev. Lett. 50 (1983) 346. Related discussions can be found in a preprint by F. Takens "Invariants Related to Dimensions and Entropy".
[16] M. Feigenbaum, J. Stat. Phys. 19 (1978) 25; 21 (1979) 669.
[17] M.C. Mackey and L. Glass, Science 197 (1977) 287.
[18] M. Hénon, Commun. Math. Phys. 50 (1976) 69.
[19] G.M. Zaslavskii, Phys. Lett. 69A (1978) 145.
[20] M.I. Rabinovich and A.L. Fabrikant, Sov. Phys. JETP 50 (1979) 311. (Zh. Exp. Theor. Fiz. 77 (1979) 617).
[21] W. Feller, An Introduction to Probability Theory and its Applications, vol. 2, 2nd ed. (Wiley, New York, 1971) p. 155.
[22] B.B. Mandelbrot, in: *Turbulence and the Navier–Stokes Equations*, R. Teman, ed., Lecture Notes in Math. 565 (Springer, Berlin, 1975). H.G.E. Hentschel and I. Procaccia, Phys. Rev. A., in press.
[23] D. Stauffer, Phys. Rep. 54C (1979) 1.
[24] T.A. Witten, Jr., and L.M. Sander, Phys. Rev. Lett. 47 (1981) 1400.
[25] N.H. Packard, J.P. Crutchfield, J.D. Farmer and R.S. Shaw, Phys. Rev. Lett. 45 (1980) 712.
[26] F. Takens, in: Proc. Warwick Symp. 1980, D. Rand and B.S. Young, eds, Lectures Notes in Math. 898 (Springer, Berlin, 1981).
[27] P. Frederickson, J.L. Kaplan, E.D. Yorke and J.A. Yorke, "The Lyapunov Dimension of Strange Attractors" (revised), to appear in J. Diff. Eq.
[28] F. Ledrappier, Commun. Math. Phys. 81 (1981) 229.
[29] L.S. Young, "Dimension, Entropy, and Lyapunov Exponents" preprint.
[30] A. Ben-Mizrachi, I. Procaccia and P. Grassberger, Phys. Rev. A, submitted.

Invariant Measures and the Variational Principle for Lozi Mappings

Marek Rychlik

Department of Mathematics
University of Arizona
Tucson, AZ 85721 USA
phone: (602) 621-6864, FAX: (602) 621-8322
rychlik@math.arizona.edu

Summary. This article contains a proof of the existence of SBR measures for the family of maps of the plane known as Lozi maps: $L(x, y) = (1-a|x|+y,\ bx)$. We also prove that the number of SBR measures is finite. Our approach also yields invariant measures for other classes of dynamical systems like piecewise expanding mappings of an interval with infinitely many pieces and Hölder derivative. Although known to specialists from my 1983 dissertation (University of California, Berkeley), the results presented in this article have never been published. I am delighted that the article found its place in the current volume, as the ideas of several papers that Jim Yorke co-authored or inspired played an important role in my approach.

1 Introduction

1.1 Preliminary Remarks

The main goal of this paper is the study of Lozi mappings introduced in (cf. [9]). Lozi noticed the chaotic behavior of the orbits of these mappings, similar to that of the Hénon mappings (cf. [5]). The formulas for Lozi mappings

$$L(x, y) = (1 - a|x| + y,\ bx) \tag{1}$$

and for Hénon mappings

$$H(x, y) = (1 - ax^2 + y,\ bx) \tag{2}$$

show some qualitative similarity. However, Lozi mappings are much easier to study because they admit uniform hyperbolic structure (cf. [10]). The only complication is the presence of the singularity $x = 0$. It makes Lozi mappings similar to the dispersed billiards of Bunimovich and Sinai. Using this analogy, Collet (cf. [3]) proved similar results to ours by constructing Markov partitions for Lozi mappings, repeating the method of Bunimovich and Sinai (cf. [2]) (however for a smaller range of parameters).

We use a different analogy. Lozi mappings are similar to maps of an interval. More precisely, if we build a factor map, utilizing the partition into stable intervals, then this factor map has properties formulated as conditions I–IV in section 2 of the paper, which are satisfied for piecewise expanding, $C^{1+\epsilon}$ mappings of an interval.

Pursuing this analogy, in section 2 we prove the existence of absolutely continuous invariant measures for a map satisfying conditions I–IV. In general, the map has more than one such measure (and this occurs for certain cases of iterations of Lozi mappings). We need a spectral decomposition theorem (Theorem 2) which gives all *ergodic* absolutely continuous measures. Actually, without much trouble (using the Pinsker partition) we could prove that the natural extension of this factor map (which for Lozi mappings coincides with L itself) has a finite number of K-components. From Pesin theory and from section 3, which gives a detailed description of invariant measures for Lozi mappings, it follows that L is Bernoulli on the K-components.

Misiurewicz in (cf. [10]) proved that L has a strange attractor for certain parameters a and b. We repeat his considerations in section 3, slightly modifying his trapping region and the methods of the proofs. In section 3 we also prove that a large class of measures (including the Lebesgue measure restricted to the trapping region) satisfy an ergodic theorem (Theorem 4).

We also prove a version of Bowen's theorem for functions (Theorem 5).

In section 4 we prove a variational principle for the measures found in section 2. This result holds in the form suggested by Bowen and Ruelle (cf. [1]). We could also say the same in Pesin's terms: For every invariant measure $\mu, h_\mu(L)$ is not larger than the positive Lyapunov exponent; $h_\mu(L)$ equals the Lyapunov exponent iff μ is one of the measures constructed in this paper.

The author hopes that the methods developed in this paper will be useful in studying other dynamical systems, in particular Hénon mappings.

There is one generalization of the results which is almost obvious: instead of L we can consider a small perturbation $L + f$ where $f \in C^{1+\epsilon}(\mathbb{R}^2)$. The results remain valid. What makes it possible is that the iterations of the singularities under the new map behave similarly to those for L.

We notice that section 2 creates a new class of examples of piecewise expanding mappings of class $C^{1+\epsilon}$ on the pieces and with infinite number of pieces. The case of finitely many pieces was considered in (cf. [7] and in one unpublished paper of the author).

2 Abstract Theorems

2.1 Abstract Reduction Theorem

Let (X, Σ, m) be a Lebesgue space with a σ-algebra Σ and a probability measure m. Let $T : X \to X$ be a measurable, nonsingular mapping. The latter means that $T_* m$ is absolutely continuous with respect to m (abbr. $T_* m \ll m$).

We define the Perron-Frobenius operator of T. We have $P_T : L^1(X, \Sigma, m) \to L^1(X, \Sigma, m)$ and:

$$P_T f = \frac{d(T_*(fm))}{dm} \qquad \text{(Radon-Nikodym derivative)}. \tag{3}$$

It is easy to see that P_T is determined by the following condition:

$$\forall h \in L^\infty(X, \Sigma, m) : \int_X (h \circ T) \cdot f dm = \int_X h \cdot P_T f dm. \tag{4}$$

This means that $T : L^\infty(X, \Sigma, m) \to L^\infty(X, \Sigma, m)$ defined by $Th = h \circ T$ is the adjoint operator of P_T.

A measurable, countable partition α of X is called regular (or T-regular) iff for every $A \in \alpha$, $T(A)$ is Σ-measurable and $T|A$ maps $(A, \Sigma|A)$ onto $(T(A), \Sigma|T(A))$ isomorphically (spaces with fixed σ-algebras, where $\Sigma|A$ and $\Sigma|T(A)$ are σ-algebras defined by the restriction of Σ to A and $T(A)$ respectively).

For any regular partition we define $g_T : X \to \mathbb{R}_+$ as follows:

$$g_T(x) = \frac{d(T_*(\chi_A \cdot m))}{dm}(Tx) \quad \text{as } x \in A \in \alpha. \tag{5}$$

In other words, $g_T = \sum_{A \in \alpha} (TP)(\chi_A)\chi_A$. One can check easily that g_T does not depend on the choice of α. Moreover, g_T is the reciprocal of the Jacobian of T. Using g_T, we can express P_T in the following way:

$$P_T f(x) = \sum_{y \in T^{-1}(x)} g_T(y) \cdot f(y), \quad x \in X. \tag{6}$$

Of course, g_T is determined up to a set of measure 0 and (6) holds m-a. e.

We study in detail the case where T is a factor of another mapping $S : Y \to Y$, where (Y, β, ν) is a Lebesgue space. We assume that S is nonsingular. By ξ we denote the measurable partition of Y which is S-invariant ($S^{-1}\xi \le \xi$) and $X = Y/\xi$ and $T = S_\xi$ (factor map). Let $\pi : Y \to X$ be the natural projection. Let $C(x)$ denote the element $\pi^{-1}(x) \in \xi$. We have $S(C(x)) \subset C(Tx)$. The situation is illustrated by following commutative diagram:

$$\begin{array}{ccc} Y & \xrightarrow{S} & Y \\ \downarrow{\scriptstyle \pi} & & \downarrow{\scriptstyle \pi} \\ X & \xrightarrow{T} & X \end{array} \tag{7}$$

Our aim is to find the connection between P_T and P_S.

Proposition 1. *For every $f \in L^1(Y, \beta, \nu)$:*

$$P_T(f|\xi) = E_\nu(P_S f|\xi) \tag{8}$$

where $E_\nu(\cdot|\xi)$ is the operator of taking conditional expectation with respect to the partition (and meant as an operator from $L^1(Y, \beta, \nu)$ to $L^1(X, \Sigma, m)$.

$$\begin{array}{ccc} L^1(\nu) & \xrightarrow{P_S} & L^1(\nu) \\ \downarrow E_\nu(\cdot|\xi) & & \downarrow E_\nu(\cdot|\xi) \\ L^1(m) & \xrightarrow{P_T} & L^1(m) \end{array} \tag{9}$$

Proof. Let $h \in L^\infty(m)$. Then:

$$\begin{aligned}
\int_X h \cdot P_T E(f|\xi)\, dm &= \int_X (h \circ T) \cdot E_\nu(f|\xi)\, dm \\
= \int_Y (h \circ T \circ \pi)(E_\nu(f|\xi) \circ \pi)\, d\nu &= \int_Y (h \circ \pi \circ S) \cdot f\, d\nu \\
= \int_Y (h \circ \pi) \cdot P_S f\, d\nu &= \int_Y (h \circ \pi)(E_\nu(P_S f|\xi) \circ \pi)\, d\nu \\
= \int_X h \cdot E_\nu(P_S f|\xi)\, dm.
\end{aligned}$$

We assume that S has a regular partition α with the following property:

$$S^{-1}\xi \vee \alpha = \xi. \tag{10}$$

Usually we will have $\xi = \alpha^- = \bigvee_{k=0}^{\infty} S^{-k}(\alpha)$ and (10) will hold automatically.

Lemma 1. *The family $\beta = \{\pi(A)\}_{A \in \alpha}$ is a T-regular partition of X.*

Proof. By (10) it follows that $\alpha \leq \xi$ and $\pi^{-1}(\pi A) = A$ for every $A \in \alpha$. Thus, β is a partition. We also check easily that $\pi^{-1}(T(\pi A)) = S(A)$. Then, $T(\pi A)$ is also measurable. Obviously $T|\pi(A) : \pi(A) \to T(\pi(A))$ is a factor of $S|A : A \to S(A)$. Moreover, (10) shows that $T|\pi(A)$ is 1:1 and $(T|\pi(A))^{-1}$ is a factor of $(S|A)^{-1}$. So, $T|\pi(A)$ is an isomorphism of $(\pi(A), \Sigma|\pi(A))$ and $(T(\pi(A)), \Sigma|T(\pi(A)))$.

By $(\nu_C)_{C \in \xi}$ we denote the family of all conditional measures of ν with respect to ξ. The following proposition joins g_S, g_T and $(\nu_C)_{C \in \xi}$ together in one formula.

Proposition 2. *For almost every $x \in X$ and for almost every $y \in C(x)$ (in the sense of $\nu_{C(x)}$):*

$$g_T(x) = g_S(y) \frac{d\left((S|A)_*^{-1} \nu_{C(Tx)}\right)}{d\nu_{C(x)}}(y) \tag{11}$$

where A is the only element of α which contains $C(x)$ (we notice that $C(x) = S^{-1}C(Tx) \cap A$). In particular, $(S|A)_*^{-1}\nu_{C(Tx)}$ is equivalent to $\nu_{C(x)}$ for almost every $x \in X$.

Proof. Let us fix some $f \in L^\infty(\nu)$. Then $E_\nu(P_S f|\xi)(x)$ can be written as $\int_{C(x)} \sum_{A\in\alpha}(f \cdot g_S) \circ (S|A)^{-1} d\nu_{C(x)}$ or $\int f\, d\sigma_{1,x}$, and $P_T E_\nu(f|\xi)(x)$ can be written as $\sum_{z\in T^{-1}(x)} \int_{C(z)} f d\nu_{C(z)} \cdot g_T(z)$ or $\int f\, d\sigma_{2,x}$, where

$$\sigma_{1,x} = \sum_{z\in T^{-1}(x)} g_S \cdot \left((S|C(z))_*^{-1} \nu_{C(x)}\right), \qquad \sigma_{2,x} = \sum_{z\in T^{-1}(x)} g_T(z) \cdot d\nu_{C(z)}.$$

Since f is arbitrary, these formulas and Proposition 1 give us $\sigma_{1,x} = \sigma_{2,x}$ for almost every $x \in X$. Since $\nu_{C(z)}$ have disjoint supports and since g_T and g_S are positive almost everywhere, this implies (11).

In many examples the elements of ξ are endowed with a measure which is different from the conditional one. We assume that $(l_C)_{C\in\xi}$ is a family of measures such that for every $C \in \xi$, l_C is equivalent to ν_C and the function

$$\rho = \frac{d\nu_C}{dl_C}$$

defined on Y is β-measurable. We notice that for a. e. $x \in X$, $(S|C(x))_*^{-1} l_{C(Tx)}$ is also equivalent to l_C. Besides, function λ, defined by

$$\lambda(y) = \frac{d\left((S|C(x))_*^{-1}\, l_{C(Tx)}\right)}{dl_{C(x)}}(y), \quad \text{as} \quad y \in C(x) \tag{12}$$

is also β-measurable.

Example 1. Let $S : Y \to Y$ be an Anosov diffeomorphism of a Riemannian manifold Y. Let ξ be an invariant measurable partition which consists of pieces of stable manifolds (one can easily construct such ξ using a Markov partition). Every element $C \in \xi$ is endowed with the induced Riemannian metric and, hence, with the volume l_C. If ν is the volume measure on the entire manifold, then (if S is at least of class $C^{1+\epsilon}$) conditional measures are equivalent to l_C.

We observe that Proposition 2 gives

$$g_T \circ \pi = g_S \frac{\rho \circ S}{\rho} \lambda. \tag{13}$$

If y and y' belong to the same C then (13) yields

$$\frac{\rho(y)}{\rho(y')} = \frac{g_S(y)}{g_S(y')} \frac{\rho(Sy)}{\rho(Sy')} \frac{\lambda(y)}{\lambda(y')}. \tag{14}$$

By induction we prove that

$$\frac{\rho(y)}{\rho(y')} = \left(\prod_{k=0}^{n-1} \frac{g_S(S^k y)\lambda(S^k y)}{g_S(S^k y')\lambda(S^k y')}\right) \frac{\rho(S^n y)}{\rho(S^n y')}. \tag{15}$$

This observation gives the following

Proposition 3. *For a. e. $x \in X$ and for $\nu_{C(x)}$ a. e. $y, y' \in C(x)$, the following conditions are equivalent:*

1.

$$\lim_{n\to\infty} \frac{\rho(S^n y)}{\rho(S^n y')} = 1$$

2.

$$\frac{\rho(y)}{\rho(y')} = \prod_{k=0}^{\infty} \frac{g_S(S^k y)\lambda(S^k y)}{g_S(S^k y')\lambda(S^k y')}$$

and the product converges.

Remark 1. If (1) holds a. e. then (2) and the condition $\int \rho dl_C = 1$ determine ρ completely.

2.2 The Existence of Invariant Measures and the Spectral Decomposition

As in the previous subsection, T is a nonsingular map of X. X is a Lebesgue space. Let β be a regular partition of X which is a generator for T. A generator means one-sided generator, i. e. $\beta^- = \bigvee_{k=0}^{\infty} T^{-k}(\beta) = \epsilon$, where ϵ is a partition into points.

We ask if there exist T-invariant measures absolutely continuous with respect to m. This problem is equivalent to that of the existence of an eigenvector of P_T corresponding to eigenvalue 1. We notice that 1 is always in the spectrum of P_T, but it may happen that it is not an eigenvalue (see cf. [8]). However, in particular situations we are able to give a positive answer to this question.

In this section we write g_n instead of g_{T^n}, g instead of $g_T = g_1$ and P instead of P_T (g_n is well defined since β^n is a regular partition for T^n); supremum (infimum) means essential supremum (infimum). For every $A \in \Sigma$ we define $\beta(A) = \{B \in \beta : m(B \cap A) > 0\}$.

The following conditions allow us to prove the existence of an eigenvector of P corresponding to eigenvalue 1.

(I) Distorsion condition: $\exists d \in (0,\infty) \forall_{n\geq 1} \forall_{B\in\beta^n} : \sup_B g_n \leq d \cdot \inf_B g_n$;

(II) Localization condition: $\exists \epsilon > 0, r \in (0,1)\ \forall_{n\geq 1}\ \forall_{B\in\beta^n}$:

$$m(T^n B) < \epsilon \implies \sum_{B'\in\beta(T^n B)} \sup_B g \leq r;$$

(III) Bounded variation condition: $\sum_{B\in\beta} \sup_B g < +\infty$.

The reader will note the order of quantifiers in these statements. Conditions (I) and (III) are often used in the theory of the existence of absolutely continuous measures (see [13]). Condition (II) is crucial in our method of treating situations with no Markov partition available.

Remark 2. Condition (II) holds automatically if for some $\epsilon' > 0$, $m(T^n B) > \epsilon'$ for all n and $B \in \beta^n$ (this is true if, for instance, β is related to a finite Markov partition). It says that if $T^n B$ is small in the sense of measure then it cannot intersect too many elements of β with a large value of g.

Let us note that if (I) and (III) hold for T, β and g, then they hold for T^N, β^n and g_N for every $N \geq 1$. Condition (I) holds with the same value of d. Besides,

$$\sum_{B \in \beta^n} \sup_B g_N \leq \left(\sum_{B \in \beta} \sup_B g \right)^N .$$

Thus we know that conditions (I)–(III) for iterations of T can be verified in the following way: we check (I) and (III) for T and then check only (II) for some iteration of T.

Theorem 1. *Let (I)–(III) be satisfied. Then the sequence $(P^n 1)_{n \geq 1}$ is bounded in $L^\infty(m)$ and the averages*

$$\frac{1}{n} \sum_{k=0}^{n-1} P^k 1 \tag{18}$$

converge in $L^1(m)$ to some $\phi \in L^\infty(m)$ such that $P\phi = \phi$.

Proof. If the averages (18) are bounded in $L^\infty(m)$ then they form a weakly sequentially conditionally compact sequence and hence, by Yosida's ergodic theorem [4] they converge to $\phi \in L^\infty(m)$ with the property $P\phi = \phi$. Therefore, we only have to check that $(P^n 1)_{n \geq 1}$ is bounded in $L^\infty(m)$.

Let $\gamma_n = \sum_{B \in \beta^n} \sup_B g_n$. It is obvious that $\|P^n 1\|_\infty \leq \gamma_n$ for every $n \geq 1$. If we prove that γ_n is bounded then the theorem would follow. We will prove by induction that

$$\gamma_{n+1} \leq r \gamma_n + C_1, \tag{19}$$

where $C_1 = (\gamma_1 \cdot d)/\epsilon$ ($\gamma_n < \infty$ because of (III)). In view of $g_{n+1} = g_n (g \circ T^n)$ we have for every $B' \in \beta^n$:

$$\sum_{\{B \in \beta^{n+1} : B \subset B'\}} \sup_B g_{n+1} \leq \sup_{B'} g_n \sum_{\{B'' \in \beta(T^n B')\}} \sup_{B''} g. \tag{20}$$

Let $\beta_1 = \{B' \in \beta^n : m(T^n B') < \epsilon\}$ and $\beta_2 = \{B' \in \beta^n : m(T^n B') \geq \epsilon\}$. If $B' \in \beta_1$ then the right-hand side of (20) is not greater than $r \cdot \sup_{B'} g_n$, by (II). If $B' \in \beta_2$ then by (I) $\sup_{B'} g_n \leq d \cdot \inf_{B'} g_n \leq d \frac{m(B')}{m(T^n B')} \leq \frac{d}{\epsilon} m(B')$.

For every $B \in \beta^{n+1}$ there is a unique $B' \in \beta^n$ containing B. We obtain $\gamma_{n+1} = \sum_{B \in \beta^{n+1}} \sup_B g_{n+1} = \sum_{B' \in \beta^n} \sum_{\{B \in \beta^{n+1} :\, B \subset B'\}} \sup_B g_{n+1} \leq \sum_{B' \in \beta_1} r \cdot \sup_{B'} g_n + \sum_{B' \in \beta_2} \gamma_1 \frac{d}{\epsilon} m(B') \leq r\gamma_n + C_1$. Hence, we have proved (19). Since $0 < r < 1$, this implies that $(\gamma_n)_{n \geq 1}$ is bounded.

Remark 3. The inequality (19) implies that $\|\phi\|_\infty \leq \frac{C_1}{1-r} = \frac{\gamma_1 \cdot d}{\epsilon(1-r)}$.

Theorem 1 states the existence of an invariant, absolutely continuous with respect to m measure $\phi \cdot m$, but it gives no information about the number and properties of such measures. One can improve this result by adding the following condition:

(IV) Expanding condition: $\exists r \in (0,1) : \sup_X g \le r$.

We can always assume that r is chosen to satisfy both (II) and (IV).

Theorem 2. *We assume (I)–(IV). Then there exists a bounded, finite dimensional projection $Q : L^1(m) \to L^\infty(m)$ such that*

1. *$Q(L^1(m)) \subset L^\infty(m)$ and Q is bounded as an operator from L^1 to L^∞.*
2. *For every $f \in L^1(m)$ the averages*

$$\frac{1}{n}\sum_{k=0}^{n-1} P^k f \tag{21}$$

   converge in $L^1(m)$ to Qf, as $n \to \infty$.
3. *The range of Q (denoted by $\mathcal{R}(Q)$) consists of all eigenvectors of P corresponding to the eigenvalue 1 (and of 0).*
4. *There exist non-negative functions $\phi_1, \phi_2, \ldots, \phi_s \in \mathcal{R}(Q)$ which span $\mathcal{R}(Q)$ and $\phi_i \wedge \phi_j = 0$ as $i \neq j$, where $\phi \wedge \psi \stackrel{\text{def}}{=} \min(\phi, \psi)$. Moreover, $m(\phi_i) = 1$ for $i = 1, 2, \ldots, s$ and if*

$$C_i = \bigcup_{n=0}^{\infty} T^{-n}\{x : \phi_i(x) > 0\}$$

   is the basin of the measure $\phi_i m$ then Q can be represented as

$$Qf = \sum_{i=1}^{s} \int_{C_i} f\, dm \cdot \phi_i. \tag{22}$$

Moreover, $\bigcup_{i=1}^{s} C_i = X$ (up to a set of measure 0) and $(\phi_i)_{i=1}^{s}$ are the only functions $\phi \in L^1(m)$ such that $\phi \cdot m$ is a T-invariant, ergodic, probabilistic measure.

Proof. Let $\mathcal{A}$ be the set of all pairs (k, B) where k is a non-negative integer and $B \in \beta^n$ for some $n \geq k$. If $a \in \mathcal{A}$, $a = (k, B)$ then for every $B' \in \beta$, aB' denotes the pair $(k+1, B \cap T^{-k}B') \in \mathcal{A}$. We notice that for $k < n$ the set $B \cap T^{-k}B'$ has positive measure for the unique B' determined by the condition $T^k B \subset B'$ and $B \cap T^{-k}B' = B$. We denote this element of $\mathcal{A}$ by $a+1$. By f_a we denote the function $P^k(\chi_B) \in L^1(m)$ (as $a = (k, B)$). Using this notation, we can write

$$Pf_a = \sum_{B' \in \beta} f_{aB'} = \begin{cases} f_{a+1} & \text{if } n > k, \\ \sum_{B' \in \beta(T^n B)} f_{aB'} & \text{if } n = k. \end{cases} \tag{23}$$

Hence, the subspace of all linear combinations of the elements of the family $(f_a)_{a\in\mathcal{A}}$ is P-invariant (but not closed in $L^1(m)$, in general). We wish to know how the coefficients in such linear combinations change under P. Therefore, we introduce the vector space V of all families $(x_a)_{a\in\mathcal{A}}$ of real numbers, only a finite number of them being different from 0. Let $(e_a)_{a\in\mathcal{A}}$ be the standard basis of V (e_a is the family of reals that has the unit on the a-th place and zeros elsewhere). We define a linear map $\Phi : V \to L^1(m)$ by putting $\Phi e_a = f_a$ for every $a \in \mathcal{A}$. If $\mathbf{x} = (x_a)_{a\in\mathcal{A}} \in V$ then

$$\Phi\mathbf{x} = \sum_{a\in\mathcal{A}} x_a f_a. \tag{24}$$

We also define the linear operator $\bar{P} : V \to V$ by

$$\bar{P}e_a = \begin{cases} e_{a+1} & \text{if } n > k \\ \sum_{B'\in\beta(T^nB)} e_{aB'} & \text{if } n = k, \end{cases} \tag{25}$$

where $a = (k, B)$ and $B \in \beta^n$, so that the following diagram commutes:

$$\begin{array}{ccc} V & \xrightarrow{\Phi} & L^1(m) \\ \downarrow{\scriptstyle \bar{P}} & & \downarrow{\scriptstyle P} \\ V & \xrightarrow{\Phi} & L^1(m) \end{array} \tag{26}$$

One can say that $\bar{P}$ acts on the coefficients of linear combinations while P acts on those combinations themselves.

By V_+ we denote the convex cone of all families of non-negative numbers. It is obvious that $W_0 = \mathcal{R}(\Phi)$ is dense in $L^1(m)$ because β is a generator. We also introduce a norm in V by

$$\|\mathbf{x}\| = \sum_{a\in\mathcal{A}} |x_a| \cdot w_a \tag{27}$$

where $(w_a)_{a\in\mathcal{A}}$ is the family of reals given by $w_a = \|f_a\|_\infty$, as $a \in \mathcal{A}$.

Let $|x|$ denote the vector $(|x_a|)_{a\in\mathcal{A}} \in V_+$ First we prove the following lemma:

Lemma 2. *There exists $R \in (0, \infty)$ such that for every $\mathbf{x} \in V_+$:*

$$\|\bar{P}\mathbf{x}\| \leq r \cdot \|\mathbf{x}\| + R\|\Phi\mathbf{x}\|_1. \tag{28}$$

Proof. Given $a = (k, B) \in \mathcal{A}$ we notice that $w_a = \sup_B g_k$. We put $C_2 = d/\epsilon$. We claim that if $\|f_a\|_\infty \geq C_2\|f_a\|_1$ then

$$\sum_{B\in\beta} w_{aB} \leq r \cdot w_a. \tag{29}$$

In fact, $\|f_a\|_1 = m(B)$ and $\|f_a\|_\infty \geq C_2\|f_a\|_1$ imply that $\sup_B g_k \geq C_2 m(B)$. But $m(B) = (m(B)/m(T^kB)) \cdot m(T^kB) \geq \inf_B g_k \cdot m(T^kB) \geq 1/d \cdot \sup_B g_k \cdot$

$m(T^kB)$. Therefore $m(T^kB) \leq C_2^{-1}d = \epsilon$. Our assertion follows by applying inequality (20) and reasoning as in the proof of Theorem 1.

Let $\mathcal{A}'$ be the set of all $a \in \mathcal{A}$ such that $\|f_a\|_\infty \geq C_2 \cdot \|f_a\|_1$ and $\mathcal{A}'' = \mathcal{A} \backslash \mathcal{A}'$. Then $\|\bar{P}\mathbf{x}\| \leq \sum_{a\in\mathcal{A}} x_a \|\bar{P}e_a\| \leq \sum_{a\in\mathcal{A}} x_a \sum_{B'\in\beta} w_{aB'} \leq \sum_{a\in\mathcal{A}'} x_a r w_a + C_2\gamma_1 \sum_{a\in\mathcal{A}''} x_a \|f_a\|_1 \leq r\|\mathbf{x}\| + C_2\gamma_1\|\Phi\mathbf{x}\|_1$. Putting $R = C_2\gamma_1$, we obtain the lemma.

Now we continue with the proof of Theorem 2. Let us fix some $x \in V_+$ and take $f = \Phi x$. We notice that $\|P^n f\|_\infty \leq \|P^n x\|$, since $\|\Phi x\|_\infty \leq \|x\|$ (see formula 26). Lemma 2 gives us $\|\bar{P}^n x\| \leq R(1 + r + \ldots + r^{n+1})\|f\|_1 + r^n\|x\| \leq \frac{R}{1-r}\|f\|_1 + r^n\|x\|$. Hence the sequence $(P^n f)_{n\geq 1}$ is bounded in $L^\infty(m)$. So, we can apply the Yosida ergodic theorem again and we get that the limit of (21) exists for at least $f \in \Phi(V_+)$. Moreover, for such f we can use the previous inequality and obtain that for every $f \in L^1(m)$, $f \geq 0$ $\|Qf\|_\infty \leq \frac{R}{1-r}\|f\|_1$. This is because $\Phi(V_+)$ is dense in the set of all nonnegative functions f, so Qf is well defined on this set (by virtue of $\|P\|_1 = 1$). Moreover, since every $f \in L^1(m)$ can be represented as $f_+ - f_-$, where $f_+ = f \vee 0$ and $f_- = (-f)\vee 0$, then this limit exists for every $f \in L^1(m)$ and $\|Qf\|_\infty \leq \|Qf_+\|_\infty + \|Qf_-\|_\infty \leq \frac{R}{1-r}\|f_+\|_1 + \frac{R}{1-r}\|f_-\|_1 = \frac{R}{1-r}\|f\|_1$. Other conclusions of Theorem 2 can be proved easily if one uses the fact that the set $\{f \in \mathcal{R}(Q) : f \geq 0, m(f) = 1\}$ has only finitely many extreme points. We define $\phi_1, \phi_2, \ldots, \phi_s$ to be the only extreme points of this set. For details, see [13].

Remark 4. In [13] it was noted that from the fact that $Q : L^1(m) \to L^\infty(m)$ is bounded with norm $\leq \frac{R}{1-r}$ follows an explicit bound on s, the number of ergodic components. If $f = \chi_{C_i}$ for some i then $\|f\|_1 = m(C_i)$ and $\|Qf\|_\infty \leq \|Q1\|_\infty = \|1\|_\infty = 1$. Hence, $1 \leq \frac{R}{1-r}m(C_i)$, and thus $m(C_i) \geq \frac{1-r}{R}$. But $\sum_{i=1}^s m(C_i) = 1$, and therefore $s\frac{1-r}{R} \geq 1$ and $s \leq \frac{R}{1-r}$.

3 Applications to Lozi Mappings

3.1 Geometrical Properties of Lozi Mappings

We consider a mapping $L : \mathbb{R}^2 \to \mathbb{R}^2$ defined by

$$L(x, y) = (1 - a|x| + by, x) \tag{30}$$

where a and b are real parameters.

For the set of parameters, which we describe later, this mapping has a strange attractor and some hyperbolicity property. These facts were proved in the paper of M. Misiurewicz [10] (we extend the set of parameters described in that paper to some negative b).

L is differentiable except for the y-axis. L is continuous for all parameters. If $p \in \mathbb{R}^2 \backslash \{x = 0\}$ then

$$DL(p) = \begin{bmatrix} \mp a & b \\ 1 & 0 \end{bmatrix} \stackrel{\text{def}}{=} A_\pm. \tag{31}$$

("+" for $x(p) > 0$ and "-" for $x(p) < 0$).

The iteration L^n is differentiable at $p \in \mathbb{R}^2 \setminus \bigcup_{k=0}^{n-1} L^k(\{x = 0\})$ and

$$DL^n(p) = DL(L^{n-1}p)\, DL(L^{n-2}p) \dots DL(p) = A_{\epsilon_{n-1}} A_{\epsilon_{n-2}} \cdots A_{\epsilon_0} \tag{32}$$

where $\epsilon_i = sgn(x(L^i p))$, $i = 0, 1, \dots, n-1$.

We want to describe some fields of invariant directions for L, or, more precisely, for DL. A convenient way to describe them is to use the projective coordinates in the tangent space of the plane. The computation $A_\pm \begin{bmatrix} 1 \\ u \end{bmatrix} = \begin{bmatrix} \mp a + bu \\ 1 \end{bmatrix} = (\mp a + bu) \begin{bmatrix} 1 \\ \frac{1}{\mp a + bu} \end{bmatrix}$ shows that the corresponding transformation on directions is $\alpha_\pm(u) = \frac{1}{\mp a + bu}$. (Each line is described by parameter u of the only point of this line which has the form $(1, u)$, except for the y-axis.)

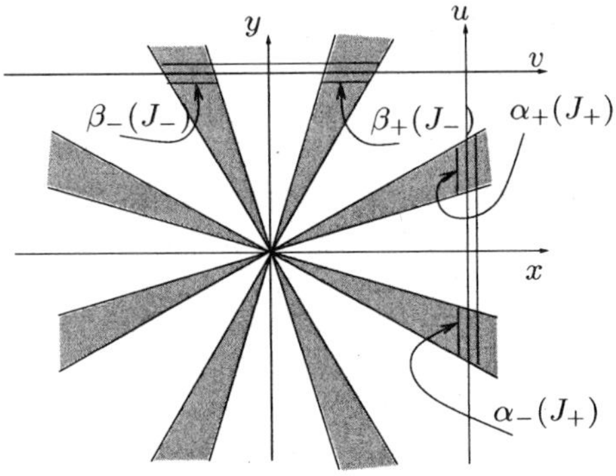

Fig. 1. Invariant cones.

If $b \neq 0$ then L^{-1} is a homeomorphism and is differentiable except for the points p on the y-axis, and

$$DL^{-1}(p) = \begin{bmatrix} 0 & 1 \\ b^{-1} & \pm ab^{-1} \end{bmatrix} \stackrel{\text{def}}{=} B_\pm. \tag{33}$$

This time "+" corresponds to $y(p) > 0$ and "−" to $y(p) < 0$. The corresponding map on directions is $\beta_\pm(v) = \frac{b}{\pm a + v}$ where any line going through 0 and point $(v, 1)$ is described by its unique point of the form $(v, 1)$.

In the sequel we always assume that $|b| < \frac{a^2}{4}$ and $a > 0$.

Lemma 3. *Let* $\theta_0 = \frac{a}{2} - \sqrt{\left(\frac{a}{2}\right)^2 - |b|}$. *Then the intervals* $J_+ = \{u \in \mathbb{R} : |bu| \leq \theta_0\}$ *and* $J_- = \{v \in \mathbb{R} : |v| \leq \theta_0\}$ *are* $\alpha_\pm$- *and* $\beta_\pm$-*invariant, respectively. Moreover, if* $\theta_1 = \frac{|b|}{a + \theta_0}$ *then* $\alpha_\pm(J_+) \subseteq \{u \in \mathbb{R} : |bu| \geq \theta_1\}$ *and* $\beta_\pm(J_-) \subseteq \{v \in \mathbb{R} : |v| \geq \theta_1\}$. *See Figure 1.*

Lemma 4. *Let* $\kappa = \frac{\frac{a}{2} - \sqrt{\left(\frac{a}{2}\right)^2 - |b|}}{\frac{a}{2} + \sqrt{\left(\frac{a}{2}\right)^2 - |b|}}$. *Then* $\sup_{J_+} |\alpha'_\pm(u)| = \sup_{J_-} |\beta'_\pm(v)| = \kappa <$

Lemmas 3 and 4 allow an easy proof of the existence of two fields of invariant directions.

For any sequence $\epsilon_0, \epsilon_1, \ldots \in \{+,-\}$ by the Mean Value Theorem we have

$$|\beta_{\epsilon_0} \circ \beta_{\epsilon_1} \circ \ldots \circ \beta_{\epsilon_{n-1}}(J)| \leq \kappa^n |J_-| \tag{34}$$

(where $|\cdot|$ stands for the length of an interval). Therefore, the set

$$\bigcap_{n=1}^{\infty} \beta_{\epsilon_0} \circ \ldots \circ \beta_{\epsilon_{n-1}}(J_-) \tag{35}$$

consists of exactly one point. This point might also be expressed as a continued fraction

$$\cfrac{b}{\epsilon_0 a + \cfrac{b}{\epsilon_1 a + \cfrac{b}{\epsilon_2 a + \cdots}}} \tag{36}$$

which converges by (34).Similarly, for any sequence $\eta_0, \eta_1, \eta_2, \ldots \in \{+,-\}$, the set $\bigcap_{n=1}^{\infty} \alpha_{\eta_0} \circ \ldots \circ \alpha_{\eta_{n-1}}(J_+)$ consists of exactly one point which can be represented as a continued fraction

$$\cfrac{1}{-\eta_0 a + \cfrac{b}{-\eta_1 a + \cfrac{b}{-\eta_2 a + \cdots}}} \tag{37}$$

The above construction gives us invariant directions for L. For example, if $p \in \mathbb{R}^2 \setminus \bigcup_{n=0}^{\infty} L^{-n}(\{x=0\})$ then we can define direction $E^s(p)$ by taking $\epsilon_i = sgn\, x(L^i p)$ for $i = 0, 1, \ldots$. Similarly, if $p \in \mathbb{R}^2 \setminus \bigcup_{n=0}^{\infty} L^n(\{y=0\})$ then by putting $\eta_i = sgn\, y(L^{-i}p)$ for $i = 1, 2, \ldots$, we can define $E^u(p)$. We also introduce two functions

$$u(p) = \cfrac{-1}{\eta_0 a + \cfrac{b}{\eta_1 a + \cfrac{b}{\eta_2 a + \cdots}}} \quad \text{and} \quad v(p) = \cfrac{b}{\epsilon_0 a + \cfrac{b}{\epsilon_1 a + \cfrac{b}{\epsilon_2 a + \cdots}}} \tag{38}$$

which correspond to these fields of directions.

These functions are discontinuous, as defined on the plane, but become Hölder if we consider them as functions of the sequences of $\pm$ only. For example, if the orbits of two points $p_1, p_2 \in \mathbb{R}^2 \setminus \bigcup_{k=0}^{n-1} L^{-k}(\{x=0\})$ fulfill the condition $sgn\, x(L^k p_1) = sgn\, x(L^k p_2)$ for $k = 0, 1, \ldots, n-1$ then $|u(p_1) - u(p_2)| \leq \kappa^n |J_+|$. A similar inequality holds for v $|v(L^n p_1) - v(L^n p_2)| \leq \kappa^n |J_-|$. We summarize these remarks in

Proposition 4. *If $a > 0$ and $|b| < (a/2)^2$ then there are two fields of directions E^s and E^u defined on $D^s = \mathbb{R}^2 \setminus \bigcup_{n=0}^{\infty} L^{-n}(\{x=0\})$ and $D^u = \mathbb{R}^2 \setminus \bigcup_{n=0}^{\infty} L^{n}(\{y=0\})$, respectively. This means that $DL(p)E^s(p) \subset E^s(Lp)$ for $p \in D^s$ and $DL^{-1}(p)E^u(p) \subset E^u(L^{-1}p)$ for $p \in D^u$.*

Remark 5. As we can see, $L(D^s) \subset D^s$ and $L^{-1}(D^u) \subset D^u$. The sets D^s and D^u are dense in the plane and have full Lebesgue measure. They are of type G_δ.

Let $\lambda^s(p)$ and $\lambda^u(p)$ be the rates of change of the length in the direction of $E^s(p)$ and $E^u(p)$ respectively (the orientation change is not reflected in the sign) under L and L^{-1}, respectively.

Lemma 5. *The following formulas hold*

$$\lambda^s(p) = |v(p)| \cdot \frac{h_1(p)}{h_1(Lp)}, \qquad \lambda^u(p) = |u(p)| \cdot \frac{h_2(p)}{h_2(L^{-1}p)} \tag{39}$$

where

$$h_1 = \frac{1}{\sqrt{1+v^2}}, \qquad h_2 = \frac{1}{\sqrt{1+u^2}}. \tag{40}$$

Proof. We have $DL(p)\begin{bmatrix} v(p) \\ 1 \end{bmatrix} = A_\pm \begin{bmatrix} v(p) \\ 1 \end{bmatrix} = \begin{bmatrix} \mp a v(p) + b \\ v(p) \end{bmatrix} = v(p)\begin{bmatrix} \mp a + b/v(p) \\ 1 \end{bmatrix} = v(p) \cdot \begin{bmatrix} v(Lp) \\ 1 \end{bmatrix}$. Hence, $\lambda^s(p) = |v(p)| \cdot \frac{\|(v(Lp),1)\|}{\|(v(p),1)\|} = |v(p)| \frac{h_1(p)}{h_1(Lp)}$. The proof of the other formula is identical.

Let $\lambda_n^s(p)$ and $\lambda_n^u(p)$ be analogous functions defined for the iterations L^n and L^{-n} respectively. We have $\lambda_n^s(p) = \lambda^s(L^{n-1}p) \cdots \lambda^s(Lp)\lambda^s(p) = |v(L^{n-1}p)| \cdots |v(Lp)| \cdot |v(p)| \frac{h_1(p)}{h_1(L^n p)}$ due to telescoping cancellations. There are constants $c_1, c_2 \in (0,\infty)$ such that $c_1 \le h_1 \le c_2$. By Lemma 3 $|v(p)| \le \theta_0$. Hence $\lambda_n^s \le c_3 \theta_0^n$ where $c_3 = c_2 c_1^{-1}$. We want L to be hyperbolic. This requires $\theta_0/|b| < 1$. After easy calculations we find that this is equivalent to $a > |b|+1$. Similarly, $\lambda_n^u \le \left| u(L^{-(n-1)}p) \cdots u(L^{-1}p)u(p) \frac{h_2(p)}{h_2(L^{-n}p)} \right| < c_4 \left(\frac{\theta_0}{|b|}\right)^n$ where c_4 is some constant $\in (0,\infty)$. We consider $|b| \le 1$ only. Thus if $a > |b|+1$ then $\theta_0 < |b|$ and we obtain

Proposition 5. *If $a < 0$, $0 < |b| \le 1$, $a > |b|+1$, and $|b| < (a/2)^2$ then L is hyperbolic in the following sense: let $\lambda_+ = \theta_0/|b|$ and $\lambda_- = \theta_0$; then $\lambda_+, \lambda_- \in (0,1)$ and there exists a constant $C \ge 0$ such that $|\lambda_n^s(p)| \le C\lambda_-^n$ as $p \in D^s$, and $|\lambda_n^u(p)| \le C\lambda_+^n$ as $p \in D^u$.*

Now we are going to describe certain invariant sets for L.

Lemma 6. *Let θ_2 be the positive root of the equation $\theta(a + |b|\theta) = 1$ (i.e. $\theta_2 = \frac{1}{|b|}\left(-\frac{a}{2} + \sqrt{\frac{a^2}{4} + |b|}\right)$. Then under the assumptions of Proposition 5 the set*

$$\Delta_0 = \{(x,y) \in \mathbb{R}^2 : |y| \le \theta_2(p-x)\},$$

where $q = (1-|b|\theta_2)^{-1}$, *is L-invariant, i.e.* $L(\Delta_0) \subset \Delta_0$.

Proof. Δ_0 is a union of rays starting from $p_0 = (q,0)$ and with directions, for which $u \in [-\theta_0, \theta_0]$. (see Figure 2(a)). Let Z be one of the rays and let u be the corresponding parameter. Hence $Z = \{(q-t, t\cdot u) \in \mathbb{R}^2 : t \in \mathbb{R}_+\}$. Z intersects the line $\{x=0\}$ at exactly one point $p_1 = (0, qu)$. We have $L(p_1) = (1+bqu, 0) \stackrel{\mathrm{def}}{=} p_2$. The calculation $1 + qbu \le 1 + \frac{|b|\theta_2}{1-|b|\theta_2} = \frac{1}{1-|b|\theta_2} = q$, implies that $p_2 \in \Delta_0$. We can see that $L(Z)$ is a union of a ray Z that starts at p_2 with the direction

$$\alpha_-(u) = (a+bu)^{-1} \in \left[\frac{1}{a+|b|\theta_0}, \frac{1}{a-|b|\theta_0}\right] \subseteq [-\theta_0, \theta_0]$$

and the interval Z_2 joining p_2 with $Lp = (1-qa, q)$. We have $Lp_0 \in \Delta_0$ because

$$\begin{aligned} \frac{q}{(1-qa)-q} &= \frac{1}{q^{-1}-(a+1)} = \frac{1}{1-|b|\theta_2-(a+1)} \\ &= \frac{1}{a+|b|\theta_2} = -\theta_2. \end{aligned} \tag{41}$$

Hence $Lp_0 \in \partial\Delta_0$. Thus $Z_1, Z_2 \subset \Delta_0$ and $L(Z) \subset \Delta_0$. In view of $|b| < 1$, L has two fixed points p_+ and p_- with $x(p_+) > 0$ and $x(p_-) < 0$. Both are hyperbolic by Proposition 5. Moreover, $p_- = (-(a+b+1)^{-1}, -(a+b-1)^{-1})$.

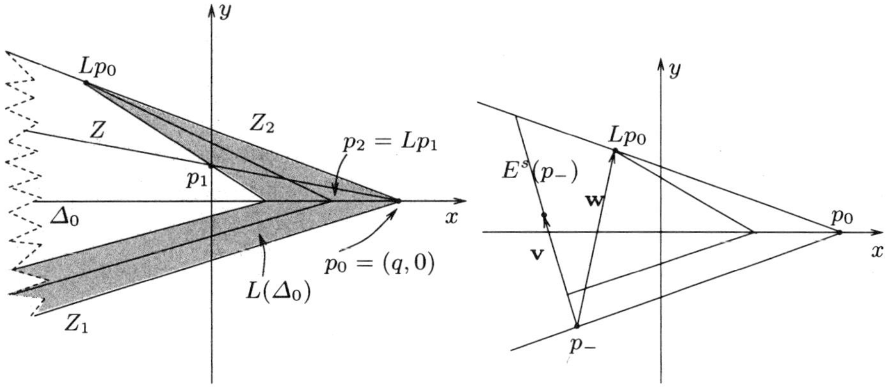

(a) The infinite trapping region.

(b) The triangular trapping region.

Fig. 2. The trapping regions for L.

Let $\mathbf{v} = (v(p_-), 1)$ and $\mathbf{w} = Lp_0 - p_-$. The condition that $L(p_0)$ is to the right of $E^s(p_-)$ is equivalent to $\mathbf{v} \times \mathbf{w} \leq 0$. This condition guarantees that the triangle

$$\Delta = \{p \in \Delta_0 : p \text{ is to the right of } E^s(p_-)\} \tag{42}$$

is invariant. The direction of $E^s(p)$ is defined by the following value of the parameter v:

$$\cfrac{b}{-a + \cfrac{b}{-a + \ldots}}. \tag{43}$$

This expression is equal to $-k$ where k is the positive root of the equation $b = k(a+k)$. From now on we always assume that the hypothesis of Proposition 5 is satisfied. We can put $\mathbf{v} = (-k, 1)$ and $\mathbf{w} = Lp_0 - p_- = (1 - aq + e, q + e)$ where $e = 1/(a + b - 1)$. Hence,

$$\mathbf{v} \times \mathbf{w} = \begin{vmatrix} -k & 1 - aq + e \\ 1 & q + e \end{vmatrix} = -k(q+e) - (1 - aq + e) \leq 0. \tag{44}$$

This condition for $b > 0$ may be rewritten as $2a + b - 4 \leq 0$, as a lengthy computation shows. In the case of $b < 0$, simplifying of this condition is more complicated.

Proposition 6. *Let a, b fulfill the conditions $a > 0$, $0 < |b| \leq 1$, $a > |b| + 1$ and (44) (as we stated, in the case of $b > 0$ this reduces to $2a + b \leq 4$). Then $L(\Delta) \subset \Delta$.*

Remark 6. If we require that $L(\Delta) \subset \operatorname{int} \Delta$ then we may use Δ shifted to the right by a sufficiently small distance. This works provided that $2a + b < 4$ ($Lp_0 \notin E^s(p_-)$).

Now we are going to give a proof of the equivalence of (44) and $2a + b - 4 \leq 0$ for $b > 1$. We notice that $k = b\theta_2$. We have a sequence of equivalent inequalities: (44) and then $1 - aq + e + k(q+e) \geq 0$ and $1 - \frac{a}{1-k} + e + \frac{k}{1-k} + ke \geq 0$. Thus $1 - k - a + e - ek + k + ke - k^2 e \geq 0$, $e(1 - k^2) \geq a - 1$ and $1 - k^2 \geq (a-1)(a+b-1)$. Because of $1 - k^2 = -b + ak + 1$ we obtain $ak + 1 - b \geq (a-1)(a+b-1) = (1-b) + a(a+b-1)$ and $k \geq a + b - 2$. Using the equation $k^2 + ak - b = 0$ we obtain $(a+b-2)^2 + a(a+b-2) - b \leq 0$. Now it is easy to check that the left-hand side is equal to $(2a+b-4)(a+b-1)$. We consider a and b for which $a + b - 1 > 0$. Thus we must have $2a + b - 4 \leq 0$.

We notice that Lemma 5 establishes "multiplicative cohomology" between λ^s and v as well as between λ^u and u. We have one more useful cohomology relation.

Lemma 7. *There is a function h_3, bounded and separated from 0, such that for all $p \in D^s \cap L^{-1}(D^u)$ we have*

$$\lambda^s(p) = \lambda^u(Lp) \cdot |b| \cdot \frac{h_3(p)}{h_3(Lp)}.$$

Proof. Let $\mathbf{v}(p) = (v(p), 1)$ and $\mathbf{u}(p) = (1, u(p))$. The linear transformation $DL(p)$ with respect to the pair of bases

$$\left(\frac{\mathbf{v}(p)}{\|\mathbf{v}(p)\|}, \frac{\mathbf{w}(p)}{\|\mathbf{u}(p)\|}\right) \text{ and } \left(\frac{\mathbf{v}(Lp)}{\|\mathbf{v}(Lp)\|}, \frac{\mathbf{w}(Lp)}{\|\mathbf{u}(Lp)\|}\right)$$

has a diagonal matrix

$$\Lambda(p) = \begin{bmatrix} \pm\lambda^s(p) & 0 \\ 0 & \pm\lambda^u(Lp)^{-1} \end{bmatrix}. \tag{45}$$

Let $S(p)$ be the transition matrix from the first basis to the standard basis. Then $\Lambda(p) = S(Lp)^{-1}DL(p)S(p)$. Taking the determinants of both sides we get $\lambda^s(p)(\lambda^u(Lp))^{-1} = |b| \cdot \frac{h_3(p)}{h_3(Lp)}$, where $h_3(p) = |\det S(p)|$.

3.2 Invariant Partitions for Lozi Mappings

The results of Misiurewicz listed in the previous subsection give the existence of stable and unstable invariant directions on sets D^s and D^u respectively.

The idea applied by Misiurewicz and coming from billiard theory, that the trajectory of almost every point does not approach the singularity too fast, gives the existence of local stable and unstable manifolds (intervals).

We use the main geometrical idea of the paper, illustrated by Lemma 8 and Proposition 7 in order to prove some stronger estimates than those used by Misiurewicz ([10]).

Let $\tilde{\alpha}$ be the cover of the plane with the half-planes $\{x \leq 0\}$ and $\{x \geq 0\}$. It is easy to see that the cover $\tilde{\alpha}^n$ consists of convex sets bounded by broken lines. Let us set $\alpha = \tilde{\alpha}|\Delta$, where Δ is the trapping region constructed in the previous subsection. Hence α^n consists of convex polygons. We will call α a partition, because it becomes a genuine partition after discarding singularities of the iterates of L from the plane (which form a set of measure 0).

Lemma 8. *For every $N \geq 1$ there is an open cover $\mathcal{U}_N$ of Δ such that every element of $\mathcal{U}_N$ intersects not more than $2N$ elements of α^n.*

Proof. We will prove by induction that not more than $2N$ elements of α^n can have a nonempty intersection. The case of $N = 1$ is obvious. Let us assume that our assertion is true for some $N \geq 1$. From the definition $\alpha^{N+1} = \alpha \vee L^{-1}(\alpha^n)$. The partition $L^{-1}(\alpha^N)$ is clearly a partition into polygons (although they may not be convex). By the induction hypothesis, not more than $2N$ segments from the boundaries of the elements of $L^{-1}(\alpha^N)$ meet at one point. Because α^{N+1} is obtained by adding the line $\{x = 0\}$ to the boundaries, the number of segments meeting at one point can increase by 2, at most. The assertion has been proved. The Lemma follows easily.

Proposition 7. *There exist constants $F \in \mathbb{R}_+$ and $r \in (0,1)$ such that for every segment I with the direction from the unstable cone J_+ we have*

$$\sum_{J \in \alpha^N | I} \frac{|J|}{|L^n(J)|} \leq F(r^n + |I|), \tag{46}$$

where $|\cdot|$ stands for the length of a segment.

Proof. We choose N in such a way that $r_0 \overset{\text{def}}{=} (2N) \cdot (C\lambda_+^N) < 1$, where C, λ_+ are as in Proposition 5.

Let ϵ_0 be the Lebesgue constant of the cover $\mathcal{U}_n$ given by Lemma 8. Let for $n = 1, 2, \ldots$

$$\gamma_n = \sum_{J \in \alpha^{nN} | I} \frac{|J|}{|L^{nN}(J)|}. \tag{47}$$

We are going to show that

$$\gamma_{n+1} \leq r_0 \gamma_n + \epsilon_0^{-1} R_0 |I|, \tag{48}$$

where $R_0 = \sup_I \gamma_1$, and I is any segment with the direction from the unstable cone.

Let $J \in \alpha^{nN}|I$. Either $|J| < \epsilon_0$ or $|J| > \epsilon_0$. In the first case $\alpha^{(n+1)N}|J$ consists of not more than $2N$ elements (Lemma 8) and

$$\begin{aligned}\sum_{J \in \alpha^{(n+1)N}|I} \frac{|J|}{|L^{(n+1)N}(J)|} &\leq 2N \cdot \max_J \frac{|J|}{|L^{(n+1)N}(J)|} \\ &\leq 2N \frac{|J|}{|L^{nN}(J)|} \cdot \max \frac{|L^{nN}(J)|}{|L^{(n+1)N}(J)|} \\ &\leq 2N \cdot \frac{|J|}{|L^{nN}(J)|} \cdot C\lambda_+^N = r_0 \frac{|J|}{|L^{nN}(J)|}. \end{aligned} \tag{49}$$

In the second case we get

$$\sum_{J' \in \alpha^{(n+1)N}|J} \frac{|J'|}{|L^{(n+1)N}(J')|} \leq R_0 \cdot \frac{|J|}{|L^{nN}(J)|} \leq R_0 \epsilon_0^{-1} |J|. \tag{50}$$

Summation over all J gives (48) easily.

To derive (46) from (48) we notice that $n = k \cdot N + l$, $0 \leq l \leq N-1$. In fact, (48) implies that $\gamma_n \leq r_0^n + \frac{\epsilon_0^{-1} R_0}{(1-r_0)\cdot|I|}$. So, we check easily that the left-hand side of (46) is not larger than $r_0^k \cdot \gamma_l + \frac{\epsilon_0^{-1} R_0}{(1-r_0)\cdot|I|}$. Let $r = r_0^{1/N}$ and let

$$F = \max\left(\max_{0 \leq l \leq N-1} \frac{\gamma_l}{r^{N-1}}, \epsilon_0^{-1} R_0 (1-r_0)^{-1} \right).$$

We check that F does not exceed the right-hand side of (46).

Remark 7. This proof was similar to that of Theorem 1.

Let $\xi = \alpha^- = \bigvee_{n=0}^{\infty} L^{-n}(\alpha)$. This partition is obviously into convex sets and the hyperbolicity of L gives that all elements of ξ are either segments with the direction from the unstable cone or points (otherwise some element of ξ would be expanded without breaking on the singularity, which contradicts the fact that Δ is bounded).

We will prove that almost every element of ξ is a segemnt of positive length.

Lemma 9. *Let*

$$D^S(\delta) = \{p \in \Delta : \ \mathit{dist}\ (L^n p, \{x = 0\}) \geq \delta\lambda_n^S(p) \ \mathit{for}\ n = 0, 1, 2, \dots \}. \quad (51)$$

For every $p \in D^S(\delta)$ the distance from p to the endpoints of $\xi(p)$ is not smaller than δ (in particular, $|\xi(p)| \geq 2\delta$).

Proof. See ([10]).

Lemma 10. *Let $\lambda = (\lambda_n)$ be a sequence of positive numbers such that $S \overset{\text{def}}{=} \sum_n \lambda_n < +\infty$. Let:*

$$D^S(\delta, \lambda) = \{p \in \Delta : \ n \in \mathbb{Z}_+ : \ \mathit{dist}(L^n p, \{x = 0\}) \leq \delta\lambda_n\}. \quad (52)$$

Let I be a segment as in Proposition 7. Then there is a constant A_1 such that

$$|I \backslash D^S(\delta, \lambda)| \leq A_1 S \cdot \delta. \quad (53)$$

Proof. See Figure 3. Let $p \in I \backslash D^S(\delta, \lambda)$.

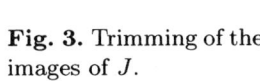

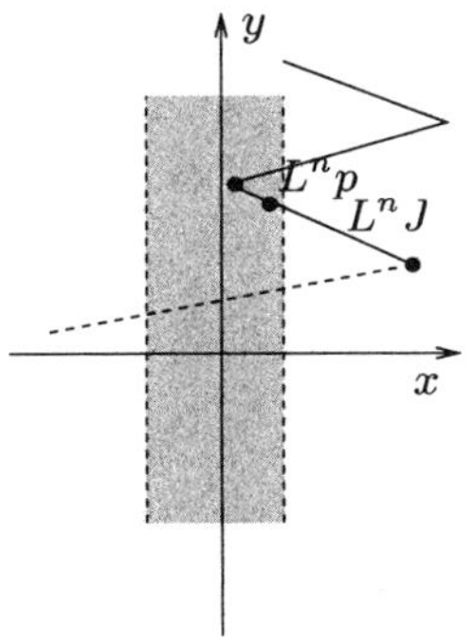

Fig. 3. Trimming of the images of J.

For some $n \in \mathbb{Z}_+$, $\text{dist}(L^n p, \{x = 0\}) < \delta\lambda_n$. Let $J \in \alpha^n | I$ be the element containing p. Thus, the image $L^n p$ belongs to an interval in $L^n J$ of length $\leq A_0 \cdot \lambda_n$.

This means that p itself belongs to the interval $J \cap L^{-n}(\{|x| \leq \delta\lambda_n\})$ which has length not larger than

$$|J \cap L^{-n}(\{|x| \leq \delta\lambda_n\})| \leq \frac{A_0 \delta\lambda_n}{|L^n J|} \cdot |J|.$$

This gives $|L^{-n}(\{|x| \leq \delta\lambda_n\}) \cap I| \leq A_0 \delta\lambda_n \cdot F(r^n + |I|) \leq A_0 F(1 + \text{diam}(\Delta))\delta\lambda_n$. Thus, by summation over all n,

$$|I \backslash D^S(\delta, \lambda)| \leq A_0 F(1 + \text{diam}(\Delta)) S\delta.$$

So, we can put $A_1 = A_0 F(1 + \text{diam}(\Delta))$.

Corollary 1. *Using the above notation,* $|I\backslash D^S(\delta)| \le A_2 \cdot \delta$ *where* $A_2 = A_1 \sum_{n=0}^{\infty} C\lambda_-^n = A_1 C/(1-\lambda_-)$.

Proof. Let $\lambda_n = C\lambda_-^n, n \in \mathbb{Z}_+$. Because $\lambda_S^n(p) \le \lambda_n$, we have $D^S(\delta) \supset D^S(\delta, \lambda)$. This gives the result.

Corollary 2. *The set* $\tilde{D}^S \stackrel{\text{def}}{=} \bigcup_{\delta>0} D^S(\delta)$ *has full Lebesgue measure in* Δ. *Moreover, there is a constant* A_2 *such that* $\nu(\Delta\backslash D^S(\delta)) \le A_2\delta$ *where* ν *is the Lebesgue measure restricted to* Δ.

We need integrability of the function $1/D$ where $D(p) = |\xi(p)|$. This cannot be derived from the above estimate, but we have:

Proposition 8. *There is a constant* $A_3 \in \mathbb{R}_+$ *such that for an arbitrary* $\delta > 0$,

$$\nu(\{p \in \Delta : \; D(p) < \delta\}) \le A_3\delta^2. \tag{54}$$

Proof. This set is contained in the set of points p such that $\text{dist}(L^n p, \{x = 0\}) < \delta\lambda_n(p)$ for at least *two* different $n \in \mathbb{Z}_+$ (because each end of $\xi(p)$ has to be trimmed). Let us consider the family of strips $\Pi_\alpha = \{p \in \mathbb{R}^2 : y(p) < \alpha\}$ $(\alpha > 0)$. The last condition means that $L^n p \in \Pi_\alpha\backslash D^S(\alpha)$ where $\alpha = \delta\lambda_n^S(p)$. By Lemma 10 and Fubini theorem, $\nu(\Pi_\alpha\backslash D^S(\alpha)) \le A_2\alpha^2 \le A_2(C\lambda_-^n)^2$. Because of the Jacobian of L is $|b|$, we have

$$\nu\left(L^{-n}(\Pi_\alpha\backslash D^S(\alpha))\right) \le A_2 C^2 \lambda_-^{2n} |b|^{-n}. \tag{55}$$

We also have $\lambda_- = \theta_0$ and $\theta_0 \le |b|$. Hence, $\lambda_-^{2n}|b|^{-n} \le \lambda_-^n$ and summing (55) over all n we get $\nu(\{p \in \Delta : \; D(p) < \delta\}) \le \sum_{n=0}^{\infty} A_2 C^2 \lambda_-^{2n} |b|^{-n} = A_2 C^2 \sum_{n=0}^{\infty} \lambda_-^u \frac{A_2 C^2}{1-\lambda_-}$. So we can put $A_3 = A_2 C^2/(1-\lambda_-)$.

Corollary 3. *The function* $\Delta \ni p \to 1/D(p) \in \mathbb{R}_+$ *is integrable with any power less than* 2.

Proof. Standard computation. Let $\beta \in (1, 2)$. Then

$$\begin{aligned} \int_\Delta \left(\frac{1}{D}\right)^\beta d\nu &= \int_0^\infty \nu(\{D^{-\beta} > \gamma\})d\gamma \le 1 + \int_1^\infty \nu(\{D^{-\beta} > \gamma\})d\gamma \\ &\le 1 + \int_1^\infty \nu(\{D < \gamma^{-1/\beta}\})d\gamma \le 1 + \int_1^\infty B\gamma^{-2/\beta}d\gamma < +\infty. \end{aligned} \tag{56}$$

Now we are going to prove that from the point of view of ergodic theory, the elements of ξ are parallel.

Let $(\nu_C)_{C\in\xi}$ denote the family of conditional measures of ν with respect to the partition ξ and let $(l_C)_{C\in\xi}$ be the family of one-dimensional Lebesgue measures on the elements of ξ.

Proposition 9. *For almost every* $C \in \xi$, v_C *is absolutely continuous with respect to* l_C *and the Radon-Nikodym derivative* $\frac{d\nu_C}{dl_C}$ *is constant on* C *(equal to* $1/|C|$*).*

Proof. Let A_n be the polygon of the partition α^n, $n \in \mathbb{Z}_+$, containing $C \in \xi$. Because A_n is convex, the projection of the measure $\frac{1}{\nu(A_n)}\nu|A_n$ onto the y-axis is a measure absolutely continuous with respect to the Lebesgue measure and the density ρ_n is a concave function on some interval (a_n, b_n) and $\rho_n \equiv 0$ outside of this interval. We have $(a_{n+1}, b_{n+1}) \subset (a_n, b_n)$ and $a_n \to a$, $b_n \to b$, where a and b are the ends of the projection of C onto the y-axis. (See Figures 4(a) and 4(b).)

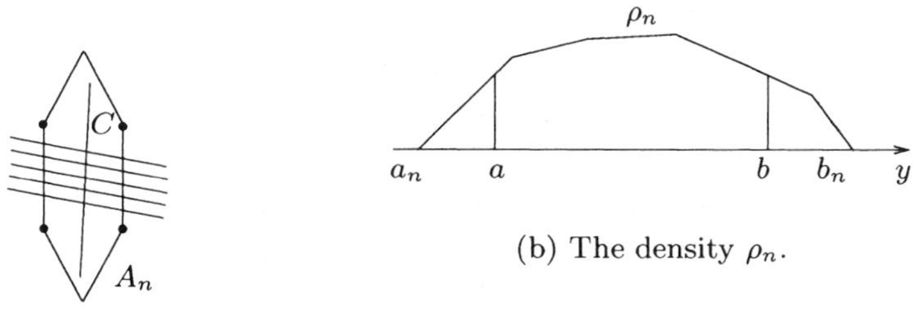

(a) A polygon of the partition α_n.

(b) The density ρ_n.

Fig. 4. Figures for the proof of Proposition 9.

Hence, ρ_{n_k} converges for some sequence $n_k \nearrow \infty$ (at least pointwise), because $\int \rho_n = 1$. Let $\rho = \lim_{n\to\infty} \rho_{n_k}$. It is easy to see that projecting ρ back onto C we get $\frac{d\nu_C}{dl_C}$ (which is concave, too). (As M. Wojtkowski noticed, even for any partition into intervals with measurable ends $\frac{d\nu_C}{dl_C}$ is linear.) Now the proposition can be derived easily from Proposition 3 and Lemma 10. Indeed, if $\rho = \frac{d\nu_C}{dl_C}$, it is enough to check that for almost every $x \in \Delta$ and almost every $y, y' \in C(x) = C$:

$$\lim_{n\to\infty} \frac{\rho(L^n y)}{\rho(L^n y')} = 1. \tag{57}$$

By Lemma 10 we assume that $\operatorname{dist}(L^n C, \{x = 0\}) > \delta\lambda^n$, where $\delta > 0$ and $\lambda \in (\lambda_-, 1)$. By concavity (see Figure 5(a)) of $\rho|\xi(L^n x)$:

$$\frac{\rho(L^n y)}{\rho(L^n y')} \le 1 + \frac{\operatorname{dist}(L^n y, L^n y')}{\operatorname{dist}(L^n C, \partial\xi(L^n x))} \le 1 + \frac{C\lambda_-^n}{\delta\lambda^n} \tag{58}$$

which goes to 1 as $n \to \infty$.

So $\overline{\lim}_{n\to\infty} \frac{\rho(L^n y)}{\rho(L^n y')} \le 1$. By symmetry we get (57).

3.3 The Reduction for Lozi Mappings

In the previous subsection we proved the basic results which allow us to apply section 2 of this paper to Lozi mappings. The purpose of this subsection is to check the assumptions of section 2 for this case.

We use the notation of subsection 2.1. So, for example, $Y = \Delta$ and β is the σ-algebra of Borel sets. First we consider the case when ν is the Lebesgue measure restricted to Δ. We put $S = L|\Delta$.

We already know that $\rho = 1/D$. Applying formula (13) and the fact that $g_S = |b|^{-1}$ we get

Lemma 11. *Under the assumptions formulated above,*

$$g_T = |b|^{-1}\lambda^S \frac{D}{D \circ T}. \tag{59}$$

What we do not like about this formula is the presence of D, which is "very discontinuous". Fortunately, it can be eliminated by replacing ν with $(1/D)\cdot\nu$ (the elimination is possible because formula (59) has a form of a cohomology relation).

Proposition 10. *If we apply the results of subsection 2.1 to the measure $\nu = 1/D\hat{\nu}$, where $\hat{\nu}$ is the Lebesgue measure restricted to Δ, then*

$$g_T = |b|^{-1}\lambda^S. \tag{60}$$

Proof. Using the definition 5 and the fact that $m = 1/D \cdot \hat{m}$ where m is the factor of the Lebesgue measure on $X = \Delta/\xi$, we get on $A \in \alpha$:

$$\begin{aligned} g_T &= \frac{d\left(T_*(\chi_A \cdot m)\right)}{dm} \circ T = \frac{d\left(T_*(\chi_A \cdot D^{-1} \cdot \hat{m})\right)}{d(D^{-1}\hat{m})} \circ T \\ &= \frac{D^{-1} \circ (T|A)^{-1} \cdot dT_*(\chi_A \hat{m})}{D^{-1} d\hat{m}} \circ T \\ &= \frac{D^{-1}}{D^{-1} \circ T} \cdot |b|^{-1}\lambda^S \cdot \frac{D}{D \circ T} = |b|^{-1}\lambda^S. \end{aligned} \tag{61}$$

(We used Lemma 11.)

This function g_T has nice properties and we can apply the results of subsection 2.2. Now using the notation of subsection 2.2 we are going to verify assumptions I–IV.

Lemma 12. *Condition I is satisfied.*

Proof. Let $n \in \mathbb{Z}_+$ and $B \in \beta^n$. Let $x_1, x_2 \in B$ ($x_1, x_2 \in X$ are also in ξ). Thus, $L^k_{x_1}$ and $L^k_{x_2}$ are on the same side of the line $x = 0$ for $k = 0, 1, \ldots, n-1$. Using formula 39 and $|v(L^k x_1) - v(L^k x_2)| \le \kappa^{n-k}|J_-|$ for $k = 0, \ldots, n-1$ we can see that for some constant $d_0 \in \mathbb{R}_+$, $\exp(-d_0\kappa^{n-k}) \le \frac{g_T(L^k x_1)}{g_T(L^k x_2)} \le \exp(d_0\kappa^{n-k})$. Because $g_n = g_T \cdot g_T \circ T \cdot \ldots \cdot g_T \circ T^{n-1}$, we get

$$\exp\left(-\sum_{k=0}^{n-1} d_0\kappa^{n-k}\right) \le \frac{g_n(x_1)}{g_n(x_2)} \le \exp\left(\sum_{k=0}^{n-1} d_0\kappa^{n-k}\right). \tag{62}$$

So, we might put $d = \exp\left(\frac{d_0}{1-\kappa}\right)$, to satisfy condition I of subsection 2.2.

Lemma 13. *Conditions II and IV are satisfied for some iteration $T^N, N \in \mathbb{Z}_+$.*

Proof. IV is satisfied because $g_n = |b|^{-n} \cdot \lambda_n^S$, so $|g_n| \le |b|^{-n}(C\theta_0^n) = C(\theta_0/|b|)^n$. For large n, $|g_n| < 1$, since $\theta_0/|b| < 1$.

The argument that shows that II is satisfied for some iteration is similar to the proof of Proposition 7. Let us choose N, r_0 and ϵ_0 as in the proof of that proposition. We notice that $\lambda_+ = \theta_0/|b|$ so $|g_n| \le C \cdot \lambda_+^n$ as $n \in \mathbb{Z}_+$.

Let $B \in \beta$ and let $A = \pi^{-1}(B)$. So $T^n(B) = \pi(S^n A)$. $S^n A$ is a convex polygon (see Figure 5(b)) such that

$$\partial(S^n A) \subset \bigcup_{k=0}^{n} L^k(\{x = 0\}). \tag{63}$$

So, the sides of $\partial(S^n A)$ are segments with the direction belonging to J_+. We have two possibilities:

1. $\operatorname{diam}(S^n A) > \epsilon_0$. Then it contains a segment I with the unstable direction (from J_+) of length $> \epsilon_0$.
   By Corollary 1, for arbitrary $\delta > 0$, $|I \backslash D^S(\delta)| \le A_2\delta$. The set $\bar{A} = \pi^{-1}\pi(S^n A) = \pi^{-1}(T^n B)$ has measure larger that $A_4^{-1}(1 - A_2\delta)$, where A_4 is a constant from the next lemma. So, we put $\delta = \frac{1}{2}A_2^{-1}$ and $\epsilon = A_4^{-1}(1 - A_2\delta) = \frac{1}{2}A_4^{-1}$. So $\nu(\bar{A}) > \epsilon$ and since $m = \pi_*\nu$, $m(T^n B) = \nu(\bar{A}) > \epsilon$.
2. $\operatorname{diam}(S^n A) < \epsilon_0$. Then $T^n B$ is contained in not more than $2N$ elements of β^N and
   $$\sum_{\beta' \in \beta^N(T^n B)} \sup_B g_N \le (2 \cdot N) \cdot (C\lambda_+^N) = r_0. \tag{64}$$

   So, in particular II holds for T^N with $r = r_0$.

Lemma 14. *Let I be a segment with the direction from J_+ and let l_I be the Lebesgue measure on I. Then the measure $\pi_*(l_I)$ is absolutely continuous with respect to m and*

$$\frac{d\pi_*(l_I)}{dm}(x) = \csc \angle(I, x) \tag{65}$$

for $x \in X$. In particular, for some $A_4 \in \mathbb{R}_+$,

$$A_4^{-1} \le \frac{d\pi_*(l_T)}{dm} \le A_4. \tag{66}$$

Proof. Fix some small $\delta > 0$. Let I_δ be a strip of width containing I. We notice that if $x \in D^S(\delta) \cap I$ and $\operatorname{dist}(x, \partial I) > \delta$, then $x \cap I_\delta$ is an interval of length $\delta \cdot \csc\omega(x)$, where $\omega(x) = \angle(I, x)$ (see Figure 5(c)). So $\nu_x(I_\delta) = \delta \csc\omega(x)/D(x)$. Let E be a subinterval of I. Thus, if $\hat{\nu}$ is the restriction of the Lebesgue measure to Δ, $\hat{\nu}(\pi^{-1}(\pi E) \cap I_\delta) = \int_{\pi(E)} \nu_x(x \cap I_\delta) d\hat{m} = \delta \int_{\pi(E)} \csc\omega(x) dm$. On the other hand, using Corollary 1 we get $\hat{\nu}(\pi^{-1}(\pi E)) = l_I(E) \cdot \delta + o(\delta)$. This implies (65) immediately.

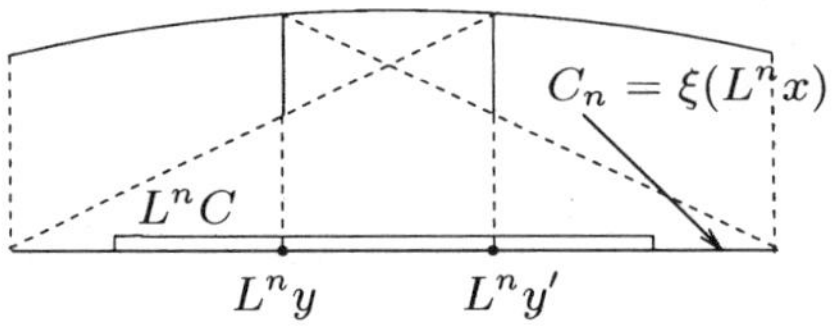

(a) Concavity of $\rho|\xi(L^n x)$.

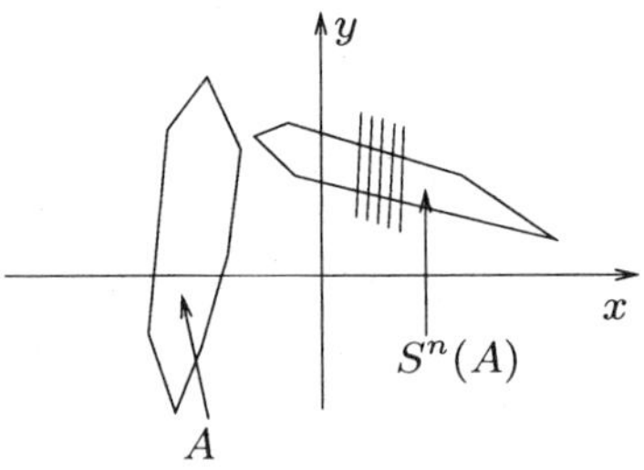

(b) The images of A.

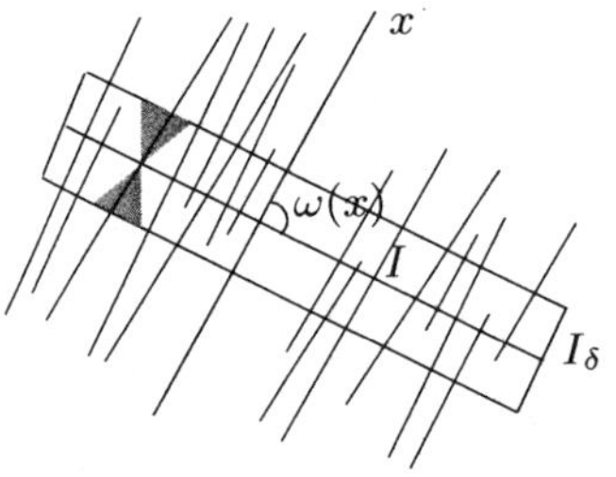

(c) The strip I_δ.

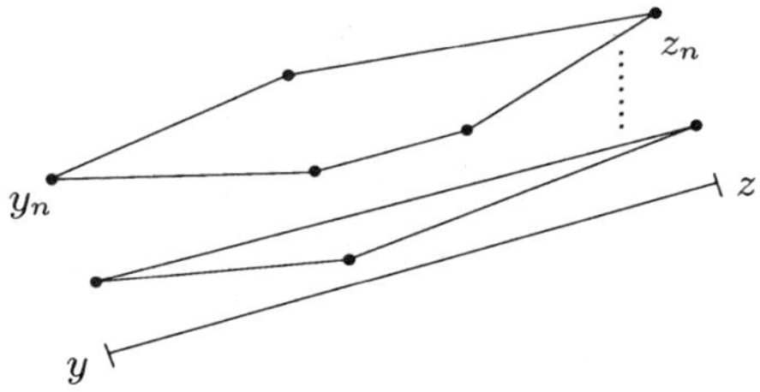

(d) An arbitrary sequence in $\mathcal{F}$.

Fig. 5. Figures illustrating convergence of partitions and densities.

Condition III of subsection 2.2 holds since β is finite.

Theorem 3. *The results of section 2, in particular Theorems 1 and 2 apply to Lozi mappings.*

3.4 Invariant Measures for Lozi Mappings

Using the general theory presented in section 2, we produced invariant measures of form $\phi \cdot m$ for the factor map T. We are now going to describe two methods of constructing the L-invariant measure μ, whose projection $\pi_*(\mu)$ onto X coincides with $\phi{\cdot}m$, $P\phi = \phi$. The first method is based on the following construction.

Let $f : \Delta \to \mathbb{R}$ be a continuous function. We have to define $\mu(f)$. Let us first define $f^<(p) = \inf_{\xi(p)} f$ and $f^>(p) = \sup_{\xi(p)} f$ where $p \in \Delta$. We put

$$\mu(f) = \lim_{n\to\infty} \tilde{\mu}\left((f \circ S^n)^<\right) \tag{67}$$

where $\tilde{\mu} = \phi \cdot m$. (We notice that $f^<$ and $f^>$ are always ξ-measurable.)

Lemma 15. *The limits* $\lim_{n\to\infty}\tilde{\mu}\left((f\circ S^n)^<\right)$ *and* $\lim_{n\to\infty}\tilde{\mu}\left((f\circ S^n)^>\right)$ *exist and are equal.*

Proof. Let $f_n^< = (f\circ S^n)^< \circ S^{-n}$ and $f_n^> = (f\circ S^n)^> \circ S^{-n}$. We have $f \geq f_n^<$ for every $n \in \mathbb{Z}_+$ and $f_1^< \leq f_2^< \leq \ldots \leq f_n^< \leq \ldots$. Similarly, $f \leq f_n^>$ and $f_1^> \geq f_2^> \geq \ldots \geq f_n^> \geq \ldots$. Besides, if $\xi_n = S^n\xi$ is a partition of $S^n\Delta$, then $f_n^>(p) - f_n^<(p) = \sup_{\xi_n(p)} f - \inf_{\xi_n(p)} f \leq \omega_{\delta_n}(f)$ where $\delta_n = \sup_p \operatorname{diam}(\xi_n(p))$ and

$$\omega_\delta(f) = \sup_{dist(x,y)<\delta} |f(x) - f(y)|. \tag{68}$$

So, $f_n^> - f_n^< \to 0$ as $n\to\infty$ and, consequently, $f_n^> \searrow f$, $f_n^< \nearrow f$ uniformly as $n\to\infty$. Since we have $(f\circ S^n)^< = f_n^< \circ S^n$, then $|\tilde{\mu}\left((f\circ S^n)^>\right) - \tilde{\mu}\left((f\circ S^n)^<\right)| \leq \sup|(f\circ S^n)^> - (f\circ S^n)^<| = \sup|f_n^> - f_n^<| \leq \omega_{\delta_n}(p)$ which goes to 0 as $n\to\infty$.

Using Theorem 2, we can construct measures $\mu_1,\ldots,\mu_S$ such that $\pi_*(\mu_i) = \phi_i m$.

Proposition 11. *Let* $\tilde{\mu}$ *be an arbitrary measure on* Λ *which is* T*-invariant and such that the sets of* Σ *are* $\tilde{\mu}$*-measurable. Then there exists a unique measure* μ *on* Δ *such that* μ *is* S*-invariant and* $\pi_*(\mu) = \tilde{\mu}$.

Proof. Let μ be constructed as the above measure and let $\bar{\mu}$ be some other measure for which (1) and (2) hold. Obviously, for every $f \in C(\Delta)$ $\bar{\mu}(f^<) \leq \bar{\mu}(f) \leq \bar{\mu}(f^>)$ and, since $\bar{\mu}(f^<) = (\pi_*\bar{\mu})(f^<) = \tilde{\mu}(f^<)$ for every f (in particular for $f\circ S^n$), we get $\tilde{\mu}\left((f\circ S^n)^<\right) \leq (\bar{\mu})(f) \leq \tilde{\mu}\left((f\circ S^n)^>\right)$ because $\bar{\mu}(f) = \bar{\mu}(f\circ S^n)$. A simple limit passage gives the proposition.

Theorem 4. *Let* μ *be Borel, regular measure on* Δ *such that* $\pi_*\mu$ *is absolutely continuous with respect to* m*. Then*

$$\frac{1}{n}\sum_{k=0}^{n-1} L_*^k\mu \overset{n\to\infty}{\longrightarrow} \sum_{i=1}^{S}\mu(\bar{C}_i)\cdot\mu_i \tag{69}$$

where $\bar{C}_i = \pi^{-1}(C_i)$ *and* $\mu_1,\ldots,\mu_s$ *are as above,* $C_1,\ldots,C_n$ *is as in Theorem 2, and the convergence is in weak-*$*$*-topology.*

Proof. Let $\rho = \frac{d(\pi_*\mu)}{dm}$. We check easily that

$$\frac{d(\pi_*(L^k\mu))}{dm} = P^k\rho, \qquad k \in \mathbb{Z}_+. \tag{70}$$

This implies

$$\pi_*\left(\frac{1}{n}\sum_{k=0}^{n-1} L_*^k\mu\right) = \left(\frac{1}{n}\sum_{k=0}^{n-1} P^k\rho\right)\cdot m. \tag{71}$$

Let $f \in C(\Delta)$ and let $N \in \mathbb{Z}_+$ be fixed. For $k \geq N$, $f_N \circ L^k$ is ξ-measurable and

$$\mu(f_N^{>} \circ L^k) = \int f_N^{>} \circ L^k \rho \, dm = \int (f_N^{>} \circ L^N) \circ T^{k-N} \rho \, dm$$
$$= \int (f_N^{>} \circ L^N) \cdot P^{k-N} \rho \, dm. \tag{72}$$

Thus by Theorem 2,

$$\frac{1}{n} \sum_{k=0}^{n-1} \mu(f_N^{>} \circ L^k) = \frac{1}{n} \sum_{k=0}^{N-1} \mu(f_N^{>} \circ L^k)$$
$$+ \int \left(\frac{1}{n} \sum_{k=N}^{n-1} P^{k-N} \rho \right) \cdot (f_N^{>} \circ L^N) dm$$
$$\to \int \left(\sum_{i=1}^{S} \mu(C_i) \cdot \phi_i \right) \cdot (f_N^{>} \circ L^N) dm$$
$$= \sum \mu(C_i) \cdot \mu_i(f_N^{>} \circ L^N), \quad \text{as} \quad n \to \infty. \tag{73}$$

Since $f_N^{>} \to f$ as $N \to \infty$ and $\mu_i(f_N^{>} \circ L^N) = \mu_i(f_N^{>})$, we get

$$\frac{1}{n} \sum_{k=0}^{n-1} \mu(f \circ L^k) \to \sum \mu(\bar{C}_i) \cdot \mu_i(f) \quad \text{as} \quad n \to \infty. \tag{74}$$

This completes the proof of the theorem.

Remark 8. The class of measures satisfying the hypothesis of Theorem 4 includes measures absolutely continuous with respect to the Lebesgue measure and measures l_I defined in Lemma 14 (as well as the convex hull of these measures).

Now we are going to describe the second construction of invariant measures. From this construction it will follow that unstable invariant intervals for L exist almost everywhere in the sense of measure $\mu_1 + \ldots + \mu_S$.

Lemma 16. *The family $\mathcal{F} = \bigcup_{n \in \mathbb{Z}_+} \bigcup_{k \leq n} L^k \alpha^n$ is precompact in the Hausdorff topology of all compact subsets of $\bar{\Delta}$. Its cluster points are segments I which are contained in some element of the partition*

$$(\alpha|\Lambda)^+ = \bigvee_{k=0}^{\infty} L^k(\alpha|\Lambda). \tag{75}$$

Proof. All compact subsets of Δ form a compact space in the Hausdorff topology. So every set, in particular $\mathcal{F}$, is precompact.

Let $A_n \in \mathcal{F}$ be an arbitrary sequence (see Figure 5(d)). We can see easily that A_n is a polygon and $\partial A_n \subset \bigcup_{j=-m(n)+k(n)}^{k(n)} L^j(\{x = 0\})$ if $A_n \in L^{k(n)}\alpha^{m(n)}$, where $k(n), m(n) \in \mathbb{Z}_+$, $k(n) \leq m(n)$. We can assume that $\operatorname{diam}(A_n) \not\to 0$ (otherwise A_n has a subsequence converging to a point of Λ). This implies that the difference $m(n) - k(n)$ is bounded by some integer N. Taking a subsequence, we may assume $m(n) - k(n) = N$ for all n. Eventually, taking a sequence $L^N A_n$ instead of A_n we may assume that $A_n \in L^{m(n)}\alpha^{m(n)}$. In this case we have $\partial A_n \subset \bigcup_{j=1}^{m(n)} L^j(\{x = 0\})$. We can also assume that $m(n) \to \infty$ as $n \to \infty$. Hence, A_n becomes "thinner and thinner". More precisely, ∂A_n is a sum of two broken lines joining the points of A_n with the largest and smallest values of the x-coordinate, which we denote by z_n and y_n respectively. Passing to a subsequence, we may assume that $y_n \to y$ and $z_n \to z_n$. Let I be the segment joining y and z. Then both the upper and the lower broken lines in ∂A_n converge to I (after a suitable reparameterization even in the Lipschitz topology). From the continuity of L it follows that $L^{-n}I \subset \Delta$ and is always contained in one of the half-planes $x \leq 0$ and $x \geq 0$ for all $n \in \mathbb{Z}_+$.

Let us use the ideas of the proof of Theorem 2.

Lemma 17. *Let $\mathcal{D} = \{$all segments $\subset \Delta$with the directions $\in J_+\}$. Then*

$$\mathcal{K} \stackrel{\text{def}}{=} \left\{ \frac{d\pi_*(l_I)}{dm} \right\}_{I \in \mathcal{D}} \tag{76}$$

is precompact in $L^1(m)$. Moreover, if $f_I = \frac{d\pi_*(l_I)}{dm}$, then

$$Pf_I = \sum_{J \in L(\alpha|I)} \lambda^u(J) f_J \tag{77}$$

where $\lambda^u(J)$ is equal to the ratio $|L^{-1}J|/|J|$.

Proof. This is an easy consequence of the fact that

$$L_*(l_I) = \sum_{J' \in \alpha|J} L_*(l_{J'}) = \sum_{J' \in \alpha|J} \frac{|J'|}{|LJ'|} LJ' \tag{78}$$

and the definition of P.

Using Proposition 1 and formula (77) we can easily derive

Lemma 18. *Let $\{t_I\}_{I \in \mathcal{D}}$ be any family of nonnegative numbers. If $f = \sum_{I \in \mathcal{D}} t_I f_I$, then $P^N f = \sum_{I \in \mathcal{D}} t_I{}' f_I$, where*

$$\left| \sum_{I \in \mathcal{D}} t_I{}' \right| \leq r_0 \left| \sum t_I \right| + R_0 \cdot \|f\|_1 \tag{79}$$

where N, r_0, R_0 are as in the proof of Proposition 1.

Corollary 4. *Let $\mathcal{L}_\gamma = (k \cdot \overline{\text{conv}(\mathcal{K} \cup \{0\})}) \cap \{\|f\|_1 \leq \gamma\}$. $\mathcal{L}_\gamma$ is a compact set in $L^1(m)$ and $P^N(\mathcal{L}_\gamma) \subset (r_0 \cdot k + R_0\gamma) \cdot \mathcal{L}_\gamma$ where N, r_0, R_0 are as in the proof of Proposition 1 and γ is an arbitrary nonnegative number. In particular, for every $k > R_0(1 - r_0)^{-1}\gamma$, $P(k\mathcal{L}_\gamma) \subset k\mathcal{L}_\gamma$.*

Thus, P^N preserves compact set $k \cdot \mathcal{L}_\gamma$ and, by Szauder fixed point theorem, it has a fixed point ϕ. This gives another proof of the existence of an invariant measure for T^N absolutely continuous with respect to m.

In the sequel we put $\gamma = 1$ and $k = R_0(1-r_0)^{-1}$ and $\mathcal{L} = k \cdot \mathcal{L}_1$.

Lemma 19. *$\bigcap_{n=0}^{\infty} P^n(\mathcal{L}) \subset k\bar{\mathcal{L}}$, where*

$$\bar{\mathcal{L}} = \left\{ \frac{d\pi_*(l_I)}{dm} \right\}_{I \in \mathcal{D}_\infty} \tag{80}$$

and $\mathcal{D}_\infty$ is the set of segments contained in some element of $(\alpha|\Lambda)^+$.

Proof. We notice that if I_n is a sequence of points of $\mathcal{D}$ and $\delta_n \stackrel{\text{def}}{=} dist(I_n, I) \to 0$ for some segment I, then by Corollary 1, $m(\pi(I_n) \div \pi(I)) \le A_2\delta_n + 2\delta_n \to 0$. This implies $\|f_{I_n} - f_I\|_1 \le (\sup f_{I_n}) \cdot m(I_n \backslash I) + \sup f_I \cdot m(I \backslash I_n) + \sup_{E_n} |f_{I_n}|E_n - f_I|E_n| \cdot m(E_n)$ where $E_n = \pi(I_n) \cap \pi(I) \subset X$. We may assume $|I| > 0$. By Lemma 14, since $\angle(I_n, x) \to \angle(I, x)$ for $x \in E_n$, we can see that the last summand tends to 0 as $n \to \infty$.

Let us fix some $N \in \mathbb{Z}_+ \cup \{\infty\}$. The set $\mathcal{L}(N) = \bigcap_{n=0}^{N} P^n(\mathcal{L})$ is convex and compact. Moreover, if $N < +\infty$, then $\mathcal{L}(N) = \overline{\text{conv}\{f_I : I \in \mathcal{D}_N\}}_N$ where $\mathcal{D}_N = \{I \subset L(\Delta) :\ x$ has the same sign on $L^{-i}(I)$ for $i = 1, \dots, N\}$. Since we proved that the map $\mathcal{D}_N \ni I \to f_I \in L^1(m)$ is continuous (and even Lipschitz), then by known theorems of functional analysis for every function $\phi \in \mathcal{L}(N)$ there is a regular, Borel, probabilistic measure τ_N on $\mathcal{D}_N$ such that $\phi = \int_{\mathcal{D}_N} f_I \cdot d\tau_N(I)$. If $\phi \in \mathcal{L}(\infty)$ then we choose a subsequence of $\{\tau_N\}_{N \in \mathbb{Z}_+}$ converging in the weak-$*$-topology to some τ. Thus, $\text{supp}(\tau) \subset \cap_{N=0}^{\infty} \mathcal{D}_N = \mathcal{D}_\infty$ and

$$\phi = \int_{\mathcal{D}_\infty} f_I d\tau(I). \tag{81}$$

Corollary 5. *For each invariant density ϕ_i, $1 \le i \le s$, there is a measure τ_i on $\mathcal{D}_\infty$ such that*

$$\phi_i = \int_{\mathcal{D}_\infty} f_I d\tau_i(I). \tag{82}$$

Let μ_i be the corresponding L-invariant measure. Then

$$\mu_i = \int_{\mathcal{D}_\infty} l_I d\tau_i(I). \tag{83}$$

(The integral makes sense, since $\mathcal{D}_\infty \ni I \to l_I \in C(\Delta)^*$ is continuous, if $\mathcal{D}$ is endowed with the Hausdorff topology and $C(\Delta)^*$ is endowed with the weak-$*$-topology set and the image of this map is a weak-$*$-compact set.)

For a while let μ_i be defined via (83).

Proof. We have $\pi_*(\mu_i) = \int_{\mathcal{D}_\infty} \pi_*(l_I) d\tau_i(I)$. Hence,

$$\frac{d\tau_*(\mu_i)}{dm} = \int_{\mathcal{D}_\infty} \frac{d\pi_*(l_I)}{dm} d\tau_i(I) = \int_{\mathcal{D}_\infty} f_I d\tau_i(I) = \phi_i. \tag{84}$$

By Proposition 11, both definitions of μ_i coincide.

Proposition 12. *For every* $\phi \in L^\infty(m)_+$ *such that* $P\phi = \phi$, *there exists a measure* τ *such that* $\text{supp}(\tau) \subset (\alpha|\Lambda)^+$ *and*

$$\phi = \int_{\mathcal{D}_\infty} f_I d\tau(I). \tag{85}$$

Moreover, if we define a measure μ *on* Δ *via*

$$\mu(f) = \int_{\mathcal{D}_\infty} l_I(f) d\tau(I), \qquad f \in C(\Delta) \tag{86}$$

then μ *is the measure satisfying* $\pi_*(\mu) = \phi$ *and the conditional measures* μ_I, *where* $I \in (\alpha|\Lambda)^+$ *are given by*

$$\mu_I = \csc\omega \cdot l_I. \tag{87}$$

Proof. The proposition is an easy consequence of formula 85. We notice that $I \in \mathcal{D}_\infty$ belongs to $(\alpha|\Lambda)^+$ iff $\partial I \subset \bigcup_{n=0}^\infty L^n(\{x=0\})$.

The idea of the proof is to iterate the measure in formula 81 in order to get more desired intervals. This is possible since $I \in \mathcal{D}_\infty$ is expanded by L, so I breaks on the singularity $x = 0$ and produces intervals $J \in (\alpha|\Lambda)^+$.

For each n and $A \in \alpha^n$, let $\Psi^{A,n} : \mathcal{D} \to \mathcal{D}$ be defined by $\Psi^{A,n}(I) = L^n(I \cap A)$. We write Ψ^A instead of $\Psi^{A,1}$. Let $U : C(\Delta)^* \to C(\Delta)^*$ be an operator defined by $U_\tau = \sum_{A\in\alpha} \lambda^u \cdot \Psi^A_*(\tau)$. We check easily that for each ϕ of the form

$$\phi = \int_{\mathcal{D}} f_I d\tau(I) \tag{88}$$

we have $P\phi = \int_{\mathcal{D}} f_I d(U\tau)(I)$. This implies $P^n\phi = \int_{\mathcal{D}_n} f_I d(U^n\tau)(I)$, $n \in \mathbb{Z}_+$, where

$$U^n\tau = \sum_{A\in\alpha^n} \lambda^u_n \Psi^{A,n}_*(\tau). \tag{89}$$

We notice that $\alpha^n|I$ consists of intervals J such that all except these containing the ends of I have the property $\partial I \subset \cup_{k=o}^{n-1} L^{-k}(\{x=0\})$. So, $L^n J \in (\alpha|\Lambda)^+$ except perhaps these two (at most) intervals. This implies that if

$$\mathcal{E}_n = \{I :\ \partial I \subset \cup_{k=1}^n L^k(\{x=0\})\} \tag{90}$$

then $U^n_\tau(\mathcal{D}\backslash\mathcal{E}_n) \le 2 \cdot \sup \lambda^n_u \cdot \|\tau\| \le 2\lambda^n_+ \cdot \|\tau\|$. Lemma 18 says that for every $\tau \in C(\Delta)^*$, $\|U^{n+N}\tau\| \le r_0\|\tau\| + R_0\tau(|I|)$. So, by considerations similar to those of Proposition 7, we get $\|U^n\tau\| \le F(r^n\|\tau\| + \tau(|I|))$ for some $F \ge 0$ and $r \in (0,1)$. In particular, $\|U^n\tau\| \le A_5\|\tau\|$ for some $A_5 \ge 0$.

If we apply (90) to $U^{m-n}\tau$ instead of τ, we get for every $m \geq n$, $U^m\tau(\mathcal{D}\backslash\mathcal{E}_n) \leq 2A_5\lambda_+^n\|\tau\|$. Let τ be a measure from (88) and let $\bar{\tau}$ be a weak accumulation point of the sequence $\tau^n = \frac{1}{n}\sum_{k=0}^{n} U^k\tau$. Then $\frac{1}{n}\sum_{k=0}^{n-1} P^k\phi \to \int_{\mathcal{D}} f_I d\,\bar{\tau}(I)$ at least for a subsequence. We have $\tau^m(\mathcal{E}_n) \geq 1 - 2\lambda_+^n A_5$ for each n and $m \geq n$. Because $\mathcal{E}_n$ is closed in $\mathcal{D}$, we get $\bar{\tau}(\mathcal{E}_n) \geq 1 - 2A_z\lambda_+^n \to 1$ as $n \to \infty$. This implies $\bar{\tau}\left(\bigcup_{n=1}^{\infty}\mathcal{E}_n\right) = 1$. Therefore, $\bar{\tau}\left((\alpha|\Lambda)^+\right) = 1$. If $P\phi = \phi$, we get $\phi = \int_{(\alpha|\Lambda)^+} f_I d\bar{\tau}(I)$.

Theorem 5. *There exist sets* $\mathcal{M}_1, \ldots, \mathcal{M}_s \subset \Delta$ *such that* $l_I(\bar{C}_i\backslash\mathcal{M}_i) = 0$ *for every* $I \in \mathcal{D}$ *and* $\mathcal{M}_i \subset \bar{C}_i$ $(i = 1, \ldots, s)$, *and for every* $p \in \mathcal{M}_i$ *and* $f \in C(\Delta)$ *we have* $\lim_{n\to\infty} \frac{1}{n}\sum_{k=0}^{n-1} f(L^k p) = \mu_i(f)$.

Proof. By the ergodic theorem, the statement about the limit holds on the set $\mathcal{M}_i^0$ of generic points of μ_i. It is well known that $\mu_i(\mathcal{M}_i^0) = 1$. Let $\mathcal{M}_i = \{p \in \bar{C}_i : \exists p' \in \mathcal{M}_i,\ j \in \mathbb{Z}_+ :\ L^j p \in \xi(p')\}$. We can see that the limit statement holds on $\mathcal{M}_i$, too. We want to prove that $l_I(\bar{C}_I\backslash\mathcal{M}_I) = 0$. Let us assume that it is not true and let us define $\mu = \chi_{\bar{C}_i\backslash\mathcal{M}_i}\cdot l_I$. By Theorem 4, $\frac{1}{n}\sum_{k=0}^{n-1} L^k(\mu) \to l_I(\bar{C}_i\backslash\mathcal{M}_i)\cdot\mu_i$. Projecting onto X, we obtain $\frac{1}{n}\sum_{k=0}^{n-1} P^k(\chi_{\bar{C}_i\backslash\mathcal{M}_i}\cdot f_I) \to l_I(\bar{C}_i\backslash\mathcal{M}_i)\cdot\phi_i$ in $L^1(m)$. The left-hand side restricted to $\pi(\mathcal{M}_i^0)$ is equal to 0, while the right-hand is not, unless $l_I(\bar{C}_I\backslash\mathcal{M}_i) = 0$.

4 The Variational Principle and the Convergence of the Iterations of the Perron-Frobenius Operator

4.1 Basic Estimates

We are going to prove that in general the convergence of the iterations of the Perron-Frobenius operator (as defined in section 2) is a sufficient condition for the measures $\mu_1, \ldots, \mu_s$ from Theorem 2 to satisfy a maximum principle, which is analogous to the variational principle of Ruelle and Bowen [1].

In order to formulate the results we have to modify the assumptions of section 2 a little.

Let (X, Σ) be a measurable space and let $T : X \to X$ be a measurable transformation. We assume that there exists a measurable partition β of X which is countable and is a one-sided generator. We also assume that for every $B \in \beta$, $T|B : (B, \Sigma|B) \to (TB, \Sigma|TB)$ is an isomorphism. (These assumptions have been copied from section 2).

We also assume that a Σ-measurable function $g : X \to \mathbb{R}_+$ is given and a measure m on X such that (X, Σ, m) is a Lebesgue space and $g = g_T$ almost everywhere. We notice that g_T is defined a. e. , while g is defined *everywhere.* We assume that $\sup_X g < +\infty$ and $\inf_X g > 0$. We also assume that the "g-operator" $Pf(x) = \sum_{y\in T^{-1}(x)} g(y)f(y)$ is well defined on the space of all bounded measurable functions $f : X \to \mathbb{R}$. For example, $\sum_{B\in\beta}\sup_B g < +\infty$ is a sufficient condition.

Let $\mathcal{M}$ be a family of T-invariant probability measures defined on Σ. We assume that for every $\mu \in \mathcal{M}$ the entropy $H_\mu(\beta) < +\infty$. Hence (see [11] and [14]), $h_\mu(T) = h_\mu(T,\beta) = \int_X \ln J\, d\mu < +\infty$. (we notice that (X, Σ, μ) is always a Lebesgue space). J, as usual, denotes the Jacobian of T.

The pressure functional $\mathcal{P} : \mathcal{M} \to \mathbb{R} \cup \{-\infty\}$ is defined by

$$\mathcal{P}(\mu) = h_\mu(T) + \int_X \ln g\, d = \int_X \ln(Jg)\, d. \tag{91}$$

We will call a measure $\mu \in \mathcal{M}$ an equilibrium state (with respect to g) iff μ realizes the maximum of $\mathcal{P}$ and $\mathcal{P}(\mu) = 0$. We sometimes write $\mathcal{P}(T, \mu)$, if it is not clear what transformation we are considering.

Lemma 20. *For every* $\mu \in \mathcal{M}$

$$\mathcal{P}(\mu) \le \inf_{\phi, n \ge 1} \frac{1}{n} \ln \int_X \frac{P^u \phi}{\phi} d\mu \tag{92}$$

where the infimum is over all positive, measurable functions $\phi : X \to \mathbb{R}_+$.

Proof. For each point $x \in X$ the set $T^{-1}(x)$ is countable, so there is a function $h : X \to \mathbb{R}_+$ such that $\mu_{T^{-1}(x)} = \sum_{y \in T^{-1}(x)} h(y) \delta_y$ where $\{\mu_{T^{-1}(x)}\}_{x \in X}$ is the family of the conditional measures with respect to the partition $\{T^{-1}(x)\}_{x \in X}$ $(= T^{-1}\epsilon$, where ϵ is the partition into points).

The condition $T_* \mu = \mu$ implies $h(y) = 1/J(y)$ almost everywhere on X (see [14]). Thus, $\mathcal{P}(\mu) = \int_X \ln(J \cdot g) d\mu = \int_X \mu_{T^{-1}(x)}(\ln(J \cdot g)) \cdot d\mu(x) = \int \sum_{y \in T^{-1}(Tx)} J^{-1}(y) \cdot \ln(Jg)(y) d\mu(x)$. By Jensen inequality

$$\sum_{y \in T^{-1}(Tx)} J^{-1}(y) \cdot \ln(J \cdot g)(y) \le \ln \left(\sum_{y \in T^{-1}(Tx)} g(y) \right) = \ln(P1 \circ T)(x).$$

Hence $\mathcal{P}(\mu) \le \int \ln(P1 \circ T) d\mu \le \ln \int P1 \circ T d\mu = \ln \int P1 d\mu$. Let $\phi : X \to \mathbb{R}_+$ be any bounded and bounded away from 0 (Σ-measurable) function. If we apply the previous inequality and definition 91 to $g \cdot \phi/\phi \circ T$ instead of g, we obtain $h_\mu(T) + \int \log(g \cdot \phi/\phi \circ T) d\mu \le \ln \int \frac{P\phi}{\phi} d\mu$. Since $\int \ln(g\phi/\phi \circ T) = \int \ln g\, d\mu + \int \ln \phi d\mu - \int \ln \phi \circ T\, d\mu = \int \ln g\, d\mu$, these last two formulas yield $\mathcal{P}(\mu) \le \ln \int \frac{P(\phi)}{\phi} d\mu$. Replacing T by T^n and using $\mathcal{P}(\mu) = \frac{1}{n} \mathcal{P}(T^n, \mu)$, we get the lemma.

Lemma 21. *Let* T, β *and* m *satisfy the assumptions of section 2. Let* $\phi \in L^\infty(m)$ *be nonnegative and* $P\phi = \phi$, $\int \phi\, dm = 1$. *Then* $\mathcal{P}(\mu) = 0$, *where* $\mu = \phi \cdot m$.

Proof. We use the following fact from [6] : Let $f : X \to \mathbb{R}$ be Σ-measurable and let $(f - f \circ T)_+$ be μ-integrable. Then $f - f \circ T$ is integrable and $\int_X (f - f \circ T) d\mu = 0$. We notice that $\phi(Tx) = P\phi(Tx) = \sum_{y \in T^{-1}(Tx)} g(y)\phi(y) \ge$

$g(x)\phi(x)$. Hence, $\phi/\phi \circ T \le g^{-1} = J$ and $\sup J = 1/\inf g$ yields $\int(\ln \phi - \ln \phi \circ T)_+ d\mu < +\infty$. Thus, $\int_X \ln \frac{\phi}{\phi \circ T} d\mu = 0$. It is not difficult to see that the Jacobian J_μ of μ is equal to $J \cdot \phi/\phi \circ T$, where $J = 1/g$. Hence $\mathcal{P}(\mu) = \int_X \ln(J_\mu \cdot g) d\mu = \int_X \ln(J \cdot g \cdot \phi/\phi \circ T) d\mu = \int_X \ln \phi/\phi \circ T d\mu = 0$.

From now on we assume that the following condition is satisfied:

(*) There is a constant $C \in \mathbb{R}_+$ such that for every $\epsilon > 0$ and $A \in \Sigma$ and for every $\mu \in \mathcal{M}$ there is a β^k-measurable set $A_1 \in \Sigma$ such that $\mu(A \div A_1) < \epsilon$ and for sufficiently large $n \in \mathbb{Z}_+$ $P^n(\chi_{A_1}) \le C \cdot m(A_1)$.

Obviously, if $\mathcal{M}$ is the family of measures $\pi_*(\nu)$, where ν is any L-invariant, Borel, probability, non-atomic measure, then this condition is satisfied by virtue of $\|P^{n^N}(\chi_{A_1})\|_\infty \le r^n \ \#\{B \in \beta^k : B \subset A_1\} + \frac{R}{1-r} m(A_1)$ following from the estimates of the proof of Theorem 2. We notice that only non-atomic measures have to be considered because g is not defined on the set $\bigcup_{k=0}^{\infty} L^{-k}(\{x = 0\})$ which could contain some atoms of μ and the integral $\int \ln g \, d\mu$ would not make sense. For nonatomic μ we have $\mu\left(\bigcup_{k=0}^{+\infty} L^k(\{x = 0\})\right) = 0$ because this set is countable.

Theorem 6. *Each equilibrium state μ is of the form $\phi \cdot m$ where $\phi \ge 0$ $\int \phi \, dm = 1$ and $P\phi = \phi$.*

Proof. Let $\mu \in \mathcal{M}$ and $\mathcal{P}(\mu) \ge 0$. We can assume that μ is ergodic, because otherwise we could work with an ergodic component of μ. We want to prove that $\mu \ll m$. Let us suppose that on the contrary, $\mu \not\ll m$. There exists a set $A \in \Sigma$ such that $m(A) = 0$ and $\mu(A) = 1$. Let $\epsilon = 1/9C$ and let $A_1 \in \Sigma$ be such that $\mu(A \div A) < \epsilon$ and $m(A_1) < \epsilon$. Let us put $\phi = t_1 \chi_{A_1} + t_2 \chi_{X \setminus A_1}$ where $t_1 = 1/3\epsilon^{-1}$ and $t_2 = \frac{1 - t_1 m(A_1)}{1 - m(A_1)}$. By condition (*) we get that for some $n \in \mathbb{Z}_+$ we have $\|P^n \phi\|_\infty \le C\,(t_1 m(A_1) + t_2(1 - m(A_1))) \le C$. Hence,

$$P^n \phi/\phi \le \begin{cases} 1/3 & \text{on } A_1 \\ 3/2C & \text{on } X \setminus A_1 . \end{cases} \tag{93}$$

(We observe that $t_2 \ge \frac{1-1/3}{1-\epsilon} \ge 2/3$). Finally, $\mu\left(\frac{P^n\phi}{\phi}\right) \le 1/3\mu(A_1) + 3/2C\mu(X \setminus A_1) \le 1/3 + 3/2C \cdot \epsilon = 1/2 < 1$. So, $\mathcal{P}(\mu) \le \frac{1}{n} \ln \mu\left(\frac{P^n\phi}{\phi}\right) \le 0$. This contradicts the assumption $\mu \ll m$.

4.2 The Variational Principle for Lozi Mappings

For every L-invariant, non-atomic, probabilistic measure μ on Δ we have $\mu\left(\bigcup_{k=-\infty}^{+\infty} L^k(\{x = 0\})\right) = 0$. This implies that both λ^u and λ^s are well-defined μ-a. e. Let $\phi^u = \log \lambda^u \circ L$. The geometrical meaning of $\lambda^u(Lp)$ is the reciprocal of the expansion coefficient along the unstable direction $E^u(p)$. So, ϕ^u is analogous to the function considered by Bowen and Ruelle [1].

Because of the cohomology relation 59 we have

$$h\mu(L) + \int_\Lambda \phi^u d\mu = h_\mu(L) + \int_\Lambda \log(\lambda^s |b|^{-1}) d\mu$$
$$+ \int \log h_3 \circ L d\mu - \int \log h_3 d\mu = h_\mu(L) + \int \log g d\mu. \tag{94}$$

The last expression is equal to $P(\pi_* \mu)$. The previous section yields:

Theorem 7. *The only measures maximizing the pressure functional $h_\mu(L) + \int_\Lambda \phi^u d\mu$ are of the form $t_1\mu_1 + \ldots + t_s\mu_s$, $\sum_i t_i = 1$, $t_i \geq 0$. Moreover, for such measures the value of the pressure functional is equal to* 0.

References

1. Bowen, R. *Equilibrium states and ergodic theory of Anosov diffeomorphisms* Lecture Notes in Math., Vol. **470**, Addison-Wesley, Berlin, Heidelberg, New York, p. 108, 1978.
2. Bunimovich, L. A., Sinai, Ya. *Markov partitions for dispersed billiards* Commun. Math. Phys. **78**, 247–280 (1980).
3. Collet, P., Levy, Y. *Ergodic properties of the Lozi Mappings* Commun. Math. Phys. **93**, 461–481 (1984).
4. Dunford, N., Schwartz, J. T. *Linear Operators*, Part 1, Wiley, New York (1959).
5. Hénon, M. *A two-dimensional mapping with a strange attractor* Commun. Math. Phys. **50**, 69–77 (1976).
6. Hofbauer, F., Keller, G. *Ergodic properties of invariant measures for piecewise monotonic transformations* Math. -Z. **180**, no. 1, 119-140 (1982).
7. Keller, G. *Generalized bounded variation and its applications to piecewise monotonic transformations* Z. Wahrsch. Verw. Gebiete **69**, no. 3, 461–478 (1985).
8. Lasota, A., Yorke, J. *On the existence of invariant measures for piecewise monotonic transformations* Trans. AMS **186**, 481–488 (1973).
9. Lozi, R. *Un attracteur étrange du type attracteur de Hénon* J. Phys. (Paris) **39**, (Coll. C5), 9–10 (1978).
10. Misiurewicz, M. *Hyperbolicity of the Lozi mappings* Annals of the New York Acad. of Sci., Vol. **357** (Nonlinear Dynamics), 348–358 (1980).
11. Rokhlin, V. A. *Lectures on the entropic theory of transformations with an invariant measure* UMN **12**, No. 5, 3–56 (1967) in Russian.
12. Rychlik, M. *Mesures invariantes et principe variationel pour les applications de Lozi* Compte Rend. Acad. Sci. Paris, t. 296 (1983).
13. Rychlik, M. *Bounded variation and invariant measures* Studia Math., t. 76, 69–80 (1983).
14. Walter, P. *Invariant measures and equilibrium states for some mappings which extend distances* Trans. AMS **236**, 121–153 (1978).

Commun. Math. Phys. 93, 461–481 (1984)

Communications in Mathematical Physics

Ergodic Properties of the Lozi Mappings

P. Collet and Y. Levy

Centre de Physique Théorique, Ecole Polytechnique, F-91128 Palaiseau Cédex, France

Abstract. In this paper, we construct the Bowen-Ruelle measure for the Lozi mapping, an almost everywhere hyperbolic diffeomorphism of the plane. We also derive some of its properties which are similar to those of an axiom A system.

I. Introduction

The Lozi mapping T is a homeomorphism of $\mathbb{R}^2$ given by

$$\begin{pmatrix} x \\ y \end{pmatrix} \rightarrow \begin{pmatrix} 1+y-a|x| \\ bx \end{pmatrix}.$$

For some values of a and b, Lozi [Lo] observed complicated behaviour for the trajectories of this system. For $b=0.5$ and $a=1.7$ one observes numerically a strange attractor, which is very similar to the attractor of the Henon map [He]. The main advantage of the Lozi map over the Henon map is that one can prove hyperbolicity without much effort. This is the main reason why so little is known for the Henon map, where hyperbolicity is believed to occur only on Cantor-like sets of parameters. Our opinion is that the Lozi mapping is an intermediate stage between the Axiom A dynamical systems and more complicated systems like the Henon map. As we shall see below, its dynamical structure is more complicated than in the Axiom A systems, although some detailed ergodic properties are the same. The Lozi map is rather similar to Sinai's billiards, and in this article, we shall use this analogy. In particular, the discontinuity of the differential allows the uniform hyperbolicity as in the billiards case. A proof of hyperbolicity for the Lozi map was first given by Misiurewicz [M]. He also derived many important consequences which will be described below.

This article is devoted to the investigation of the metric properties of the Lozi map. In the next paragraph, we briefly describe some properties of the map which will be needed later on. Most of them were known before. In the third paragraph we construct an invariant measure; its ergodic properties (absolute continuity with respect to the Lebesgue measure in the unstable direction, ergodicity, K-property, Bowen-Ruelle property, Bernoulli character) are derived in Sect. IV.

Similar results were obtained by Rychlik [Ry] using a different proof. We learned that Young [Y1] has proven similar results for piecewise C^2 hyperbolic maps.

II. Notations – General Properties

We set some notations which will be of constant use in this paper. $\mathcal{M}$ is the Borel σ-algebra of $\mathbb{R}$. If A is a measurable subset of $\mathbb{R}^2$, $\mathcal{M}_A$ is the corresponding factor sub σ-algebra. l and m are the 1-dim and 2-dim Lebesgue measure. d is the Euclidean distance in $\mathbb{R}^2$. We shall sometimes use the notations $\pm$ and $\mp$. We adopt the values of [M] for a and small values for b. For a partition ξ of the space X, $\xi(x)$, $x \in X$ is the atom of ξ containing x. Let $S_0^- = Oy$, $S_0^+ = Ox = TS_0^-$; $S_n^\pm = T^{\pm n} S_0^\pm$, $n \in \mathbb{N}$, are finite broken lines. Note that $T^{\pm n}$ is singular on $S_n^\mp$, $n \in \mathbb{N}$.

The fields of stable (unstable) directions $E^s(\cdot)$ $[E^u(\cdot)]$ are defined outside $\bigcup_{n \geqq 0} S_n^- \left(\bigcup_{n \geqq 0} S_n^+ \right)$ (cf. [M]). If $\theta^s(x)$ and $\theta^u(x)$ are their angles with respect to the x axis, we have continuous fraction expansions for $\tan\theta^s(x)$ and $\tan\theta^u(x)$ which are given by

$$\tan\theta^u(x) = \frac{-b}{a \cdot \varepsilon(T^{-1}x) - \tan\theta^u(T^{-1}x)}, \qquad \tan\theta^s(x) = a \cdot \varepsilon(x) + \frac{b}{\tan\theta^s(Tx)},$$

where $\varepsilon(x)$ is the sign of x. These formulae express the fact that $E^s(\cdot)$ and $E^u(\cdot)$ are invariant fields. If b/a is small enough, the above continuous fraction expansions are convergent. We shall denote by $\lambda(x)$ the expansion factor in the direction $E^u(x)$, i.e. this is the length of the image by DT_x of the unit vector in the direction $E^u(x)$. It is easy to verify that

$$\lambda(x) = \frac{[a^2 + b^2 + \tan^2(\theta^u(x)) - 2a\varepsilon(x)\tan\theta^u(x)]^{1/2}}{(1 + \tan^2\theta^u(x))^{1/2}}.$$

Let $W^s(x)$ [respectively $W^u(x)$] be the global stable (respectively unstable) manifold of x. The maximal smooth component $W^s_{\text{loc}}(x)$ $[W^u_{\text{loc}}(x)]$ of $W^s(x)$ $[W^u(x)]$ containing x will be called the local stable (local unstable) manifold of x. x splits $W^u_{\text{loc}}(x)$ into two "semi local unstable manifolds."

Let X be the fixed point of T with positive coordinates. Z is the intersection of the positive x axis with $\overline{W^u_{\text{loc}}(X)}$. We shall denote by F the triangle defined by the points Z, $T(Z)$, $T^2(Z)$ (see [M]). Ω will denote the strange attractor of T which is equal to $\bigcap_0^\infty T^n(F)$ [M]. We shall denote by λ and λ_+ the infimum and the supremum of $\lambda(x)$ for x in the set $F \Big\backslash \bigcup_{-\infty}^{+\infty} S_l^\pm : \lambda > 0$.

It is now easy, using the continuous fraction expansion, to verify the following lemma

Lemma II.1. *Let x and y in F be such that for $0 \leqq j \leqq q$, $T^j(x)$ and $T^j(y)$ are on the same side of S_0. Then*

i) *The angle between* $W^s_{\mathrm{loc}}(x)$ *and* $W^s_{\mathrm{loc}}(y)$ *(if they are defined) is bounded by* $2\left(\frac{b}{\lambda}\right)^{q-1}$.

ii) $\left|\frac{\lambda(x)}{\lambda(y)}-1\right| \leqq 2\left(\frac{b}{\lambda}\right)^{q-1}$.

We also observe that the angle between two local stable manifolds is between $\frac{2\pi}{5}$ and $-\frac{2\pi}{5}$ if $a<2$, and the angle between a local stable and a local unstable manifold is greater than $\frac{\pi}{5}$.

We now introduce a set H which is more convenient for our purposes than the set F. This set was already used by Misiurewicz but we recall here its properties.

Lemma II.2. *For b small enough, there is a polygon H such that*

i) $TH \subset H$,

ii) $\Omega \subset H$,

iii) *The boundary of H is contained in* $T^q W^u_{\mathrm{loc}}(X) \cup W^s_{\mathrm{loc}}(X)$ *for some* $q>0$.

iv) $T^r F \subset H$ *for some positive r.*

Proof. We adopt the notations of [M]. H_0 is the triangle XZP, where $P = W^s_{\mathrm{loc}}(X) \cap (Z, T(Z))$ and $H = \bigcup_{n=0}^{\infty} T^n(H_0)$. ($T$ is denoted f in [M].)

In Proposition 2 of [M], it is shown that there is a positive integer p such that

$$H = \bigcup_{n=0}^{p} T^n(H_0).$$

Moreover, in the same Proposition 2, it is established that

$$T^p F \subset H \quad \text{and} \quad TH \subset H. \quad \text{Q.E.D.}$$

We shall use the following mixing property derived by [M].

Proposition II.3. (Ω, T) *is topologically mixing, i.e.: if* $A, B \subset \mathbb{R}^2$ *are open, then*

$$(A \cap \Omega \neq \emptyset,\ B \cap \Omega \neq \emptyset) \Rightarrow (\exists N, n>N \Rightarrow T^n A \cap B \cap \Omega \neq \emptyset).$$

We investigate now the absolute continuity of the unstable foliation. Let W^1 and W^2 be two local unstable manifolds in Ω. We define a map $P = P_{W^1 W^2}$ from W^1 to W^2 by

$$x \in W^1 \to P(x) = W^s_{\mathrm{loc}}(x) \cap W^2 \quad \text{if} \quad W^s_{\mathrm{loc}}(x) \cap W^2 \neq \emptyset.$$

P is defined on $\mathcal{D}(P) = \{x \in W^1 \mid W^s_{\mathrm{loc}}(x) \cap W^2 \neq \emptyset\}$.

Proposition II.4. *Given* W^1 *and* W^2 *as above there is a constant* $L_1 > 0$ *which is* $\mathcal{O}(1)$ *such that for any Borel subset A of* W^1, $A \subset \mathcal{D}(P)$,

$$(1 - L_1(d(W^1, W^2))^{1/3}) l(A) \leqq l(P(A)) \leqq (1 + L_1(d(W^1, W^2))^{1/3}) l(A).$$

The proof is given in the appendix.

III. Construction of Invariant Measures

Let $L_0 = W^u_{\text{loc}}(X)$. We know from Misiurewicz's work [M] that Ω is the closure of $\bigcup_{n\geqq 0} T^n L_0$. Therefore, it is natural to try to obtain an invariant measure by iterating the Lebesgue measure supported by L_0. We define a sequence $(\mu'_n)_{n\in\mathbb{Z}}$ of probability measures on $\mathcal{M}_\Omega$ (μ'_n has support in $T^n L_0$) by $\mu'_n(A) = l(T^{-n}A \cap L_0)/l(L_0)$ for $A \in \mathcal{M}_\Omega$. The sequence $(\mu_n)_{n\in\mathbb{N}}$ defined by

$$\mu_n = \frac{1}{n}\sum_{j=0}^{n-1} \mu'_j$$

is a sequence of probability measures on $\mathcal{M}_\Omega$. As Ω is compact one can extract a subsequence which vaguely converges to an invariant probability measure μ (cf. [B]). Let $A^\pm_\varepsilon$ denote the stripe of width ε around $S^\pm_0$.

In order to obtain some properties of the measure μ, we shall first estimate the μ_n-measures of $A^\pm_\varepsilon$.

Proposition III.1. *For any positive number* $\tau < \left(1 - \frac{1}{K\log\lambda}\right)$ *(K is the integer appearing in Lemma III.2), there is a positive real number* ε_0, *such that for* $0 < \varepsilon \leqq \varepsilon_0$, *we have* $\mu_n(A^\pm_\varepsilon) \leqq \varepsilon^\tau$, $\forall n \in \mathbb{N}$.

We give the proof for $A_\varepsilon = A^-_\varepsilon$. We shall first give some geometrical considerations. For $n \in \mathbb{N}$, $T^n L_0$ is a segment or a broken line. Let J_n be the set of maximal smooth components contained in $T^n L_0$. For $M \in J_n$, the endpoints of M belong to $S_p \cup S_q$ for some p, q $1 \leqq p, q \leqq n+2$. We define $k(M)$ by $k(M) = \inf(p,q)$. For M belonging to J_n, we shall denote by $R_p(M)$ the element of J_{n-p} containing $T^{-p}M$ for $p \in \mathbb{Z}$. We now define recursively a finite sequence of integers $k_i(M)$ by

$$k_0(M) = 0,$$

$$k_1(M) = k(M),$$

$$k_{i+1}(M) = k_i(M) + k(R_{k_i(M)}(M)) \quad \text{as long as} \quad R_{k_i(M)}(M) \neq L_0.$$

We shall write k_i instead of $k_i(M)$ when there is no ambiguity. We now prove a lower bound on $k(\cdot)$.

Lemma III.2. *Assume b is small enough, there is an integer* $K > 4$ *and* $\theta > 0$ *such that if* $M \in J_n$, $l(M) < \theta$, *and* $M \cap S^-_0 \neq \emptyset$, *then* $k(M) > K$.

Proof. For b small enough, there is an integer $K > 2$ such that $0 < p \leqq K$ implies $(S^-_0 \cap F) \cap S^-_p = \emptyset$.

Let $\theta = \inf_{0 < p \leqq K} d(S^-_0 \cap F, S^-_p \cap F)$, then if $l(M) < \theta$, we have $k(M) > K$. Q.E.D.

We now come to the basic estimate. It is enough to prove the assertion for μ'_n, $n \in \mathbb{Z}$. The proof is recursive. Let τ be a number such that $0 < \tau < 1 - \frac{1}{K\log\lambda}$. The estimate is obvious for μ'_{-1}, since $l(L_0) > 1$; for μ'_n, $n \leqq -2$ and ε small enough, $\mu'_n(A_\varepsilon) = 0$. From now on, we assume the bound has been already proven for μ'_p, $-1 \leqq p < n$.

Let ϱ, $0<\varrho<1$ be such that

$$(1-\varrho)\left(1-\frac{1}{K\log\lambda}\right)>\tau.$$

Note that such a ϱ exists since $\tau<1-\frac{1}{K\log\lambda}$.

Let $E_1=\{M\in J_n|l(M)>4\varepsilon^\varrho\}$. If $M\in E_1$, M is a straight segment of T^nL_0, therefore

$$\frac{l(T^{-n}(A_\varepsilon\cap M))}{l(T^{-n}M)}=\frac{l(A_\varepsilon\cap M)}{l(M)}\leqq\frac{4\varepsilon}{4\varepsilon^\varrho}=\varepsilon^{1-\varrho},$$

since $l(A_\varepsilon\cap M)\leqq 4\varepsilon$ (we have used the fact that the contraction coefficient by T^{-n} is constant along M). Therefore,

$$\sum_{M\in E_1} l(T^{-n}(A_\varepsilon\cap M))\leqq\varepsilon^{1-\varrho}\sum_{M\in E_1} l(T^{-n}M)\leqq\varepsilon^{1-\varrho}l(L_0).$$

We now define a subset E_2 of J_n by

$$E_2=\{M\in J_n|\exists i\in\mathbb{N},\ k_i(M)<c|\log\varepsilon|,\ l(R_{k_i}(M))>4\varepsilon^\varrho\lambda^{-k_i(M)}\},$$

where c is a fixed number satisfying

$$K(1-\varrho)/\log 2>c>(1-\varrho)/\log\lambda\quad\left(\text{note that }\frac{K\log\lambda}{\log 2}>1\right).$$

For $M\in E_2$ let $\sigma(M)$ be the smallest integer i such that

$$l(R_{k_i}(M))>4\varepsilon^\varrho\lambda^{-k_i(M)}.$$

Let p be a positive integer, and let $\tilde{M}$ be an element of J_{n-p} such that $\tilde{M}\cap S_0^-\neq\emptyset$. We shall denote by $N(\tilde{M},p)$ the number of elements M of J_n such that $T^{-p}M$ is included in $\tilde{M}$, and $k_{\sigma(M)}\geqq p$.

Let

$$L_p=\sup_{0\leqq k\leqq p}\ \sup_{\substack{\tilde{M}\in J_{n-k}\\ \tilde{M}\cap S_0\neq\emptyset}} N(\tilde{M},k).$$

Note that L_p is a non-decreasing sequence. We shall now give an upper bound for L_p. For $\tilde{M}$ as above let $\tilde{M}_L$ (respectively $\tilde{M}_R$) be the segment $\tilde{M}\cap\{(x,y)|x\leqq 0\}$ [respectively $\tilde{M}\cap\{(x,y)|x\geqq 0\}$]. Assume moreover that the subset $E_L(p)$ of E_2 defined by $E_L(p)=\{M\in E_2|T^{-p}M\subset\tilde{M}_L,\ k_{\sigma(M)}\geqq p\}$ is non-empty. [If $E_L(p)$ is empty, then $E_R(p)$ is not empty or $L_p=0$.] Let q be the smallest positive integer such that $T^q\tilde{M}_L\cap S_0^-\neq\emptyset$. There are two cases.

Case 1. $q<p$. Then $T^q\tilde{M}_L\in J_{n-p+q}$, and for any $M\in E_L(p)$, we have $R_{k_i}(M)=T^q\tilde{M}_L$ for some integer i. Moreover, $k_i(M)=p-q<p\leqq k_{\sigma(M)}$. This implies $i<\sigma(M)$, and

$$l(T^q\tilde{M}_L)\leqq 4\varepsilon^\varrho\lambda^{-k_i(M)}.$$

If ε_0 is small enough, we have $4\varepsilon^\varrho < \theta$, and we can apply Lemma III.2 to conclude that $q > K$. Therefore,

$$\begin{aligned}\operatorname{card}(E_L(p)) &= \operatorname{card}\{M \in E_2 | T^{-p+q}M \subset T^q\tilde{M}_L,\ k_{\sigma(M)} \geqq p\} \\ &\leqq \operatorname{card}\{M \in E_L(p) | T^{-p+q}M \subset T^q\tilde{M}_L,\ k_{\sigma(M)} \geqq p-K\} \\ &\leqq \sup_{0 \leqq k \leqq p-K} \ \sup_{\substack{\tilde{M}' \in J_{n-k} \\ \tilde{M}' \cap S_0 \neq \emptyset}} N(\tilde{M}', k) = L_{p-K}.\end{aligned}$$

Case 2. $q \geqq p$. In this case we have $T^p\tilde{M}_L \in J_n$, and $\operatorname{card} E_L(p) = 1$. Similarly, we define $E_R(p)$ by

$$E_R(p) = \{M \in E_2 | T^{-p}M \subset \tilde{M}_R,\ k_{\sigma(M)} \geqq p\},$$

and we obtain as before $\operatorname{card} E_R(p) = 1$ or $\operatorname{card} E_R(p) \leqq L_{p-K}$. Therefore, $N(\tilde{M}, p) = \operatorname{card} E_L(p) + \operatorname{card} E_R(p) \leqq \sup(2, 2L_{p-K})$, and we obtain the bound

$$L_p \leqq 2^{1+p/K}.$$

Let now p be an integer such that $0 \leqq p \leqq c|\log\varepsilon|$. Let $\tilde{M} \in J_{n-p}$. We note that if there is an $M \in J_n$ for which $T^{-p}M \subset \tilde{M}$, and $k_{\sigma(M)} = p$, we have $\tilde{M} \cap S_0 \neq \emptyset$. For such an M, we have

$$l(T^{-p}(A_\varepsilon \cap M)) \leqq \lambda^{-p}, \qquad l(A_\varepsilon \cap M) < 4\varepsilon\lambda^{-p},$$

and from $l(\tilde{M}) > 4\varepsilon^\varrho\lambda^{-p}$ [since $p = k_{\sigma(M)}(M)$], we deduce

$$\begin{aligned}\sum_{\substack{M \in E_2 \\ T^{-p}M \subset \tilde{M} \\ k_{\sigma(M)}(M) = p}} l(T^{-p}(A_\varepsilon \cap M)) &< 4\varepsilon\lambda^{-p}L_p \leqq 4\varepsilon\lambda^{-p}2^{1+p/K} \\ &< 2\varepsilon^{1-\varrho-\frac{c}{K}\log 2}\, l(\tilde{M}).\end{aligned}$$

If $M \in J_n$ and $T^{-p}M \subset \tilde{M}$, $\tilde{M} \in J_{n-p}$, we have

$$\frac{l(T^{-n}(A_\varepsilon \cap M))}{l(T^{-n+p}\tilde{M})} = \frac{l(T^{-p}(A_\varepsilon \cap M))}{l(\tilde{M})}.$$

Therefore,

$$\sum_{\substack{M \in E_2 \\ T^{-p}M \subset \tilde{M} \\ k_{\sigma(M)}(M) = p}} l(T^{-n}(A_\varepsilon \cap M)) < 2\varepsilon^{1-\varrho-\frac{c}{K}\log 2}\, l(T^{-(n-p)}\tilde{M}),$$

and

$$\begin{aligned}\sum_{M \in E_2} l(T^{-n}(A_\varepsilon \cap M)) &= \sum_{p=0}^{E(c|\log\varepsilon|)} \sum_{\tilde{M} \in J_{n-p}} \sum_{\substack{M \in E_2 \\ T^{-p}M \subset \tilde{M} \\ k_{\sigma(M)}(M) = p}} l(T^{-n}(A_\varepsilon \cap M)) \\ &< 2\varepsilon^{1-\varrho-\frac{c}{K}\log 2} \sum_{p=0}^{E(c|\log\varepsilon|)} \sum_{\tilde{M} \in J_{n-p}} l(T^{-(n-p)}\tilde{M}) \\ &< 4\varepsilon^{1-\varrho-\frac{c}{K}\log 2} c|\log\varepsilon|\, l(L_0).\end{aligned}$$

Let now E_2 be defined by $E_3 = J_n \backslash (E_1 \cup E_2)$. We shall assume $E_3 \neq \emptyset$, otherwise, the proof is finished. This implies $n > c|\log\varepsilon|$. For $M \in E_3$, let i be the unique integer such that

$$k_i(M) < c|\log\varepsilon| \leqq k_{i+1}(M).$$

Note that $k_{i+1}(M)$ exists if ε_0 is small enough since $l(L_0) > 1$. We now observe that $T^{-(k_{i+1}-k_i)}R_{k_i}(M) \cap S_0^- \neq \emptyset$, and

$$l(T^{-(k_{i+1}-k_i)}R_{k_i}(M)) < \lambda^{-(k_{i+1}-k_i)}l(R_{k_i}(M)) \leqq 4\varepsilon^{\varrho}\lambda^{-k_{i+1}}.$$

Therefore, $T^{-(k_{i+1}-k_i)}R_{k_i}(M) \subset A_{4\varepsilon^{\varrho}\lambda^{-k_{i+1}}}$. We shall now use the recursive assumption. We have

$$\sum_{\substack{M\in E_3\\ k_{i+1}(M)=p}} l(T^{-n}(A_\varepsilon \cap M)) = \mu'_{n-p}\Bigg(\bigcup_{\substack{M\in E_3\\ k_{i+1}(M)=p}} T^{-p}(A_\varepsilon \cap M)\Bigg)\cdot l(L_0)$$

$$\leqq \mu'_{n-p}(A_{4\varepsilon^{\varrho}\lambda^{-p}})\cdot l(L_0).$$

Therefore,

$$\sum_{M\in E_3} l(T^{-n}(A_\varepsilon \cap M)) \leqq \sum_{p=c|\log\varepsilon|}^{n} \mu'_{n-p}(A_{4\varepsilon^{\varrho}\lambda^{-p}})$$

$$\leqq \frac{4^{\tau}\varepsilon^{\varrho\tau}\lambda^{-c\tau|\log\varepsilon|}}{1-\lambda^{-\tau}}.$$

From $\varrho + c\log\lambda > 1$, $1-\varrho > \tau$, and $1-\varrho-\frac{c}{K}\log 2 > \tau$, we obtain $\mu'_n(A_\varepsilon) \leqq \varepsilon^{\tau}$ if $\varepsilon < \varepsilon_0$ for ε_0 small enough but independent of n. Q.E.D.

We fix now τ as in Proposition III.1 and denote by $(\mu_n)_{n\in\mathbb{N}}$ a subsequence of the previous sequence which converges weakly to μ.

Corollary III.3. i) *For ε small, $n\in\mathbb{N}$, $k\in\mathbb{Z}$, $\mu_n(T^kA_\varepsilon) < \varepsilon^{\tau}$, $\mu(T^kA_\varepsilon) < \varepsilon^{\tau}$.*

ii) *For $N\in\mathbb{N}$, we define $H^N = \Big\{x\in\Omega$: one (at least) of the endpoints of $W^u_{\mathrm{loc}}(x)$ does not lie on $\bigcup_{k=0}^{N} S_k^+\Big\}$. There exist $c>0$ and α, $0<\alpha<1$, such that for $n\in\mathbb{N}$, $\mu_n(H^N) < c\alpha^N$, and $\mu(H^N) < c\alpha^N$.*

iii) *For $\varepsilon\in\mathbb{R}^+_*$, we define $H_\varepsilon = \{x\in\Omega$: one (at least) of the semi-loc unstable manifolds is shorter than $\varepsilon\}$. For ε small, $\mu_n(H_\varepsilon) < \varepsilon^{\tau}$, $n\in\mathbb{N}$, and $\mu(H_\varepsilon) < \varepsilon^{\tau}$.*

iv) *For $x\in\Omega$, $\mu(\{x\}) = 0$.*

v) *For $x\in\Omega$, $\mu(\{W^u(x)\}) = \mu(\{W^s(x)\}) = 0$.*

Proof. The first part of i) follows from Proposition III.1, by the definition of μ_n; the second part is a consequence of the weak convergence of (μ_n), for A_ε has smooth boundary.

To prove ii) consider the line Δ passing through $x\in\Omega$ in the unstable direction, and let y be one of the endpoints of $W^u_{\mathrm{loc}}(x)$.

By construction, either $y\in S_k^+$ for some $k\in\mathbb{N}$ or $y\notin \bigcup_{k\geqq 0} S_k^+$, but y is an accumulation point of $\bigcup_{k\geqq 0} S_k^+ \cap \Delta$. In the latter case, one can construct a sequence $(y_n)_{n\in\mathbb{N}}$, $y_n\in\Delta$, such that, for some strictly increasing function $k(\cdot)$, $y_n\in S^+_{k(n)}$, $n\in\mathbb{N}$, and moreover, $\bigcup_{k=0}^{k(n)} S_k^+$ does not cross Δ between y_n and x. Then, $T^{-k(n)}$ is linear on the segment $[x, y_n] \subset \Delta$. As $\lim_{n\to\infty} y_n = y \in \Omega$ and Ω is bounded, we can suppose $d(x, y_n) < 10$, so that $d(T^{-k(n)}x, S_0^+) < 10/\lambda^{k(n)}$, $\forall n \in \mathbb{N}$. Thus, $x \in \bigcap_{n\geq 0} T^{k(n)} A_{10/\lambda^{k(n)}}$, which is a set of μ-measure zero, by i): almost surely, the endpoints of $W^u_{\text{loc}}(x)$, $x \in \Omega$, lie in $\bigcup_{k\geq 0} S_k^+$. Suppose now $y \in S_k^+$, $k \geqq N$. Then $x \in T^k A_{10/\lambda^k}$ whence ii) by i) with $\alpha = \lambda^{-\tau}$.

With the same notations, suppose $x \in H_\varepsilon$; then if $y \in S_k^+$, $k \in \mathbb{N}$, $x \in T^k A_{\varepsilon/\lambda^k}$, whence iii) by i) and the convergence of $\sum_{k\geq 0} \lambda^{-k\tau}$.

The proof of iv) is similar to the proof of Proposition III.1: Let $x \in \Omega$, and let ε be a positive number. Let B be the ball of radius ε centered at x. Let J_n be as above. We have

$$\mu_n(B) = \sum_{M\in J_n} l(T^{-n}(B\cap M))/l(L_0)\,.$$

Let $J_n^+ = \left\{M \in J_n \,|\, l(M) > \varepsilon^{\frac{1}{1+\tau}}\right\}$, $J_n^- = J_n \setminus J_n^+$. If $M \in J_n^+$, from $l(B\cap M) \leqq 2\varepsilon$, we obtain

$$l(T^{-n}(B\cap M)) = l(T^{-n}M)\frac{l(B\cap M)}{l(M)} \leqq \varepsilon^{\frac{\tau}{1+\tau}} l(T^{-n}M)\,.$$

Therefore,

$$\sum_{M\in J_n^+} l(T^{-n}(B\cap M)) \leqq \varepsilon^{\frac{\tau}{1+\tau}} \sum_{M\in J_n^+} l(T^{-n}M) \leqq \varepsilon^{\frac{\tau}{1+\tau}} l(L_0)\,.$$

For J_n^-, we have

$$\sum_{M\in J_n^-} l(T^{-n}(B\cap M)) \leqq \sum_{M\in J_n^-} l(T^{-n}M) = \mu_n\Big(\bigcup_{M\in J_n^-} M\Big)\cdot l(L_0)$$
$$\leqq \mu_n(H\varepsilon^{1/1+\tau})\cdot l(L_0) \leqq c\cdot l(L_0)\cdot \varepsilon^{\frac{\tau}{1+\tau}} \quad \text{by iii)}\,.$$

We obtain

$$\mu(B) \leqq l(L_0)(1+c)\cdot \varepsilon^{\frac{\tau}{1+\tau}}$$

and get iv) if we let $\varepsilon \to 0$.

We shall prove v) for $W^s(x)$, the proof for $W^u(x)$ is similar. It is enough to prove that $(W^s_{\text{loc}}(x)) = 0$, since $W^s(x) = \bigcup_{j=0}^{\infty} T^{-j} W^s_{\text{loc}}(T^j x)$. Assume $x \in \Omega$ satisfies $\mu(W^s_{\text{loc}}(x)) > \beta > 0$. Then, there is a couple of integers i and j, $i \neq j$ such that $T^i W^s_{\text{loc}}(x) \cap T^j W^s_{\text{loc}}(x) \neq \emptyset$, otherwise, we get a contradiction from $\mu(T^l W^s_{\text{loc}}(x)) > \beta$, $\forall l \geqq 0$. Assume $i > j$ and let $k = i - j$. Since T is a bijection, we have $T^k W^s_{\text{loc}}(x) \cap W^s_{\text{loc}}(x) \neq \emptyset$, and therefore, $T^k W^s_{\text{loc}}(x) \subset W^s_{\text{loc}}(x)$. Since $T^k|_{W^s_{\text{loc}}(x)}$ is a contraction, this means that $W^s_{\text{loc}}(x)$ contains a k periodic point P. Moreover, $P = \bigcap_{l=0}^{\infty} T^{kl} W^s_{\text{loc}}(x)$, and therefore,

$$\mu(\{P\}) = \inf_l \mu(T^{kl} W^s_{\text{loc}}(x)) = \mu(W^s_{\text{loc}}(x)) > \beta\,.$$

a contradiction with iv).

IV. Ergodic Properties of the Invariant Measure

So far, the measure μ is not unique. The uniqueness will be proven by showing that μ is the Bowen-Ruelle measure. We first investigate the properties of the conditional expectations of μ on the unstable foliation, making use of the sequence $(\mu_n)_{n\in\mathbb{N}}$. In order to investigate and use these properties, we define two countable partitions α and β which decompose Ω.

Let ζ^+ (ζ^-) be the decomposition of μ.a.a. Ω into local unstable (stable) manifolds. As ζ^+ is a partition generated by $\bigcup_{n\geq 0} S_n^+$, it is measurable. We can define the restriction μ^+ of μ to the sub σ algebra $\mathcal{M}^+ \subset \mathcal{M}_\Omega$ of the sets ζ^+-saturate. By Corollary III.3ii), for μ.a.e. $W_0 \in \zeta^+$, there are two maximal smooth components I_0 and J_0, contained in $\bigcup_{k=0}^{N_0} S_k^+$ for some $N_0 \geq 0$ such that the endpoints of W_0 lie on I_0 and J_0. We then define the partition α by $\alpha(W_0) = \{W \in \zeta^+$ with endpoints on I_0 and $J_0\}$. Let $\zeta_N^+ \subset \zeta^+$ be the union of the elements of ζ^+ with endpoints on $\bigcup_{k=0}^{N} S_k^+$, $N \in \mathbb{N}$. ζ_N^+ is a finite union of atoms of α; since, by Corollary III.3ii), $\lim_{N\to\infty} \mu^+(\zeta_N^+) = 1$, α is a countable partition of ζ^+.

We now look for a partition of Ω into parallelograms. Let P_N^+ (P_N^-) be the partitions generated by $\bigcup_{n=0}^{N} S_n^+ \left(\bigcup_{n=0}^{N} S_n^-\right)$. As S_n^+ is a broken line, folded only on $\bigcup_{i=0}^{n-1} S_i^+$, a simple recursion argument shows that the $P_N^\pm(x)$, $x \in \Omega$, are convex sets. Note that $\zeta^\pm = \lim_{N\to\infty} P_N^\pm$. Let $x \in \Omega$ such that $W^+ = W_{\mathrm{loc}}^u(x)$ and $W^- = W_{\mathrm{loc}}^s(x)$ have positive length. Let $N_\pm(x)$ be the smallest integers such that the endpoints of W^- (W^+) lie outside $P_{N^+(x)}^+(x)$ [$P_{N^-(x)}^-(x)$]. It is easy to see that $N_\pm$ are finite; $N_\pm$ are obviously measurable functions, so that we can define, for such an element $x \in \Omega$ the atom $\beta(x)$ by:

$$\beta(x) = N_x^{-1}(N_+(x)) \cap N_-^{-1}(N_-(x)) \cap P_{N^+(x)}^+(x) \cap P_{N^-(x)}^-(x).$$

As usual, we define the unstable and stable fibers $\gamma^+(x)$, $\gamma^-(x)$ by

$$\gamma^+(x) = \beta(x) \cap W_{\mathrm{loc}}^u(x),$$

$$\gamma^-(x) = \beta(x) \cap W_{\mathrm{loc}}^s(x), \quad \text{i.e.} \quad \gamma^\pm = \beta \vee \zeta^\pm.$$

The set $\beta(y)$ is a parallelogram in the following sense

$$x \in \beta(y) \Rightarrow \begin{cases} \exists!\, z \in \beta(y), & z = \gamma^+(x) \cap \gamma^-(y), \\ \exists!\, z' \in \beta(y), & z' = \gamma^-(x) \cap \gamma^+(y). \end{cases}$$

This is easily checked by using the convexity of $P_{N^\pm(x)}^\pm(x)$ and the fact that $N^\pm$ are constant on atoms of $P_{N^\pm(x)}^\pm \vee \zeta^\mp$. By definition β is a countable partition of Ω into parallelograms.

We now come back to the properties of μ. As ζ^+ is a measurable partition, we can apply the usual theorem on disintegration (see [Ro] e.g.).

Namely, there is a μ-a.s. unique family $\{\mu_W, W \in \zeta^+\}$ of probability measures on Ω such that:

a) μ_W has μ-a.s. support on W,

b) for $A \in \mathcal{M}_\Omega$ the map $W \to \mu_W(A)$ is in $L^1(\zeta^+, d\mu^+)$, and $\mu(A) = \int_{\zeta^+} \mu_W(A) d\mu^+(W)$, denoted $\mu^+(\mu.(A))$.

For $W \in \zeta^+$, let l_W denote the normalized $1-d$ Lebesgue measure on W.

Proposition IV.1. *The conditional expectations of μ on the local unstable manifolds are the corresponding $1-d$ Lebesgue probabilities, i.e.: $\mu_W = l_W$ for μ.a.e. $W \in \zeta^+$.*

Proof. As stated before, some subsequence $(\mu_{n_i})_{i \in \mathbb{N}}$ of $(\mu_n)_{n \in \mathbb{N}}$ converges weakly to μ; we shall still denote it $(\mu_n)_{n \in \mathbb{N}}$. Let μ_n^+ denote the restriction of μ_n to $\mathcal{M}^+$, $n \in \mathbb{N}$. By the geometrical properties of T, it is easy to show that for $A \in \mathcal{M}_\Omega$, $\mu_n(A) = \int_{\zeta^+} l_W(A) d\mu_n^+(W) = \mu_n^+(l.(A))$. It is enough to prove $\lim_{n \to \infty} \mu_n^+(l.(f)) = \mu^+(l.(f))$ for $f \in C^0(\Omega)$. Let $f \in C^0(\Omega)$. With respect to the Hausdorff topology τ_H on ζ^+, the map $\zeta^+ \to \mathbb{R} : W \to l_W(f)$ is continuous. We shall show that μ_n^+ converges weakly to μ^+ in the sense of τ_H.

By Corollary III.3ii), $\mu_k^+(\zeta^+ \backslash \zeta_N^+)$, $k \in \mathbb{N}$ and $\mu^+(\zeta^+ \backslash \zeta_N^+)$ are simultaneously bounded by $C \cdot \varrho^N$ for some $\varrho \in]0, 1[$ and some positive constant C. Let $Q \subset \zeta^+$ be an atom of α. Since ζ_N^+ is a finite union of atoms of α the compactness of ζ_N^+ would follow from the compactness of Q. As two elements of ζ^+ cannot intersect, the elements of Q depend continuously on, e.g. the vertical coordinate of their rightmost endpoint. Since Ω is compact, the limit of a convergent sequence of elements of Q is a segment contained in Ω, thus, it belongs to Q : Q is compact. As Ω is totally regular, the hypothesis of Prokhorov's theorem are fulfilled (cf. [B]), so that $(\mu_n^+)_{n \in \mathbb{N}}$ converges weakly to some probability measure on ζ^+, which has to be the restriction μ^+ of μ. Q.E.D.

We are now able to derive the ergodic properties of T.

Proposition IV.2. *(Ω, T, μ) is ergodic.*

Proof. Take $f \in C^0(\Omega)$. By Birkhoff's ergodic theorem, there is a set $B \subset \Omega$, $\mu(B) = 0$ and a function $\bar{f} \in L^1(d\mu)$ such that, if $x \in \Omega \backslash B$ the limits $f^\pm(x) = \lim_{N \to \infty} \frac{1}{N} \sum_{n=0}^{N-1} f(T^{\pm n} x)$ exist and $f^+(x) = f^-(x) = \bar{f}(x)$. What we shall prove is that $\bar{f}$ is almost surely constant. Let $x, y \in \Omega \backslash B$. As $\mu(B) = \mu\left(\bigcap_\varepsilon H_\varepsilon\right) = 0$, the conditional measures of $B \cup \left(\bigcap_\varepsilon H_\varepsilon\right)$ (i.e. the corresponding normalized lengths) are zero on $W^u_{\text{loc}}(x)$ and $W^u_{\text{loc}}(y)$ for $\mu \times \mu$-a.e. (x, y). Thus, by Proposition II.4, it is enough to find a subset $A \subset W^u_{\text{loc}}(x)$ of positive length and an integer N such that for $z \in A$, $W^s_{\text{loc}}(z)$ crosses $T^N W^u_{\text{loc}}(y)$, because f^+ (f^-) is obviously constant on stable (unstable) manifolds. Consider now $\beta(x)$. Almost surely $\mu(\beta(x)) > 0$, and thus, by Proposition II.4, $l(\gamma^+(x)) > 0$. By Corollary III.3v), we can define $x_1, x_2 \in \beta(x)$ by demanding that the quadrilateral Q defined by $\zeta^\pm(x_1)$, $\zeta^\pm(x_2)$ be the smallest such that $\mu(Q \cap \beta(x)) = \mu(\beta(x))$.

For some N, by Proposition II.3, $T^N W^u_{\text{loc}}(y)$ "enters" Q, and one of its smooth components, that we shall call $\tilde{W}$, crosses $\zeta^-(x_1)$ or $\zeta^-(x_2)$ or both. Thus, we can define the canonical isomorphism P along stable fibers from $W^u_{\text{loc}}(x)$ into $\tilde{W} \subset T^N W^u_{\text{loc}}(y)$, with a domain $\mathscr{D}(P)$ of positive length. The proof is completed by taking $A = \mathscr{D}(P)$.

Proposition IV.3. i) *For* $n \in \mathbb{N}^*$, (Ω, T^n, μ) *is ergodic.*
ii) (Ω, T, μ) *is a K-system.*

Proof. The proof of i) is similar to the proof of Proposition IV.2 ii) follows from the fact the Pinsker σ-algebra Π of T is smaller than the σ-algebra of the measurable sets saturate by stable and unstable manifolds (cf. [P]). The proof of Proposition IV.2 shows that if $f \in L^1(d\mu)$ is constant along stable and unstable manifolds then f is μ-a.s. constant; thus, Π is trivial for μ, whence ii).

In order to prove the Bernoullian property, we introduce some notations. We define decreasing sequences $(\mathscr{M}_n^{\pm})_{n \in \mathbb{N}}$, where $\mathscr{M}_n^{\pm}$ is the sub σ-algebra of the elements of $\mathscr{M}_\Omega$ which are $T^{\pm n}\zeta^{\pm}$-saturate. By Proposition IV.3ii) the σ-algebras $\lim_n \mathscr{M}_n^{\pm} = \bigcap_n \mathscr{M}_n^{\pm}$ are both trivial for μ. We note $\mu_{\pm}$ the restriction of μ to $\mathscr{M}_0^{\pm}$.

Proposition IV.1 about the conditional expectation of μ with respect to $\mathscr{M}_0^+$ will allow us to prove the following:

Proposition IV.4. (Ω, T, μ) *is isomorphic to a Bernoulli shift.*

Proof. The proof is similar to [L1].

The canonical map $(\Omega, \mathscr{M}_0^+ \vee \mathscr{M}_0^-) \xrightarrow{P} (\zeta^+ \times \zeta^-, \mathscr{M}_0^+ \times \mathscr{M}_0^-)$ given by $P(x) = (\zeta^+(x), \zeta^-(x))$ is a.e. defined. Let $\nu = \mu \circ P^{-1}$ denote the image of μ through P. We have:

Lemma IV.5. ν *is absolutely continuous with respect to* $\mu^+ \otimes \mu^-$, i.e. $\nu \ll \mu^+ \otimes \mu^-$.

Proof. Let $A \in \mathscr{M}_0^+ \vee \mathscr{M}_0^-$ such that $P(A) \in \mathscr{M}_0^+ \times \mathscr{M}_0^-$. By definition, we have:

$$\mu^+ \otimes \mu^-(P(A)) = \int_{W \in \zeta^+} d\mu^+(W) \cdot \mu^-\left(\bigcup_{x \in A \cap W} \zeta^-(x)\right).$$

Suppose $\mu^+ \otimes \mu^-(P(A)) = 0$. Then, for some $\mathscr{W}_1 \subset \zeta^+$ of full measure:

$$W \in \mathscr{W}_1 \Rightarrow \mu^-\left(\bigcup_{x \in A \cap W} \zeta^-(x)\right) = 0.$$

Suppose that, for some $W' \in \gamma^+$, included in an element C of β of positive measure, we have $\mu(A|W') = 0$. Then, by Proposition II.4 and Proposition IV.1, we have:

$$\mu\left(C \cap \bigcup_{x \in A \cap W'} \zeta^-(x)\right) > 0,$$

so that if W is the ζ^+-saturate of W', we have:

$$\mu^-\left(\bigcup_{x \in A \cap W} \zeta^-(x)\right) > 0,$$

so that $W \notin \mathscr{W}_1$. Thus, $\mu(A|W') = 0$ almost surely and $\mu(A) = 0$. Q.E.D.

We shall denote the measure $\mu^+ \otimes \mu^-$ by $\mu^\times$. By Radon-Nikodym's theorem, there is a $\mu^\times$-integrable function $h: \zeta^+ \times \zeta^- \to \mathbb{R}^+$, such that $d\nu(x) = h(x) \cdot d\mu^\times(x)$. If $A \in \mathscr{M}_0^+ \times \mathscr{M}_0^-$ and $\nu(A) > 0$, the conditional probability $\nu_A = \nu(\cdot|A)$ is given, for a function f, by:

$$\nu_A(f) = \nu(f \cdot \chi_A)/\nu(\chi_A) = \mu^\times(f \cdot h \cdot \chi_A)/\mu^\times(h \cdot \chi_A)$$
$$= \mu^\times(f \cdot h)/\mu_A^\times(h)$$

or, more briefly:

$$\nu_A = \mu_A^\times(h \cdot)/\mu_A^\times(h).$$

We are now able to prove the weak Bernoulli property.

Lemma IV.5. *$(\Omega, \mathcal{M}, \mu, T)$ is weak Bernoulli, that is μ and $\mu^\times \circ P$ coïncide on $\bigwedge_n (\mathcal{M}_n^+ \vee \mathcal{M}_n^-)$.*

Proof. As $(\mathcal{M}_n^+ \vee \mathcal{M}_n^-)_{n \in \mathbb{N}}$ is a decreasing sequence, the conditional expectation on A of μ with respect to $\bigwedge_n (\mathcal{M}_n^+ \vee \mathcal{M}_n^-)$ is given almost everywhere by:

$$\mu_A^{\bigwedge_n (\mathcal{M}_n^+ \vee \mathcal{M}_n^-)} = \lim_{n\to\infty} \mu_A^{\mathcal{M}_n^+ \vee \mathcal{M}_n^-}$$
$$= \lim_{n\to\infty} \nu_{P(A)}^{\mathcal{M}_n^+ \times \mathcal{M}_n^-},$$

by definition of ν. Therefore,

$$\mu_A^{\bigwedge_n (\mathcal{M}_n^+ \vee \mathcal{M}_n^-)} = \lim_{n\to\infty} (\mu^\times)_{P(A)}^{\mathcal{M}_n^+ \times \mathcal{M}_n^-}(h \cdot)/(\mu^\times)_{P(A)}^{\mathcal{M}_n^+ \times \mathcal{M}_n^-}(h).$$

As $(\Omega, \mathcal{M}_\Omega, \mu, T)$ is a K-system, the σ-algebras $\bigwedge_{n \in \mathbb{N}} \mathcal{M}_n^+$ and $\bigwedge_{n \in \mathbb{N}} \mathcal{M}_n^-$ both coïncide with the trivial algebra mod μ: the conditional expectations

$$\mu^{\bigwedge_{n \in \mathbb{N}} \mathcal{M}_n^\pm} = \lim_{n\to\infty} \mu_\pm^{\mathcal{M}_n^\pm}$$

are constant μ-almost everywhere.

Thus, both $\mu^{\bigwedge_n (\mathcal{M}_n^+ \vee \mathcal{M}_n^-)}$ and $(\mu^\times \circ P)^{\bigwedge_n (\mathcal{M}_n^+ \vee \mathcal{M}_n^-)}$ are μ-a.e. constant and thus coïncide. Q.E.D.

The statement of Proposition IV.4 follows from Lemma IV.5 (cf. [L2] for instance).

We prove now that μ is the (unique) Bowen-Ruelle measure, id est:

Proposition IV.6. *For $g \in C^0(F)$ and m-almost any $x \in F$,*

$$\lim_{n\to\infty} \frac{1}{n} \sum_{k=0}^{n-1} g(T^k x) = \mu(g).$$

The proof follows three steps: We consider the points x of $\bigcup_{n \in \mathbb{N}} T^n L_0$ such that, for some $g \in C^0(F)$, we have not $\lim_{N\to\infty} \frac{1}{N} \sum_{n=0}^{N-1} g(T^n x) = \mu(g)$, and prove that the length of this set is zero (Lemma IV.7). Then we notice that, for m-a.e. $x \in F$ one can find positive integers p, n such that $W_{\text{loc}}^s(T^n x)$ crosses $T^p L_0$ (Lemma IV.8). Thus, what we have to show is that, m-almost surely in F, this intersection does not fall in the exceptional set estimated in Lemma IV.7.

Let $\mathscr{A}=\left\{x\in \bigcup_{n\geqq 0} T^nL_0 | \exists g\in C^0(F) \text{ and } \lim_{N\to+\infty} \frac{1}{N}\sum_{n=0}^{N-1} g(T^nx) \text{ does not exist or is not equal to } \mu(g)\right\}$.

Lemma IV.7. $\forall n\in \mathbb{N}$, $\mu_n(\mathscr{A})=l(\mathscr{A}\cap L_0)=0$.

Proof. We first observe that $\mathscr{A}=T\mathscr{A}$, and if $x, y\in T^nL_0$, $y\in W^s_{\text{loc}}(x)$, then $x\in\mathscr{A}$ is equivalent to $y\in\mathscr{A}$. This last property implies using Propositions II.4 and IV.1, that $W\to\mu(\mathscr{A}|W)$ is a continuous function on (ζ^+, τ_H). From $\mu\left(\bigcup_{n\geqq 0} T^nL_0\right)=0$, we deduce $\mu(\mathscr{A})=0$. Therefore, using the Birkhoff ergodic Theorem, we have

$$0=\mu(\mathscr{A})=\int_{\zeta^+} d\nu(W)\mu(\mathscr{A}|W)=\lim_{n\to+\infty}\int_{\zeta^+} d\nu_n(W)\mu(\mathscr{A}|W)=\lim_{n\to+\infty}\mu_n(\mathscr{A}).$$

However, $\mu_n(\mathscr{A})=\dfrac{l(T^{-n}\mathscr{A}\cap L_0)}{l(L_0)}=\dfrac{l(\mathscr{A}\cap L_0)}{l(L_0)}$, hence $l(\mathscr{A}\cap L_0)=0$. Q.E.D.

Let $\mathscr{L}=\{x\in F | \forall n, p\in\mathbb{N},\ W^s_{\text{loc}}(T^nx)\cap T^pL_0=\emptyset\}$.

Lemma IV.8. $m(\mathscr{L})=0$.

Proof. Let H and r be as in Lemma II.2. This lemma implies $\mathscr{L}\subset\bigcup_{n>0} T^{-n}(\mathscr{L}\cap H)$. Therefore, it is enough to show that $m(\mathscr{L}\cap H)=0$ since T^{-1} is absolutely continuous. For $\alpha>0$ and $n\in\mathbb{N}$, let $\mathscr{M}^\alpha_n=\{x\in H | l(W^s_{\text{loc}}(T^nx))<\alpha^n\}$. We have

$$\mathscr{M}^\alpha_n=T^{-n}\{y\in T^nH | l(W^s_{\text{loc}}(y))<\alpha^n\}$$

and using [M], we derive

$$m(\mathscr{M}^\alpha_n)\leqq\delta\left(\frac{\alpha}{b}\right)^n,$$

where δ is a positive constant. Therefore, if $\alpha<b$, we have with $\mathscr{M}^\alpha=\bigcap_{n=0}^{\infty}\mathscr{M}^\alpha_n$,

$$m(\mathscr{M}^\alpha)=0.$$

Let $x\in(H\backslash\mathscr{M}^\alpha)\cap\mathscr{L}$. We have $x\notin\mathscr{M}^\alpha_n$ for n large enough, but

$$W^s_{\text{loc}}(T^nx)\cap T^{n+r}L_0=\emptyset,$$

therefore,

$$W^s_{\text{loc}}(T^nx)\subset T^nH.$$

$T^{-n}W^s_{\text{loc}}(T^nx)$ is a broken line of total length bounded below by $\left(\dfrac{\lambda\alpha}{b}\right)^n$. However, for b small enough, it is easy to verify that for $y\in H$ such that $W^s_{\text{loc}}(y)\subset H$, then $\overset{0}{\widehat{W^s_{\text{loc}}(y)}}$ does not intersect both S^-_0 and S^+_0. This implies that $T^{-n}W^s_{\text{loc}}(T^nx)$ is composed of at most $2^{[n/2]+1}$ straight segments. Therefore, since each segment has a length at most one, we have

$$l(T^{-n}(W^s_{\mathrm{loc}}(T^n x))) \leqq 2^{[n/2]+1}.$$

We now choose $\alpha < b$, such that $\frac{\sqrt{2}}{\lambda}\frac{b}{\alpha} < 1$ (this is possible since $\lambda > \sqrt{2}$). We have

$$2^{[n/2]+1} \geqq l(T^{-n}(W^s_{\mathrm{loc}}(T^n x))) \geqq \left(\frac{\lambda\alpha}{b}\right)^n,$$

which is a contradiction if n is large enough. Therefore, $\mathscr{L} \cap H \subset \mathscr{M}^\alpha$, and we have

$$m(\mathscr{L} \cap H) \leqq m(\mathscr{M}^\alpha) = 0. \quad \text{Q.E.D.}$$

We now come to the proof of Proposition IV.6.

We first observe that if $x \in F\backslash\mathscr{L}$, one can find an integer n, and $p \geqq n$ such that $W^s_{\mathrm{loc}}(T^n x) \cap T^p L_0 \neq \emptyset$. We define a new set $\mathscr{B}$ by

$$\mathscr{B} = \{x \in F\backslash\mathscr{L} \mid \forall n, p, p \geqq n,\ W^s_{\mathrm{loc}}(T^n x) \cap T^p L_0 \subset \mathscr{A}\}.$$

If $x \in F\backslash(\mathscr{L} \cup \mathscr{B})$, there is an $n \in \mathbb{N}$, and an integer $p \geqq n$ such that some point y of $W^s_{\mathrm{loc}}(T^n x) \cap T^p L_0$ does not belong to $\mathscr{A}$. Therefore, if g belongs to $C^0(\mathbb{R}^2)$, we have

$$\begin{aligned}\lim_{m \to +\infty} \frac{1}{m} \sum_{j=0}^{m-1} g(T^j x) &= \lim_{m \to \infty} \frac{1}{m} \sum_{j=0}^{m-1} g(T^{j+n} x) \\ &= \lim_{m \to \infty} \frac{1}{m} \sum_{j=0}^{m-1} g(T^j y) = \mu(g).\end{aligned}$$

We shall now show that $m(\mathscr{B}) = 0$. Let $\mathscr{B}_p$ be defined by

$$\mathscr{B}_p = \{y \in F\backslash\mathscr{L} \mid W^s_{\mathrm{loc}}(y) \cap T^p L_0 \subset \mathscr{A}\}.$$

From $\mathscr{B} = \bigcap_{p > n \geqq 0} T^{-n}\mathscr{B}_p$, it is enough to show that $m(\mathscr{B}_p) = 0$ for every integer p. From the definition of $\mathscr{B}_p$, we have

$$\mathscr{B}_p \subset \bigcup_{x \in \mathscr{B} \cap T^p L_0} (W^s_{\mathrm{loc}}(x) \cap F).$$

Thus, an unstable segment W being chosen, we consider $\mathscr{B}_W = \bigcup_{x \in \mathscr{A} \cap W} W^s_{\mathrm{loc}}(x)$, and prove $m(\mathscr{B}_W) = 0$, using $l(\mathscr{A} \cap W) = 0$.

Let $\mathscr{A}_W = \{x \in \mathscr{A} \cap W \mid W^s_{\mathrm{loc}}(x) \neq \{x\}\}$. It is enough to prove $m\left(\bigcup_{x \in \mathscr{A}_W} W^s_{\mathrm{loc}}(x)\right) = 0$. If the W^s_{loc}'s were depending smoothly on x, this would be a consequence of Fubini's theorem. Instead we use Lemma II.1. Let $\varepsilon > 0$ be given sufficiently small. As $l(\mathscr{A} \cap W) = 0$, $\mathscr{A} \cap W$ can be covered by a countable union of open disjoint intervals of total length smaller than ε. Consider one of these intervals, say I; let $\varepsilon' < \varepsilon$ be its length. Let now $n \in \mathbb{N}^*$; $\bigcup_0^{n+1} S_k^-$ splits I into at most 2^{n+2} segments; if J is such a segment, by Lemma II.1, the dispersion of the angles of $\{W^s_{\mathrm{loc}}(x),\ x \in J\}$ is bounded by $2(b/\lambda)^n$, so we obtain

$$m\left(\bigcup_{x \in J} W^s_{\mathrm{loc}}(x)\right) \leqq 4(b/\lambda)^n + l(J).$$

Thus, $m\left(\bigcup_{x\in I} W^s_{\text{loc}}(x)\right) \leqq 16(2b/\lambda)^n + l(I) \leqq 16(2b/\lambda)^n + \varepsilon'$. As n is arbitrary and $\lambda > 2b$, we get

$$m\left(\bigcup_{x\in I} W^s_{\text{loc}}(x)\right) \leqq 2\varepsilon',$$

so that

$$m\left(\bigcup_{x\in \mathscr{A}W} W^s_{\text{loc}}(x)\right) \leqq 2\varepsilon,$$

which proves that $m(\mathscr{B}_W)=0$.

V. The Hausdorff Dimension

Let us denote by χ_+ and χ_- the characteristic exponents of (Ω, T, μ): $\chi_\pm = \lim_{n\to\pm\infty} \frac{1}{n}\log\|D_x T^n\|$ for μ.a.e. $x\in\Omega$. Let h denote the μ-entropy of T. In [Y2] Young proves that, if M is a compact surface, (M, f, m) an ergodic C^2 dynamical system with characteristic exponents $\chi_1 \geqq 0 \geqq \chi_2$, the Hausdorff dimension of m is given by $HD(m) = h_m(f)\left(\frac{1}{\chi_1} - \frac{1}{\chi_2}\right)$, where $h_m(f)$ is the m-entropy of f. If the limit exists almost everywhere, $HD(m)$ is defined by $HD(m) = \lim_{\varepsilon\to 0} \log m(\mathscr{B}(x,\varepsilon))/\log\varepsilon$, where$\mathscr{B}(x,\varepsilon) = \{y\in M \mid d(x,y) < \varepsilon\}$. As we shall show this result remains valid in our case, despite the fact that T is only almost everywhere C^∞. We have the following theorem.

Proposition V.1. *For μ-a.e. $x\in\Omega$,*

$$\lim_{\alpha\to 0} \log\mu(\mathscr{B}(x,\alpha))/\log\alpha = h\left(\frac{1}{\chi_+} - \frac{1}{\chi_-}\right).$$

In order to prove this, we merely adapt ideas of Ledrappier [L3] to our case. We prove separately two inequalities which lead to Proposition V.1.

Lemma V.2. $\liminf_{\alpha\to 0} \log m(\mathscr{B}(x,\alpha))/\log\alpha \geqq h\left(\frac{1}{\chi_+} - \frac{1}{\chi_-}\right)$ *μ-a.e.*

Proof. Let P denote the partition of Ω defined by $S = S_0^+ \cup S_0^-$. For simplicity, let $F_\pm(x)$ denote the quantities $\|D_x T^{\pm 1}\|$, for $x\in\Omega$. Note that $F_\pm$ is constant on each side of $S_0^\mp$.

We first prove that, $\varepsilon>0$ being given, we can find for μ-a.e. $x\in\Omega$ an integer $N(x)$ and a number $C(x,\varepsilon)>0$ such that if $n > N(x)$ and $d(x,y) \leqq C(x,\varepsilon)\cdot e^{-n}$, then x and y belong to the same atom of $\bigvee_{-n_-}^{n_+} T^k P$, where $n_\pm = [n/(\mu(F_\pm)+2\varepsilon)]$. Suppose $d(x,y) \leqq d(x,S)$. Then x and y belong to the same atom of P, and $d(Tx,Ty) \leqq F_+(x)\cdot d(x,y)$. If, moreover, $d(x,y) \leqq d(Tx,S)/F_+(x)$, we deduce $d(Tx,Ty) \leqq d(Tx,S)$: x and y belong to the same atom of $T^{-1}P$, and $d(T^2x\cdot T^2y) \leqq F_+(Tx)\cdot F_+(x)\cdot d(x,y)$. Similar arguments are valid for T^{-1} and F_-. Thus, we see that a sufficient set of conditions to insure that x and y lie in the same atom of $\bigvee_{-n_-}^{n_+} T^k P$ is that

$$\begin{cases} d(x,y) \leqq d(T^k x, S) \Big/ \prod_{j=0}^{k-1} F_+(T^j x), & 0 \leqq k \leqq n_-, \\ d(x,y) \leqq d(T^{-k} x, S) \Big/ \prod_{j=0}^{k-1} F_-(T^{-j} x), & 0 \leqq k \leqq n_+ \end{cases}$$

(where the product is taken equal to 1 if $k=0$).

We have the two following estimates:

1) Let $B(\alpha^k)$ be the neighborhood of S of diameter α^k. We recall that $\mu(B(\alpha^k)) < K \cdot \alpha^{\tau k}$, where K, τ are two positive constants. Thus, if $\alpha < 1$, the series $\sum_{k=0}^{\infty} \mu(B(\alpha^k))$ converges. This allows us, through a measure theoretic result (see [L3] e.g.) to derive the existence a.e. of a measurable function $C(x, \varepsilon)$, $0 < C(x, \varepsilon) \leqq 1$ for μ-a.e. $x \in \Omega$ such that for $k \in \mathbb{Z}$ $d(T^k x, S) \geqq C(x, \varepsilon) e^{-|k|\varepsilon}$. In particular, we get

$$\begin{cases} d(T^k x, S) \geqq C(x,\varepsilon) e^{-n_- \cdot \varepsilon}, & 0 \leqq k \leqq n_-, \\ d(T^{-k} x, S) \geqq C(x,\varepsilon) e^{-n_+ \cdot \varepsilon}, & 0 \leqq k \leqq n_+. \end{cases}$$

2) By the ergodic theorem, for μ-a.e. $x \in \Omega$, $\exists \bar{N}(x)$ such that if $n_- > \bar{N}(x)$, $n_+ > \bar{N}(x)$, we have

$$\begin{cases} \prod_{j=0}^{k-1} F_+(T^j x) \leqq \exp(n_-(\mu(F_+) + \varepsilon)), & 0 \leqq k \leqq n_-, \\ \prod_{j=0}^{k-1} F_-(T^{-j} x) \leqq \exp(n_+(\mu(F_-) + \varepsilon)), & 0 \leqq k \leqq n_+. \end{cases}$$

Assuming ε is sufficiently small and setting $N(x) = \bar{N}(x) \cdot \mu(F_+ + F_-)$, we get, for $n > N(x)$,

$$\begin{cases} d(T^k x, S) \Big/ \prod_{j=0}^{k-1} F_+(T^j x) \geqq C(x,\varepsilon) \cdot \exp[-n_-(\mu(F_+) + 2\varepsilon)] = C(x,\varepsilon) \cdot e^{-n}, \\ \qquad 0 \leqq k \leqq n_-, \\ d(T^{-k} x, S) \Big/ \prod_{j=0}^{k-1} F_-(T^{-j} x) \geqq C(x,\varepsilon) \cdot \exp[-n_+(\mu(F_-) + 2\varepsilon)] = C(x,\varepsilon) \cdot e^{-n}, \\ \qquad 0 \leqq k \leqq n_+. \end{cases}$$

Then, $B(x, C(x,\varepsilon) \cdot e^{-n}) \subset \left(\bigvee_{-n_-}^{n_+} T^k P \right)(x)$ for $n > N(x)$.

Using Shannon-Mac Millan-Breimann's theorem (see [Bi]) we obtain for μ-a.e. $x \in \Omega$,

$$\liminf_{\alpha \to 0} \log \mu(B(x,\alpha)) / \log \alpha \geqq h \cdot \left(\frac{1}{\mu(F_+) + 2\varepsilon} + \frac{1}{\mu(F_-) + 2\varepsilon} \right).$$

As ε is arbitrarily small, we have μ-a.e.:

$$\liminf_{\alpha \to 0} \log \mu(B(x,\alpha)) / \log \alpha \geqq h \cdot \left(\frac{1}{\mu(F_+)} + \frac{1}{\mu(F_-)} \right).$$

We can reproduce the same arguments replacing T by T^n, $S_0^\pm$ by $\bigcup_{k=0}^{n-1} T^{\pm k}S_0^\pm$, P by $\bigvee_{|k|<n} T^kP$, $F_\pm$ by $\|DT^{\pm n}\|$.

As $\lim_{n\to\infty} \int \|D_xT^{\pm n}\| d\mu(x) = \chi_\pm$, we get

$$\liminf_{\alpha\to 0} \log\mu(B(x,\alpha))/\log\alpha \geqq h\cdot\left(\frac{1}{\chi_+} - \frac{1}{\chi_-}\right).$$

Lemma V.3. $\limsup_{\alpha\to 0} \log\mu(B(x,\alpha))/\log\alpha \leqq h\cdot\left(\frac{1}{\chi_+} - \frac{1}{\chi_-}\right).$

Proof. Let P still denote the partition defined by S. We first prove that, $\varepsilon>0$ being given, there is a constant K such that, for μ-a.e. $x\in\Omega$, we can find $N(x)$ such that

$$n>N(x) \Rightarrow \left(\bigvee_{-n_-}^{n_+} T^kP\right)(x) \subset B(x, Ke^{-n}),$$

where $n_+ = n/[-\chi_- - \varepsilon]$, $n_- = n/[\chi_+ - \varepsilon]$.

Let $x\in\Omega$ and let $Q = \left(\bigvee_{-n_-}^{n_+} T^kP\right)(x)$.

We first exhibit "unstable and stable widths" $w_\pm$ of Q. For $y\in Q$, we draw the line $\Delta = \Delta(y)$ passing through y in the unstable direction, and note $w_+(y)$ the length of $\Delta\cap Q$; we recall that Q is a convex set. We set $w_+ = \sup_{y\in Q} w_+(y)$ and define w_- similarly. As the unstable and stable directions are transverse, and as the S_p^+, $p\in\mathbb{N}$, and the S_q^-, $q\in\mathbb{N}$, are also transverse, the minimum ball constaining Q has a radius smaller than $C\cdot(w_+ + w_-)$, where C is a constant. We now estimate w_+; we fix $y\in Q$ such that $w_+(y) > w_+/2$. T^{n_-} is linear on $\Delta(y)\cap Q$. Thus, the usual argument shows that $10 > l(T^{n_-}(\Delta(y)\cap Q)) = w_+(y)\cdot \prod_{i=0}^{n_- -1} J^+(T^iy)$, where J^+ is the jacobian in the unstable direction. By the ergodic theorem, $\prod_{i=0}^{n_- -1} J^+(T^iy) = e^{n_-(\chi_+ - \varepsilon(n_-))}$, where $\varepsilon(\cdot)$ is a function depending on y such that $\lim_{m\to\infty} \varepsilon(m) = 0$. As y is here fixed, for n large enough, n_- is large enough and $\varepsilon(n_-) < \varepsilon$. We get $10 > w_+(y)\cdot e^n$; similarly, $w_-(y) < 10\cdot e^{-n}$, and we get $\left(\bigvee_{-n_-}^{n_+} T^kP\right)(x) \subset B(x, Ke^{-n})$ with $K = 20\cdot C$. Letting n go to infinity, we get

$$\limsup_{\alpha\to 0} \frac{\log\mu(B(x,\alpha))}{\log\alpha} \leqq h\left(\frac{1}{\chi_+ - \varepsilon} + \frac{1}{-\chi_- - \varepsilon}\right).$$

Lemma V.3 now follows from the fact that ε is arbitrarily small, Lemmata V.2 and V.3 end the proof of Proposition V.1.

We add the following result, which is the mere consequence of [L, S].

Proposition V.4. *With the above notations,* $h = \chi_+$.

Proof. One can check that hypothesis of [K, S] are fulfilled, so that the result of [L, S] applies.

Appendix. Proof of Proposition II.4

Lemma A.1. *There is a positive constant L'_1 such that for any pair W_1, W_2 of local unstable manifolds in Ω, there is an ε_0, $0<\varepsilon_0<1$ such that if $0<\varepsilon<\varepsilon_0$, one can find an open subset $\mathscr{A}_\varepsilon$ of Ω with $l(\mathscr{A}_\varepsilon \cap W_1) < \frac{2\varepsilon}{1-4\sqrt{b}/\lambda}$, and if $x, x' \in \mathscr{D}(P_{W^1W^2}) \backslash \mathscr{A}_\varepsilon$, $d(x,x')<\varepsilon^2$, then*

$$(1-L'_1 D^{1/3}) d(x,x') \leqq d(P_{W^1W^2}(x), P_{W^1W^2}(x')) \leqq d(x,x')(1+L'_1 D^{1/3}),$$

where $D = d(W_1, W_2)$.

Proof. Let K be a real number such that $\frac{2}{\lambda} < K < \frac{1}{\sqrt{b}}$. We define $\mathscr{A}_\varepsilon$ by

$$\mathscr{A}_\varepsilon = \bigcup_{j=0}^{\infty} (A^{-j}_{2\varepsilon/K^j} \cap \Omega), \quad \text{where} \quad A^{-j}_\alpha = T^{-j} A^-_\alpha .$$

Every $A^{-j}_{2\varepsilon/K^j}$ crosses W_1 at most 2^j times, therefore,

$$l(\mathscr{A}_\varepsilon \cap W_1) \leqq \sum_{j=0}^{\infty} l(A_{2\varepsilon/K^j} \cap W_1)$$

$$\leqq 2 \sum_{j=0}^{\infty} 2^j \frac{1}{(K\lambda)^j} 2\varepsilon = \frac{4\varepsilon}{1-2/K\lambda},$$

because the stable and unstable manifolds are transverse. We note $p = P_{W^1W^2}$. Assume $x, x' \in \mathscr{D}(p)$, and $x \notin \mathscr{A}_\varepsilon$ and $d(x,x') \leqq \varepsilon^2$. Let $\delta = d(x,x')$. Let N be the integer such that $2\varepsilon \left(\frac{1}{K\lambda_+}\right)^{N+1} \leqq \delta < 2\varepsilon \left(\frac{1}{K\lambda_+}\right)^N$. Assume that for some k, $0 \leqq k \leqq N$, S^-_k crosses W_1 at some point y between x and x'. From $d(x,y)<\delta$, we deduce $d(T^k x, T^k y) \leqq \delta \lambda^k_+$. Therefore,

$$d(T^k x, S_0) \leqq \delta \lambda^k_+ \leqq \delta \lambda^N_+ \leqq 2\varepsilon K^{-N} \leqq 2\varepsilon K^{-k},$$

which contradicts $x \notin \mathscr{A}_\varepsilon$. Assume now that S^-_k crosses W_2 for some k, $0 \leqq k \leqq N$, at some point which is on the segment $(p(x), p(x'))$. Since S^-_k cannot cross a local stable manifold, S^-_k must have a corner inside the parallelogram $Q = (x, x', p(x), p(x'))$. This implies that there is a point z belonging to this parallelogram such that $z \in S^-_{k'}$, for some k', $0 \leqq k' < k$. By induction we obtain that Q must intersect S^-_0, a contradiction since S^-_0 is a straight line. Let j be an integer such that $0 \leqq j \leqq N$. The above argument implies that $T^{N-j} Q$ is a parallelogram. Moreover, we have

$$d(T^{N-j}x, T^{N-j}x') \geqq \delta \lambda^{N-j}, d(T^{N-j}x, T^{N-j}(p(x))) < 2\left(\frac{b}{\lambda}\right)^{N-j} D,$$

and

$$d(T^{N-j}x', T^{N-j}(p(x'))) \leqq 2\left(\frac{b}{\lambda}\right)^{N-j} D,$$

since the angle between a local stable and a local unstable manifold is greater than $\frac{\pi}{5}$. We have, therefore,

$$\left|\frac{d(T^{N-j}(p(x)), T^{N-j}(p(x')))}{d(T^{N-j}(x), T^{N-j}(x'))} - 1\right| < 10\left(\frac{b}{\lambda^2}\right)^{N-j} \frac{D}{\delta},$$

which implies

$$\left|\frac{d(T^{N-j}(p(x)), T^{N-j}(p(x')))}{d(T^{N-j}(x), T^{N-j}(x'))} - 1\right| < \frac{1}{10}\sqrt{D} \quad \text{by our choice of } N,$$

provided ε_0 is chosen small enough (independently of N and δ), and $j \leqq \frac{1}{2}\frac{|\log D|}{\log(\lambda^2/b)}$. We also have

$$\frac{d(T^{N-j}(p(x)), T^{N-j}(p(x')))}{d(T^{N-j}(x), T^{N-j}(x'))} \Big/ \frac{d(p(x), p(x'))}{d(x, x')} = \prod_{l=1}^{N-j} \Delta_l,$$

where

$$\Delta_l = \frac{d(T^l(p(x)), T^l(p(x')))}{d(T^l(x), T^l(x'))} \Big/ \frac{d(T^{l-1}(p(x)), T^{l-1}(p(x')))}{d(T^{l-1}(x), T^{l-1}(x'))}$$

satisfies

$$1 - 2\left(\frac{b}{\lambda}\right)^{N-l} \leqq \Delta_l \leqq 1 + 2\left(\frac{b}{\lambda}\right)^{N-l},$$

according to Lemma I.1. Therefore, if $j = \left[\frac{1}{2}\frac{|\log D|}{\log(\lambda^2/b)}\right] - \log 8$, we have

$$\left|\prod_{l=1}^{N-j} \Delta_l - 1\right| \leqq \frac{\sqrt{D}}{2}, \quad \text{if } D \text{ is smaller than } \frac{b}{64\lambda^2}.$$

Combining the two estimates, we obtain the result for $D < \frac{b}{64\lambda^2}$ with $L'_1 = 1$. For $D > \frac{b}{64\lambda^2}$, we apply T one time [this is enough since $D \leqq \mathcal{O}(1)b$],and we can apply the estimate unless W^1 and W^2 cross S_0, in which case the estimation is performed with respect to S_0. Q.E.D.

We still note $p = P_{W^1W^2}$.

Proof of Proposition II.4. It is enough to prove the proposition with $A = \bar{A} \cap \mathcal{D}(p)$, $\bar{A}$ a closed interval in W^1 and $l(A) > 0$. Let ε_0 be as in Lemma A.1, and choose ε such that $0 < \varepsilon < \varepsilon_0$ and $\varepsilon < l(A)D^{1/3}$. We observe that $W_1 \backslash \mathcal{D}(p)$ is an open subset. Therefore, there is a sequence $(U_k)_{k \in \mathbb{N}}$ of disjoint open intervals such that

$$W_1 \backslash \mathcal{D}(p) = \bigcup_{k=0}^{\infty} U_k.$$

In the following, for $a, b \in \mathscr{D}(p)$ we shall denote by $\tilde{p}(]a,b[)$ the segment $]p(a), p(b)[$ (although p is not everywhere defined on $]a,b[$, this definition makes sense because p is order preserving).

Let $V_k = \tilde{p}(U_k)$. From $l(W_1) < \infty$ and $l(W_2) < \infty$, we deduce that there is an integer N_1 such that

$$\sum_{k=N_1}^{\infty} l(U_k) < \varepsilon\,, \qquad \sum_{k=N_1}^{\infty} l(V_k) < \varepsilon\,, \qquad l(V_k) < \varepsilon^2 \quad \text{if} \quad k \geqq N_1\,.$$

Let $A' = \bar{A} \Big\backslash \bigcup_{k=0}^{N_1-1} U_k$. We have $l(A) \leqq l(A') \leqq l(A) + \varepsilon$ and $l(p(A)) \leqq l(\tilde{p}(A'))$. The set $\mathscr{A}' = W_2 \cap \mathscr{A}_\varepsilon \cup \left(\bigcup_{k=N_1}^{\infty} V_k \right)$ is an open subset of W_2 such that

$$l(\mathscr{A}') \leqq \left(1 + \frac{2}{1 - 4\sqrt{b/\lambda}} \right) \varepsilon \leqq 4\varepsilon \quad \text{if } b \text{ is small enough}\,.$$

We can find a sequence $(V_k')_{k \in \mathbb{N}}$ of disjoint open intervals of W_2 such that

$$\mathscr{A}' = \bigcup_{k=0}^{\infty} V_k'\,.$$

Let N_2 be an integer such that

$$l(V_k') < \varepsilon^2 \quad \text{if} \quad k \geqq N_2 \quad \text{and} \quad \sum_{k=N_2}^{\infty} l(V_k') < \varepsilon\,.$$

Let

$$A'' = \tilde{P}^{-1} \left(\tilde{P}(A') \Big\backslash \bigcup_{k=0}^{N_2-1} V_k' \right),$$

we have $l(A'') \leqq l(A')$ and $l(\tilde{p}(A'')) \leqq l(\tilde{p}(A')) \leqq l(\tilde{p}(A'')) + 4\varepsilon$. $\tilde{p}(A'')$ is a finite union of closed intervals of W_2 whose endpoints are in $\mathscr{D}(p^{-1}) \backslash \mathscr{A}_\varepsilon$. Let $I = [u, v]$ be such an interval. We claim that there is a finite sequence $(u_s)_{s=0,\ldots,q}$ such that

i) $u_0 = u$, $u_q = v$,

ii) $u_j \in \mathscr{D}(p^{-1} \backslash \mathscr{A}_\varepsilon)$,

iii) $d(u_j, u_{j+1}) \leqq \varepsilon^2$.

This is obvious from the above construction. From Lemma A.2.1, we have

$$l(\tilde{p}^{-1}[u_j, u_{j+1}]) \leqq (1 + L_1' D^{1/3}) l[u_j, u_{j+1}] \forall j, \qquad 0 \leqq j \leqq q-1\,,$$

which implies

$$l(\tilde{p}^{-1}(I)) \leqq (1 + L_1' D^{1/3}) l(I) \quad \text{and} \quad l(A'') \leqq (1 + L_1' D^{1/3}) l(\tilde{p}(A''))\,.$$

Therefore,

$$\begin{aligned} l(p(A)) &\leqq l(\tilde{p}(A')) \leqq l(\tilde{p}(A'')) + 4\varepsilon \leqq (1 + L_1' D^{1/3}) l(A'') + 4\varepsilon \\ &\leqq (1 + L_1' D^{1/3}) l(A') + 4\varepsilon \leqq (1 + L_1' D^{1/3}) l(A) + (5 + L_1' D^{1/3}) \varepsilon \\ &\leqq (1 + L_1 D^{1/3}) l(A)\,, \quad \text{where} \quad L_1 = 2L_1' + 5\,. \end{aligned}$$

The inequality in the other direction is obtained by interchanging W_1 and W_2. Q.E.D.

Acknowledgement. We would like to thank J. Lascoux and F. Ledrappier for many helpful discussions.

References

[B] Bourbaki, M.: Eléments de Mathématiques, Integration chapter IX. Paris: Hermann 1969
[Bi] Billingsley, P.: Ergodic theory and information. New York: Wiley 1965
[He] Henon, M.: A two-dimensional mapping with a strange attractor. Commun. Math. Phys. **50**, 69–77 (1976)
[K, S] Katok, A., Strelcyn, J.M.: Invariant manifolds for smooth maps with singularities, Part I: Existence. Preprint (1980)
[L1] Ledrappier, F.: Propriétés ergodiques des mesures de Sinaï. Preprint (1982)
[L2] Ledrappier, F.: Sur la condition de Bernoulli faible et ses applications. Théorie ergodique Rennes 73/74. In: Lecture Notes in Mathematics, Vol. 532. Berlin, Heidelberg, New York: Springer 1975
[L3] Ledrappier, F.: Quelques propriétés des exposants caractéristiques. Ecole d'été de probabilités de St. Flour 1982. In: Lecture Notes in Mathematics. Berlin, Heidelberg, New York: Springer 1983 (to appear)
[L, S] Ledrappier, F., Strelcyn, J.M.: A proof of the estimation from below in Pesin entropy formula. Preprint. Paris-Nord (1982)
[Lo] Lozi, R.: Un attracteur étrange du type attracteur de Henon. J. Phys. (Paris) **39** (Coll. C5), 9–10 (1978)
[M] Misiurewicz, M.: Strange attractors for the Lozi mappings. In: Nonlinear dynamics, R. G. Helleman (ed.). New York: The New York Academy of Sciences 1980
[P] Parry, W.: Entropy and generators in ergodic theory. New York: Benjamin 1969
[Ro] Rohlin, V.A.: On the fundamental ideas of measure theory. Math. Sbornik **25** (67), 107–150 (1949)
[Ry] Rychlik, M.: Théorie ergodique, mesures invariantes et principe variationnel pour les applications de Lozi. C. R. Acad. Sci. Paris (1983) (to appear)
[Y1] Young, Lai-Sang: Bowen-Ruelle measures for certain piecewise hyperbolic maps. Preprint, University of Wisconsin (1982)
[Y2] Young, Lai-Sang: Dimension, entropy, and Lyapunov exponents. Ergod. Theor. Dynam. Syst. **2**, 109–124 (1982)

Communicated by J. Lascoux

Received June 13, 1983; in revised form November 1, 1983

Commun. Math. Phys. 99, 177–195 (1985)

Communications in Mathematical Physics

On the Concept of Attractor

John Milnor
Institute for Advanced Study, Princeton, NJ 08540, USA

Abstract. This note proposes a definition for the concept of "attractor," based on the probable asymptotic behavior of orbits. The definition is sufficiently broad so that every smooth compact dynamical system has at least one attractor.

Attractors have played an increasingly important role in thinking about dynamical systems since their introduction some twenty years ago; yet there is no agreement as to the most useful definition. Section 1 of this note compares several definitions from the literature, and Sect. 2 proposes an alternative definition based on asymptotic behavior for almost every choice of initial point. The remaining two sections illustrate this definition by a number of examples, and discuss the stability and robustness of these attractors. There are three appendices. The first compares a closely related purely topological definition, the second studies real quadratic maps of their interval as a test case, and the third discusses strange attractors.

1. History

The following is quoted from Auslander, Bhatia, and Seibert (1964):

> "In the study of topological properties of ordinary differential equations, the stability theory of compact invariant sets (which may be regarded as generalizations of critical points and limit cycles) plays a central role. ... By Liapunov stability (or just stability) of the compact invariant set M, we mean that every orbit starting sufficiently close to M will remain in a given neighborhood of M. The set M is asymptotically stable if it is stable and is also an 'attractor' – that is, all orbits in a neighborhood ... of M approach M."

(Compare La Salle and Lefschetz, p. 31.) The word attractor, applied to a single invariant point for a smooth flow, had been used earlier by Coddington and

Levinson (1955) and by Mendelson (1960). However, attractors consisting of more than one point seem to have been first studied in this Auslander-Bhatia-Seibert paper. They explicitly considered the unstable case, in which an orbit which starts arbitrarily close to the attractor may wander far away before converging back towards the attractor. Compare Example 1 in Sect. 4 below. Although their definition is occasionally used by other authors (see La Salle, Hirsch, and perhaps Smale 1977), it has not been widely accepted, and most subsequent authors have required some form of stability as part of the definition. Smale (1967), in a widely read survey article, defined a more complicated object which might better be called an axiom A attractor. Smale's attractors, for a smooth map f from a compact manifold to itself, were "hyperbolic sets," containing a dense orbit, with periodic points everywhere dense, and satisfying the following rather awkward stability condition: The attractor A must have a neighborhood U so that A is equal to the intersection of the images $f^m(U)$ for $m>0$. Williams (1968) gave a related but simpler definition:

> "A subset A of $\Omega(f)$ is an *attractor* of f, provided it is indecomposable and has a neighborhood U such that $f(U)\subset U$ and $\bigcap_{i>0} f^i(U)=A$",

where $\Omega(f)$ is the non-wandering set of f, and where a closed f-invariant set is *indecomposable* if is not the union of two disjoint closed invariant subsets. The concept of attractor acquired great interest when Ruelle and Takens (1971) suggested that turbulent behavior in fluids might be caused by the presence of "strange" attractors. In fact, an explicit example in support of this idea had been worked out much earlier by Lorenz (1963). The definition of attractor used by Ruelle and Takens, like that of Smale, used an awkward form of stability:

> "A closed subset A of the non-wandering set Ω is an attractor if it has a neighborhood U such that $\bigcap_{t>0} D_{X,t}(U)=A$",

where the notation $D_{X,t}$ stands for the flow on a smooth manifold generated by a vector field X. (Note that this condition does not imply asymptotic stability. For example, according to Besicovitch, there is a homeomorphism of the plane, fixing the origin, such that the intersection of the successive images of the unit disk is the origin, even though every nonzero orbit is everywhere dense. See also Anosov and Katok.)

It is my contention that all of these definitions are too restrictive, since they exclude many interesting examples. Furthermore, in many cases they lead to an awkward situation in which there is no convenient language to describe where most points actually go when one follows a flow or iterates a mapping. In order to illustrate this point, Sect. 3 will describe a quadratic map of the interval which has no attractor at all according to the above definitions, although almost every orbit converges to a single uniquely defined compact set. A less restrictive definition has been given by Guckenheimer (1976), who requires only that an attractor must have:

> "... a fundamental system of neighborhoods, each of which is forward invariant under the flow generated by X".

In the language of Auslander, Bhatia, and Seibert, this is the condition of Liapunov stability (cf. Sect. 4). Sets having this property are important and useful; however, I believe that not every Liapunov stable set should be called an attractor. Here is an example. Consider a diffeomorphism or a flow on the plane which reduces to the identity map on a collection of concentric circles converging to the origin, but which pushes points slightly away from the origin otherwise. Then the origin is Liapunov stable, and hence would be called an attractor by Guckenheimer's definition, although it does not attract any other point.

Many other definitions of attractor can be found in the literature. The author's favorite reference is Collet and Eckmann (1980), which informally defines the attractor of a map f as

> "the set of points to which most points evolve under iterates of f".

This idea forms the basis for the present paper. For further discussion and other definitions the reader is referred to Conley, Guckenheimer and Holmes (Sect. 5.4), Kan, Ruelle (1981, 1983), as well as Zeeman.

2. Attractors, Minimal Attractors, and the Likely Limit Set

Let M be a smooth compact manifold, possibly with boundary, and let f be a continuous map from M into itself. The notation $f^n = f \circ \ldots \circ f$ will stand for the nth iterate of f. Recall that the *omega limit set* $\omega(x)$ of a point $x \in M$ is the collection of all accumulation points for the sequence $x, f(x), f^2(x), \ldots$ of successive images of x. If we choose some metric for the topological space M, then $\omega(x)$ can also be described as the smallest closed set S such that the distance from $f^n(x)$ to the nearest point of S tends to zero as $n \to \infty$. The definition of omega limit set in the case of a continuous flow on M is completely analogous. Note that $\omega(x)$ is always closed and nonvacuous, with $f(\omega(x)) = \omega(x)$. Furthermore, $\omega(x)$ is always contained in the nonwandering set $\Omega(f)$.

Choose some measure μ on M which is equivalent to Lebesgue measure when restricted to any coordinate neighborhood. This can be constructed using a partition of unity, or using the volume form associated with a Riemannian metric. It doesn't really matter which particular measure we use, since we will usually only distinguish between sets of measure zero and sets of positive measure.

Definition. A closed subset $A \subset M$ will be called an *attractor* if it satisfies two conditions:

(1) the *realm of attraction* $\varrho(A)$, consisting of all points $x \in M$ for which $\omega(x) \subset A$, must have strictly positive measure; and

(2) there is no strictly smaller closed set $A' \subset A$ so that $\varrho(A')$ coincides with $\varrho(A)$ up to a set of measure zero.

The first condition says that there is some positive possibility that a randomly chosen point will be attracted to A, and the second says that every part of A plays an essential role.

Note. In the literature, the set $\varrho(A)$ is usually called the "basin of attraction" if it is an open set, and the "stable manifold" if it is a lower dimensional smooth manifold. I have avoided both terminologies since our sets $\varrho(A)$ are not open in general (cf. Sect. 3), and are certainly not lower dimensional manifolds. For any closed set A, it is not difficult to check that $\varrho(A)$ is necessarily a Borel set (or more precisely a countable intersection of σ-compact sets), and hence is measurable.

Basic properties of attractors are the following. An attractor A is necessarily closed, nonvacuous, and contained in the nonwandering set $\Omega(f)$, with $f(A)=A$. Any finite union of attractors is again an attractor; and more generally the closure of an arbitrary union of attractors is an attractor. Proofs are easily supplied. In order to show that attractors always exist, we will first construct one particularly important attractor.

Definition. The *likely limit set* $\Lambda=\Lambda(f)$ is the smallest closed subset of M with the property that $\omega(x)\subset\Lambda$ for every point $x\in M$ outside of a set of measure zero.

Lemma 1. *This likely limit set Λ is well defined and is an attractor for f. In fact, Λ is the unique maximal attractor, which contains all others.*

Sketch of Proof. Let $\{U_i\}$ be a countable basis for the open subsets of M, and let U be the union of those U_i such that $U_i\cap\omega(x)=\emptyset$ for almost every x. Then it follows that $U\cap\omega(x)=\emptyset$ for almost every x. The complement of U is the required likely limit set Λ. Further details of the argument are straightforward. □

If f is a measure preserving transformation, that is if $\mu(S)=\mu(f^{-1}(S))$ for every measurable set S, then $\Lambda(f)$ will be equal to the entire manifold M. Furthermore, every attractor will coincide with its realm of attraction, up to a set of measure zero. In this case, these constructions are probably not too interesting. However, when Λ is a proper subset of M, it provides a usefull tool for studying asymptotic behavior for almost all orbits.

Here is a more general construction for attractors. If $S\subset M$ is any subset of positive measure, then we define $\Lambda(f,S)$ to be the smallest closed subset of M which contains $\omega(x)$ for almost every point x of S. It is easy to check that $\Lambda(f,S)$ is well defined, and is an attractor. Note that its realm of attraction necessarily contains S, up to a set of measure zero. Evidently every attractor can be obtained by this construction. One special case is of particular interest.

Lemma 2. *If S is a compact set of positive measure with the property that $f(S)\subset S$, then S necessarily contains at least one attractor.*

For $\Lambda(f,S)$ is an attractor, and is clearly contained in S. □

We will be particularly interested in *minimal attractors*, that is, attractors for which no proper subset is an attractor. Evidently a closed set $A\subset M$ is a minimal attractor if and only if

(1′) its realm of attraction $\varrho(A)$ has positive measure, and

(2′) there is no strictly smaller closed set $A'\subset A$ for which $\varrho(A')$ has positive measure.

The number of distinct minimal attractors for f is at most countably infinite. If A is a minimal attractor, note that $\omega(x)$ is precisely equal to A for almost every x in $\varrho(A)$. There are many interesting cases in which the union of the minimal attractors for f is equal to the entire likely limit set Λ, or at least is everywhere dense in Λ.

Lemma 3. *Suppose that the likely limit set Λ is a union of finitely many disjoint minimal attractors $A_1, \ldots, A_n$. Then the corresponding realms of attraction $\varrho(A_i)$ form a partition of M into disjoint sets of positive measure, up to a set of measure zero. Every attractor is a union of minimal attractors, and for almost every point x of M the limit set $\omega(x)$ is precisely equal to some A_i.*

Proof. Choose neighborhoods U_i of the A_i so that $f(U_i)$ is disjoint from U_j for $i \neq j$, and let U be the union of the U_i. Then almost every orbit $x_0 \mapsto x_1 \mapsto \ldots$ will satisfy $x_t \in U$ for large t; hence x_t must belong to just one U_i for large t. The proof is now straightforward. □

In cases where this lemma applies, the collection of minimal attractors forms a very useful tool. However, the minimality condition may be very hard to verify, and there is no guarantee that a map f has any minimal attractors at all. Counterexamples are provided by the identity map, the concentric circle example of Sect. 1, or by more interesting examples such as the polynomial map

$$(c, y) \mapsto (c, y^2 - c)$$

in two variables (see Fig. 1). In such cases where there are not enough minimal attractors, or in cases where the minimal attractors are not known, it is necessary to fall back on the more general concept of attractor.

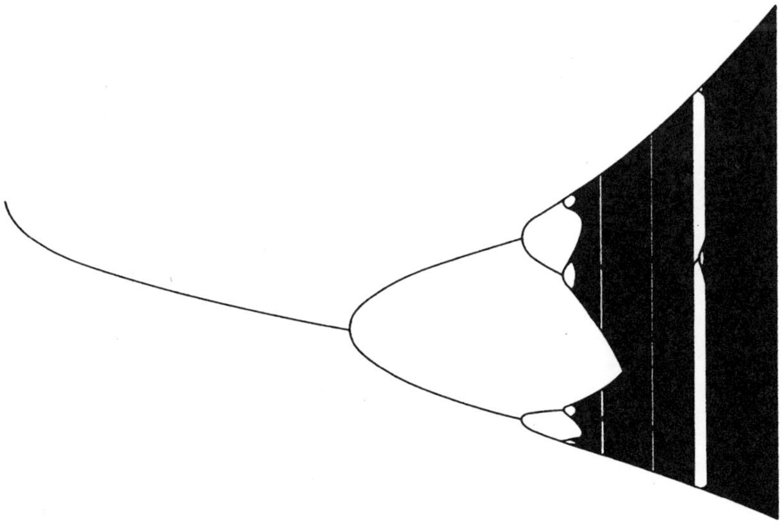

Fig. 1. Likely limit set (in the finite plane) for $(c, y) \mapsto (c, y^2 - c)$

3. The Feigenbaum Attractor

This section will illustrate these ideas by describing a smooth map which has a unique attractor according to the definitions in Sect. 2, but no attractor at all according to most of the definitions in Sect. 1. Let $c=1.401155189\ldots$ be the smallest real number for which the real quadratic map $x \mapsto x^2-c$ has infinitely many distinct periodic orbits. This map $f(x)=x^2-c$ has been studied by Feigenbaum, since it represents the point of transition from stable periodic behavior to chaotic behavior. In order to have a compact domain of definition, let us choose some interval I containing the origin with $f(I)\subset I$, and think of f as a map from I into itself (Fig. 2).

The orbit of zero under f is an almost periodic sequence which can be described as follows (cf. Misiurewicz). The numbers $0=a_0 \mapsto a_1 \mapsto a_2 \ldots$ are all distinct. However, the difference a_m-a_n is very small whenever $m-n$ is divisible by a high power of two. Thus the closure A of this orbit is a Cantor set, homeomorphic to the ring $\varprojlim(Z/2^kZ)$ of 2-adic integers in such a way that each $a_n=f^n(0)$ corresponds to the 2-adic integer n. Note that f restricted to A corresponds to the homeomorphism which adds one to each 2-adic integer. It follows that there are no periodic points in A. However, it can be shown that there are periodic points arbitrarily close to every point of A. These periodic points are all unstable, with period equal to a power of two.

The dynamic structure of $f: I\to I$ is as follows. For almost every initial point x_0 in I the successive images $x^n=f^n(x_0)$ converge towards the Cantor set A. Hence $A=\Lambda(f)$ is the unique attractor for f. But there are a countable infinity of exceptional points whose successive images do not converge towards A. Furthermore, these exceptional points are everywhere dense in I. For according to Guckenheimer (1979), since f has no periodic attractor, any arbitrarily small open interval $U\subset I$ has some forward image $f^n(U)$ which contains the origin. Choosing a point x in U which maps to a periodic point in this neighborhood of the origin, we see that $\omega(x)$ is a periodic orbit disjoint from A. Thus A is certainly not asymptotically stable.

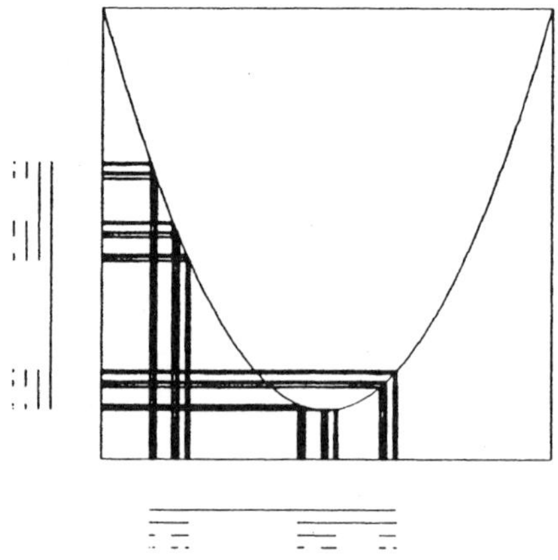

Fig. 2. The unique attractor for the map $x \mapsto x^2-1.4011\ldots$ on an appropriate interval is the indicated Cantor set

However, A is Liapunov stable, and in fact is equal to the intersection of a nested sequence of asymptotically stable sets. In spite of the dense set of exceptional points, the tendency to converge towards A is extremely strong. If U is an arbitrarily small neighborhood of A, let us choose a smaller neighborhood V so that $f(V) \subset V$. Then if we start at any point x_0 in I and follow the successive images $x_n = f^n(x_0)$ there are just two possibilities. Either this orbit eventually hits the open set V and remains trapped within $V \subset U$ forever after, or else it manages to precisely hit one of the finitely many periodic points which lie outside of V, and remains trapped in a finite periodic orbit outside of V thereafter. The first possibility occurs for a dense open set of initial points x_0, while the second possibility occurs only for a countable nowhere dense set of x_0.

Further details of this argument will be given in Appendix 2. For readers who prefer to work with flows, Kan has described somewhat analogous examples of smooth flows on the 3-dimensional sphere so that almost every orbit converges towards a "solenoid," locally homeomorphic to the product of a Cantor set and a 1-manifold, which contains no periodic orbits, but can be approximated arbitrarily closely by periodic orbits. These examples are also Liapunov stable, but not asymptotically stable. (See also Grebogi, Ott, Pelikan, and Yorke. Related but less differentiable examples have been described by Bowen and Franks, and by Franks and Young.) It seems likely that this lack of asymptotic stability may be shared by many other fractal attractors.

4. Stability and Robustness

Let $f: M \to M$ be a fixed smooth map. As in previous sections, we will say that a closed subset $A \subset M$ with $f(A) = A$ is *Liapunov stable* (also called "orbitally stable") if it has arbitrarily small neighborhoods U with $f(U) \subset U$; and *asymptotically stable* if it is Liapunov stable and also satisfies the Auslander-Bhatia-Seibert condition that its realm of attraction $\varrho(A)$ is an open set. In the asymptotically stable case, if we choose U with closure contained in $\varrho(A)$, then it follows easily that A is equal to the intersection of the sequence of forward images $U \supset f(U) \supset f^2(U) \supset \dots$.

Note that we have not imposed any stability requirement at all as part of the definition of attractor. Of course, many interesting attractors are asymptotically stable, or at least Liapunov stable. However, if we required some form of stability as part of the definition, then it would be awkward to talk about transitional case, and it would no longer be true that every smooth map on a compact manifold has at least one attractor. Here is a simple minded example to illustrate this point.

Example 1. Let M be the circle of real numbers modulo 2π, and let

$$f(\theta) \equiv \theta + 1 - \cos(\theta) \pmod{2\pi}.$$

Then the fixed point $\theta = 0$ is the unique attractor. This attractor is not Liapunov stable, since no nontrivial neighborhood is mapped into itself by f. (Compare Fig. 3. We will describe such a fixed point as "one-sided stable.") A completely analogous example on the real projective line $R \cup \infty$, is the map $f(x) = x + 1$, with

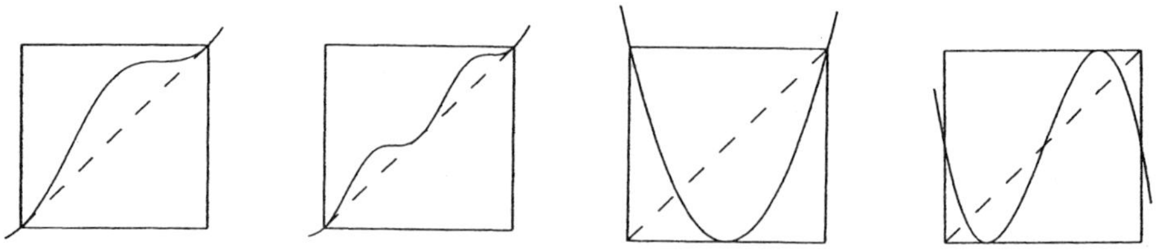

Fig. 3. Graphs of Examples 1–4

the point at infinity as unique attractor. There is a well known classical example of a fractal attractor with similar stability properties, namely the Denjoy C^1-diffeomorphism of the circle, which has a Cantor set as unique attractor (cf. Schweitzer). Here is a really pathological example. Let $f(x)=4x(1-x^2)^2$ for $|x|\leqq 1$, with $f(x)=0$ otherwise. Then surely the only attractor is the origin, which is a strictly repulsive fixed point.

Here is a simple variant of Example 1.

Example 2. On the circle of real numbers modulo 2π, let $f(\theta)=\theta+\sin^2(\theta)$. Then A consists of the two points 0 and π. Each is a minimal attractor, but neither is stable since there are points arbitrarily close to either one whose successive images converge to the other.

Note that an attractor which has positive measure need not attract anything outside of itself. For example an irrational rotation of the circle has the entire circle as minimal attractor.

Example 3. The Ulam von Neumann Attractor. Consider the quadratic map $f(x)=2x^2-1$ from $R\cup\infty$ to itself. Then for almost every point x in the interval $I=[-1,1]$ the limit set $\omega(x)$ is equal to the entire interval I. A proof can be based on the corresponding property for the squaring map $z\mapsto z^2$ on the unit circle, making use of the observation that f restricted to I is equal to the correspondence $\mathrm{Re}(z)\mapsto\mathrm{Re}(z^2)$, where z ranges over the unit circle. On the other hand, for every point x outside of I the orbit of x diverges to infinity. It follows that $A(f)$ consists of the interval I, which is an unstable minimal attractor, together with the point at infinity which is a stable attractor. Every point outside of I is repelled by I.

By a linear change of coordinate, this map can be put into the form $x\mapsto x^2-2$. More generally, for any parameter c we can consider the map $x\mapsto x^2-c$. If c belongs to the interval $[-1/4,2]$, this map has at least one bounded attractor. It may be true that it has only one bounded attractor, which is either a periodic orbit, a finite union of intervals, or (in uncountably many exceptional cases) an almost periodic Cantor set as in Sect. 3 (cf. Lemma 4 in Appendix 2). Presumably, for almost all parameter values this bounded attractor is asymptotically stable. However, in the almost periodic case it is only Liapunov stable, and there are at least a countable infinity of cases which are not even Liapunov stable. These include the two extreme values $c=-1/4$ and $c=2$, and the value $c=7/4$ which corresponds to a one-sided stable orbit of period 3.

Note that two distinct minimal attractors need not be disjoint from each other. However, the intersection of two minimal attractors must be too small to attract points from any set of positive measure.

Example 4. The likely limit set for the cubic map $f(x)=3\sqrt{3}(x-x^3)/2$ consists of two unstable minimal attractors $[-1,0]$ and $[0,1]$, together with the stable attractor $\{\infty\}$. Here the coefficient $3\sqrt{3}/2$ is chosen so that each of these intervals will map precisely onto itself. In this case, it is interesting to note that the union $[-1,1]$ of the two unstable attractors is asymptotically stable. (The proof that $[0,1]$ is a minimal attractor can be outlined as follows. One can first check that the change of variable $x=\sin^2(\theta)$ converts f, restricted to this interval, to an expanding map. It then follows from Li and Yorke that it has an ergodic invariant measure which is equivalent to Lebesgue measure.)

It is often convenient to describe instability properties of such examples by constructing an associated graph.

Definition. The *stability diagram* associated with a map f is the 1-dimensional complex with one vertex a_i for each minimal attractor A_i, and with edges as follows. There is a directed edge from vertex a_i to the distinct vertex a_j whenever there exist points arbitrarily close to A_i whose limit set $\omega(x)$ intersects A_j; and also there is a loop joining a_i to itself whenever there exist points arbitrarily close to A_i whose successive images temporarily wander away, outside of some fixed neighborhood of A_i, and yet have limit set $\omega(x)$ which intersects A_i.

If A_i is Liapunov stable, note that no edge can lead away from the corresponding vertex. The stability diagrams corresponding to the four examples are shown in Fig. 4.

Next let us study the behavior of the set $\Lambda=\Lambda(f)$ as we perturb the map f. Recall that the *Hausdorff distance* between two closed sets is the smallest number δ such that each closed ball of radius δ centered at a point of either set necessarily contains a point of the other set.

Definition. We will say that the likely limit set Λ for a given map f is *robust* if the Hausdorff distance between $\Lambda(f)$ and $\Lambda(g)$ tends to zero whenever g tends to f in the C^∞-topology (roughly speaking, whenever the kth derivative of g converges uniformly to the kth derivative of f for every $k \geqq 0$). Similarly, an attractor A for f is *robust* if any g which is C^∞-close to f has an attractor A' which is Hausdorff-close to A.

Note that this attractor A' may be qualitatively quite different from A. The condition of Hausdorff closeness guarantees only that these two sets will be rather hard to distinguish by a computer experiment. For example, it follows from Lemma 2 that any asymptotically stable periodic orbit is necessarily robust.

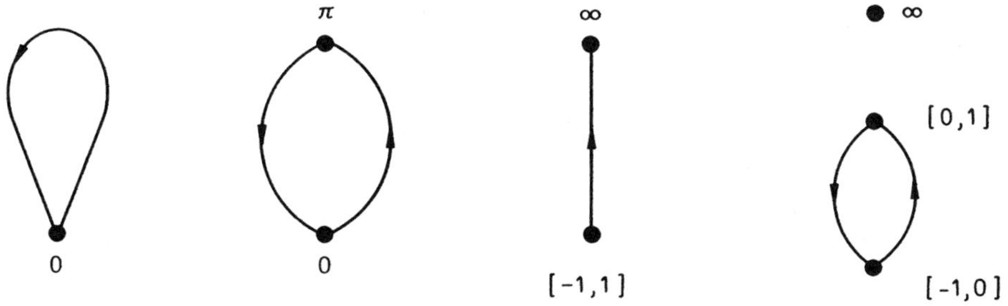

Fig. 4. Stability diagrams for Examples 1–4

However, the attractor A' which Hausdorff approximates this orbit A may have period equal to some multiple of the period of A, or may even be infinite. The most common possibility is bifurcation. For example, each of the maps $x \mapsto x - x^3$ and $x \mapsto x^2 - 3/4$ has a unique stable finite fixed point, which bifurcates under perturbation. The unique attractor $\Lambda(f)$ for the Feigenbaum map f of Sect. 3 is also robust. Again, for g close to f the likely limit set $\Lambda(g)$ may be qualitatively quite different from $\Lambda(f)$. This Cantor set may simplify into a stable periodic orbit, or complicate into a chaotic attractor.

It is conjectured that any robust attractor must be Liapunov stable. To illustrate this conjecture, we have noted that Examples 1, 2, 3, and 4 are not stable, and it is easy to show that they also are not robust.

An attractor may fail to be robust in at least three different ways. It may simply evaporate, or explode into a much bigger set, or implode into a much smaller set. (In the case of an asymptotically stable attractor, only the last possibility can occur, by Lemma 2.) Here are examples to illustrate these three possibilities. The map $x \mapsto x + x^2$, with a one-sided stable attractor at the origin, can be approximated arbitrarily closely by a map for which the only attractor is the point at infinity. The translation $x \mapsto x + 1$ of Example 1, with the infinite point as unique attractor, can be approximated arbitrarily closely by a fractional linear transformation $x \mapsto (1+x)/(1-\varepsilon x)$, for which every orbit is dense, so that the only attractor is the entire set $R \cup \infty$. Finally, let $f(x) = x^2 - c$, where $c = 1.543689\ldots$ is the real root of the equation $c^3 - 2c^2 + 2c - 2 = 0$, so that $f^4(0) = f^3(0) = -f^2(0)$. Then f has a unique finite attractor $A = [f(0), f^2(0)]$, which is asymptotically stable. (The proof is similar to that in Example 4.) Yet f is equal to the limit of a sequence $f_3, f_4, \ldots$ of quadratic maps, where the unique finite attractor A_n for f_n is a superstable orbit of period n consisting of zero and $n-2$ negative numbers, but only one positive number $f_n^2(0)$. Thus A_n is substantially smaller than A for all n.

Note that the likely limit set may fail to be robust even if every attractor is robust, since new attractors may suddenly appear. As an example, the unique attractor for the map $f(x) = x + x^3$ is the stable fixed point at infinity; yet maps arbitrarily close to f have an additional stable fixed point at the origin.

Even when an attractor disappears under perturbation of a map, it leaves behind a "resonance," that is a region where successive images of a point may get trapped for many iterations before escaping (cf. Grebogi, Ott, and Yorke). In practical applications, if the perturbation is very small so that the expected escape time is very large, it may be difficult to tell the difference between such a resonance and an actual attractor.

Appendix 1. The Generic Limit Set

The following variation on the likely limit set Λ of Sect. 2 is sometimes easier to work with. Let us say that a property of a point $x \in M$ is true for *generic* x if it is true for almost all x in the sense of Baire category theory, i.e., for all x outside of a countable union of nowhere dense subsets of M. Define the *generic limit set* $\Gamma \subset M$ to be the smallest closed subset of M with the property that, for generic x, the limits set $\omega(x)$ is contained in Γ. There is an analogous concept of "generic-attractor." The definition will be left to the reader.

In many cases the two limit sets Γ and Λ are equal to each other. However, this is not always true, and I do not know any good criterion for equality. The set Γ is likely to be easier to compute since its definition is purely topological, whereas the definition of Λ uses both topology and measure theory. On the other hand, the set Λ is more closely related to the probable asymptotic behavior which one would like to study in applications. Also, the set Λ is closer to something which one can actually draw pictures of in numerical experiments.

The only general relation between these two sets which is known to me is the statement that the intersection $\Gamma \cap \Lambda$ is always non-vacuous. This follows from the more precise statement that, for generic x, the intersection $\omega(x) \cap \Lambda$ is non-vacuous. In fact, for any neighborhood U of Λ the set of x whose full orbit is disjoint from U is closed and of measure zero, hence nowhere dense. Therefore, the orbit of a generic x intersects every such neighborhood, and hence satisfies $\omega(x) \cap \Lambda \neq \emptyset$.

Following are two rather pathological examples.

Example 5. The likely limit set Λ can be strictly larger than Γ. Consider the gradient flow associated with a smooth real valued function on the 2-dimensional sphere which takes its minimum along a Cantor set of positive measure, but has only non-degenerate critical points otherwise, with just one local maximum x_0. Then $\Gamma = \{x_0\}$, yet Λ contains an entire Cantor set. A similar example involving a C^1-horseshoe has been given by Bowen (1975).

Example 6. The set Λ can be strictly smaller than Γ, at least in the case of a non-differentiable map f. More precisely, there is a continuous map of the interval for which a generic orbit is dense, so that Γ is the entire interval, and for which Λ is a Cantor set. In fact, with suitable modification of the example, Λ will split up into two or more minimal attractors, which may be disjoint or not according to taste. I don't know whether there exists a smooth map with these properties.

Let us start with the continuous measure-preserving map $F(x) = 2 \cdot |x| - 1$ on the interval $[-1, 1]$. Then it is not difficult to show that, for generic x, the limit set $\omega(x)$ is equal to the entire interval $[-1, 1]$ (cf. Appendix 2). Hence $\Gamma = [-1, 1]$. However, we will construct a Borel probability measure μ on M so that, with μ-probability equal to one, the limit set $\omega(x)$ avoids some fixed open subset of $[-1, 1]$. Let C be the closed set consisting of all points x whose full orbit is disjoint from the interval $(1/2, 1]$. It is not difficult to check that C is a Cantor set. In fact, the "kneading sequence" $\operatorname{sgn}(x_0)$, $\operatorname{sgn}(x_1), \ldots$ associated with an orbit $x_0 \mapsto x_1 \mapsto \ldots$ in C can be any sequence of signs which does not have two $+1$'s in a row. Therefore, we can choose a probability measure μ_0 with support equal to C, and so that points have measure zero. Now define a sequence of probability measures inductively by setting

$$2\mu_{n+1}(S) = \mu_n(F(S \cap [-1, 0])) + \mu_n(F(S \cap [0, 1])),$$

so that the support of μ_n is equal to the full t-fold inverse image $F^{-n}(C)$. Then the measure $\sum \mu_n / 2^{n+1}$ has support equal to the entire interval $[-1, 1]$. Yet, with μ-probability equal to one, the limit set $\omega(x)$ is contained in C. [If μ_0 is constructed with more care, then $\omega(x)$ will actually be equal to C with probability one.] Note that the continuous change of coordinate $y = \int_{-1}^{x} d\mu$ transform μ to the standard

Lebesgue measure on the unit interval. Therefore, F is topologically conjugate to a continuous map for which Γ is strictly larger than Λ.

Similarly, let $\hat{C}$ be the Cantor set consisting of all points whose orbits are disjoint from the interval $(-1/2,0)$, and let $\hat{\mu}$ be an associated measure, so that with $\hat{\mu}$-probability equal to one the limit set $\omega(x)$ must be equal to $\hat{C}$. If we combine these two constructions by using the probability measure $(\mu+\hat{\mu})/2$, then $\omega(x)$ will be equal to C with probability 1/2, and to $\hat{C}$ with probability 1/2. Thus in this case there are two minimal attractors, which intersect only along the periodic orbit $C\cap\hat{C}=\{-3/5,1/5\}$. We can modify the example so that the two minimal attractors will be disjoint, by using the slightly larger interval $(-3/4,0)$ for the construction of $\hat{C}$.

Appendix 2. Maps of the Interval

The following discussion will depend strongly on Guckenheimer [17].

Let $x\mapsto f(x)$ be a real quadratic mapping which carries some finite interval I into itself. If we choose this interval I to be maximal, then $f:I\to I$ will belong to the class of maps studied by Guckenheimer. That is: f is smooth of class C^3, with a single interior critical point, with negative Schwarzian derivative

$$2(Df)(D^3f)-3(D^2f)^2<0,$$

and f maps both endpoints of I to a single endpoint x_0 at which the derivative satisfies $Df(x_0)\geqq 1$. (This last condition is essential, but was inadvertently omitted in [17].)

Lemma 4. *For $f:I\to I$ satisfying these conditions, the generic limit set Γ is either a finite periodic orbit, a Cantor set equal to the closure of the orbit of the critical point, or a finite union of say n intervals, bounded by the first $2n$ forward images of the critical point, and containing the critical point in its interior. Furthermore, for generic x, the omega limit set $\omega(x)$ is precisely equal to Γ.*

In the first two cases, for every x outside of a set of measure zero we have $\omega(x)=\Gamma$. Hence the likely limit set Λ is equal to Γ, and is a minimal attractor. It is quite possible that this statement is true in the third case also, but I have not been able to decide this question (cf. Example 6 above).

Proof. First suppose that f has a periodic attractor, that is either a stable or a one-sided stable periodic orbit A. Then there exists a neighborhood (or one-sided neighborhood) U consisting of points whose orbits converge uniformly to A. According to [17, p. 135 and 2.8, 3.1], for every point x outside of a closed set of measure zero the orbit of x eventually falls into this neighborhood U, and hence converges to A. Clearly, it follows that $\Gamma=\Lambda=A$.

Henceforth, let us assume that f has no stable or one-sided stable periodic orbit. In order to handle this case we will need the following.

Definition. We will say that the map f is *reducible* if there exists a closed interval J about the origin and an integer $n\geqq 2$ so that the successive images $J,f(J),f^2(J),\dots,f^{n-1}(J)$ have disjoint interiors, and so that $f^n(J)\subset J$.

The idea is that whenever this is the case we can reduce the study of f to the study of the n-fold iterate $g=f^n$, considered a map from the subinterval J into itself. In fact, if f is reducible, [17, 2.6 and 3.1] proves that, for every x outside of a closed set of measure zero, the orbit of x eventually meets the interior of J. It follows that the likely limit set $\Lambda=\Lambda(f)$ is equal to the n-fold union $\Lambda(g)\cup f(\Lambda(g)) \cup \ldots f^{n-1}(\Lambda(g))$; with a similar description for $\Gamma(f)$.

We may always assume that the interval J is chosen so as to be maximal. It then follows easily that $g=f^n$ maps the boundary of J to one boundary point. Since there are no stable periodic orbits, the derivative of g at this boundary fixed point, must be at least one. Since the condition of negative Schwarzian derivative is preserved under composition, we see that this new map $g: J\rightarrow J$ satisfies all of Guckenheimer's conditions.

If this new map $g: J\rightarrow J$ is also reducible, then we iterate the construction. We must distinguish between the "finitely reducible case" in which the reduction process terminates with a non-reducible map after finitely many steps, and the "infinitely reducible case" in which it continues ad infinitum (Fig. 2). Evidently, in order to understand the finitely reducible case, we need only understand the case of a non-reducible map.

Suppose then that f is not reducible. We will prove that f is topologically conjugate to a piecewise linear map of the form

$$F(y)=s\cdot|y|-1 \quad \text{for} \quad |y|\leqq 1/(s-1),$$

where $\log(s)>0$ is the topological entropy [17, 4.5]). Since we have excluded the trivial case where f has a stable or one-sided stable fixed point, it is not difficult to check that f must have strictly positive topological entropy. Hence, by Milnor and Thurston, f is topologically semi-conjugate to F, say $h\circ f=F\circ h$, where h is a monotone map. Thus every inverse image $h^{-1}(y)$ is either a point or an interval J. If some $h^{-1}(y)$ were an interval, then the orbit of y would have to contain the critical point 0, say $F^m(y)=0$. For otherwise any two points of J would have the same kneading sequence, which is impossible by [17, 2.6]. In fact, this orbit would have to contain 0 more than once, since the interval $f^{m+1}(J)$ also maps to a single point under h. Hence $F^n(0)$ would be 0 for some n, necessarily greater than one. Therefore, f^n would map the interval $h^{-1}(0)$ into itself, and hence f would be reducible, contradicting our hypothesis. This contradiction proves that h must in fact be a homeomorphism, so that f is topologically conjugate to F.

Thus we are reduced to studying the piecewise linear map F. We may assume that the growth number s is greater than $\sqrt{2}$, since otherwise it is easy to check that F is reducible. Then we will prove that the generic limit set of F is equal to the interval

$$A=[-1,s-1]=[F(0),F^2(0)].$$

In fact, it is easy to check that $F(A)=A$, and that the orbit of any y in the open interval $|y|<1/(s-1)$ eventually gets trapped in this subinterval A.

The map F from A to itself has a strong transitivity property as follows. *For any subinterval $J\subset A$ there exists a positive integer n so that $F^n(J)$ is equal to the entire interval A.* For if J is small enough so that the two sets J and $F(J)$ do not both contain zero, then $F^2(J)$ is longer than J by a factor of at least $s^2/2>1$. Thus the successive images of J expand exponentially until say $F^{m-1}(J)$ and $F^m(J)$ both contain zero; and hence $F^{m+1}(J)=A$.

It follows from this property that, for a generic point y, the limit set $\omega(y)$ is equal to the entire set A. For if $\{U_i\}$ is a basis for the open subsets of A, then the set of y whose orbit intersects U_i is dense and open. Thus the orbit of a generic y intersects every U_i, and hence is dense in A. Therefore, the generic limit set $\Gamma(F)$ equals A, and it follows that the generic limit set $\Gamma(f)$ for the smooth map f is the corresponding interval $h^{-1}(A)$. More explicitly, if c is the critical point of the non-reducible map f, then $\Gamma(f)$ is equal to the closed interval bounded by $f(c)$ and $f^2(c)$. Similarly, the generic limit set for a finitely reducible map f is a finite union of intervals, bounded by appropriate forward images of the critical point.

The likely limit set Λ seems much harder to determine. It follows easily from this discussion that Λ is a subset of Γ, but I have not been able to prove that $\Lambda=\Gamma$ in the non-reducible or finitely reducible case.

Finally, we must consider the case where the reduction process does not terminate after finitely many steps, but rather continues ad infinitum. In this infinitely reducible case it is not difficult to check that there is a unique attractor $A=\Lambda=\Gamma$ which is a Cantor set, homeomorphic to an inverse limit of finite cyclic groups Z/nZ. Furthermore, the map f is almost periodic on A, and corresponds to the adding machine map $\alpha\mapsto\alpha+1$ from this limit group to itself. Any such inverse limit of finite cyclic groups can occur, so there are uncountably many distinct examples. Details will be omitted. □

It follows easily from this discussion that f has sensitive dependence on initial conditions (as defined in [17]) if and only if Γ is a finite union of intervals. For the class of maps studied by Guckenheimer, there are just two qualitatively different forms of behavior. If f has a periodic or almost periodic attractor A, then $A=\Lambda(f)=\Gamma(f)$ is a set of measure zero, f does not have sensitive dependence on initial conditions, and the topological entropy of $f|A$ is zero. In all other cases, the generic limit set is an interval or union of intervals, with positive measure, with sensitive dependence on initial conditions, and the entropy of $f|\Gamma(f)$ is strictly positive. In the latter case, the dynamics of f must be described as *chaotic*.

Concluding Remarks. There are many unsolved problems, even for real quadratic maps of the interval. For example, in the chaotic case, I don't know how to decide whether $\Lambda=\Gamma$, or to decide whether there can be more than one attractor. Equivalently, is it true that almost every orbit in the union of intervals Γ is everywhere dense in Γ? More generally, I don't know how to answer the following basic question for maps of $R\cup\infty$ to itself. Suppose either that $f(x)$ is a rational function of degree at least two, or that f has negative Schwarzian derivative. Does it follow that the likely limit set $\Lambda(f)$ is equal to $\Gamma(f)$, and can be expressed as the union of finitely many minimal attractors? Similarly, I have no idea how to decide when the set $\Gamma(f)$ [much less $\Lambda(f)$] is robust. Here is an explicit guess for the class of maps considered by Guckenheimer: The interval $[f(0), f^2(0)]$ is conjectured to be a robust necessarily minimal attractor if and only if the orbit of the critical point is everywhere dense in this interval. (Compare the discussion of $x\mapsto x^2-1.543689\ldots$ in Sect. 4.)

Appendix 3. Strange Attractors

The concept of a *strange attractor*, as described by Ruelle and Takens, is perhaps intentionally not precisely defined. The original discussion emphasized wild topology. An excellent introduction to this subject can be found in Ruelle's survey article (1979–80), which rather emphasizes chaotic dynamical behavior. Guckenheimer and Holmes define a strange attractor as one which has a transversal homoclinic point, and hence has chaotic behavior.

It is important to note that there is no necessary connection between wild topology or fractal geometry and strange dynamical behavior. For example, the fractal attractor of Sect. 3 is topologically wild, but has very sedate almost periodic dynamics, with topological entropy equal to zero. On the other hand, the map $(\theta, t) \mapsto (2\theta, t/2)$ from the cylinder $S^1 \times [-1, 1]$ to itself has a smooth manifold $S^1 \times 0$ as unique attractor; yet the dynamical behavior is completely chaotic. In fact, this map is ergodic and mixing with respect to the standard invariant measure on the circle, and has positive entropy. The behavior of this example under perturbation will be studied below. For an analogous example using a diffeomorphism or flow in place of an iterated map, we could substitute an Anosov diffeomorphism of the torus, or the geodesic flow on the unit tangent bundle of a manifold of negative curvature, in place of the squaring map on the circle.

Here is an example of an attractor which is chaotic and topologically wild, yet simple enough to analyze almost completely (cf. Kaplan and Yorke). Let us start with the map $f_0(e^{i\theta}, x) = (e^{2i\theta}, cx)$ from the cylinder $S^1 \times R$ to itself, where c is a constant in the range $0 < c < 1/2$. Clearly, f_0 has a unique attractor which is asymptotically stable, since all orbits tend towards the circle $S^1 \times 0$, and since almost very orbit on the circle is everywhere dense. This attractor $S^1 \times 0$ is a smooth manifold, but is dynamically chaotic. Now make a small perturbation of f_0 to the map

$$f_\varepsilon(e^{i\theta}, x) = (e^{2i\theta}, cx + \varepsilon \cos\theta).$$

Then the smooth attractor $S^1 \times 0$ will be replaced by a fractal attractor (Fig. 5).

To see this, first note that a scale change in the x-coordinate will replace f_ε by the map

$$f_1(e^{i\theta}, x) = (e^{2i\theta}, cx + \cos\theta).$$

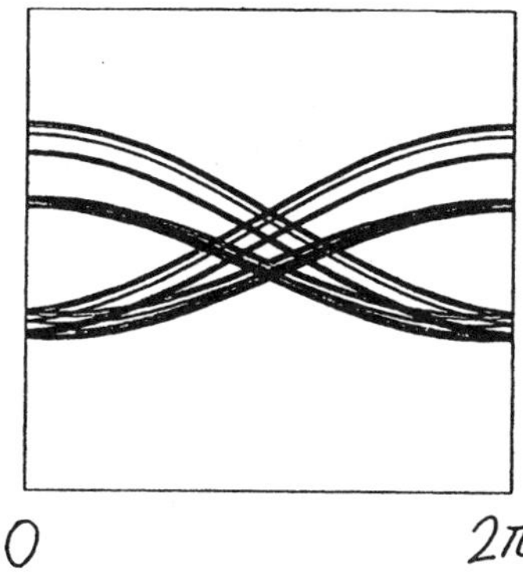

Fig. 5. The unique attractor for the map $(\theta, x) \mapsto (2\theta, 0.4x + \cos\theta)$

We can further simplify by introducing a third coordinate y, and extending f_1 to the smooth map

$$(e^{i\theta}, x, y) \mapsto (e^{2i\theta}, cx + \cos\theta, cy + \sin\theta)$$

from $S^1 \times R \times R$ to itself. Setting $w = e^{i\theta}$ and $z = x + iy$, we can write this map as

$$F(w, z) = (w^2, cz + w)$$

from the 3-dimensional manifold $S^1 \times C$ to itself.

Lemma 5. *There is a unique attractor $A \subset S^1 \times C$, homeomorphic to the dyadic solenoid, which is asymptotically stable, and in fact, attracts every orbit. More precisely, for any compact neighborhood K of A the successive images $F^n(K)$ converge towards A, and for almost every point (w, z) of $S^1 \times C$ the omega limit set $\omega(x)$ is equal to A. This set A supports a measure so that F restricted to A is measure theoretically equivalent to the full Bernoulli shift of entropy* log2.

Here by the dyadic solenoid $\hat{S}$ we mean the inverse limit of the sequence of squaring homomorphisms

$$S^1 \leftarrow S^1 \leftarrow S^1 \leftarrow \dots .$$

Thus, by definition, a point of $\hat{S}$ is a sequence $\hat{w} = (w_0, w_1, w_2, \dots)$ of points on the unit circle so that

$$w_1 = \pm\sqrt{w_0}, \qquad w_2 = \pm\sqrt{w_1}, \dots .$$

Define $A = g(\hat{S})$ to be the image of $\hat{S}$ under the continuous map

$$g(\hat{w}) = (w_0, w_1 + cw_2 + c^2 w_3 + \dots).$$

Since we have assumed that $|c| < 1/2$, it is quite easy to check that g is a topological embedding. (With more work, one could check this for somewhat larger c.) Note that F maps A homeomorphically onto itself. In fact,

$$F(w_0, w_1 + cw_2 + c^2 w_3) = (w_0^2, w_0 + cw_1 + c^2 w_2 \dots),$$

so F restricted to A corresponds to the "shift automorphism"

$$(w_0, w_1, w_2, \dots) \mapsto (w_0^2, w_0, w_1, \dots)$$

from the topological group $\hat{S}$ to itself.

Next we will see that, for any starting point (w, z) in $S^1 \times C$, the distance between $F^n(w, z)$ and the set A converges geometrically (and uniformly on any compact set) to zero. In fact, choose any point $\hat{w} = (w_0, w_1, \dots)$ in $\hat{S}$ with $w_0 = w$, and let $g(\hat{w}) = (w_0, z_0) \in A$. Then by inspecting the definition of F we see that the distance between $F^n(w, z)$ and the point $F^n(w_0, z_0) \in A$ is exactly $c^n|z - z_0|$, which converges to zero as $n \to \infty$.

More geometrically, let k be some constant in the open interval from $1/(1-c)$ to $1/c$, and let T be the solid torus consisting of all (w, z) with $|z| \leqq k$. Then it is not hard to see that F maps T homeomorphically into its own interior, with degree two, and that A is equal to the intersection of the forward images $F^n(T)$. In this form, the example has been studied by Smale (1967), by Ruelle and Takens, and by Ruelle (1979–80).

To study the ergodic behavior for F restricted to A, or equivalently for the shift automorphism of $\hat{S}$, write each point of $\hat{S}$ as $(w_0, w_1, \ldots)$, with $w_n = \exp(\pi i x_n)$. Expressing the number x_0 in binary notation as $a_0 + a_1/2 + a_2/4 + \ldots$, it follows that x_1 has the form $a_{-1} + a_0/2 + a_1/4 \ldots$, with similar expressions for the higher x_n. Thus the shift automorphism of $\hat{S}$ corresponds to the Bernoulli shift on a doubly infinite sequence $\ldots a_{-1}, a_0, a_1, a_2, \ldots$ of zeros and ones.

The Hausdorff dimension of $A \subset S^1 \times C$ can be computed as $1 - \log(2)/\log(c)$. This lies strictly between 1 and 2, so A is a fractal set, as defined by Mandelbrodt.

Now let A' be the image of A under the projection map from $S^1 \times C$ to $S^1 \times R$. Then it follows easily from the discussion above that A' is an attractor for the map f. In fact, for any compact neighborhood N of A', the successive images $f^k(N)$ converge to A'. Furthermore, A' supports an invariant measure, with entropy $\log 2$, so that almost every orbit in A' is dense. This attractor A' is nowhere dense, since it is a compact set of Hausdorff dimension $d \leqq 1 - \log(2)/\log(c) < 2$. In fact, A' can be described as the closure of the union of the graphs of the smooth almost periodic functions

$$g_n(\theta) = \cos((\theta + 2\pi n)/2) + c\cos((\theta + 2\pi n)/4) + c^2\cos((\theta + 2\pi n)/8) + \ldots,$$

where n ranges over all integers. Since g_n is uniformly close to g_m whenever n is close to m in the 2-adic topology, the function g_N is also defined and real analytic for any 2-adic integer N. With this convention, we can say simply that A' is the union of the graphs of the uncountably many distinct functions g_N. Thus A' is certainly not a manifold. It seems likely that its Hausdorff dimension is equal to $1 - \log(2)/\log(c) > 1$. (Compare Farmer, Ott, and Yorke.) On the other hand, if we modify this example by choosing c sufficiently close to 1, then it can be shown that the corresponding attractor A' contains an entire open subset of the cylinder. (See Frederickson, Kaplan, Yorke, and Yorke.)

If we carry out an analogous construction starting out with the Anosov diffeomorphism $(\phi, \psi) \mapsto (2\phi + \psi, \phi + \psi)$ of the torus in place of the squaring map on the circle, then we obtain an attractor which is topologically a torus $S^1 \times S^1$. Let $\lambda = (\sqrt{5} - 1)/2$ be the smaller eigenvalue of the associated integer matrix. Then this torus is embedded in a nowhere differentiable manner into the manifold $S^1 \times S^1 \times R$ if $\lambda < c < 1$, but is n times continuously differentiable if $0 < c < \lambda^n$ (see Kaplan, Mallet-Paret, and Yorke).

Acknowledgements. I am grateful to R. Butler, J. Guckenheimer, W. Thurston, R. F. Williams, and J. Yorke for a number of helpful suggestions.

References

Anosov, D.V., Katok, A.B.: New examples in smooth ergodic theory. Trans. Moscow Math. Soc. **23**, 1–35 (1970)

Auslander, J., Bhatia, N.P., Seibert, P.: Attractors in dynamical systems. Bol. Soc. Mat. Mex. **9**, 55–66 (1964)

Besicovitch, A.S.: A problem on topological transformation of the plane. Fundam. Math. **27**, 61–65 (1937)

Bowen, R.: A horseshoe with positive measure. Invent. Math. **28**, 203–204 (1975)

Bowen, R.: Invariant measures for Markov maps of the interval. Commun. Math. Phys. **69**, 1–17 (1979)

Bowen, R., Franks, J.: The periodic points of maps of the disk and the interval. Topology **15**, 337–342 (1976)

Coddington, E., Levinson, N.: Theory of ordinary differential equations. New York: McGraw-Hill 1955

Collet, P., Eckmann, J.-P.: Iterated maps on the interval as dynamical systems. Boston: Birkhäuser 1980

Conley, C.: Isolated invariant sets and the morse index. C.B.M.S. Regional Lect. **38**, A.M.S. 1978

Farmer, J.D., Ott, E., Yorke, J.A.: The dimension of chaotic attractors. Physica **7** D, 153–180 (1983)

Feigenbaum, M.: Quantitative universality for a class of nonlinear transformations. J. Stat. Phys. **19**, 25–52 (1978); **21**, 669–706 (1979)

Franks, J., Young, L.-S.: A C^2 Kupka-Smale diffeomorphism of the disk with no sources or sinks (preprint)

Frederickson, P., Kaplan, J., Yorke, E., Yorke, J.: The Liapunov dimension of strange attractors. J. Diff. Eq. **49**, 185–207 (1983)

Grebogi, C., Ott, E., Pelikan, S., Yorke, J.: Strange attractors that are not chaotic. Physica **13** D, 261–268 (1984)

Grebogi, C., Ott, E., Yorke, J.: Crises, sudden changes in chaotic attractors, and transient chaos. Physica **7** D, 181–200 (1983)

Guckenheimer, J.: A strange attractor, pp. 368–381. In: The Hopf bifurcation and applications. Marsden and McCracken (eds.). Berlin, Heidelberg, New York: Springer 1976

Guckenheimer, J.: Sensitive dependence on initial conditions for one-dimensional maps. Commun. Math. Phys. **70**, 133–160 (1979)

Guckenheimer, J., Williams, R.F.: Structural stability of Lorenz attractors. Pub. Math. I.H.E.S. **50**, 59–72 (1979)

Guckenheimer, J., Holmes, P.: Nonlinear oscillations, dynamical systems and bifurcation of vector fields. Berlin, Heidelberg, New York: Springer 1983

Hénon, M.: A two-dimensional mapping with a strange attractor. Commun. Math. Phys. **50**, 69–77 (1976)

Hirsch, M.: The dynamical systems approach to differential equations. Bull. Am. Math. Soc. **11**, 1–64 (1984)

Jakobson, M.V.: Absolutely continuous invariant measures for one-parameter families of one-dimensional maps. Commun. Math. Phys. **81**, 39–88 (1981)

Jonker, L., Rand, D.: Bifurcations in one dimension. Invent. Math. **62**, 347–365 (1981); **63**, 1–15 (1981)

Kan, I.: Strange attractors of uniform flows. Thesis, Univ. Ill. Urbana 1984

Kaplan, J.L., Mallet-Paret, J., Yorke, J.A.: The Lyapunov dimension of a nowhere differentiable attracting torus. Erg. Th. Dyn. Syst. (to appear)

Kaplan, J.L., Yorke, J.A.: Chaotic behavior of multidimensional difference equations, pp. 204–227. In: Functional differential equations and the approximation of fixed points. Peitgen and Walther (eds.). Lecture Notes in Mathematics, Vol. 730. Berlin, Heidelberg, New York: Springer 1970

Lanford, O.: A computer assisted proof of the Feigenbaum conjecture. Bull. Am. Math. Soc. **6**, 427–434 (1982)

La Salle, J.P.: The stability of dynamical systems. Reg. Conf. App. Math. **25**, S.I.A.M. 1976

La Salle, Lefschetz, S.: Stability by Liapunov's direct method. New York: Academic Press 1961

Li, T., Yorke, J.: Ergodic transformations from an interval into itself. Trans. Am. Math. Soc. **235**, 183–192 (1978)

Liapunoff, M.A.: Problème général de la stabilité du mouvement. Annal. Math. Stud. **17**, Princeton Univ. Press 1947 (1907 translation from the 1892 Russian original)

Lorenz, E.N.: Deterministic nonperiodic flow. J. Atmosph. Sci. **20**, 130–141 (1963)

Mandelbrodt, B.: The Fractal Geometry of Nature. San Francisco: Freeman 1982
Mendelson, P.: On unstable attractors. Bol. Soc. Mat. Mex. **5**, 270–276 (1960)
Milnor, J., Thurston, W.: On iterated maps of the interval (in preparation)
Misiurewicz, M.: Invariant measures for continuous transformations of [0,1] with zero topological entropy, pp. 144–152 of Ergodic Theory, Denker and Jacobs, Lecture Notes in Mathematics, Vol. 729. Berlin, Heidelberg, New York: Springer 1979
Misiurewicz, M.: Absolutely continuous measures for certain maps of an interval. Pub. Math. I.H.E.S. **53**, 17–51 (1981)
Nusse, L.: Chaos, yet no chance to get lost. Thesis, Utrecht 1983
Ruelle, D., Takens, F.: On the nature of turbulence. Commun. Math. Phys. **20**, 343–344 (1971)
Ruelle, D.: Strange attractors. Math. Intell. **2**, 126–137 (1979–80)
Ruelle, D.: Small random perturbations of dynamical systems and the definition of attractor. Commun. Math. Phys. **82**, 137–151 (1981)
Ruelle, D.: Small random perturbations and the definition of attractor, pp. 663–676 of Geometric Dynamics, Palis (ed.). Lecture Notes in Mathematics, Vol. 1007. Berlin, Heidelberg, New York: Springer 1983
Schweitzer, P.A.: Counterexamples to the Seifert conjecture and opening closed leaves of foliations. Annal. Math. **100**, 386–400 (1974)
Shaw, R.: Strange attractors, chaotic behavior and information flow. Z. Naturforsch. A**36**, 80–112 (1981)
Smale, S.: Differential dynamical systems. Bull. Am. Math. Soc. **73**, 747–817 (1967)
Smale, S.: Dynamical systems and turbulence, pp. 71–82 of Turbulence seminar, Chorin, Marsden, Smale (eds.). Lecture Notes in Mathematics, Vol. 615. Berlin, Heidelberg, New York: Springer 1977
Ulam, S., v. Neumann, J.: On combination of stochastic and deterministic processes. Bull. Am. Math. Soc. **53**, 1120 (1947)
Williams, R.F.: The zeta function of an attractor, pp. 155–161 of Conference on the Topology of Manifolds, J. Hocking (ed.). Boston: Prindle, Weber & Schmidt 1968
Williams, R.F.: Expanding attractors. Pub. Math. I.H.E.S. **43**, 169–203 (1974)
Williams, R.F.: The structure of Lorenz attractors, pp. 94–112 of Turbulence Seminar, Chorin, Marsden, Smale (eds.). Lecture Notes in Mathematics, Vol. 615. Berlin, Heidelberg, New York: Springer 1977
Williams, R.F.: The structure of Lorenz attractors. Pub. Math. I.H.E.S. **50**, 73–100 (1979)
Williams, R.F.: Attractors, strange and perverse (to appear)
Zeeman, E.C.: Catastrophe theory and applications. Summer school on dynamical systems. I.C.T.P. Trieste 1983

Communicated by O. E. Lanford

Received December 5, 1984; in revised form December 26, 1984

Commun. Math. Phys. 102, 517–519 (1985)

Communications in Mathematical Physics

Comments

On the Concept of Attractor: Correction and Remarks

John Milnor
Institute for Advanced Study, Princeton, NJ 08540, USA

The following consists of three unrelated comments on the author's paper [1].

1. Correction

Let f be a continuous map from a compact metric space X to itself, with n^{th} iterate denoted by f^n, and let $A \subset X$ be a closed non-vacuous subset with $f(A)=A$. Consider the following two properties of A.

(I) *For any sufficiently small neighborhood U of A, the intersection of the images $f^n(U)$ for $n \geqq 0$ is equal to A (compare Smale* [2, p. 786]*).*

(II) (Asymptotic stability) *For any sufficiently small neighborhood U, the successive images $f^n(U)$ converge to A, in the sense that for any neighborhood V there exists n_0 so that $f^n(U) \subset V$, for $n \geqq n_0$.*

In [1, Sect. 1] the author mistakenly described an example satisfying (I) but not (II). (The example was based on a remark of Besicovitch [3], which was corrected in a later paper [4].) In fact, (I) implies (II). The following proof is a minor modification of Hurley [8, Lemma 1.6], which demonstrates a corresponding statement for flows on a compact manifold. The proof shows also that (I) implies the existence of arbitrarily small neighborhoods $W \supset A$ with $f(W) \subset W$.

Proof that (I) *implies* (II). Let U be an open neighborhood which is small enough so that the intersection of the forward images of the closure $\bar{U}$ is equal to A. Let U_n be the open neighborhood consisting of all points x such that $f^i(x) \in U$ for $0 \leqq i \leqq n$. Thus $U = U_0 \supset U_1 \supset \ldots \supset A$ and $f(U_n) \subset U_{n-1}$. Hence the intersection W of the U_n satisfies $f(W) \subset W$. We will show that W is equal to U_n for n sufficiently large, and hence that W is an open set. Otherwise, for infinitely many integers n there must exist a point x_n which belongs to U_n but not U_{n+1}. Let $y_n = f^n(x_n) \in U$. Then we can choose some subsequence of these points y_n which converges to a point $y \in \bar{U}$. Since y_n belongs to the intersection of the sets $f^i(\bar{U})$ for $0 \leqq i \leqq n$, it follows that y belongs to the intersection of all of the $f^i(\bar{U})$, which is equal to A by hypothesis. But $f(y_n) \notin U$, hence $f(y) \notin U$, contradicting the hypothesis that $f(A) = A \subset U$. This proves that W is open. Hence the compact set $\bar{W} \subset \bar{U}$ is a neighborhood of A with $f(\bar{W}) \subset \bar{W}$. It follows easily from compactness that the successive images $\bar{W} \supset f(\bar{W}) \supset f^2(\bar{W}) \supset \ldots$ with intersection A actually converge to A in the sense described in (II). □

Here is an example. Let $U \subset X$ be any open set satisfying $f^n(\bar{U}) \subset U$ for some n, and suppose that the intersection $B = \bar{U} \cap f(\bar{U}) \cap \ldots \cap f^{n-1}(\bar{U})$ is non-vacuous. Then the intersection of the forward images of B is a non-vacuous, compact, f-invariant set which satisfies (I) and (II). The proof is not difficult.

In the context of a smooth flow, Conley [5] uses the term *attractor* for a compact, non-vacuous invariant set satisfying the analogues of conditions (I) and (II), while Auslander et al. [6] call such a set an [asymptotically] *stable attractor*. Note, however, that the word attractor is used in a quite different sense in [1]. Here is an example: Let f be the map of the unit square given by $f(x, y) = (1-x)(x, y)$. Then the edge $x=0$ is the unique compact invariant set satisfying (I) and (II). However, the omega limit set $\omega(x, y)$ is equal to the origin for almost every point (x, y) in the square, hence by the definitions of [1] the origin is the unique attractor.

2. Remark

In order to know that the concept of "minimal attractor," as defined in [1], is reasonable and useful, one would at least like an affirmative answer to the following. *For a C^k-generic map or flow on a compact manifold, is it true that almost every point belongs to the realm of attraction of some "minimal attractor"?* Using standard definitions, a related question would be the following: *Does the chain-recurrent set of a C^k-generic map or flow have at most a countable number of chain components?* Whenever this is true, one can at least say that almost every point belongs to the realm of attraction $\varrho(A)$ for some chain component A which is "attractive" in the sense that $\varrho(A)$ has strictly positive measure.

For some of the known generic and non-generic properties of maps or flows, see [7–12] below.

3. Addendum

The discussion of an attractor related to the solenoid in [1, Appendix 3] should have included a reference to Mayer and Roepstorff [13], which contains a detailed discussion of a similar example.

References

1. Milnor, J.: On the concept of attractor. Commun. Math. Phys. **99**, 177–195 (1985)
2. Smale, S.: Differentiable dynamical systems. Bull. Am. Math. Soc. **73**, 747–817 (1967)
3. Besicovitch, A.S.: A problem on topological transformation of the plane. Fundam. Math. **27**, 61–65 (1937) (compare [4])
4. Besicovitch, A.S.: A problem on topological transformations of the plane. II. Proc. Camb. Phil. Soc. **47**, 38–45 (1951)
5. Conley, C.: Isolated invariant sets and the Morse index. CBMS-NSF Reg. Conf. **38**, Am. Math. Soc. 1978
6. Auslander, J., Bhatia, N.P., Seibert, P.: Attractors in dynamical systems. Bol. Soc. Mat. Mex. **9**, 55–66 (1964)

7. Dobrynskii, V.A., Sarkovskii, A.N.: Genericity of dynamical systems almost all orbits of which are stable under sustained perturbations. Sov. Math. Dokl. **14**, 997–1005 (1973)
8. Hurley, M.: Attractors: persistence, and density of their basins. Trans. Am. Math. Soc. **269**, 247–271 (1982) (compare [9])
9. Hurley, M.: Bifurcation and chain recurrence. Ergodic Theory Dyn. Syst. **3**, 231–240 (1983)
10. Newhouse, S.: Lectures on dynamical systems, pp. 1–114 of "Dynamical Systems". Guckenheimer, Moser, Newhouse (ed.). Boston, Basel, Stuttgart: Birkhäuser 1980
11. Robinson, R.C., Williams, R.F.: Finite stability is not generic, pp. 451–462 of "Dynamical Systems". Peixoto, M. (ed.). New York: Academic Press 1973
12. Takens, F.: Tolerance stability, pp. 293–304 of "Dynamical Systems". Lecture Notes in Mathematics, Vol. 468, Manning, A. (ed.). Berlin, Heidelberg, New York: Springer 1975
13. Mayer, D., Roepstorff, G.: Strange attractors and asymptotic measures of discrete-time dissipative systems. J. Stat. Phys. **31**, 309–326 (1983)

Communicated by A. Jaffe

Received September 3, 1985

TRANSACTIONS OF THE
AMERICAN MATHEMATICAL SOCIETY
Volume 287, Number 1, January 1985

BOWEN-RUELLE MEASURES FOR CERTAIN PIECEWISE HYPERBOLIC MAPS

BY
LAI-SANG YOUNG[1]

ABSTRACT. We consider a class of piecewise C^2 Lozi-like maps and prove the existence of invariant measures with absolutely continuous conditional measures on unstable manifolds

This work is a small step forward in the following program: Suppose a compact neighborhood is mapped into itself and the map displays some chaotic behavior. Is there a strange attractor, or more specifically, is there a Bowen-Ruelle measure? Axiom A systems aside, current techniques have hardly begun to provide answers to these questions. One of the purposes of this note is to demonstate, in a very limited way, how certain 1-dimensional results can sometimes be useful in handling dissipative systems in 2-dimension.

The 1-dimensional result we alluded to says that piecewise expanding endomorphisms of the unit interval have smooth invariant measures [**LY**]. What we prove here is an analogous statement for certain piecewise hyperbolic attractors in 2-dimension. The prime examples that motivate this study are the Lozi maps, though our analysis has little to do with the precise nature of the equations studied by Lozi or Misiurewicz [**M**]. Our main result is that these "generalized Lozi maps" have invariant measures with absolutely continuous conditional measures on unstable manifolds. As a consequence they have Bowen-Ruelle measures.

We begin by isolating several properties of the Lozi maps. They are essentially the properties upon which our proof depends. Maps satisfying these hypotheses will henceforth be called "generalized Lozi maps". (See Figure 1.)

Let $R = [0,1]\times[0,1]$ and let $f\colon R \to R$ be a continuous injective map. We assume that f or some iterate of f takes R into its interior. Let $0 < a_1 < \cdots < a_q < 1$ and let $S = \{a_1, \ldots, a_q\} \times [0,1]$. We assume that $f|(R-S)$ is a C^2 diffeomorphism onto its image with $|\mathrm{Jac}(f)| < 1$ and that both $f|(R-S)$ and $f^{-1}|f(R-S)$ have bounded second derivative. We further impose the following conditions on $f|(R-S)$ (geometric interpretations are given in parentheses).

(H1) $$\operatorname{Inf}\left\{\left(\left|\frac{\partial f_1}{\partial x}\right| - \left|\frac{\partial f_1}{\partial y}\right|\right) - \left(\left|\frac{\partial f_2}{\partial x}\right| + \left|\frac{\partial f_2}{\partial y}\right|\right)\right\} \geq 0$$

(Df preserves cones making $< 45^\circ$ with the x-axis),

(H2) $$\operatorname{Inf}\left\{\left|\frac{\partial f_1}{\partial x}\right| - \left|\frac{\partial f_1}{\partial y}\right|\right\} = u > 1$$

Received by the editors November 15, 1982.
1980 *Mathematics Subject Classification*. Primary 58F15.
Key words and phrases. Hyperbolicity, unstable manifolds, invariant measures.
[1]Research partially supported by NSF.

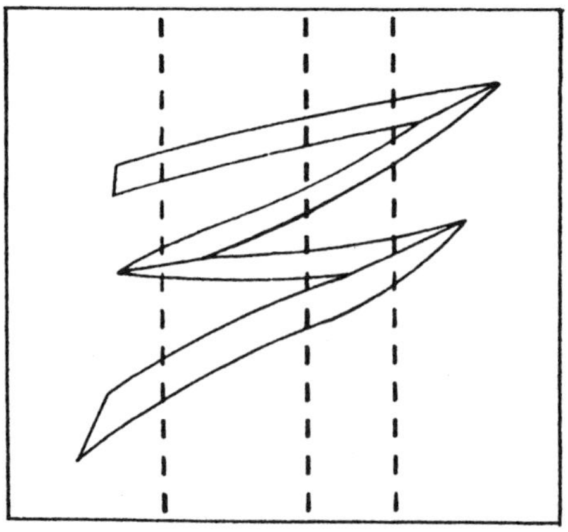

FIGURE 1. Example of a "generalized Lozi map"

(restricted to these cones the action of Df when projected onto the x-axis is uniformly expanding),

(H3) $$\operatorname{Sup}\left\{\frac{|\partial f_1/\partial y| + |\partial f_2/\partial y|}{(|\partial f_1/\partial x| - |\partial f_1/\partial y|)^2}\right\} < 1$$

(horizontal expansion dominates the action of Df on vertical vectors), and

(H4) $$\exists N \in \mathbf{Z}^+ \text{ s.t. } u^N > 2 \text{ and } f^k S \cap S = \emptyset \text{ for } 1 \leq k \leq N$$

(f expands horizontally more than it folds).

Note that (H4) is vacuous when $u > 2$. (Hence Figure 1 is legitimate.) Note also that these hypotheses are indeed satisfied by many Lozi maps. Lozi maps are usually given by

$$L(x, y) = (1 + y - a|x|, bx).$$

Changing coordinates this can be written as

$$f(x, y) = (1 + by - a|x|, x).$$

It is easy to verify that for open intervals of a and b, f takes some square $[c, c] \times [c, c]$ into itself and satisfies (H1)–(H4).[2]

DEFINITION 1. A Borel probability measure μ on R is said to have absolutely continuous conditional measures on unstable manifolds if there exist measurable partitions $\mathcal{P}_1 \subset \mathcal{P}_2 \subset \cdots$ of R and measurable sets $V_1 \subset V_2 \subset \cdots$ s.t.

1. $\mu V_n \uparrow 1$ as $n \to \infty$,
2. each element of $\mathcal{P}_n | V_n$ is an open subset of some unstable manifold and
3. if $\{\mu_c \colon c \in \mathcal{P}_n | V_n\}$ denotes the system of conditional measures on elements of $\mathcal{P}_n | V_n$, and m_c denotes Riemannian measure on c, then for almost every $c \in \mathcal{P}_n | V_n$, we have $\mu_c \ll m_c$.

We now state our main result.

[2]As this manuscript was being written I learned that P. Collet and Y. Levy had jointly obtained a result similar to ours for certain parameter values of the Lozi map [CL].

THEOREM. *If f: $R \to R$ is a generalized Lozi map, then f has an invariant Borel probability measure μ s.t.*

1. *Local unstable manifolds exist at μ-a.e. point and*
2. *μ has absolutely continuous conditional measures on unstable manifolds.*

The idea of our proof is as follows: Observe that unstable manifolds (when they exist) are piecewise smooth curves zigzagging across R (turning around at random places. Our strategy is first to construct an invariant measure μ that behaves nicely on neighborhoods of the singularity set S. This is done following a combination of the methods used by Sinai [S] and Lasota and Yorke [**LY**]. Since all turns are created by passing through S, we now have control over their impact as well. In particular, this allows us to construct a noninvariant measure $\tilde{\mu}$ equivalent to μ on an arbitrarily large set and having the property that its conditional measures on unstable manifolds are (obviously) absolutely continuous. This finishes the proof.

The following notations will be used: The map p: $R \to [0,1]$ denotes projection onto the first factor. Lebesgue measure on $[0,1]$ is denoted by m. For g: $[a,b] \to \mathbf{R}$, $\bigvee_a^b g$ denotes the total variation of g on $[a,b]$. If μ is a measure on R, then $f_*\mu$ is given by $f_*\mu(E) = \mu(f^{-1}E)$. If $J \subset [0,1]$ is a closed interval and α: $J \to [0,1]$ is a C^2 function with $|\alpha'| \leq 1$, then $f(\text{graph}(\alpha))$ is a union of finitely many smooth curves (H1). We denote them by $\{L_i(\alpha)\}$, dropping the α whenever no ambiguity arises. For $k > 1$, denote the smooth segments of $f^k\text{graph}(\alpha)$ by $\{L_{i_1\cdots i_k}\}$, where the indices are chosen so that $fL_{i_1\cdots i_k} = \bigcup_j L_{i_1\cdots i_k j}$. Let μ_0 be the measure on graph(α) s.t. $p_*\mu_0 =$ normalized Lebesgue measure on J. For $k = 1, 2, \ldots$, define $\mu_k = (f^k)_*\mu_0$. From (H1) and (H2) we know that for each $i_1\cdots i_k$, $p_*\mu_k|L_{i_1\cdots i_k}$ is absolutely continuous to m. We denote its density by $g_{i_1\cdots i_k}$ and the density of $p_*\mu_k$ by $\hat{g}_k$. That is, we have $\sum_{i_1\cdots i_k} g_{i_1\cdots i_k} = \hat{g}_k$. When it is convenient to consider f^N instead of f, we will write

$$f^{Nk}\text{graph}(\alpha) = \bigcup_{i_1\cdots i_k} L^{(N)}_{i_1\cdots i_k}$$

and

$$d(p_*\mu_{Nk}|L^{(N)}_{i_1\cdots i_k}) = g^{(N)}_{i_1\cdots i_k}\, dm,$$

etc.

LEMMA. *Under the hypotheses of the theorem, there exists an invariant Borel probability measure μ and a function g: $[0,1] \to [0,\infty)$ of bounded variation s.t. $d(p_*\mu) = gdm$.*

PROOF. Fix $J \subset [0,1]$ and a C^2 function α: $J \to [0,1]$ with $|\alpha'| \leq 1$. We will show that $\exists M$ s.t. $\bigvee_0^1 \hat{g}_k \leq M$ for all k. This will imply that $\bigvee_0^1(n^{-1}\sum_{k=1}^n \hat{g}_k) \leq M$, and since $\int_0^1(n^{-1}\sum_{k=1}^n \hat{g}_k)\,dm = 1$ the sequence $\{n^{-1}\sum_{k=1}^n \hat{g}_k\}_{n=1,2,\cdots}$ is precompact in $L^1([0,1],m)$. Choose a subsequence $\{n_i\}$ s.t. as $i \to \infty$, $(n_i)^{-1}\sum_{k=1}^{n_i} \mu_k$ converges in the weak star topology to a Borel probability measure μ and

$$\frac{1}{n_i}\sum_{k=1}^{n_i} \hat{g}_k \xrightarrow{L^1} \text{some function } g.$$

It follows immediately that μ is invariant, $d(p_*\mu) = gdm$ and that $\bigvee_0^1 g \leq M$.

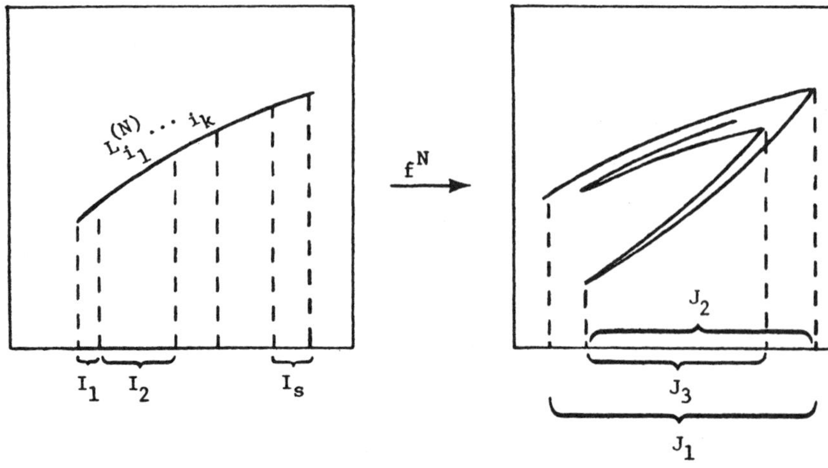

FIGURE 2

Let $N \in \mathbf{Z}^+$ satisfy (H4). It suffices to show the uniform boundedness of $\bigvee_0^1 \hat{g}_{Nk}$ for $k = 1, 2, \ldots$. We will prove that $\exists A$ s.t. for every $i_1 \cdots i_k$,

$$(*) \qquad \sum_j \bigvee_0^1 g^{(N)}_{i_1\cdots i_k j} \le A \int_0^1 g^{(N)}_{i_1\cdots i_k}\, dm + \frac{2}{u^N} \bigvee_0^1 g^{(N)}_{i_1\cdots i_k}.$$

Then if $\beta_k = \sum_{i_1\cdots i_k} \bigvee_0^1 g^{(N)}_{i_1\cdots i_k}$, we would have

$$\beta_{k+1} = \sum_{i_1\cdots i_k} \left(\sum_j \bigvee_0^1 g^{(N)}_{i_1\cdots i_k j} \right) \le A + \frac{2}{u^N}\beta_k$$

so that for all k,

$$\bigvee_0^1 g_{Nk} \le \beta_k \le A \sum_{i=0}^{\infty} \left(\frac{2}{u^N} \right)^i < \infty.$$

We now fix $i_1 \cdots i_k$ and prove $(*)$. Let $pL^{(N)}_{i_1\cdots i_k} = \bigcup_{j=1}^s I_j$, where $\{f^N(x,y)\colon x \in I_j\} = L^{(N)}_{i_1\cdots i_k j}$ and $J_j = pL^{(N)}_{i_1\cdots i_k j}$. Define $\tilde{f}_j\colon I_j \to J_j$ by $\tilde{f}_j(x) = p \circ f^N(x,y)$ for $(x,y) \in L^{(N)}_{i_1\cdots i_k}$. Then each $\tilde{f}_j$ is a C^2 diffeomorphism between I_j and J_j with $|\tilde{f}_j'| \ge u^N$ ((H1), (H2)) and $|(\tilde{f}_j^{-1})''| \le Q$ for some universal Q. (This follows from (H3) by a direct computation.) Now

$$\sum_j \bigvee_0^1 g^{(N)}_{i_1\cdots i_k j} = \underbrace{\sum_j \bigvee_{J_j} \left[(g^{(N)}_{i_1\cdots i_k} \circ \tilde{f}_j^{-1}) |(\tilde{f}_j^{-1})'| \right]}_{①}$$

$$+ \underbrace{\sum_j \left[|\tilde{f}_j'(l_j)|^{-1} g^{(N)}_{i_1\cdots i_k}(l_j) + |\tilde{f}_j'(r_j)|^{-1} g^{(N)}_{i_1\cdots i_k}(r_j) \right]}_{②},$$

where l_j and r_j denote the left and right end points of I_j respectively. Consider one j at a time.

$$\begin{aligned}\bigvee_{J_j}(g^{(N)}_{i_1\cdots i_k}\circ\tilde{f}_j^{-1})|(\tilde{f}_j^{-1})'| &= \int_{J_j}|((g^{(N)}_{i_1\cdots i_k}\circ\tilde{f}_j^{-1})(\tilde{f}_j^{-1})')'|\,dm\\ &= \int_{J_j}|(g^{(N)}_{i_1\cdots i_k}\circ\tilde{f}_j^{-1})'|\,|(\tilde{f}_j^{-1})'|\,dm+\int_{J_j}(g^{(N)}_{i_1\cdots i_k}\circ\tilde{f}_j^{-1})|(\tilde{f}_j^{-1})''|\,dm\\ &\le \frac{1}{u^N}\bigvee_{I_j} g^{(N)}_{i_1\cdots i_k}+\frac{Q}{u^N}\int_{I_j} g^{(N)}_{i_1\cdots i_k}\,dm\end{aligned}$$

so that

$$(**)\qquad \begin{aligned}&①+|\tilde{f}_1'(l_1)|^{-1}g^{(N)}_{i_1\cdots i_k(l_1)}+|\tilde{f}_s'(r_s)|^{-1}g^{(N)}_{i_1\cdots i_k}(r_s)\\ &\le \frac{1}{u^N}\bigvee_0^1 g^{(N)}_{i_1\cdots i_k}+\frac{Q}{u^N}\int_0^1 g^{(N)}_{i_1\cdots i_k}\,dm.\end{aligned}$$

Now we claim that there is a universal $d>0$ s.t. the points $0, l_2, \ldots, l_s, 1$ are pairwise at least d apart. We have $|0-l_2|, |l_s-1|\ge d$ because we may assume that $f^{Nk}R\subset \operatorname{int} R$. That $|r_j-l_j|\ge$ some d for $j=2,\ldots,s-1$ follows from (H4). Thus if $I=[l,r]$ is either $[0,l_1]$ or $[l_s,1]$ or $[l_j,r_j]$, $j=2,\ldots,s-1$, then

$$\begin{aligned}g^{(N)}_{i_1\cdots i_k}(l)+g^{(N)}_{i_1\cdots i_k}(r)&\le 2\left(\min_I g^{(N)}_{i_1\cdots i_k}\right)+\bigvee_I g^{(N)}_{i_1\cdots i_k}\\ &\le \frac{2}{d}\int_I g^{(N)}_{i_1\cdots i_k}\,dm+\bigvee_I g^{(N)}_{i_1\cdots i_k}.\end{aligned}$$

Note that $g^{(N)}_{i_1\cdots i_k}(0)=g^{(N)}_{i_1\cdots i_k}(1)=0$, so that the terms in ② not accounted for in $(**)$

$$\le \frac{1}{u^N}\frac{2}{d}\int_0^1 g(N)_{i_1\cdots i_k}\,dm+\frac{1}{u^N}\bigvee_0^1 g^{(N)}_{i_1\cdots i_k}.$$

Thus

$$①+②\le\left(Q+\frac{2}{d}\right)\frac{1}{u^N}\int_0^1 g^{(N)}_{i_1\cdots i_k}\,dm+\frac{2}{u^N}\bigvee_0^1 g^{(N)}_{i_1\cdots i_k},$$

which completes the proof of $(*)$ and the lemma. □

Let $W^u_\delta(x)$ denote the local unstable manifold at x (assuming it exists) s.t. $pW^u_\delta(x)=[p(x)-\delta, p(x)+\delta]$. Implicit in this notation is the assertion that $W^u_\delta(x)$ contains no cusps. Let $D(S,\delta)$ denote the δ-neighborhood of S.

PROOF OF THEOREM. If μ is any invariant probability measure with $d(p_*\mu)=g\,dm$ for some bounded g, say $g\le M_0$, then for $\delta>0$,

$$\sum_{k=0}^{\infty}\mu(f^kD(S,\delta u^{-k}))=\sum_{k=0}^{\infty}\mu D(S,\delta u^{-k})\le 2\delta M_0\sum_{k=0}^{\infty}u^{-k}<\infty.$$

By the Borel-Cantelli lemma, μ-a.e. x is in $f^kD(S,\delta u^{-k})$ for at most finitely many k. That is, for μ-a.e. x, $\exists\delta(x)>0$ s.t. $f^{-k}x\notin D(S,\delta(x)u^{-k})$ for all $k>0$. This implies the existence of $W^u_{\delta(x)}(x)$. (For more details see [**KS**].) Let us fix one measure μ constructed as in the lemma with graph$(\alpha) = W^u_{\delta(x)}$ for some x. This guarantees that the $L_{i_1\cdots i_k}(\alpha)$ will not cross other unstable manifolds. For $\delta > 0$, let $\Lambda_\delta = \{x \in R\colon d(f^{-k}x, S) \geq \delta u^{-k}\ \forall k \geq 0\}$. Then each Λ_0 is closed and $\lim_{\delta\to 0} \mu\Lambda_\delta = 1$.

We now define a sequence of measurable partitions $\mathcal{P}_1 \subset \mathcal{P}_2 \subset \cdots$. All notations will be as in Definition 1. For $n \in \mathbf{Z}^+$, let $\{U_1, \ldots, U_{2^n}\}$ be the partition of R into 2^n vertical columns of width 2^{-n}. For $x \in U_i \cap \Lambda_{2^{-n}}$, let $c(x) = W^u_{2^{-n}}(x) \cap U_i$. Let $V_n = \bigcup_{x\in\Lambda_{2^{-n}}} c(x)$ and $\mathcal{P}_n = \{c(x)\colon x \in V_n\} \cup \{R - V_n\}$. It suffices to show that for every $\varepsilon > 0$ and every n, there is a set $\tilde{V}_n \subset V_n$ s.t. $\mu\tilde{V}_n > \mu V_n - \varepsilon$ and condition 3 in Definition 1 is satisfied when μ is replaced by $\chi_{\tilde{V}_n}\mu$. In fact, for given ε and n, we will construct a noninvariant measure $\tilde{\mu}$ and a set $\tilde{V}$ with $\mu\tilde{V} > 1 - \varepsilon$ s.t. $\tilde{\mu} \ll \mu$, is equivalent to μ on $\tilde{V}$, and satisfies condition 3 with respect to $\mathcal{P}_{n'}$ for some $n' \geq n$. It is straightforward to verify that this implies the desired result.

Now let $n \in \mathbf{Z}^+$ and $\varepsilon > 0$ be given. Let n' be a large number to be determined later and let $\tilde{U}_1, \ldots, \tilde{U}_{2^{n'}}$ be pairwise disjoint vertical columns of width $2^{-n'}$. We define a sequence of measures $\{\tilde{\mu}_k\}_{k=1,2,\ldots}$ as follows: Recall that in the definition of μ_k, μ_k is carried by $f^k(\text{graph}(\alpha))$, where $f^k(\text{graph}(\alpha))$ is a finite union of smooth curve segments. Let $\tilde{\mu}_k$ be μ_k annihilated on those parts of its support that only partially cross some $\tilde{U}_i$ (See Figure 3.) $\tilde{\mu}_k(R)$ is probably < 1.

We claim that given $\delta > 0$, $\exists n'$ such that for sufficiently large k, $\tilde{\mu}_k(R) > 1 - \delta$. Recall that $\exists M_0$ s.t. $\hat{g}_k \leq M_0\ \forall k$. If $x \in (\operatorname{supp}\mu_k - \operatorname{supp}\tilde{\mu}_k)$, then either x lies in one of the two end pieces of $f^k(\text{graph}(\alpha))$ that only partially crosses some $\tilde{U}_i$, or the horizontal distance between x and a cusp in $f^k(\text{graph}(\alpha))$ is $< 2^{-n'}$, which says that $d(f^{-i}x, S) \leq 2^{-n'}u^{-i}$ for some $1 \leq i \leq k$. Thus

$$\begin{aligned}
1 - \tilde{\mu}_k(R) &= \mu_k(\operatorname{supp}\mu_k - \operatorname{supp}\tilde{\mu}_k) \\
&= \sum_{i=1}^{k} \mu_k\{x\colon d(f^{-i}x, S) \leq 2^{-n'}u^{-i}\} + \mu_k\{2 \text{ end pieces}\} \\
&= \sum_{i=1}^{k} \mu_i\{x\colon d(x, S) \leq 2^{-n'}u^{-i}\} + \mu_k\{2 \text{ end pieces}\} \\
&\leq 2M_0 q 2^{-n'} \sum_{i=1}^{k} u^{-i} + \mu_k\{2 \text{ end pieces}\}.
\end{aligned}$$

The second term $\to 0$ as $k \to \infty$. The first term becomes arbitrarily small as $n' \uparrow \infty$. Recall also that $(n_i)^{-1}\sum_{k=1}^{n_i} \mu_k \to \mu$. Choose a subsequence $\{n_i'\}$ of $\{n_i\}$ s.t. $(n_i')^{-1}\sum_{k=1}^{n_i'} \tilde{\mu}_k \to \tilde{\mu}$ for some $\tilde{\mu}$. It is easy to verify that $\tilde{\mu}E \leq \mu E$ for every Borel set E and hence we have $\tilde{\mu} \ll \mu$ with $0 \leq d\tilde{\mu}/d\mu \leq 1$. But since $\tilde{\mu}(R)$ can be made arbitrarily near 1, we can choose $\tilde{\mu}$ s.t. $\tilde{\mu}$ is equivalent to μ except on a set of μ-measure $< \varepsilon$.

It remains to show that if $\tilde{\mu}_T$ is the transverse measure on $\mathcal{P}_{n'}$ induced by $\tilde{\mu}$, then for $\tilde{\mu}_T$-a.e. $c \in \mathcal{P}_{n'}|V_{n'}$, $\tilde{\mu}_c \ll m_c$. Recall that $g_{i_1\cdots i_k}$ = density of $p_*(\mu_k|L_{i_1\cdots i_k})$. Let $\tilde{g}_{i_1\cdots i_k}$ = density of $p_*(\tilde{\mu}_k|L_{i_1\cdots i_k})$. We claim that $\exists B > 0$ s.t. for any $1 \leq i \leq 2^{n'}$

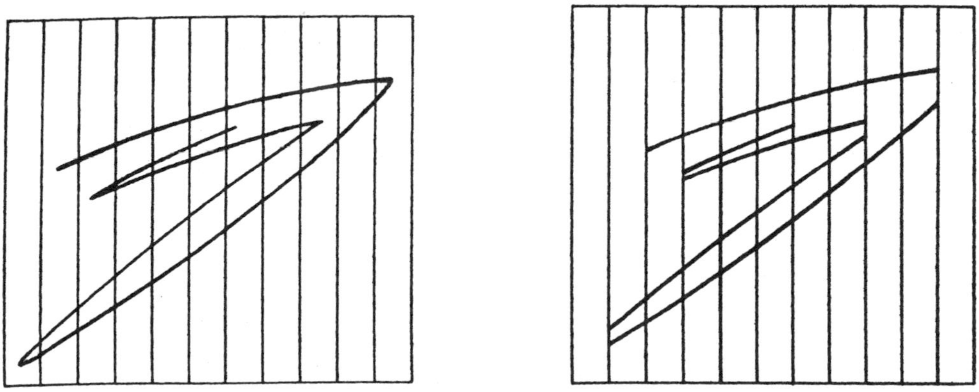

supp μ_k supp $\tilde{\mu}_k$

FIGURE 3

and any $i_1 \cdots i_k$, we have either

$$\tilde{g}_{i_1 \cdots i_k} \equiv 0 \quad \text{on } p\tilde{U}_i$$

or

$$\frac{\tilde{g}_{i_1 \cdots i_k}(x)}{\tilde{g}_{i_1 \cdots i_k}(y)} \leq B \quad \forall x, y \in p\tilde{U}_i.$$

This would prove that for almost every $c \in \mathcal{P}_{n'}|V_{n'}$, $d\tilde{\mu}_c = g_c dm_c$ where $\forall x, y \in c$, $g_c(x)/g_c(y) \leq B'$ for some B'. To prove the claim fix $1 \leq i \leq 2^{n'}$ and $i_1 \cdots i_k$. If $L_{i_1 \cdots i_k}$ does not cross the full width of $\tilde{U}_i$, then $g_{i_1 \cdots i_k}|p\tilde{U}_i \equiv 0$. Otherwise there are intervals $E_0, E_1, \ldots, E_k \subset [0,1]$, $E_k = p\tilde{U}_i$, and C^2 diffeomorphisms $h_j \colon E_{j-1} \to E_j$ s.t. $\tilde{g}_{i_1 \cdots i_k}|p\tilde{U}_i = |((h_k \circ \cdots \circ h_1)^{-1})'|$ (see lemma). While these h_j's depend strictly on i and $i_1 \cdots i_k$, they satisfy $|h_j'| \geq u > 1$ and $|h_j''| \leq Q$, where u and Q are universal constants.

The reader can easily verify that

$$\begin{aligned} &|\log|((h_k \circ \cdots \circ h_1)^{-1})'(x)| - \log|((h_k \cdots h_1)^{-1})'(y)||\\ &\qquad \leq \text{ some constant depending only on } u,\ Q \text{ and } n'. \end{aligned}$$

This completes the proof. □

We have proved our result for a specific class of maps of the square into itself. It is obvious that our hypotheses are more stringent than necessary and that it is easy to make slight generalizations. We do not attempt to do that here, because we do not know what a natural general statement ought to be.

We mention a few corollaries. Since the proofs are standard, we will provide only references. The standing hypothesis for the rest of this article is that f is a generalized Lozi map.

COROLLARY 1. *Let μ be constructed as in our lemma. Then there are measurable sets $E_1, E_2, \cdots \subset R$ s.t. $f^{-1}E_i = E_i$, $\mu E_i > 0\ \forall i$, $\mu(\bigcup E_i) = 1$ and for each i, $f|E_i \colon (E_i, \mu|E_i) \to (E_i, \mu|E_i)$ is ergodic.*

PROOF. The proof follows that in [**P**]. It uses the absolute continuity of stable manifolds [**KS**] and the fact that on most $c \in \mathcal{P}_n|V_n$, μ_c is equivalent to m_c. □

DEFINITION 2. A Borel probability measure μ on R is called a *Bowen-Ruelle measure* [**B**] if there is a set $U \subset R$ of positive Lebesgue measure s.t. for every continuous function φ: $\mathbf{R} \to \mathbf{R}$,

$$\frac{1}{n}\sum_{i=0}^{n-1} \varphi f^i x \to \int \varphi \, d\mu$$

for Lebesgue-a.e. $x \in U$.

COROLLARY 2. *Let μ be constructed as in our lemma. Then corresponding to each E_i in Corollary* 1, *$\mu|E_i$ normalized is a Bowen-Ruelle measure.*

PROOF. This follows from [**KS**]. □

Unlike the case of Axiom A attractors, assuming only (H1)–(H4) there can easily be more than one Bowen-Ruelle measure.

COROLLARY 3. *Let f: $R \to R$ be a generalized Lozi map. Then f has an invariant Borel probability measure μ s.t.*

$$h_\mu(f) = \int \lambda_1(x) \, d\mu(x),$$

where $\lambda_1(x)$ is the positive μ-exponent of f at x.

PROOF. See [**LS**]. □

REFERENCES

[**B**] R. Bowen, *Equilibrium states and the ergodic theory of Anosov diffeomorphisms,* Lecture Notes in Math., vol. 47, Springer, Berlin and New York, 1975.

[**CL**] P. Collet and Y. Levy, *Ergodic properties of the Lozi mappings,* Comm. Math. Phys. **93** (1984), 461–482.

[**KS**] A. Katok and J. M. Strelcyn, *Invariant manifolds for smooth maps with singularities, Part* I: *Existence* (preprint).

[**LS**] F. Ledrappier and J. M. Strelcyn, *A proof of the estimation from below in Pesin entropy formula,* J. Ergodic Theory and Dynam. Syst. (to appear).

[**LY**] A. Lasota and J. Yorke, *On the existence of invariant measures for piecewise monotonic transformations,* Trans. Amer. Math. Soc. **186** (1973), 481–488.

[**M**] M. Misiurewicz, *Strange attractors for the Lozi mappings,* Proc. N.Y. Acad. Sci. (1980), 348–358.

[**P**] Ja. Pesin, *Characteristic Lyapunov exponents and smooth ergodic theory,* Russian Math. Surveys **32** (1977), 55–114.

[**S**] Ya. G. Sinai, *Markov partitions and C-diffeomorphisms,* Functional Anal. Appl. **2** (1968), 61–82.

DEPARTMENT OF MATHEMATICS, MICHIGAN STATE UNIVERSITY, EAST LANSING, MICHIGAN 48824 (Current address)

DEPARTMENT OF MATHEMATICS, UNIVERSITY OF NORTH CAROLINA, CHAPEL HILL, NORTH CAROLINA 27514

Ergodic theory of chaos and strange attractors

J.-P. Eckmann

Université de Genève, 1211 Genève 4, Switzerland

D. Ruelle

Institut des Hautes Etudes Scientifiques, 91440 Bures-sur-Yvette, France

Physical and numerical experiments show that deterministic noise, or chaos, is ubiquitous. While a good understanding of the onset of chaos has been achieved, using as a mathematical tool the geometric theory of differentiable dynamical systems, moderately excited chaotic systems require new tools, which are provided by the *ergodic* theory of dynamical systems. This theory has reached a stage where fruitful contact and exchange with physical experiments has become widespread. The present review is an account of the main mathematical ideas and their concrete implementation in analyzing experiments. The main subjects are the theory of *dimensions* (number of excited degrees of freedom), *entropy* (production of information), and *characteristic exponents* (describing sensitivity to initial conditions). The relations between these quantities, as well as their experimental determination, are discussed. The systematic investigation of these quantities provides us for the first time with a reasonable understanding of dynamical systems, excited well beyond the quasiperiodic regimes. This is another step towards understanding highly turbulent fluids.

CONTENTS

*Sections marked with * contain supplementary material which can be omitted at first reading.

I. INTRODUCTION

In recent years, the ideas of differentiable dynamics have considerably improved our understanding of irregular behavior of physical, chemical, and other natural phenomena. In particular, these ideas have helped us to understand the onset of turbulence in fluid mechanics. There is now ample experimental and theoretical evidence that the qualitative features of the time evolution of many physical systems are the same as those of the solution of a typical evolution equation of the form

$$\dot{x}(t)=F_\mu(x(t)), \quad x\in \mathbf{R}^m \tag{1.1}$$

in a space of small dimension m. Here, x is a set of coordinates describing the system (typically, mode amplitudes, concentrations, etc.), and F_μ determines the nonlinear time evolution of these modes. The subscript μ corresponds to an experimental *control parameter*, which is kept constant in each run of the experiment. (Typically, μ is the intensity of the force driving the system.) We write

$$x(t)=f_\mu^t(x(0)) . \tag{1.2}$$

We usually *assume* that there is a parameter value, say $\mu=0$, for which the equation is well understood and leads to a motion in phase space which, after some transients, settles down to be stationary or periodic.

As the parameter μ is varied, the nature of the asymptotic motion may change.[1] The values μ for which this change of asymptotic regime happens are called bifurcation points. As the parameter increases through successive bifurcations, the asymptotic motion of the system typically gets more complicated. For special sequences of these bifurcations a lot is known, and even quantitative features are predicted, as in the case of the period-doubling cascades ("Feigenbaum scenario"). We do not, however, possess a complete classification of the possible transitions to more complicated behavior, leading eventually to turbulence. *Geometrically*, the asymptotic motion follows an *attractor* in phase space, which will become more and more complicated as μ increases.

The aim of the present review is to describe the current state of the theory of *statistical* properties of dynamical systems. This theory becomes relevant as soon as the system is "excited" beyond the simplest bifurcations, so that precise geometrical information about the shape of the attractor or the motion on it is no longer available. See Eckmann (1981) for a review of the *geometrical* aspects of dynamical systems. The statistical theory is still capable of distinguishing different degrees of complexity of attractors and motions, and presents thus a further step in bridging the gap between simple systems and fully developed turbulence. In particular, the present treatment does not exclude the description of space-time patterns.

After introducing precise dynamical concepts in Sec. II, we address the theory of characteristic exponents in Sec. III and the theory of entropy and information dimension in Sec. IV. In Sec. V we discuss the extraction of dynamical quantities from experimental time series.

It is necessary at this point to clarify the role of the physical concept of *mode*, which appears naturally in simple theories (for instance, Hamiltonian theories with quadratic Hamiltonians), but which loses its importance in nonlinear dynamical systems. The usual idea is to represent a physical system by an appropriate change of variables as a collection of independent oscillators or modes. Each mode is periodic, and its state is represented by an angular variable. The global system is *quasiperiodic* (i.e., a superposition of periodic motions). From this perspective, a dissipative system becomes more and more turbulent as the number of *excited modes* grows, that is, as the number of independent oscillators needed to describe the system progressively increases. This point of view is very widespread; it has been extremely useful in physics and can be formulated quite coherently (see, for example, Haken, 1983). However, this philosophy and the corresponding intuition about the use of Fourier modes have to be completely modified when nonlinearities are important: *even a finite-dimensional motion need not be quasiperiodic in general.* In particular, the concept of "number of excited modes" will have to be replaced by new concepts, such as "number of non-negative characteristic exponents" or "information dimension." These new concepts come from a statistical analysis of the motion and will be discussed in detail below.

In order to talk about a statistical theory, one needs to say what is being averaged and in which sample space the measurements are being made. The theory we are about to describe treats *time averages.* This implies and has the advantage that *transients* become irrelevant. (Of course, there may be formidable experimental problems if the transients become too long.) Once transients are over, the motion of the solution x of Eq. (1.1) settles typically near a subset of $\mathbf{R}^m$, called an *attractor* (mathematical definitions will be given later). In particular, in the case of dissipative systems, on which we focus our attention, the volume occupied by the attractor is in general very small relative to the volume of phase space. We shall not talk about attractors for conservative systems, where the volume in phase space is conserved. For dissipative systems we may assume that phase-space volumes are contracted by the time evolution (if phase space is finite dimensional). Even if a system contracts volumes, this does not mean that it contracts lengths in *all* directions. Intuitively, some directions may be stretched, provided some others are so much contracted that the final volume is smaller than the initial volume (Fig. 1). This seemingly trivial remark has profound consequences. It implies that, even in a dissipative system, the final motion may be unstable *within the attractor.* This instability usually manifests itself by an exponential separation of orbits (as time goes on) of points which initially are very close to each other (on the attractor). The exponential separation takes place in the direction of stretching, and an attractor having this stretching property will be called a *strange attractor.* We shall also say that a system with a strange attractor is *chaotic* or has *sensitive dependence on initial conditions.* Of course, since the attractor is in general bounded, exponential separation can only hold as long as distances are small.

Fourier analysis of the motion on a strange attractor (say, of one of its coordinate components) in general reveals a *continuous power spectrum.* We are used to interpreting this as corresponding to an infinite number of modes. However, as we have indicated before, this

[1] It is to be understood that the experiment is performed with a *fixed* value of the parameter.

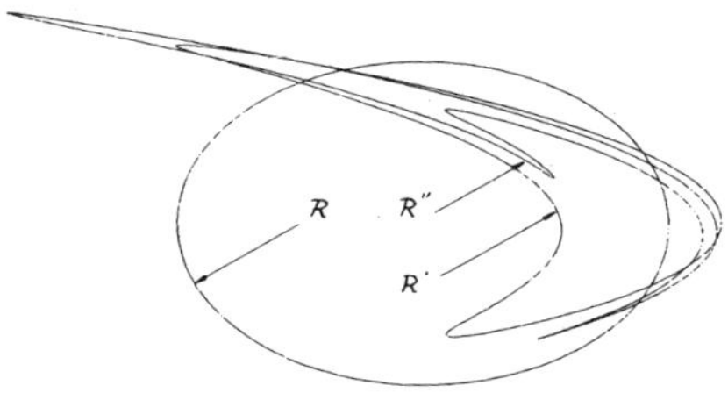

FIG. 1. The Hénon map $x_1' = 1 - 1.4x_1^2 + x_2$, $x_2' = 0.3x_1$ contracts volumes but stretches distances. Shown are a region R, and its first and second images R' and R'' under the Hénon map.

reasoning is only valid in a "linear" theory, which then has to take place in an infinite-dimensional phase space. Thus, if we are confronted experimentally with a continuous power spectrum, there are two possibilities: We are either in the presence of a system that "explores" an infinite number of dimensions in phase space, or we have a system that evolves nonlinearly on a finite-dimensional attractor. Both alternatives are possible, and the second appears frequently in practice. We shall give below an algorithm which, starting from measurements, gives information on the effective dimension. This algorithm has been successfully used in several experiments, e.g., Malraison *et al.* (1983), Abraham *et al.* (1984), Grassberger and Procaccia (1983b); it has indicated *finite dimensions* in hydrodynamic systems, even though the phase space is infinite dimensional and the system therefore could potentially excite an infinite number of degrees of freedom.

The tool with which we want to measure the dimension and other dynamical quantities of the system is *ergodic theory.* Ergodic theory says that a time average equals a space average. The weight with which the space average has to be taken is an *invariant measure.* An invariant measure ρ satisfies the equation

$$\rho[f^{-t}(E)] = \rho(E), \quad t > 0 , \tag{1.3}$$

where E is a subset of points of $\mathbf{R}^m$ and $f^{-t}(E)$ is the set obtained by evolving each of the points in E backwards during time t. There are in general many invariant measures in a dynamical system, but not all of them are physically relevant. For example, if x is an unstable fixed point of the evolution, then the δ function at x is an invariant measure, but it is not observed. From an experimental point of view, a reasonable measure is obtained according to the following idea of Kolmogorov (see Sec. II.F). Consider Eq. (1.1) with an external noise term added,

$$\dot{x}(t) = F_\mu(x(t)) + \varepsilon\omega(t) , \tag{1.4}$$

where ω is some noise and $\varepsilon > 0$ is a parameter. For suitable noise and $\varepsilon > 0$, the stochastic time evolution (1.3) has a unique stationary measure ρ_ε, and the measure we propose as "reasonable" is $\rho = \lim_{\varepsilon \to 0} \rho_\varepsilon$. We shall come back later to the problem of choosing a reasonable measure ρ.

For the moment we assume that such a measure exists, and call it the *physical measure.* Physically, we assume that it represents experimental time averages. Mathematically, we only require (for the moment) that it be invariant under time evolution.

A basic virtue of the ergodic theory of dynamical systems is that it allows us to consider only the long-term behavior of a system and not to worry about transients. In this way, the problems are at least somewhat simplified. The physical long-term behavior is on attractors, as we have already noted, but the geometric study of attractors presents great mathematical difficulties. Shifting attention from attractors to invariant measures turns out to make life much simpler.

An invariant probability measure ρ may be *decomposable* into several different pieces, each of which is again invariant. If not, ρ is said to be *indecomposable* or *ergodic.* In general, an invariant measure can be uniquely represented as a superposition of ergodic measures. In view of this, it is natural to assume that the physical measure is not only invariant, but also ergodic. If ρ is ergodic, then the *ergodic theorem* asserts that for every continuous function φ,

$$\lim_{T\to\infty} \frac{1}{T} \int_0^T \varphi[f^t x(0)]dt = \int \rho(dx)\varphi(x) \tag{1.5}$$

for almost all initial conditions $x(0)$ with respect to the measure ρ. Since the measure ρ might be singular, for instance concentrated on a fractal set, it would be better if we could say something for almost all $x(0)$ with respect to the ordinary (Lebesgue) measure on some set $S \subset \mathbf{R}^m$. We shall see below that this is sometimes possible.

One crucial decision in our study of dynamics is to concentrate on the analysis of the separation in time of two infinitely close initial points. Let us illustrate the basic idea with an example in which time is discrete [rather than continuous as in (1.1)]. Consider the evolution equation

$$x(n+1) = f(x(n)), \quad x(i) \in \mathbf{R} , \tag{1.6}$$

where n is the discrete time. The separation of two initial points $x(0)$ and $x(0)'$ after time N is then

$$x(N) - x(N)' = f^N(x(0)) - f^N(x(0)')$$
$$\approx \left| \frac{d}{dx}(f^N)(x(0)) \right| [x(0) - x(0)'] , \tag{1.7}$$

where $f^N(x) = f(f(\cdots f(x)\cdots))$, N times. By the chain rule of differentiation,

$$\frac{d}{dx}(f^N)(x(0)) = \frac{d}{dx} f(x(N-1))$$
$$\times \frac{d}{dx} f(x(N-2)) \cdots \frac{d}{dx} f(x(0)) . \tag{1.8}$$

[In the case of m variables, i.e., $x \in \mathbf{R}^m$, we replace the derivative $(d/dx)f$ by the Jacobian matrix, evaluated at x: $D_x f = (\partial f_i / \partial x_j)$.] Assuming that all factors in the above expression are of comparable size, it seems plausible that df^N/dx grows (or decays) exponentially with N.

The same is true for $x(N)-x(N)'$, and we can define the average rate of growth as

$$\lambda=\lim_{N\to\infty}\frac{1}{N}\log|D_{x(0)}f^N\delta x(0)| \ . \tag{1.9}$$

By the theorem of Oseledec (1968), this limit exists for almost all $x(0)$ (with respect to the invariant measure ρ). The average expansion value depends on the direction of the initial perturbation $\delta x(0)$, as well as on $x(0)$. However, if ρ is ergodic, the largest λ [with respect to changes of $\delta x(0)$] is independent of $x(0)$, ρ-almost everywhere. This number λ_1 is called the *largest Liapunov exponent* of the map f with respect to the measure ρ. Most choices of $\delta x(0)$ will produce the largest Liapunov exponent λ_1. However, certain directions will produce smaller exponents $\lambda_2,\lambda_3,\ldots$ with $\lambda_1\geq\lambda_2\geq\lambda_3\geq\cdots$ (see Sec. III.A for details).

In the continuous-time case, one can similarly define

$$\lambda(x,\delta x)=\lim_{T\to\infty}\frac{1}{T}\log|(D_xf^T)\delta x| \ . \tag{1.10}$$

We shall see that the Liapunov exponents (i.e., characteristic exponents) and quantities derived from them give useful bounds on the *dimensions of attractors,* and on the *production of information* by the system (i.e., *entropy* or Kolmogorov-Sinai invariant). It is thus very fortunate that λ and related quantities are experimentally accessible. [We shall see below how they can be estimated. See also the paper by Grassberger and Procaccia (1983a).]

The Liapunov exponents, the entropy, and the Hausdorff dimension associated with an attractor or an ergodic measure ρ all are related to how excited and how chaotic a system is (how many degrees of freedom play a role, and how much sensitivity to initial conditions is present). Let us see by an example how entropy (information production) is related to sensitive dependence on initial conditions.

We consider the dynamical system given by $f(x)=2x \bmod 1$ for $x\in[0,1)$. (This is "left-shift with leading digit truncation" in binary notation.) This map has sensitivity to initial conditions, and $\lambda=\log 2$. Assume now that our measuring apparatus can only distinguish between $x<\frac{1}{2}$ and $x>\frac{1}{2}$. Repeated measurements in time will nevertheless yield eventually all binary digits of the initial point, and it is in this sense that information is produced as "time" (i.e., the number of iterations) goes on. Thus changes of initial condition may be unobservable at time zero, but become observable at some later time. If we denote by ρ the Lebesgue measure on [0,1), then ρ is an invariant measure, and the corresponding mean information produced per unit time is exactly one bit. More generally, the average rate $h(\rho)$ of information production in an ergodic state ρ is related to sensitive dependence on initial conditions. [The quantity $h(\rho)$ is called the entropy of the measure ρ; see Sec. IV.] It may be bounded in terms of the characteristic exponents, and one finds

$$h(\rho)\leq\sum \text{postive characteristic exponents} \ . \tag{1.11}$$

In fact, in many cases (but not all), when a physical measure ρ may be identified, we have Pesin's formula (Pesin, 1977):

$$h(\rho)=\sum \text{positive characteristic exponents} \ . \tag{1.12}$$

Another quantity of interest is the *Hausdorff dimension.* [This quantity has been brought very much to the attention of physicists by Mandelbrot (1982), who uses the term *fractal dimension.* This is also used as a sort of generic name for different mathematical definitions of dimension for "fractal" sets.] The dimension of a set is roughly the amount of information needed to specify points on it accurately. For instance, let S be a compact set and assume that $N(\varepsilon)$ balls of radius ε are needed to cover S. Then a dimension $\dim_K S$, the "capacity" of S, is defined by

$$\dim_K S=\limsup_{\varepsilon\to 0}\log N(\varepsilon)/|\log\varepsilon| \ .$$

[This is a little less than requiring $N(\varepsilon)\varepsilon^d\to$finite, which means that the "volume" of the set S is finite in dimension d.] Mañé (1981) has shown that the points of S can be parametrized by m real coordinates as soon as $m\geq 2\dim_K S+1$.

The definition of the Hausdorff dimension $\dim_H S$ is slightly more complicated than that of $\dim_K S$; it does not assume that S is compact (see Sec. II.J). We next define the *information dimension* $\dim_H\rho$ of a probability measure ρ as the minimum of the Hausdorff dimensions of the sets S for which $\rho(S)=1$. It is not *a priori* clear that sets defined by dynamical systems have *locally* the *same* Hausdorff dimension everywhere, but this follows from the ergodicity of ρ in the case of $\dim_H\rho$. A result of Young [see Eq. (1.13) below] permits in many cases the evaluation of $\dim_H\rho$. Starting from different ideas, Grassberger and Procaccia (1983a,1983b) have arrived at a very similar way of computing the information dimension $\dim_H\rho$ of the measure ρ. Their proposal has been extremely successful, and has been used to measure reproducibly dimensions of the order of 3–10 in hydrodynamical experiments (see, for example, Malraison *et al.*, 1983).

We present some details of the method. Let $\rho[B_x(r)]$ be the mass of the measure ρ contained in a ball of radius r centered at x, and assume that the limit

$$\lim_{r\to 0}\frac{\log\rho[B_x(r)]}{\log r}=\alpha \tag{1.13}$$

exists for ρ-almost all x. The existence of the limit implies that it is constant, by the ergodicity of ρ. Under these conditions, α is equal to the information dimension $\dim_H\rho$, as noted by Young. In an experimental situation, one takes N points $x(i)$, regularly spaced in time, on an orbit of the dynamical system, and estimates $\rho[B_{x(i)}(r)]$ by

$$\frac{1}{N}\sum_{j=1}^{N}\Theta[r-|x(j)-x(i)|] \quad (N \text{ large}) \ , \tag{1.14}$$

where $\Theta(u)=(1+\operatorname{sgn}u)/2$. This permits us in principle to test the existence of the limit. In practice (Grassberger and Procaccia, 1983a,1983b) one defines

$$C(r)=\frac{1}{N^2}\sum_{i,j}\Theta[r-|x(j)-x(i)|] \quad (N \text{ large}) , \tag{1.15}$$

$$\text{information dimension}=\lim_{r\to 0}\frac{\log C(r)}{|\log r|} . \tag{1.16}$$

The problem of associating an orbit in $\mathbf{R}^m$ with experimental results will be discussed later. We also postpone discussion of relations between the Hausdorff dimension and characteristic exponents [such relations are described in the work of Frederickson, Kaplan, Yorke, and Yorke (1983); Douady and Oesterlé (1980); and Ledrappier (1981a)].

One may ask to what extent the definition of the above quantities is more than wishful thinking: is there any chance that the dimensions, exponents, and entropies about which we have been talking are finite numbers? For the case of the Navier-Stokes equation,

$$\frac{\partial v_i}{\partial t}=-\sum_j v_j\partial_j v_i+\nu\Delta v_i-\frac{1}{d}\partial_i p+g_i , \tag{1.17}$$

with the incompressibility conditions $\sum\partial_j v_j=0$, one has some comforting results given below. [Note that, in the case of two-dimensional hydrodynamics, one has good existence and uniqueness results for the solutions to Eq. (1.17). Assuming the same to be true in three dimensions (for reasonable physical situations), the conclusions given below for the two-dimensional case will carry over.]

Consider the Navier-Stokes equation in a bounded domain $\Omega\subset\mathbf{R}^d$, where $d=2$ or 3 is the spatial dimension. For *every* invariant measure ρ one has the following relations between the energy dissipation ε (per unit volume and time) and the ergodic quantities described earlier:

$$h(\rho)\le\sum_{\lambda_i\ge 0}\lambda_i\le\frac{B_d}{\nu^{1+d}}\left\langle\int_\Omega\varepsilon^{(2+d)/4}\right\rangle , \tag{1.18}$$

$$\dim_H\rho\le B_d'\frac{|\Omega|^{2/(d+2)}}{\nu^{d/2}}\left\langle\int_\Omega\varepsilon^{(2+d)/4}\right\rangle^{d/(d+2)} , \tag{1.19}$$

where B_d,B_d' are universal constants (see Ruelle, 1982b,1984, and Lieb, 1984, for a detailed discussion of these inequalities). Thus, if some average dissipation is finite, then all of these quantities are finite. In two dimensions, if the average dissipation is finite, i.e., if the *power* pumped into the system is finite, then $h(\rho)$ and $\dim_H\rho$ are also finite. In three dimensions, the situation is less clear because the average of $\int\varepsilon^{5/4}$ occurs instead of the average of $\int\varepsilon$. The lack of an existence and uniqueness theorem is in fact related to this difficulty. Experimentally, however, one finds that $\dim_H\rho$ is finite (implying that there are only finitely many $\lambda_i>0$).

To conclude, let us remark that the dynamical theory of physical systems is a rather mathematical subject, in the sense that it appeals to difficult mathematical theories and results. On the other hand, these mathematical theories still have many loose ends. One might thus be tempted either to disregard rigorous mathematics and go ahead with the physics, or on the contrary to wait until the mathematical situation is sufficiently clarified before going ahead with the physics. Both attitudes would be unfortunate. We believe in the value of the interplay between mathematics and physics, although either discipline offers only incomplete results. A mathematical theorem can prevent us from making "intuitive" assumptions that are already proved to be invalid. On the other hand, the relation between the two disciplines can help us to formulate mathematical conjectures which are made plausible on the basis of our experience as physicists. We are fortunate that the theory of dynamical systems has reached a stage where this kind of attitude seems especially fruitful.

The following are a few general references which are of interest in relation to the topics discussed in the present paper. (These references include books, conference proceedings, and reviews.)

Abraham, Gollub, and Swinney (1984): An overview of the experimental situation.

Bergé, Pomeau, and Vidal (1984): A very nice physics-oriented introduction, to be translated into English.

Bowen (1975): A more advanced introduction, stressing the ergodic theory of hyperbolic systems.

Campbell and Rose (1983): Los Alamos conference.

Collet and Eckmann (1980): A monograph, mostly on maps of the interval.

Cvitanović (1984): A very useful reprint collection.

Eckmann (1981): Review article on the geometric aspects of dynamical systems theory.

Ghil, Benzi, and Parisi (1985): Summer school proceedings on turbulence and predictability in geophysics.

Guckenheimer and Holmes (1983): An easy introduction to differential dynamical systems, oriented towards chaos.

Gurel and Rössler (1979): N.Y. Academy Conference.

Helleman (1980): N.Y. Academy Conference. These two conferences played an important historical role.

Iooss, Helleman, and Stora (1981): Proceedings of a summer school in Les Houches, 1981, with many interesting lectures.

Nobel symposium on chaos (1985).

Shaw (1981): A nice intuitive introduction to the information aspects of chaos.

Vidal and Pacault (1981): Conference proceedings on chemical turbulence.

Young (1984): A brief, but excellent, exposition of the inequalities for entropy and dimension.

II. DIFFERENTIABLE DYNAMICS AND THE RECONSTRUCTION OF DYNAMICS FROM AN EXPERIMENTAL SIGNAL

A. What is a differentiable dynamical system?

A differentiable dynamical system is simply a time evolution defined by an evolution equation

$$\frac{dx}{dt}=F(x) \tag{2.1}$$

(continuous-time case) or by a map

$$x(n+1)=f(x(n)) \tag{2.2}$$

(discrete-time case), where f or F are *differentiable* functions. In other words, f or F have continuous first-order derivatives. We may require f or F to be twice differentiable or more, i.e., to have continuous derivatives of second or higher order. Differentiability (possibly of higher order) is also referred to as *smoothness.* The physical justification for the assumed continuity of the derivatives of f or F is simply that physical quantities are usually continuous (small causes produce small effects). This philosophy, however, should, not be adhered to blindly (see Sec. III.D.2).

One introduces the nonlinear time-evolution operators f^t, t real or integer, requiring sometimes $t\geq 0$. They have the property

$$f^0=\text{identity}, \quad f^s f^t=f^{s+t} .$$

The variable x varies over the *phase space* M, which is $\mathbf{R}^m$, or a manifold like a sphere or a torus, or infinite dimensional (Banach spaces, in particular Hilbert spaces, are important in hydrodynamics). If M is a linear space, we define the linear operator $D_x f^t$ (matrix of partial derivatives of f at x, or a bounded operator if M is a Banach space). Writing $f^1=f$, we have

$$D_x f^n=D_{f^{n-1}x} f \cdots D_{fx} f D_x f \tag{2.3}$$

by the chain rule.

Example.

A viscous fluid in a bounded container $\Omega\subset\mathbf{R}^2$ or $\mathbf{R}^3$ is described by the Navier-Stokes equation

$$\frac{\partial v_i}{\partial t}=-\sum_j v_j \partial_j v_i+\nu\Delta v_i-\frac{1}{d}\partial_i p+g_i , \tag{2.4}$$

where (v_i) is the velocity field in Ω, ν a constant (the kinematic viscosity), d the (constant) density, p the pressure, and g an external force field. We add to Eq. (2.4) the incompressiblity condition

$$\sum_j \partial_j v_j=0 , \tag{2.5}$$

which expresses that v_i is divergence free, and we impose $v_i=0$ on $\partial\Omega$ (the fluid sticks to the boundary). Note that the divergence-free vector fields are orthogonal to gradients, so that one can eliminate the pressure from Eq. (2.4) by orthogonal projection of the equation on the divergence-free fields. One obtains thus an equation of the type (2.1) where M is the Hilbert space of square-integrable vector fields which are orthogonal to gradients. In two dimensions (i.e., for $\Omega\subset\mathbf{R}^2$), one has a good existence and uniqueness theorem for solutions of Eqs. (2.4) and (2.5), so that f^t is defined for t real ≥ 0 (Ladyzhenskaya, 1969; Foias and Temam, 1979; Temam, 1979). In three dimensions one has only partial results (Caffarelli, Kohn, and Nirenberg, 1982).

B. Dissipation and attracting sets

For a *conservative system* (Hamiltonian time evolution), Liouville's theorem says that the volume in phase space M is conserved by the time evolution. We shall be mainly interested in *dissipative systems,* for which this is not the case and for which the volume is usually contracted. Let us therefore assume that there is an open set U in M which is contracted by time evolution asymptotically to a compact set A. To be precise, we say that A is an *attracting set* with *fundamental neighborhood* U if (a) for every open set $V\supset A$ we have $f^tU\subset V$ when t is large enough, and (b) $f^tA=A$ for all t. (See Fig. 2.) The open set $\bigcup_{t>0}(f^t)^{-1}U$ is the *basin of attraction* of A. If the basin of attraction of A is the whole of M, we say that A is the *universal attracting set.*

Examples.

(a) If U is an open set in M, and the closure of f^tU is compact and contained in U for all sufficiently large t, then the set $A=\bigcap_{t\geq 0}f^tU$ is a (compact) attracting set with fundamental neighborhood U (see Ruelle, 1981).

(b) The Lorenz time evolution in $\mathbf{R}^3$ is defined by the equation

$$\frac{d}{dt}\begin{bmatrix} x_1 \\ x_2 \\ x_2 \end{bmatrix}=\begin{bmatrix} -\sigma x_1+\sigma x_2 \\ -x_1x_3+rx_1-x_2 \\ x_1x_2-bx_3 \end{bmatrix}, \tag{2.6}$$

with $\sigma=10$, $b=\frac{8}{3}$, $r=28$ (see Lorenz, 1963). If U is a sufficiently large ball, [i.e., $U=\{(x_1,x_2,x_3):\sum x_i^2\leq R^2\}$ with large R], then U is mapped into itself by time evolution. It contains thus an attracting set A, and A is universal (see Fig. 3).

(c) The Navier-Stokes time evolution in two dimensions also gives rise to a universal attracting set A, because one can again apply (a) to a sufficiently large ball (in a suitable Hilbert space). It can be shown that A has finite dimension (see Mallet-Paret, 1976).

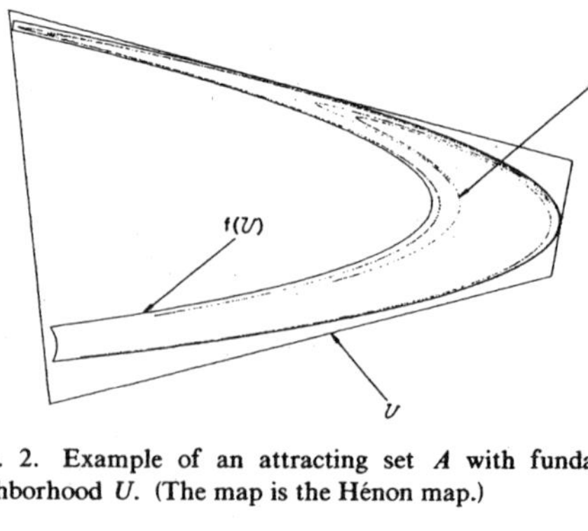

FIG. 2. Example of an attracting set A with fundamental neighborhood U. (The map is the Hénon map.)

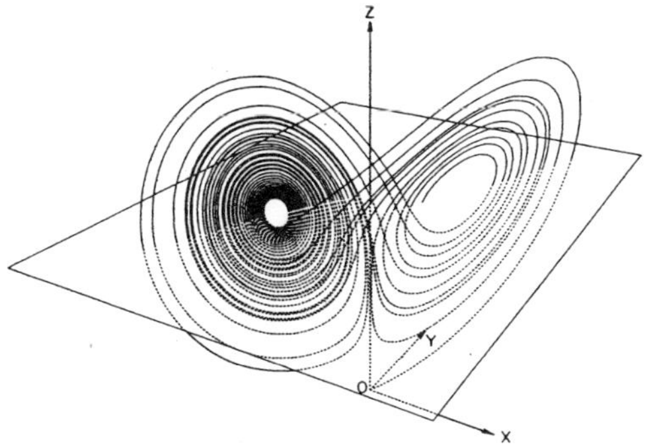

FIG. 3. The Lorenz attractor. From Lanford (1977).

C. Attractors

Physical experiments and computer experiments with dynamical systems usually exhibit transient behavior followed by what seems to be an asymptotic regime. Therefore the point $f^t x$ representing the system should eventually lie on an attracting set (or near it). However, in practice *smaller* sets, which we call *attractors,* will be obtained (they should be carefully distinguished from attracting sets). This is because some parts of an attracting set may not be attracting (Fig. 4).

We should also like to include in the mathematical definition of an attractor A the requirement of irreducibility (i.e., the union of two disjoint attractors is not considered to be an attractor). This (unfortunately) implies that one can no longer impose the requirement that there be an *open* fundamental neighborhood U of A such that $f^t U \to A$ when $t\to\infty$. Instead of trying to give a precise mathematical definition of an attractor, we shall use here the *operational* definition, that it is a set on which *experimental points* $f^t x$ accumulate for large t. We shall come back later to the significance of this operational definition and its relation to more mathematical concepts.

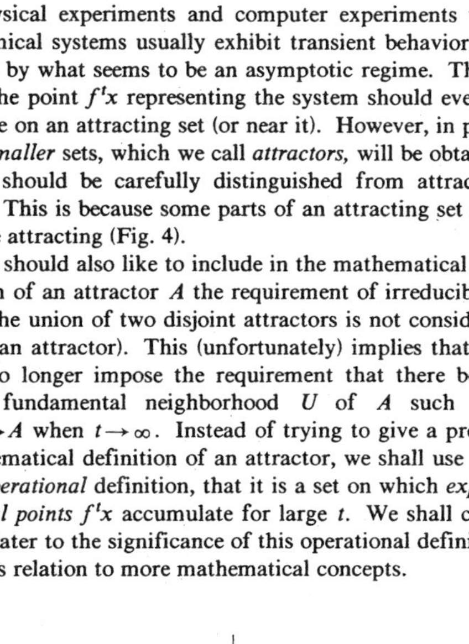

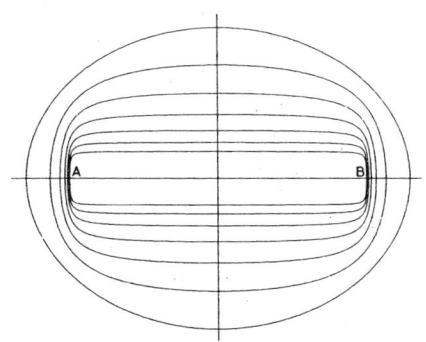

FIG. 4. The dynamical system is $\dot{x}_1 = x_1 - x_1^3$, $\dot{x}_2 = -x_2$. The segment A,B is the universal attracting set, but only the points A,B are attractors. In other words, the whole space is attracted to the segment A,B but only A and B are attractors.

Examples.

(a) Attracting fixed point. Let P be a *fixed point* for our dynamical system, i.e., $f^t P = P$ for all t. The derivative $D_P f^1$ of f^1 (time-one map) at the fixed point is an $m\times m$ matrix or an operator in Hilbert space. If its spectrum is in a disk $\{z: |z| < \alpha\}$ with $\alpha<1$, then P is an attracting fixed point. It is an attracting set (and an attractor). When the time evolution is defined by the differential equation (1.1) in $\mathbb{R}^m$, the attractiveness condition is that the eigenvalues of $D_P F_\mu$ all have a negative real part. For a discrete-time dynamical system, we say that $(P_1,\ldots,P_n)$ is an attracting periodic orbit, of period n, if $fP_1 = P_2, \ldots, fP_n = P_1$, and P_i is an attracting fixed point for f^n.

(b) Attracting periodic orbit for continuous time. For a continuous-time dynamical system, suppose that there are a point a and a $T>0$, such that $f^T a = a$ but $f^t a \neq a$ when $0<t<T$. Then a is a periodic point of period T, and $\Gamma = \{f^t a : 0 \le t < T\}$ is the corresponding *periodic orbit* (or closed orbit). The derivative $D_a f^T$ has an eigenvalue 1 corresponding to the direction tangent to Γ at a. If the rest of the spectrum is in $\{z: |z| < \alpha\}$ with $\alpha<1$, then Γ is an attracting periodic orbit. It is again an attracting set and an attractor. The attracting character of a periodic orbit may also be studied with the help of a Poincaré section (see Sec. II.H).

(c) Quasiperiodic attractor. A periodic orbit for a continuous system is really a circle, and the motion on it (by proper choice of coordinate φ) may be written

$$\varphi(t) = \varphi(0) + \omega t \pmod{2\pi}, \tag{2.7}$$

where $\omega = 2\pi/T$. This may be thought of as representing the time evolution of a simple oscillator. Consider now a collection of k oscillators with frequencies $\omega_1,\ldots,\omega_k$ (without rational relations between the ω_i: no linear combination with nonzero integer coefficients vanishes). The motion of the oscillators is described by

$$\varphi_i(t) = \varphi_i(0) + \omega_i t \pmod{2\pi},\ i=1,\ldots,k, \tag{2.8}$$

and this motion takes place on the product of k circles, $(k>1)$, which is a k-dimensional torus T^k. Suppose that the torus T^k is embedded in $\mathbb{R}^m$, $m \ge k$ (or in Hilbert space), as the periodic orbit Γ was in the previous example; suppose, furthermore, that this torus is an attracting set. Then we say that T^k is a quasiperiodic attractor. Asymptotically, the dynamical system will thus be described by

$$x(t) = f^t x = \Phi[\varphi_1(t),\ldots,\varphi_k(t)] \tag{2.9}$$

$$= \Psi(\omega_1 t, \ldots, \omega_k t), \tag{2.10}$$

where Ψ is periodic, of period 2π, in each argument. A function of the form $t \to \Psi(\omega_1 t,\ldots,\omega_k t)$ is known as a *quasiperiodic function* (with k different periods). Quasiperiodic attractors are a natural generalization of periodic orbits, and they occur fairly frequently in the description of moderately excited physical systems.

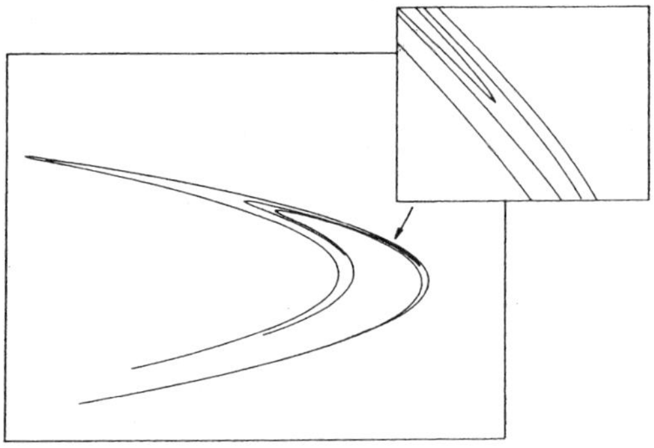

FIG. 5. The Hénon attractor for $a=1.4$, $b=0.3$. The successive iterates f^k of f have been applied to the point (0,0), producing a sequence asymptotic to the attractor. Here, 30 000 points of this sequence are plotted, starting with $f^{20}(0,0)$.

D. Strange attractors

The attractors discussed under (a), (b), and (c) above are also attracting sets. They are nice manifolds (point, circle, torus). Notice also that, if a small change $\delta x(0)$ is made to the initial conditions, then $\delta x(t)=(D_x f^t)\delta x(0)$ remains small when $t \to \infty$. [In fact, for a quasiperiodic motion, Eq. (2.7) gives $\delta\varphi(t)=\delta\varphi(0)$.] We shall now discuss more complicated situations.

Examples.

(a) Hénon attractor (Hénon, 1976; Feit, 1978; Curry, 1979). Consider the discrete-time dynamical system defined by

$$f\begin{pmatrix} x_1 \\ x_2 \end{pmatrix} = \begin{pmatrix} 1+x_2-ax_1^2 \\ bx_1 \end{pmatrix} \tag{2.11}$$

and the corresponding attractor, for $a=1.4$, $b=0.3$ (see Fig. 5). One finds here numerically that

$$\delta x(t) \approx \delta x(0) e^{\lambda t}, \quad \lambda=0.42 ,$$

i.e., the errors grow exponentially. This is the phenomenon of *sensitive dependence on initial conditions.* In fact (Curry, 1979), computing the successive points $f^n x$ for $n=1,2,\ldots$, with 14 digits' accuracy, one finds that the error of the sixtieth point is of order 1. Sensitive dependence on initial conditions is also expressed by saying that the system is *chaotic* [this is now the accepted use of the word *chaos,* even though the original use by Li and Yorke (1975) was somewhat different].

(b) Feigenbaum attractor (Feigenbaum, 1978,1979,1980; Misiurewicz, 1981; Collet, Eckmann, and Lanford, 1980). A map of the interval [0,1] to itself is defined by

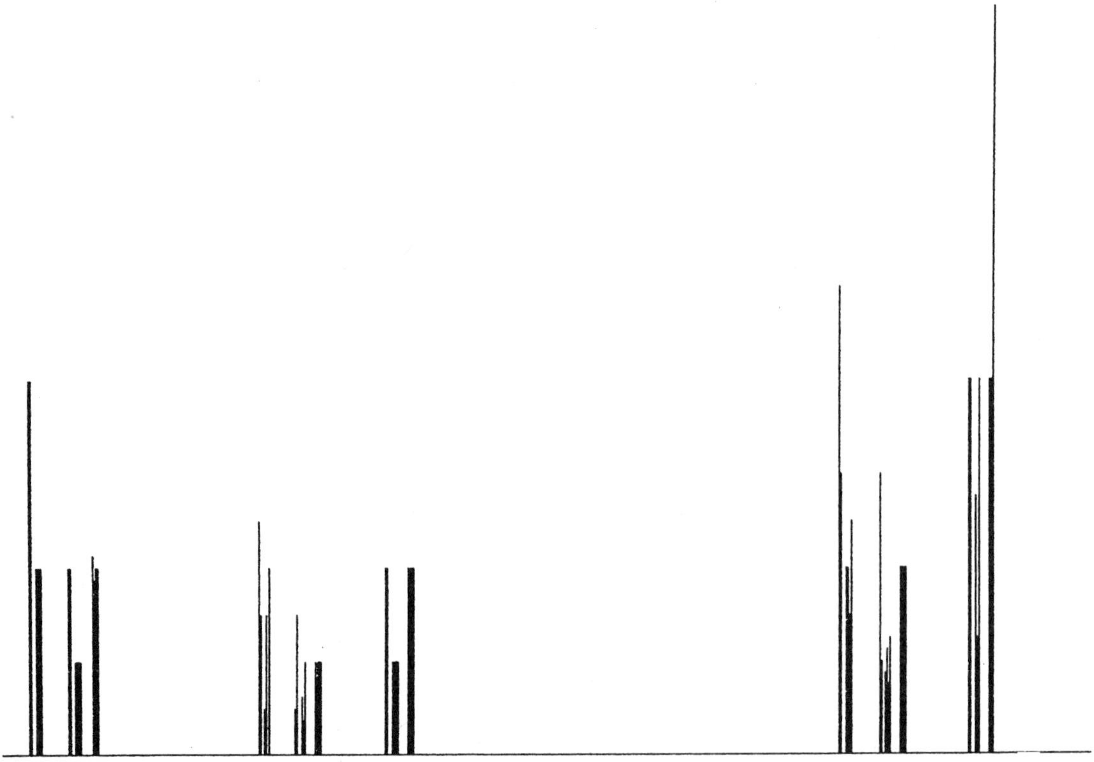

FIG. 6. The Feigenbaum attractor. Histogram of 50 000 points in 1024 bins. This histogram shows the unique ergodic measure, which is clearly singular.

$$f_\mu(x) = \mu x(1-x) \tag{2.12}$$

when $\mu \in [0,4]$. It has attracting periodic orbits of period 2^n, with n tending to infinity as μ tends to 3.57... through lower values. For the limiting value $\mu = 3.57\ldots$, there is a very special attractor A shown in Fig. 6. We shall call it the Feigenbaum attractor (although it was known earlier to many authors). Note that interspersed with this attractor, and arbitrarily close to it, there are repelling periodic orbits of period 2^n, for all n. Therefore the attractor A cannot be an attracting set. One can show, moreover, that, for this very special attractor, there is no sensitive dependence on initial conditions (no exponential growth of errors): the Feigenbaum attractor is not chaotic.

The Hénon and Feigenbaum attractors, as depicted in Figs. 5 and 6, have a complicated aspect typical of *fractal* objects. In general, a fractal set is a set for which the Hausdorff dimension is different from the topological dimension, and usually not an integer. (The exact definition of the Hausdorff dimension is given in Sec. II.J.) The name *fractal* was coined by Mandelbrot. For the rich lore of fractal objects, see Mandelbrot (1982). While many attractors are fractals, and therefore complicated objects, they are by no means featureless. They are unions of *unstable manifolds* (to be defined in Sec. III.E) and often have a Cantor-set structure in the direction transversal to the unstable manifolds. (For the Feigenbaum attractor the unstable manifolds have dimension 0, and only a Cantor set is visible; for the Hénon attractor the unstable manifolds have dimension 1.) An attractor is by definition invariant under a dynamical evolution, and this creates a self-similarity that is often strikingly visible.

In view of both its chaotic and fractal characters, the Hénon attractor deserves to be called a *strange attractor* (this name was introduced by Ruelle and Takens, 1971). The property of being chaotic is actually a more important dynamical concept than that of being fractal, and we shall therefore say that the Feigenbaum attractor is *not* a strange attractor (this differs somewhat from the point of view in Ruelle and Takens). We therefore define a strange attractor to be an attractor with *sensitive dependence on initial conditions.* The notion of strangeness refers thus to the *dynamics* on the attractor, and *not just to its geometry;* it applies whether the time is discrete or continuous. This is again an operational definition rather than a mathematical one. We shall see in Sec. III what should be clarified mathematically. For physics, however, the above operational concept of strange attractors has served well and deserves to be kept.

Example.

(c) Thom's toral automorphisms and Arnold's cat map. Let x_1 (mod1) and x_2 (mod1) be coordinates on the 2-torus T^2; a map $f: T^2 \to T^2$ is defined by

$$f\begin{pmatrix} x_1 \\ x_2 \end{pmatrix} = \begin{pmatrix} x_1 + x_2 \\ x_1 + 2x_2 \end{pmatrix} \pmod 1 . \tag{2.13}$$

[Because $\det\binom{1\,1}{1\,2} = 1$, the map $\mathbb{R}^2 \to \mathbb{R}^2$ defined by the matrix $\binom{1\,1}{1\,2}$ maps Z^2 to Z^2 and therefore, going to the quotient $T^2 = \mathbb{R}^2/Z^2$, a map $f: T^2 \to T^2$ of the 2-torus to itself is defined. The system is area preserving, and the whole torus is an "attractor."] This is Arnold's celebrated cat map (Arnold and Avez, 1967), well known to be chaotic (see Fig. 7). In fact we have here

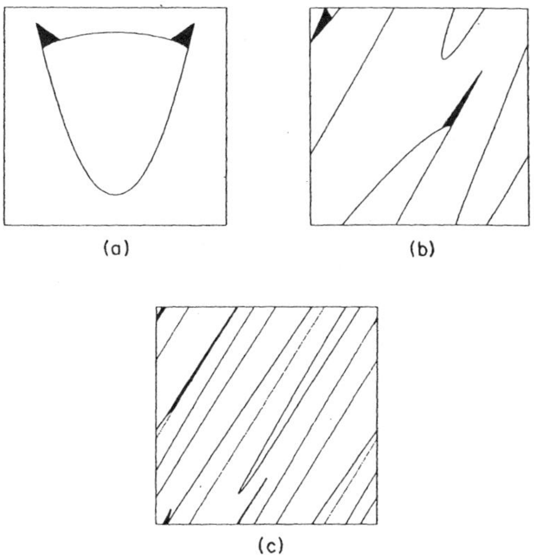

FIG. 7. Arnold's "cat map": (a) the cat; (b) its image under the first iterate; (c) image under the second iterate of the cat map.

$$\delta x(t) \approx \delta x(0) e^{\lambda t} , \tag{2.14}$$

with

$$\lambda = \log \frac{3+\sqrt{5}}{2} ,$$

$(3+\sqrt{5})/2$ being the eigenvalue larger than 1 of the matrix $\binom{1\,1}{1\,2}$.

More generally, if V is an $m \times m$ matrix with integer entries and determinant ± 1, it defines a toral automorphism $T^m \to T^m$, and Thom noted that these automorphisms have sensitive dependence on initial conditions if V has an eigenvalue α with $|\alpha| > 1$.

Returning to Arnold's cat map, we may imbed T^2 as an attractor A in a higher-dimensional Euclidean space. In this case A is chaotic, but not fractal.

Our examples clearly show that the notions of fractal attractor and chaotic (i.e., strange) attractor are independent. A periodic orbit is neither strange nor chaotic, Arnold's map is strange but not fractal, Feigenbaum's attractor is fractal but not strange, the Lorenz and Hénon attractors are both strange and fractal. [Another strange and fractal attractor with a simple equation has been introduced by Rössler (1976).]

E. Invariant probability measures

An attractor A, be it strange or not, gives a global picture of the long-term behavior of a dynamical system. A more detailed picture is given by the probability measure ρ on A, which describes how frequently various parts of A are visited by the orbit $t \to x(t)$ describing the system (see Fig. 8). Operationally, ρ is defined as the time average of Dirac deltas at the points $x(t)$,

$$\rho = \lim_{t\to\infty} \frac{1}{T} \int_0^T dt\, \delta_{x(t)} \,. \tag{2.15}$$

Similarly, if a continuous function φ is given, then we define

$$\rho(\varphi) \equiv \int \rho(dx)\varphi(x) = \lim_{T\to\infty} \frac{1}{T} \int_0^T dt\, \varphi[x(t)] \,. \tag{2.16}$$

The measure is *invariant* under the dynamical system, i.e., invariant under time evolution. This invariance may be expressed as follows: For all φ one has

$$\rho(\varphi \circ f^t) = \rho(\varphi) \,. \tag{2.17}$$

Suppose that the invariant probability measure ρ cannot be written as $\frac{1}{2}\rho_1 + \frac{1}{2}\rho_2$ where ρ_1, ρ_2 are again invariant probability measures and $\rho_1 \neq \rho_2$. Then ρ is called *indecomposable,* or equivalently, *ergodic.*

Theorem. If the compact set A is invariant under the dynamical system (f^t), then there is a probability measure ρ invariant under (f^t) and with support contained in A. One may choose ρ to be ergodic.

[The important assumptions are that the f^t commute and are continuous $A \to A$ (A compact). The theorem results from the Markov-Kakutani fixed-point theorem (see Dunford and Schwartz, 1958, Vol. I).] This is not a very detailed result; it is more in the class referred to as "general nonsense" by mathematicians. But since we shall talk a lot about ergodic measures in what follows, it is good to know that such measures are indeed present.

Theorem (Ergodic theorem). If ρ is ergodic, then for ρ almost all initial $x(0)$ the time averages (2.15) and (2.16) reproduce ρ.

The above theorems show that there are invariant (ergodic) measures defined by time averages. Unfortunately,

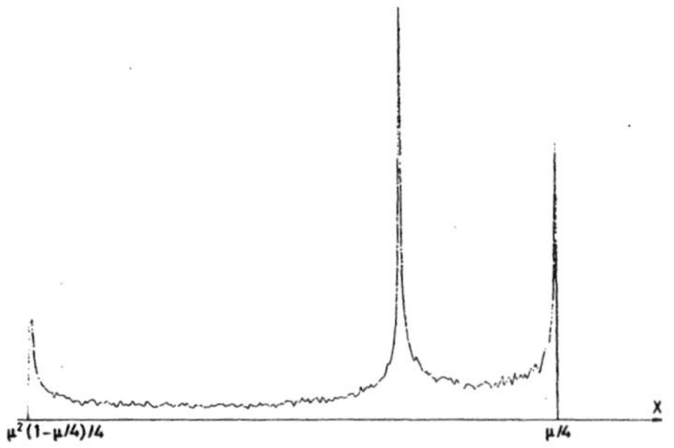

FIG. 8. Histogram of 50 000 iterates of the map $x \to \mu x(1-x)$, in 400 bins. The parameter $\mu = 3.678\,57\ldots$ is the real solution of the equation $(\mu-2)^2(\mu+2) = 16$. It is known that the invariant density is smooth with square-root singularities.

a strange attractor typically carries *uncountably many* distinct ergodic measures. Which one do we choose? We shall propose natural definitions in the next section.

Example.

The points of the circle T^1 may be parametrized by numbers in [0,1), and each such number has a binary expansion $0.a_1a_2a_3\cdots$, where, for each i, $a_i = 0$ or 1 (this coding introduces a little ambiguity, of no importance for what follows). We define a map $f: T^1 \to T^1$ by

$$f(x) = 2(x) \pmod 1 \,. \tag{2.18}$$

Clearly, f replaces $0.a_1a_2a_3\cdots$ by $0.a_2a_3\cdots$ (an operation called a *shift).* We now choose p between 0 and 1. A probability distribution ρ_p on binary expansions $0.a_1a_2a_3\cdots$ is then defined by requiring that a_i be 0 with probability p, and 1 with probability $1-p$ (independently for each i). One can check that ρ_p is invariant under the shift, and in fact ergodic. It thus defines an ergodic measure for the differentiable dynamical system (2.18), $f: T^1 \to T^1$, and there are uncountably many such measures, corresponding to the different values of p in (0,1).

F. Physical measures

Operationally, it appears that (in many cases, at least) the time evolution of physical systems produces well-defined time averages. The same applies to computer-generated time evolutions. There is thus a selection process of a particular measure ρ which we shall call *physical measure* (another operational definition).

One selection process was discussed by Kolmogorov (we are not aware of a published reference) a long time ago. A physical system will normally have a small level ε of random noise, so that it can be considered as a stochastic process rather than a deterministic one. In a computer study, roundoff errors should play the role of the random noise. Due to sensitive dependence on initial conditions, even a very small level ε of noise has important effects, as we saw in Sec. II.D for the Hénon attractor. On the other hand, a stochastic process such as the one described above normally has only one stationary measure ρ_ε, and we may hope that ρ_ε tends to a specific measure (the *Kolmogorov measure)* when $\varepsilon \to 0$. As we shall see below, this hope is substantiated in the case of *Axiom-A* dynamical systems. However, this approach may have difficulties in general, because an attractor A does not always have an open basin of attraction, and thus the added noise may force the system to jump around on several attractors.

Another possibility is the following: Suppose that M is finite dimensional, and that there is a set $S \subset M$ with Lebesgue measure $\mu(S) > 0$ such that ρ is given by the time averages (2.15) and (2.16) when $x(0) \in S$. This property holds if ρ is an *SRB measure* (to be defined and studied in Sec. IV.C; Sinai, 1972; Bowen and Ruelle, 1975; Ruelle, 1976). For Axiom-A systems, the Kolmogorov and SRB measures coincide, but in general SRB measures are easier to study.

Clearly, Kolmogorov measures and SRB measures are candidates for the description of physical time averages, but they are not always easy to define. Fortunately, many important results hold for an arbitrary invariant measure ρ. Results of this type, which constitute a large part of the ergodic theory of differentiable dynamical systems, will be discussed in Secs. III and IV of this paper.

G. Reconstruction of the dynamics from an experimental signal

In a computer study of a dynamical system in m dimensions, we have an m-dimensional signal $x(t)$, which can be submitted to analysis. By contrast, in a physical experiment one monitors typically only one scalar variable, say $u(t)$, for a system that usually has an infinite-dimensional phase space M. How can we hope to understand the system by analyzing the single scalar signal $u(t)$? The enterprise seems at first impossible, but turns out to be quite doable. This is basically because (a) we restrict our attention to the dynamics on a finite-dimensional attractor A in M, and (b) we can generate several different scalar signals $x_i(t)$ from the original $u(t)$. We have already mentioned that the universal attracting set (which contains all attractors) has finite dimension in two-dimensional hydrodynamics, and we shall come back later to this question of finite dimensionality.

The easiest, and probably the best way of obtaining several signals from a single one is to use *time delays*. One chooses different delays $T_1=0,T_2,\ldots,T_N$ and writes $x_k(t)=u(t+T_k)$. In this manner an N-dimensional signal is generated. The experimental points in Fig. 9 below have been obtained by this method. Successive time derivatives of the signal have also been used: $x_{k+1}(t)=d^k x_1(t)/dt^k$, but the numerical differentiations tend to produce high levels of noise. Of course one should measure several experimental signals instead of only one whenever possible.

The reconstruction just outlined will provide an N-dimensional image (or projection) πA of an attractor A which has finite Hausdorff dimension, but lives in a usually infinite-dimensional space M. Depending on the choice of variables (in particular on the time delays), the projection will look different. In particular, if we use fewer variables than the dimension of A, the projection πA will be bad, with trajectories crossing each other. There are some theorems (Takens, 1981; Mañé, 1981) which state that if we use enough variables, typically about twice the Hausdorff dimension, we shall generally get a good projection.

Theorem (Mañé). Let A be a compact set in a Banach space B, and E a subspace of finite dimension such that

$$\dim E > \dim_H(A\times A)+1 ,$$

or let A be compact and

$$\dim E > 2\dim_K(A)+1 ,$$

where $\dim_H$ is the Hausdorff dimension and $\dim_K$ is the capacity. Then the set of projections $\pi:B\to E$ such that π restricted to A is injective (i.e., one to one into E) is dense among all projections $B\to E$ with respect to the norm operator topology.

[More precisely, the injective projections are "residual," i.e., contain a countable intersection of dense sets. As noticed by Mañé, his original statement of the theorem needs a slight correction, which is made in the above formulation.]

The choice of variables for the reconstruction of a dynamical system has to be made carefully (by trial and error). This is discussed in Roux, Simoyi, and Swinney (1983).

H. Poincaré sections

The reconstruction process described above yields a line $(f^t x)_0^T$ that may look like a heap of spaghetti and may be difficult to interpret. It is often possible and useful to make a transverse cut through this mess, so that instead of a long curve in N dimensions one now has a set of points S in $N-1$ dimensions (Poincaré section). Figure 9 gives an experimental example corresponding to the Beloussov-Zhabotinski chemical reaction. Given a point x of the Poincaré section, the *first return map* will bring it to Px, which is again in the Poincaré section. When a good model of S and P can be deduced from the experiment, one has essentially understood the dynamical system. This is, however, possible only for low-dimensional attractors.

Notice that the use of a Poincaré section is different from a *stroboscopic* study, where one looks at the system at integer multiples of a *fixed* time interval. By contrast, the time of first return to the Poincaré section is variable

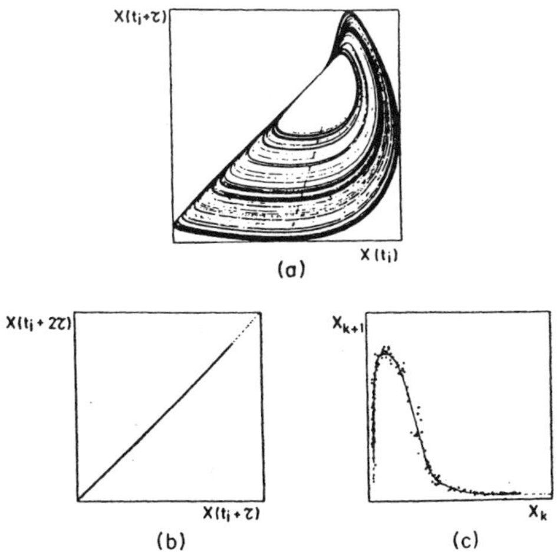

FIG. 9. Experimental plot of a Poincaré section in the Beloussov-Zhabotinski reaction, after Roux and Swinney (1981): (a) the attractor and the plane of Poincaré section; (b) the Poincaré section; (c) the corresponding first return map.

(and has to be determined numerically by interpolation). Sometimes, as for quasiperiodic motions, there is a *natural frequency* (or several) that will stabilize the stroboscopic image. But in general this is not the case, and therefore the stroboscopic study is useless.

I. Power spectra

The power spectrum $S(\omega)$ of a scalar signal $u(t)$ is defined as the square of its Fourier amplitude per unit time. Typically, it measures the amount of energy per unit time (i.e., the power) contained in the signal as a function of the frequency ω. One can also define $S(\omega)$ as the Fourier transform of the time correlation function $\langle u(0)u(t)\rangle$ equal to the average over τ of $u(\tau)u(t+\tau)$. If the correlations of u decay sufficiently rapidly in time, the two definitions coincide, and one has (Wiener-Khinchin theorem; see Feller, 1966)

$$S(\omega)=(\text{const})\lim_{T\to\infty}\frac{1}{T}\left|\int_0^T dt\, e^{i\omega t}u(t)\right|^2$$
$$=(\text{const})\int_{-\infty}^{\infty}dt\, e^{i\omega t}\lim_{T\to\infty}\frac{1}{T}\int_0^T d\tau' u(\tau')u(t+\tau')\,. \tag{2.19}$$

Note that the above limit (2.19) makes sense only after averaging over small intervals of ω. Without this averaging, the quantity

$$\frac{1}{T}\left|\int_0^T dt\, e^{i\omega t}u(t)\right|^2$$

fluctuates considerably, i.e., it is very noisy. (Instead of averaging over intervals of ω, one may average over many runs).

The power spectrum indicates whether the system is periodic or quasiperiodic. The power spectrum of a periodic system with frequency ω has Dirac δ's at ω and its harmonics $2\omega, 3\omega, \ldots$. A quasiperiodic system with basic frequencies $\omega_1, \ldots, \omega_k$ has δ's at these positions and also at all linear combinations with integer coefficients. (The choice of basic frequencies is somewhat arbitrary, but the number k of independent frequencies is well defined.) In experimental power spectra, the Dirac δ's are not infinitely sharp; they have at least an "instrumental width" $2\pi/T$, where T is the length of the time series used. The linear combinations of the basic frequencies $\omega_1, \ldots, \omega_k$ are dense in the reals if $k>1$, but the amplitudes corresponding to complicated linear combinations are experimentally found to be small. (A mathematical theory for this does not seem to exist.) A careful experi-

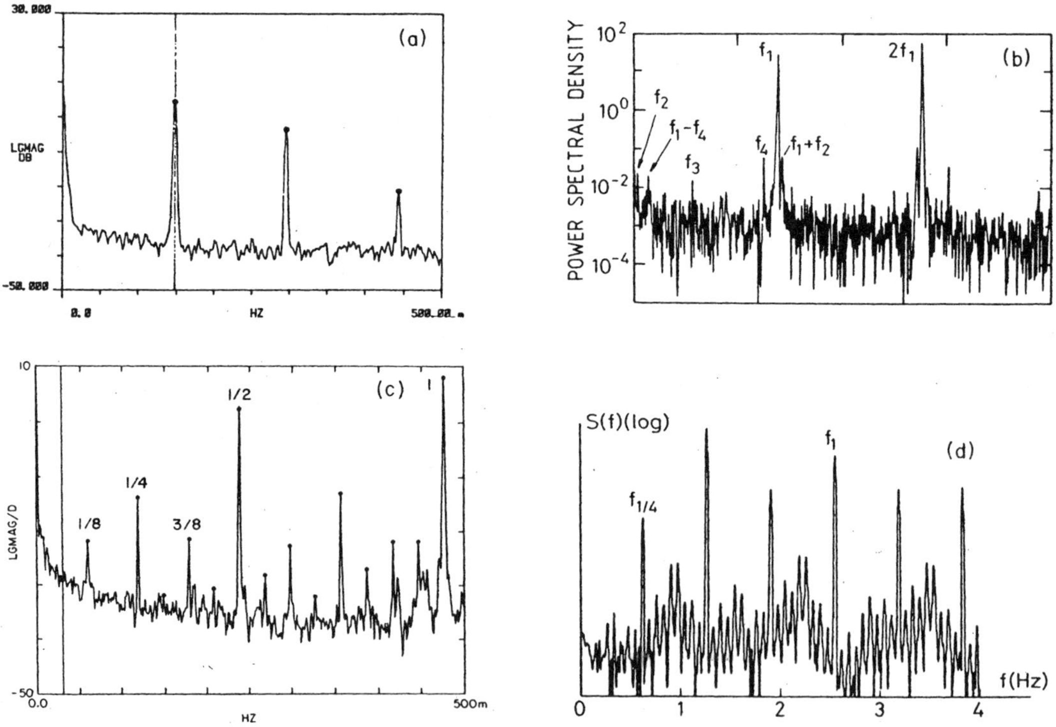

FIG. 10. Some spectra: (a) The power spectrum of a periodic signal shows the fundamental frequency and a few harmonics. Fauve and Libchaber (1982). (b) A quasiperiodic spectrum with four fundamental frequencies. Walden, Kolodner, Passner, and Surko (1984). (c) A spectrum after four period doublings. Libchaber and Maurer (1979). (d) Broadband spectrum invades the subharmonic cascade. The fundamental frequency and the first two subharmonics are still visible. Croquette (1982).

ment may show very convincing examples of quasiperiodic systems with two, three, or more basic frequencies. In fact, $k=2$ is common, and higher k's are increasingly rare, because the nonlinear couplings between the modes corresponding to the different frequencies tend to destroy quasiperiodicity and replace it by chaos (see Ruelle and Takens, 1971; Newhouse, Ruelle, and Takens, 1978). However, for weakly coupled modes, corresponding for instance to oscillators localized in different regions of space, the number of observable frequencies may become large (see, for instance, Grebogi, Ott, and Yorke, 1983a; Walden, Kolodner, Passner, and Surko, 1984). Nonquasiperiodic systems are usually chaotic. Although their power spectra still may contain peaks, those are more or less broadened (they are no longer instrumentally sharp). Furthermore, a noisy background of *broadband spectrum* is present. For this it is *not* necessary that the system be infinite dimensional [Figs. 10(a)–10(d)].

In general, power spectra are very good for the visualization of periodic and quasiperiodic phenomena and their separation from chaotic time evolutions. However, the analysis of the chaotic motions themselves does not benefit much from the power spectra, because (being squares of absolute values) they lose phase information, which is essential for the understanding of what happens on a strange attractor. In the latter case, as already remarked, the dimension of the attractor is no longer related to the number of independent frequencies in the power spectrum, and the notion of "number of modes" has to be replaced by other concepts, which we shall develop below.

J. Hausdorff dimension and related concepts*

Most concepts of dimension make use of a metric. Our applications are to subsets of $\mathbb{R}^m$ or Banach spaces, and the natural metric to use is the one defined by the norm.

Let A be a compact metric space and $N(r,A)$ the minimum number of open balls of radius r needed to cover A. Then we define

$$\dim_K A = \limsup_{r\to 0} \frac{\log N(r,A)}{\log(1/r)} .$$

This is the *capacity* of A (this concept is related to Kolmogorov's ε *entropy and has nothing to do with Newtonian capacity*). *If A and B are compact metric spaces, their product $A\times B$ satisfies*

$$\dim_K(A\times B)\le \dim_K A+\dim_K B . \tag{2.20}$$

Given a nonempty set A, with a metric, and $r>0$, we denote by σ a covering of A by a (countable) family of sets σ_k with diameter $d_k=\operatorname{diam}\sigma_k\le r$. Given $\alpha\ge 0$, we write

$$m_r^\alpha(A)=\inf_\sigma \sum_k (d_k)^\alpha .$$

When $r\downarrow 0$, $m_r^\alpha(A)$ increases to a (possibly infinite) limit $m^\alpha(A)$ called the *Hausdorff measure of A in dimension α*. We write

$$\dim_H A=\sup\{\alpha : m^\alpha(A)>0\}$$

and call this quantity the *Hausdorff dimension* of A. Note that $m^\alpha(A)=+\infty$ for $\alpha<\dim_H A$, and $m^\alpha(A)=0$ for $\alpha>\dim_H A$. The Hausdorff dimension of a set A is, in general, strictly smaller than the Hausdorff dimension of its closure. Furthermore, the inequality (2.20) on the dimension of a product does not extend to Hausdorff dimensions. It is easily seen that for every compact set A, one has $\dim_H A\le\dim_K A$.

If A and B are compact sets satisfying $\dim_H A=\dim_K A$, $\dim_H B=\dim_K B$, then

$$\dim_H(A\times B)=\dim_K(A\times B)=\dim_H A+\dim_H B .$$

We finally introduce a *topological* dimension $\dim_L A$. It is defined as the smallest integer n (or $+\infty$) for which the following is true: For every finite covering of A by open sets $\sigma_1,\dots,\sigma_N$ one can find another covering $\sigma_1',\dots,\sigma_N'$ such that $\sigma_i'\subset\sigma_i$ for $i=1,\dots,N$ and any $n+2$ of the σ_i' will have an empty intersection:

$$\sigma_{i_0}'\cap\sigma_{i_1}'\cap\cdots\cap\sigma_{i_{n+1}}'=\varnothing .$$

The quantity $\dim_L A$ is also called the *Lebesgue* or *covering dimension* of A.

For more details on dimension theory, see Hurewicz and Wallman (1948) and Billingsley (1965).

III. CHARACTERISTIC EXPONENTS

In this section we review the ergodic theory of differentiable dynamical systems. This means that we study invariant probability measures (corresponding to time averages). Let ρ be such a measure, and assume that it is ergodic (indecomposable). The present section is devoted to the *characteristic exponents of ρ* (also called *Liapunov exponents*) and related questions. We postpone until Sec. V the discussion of how these characteristic exponents can be measured in physical or computer experiments.

A. The multiplicative ergodic theorem of Oseledec

If the initial state of a time evolution is slightly perturbed, the exponential rate at which the perturbation $\delta x(t)$ increases (or decreases) with time is called a *characteristic exponent*. Before defining characteristic exponents for differentiable dynamics, we introduce them in an abstract setting. Therefore, we speak of measurable maps f and T, but the application intended is to continuous maps.

Theorem (multiplicative ergodic theorem of Oseledec). Let ρ be a probability measure on a space M, and $f:M\to M$ a measure preserving map such that ρ is ergodic. Let also $T:M\to$ the $m\times m$ matrices be a measurable map such that

$$\int \rho(dx)\log^+\|T(x)\|<\infty ,$$

where $\log^+ u=\max(0,\log u)$. Define the matrix

$T_x^n = T(f^{n-1}x) \cdots T(fx)T(x)$. Then, for ρ-almost all x, the following limit exists:

$$\lim_{n\to\infty} (T_x^{n*} T_x^n)^{1/2n} = \Lambda_x \ . \tag{3.1}$$

(We have denoted by T_x^{n*} the adjoint of T_x^n, and taken the $2n$th root of the positive matrix $T_x^{n*} T_x^n$.)

The logarithms of the eigenvalues of Λ_x are called *characteristic exponents.* We denote them by $\lambda_1 \geq \lambda_2 \geq \cdots$. They are ρ-almost everywhere constant. (This is because we have assumed ρ ergodic. Of course, the λ_i depend on ρ.) Let $\lambda^{(1)} > \lambda^{(2)} > \cdots$ be the characteristic exponents again, but no longer repeated by multiplicity; we call $m^{(i)}$ the multiplicity of $\lambda^{(i)}$. Let $E_x^{(i)}$ be the subspace of $\mathbb{R}^m$ corresponding to the eigenvalues $\leq \exp\lambda^{(i)}$ of Λ_x. Then $\mathbb{R}^m = E_x^{(1)} \supset E_x^{(2)} \supset \cdots$ and the following holds

Theorem. For ρ-almost all x,

$$\lim_{n\to\infty} \frac{1}{n} \log \| T_x^n u \| = \lambda^{(i)} \tag{3.2}$$

if $u \in E_x^{(i)} \setminus E_x^{(i+1)}$. In particular, for all vectors u that are not in the subspace $E_x^{(2)}$ (viz., almost all u), the limit is the largest characteristic exponent $\lambda^{(1)}$.

The above remarkable theorem dates back only to 1968, when the proof of a somewhat different version was published by Oseledec (1968). For different proofs see Raghunathan (1979), Ruelle (1979), Johnson, Palmer, and Sell (1984). What does the theorem say for $m=1$? The 1×1 matrices are just ordinary numbers. Assuming them to be positive and taking the log, the reader will verify that the multiplicative ergodic theorem reduces to the ordinary ergodic theorem of Sec. II.E. The novelty and difficulty of the multiplicative ergodic theorem is that for $m>1$ it deals with *noncommuting* matrices.

In some applications we shall need an extension, where $\mathbb{R}^m$ is replaced by an infinite-dimensional Banach or Hilbert space E and the $T(x)$ are bounded operators. Such an extension has been proved under the condition that the $T(x)$ are *compact* operators. In the Hilbert case this means that the spectrum of $T(x)^* T(x)$ is discrete, that the eigenvalues have finite multiplicities, and that they accumulate only at 0.

Theorem (multiplicative ergodic theorem—compact operators in Hilbert space). All the assertions of the multiplicative ergodic theorem remain true if $\mathbb{R}^m$ is replaced by a separable Hilbert space E, and T maps M to compact operators in E. The characteristic exponents form a sequence tending to $-\infty$ (it may happen that only finitely many characteristic exponents are finite).

See Ruelle (1982a) for a proof. For compact operators on a Banach space, Eq. (3.1) no longer makes sense, but there are subspaces $E_x^{(1)} \supset E_x^{(2)} \supset \cdots$ such that (3.2) holds. This was shown first by Mañé (1983), with an unnecessary injectivity assumption, and then by Thieullen (1985) in full generality. (Thieullen's result applies in fact also to noncompact situations.)

B. Characteristic exponents for differentiable dynamical systems

1. Discrete-time dynamical systems on $\mathbb{R}^m$

We consider the time evolution

$$x(n+1) = f(x(n)) \ , \tag{3.3}$$

where $f: \mathbb{R}^m \to \mathbb{R}^m$ is a differentiable vector function. We denote by $T(x)$ the matrix $(\partial f_i / \partial x_j)$ of partial derivatives of the components f_i at x. For the nth iterate f^n of f, the corresponding matrix of partial derivatives is given by the chain rule:

$$\partial (f^n)_i / \partial x_j = T(f^{n-1}x) \cdots T(fx)T(x) \ . \tag{3.4}$$

Now, if ρ is an ergodic measure for f, with compact support, the conditions of the multiplicative ergodic theorem are all satisfied and the characteristic exponents are thus defined.

In particular, if $\delta x(0)$ is a small change in initial condition (considered as infinitesimally small), the change at time n is given by

$$\delta x(n) = T_x^n \delta x(0)$$
$$= T(f^{n-1}x) \cdots T(x) \delta x(0) \ . \tag{3.5}$$

For most $\delta x(0)$ [i.e., for $\delta x(0) \notin E_{x(0)}^{(2)}$] we have $\delta x(n) \approx \delta x(0) e^{n\lambda_1}$, and sensitive dependence on initial conditions corresponds to $\lambda_1 > 0$. Note that if $\delta x(0)$ is finite rather than infinitely small, the growth of $\delta x(n)$ may not go on indefinitely: if $x(0)$ is in a bounded attractor, $\delta x(n)$ cannot be larger than the diameter of the attractor.

2. Continuous-time dynamical systems on $\mathbb{R}^m$

If the time is continuous, we apply the multiplicative ergodic theorem to the time-one map $f = f^1$. The limits defining the characteristic exponents hold again, with $t \to \infty$ replacing $n \to \infty$ (because of continuity it is not necessary to restrict t to integer values). To be specific, we define

$$T_x^t = \text{matrix}\,(\partial f_i^t / \partial x_j) \ . \tag{3.6}$$

If ρ is an ergodic measure with compact support for the time evolution, then, for ρ-almost all x, the following limits exist:

$$\lim_{t\to\infty} (T_x^{t*} T_x^t)^{1/2t} = \Lambda_x \ , \tag{3.7}$$

$$\lim_{t\to\infty} \frac{1}{t} \log \| T_x^t u \| = \lambda^{(i)} \ \text{ if } u \in E_x^{(i)} \setminus E_x^{(i+1)} \ , \tag{3.8}$$

where $\lambda^{(1)} > \lambda^{(2)} > \cdots$ are the logarithms of the eigenvalues of Λ_x, and $E_x^{(i)}$ is the subspace of $\mathbb{R}^m$ corresponding to the eigenvalues $\leq \exp\lambda^{(i)}$. Notice, incidentally, that if the Euclidean norm $\|\ \|$ is replaced by some other norm on $\mathbf{R}^m$, *the characteristic exponents and the* $E_x^{(i)}$ *do not change.*

3. Dynamical systems in Hilbert space

We assume that E is a (real) Hilbert space, ρ a probability measure with compact support in E, and f^t a time evolution such that the linear operators $T_x^t = D_x f^t$ (derivative of f^t at x) are compact linear operators for $t > 0$. This situation prevails, for instance, for the Navier-Stokes time evolution in two dimensions (as well as in three dimensions, so long as the solution has no singularities). The definition of characteristic exponents is the same here as for dynamical systems in $\mathbf{R}^m$.

4. Dynamical systems on a manifold M

For definiteness, let M be a compact manifold like a sphere or a torus; ρ is a probability measure on M, invariant under the dynamical system. If M is m dimensional, we may cut M into a finite number of pieces which are smoothly parametrized by subsets of $\mathbf{R}^m$ (see Fig. 11). In terms of this new parametrization, the map f is continuous except at the cuts, and so is the matrix of partial derivatives. Since only measurability is needed for the abstract multiplicative ergodic theorem, we can again define characteristic exponents. This definition is independent of the partition of the manifold M that has been used, and of the choice of parametrization for the pieces. The reason is that, for any other choice, the norm used would differ from the original norm by a bounded factor, which disappears in the limit. One could alternatively use a Riemann metric on the manifold and define the characteristic exponents in terms of this metric. If $\mathscr{T}_x M$ denotes the tangent space at x, we now have $\mathscr{T}_x M = E_x^{(1)} \supset E_x^{(2)} \supset \cdots$.

C. Steady, periodic, and quasiperiodic motions

1. Examples and parameter dependence

Before proceeding with the general theory, we pause to discuss illustrations of the preceding results.

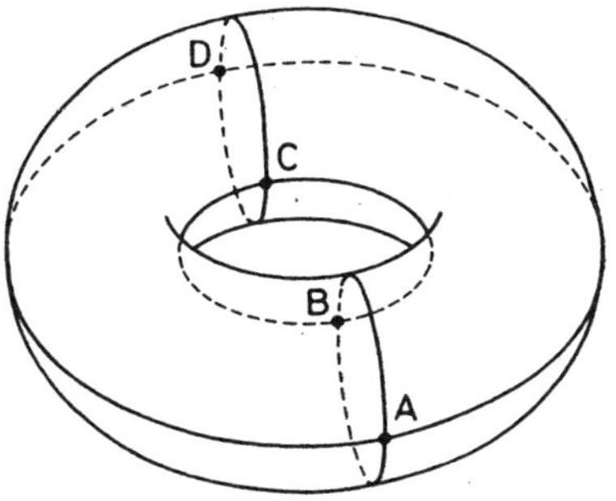

FIG. 11. A two-dimensional torus cut into four rectangular pieces by two horizontal and two vertical circles.

A *steady state* of a physical time evolution is associated with a fixed point P of the corresponding dynamical system. The steady state is thus described by the probability measure $\rho = \delta_P$ (Dirac's delta at P), which is of course invariant and ergodic. We denote by $\alpha_1, \alpha_2, \ldots,$ the eigenvalues of the operator $D_P f^1$ (derivative of the time-one map f^1 at P), in decreasing order of absolute values, and repeated according to multiplicity. Then the characteristic exponents are

$$\lambda_1 = \log|\alpha_1|, \quad \lambda_2 = \log|\alpha_2|, \quad \ldots . \tag{3.9}$$

In particular, a stable steady state associated with an attracting fixed point (see Sec. III.C.2) has negative characteristic exponents. If the dynamical system depends continuously on a bifurcation parameter μ, the $\lambda_i = \log|\alpha_i|$ depend continuously on μ, but we shall see in Sec. III.D that this situation is rather exceptional.

A *periodic state* of a physical time evolution is associated with a periodic orbit $\Gamma = \{f^t a : 0 \le t < T\}$ of the corresponding continuous-time dynamical system. It is thus described by the ergodic probability measure

$$\rho = \delta_\Gamma = \frac{1}{T}\int_0^T dt\, \delta_{f^t a} . \tag{3.10}$$

We denote by α_i^T the eigenvalues of $D_a f^T$; then one of these eigenvalues is 1 (corresponding to the direction tangent to Γ at a). The characteristic exponents are the numbers

$$\lambda_i = \frac{1}{T}\log|\alpha_i^T| ,$$

and one of them is thus 0. In particular, a stable periodic state, associated with an attracting periodic orbit (see Sec. II.C.2), has one characteristic exponent equal to zero and the others negative. Here again, if there is a bifurcation parameter μ, the λ_i depend continuously on μ.

Consider now a *quasiperiodic state with k frequencies*, stable for simplicity. This is represented by a quasiperiodic attractor (Sec. II.C.3), i.e., an attracting invariant torus T^k on which the time evolution is described by translations (2.8) in terms of suitable angular variables $\varphi_1, \ldots, \varphi_k$. There is only one invariant probability measure here: the Haar measure ρ on T^k, defined in terms of the angular variables by

$$(2\pi)^{-k} d\varphi_1 \cdots d\varphi_k .$$

Here, k characteristic exponents are equal to zero, and the others are negative. If the dynamical system depends continuously on a bifurcation parameter μ and has a quasiperiodic attractor for $\mu = \mu_0$, it will still have an attracting k torus for μ close to μ_0, but the motion *on* this k torus may no longer be quasiperiodic. For $k \ge 2$, frequency locking may lead to attracting periodic orbits (and negative characteristic exponents). For $k \ge 3$, strange attractors and positive characteristic exponents may be present for μ arbitrarily close to μ_0 (see Ruelle and Takens, 1971; Newhouse, Ruelle, and Takens, 1978). Nevertheless, we have continuity at $\mu = \mu_0$: the characteristic exponents for μ close to μ_0 are close to their values at μ_0.

2. Characteristic exponents as indicators of periodic motion

The examples of the preceding section are typical for the case of negative characteristic exponents. We now point out that, conversely, it is possible to deduce from the negativity of the characteristic exponents that the ergodic measure ρ describes a steady or a period state.

Theorem (continuous-time fixed point). Consider a continuous-time dynamical system and assume that all the characteristic exponents are different from zero. Then $\rho=\delta_P$, where P is a fixed point. (In particular, if all characteristic exponents are negative, P is an attracting fixed point.)

Another formulation: If the support of ρ does not reduce to a fixed point, then one of the characteristic exponents vanishes.

Sketch of proof. One considers the vector function F,

$$F(x)=\left.\frac{d}{d\tau}f^{\tau}x\right|_{\tau=0} . \tag{3.11}$$

If the support of ρ is not reduced to a fixed point, we have $F(x)\neq 0$ for ρ-almost all x. Furthermore, Eq. (3.11) yields

$$T_x^t F(x)=(D_x f^t)F(x)=F(f^t x) .$$

Since ρ is ergodic, $f^t x$ comes close to x again and again, and we find for the limit (3.8)

$$\lim_{t\to\infty}\frac{1}{t}\log\|T_x^t F(x)\|=0 .$$

Thus there is a characteristic exponent equal to 0.

In the next two theorems we assume that the dynamical system is defined by functions that have continuous second-order derivatives. (The proofs use the stable manifolds of Sec. III.E.)

Theorem (discrete-time periodic orbit). Consider a discrete-time dynamical system and assume that all the characteristic exponents of ρ are negative. Then

$$\rho=\frac{1}{N}\sum_{1}^{N}\delta_{f^k a} ,$$

where $\{a, fa, \ldots, f^{N-1}a\}$ is an attracting periodic orbit, of period N.

Proof. See Ruelle (1979).

Theorem (continuous-time periodic orbit). Consider a continuous-time dynamical system and assume that all the characteristic exponents of ρ are negative, except λ_1. There are then two possibilities: (a) $\rho=\delta_P$, where P is a fixed point, (b) ρ is the measure (3.9) on an attracting periodic orbit (and $\lambda_1=0$).

Proof. See Campanino (1980).

As an application of these results, consider the time evolution given by a differential equation (2.1) in two dimensions. We have the following possibilities for an ergodic measure ρ:

$\lambda_1=\lambda_2=0$,

$\lambda_1=0$, $\lambda_2<0$: ρ is associated with a fixed point or an attracting period orbit,

$\lambda_1>0$, $\lambda_2=0$: this reduces to the previous case by changing the direction of time, and therefore ρ is associated with a fixed point or a repelling periodic orbit,

λ_1 and λ_2 are nonvanishing: ρ is associated with a fixed point.

None of these possibilities corresponds to an *attractor* with a positive characteristic exponent. Therefore, an evolution (2.1) can be chaotic only in three or more dimensions.

D. General remarks on characteristic exponents

We now fix an ergodic measure ρ, and the characteristic exponents that occur in what follows are with respect to this measure.

1. The growth of volume elements

The rate of exponential growth of an infinitesimal vector $\delta x(t)$ is given in general by the largest characteristic exponent λ_1. The rate of growth of a surface element $\delta\sigma(t)=\delta_1 x(t)\wedge\delta_2 x(t)$ is similarly given in general by the sum of the largest two characteristic exponents $\lambda_1+\lambda_2$. In general for a k-volume element $\delta_1 x(t)\wedge\cdots\wedge\delta_k x(t)$ the rate of growth is $\lambda_1+\cdots+\lambda_k$. (Of course, if this sum is negative, the volume is contracted.) The construction above gives computational access to the lower characteristic exponents (and is used in the proof of the multiplicative ergodic theorem). For instance, for a dynamical system in $\mathbb{R}^m$, the rate of growth of the m-volume element is the rate of growth of the *Jacobian determinant* $|J_x^t|=|\det(\partial f_i^t/\partial x_j)|$, and is given by $\lambda_1+\cdots+\lambda_m$. For a volume-preserving transformation we have thus $\lambda_1+\cdots+\lambda_m=0$. For a map f with constant Jacobian J, we have $\lambda_1+\cdots+\lambda_m=\log|J|$.

Examples.

In the case of the Hénon map [example (a) of Sec. II.D] we have $J=-b=-0.3$, hence $\lambda_2=\log|J|-\lambda_1\approx-1.20-0.42=-1.62$.

In the case of the Lorenz equation [example (b) of Sec. II.B] we have $dJ^t/dt=-(\sigma+1+b)$. Therefore, if we know $\lambda_1>0$ we know all characteristic exponents, since $\lambda_2=0$ and $\lambda_3=-(\sigma+1+b)-\lambda_1$.

2. Lack of explicit expressions, lack of continuity

The ordinary ergodic theorem states that the time average of a function φ tends to a limit (ρ-almost everywhere) and asserts that this limit is $\int\varphi(x)\rho(dx)$. By contrast, the multiplicative ergodic theorem gives no explicit expression for the characteristic exponents. It is true that in the proof of the theorem as given by Johnson *et al.* (1984) there is an integral representation of characteristic exponents in terms of a measure on the space of points (x,Q), where x is a point of our m-dimensional manifold, and Q is an $m\times m$ orthogonal matrix. However, this measure is not constructively given. This situation is similar to that in statistical mechanics where, for example, there is in general no explicit expression for the pressure in terms of the interparticle forces.

For a dynamical problem depending on a bifurcation parameter μ, one would like at least to know some continuity properties of the λ_i as functions of μ. The situation there is unfortunately quite bad (with some exceptions—see Sec. III.C). For each μ there may be several attractors A_α^μ, each having at least one physical measure ρ_α^μ. The dependence of the attractors on μ need not be continuous, because of captures and "explosions," and we do not know that ρ_α^μ depends continuously on A_α^μ. Finally, even if ρ_α^μ depends continuously on μ, it is not true in general that the characteristic exponents do the same. To summarize: the characteristic exponents are in general discontinuous functions of the bifurcation parameter μ.

Example.

The interval [0,1] is mapped into itself by $x\to\mu x(1-x)$ when $0\le\mu\le 4$, and Fig. 12 shows λ_1 as a function of μ. There are intervals of values of μ where λ_1 is negative, corresponding to an attracting periodic orbit. It is believed that these intervals are dense in [0,4]. If this is so, λ_1 is necessarily a discontinuous function of μ wherever it is positive. It is believed that $\{\mu\in[0,4]:\lambda_1>0\}$ has positive Lebesgue measure (this result has been announced by Jakobson, but no complete proof has appeared). For some positive results on these difficult problems see Jakobson (1981), Collet and Eckmann (1980a,1983), and Benedicks and Carleson (1984).

The wild discontinuity of characteristic exponents raises a philosophical question: should there not be at least a piecewise continuous dependence of physical quantities on parameters such as one sees, for example, in the solution of the Ising model? Yet we obtain here discontinuous predictions. Part of the resolution of this paradox lies in the fact that our mathematical predictions are *measurable functions* if not continuous, and that measurable functions have much more controllable discontinuities (cf. Luzin's theorem, for instance) than those one could construct with help of the axiom of choice. Another fact is that physical measurements are smoothed by the instrumental procedure. In particular, the definition of characteristic exponents involves a limit $t\to\infty$ [see Eqs. (3.7) and (3.8)], and the great complexity of a curve $\mu\to\lambda_1(\mu)$ will only appear progressively as t is made larger and larger. The presence of noise also smooths out experimental results. At a given level of precision one may find, for instance, that there is one positive characteristic exponent $\lambda_1(\mu)$ in the interval $[\mu_1,\mu_2]$. *This is a meaningful statement,* even though it probably will have to be revised when higher-precision measurements are made; those may introduce *small* subintervals of $[\mu_1,\mu_2]$ where all characteristic exponents are negative. Let us also mention the possibility that for a *large* chaotic system (like a fully turbulent fluid) the distribution of characteristic exponents could again be a smooth function of bifurcation parameters.

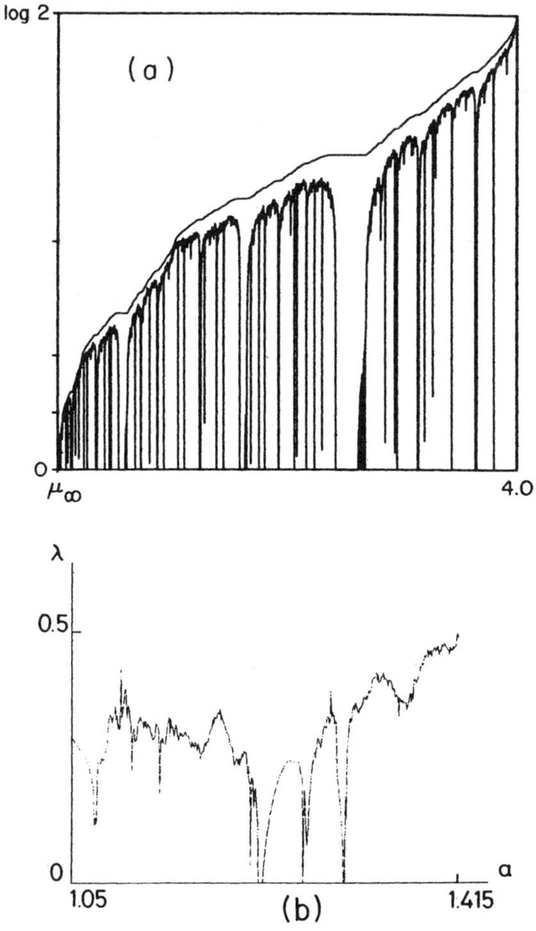

FIG. 12. (a) Topological entropy (upper curve) and characteristic exponent (lower curve) as a function of μ for the family $x\to\mu x(1-x)$. (Graph by J. Crutchfield.) Note the discontinuity of the lower curve. (b) Similar figure for the Hénon map, with $b=0.3$, after Feit (1978).

3. Time reflection

Let us assume that the time-evolution maps f^t are defined for t negative as well as positive. In the discrete-time case this means that f has an inverse f^{-1} which is a smooth map (i.e., f is a *diffeomorphism*). We may consider the time-reversed dynamical system, with time-evolution map $\bar{f}^t=f^{-t}$. If ρ is an invariant (or ergodic) probability measure for the original system, it is also invariant (or ergodic) for the time-reversed system. Furthermore, *the characteristic exponents of an ergodic measure ρ for the time-reversed system are those of the original* *system, but with opposite sign.* We have correspondingly a sequence of subspaces $\bar{E}_x^{(1)} \subset \bar{E}_x^{(2)} \subset \cdots$ for almost all x, such that

$$\lim_{t \to -\infty} \frac{1}{|t|} \log \| T_x^t u \| = -\lambda^{(i)} \quad \text{if } u \in \bar{E}_x^{(i)} \setminus \bar{E}_x^{(i-1)} .$$

Define $F_x^{(i)} = E_x^{(i)} \cap \bar{E}_x^{(i)}$. Then, for ρ-almost all x, the subspaces $F_x^{(i)}$ span $\mathbf{R}^m$ (or the tangent space to the manifold M, as the case may be; compact operators in infinite-dimensional Hilbert space are excluded here because they are not compatible with $t<0$). Furthermore, if T_x^t is the derivative matrix or operator corresponding to $f^{|t|}$ when $t<0$, we have

$$\lim_{|t| \to \infty} \frac{1}{t} \log \| T_x^t u \| = \lambda^{(i)} \quad \text{if } u \in F_x^{(i)} ,$$

where t may go to $+\infty$ or $-\infty$. (For details see Ruelle, 1979.)

4. Relations between continuous-time and discrete-time dynamical systems

We have defined the characteristic exponents for a continuous-time dynamical system [see Eqs. (3.7) and (3.8)] so that they are the same as the characteristic exponents for the discrete-time dynamical system generated by the time-one map $f = f^1$.

Given a Poincaré section (see Sec. II.H), we want to relate the characteristic exponents λ_i for a continuous-time dynamical system with the characteristic exponents $\tilde{\lambda}_i$ corresponding to the first return map P. Note that one of the λ_i is zero (first theorem in Sec. III.C); we claim that the other λ_i are given by

$$\lambda_i = \tilde{\lambda}_i / \langle \tau \rangle_\sigma , \tag{3.12}$$

where $\langle \tau \rangle_\sigma$ is the average time between two crossings of the Poincaré section Σ, computed with respect to the probability measure σ on Σ naturally associated with ρ. (The measure σ gives the density of intersections of orbits with Σ.) The proof is not hard and is left to the reader.

5. Hamiltonian systems

Consider a Hamiltonian (i.e., conservative) system with m degrees of freedom. This is a continuous-time dynamical system in $2m$ dimensions. We claim that the set of λ_i's is symmetric with respect to 0. This is readily checked from Eq. (3.7) and the fact that T_x^t is a symplectic matrix. Actually, two of the λ_i vanish; we get rid of one by going to a $(2m-1)$-dimensional energy surface, and one zero characteristic exponent survives in accordance with the first theorem of Sec. III.C.2.

E. Stable and unstable manifolds

The multiplicative ergodic theorem asserts the existence of *linear* spaces $E_x^{(1)} \supset E_x^{(2)} \supset \cdots$ such that

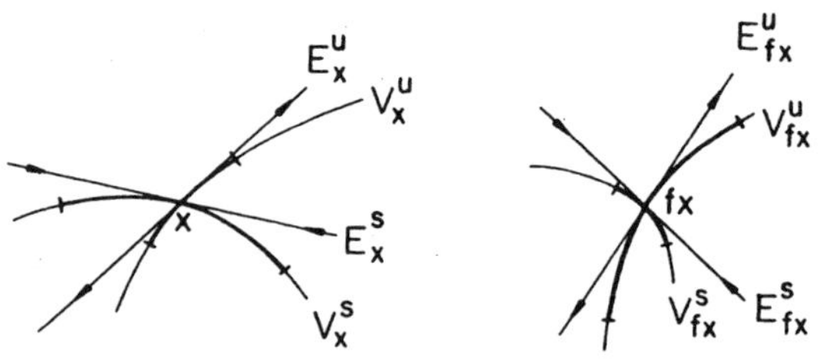

FIG. 13. Stable and unstable manifolds can be defined for points that are neither fixed nor periodic. The stable and unstable directions E_x^s and E_x^u are tangent to the stable and unstable manifolds V_x^s and V_x^u, respectively. They are mapped by f onto the corresponding objects at fx.

$$\lim_{t \to \infty} \frac{1}{t} \log \| T_x^t u \| \le \lambda^{(i)} \quad \text{if } u \in E_x^{(i)} .$$

This means that there exist subspaces $E_x^{(i)}$ such that the vectors in $E_x^{(i)} \setminus E_x^{(i+1)}$ are expanded exponentially by time evolution with the rate $\lambda^{(i)}$. (This expansion is of course a contraction if $\lambda^{(i)} < 0$.) See Fig. 13.

One can define a nonlinear analog of those $E_x^{(i)}$ which correspond to negative characteristic exponents. Let $\lambda < 0$, $\varepsilon > 0$, and write

$$V_x^s(\lambda, \varepsilon) = \{ y : d(f^t x, f^t y) \le \varepsilon e^{\lambda t} \text{ for all } t \ge 0 \} ,$$

where $d(x,y)$ is the distance of x and y (Euclidean distance in $\mathbf{R}^m$, norm distance in Hilbert space, or Riemann distance on a manifold). We shall assume from now on that the time-one map f^1 has continuous derivatives of second as well as first order. If $\lambda^{(i-1)} > \lambda > \lambda^{(i)}$, the set $V_x^s(\lambda, \varepsilon)$ is in fact, for ρ-almost all x and small ε, a piece of *differentiable* manifold, called a *local stable manifold* at x; it is tangent at x to the linear space $E_x^{(i)}$ (and has the

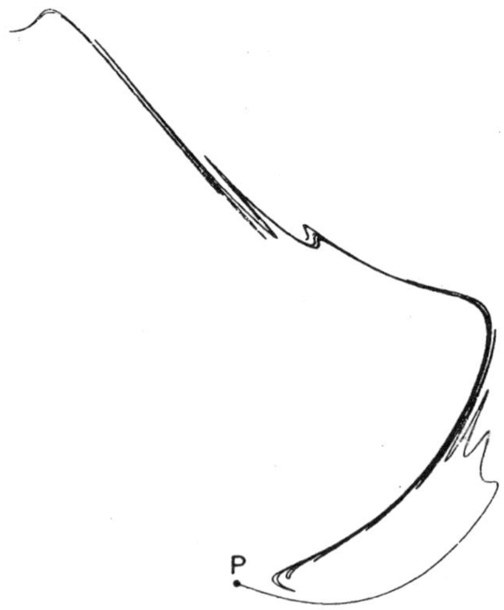

FIG. 14. The stable manifold of a hyperbolic fixed point folds up on itself. (The map is after Hénon and Heiles, 1964.)

same dimension). One shows that $V_x^s(\lambda,\varepsilon)$ is differentiable as many times as f^1.

If we assume that our dynamical system is defined for negative as well as positive times, we can define *global stable manifolds* such that

$$V_x^{(i)s} = \left\{ y : \lim_{t\to\infty} \frac{1}{t} \log d(f^t x, f^t y) \le \lambda^{(i)} \right\}$$

$$= \bigcup_{t>0} f^{-t} V_x^s(\lambda,\varepsilon) ,$$

with negative λ between $\lambda^{(i-1)}$ and $\lambda^{(i)}$ as above.

These global manifolds have the somewhat annoying feature that, while they are locally smooth, they tend to fold and accumulate in a very complicated manner, as suggested by Fig. 14. We can also define *the* stable manifold of x by

$$V_x^s = \left\{ y : \lim_{t\to\infty} \frac{1}{t} \log d(f^t x, f^t y) < 0 \right\}$$

(it is the largest of the stable manifolds, equal to $V_x^{(i)s}$ where $\lambda^{(i)}$ is the largest negative characteristic exponent).

For a dynamical system where negative times are allowed, we obtain *unstable manifolds* V^u instead of stable manifolds simply through replacement of t by $-t$ in the definitions. Instead of assuming that f^t is defined for $t<0$, we find it desirable to make the weaker assumption that f^t and Df^t (defined for $t \ge 0$) are *injective*. This means that $f^t x = f^t y$ implies $x=y$ and $D_x f^t u = D_x f^t v$ implies $u=v$. This injectivity assumption is satisfied when the dynamical system is defined for negative as well as positive times, but also in the case of the Navier-Stokes time evolution. The *global unstable manifold* V_x^u is then defined, provided that for every $t>0$ there is x_{-t} such that $f^t x_{-t} = x$; the definition is

$$V_x^u = \left\{ y : \text{there exists } y_{-t} \text{ such that } f^t y_{-t} = y \text{ and } \lim_{t\to\infty} \frac{1}{t} \log d(x_{-t}, y_{-t}) < 0 \right\} .$$

If $\lambda > 0$ we define similarly

$$V_x^u(\lambda,\varepsilon) = \{ y : \text{there exists } y_{-t} \text{ such that } f^t y_{-t} = y \text{ and } d(x_{-t}, y_{-t}) \le \varepsilon e^{-\lambda t} \text{ for all } t \ge 0 \} .$$

and if $\lambda > 0$ and $\lambda^{(i+1)} < \lambda < \lambda^{(i)}$, we write

$$V_x^{(i)u} = \left\{ y : \text{there exists } y_{-t} \text{ such that } f^t y_{-t} = y \text{ and } \lim_{t\to\infty} \frac{1}{t} d(x_{-t}, y_{-t}) \le -\lambda^{(i)} \right\}$$

$$= \bigcup_{t>0} f^t V_x^u(\lambda,\varepsilon) .$$

The global unstable manifold V_x^u is the largest of the $V_x^{(i)u}$, corresponding to the smallest positive characteristic exponent $\lambda^{(i)}$. Here again one shows that the local unstable manifolds $V_x^u(\lambda,\varepsilon)$ are differentiable (as many times, in fact, as f^1), while the global unstable manifolds $V_x^{(i)u}$ and V_x^u are locally differentiable, but may accumulate on themselves in a complicated manner globally.

The theory of stable and unstable manifolds is part of Pesin theory (for some details, see Sec. III.G).

Examples.

(a) *Fixed points.* If P is a fixed point for a dynamical system (with discrete or continuous time), the characteristic exponents of the δ-measure δ_P at P are called characteristic exponents of the fixed point. They are given explicitly by Eq. (3.9). The fixed point P is said to be *hyperbolic* if all characteristic exponents λ_i are nonzero. When all $\lambda_i < 0$, P is *attracting.* When all $\lambda_i > 0$, P is *repelling.* When some λ_i are >0 and some <0, P is of *saddle type.* The *stable* and *unstable manifolds* of the hyperbolic fixed point P are defined to be the stable and unstable manifolds of δ_P. One has

$$V_x^s = \left\{ y : \lim_{t\to+\infty} f^t y = x \right\} ,$$

$$V_x^u = \left\{ y : \lim_{t\to-\infty} f^t y = x \right\} .$$

(b) *Periodic orbits.* Let Γ be a closed orbit for a continuous-time dynamical system. There is only one invariant measure with support Γ, namely δ_Γ given by Eq. (3.10); it is ergodic. If u is a vector tangent to Γ at x, the corresponding characteristic exponent is zero as one may easily check. If all other characteristic exponents are nonzero, Γ is a *hyperbolic periodic orbit.* The *attracting, repelling,* and *saddle-type* periodic orbits are similarly defined. If $x \in \Gamma$ we have

$$V_x^s = \left\{ y : \lim_{t\to+\infty} d(f^t x, f^t y) = 0 \right\} .$$

This is also called the *strong stable manifold* of x, and a stable manifold of Γ is defined by

$$V_\Gamma^{cs} = \bigcup_{x\in\Gamma} V_x^s$$

$$= \bigcup_{t>0} f^{-t} V_\Gamma^{cs}(\varepsilon) ,$$

where the local stable manifold $V_\Gamma^{cs}(\varepsilon)$ is defined for small ε by

$$V_\Gamma^{cs}(\varepsilon)=\{y:d(f^t x,f^t y)<\varepsilon \text{ for all } t\geq 0\} .$$

Theorem. If A is an attracting set, and $x\in A$, then $V_x^u\subset A$ i.e., the unstable manifold of x is contained in A.

Proof. If U is a fundamental neighborhood of A, and $y\in V_x^u$, then $f^{-\tau}y\in U$ for sufficiently large τ (because $f^{-\tau}y$ is close to $f^{-\tau}x\in A$). Therefore $y\in\bigcap_{\tau>T}f^\tau U=A$.

Corollary. Let A be an attracting set. The number of characteristic exponents $\lambda_i>0$ for any ergodic measure with support in A is a lower bound to the dimension of A.

Proof. The dimension of A is at least that of V_x^u, which is equal to the dimension of $\overline{E}_x^{(k)}$, where $\lambda^{(k)}$ is the smallest positive characteristic exponent. But $\dim\overline{E}_x^{(k)}$ is the sum of the multiplicities of the positive $\lambda^{(i)}$, i.e., the number of positive characteristic exponents λ_i.

(c) *Visualization of the unstable manifolds.* The Hénon attractor has a characteristic appearance of a line folded over many times (see Fig. 5). A similar picture appears for attractors of other two-dimensional dynamical systems generated by a diffeomorphism (differentiable map with differentiable inverse). The theorem stated above suggests that the convoluted lines seen in such attractors are in fact *unstable manifolds.* This suggestion is confirmed by the fact that in many cases the *physical measure* on an attractor is *absolutely continuous on unstable manifolds,* as we shall discuss below.

In higher dimensions, the unstable manifolds forming an attractor may be lines (one dimension), veils (two dimensions), etc. Attractors corresponding to noninvertible maps in two dimensions often have the characteristic appearance of folded veils or drapes, and it is thus immediately apparent that they do not come from a diffeomorphism (Fig. 15).

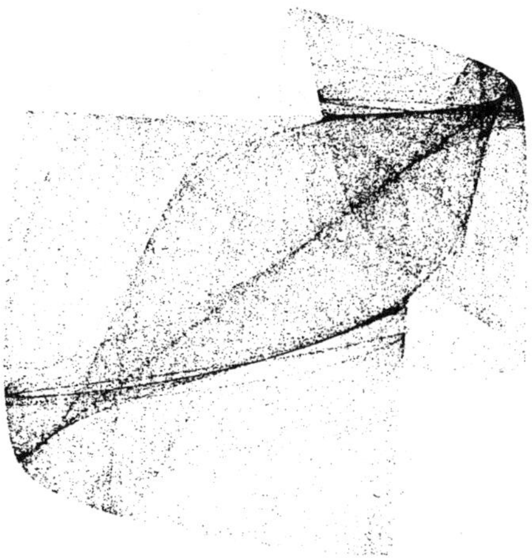

FIG. 15. The map $x'=(A-x-B_1y)x$, $y'=(A-B_2x-y)y$ with $A=3.7$, $B_1=0.1$, $B_2=0.15$ is not invertible. Shown are 50 000 iterates. The map was described in Ushiki *et al.* (1980).

F. Axiom-A dynamical systems*

We discuss here some concepts of *hyperbolicity* which will be referred to in Sec. IV. The hurried reader may skip this discussion without too much disadvantage. In this section, M will always be a compact manifold of dimension m. We shall denote by T_xM the tangent space to M at x. If $f:M\to M$ is a differentiable map, we shall denote by $T_xf:T_xM\to T_{fx}M$ the corresponding tangent map. (We refer the reader to standard texts on differential geometry for the definitions.) If a Riemann metric is given on M, the vector spaces T_xM acquire norms $\|\ \|_x$.

1. Diffeomorphisms

Let $f:M\to M$ be a diffeomorphism, i.e., a differentiable map with differentiable inverse f^{-1}.

We say that a point a of M is *wandering* if there is an open set B containing a [say a ball $B_a(\varepsilon)$] such that $B\cap f^kB=\emptyset$ for all $k>0$ (or we might equivalently require this only for all k large enough). The set of points that are not wandering is the *nonwandering set* Ω. It is a closed, f-invariant subset of M.

Let Λ be a closed f-invariant subset of M, and assume that we have linear subspaces E_x^-,E_x^+ of T_xM for each $x\in\Lambda$, depending continuously on x and such that

$$T_xM=E_x^+ + E_x^-,\quad \dim E_x^+ +\dim E_x^- =m .$$

Assume also that $T_xfE_x^-=E_{fx}^-$ and $T_xfE_x^+=E_{fx}^+$ (i.e., E^-,E^+ form a continuous invariant splitting of TM over Λ). One says that Λ is a *hyperbolic set* if one may choose E^- and E^+ as above, and constants $C>0$, $\Theta>1$ such that, for all $n\geq 0$,

$$\|T_xf^n u\|_{f^n x}\leq C\Theta^{-n}\|u\|_x \quad \text{if } u\in E_x^- ,$$

$$\|T_xf^{-n} v\|_{f^{-n}x}\leq C\Theta^{-n}\|v\|_x \quad \text{if } v\in E_x^+ .$$

[Note that, as a consequence, no ergodic measure with support in Λ has characteristic exponents in the interval $(-\Theta^{-1},\Theta^{-1})$.]

If the whole manifold M is hyperbolic, f is called an *Anosov diffeomorphism.* [Arnold's cat map, Sec. II.D, example (c), is an Anosov diffeomorphism.]

If the nonwandering set Ω is hyperbolic, and if the periodic points are dense in Ω, f is called an *Axiom-A diffeomorphism.* (Every Anosov diffeomorphism is an Axiom-A diffeomorphism.)

2. Flows

Consider a continuous-time dynamical system (f^t) on M, where f^t is defined for all $t\in R$; (f^t) is then also called a *flow.*

We say that a point a of M is *wandering* if there is an open set B containing a [say a ball $B_a(\varepsilon)$] such that $B\cap f^tB=\emptyset$ for all sufficiently large t. The set of points that are not wandering is the *nonwandering set* Ω. It is a closed, (f^t)-invariant subset of M.

Let Λ be a closed invariant subset of M containing no fixed point. Assume that we have linear subspaces E_x^-, E_x^0, E_x^+ of T_xM for each $x \in \Lambda$, depending continuously on x, and such that

$$T_xM = E_x^- + E_x^0 + E_x^+ ,$$

$$\dim E_x^0 = 1, \quad \dim E_x^- + \dim E_x^+ = m - 1 .$$

Assume also that E_x^0 is spanned by

$$\left. \frac{d}{dt} f^t x \right|_{t=0} ,$$

i.e., E_x^0 is in the direction of the flow, and that

$$T_x f^t E_x^- = E_{f^t x}^- ,$$

$$T_x f^t E_x^+ = E_{f^t x}^+ .$$

One says that Λ is a *hyperbolic set* if one may choose E^-, E^0, and E^+ as above, and constants $C>0$, $\Theta>1$ such that, for all $t \geq 0$

$$\| T_x f^t u \|_{f^t x} \leq C \Theta^{-t} \| u \|_x \quad \text{if } u \in E_x^- ,$$

$$\| T_x f^{-t} v \|_{f^{-t} x} \leq C \Theta^{-t} \| v \|_x \quad \text{if } v \in E_x^+ .$$

More generally, we shall also say that Λ^* is a *hyperbolic set* if Λ^* is the union of Λ as above and of a finite number of hyperbolic fixed points [Sec. III.E, example (a)].

If the whole manifold M is a hyperbolic, (f^t) is called an *Anosov flow.*

If the nonwandering set Ω is hyperbolic, and if the periodic orbits and fixed points are dense in the Ω, then (f^t) is called an *Axiom-A flow.*

3. Properties of Axiom-*A* dynamical systems

Axiom-*A* dynamical systems were introduced by Smale [for reviews, see Smale's original paper (1967) and Bowen (1978)]. Smale proved the following "*spectral theorem*" valid both for diffeomorphism and flows.

Theorem. Ω is the union of finitely many disjoint closed invariant sets $\Omega_1, \ldots, \Omega_s$, and for each Ω_i there is $x \in \Omega_i$ such that the orbit $\{f^t x\}$ is dense in Ω_i. The decomposition $\Omega = \Omega_i \cup \cdots \cup \Omega_s$ is unique with these properties.

The sets Ω_i are called *basic sets,* while those which are attracting sets are called *attractors* (there is always at least one attractor among the basic sets).

Some of the ergodic properties of Axiom-*A* attractors will be discussed in Sec. IV. The great virtue of these systems is that they can be analyzed mathematically in detail, while many properties of a map apparently as simple as the Hénon diffeomorphism [Sec. II.D, example (a)] remain conjectural.

It should be pointed out that there is a vast literature on the Axiom-*A* systems, concerned in particular with *structural stability.*

G. Pesin theory*

We have seen above that the stable and unstable manifolds (defined almost everywhere with respect to an ergodic measure ρ) are differentiable. This is part of a theory developed by Pesin (1976,1977).[2] Pesin assumes that ρ has differentiable density with respect to Lebesgue measure, but this assumption is not necessary for the study of stable and unstable manifolds [see Ruelle (1979), and for the infinite-dimensional case Ruelle (1982a) and Mañé (1983)].

The earlier results on differentiable dynamical systems had been mostly geometric and restricted to *hyperbolic* (Anosov, 1967) or *Axiom-A* systems (Smale, 1967). Pesin's theory extends a good part of these geometric results to arbitrary differentiable dynamical systems, but working now *almost everywhere* with respect to some ergodic measure ρ. (The results are most complete when all characteristic exponents are different from zero.) The original contribution of Pesin has been extended by many workers, notably Katok (1980) and Ledrappier and Young (1984). Many of the results quoted in Sec. IV below depend on Pesin theory, and we shall give an idea of the present aspect of the theory in that section. Here we mention only one of Pesin's original contributions, a striking result concerning area-preserving diffeomorphisms (in two dimensions).

Theorem (Pesin). Let f be an area-preserving diffeomorphism, and f be twice differentiable. Suppose $fS = S$ for some bounded region S, and let S' consist of the points of S which have nonzero characteristic exponents. Then (up to a set of measure 0) S' is a countable union of ergodic components.

In this theorem the area defines an invariant measure on S, which is not ergodic in general, and S can therefore be decomposed into further invariant sets. This may be a continuous decomposition (like that of a disk into circles). The theorem states that *where the characteristic exponents are nonzero, the decomposition is discrete.*

IV. ENTROPY AND INFORMATION DIMENSION

In this section we introduce two more ergodic quantities: the *entropy* (or *Kolmogorov-Sinai invariant*) and the *information dimension.* We discuss how these quantities are related to the characteristic exponents. The measurement of the entropy and information dimension in physical and computer experiments will be discussed in Sec. V.

A. Entropy

As we have noted already in the Introduction, a system with sensitive dependence on initial conditions produces

---

[2]For a systematic exposition see Fathi, Herman, and Yoccoz (1983).

information. This is because two initial conditions that are different but indistinguishable at a certain experimental precision will evolve into distinguishable states after a finite time. If ρ is an ergodic probability measure for a dynamical system, we introduce the concept of *mean rate of creation of information* $h(\rho)$, also known as *measure-theoretic entropy* or the *Kolmogorov-Sinai invariant* or simply *entropy*. When we study the dynamics of a dissipative physicochemical system, it should be noted that the Kolmogorov-Sinai entropy is not the same thing as the thermodynamic entropy of the system. To define $h(\rho)$ we shall assume that the support of ρ is a compact set with a given metric. (More general cases can be dealt with, but in our applications supp ρ is indeed a compact metric space.) Let $\mathscr{A}=(\mathscr{A}_1,\ldots,\mathscr{A}_\alpha)$ be a finite (ρ-measurable) partition of the support of ρ. For every piece $\mathscr{A}_j$ we write $f^{-k}\mathscr{A}_j$ for the set of points mapped by f^k to $\mathscr{A}_j$. We then denote by $f^{-k}\mathscr{A}$ the partition $(f^{-k}\mathscr{A}_1,\ldots,f^{-k}\mathscr{A}_\alpha)$. Finally, $\mathscr{A}^{(n)}$ is defined as

$$\mathscr{A}^{(n)}=\mathscr{A}\vee f^{-1}\mathscr{A}\vee\cdots\vee f^{-n+1}\mathscr{A}\,,$$

which is the partition whose pieces are

$$\mathscr{A}_{i_1}\cap f^{-1}\mathscr{A}_{i_2}\cap\cdots\cap f^{-n+1}\mathscr{A}_{i_n}$$

with $i_j\in\{1,2,\ldots,\alpha\}$. What is the significance of these partitions? The partition $f^{-k}\mathscr{A}$ is deduced from $\mathscr{A}$ by time evolution (note that $f^k\mathscr{A}$ need not be a partition, since f might be many-to-one; this is why we use $f^{-k}\mathscr{A}$). The partition $\mathscr{A}^{(n)}$ is the partition generated by $\mathscr{A}$ in a time interval of length n. We write

$$H(\mathscr{A})=-\sum_{i=1}^{\alpha}\rho(\mathscr{A}_i)\log\rho(\mathscr{A}_i)\,,\tag{4.1}$$

with the understanding that $u\log u=0$ when $u=0$. (We strongly advise using natural logarithms, but $\log_{10}$ and $\log_2$ have their enthusiasts.) Thus $H(\mathscr{A})$ is the information content of the partition $\mathscr{A}$ with respect to the state ρ, and $H(\mathscr{A}^{(n)})$ is the same, over an interval of time of length n. The following limits are asserted to exist, defining $h(\rho,\mathscr{A})$ and $h(\rho)$:

$$h(\rho,\mathscr{A})=\lim_{n\to\infty}[H(\mathscr{A}^{(n+1)})-H(\mathscr{A}^{(n)})]$$

$$=\lim_{n\to\infty}\frac{1}{n}H(\mathscr{A}^{(n)})\,,\tag{4.2}$$

$$h(\rho)=\lim_{\text{diam}\mathscr{A}\to 0}h(\rho,\mathscr{A})\,,\tag{4.3}$$

where $\text{diam}\mathscr{A}=\max_i\{\text{diameter of }\mathscr{A}_i\}$. Clearly, $h(\rho,\mathscr{A})$ is the rate of information creation with respect to the partition $\mathscr{A}$, and $h(\rho)$ its limit for finer and finer partitions. This last limit may sometimes be avoided [i.e., $h(\rho,\mathscr{A})=h(\rho)$]; this is the case when $\mathscr{A}$ is a *generating partition*. This holds in particular if $\text{diam}\mathscr{A}^{(n)}\to 0$ when $n\to\infty$, or if f is invertible and $\text{diam}f^n\mathscr{A}^{(2n)}\to 0$ when $n\to\infty$. For example, for the map of Fig. 8, a generating partition is obtained by dividing the interval at the singularity in the middle. For more details we must refer the reader to the literature, for instance the excellent book by Billingsley (1965).

The above definition of the entropy applies to continuous as well as discrete-time systems. In fact, the entropy in the continuous-time case is just the entropy $h(\rho,f^1)$ corresponding to the time-one map. We also have the formula

$$h(\rho,f^T)=|T|\,h(\rho,f^1)\,.$$

Note that the definition of the entropy in the continuous-time case *does not involve a time step t tending to zero,* contrary to what is sometimes found in the literature. Note also that the entropy does not change if f is replaced by f^{-1}.

If (f^t) has a Poincaré section Σ, we let σ be the probability measure on Σ, invariant under the Poincaré map P and corresponding to ρ (i.e., σ is the density of intersection of orbits of the continuous dynamical system with Σ). If we also let τ be the first return time, then we have *Abramov's formula,*

$$h(\rho)=\frac{h(\sigma)}{\langle\tau\rangle_\sigma}\,,$$

which is analogous to Eq. (3.12) for the characteristic exponents.

The relationship of entropy to characteristic exponents is very interesting. First we have a general inequality.

Theorem (Ruelle, 1978). Let f be a differentiable map of a finite-dimensional manifold and ρ an ergodic measure with compact support. Then

$$h(\rho)\le\Sigma\text{ positive }\lambda_i\,.\tag{4.4}$$

The result is believed to hold in infinite dimensions as well, but no proof has been published yet.

It is of considerable interest that the *equality* corresponding to Eq. (4.4) seems to hold often (but not always) for the *physical measures* (Sec. II.F) in which we are mainly interested. This equality is called the *Pesin identity:*

$$h(\rho)=\Sigma\text{ positive }\lambda_i\,.$$

Pesin proved that it holds if ρ is invariant under the diffeomorphism f, and ρ has smooth density with respect to Lebesgue measure. More generally, the Pesin identity holds for the *SRB measures* to be studied in Sec. IV.B.

In Sec. V we shall use in addition an entropy concept different from that of Eqs. (4.1)—(4.3). It is given by

$$H_2(\mathscr{A})=-\log\sum_{i=1}^{\alpha}\rho(\mathscr{A}_i)^2\,,$$

$$K_2(\rho)=\lim_{\text{diam}\mathscr{A}\to 0}\lim_{n\to\infty}\frac{1}{n}H_2(\mathscr{A}^{(n)})\,,\tag{4.5}$$

if these limits exist (see Grassberger and Procaccia, 1983a). It can be shown that *the K_2 entropy is a lower bound to the entropy $h(\rho)$*:

$$K_2(\rho)\le h(\rho)\,.\tag{4.6}$$

B. SRB measures

We have seen in Sec. III.E that attracting sets are unions of unstable manifolds. Transversally to these, one often finds a discontinuous structure corresponding to the complicated piling up of the unstable manifolds upon themselves. This suggests that invariant measures may have very rough densities in the directions transversal to the foliations of the unstable manifolds. On the other hand, we may expect that—due to stretching in the unstable direction—the measure is smooth when viewed along these directions. We shall call SRB measures (for Sinai, Ruelle, Bowen) those measures that are smooth along unstable directions. They turn out to be a natural and useful tool in the study of physical dynamical systems.

Much of this section is concerned with consequences of the existence of SRB measures. These are mostly relations between entropy, dimensions, and characteristic exponents. To prove the existence of SRB measures for a given system is a hard task, and whether they exist is not known in general. Sometimes no SRB measures exist, but it is unclear how frequently this happens. On the other hand, we do not have much of physical relevance to say about systems without SRB measures.

To repeat, we should like to define, intuitively, SRB measures as measures with smooth density in the stretching, or *unstable,* directions of the dynamical system defined by f. The geometric complexities described above make a rather technical definition necessary. Before going into these technicalities, we discuss the framework in which we shall work.

(a) In the ergodic theory of differentiable dynamical systems, there is no essential difference between discrete-time and continuous-time systems. In fact, if we discretize a continuous-time dynamical system by restricting t to integer values (i.e., use the time-one map $f=f^1$ as a generator), then the characteristic exponents, the stable and unstable manifolds, and the entropy are unchanged. (The information dimension to be defined in Sec. IV.C also remains the same.) We may thus, for simplicity, *consider only discrete-time systems.*

(b) If f is a diffeomorphism (i.e., a differentiable map with differential inverse), then our dynamical system is defined for negative as well as positive times. If, in addition, f is twice differentiable, then the inverse map is also twice differentiable. We shall assume a little less, namely, that f is *twice differentiable* and either a *diffeomorphism* or at least such that f and Df are *injective* (i.e., $fx=fy$ implies $x=y$, and $D_xfu=D_xfv$ implies $u=v$; these conditions hold for the Navier-Stokes time evolution).

Given an ergodic measure ρ (with compact support as usual), unstable manifolds V_x^u are defined for almost all x according to Eq. (3.13). Notice that $y\in V_x^u$ is the same thing as $x\in V_y^u$, so that the unstable manifolds V^u partition the space into equivalence classes. It might seem natural to define SRB measures by using this partition for a decomposition of ρ into pieces ρ_α, carried by different unstable manifolds:

$$\rho=\int \rho_\alpha m(d\alpha)\,, \tag{4.7}$$

where α parametrizes the V^u's, and m is a measure on the "space of equivalence classes." In reality, this space of equivalence classes does not exist in general (as a measurable space) because of the folding and accumulation of the global unstable manifolds [and the existence of a nontrivial decomposition (4.7) would contradict ergodicity].

The correct approach is as follows. Let S be a ρ-measurable set of the form $S=\cup_{\alpha\in A}S_\alpha$, where the S_α are disjoint small open pieces of the V^u's (say each S_α is contained in a *local* unstable manifold). If this decomposition is ρ measurable, then one has

$$\rho \text{ restricted to } S=\int \rho_\alpha m(d\alpha)\,,$$

where m is a measure on A, and ρ_α is a probability measure on S_α called the *conditional probability measure* associated with the decomposition $S=\cup_{\alpha\in A}S_\alpha$. The ρ_α are defined m-almost everywhere. See Fig. 16. The situation of interest for the definition of SRB measures occurs when *the conditional probabilities ρ_α are absolutely continuous with respect to Lebesgue measure* on the V^u's. This means that

$$\rho_\alpha(d\xi)=\varphi_\alpha(\xi)d\xi \quad \text{on } S_\alpha\,, \tag{4.8}$$

where $d\xi$ denotes the volume element when S_α is smoothly parametrized by a piece of $\mathbb{R}^{m_+}$ and φ_α is an integrable function. The *unstable dimension* m_+ of S_α or V^u is the sum of the multiplicities of the positive characteristic exponents. It is finite even for the case of the Navier-Stokes equation discussed earlier (because $\lambda_i\to-\infty$ when $i\to\infty$, as we have noted).

We say that the ergodic measure ρ is an *SRB measure* if its conditional probabilities ρ_α are absolutely continuous with respect to Lebesgue measure for some choice of S with $\rho(S)>0$, and a decomposition $S=\cup_\alpha S_\alpha$ as above. The definition is independent of the choice of S and its decomposition (this is an easy exercise in ergodic theory). We shall also say that ρ *is absolutely continuous along unstable manifolds.*

Theorem (Ledrappier and Young, 1984). Let f be a twice differentiable diffeomorphism of an m-dimensional manifold M and ρ an ergodic measure with compact support. The following conditions are then equivalent: (a) The measure ρ is an SRB measure, i.e., ρ is absolutely

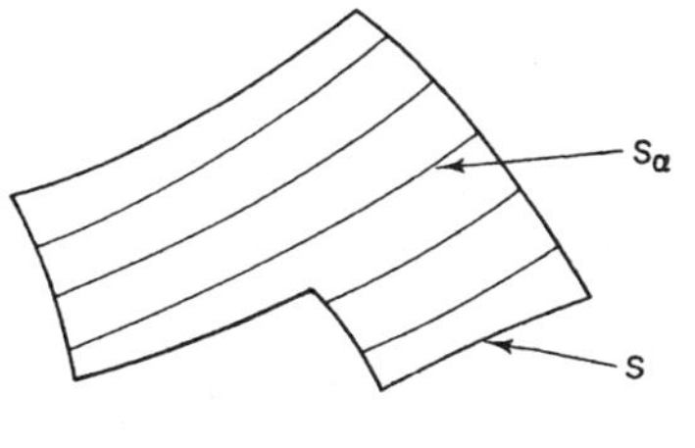

FIG. 16. A decomposition of the set S into smooth leaves S_α, each of which is contained in the unstable manifold.

continuous along unstable manifolds. (b) The measure ρ satisfies Pesin's identity,

$$h(\rho)=\Sigma \text{ positive characteristic exponents} .$$

Furthermore, if these conditions are satisfied, the density functions φ_α in Eq. (4.8) are differentiable.

The theorem says that if ρ is absolutely continuous along unstable manifolds, then the rate of creation of information is the mean rate of expansion of m_+-dimensional volume elements. If, however, ρ is singular along unstable manifolds, then this rate is strictly less than the rate of expansion. These assertions are intuitively quite reasonable, but in fact quite hard to prove. The first proofs have been given for Axiom-A systems (see Sec. III.F) by Sinai (1972; Anosov systems), Ruelle (1976; Axiom-A diffeomorphisms), and Bowen and Ruelle (1975; Axiom-A flows). The general importance of (a) and (b) was stressed by Ruelle (1980).

One hopes that there is an infinite-dimensional extension applying to Navier-Stokes, but such an extension has not yet been proved. Ledrappier (1981b) has obtained a version of the above theorem that is valid for noninvertible maps in one dimension.

The SRB measures are of particular interest for physics because one can show—in a number of cases—that the ergodic averages

$$\frac{1}{n}\sum_{k=0}^{n-1}\delta_{f^k x}$$

tend to the SRB measure ρ when $n\to\infty$, not just for ρ-almost all x, but for x in a set of positive *Lebesgue* measure. Lebesgue measure corresponds to a more *natural notion of sampling* than the measure ρ (which is carried by an attractor and usually singular). The above property is thus both strong and natural.

To formulate this result as a theorem, we need the notion of a *subset of Lebesgue measure zero* on an m-dimensional manifold. We say that a set $S\subset M$ is Lebesgue measurable (has zero Lebesgue measure or positive Lebesgue measure) if for a smooth parametrization of M by patches of $\mathbf{R}^m$ one finds that S is Lebesgue measurable (has zero Lebesgue measure or positive Lebesgue measure). These definitions are independent of the choice of parametrization (in contrast to the *value* of the measure).

Theorem (SRB measures for Axiom-A systems). Consider a dynamical system determined by a twice differentiable diffeomorphism f (discrete time) or a twice differentiable vector field (continuous time) on an m-dimensional manifold M. Suppose that A is an Axiom-A attractor, with basin of attraction U. (a) There is one and only one SRB measure with support in A. (b) There is a set $S\subset U$ such that $U\setminus S$ has zero Lebesgue measure, and

$$\lim_{n\to\infty}\frac{1}{n}\sum_{k=0}^{n-1}\delta_{f^k x}=\rho \quad \text{(discrete time)} ,$$

or

$$\lim_{T\to\infty}\frac{1}{T}\int_0^T dt\,\delta_{f^t x}=\rho \quad \text{(continuous time)} ,$$

whenever $x\in S$.

For a proof, see Sinai (1972; Anosov systems), Ruelle (1976; Axiom-A diffeomorphisms), or Bowen and Ruelle (1975; Axiom-A flows). The "geometric Lorenz attractor" can be treated similarly.

The following theorem shows that the requirement of Axiom A can be replaced by weaker information about the characteristic exponents.

Theorem (Pugh and Shub, 1984). Let f be a twice differentiable diffeomorphism of an m-dimensional manifold M and ρ an SRB measure such that all characteristic exponents are different from zero. Then there is a set $S\subset M$ with positive Lebesgue measure such that

$$\lim_{n\to\infty}\frac{1}{n}\sum_{k=0}^{n-1}\delta_{f^k x}=\rho \tag{4.9}$$

for all $x\in S$.

This theorem is in the spirit of the "absolute continuity" results of Pesin. An infinite-dimensional generalization has been promised by Brin and Nitecki (1985). The theorem fails if 0 is a characteristic exponent, as the following example shows.

Counterexample. A dynamical system is defined by the differential equation

$$\frac{dx}{dt}=x^3$$

on $\mathbf{R}$. Its time-one map has δ_0 as an ergodic measure, with $\lambda_1=0$. However, 0 is (weakly) repelling, so that Eq. (4.9) cannot hold for $x\neq 0$. In fact, if $x\neq 0$, $f^t x$ goes to infinity in a finite time.

We give now an example showing that there is not always an SRB measure lying around, and that there are physical measures that are not SRB.

Counterexample (Bowen, and also Katok, 1980). Consider a continuous-time dynamical system (flow) in $\mathbf{R}^2$ with three fixed points A,B,C where A,C are repelling and B of saddle type, as shown in Fig. 17. The system has an invariant curve in the shape of a "figure 8" (or rather, figure ∞), which is attracting. It can be seen that any point different from A or C yields an ergodic average corresponding to a Dirac δ at B. Therefore δ_B is the physical measure for our system. Clearly it has zero entropy, one strictly positive characteristic exponent and the other strictly negative (and thus, in particular, not zero),

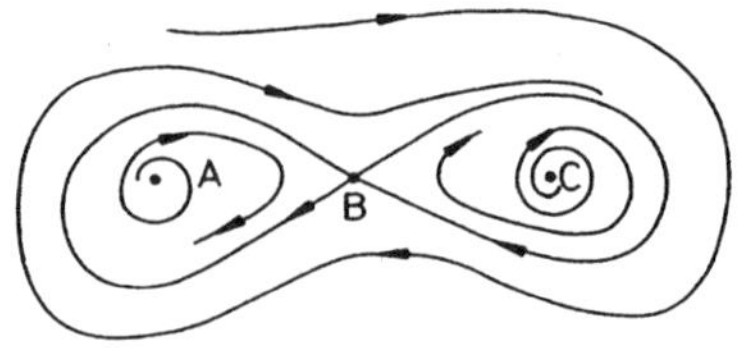

FIG. 17. The figure ∞ counterexample of Bowen (1975).

and is not absolutely continuous with respect to Lebesgue measure on the unstable manifold. (Note that the unstable manifold at B consists of the "figure 8.") On the other hand, it is not hard to see that δ_B is a Kolmogorov measure; i.e., a system perturbed with a little noise ε will spend most of its time near B, and as $\varepsilon \to 0$ the fraction of time spent near B goes to 1.

C. Information dimension

Given a probability measure ρ, we know that its *information dimension* $\dim_H \rho$ is the smallest Hausdorff dimension of a set S of ρ measure 1. Note that the set S is not closed in general, and therefore the Hausdorff dimension $\dim_H(\text{supp}\rho)$ of the support of ρ may be strictly larger than $\dim_H \rho$.

Example. The rational numbers of the interval [0,1], i.e., the fractions p/q with p,q integers, form a countable set. This means that they can be ordered in a sequence $(a_n)_1^\infty$. Consider the probability measure

$$\rho = \frac{1}{e} \sum_{n=1}^{\infty} \frac{1}{n!} \delta_{a_n} ,$$

where δ_x is the δ measure at x. Then ρ is carried by the set S of rational numbers of [0,1], and since this is a countable set we have $\dim_H \rho = 0$. On the other hand, $\text{supp}\rho = [0,1]$, so that $\dim_H(\text{supp}\rho) = 1$.

It turns out that the information dimension of a physical measure ρ is a more interesting quantity than the Hausdorff dimension of the attractor or attracting set A which carries ρ. This is both because $\dim_H \rho$ is more accessible experimentally and because it has simple mathematical relations with the characteristic exponents. In any case, we have $\text{supp}\rho \subset A$ and therefore

$$\dim_H \rho \le \dim_H(\text{supp}\rho) \le \dim_H A .$$

The next theorem shows that the information dimension is naturally related to the measure of small balls in phase space.

Theorem (Young, 1982). Let ρ be a probability measure on a finite-dimensional manifold M. Assume that

$$\lim_{r \to 0} \frac{\log \rho[B_x(r)]}{\log r} = \alpha \tag{4.10}$$

for ρ-almost all x. Then $\dim_H \rho = \alpha$.

Young shows that α is also equal to several other "fractal dimensions" (in particular, the "Rényi dimension").

We are of course mostly interested in the case when ρ is ergodic for a differentiable dynamical system. In that situation, the requirement that

$$\lim_{r \to 0} \frac{\log \rho[B_x(r)]}{\log r}$$

exists ρ-almost everywhere already implies that the limit is almost everywhere constant, and therefore equal to $\dim_H \rho$. [The above limit does not always exist, as Ledrappier and Misiurewicz (1984) have shown for certain maps of the interval.]

An interesting relation between $\dim_H \rho$ and the characteristic exponents λ_i has been conjectured by Yorke and others (see references below). We denote by

$$c_\rho(k) = \sum_{i=1}^{k} \lambda_i$$

the sum of the k largest characteristic exponents, and extend this definition by linearity between integers (see Fig. 18):

$$c_\rho(s) = \sum_{i=1}^{k} \lambda_i + (s-k)\lambda_{k+1} \quad \text{if } k \le s < k+1 .$$

The function c_ρ is defined on the interval $[0, +\infty)$ for a dynamical system on a Hilbert space, and on the interval $[0,m]$ for a system on $\mathbb{R}^m$ or an m-dimensional manifold. In the latter cases we write $c_\rho(s) = -\infty$ for $s > m$, so that c_ρ is now in all cases a *concave* function on $[0, +\infty)$, as in Fig. 18. Notice that $c_\rho(0) = 0$, that the maximum of $c_\rho(s)$ is the sum of the positive characteristic exponents, and that $c_\rho(s)$ becomes negative for sufficiently large s. (This is because, in the Hilbert case, the λ_i tend to $-\infty$.)

The *Liapunov dimension* of ρ is now defined as

$$\dim_\Lambda \rho = \max\{s : c_\rho(s) \ge 0\} .$$

Notice that when $c_\rho(k) \ge 0$ and $c_\rho(k+1) < 0$ we have

$$\dim_\Lambda \rho = k + \frac{c_\rho(k)}{|\lambda_{k+1}|} .$$

The $(k+1)$-volume elements are thus contracted by time evolution, and this suggests that the dimension of ρ must be less than $k+1$, a result made rigorous by Ilyashenko (1983). Yorke and collaborators have gone further and made the following guess.

Conjecture (Kaplan and Yorke, 1979; Frederickson, Kaplan, Yorke, and Yorke, 1983; Alexander and Yorke, 1984). If ρ is an SRB measure, then generically

$$\dim_H \rho = \dim_\Lambda \rho . \tag{4.11}$$

The SRB measures have been defined in Sec. IV.B. and "genericity" means here "in general." What concept of

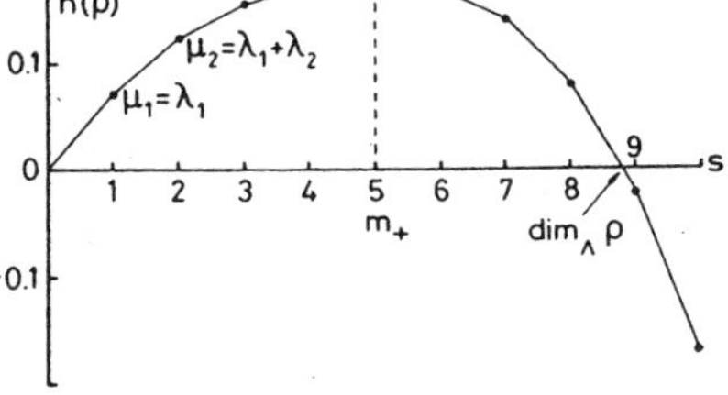

FIG. 18. Determination of the Liapunov dimension $\dim_\Lambda(\rho)$. The number of positive Liapunov exponents (unstable dimension) is m_+. The graph is from Manneville (1985), for the Kuramoto-Sivashinsky model.

genericity is adequate is a very difficult question; we do not know—among other things—how frequently a dynamical system has an SRB measure. An *inequality* is, however, available in full generality.

Theorem. Let f be a twice continuously differentiable map and let ρ be an ergodic measure with compact support. Then

$$\dim_H \rho \le \dim_\Lambda \rho \ . \tag{4.12}$$

The basic result was proved by Douady and Oesterlé (1980), and from this Ledrappier (1981a) derived the theorem as stated. (It holds for Hilbert spaces as well as in finite dimensions.)

An *equality* is also known in special cases, notably the following.

Theorem (Young, 1982). Let f be a twice differentiable diffeomorphism of a two-dimensional manifold, and let ρ be an ergodic measure with compact support. Then the limit (4.10) exists ρ-almost everywhere, and we have

$$\dim_H \rho = h(\rho)\left[\frac{1}{\lambda_1}+\frac{1}{|\lambda_2|}\right], \tag{4.13}$$

where $\lambda_1>0$ and $\lambda_2<0$ are the characteristic exponents of ρ.

Note, incidentally, that if f is replaced by f^{-1}, the characteristic exponents change sign, the entropy remains the same, and the formula remains correct, as it should. Note also that the cases where λ_1 and λ_2 are not of opposite sign are relatively trivial. From the inequality (4.4) applied for f or f^{-1} we see that $\lambda_1=0$ or $\lambda_2=0$ implies $h(\rho)=0$, so that the right-hand side of Eq. (4.13) becomes indeterminate. If $\lambda_1\ge\lambda_2>0$ or $0>\lambda_1\ge\lambda_2$, a theorem of Sec. III.C.2 applied to f or f^{-1} shows that ρ is carried by a periodic orbit, so that Eq. (4.13) holds with $\dim_H\rho=h(\rho)=0$.

Variants of the above theorem that do not assume the invertibility of f are known (for one dimension see Ledrappier, 1981b, Proposition 4; for holomorphic functions see Manning, 1984).

If ρ is an SRB measure, then Eq. (4.13) becomes $\dim_H\rho=1+\lambda_1/|\lambda_2|$, which is just the conjecture (4.11). The next example shows that the conjecture does not always hold.

Counterexample. Notice first that if a measure ρ has no positive characteristic exponent, then $h(\rho)=0$ by a theorem in Sec. IV.A, and therefore ρ is an SRB measure. If $\dim_H\rho$ is strictly between 0 and 1, then Eq. (4.11) cannot hold (because $\dim_\Lambda\rho$ can only have the value 0 or a value ≥ 1). In particular, the Feigenbaum attractor [example (b) of Sec. II.D] carries a unique probability measure ρ with $\lambda_1=0$ and $\dim_H\rho=0.538\ldots$ so that Eq. (4.11) is violated here. [For the Hausdorff dimension of the Feigenbaum measure see Grassberger (1981); Vul, Sinai, and Khanin (1984); Ledrappier and Misiurewicz (1984).]

Finally, let us mention *lower* bounds on $\dim_H\rho$.

Theorem. If ρ is an SRB measure, then $\dim_H\rho\ge m_+$, where m_+ is the sum of the multiplicities of the positive characteristic exponents (unstable dimension).

This follows readily from the definitions.

D. Partial dimensions

Given an ergodic measure ρ, we can associate with each characteristic exponent $\lambda^{(i)}$ a *partial dimension* $D^{(i)}$. Roughly speaking, $D^{(i)}$ is the Hausdorff dimension in the direction of $\lambda^{(i)}$. The entropy inequality (4.4) and the dimension inequality (4.12) will be natural consequences of the existence of the $D^{(i)}$.

In order to give a precise definition, we assume that f is a twice differentiable diffeomorphism of a compact manifold M (in the case of a continuous-time dynamical system we take for f the time-one map). If $\lambda^{(i)}$ is a positive characteristic exponent and $\lambda^{(i)}>\lambda>\max(0,\lambda^{(i+1)})$, we have defined in Sec. III.E the local *unstable manifolds* $V_x^u(\lambda,\varepsilon)$, which we simply denote here by $V_{\rm loc}^{(i)}$, and $S=\cup_{\alpha\in A}S_\alpha$, where the S_α are open pieces of the $V_{\rm loc}^{(i)}$. Suppose $S=\cup_{\alpha\in A}S_\alpha$, where the S_α are open pieces of the $V_{\rm loc}^{(i)}$, and define *conditional probability measures* $\rho_\alpha^{(i)}$ on S_α such that

$$\rho \text{ restricted to } S=\int \rho_\alpha^{(i)} m(d\alpha) \ ,$$

where m is some measure on A. This definition is a bit more general than that given in Sec. IV.B. There $\lambda^{(i)}$ was the smallest positive characteristic exponent. We define

$$\delta^{(i)}=\dim_H\rho_\alpha^{(i)}$$

(this is a constant almost everywhere) and write

$$D^{(1)}=\delta^{(1)} \quad \text{if } \lambda^{(1)}>0 \ ,$$

and

$$D^{(i)}=\delta^{(i)}-\delta^{(i-1)} \quad \text{if } i>1 \text{ and } \lambda^{(i)}>0 \ .$$

Similarly, if $\lambda^{(i)}<0$, we define conditional probabilities $\rho_\alpha^{(j)}$ on pieces of *stable manifolds* and let

$$\delta^{(j)}=\dim_H\rho_\alpha^{(j)} \ .$$

We then write

$$D^{(r)}=\delta^{(r)}$$

if the smallest characteristic exponent $\lambda^{(r)}$ is negative, and

$$D^{(j)}=\delta^{(j)}-\delta^{(j+1)} \quad \text{if } j<r,\ \lambda^{(j)}<0 \ .$$

The definition of $D^{(k)}$ for $\lambda^{(k)}=0$ is somewhat arbitrary [between 0 and the multiplicity $m^{(k)}$ of $\lambda^{(k)}=0$]; we take $D^{(k)}=m^{(k)}$.

Theorem. The partial dimensions $D^{(1)},\ldots,D^{(r)}$ satisfy

$$0\le D^{(i)}\le m^{(i)} \quad \text{for } i=1,\ldots,r \ ,$$

where $m^{(i)}$ is the multiplicity of $\lambda^{(i)}$. The entropy is given by

$$h(\rho)=\sum_i{}^{+}\lambda(i)D^{(i)}=-\sum_i{}^{-}\lambda^{(i)}D^{(i)} \ , \tag{4.14}$$

where $\sum^+$ ($\sum^-$) is the sum over positive (negative) characteristic exponents, in particular

$$\sum_i \lambda^{(i)}D^{(i)}=0 \ . \tag{4.15}$$

The Hausdorff dimension satisfies

$$\dim_H \rho \le \sum_i D^{(i)} . \tag{4.16}$$

(It is not known if there is equality when no characteristic exponent vanishes.)

The proof of this theorem by Ledrappier and Young (1984) is not easy, but brings further dividends, in particular an interpretation of the numbers $|\lambda^{(i)}| D^{(i)}$ as *partial entropies.*

Some earlier theorems on entropy and Hausdorff dimension are recovered as corollaries of the above theorem, as we now indicate. [We follow Ledrappier and Young (1984); Grassberger (1984); Procaccia (1984).]

(a) First we recover from

$$h(\rho) = \sum_i{}^{+} \lambda^{(i)} D^{(i)}$$

the entropy inequality (4.4),

$$h(\rho) \le \sum_i{}^{+} \lambda^{(i)} m^{(i)} .$$

(b) The above inequality is in fact an equality (i.e., *ρ is an SRB measure) if and only if $D^{(i)} = m^{(i)}$ for all positive $\lambda^{(i)}$.*

Next we check that we recover the dimension inequality (4.12):

$$\dim_H \rho \le \dim_\Lambda \rho \tag{4.17}$$

from

$$\dim_H \rho \le \sum_i D^{(i)} \le \max \left\{ \sum_i d^{(i)} : 0 \le d^{(i)} \le m^{(i)} \text{ and } \sum_i d^{(i)} \lambda^{(i)} = 0 \right\} .$$

Proof. Let k be such that

$$\sum_1^k \lambda^{(i)} \ge 0 > \sum_i^{k+1} \lambda^{(i)} .$$

We then have $\lambda^{(k+1)} < 0$ and

$$\begin{aligned}
\dim_H \rho \le \sum_i D^{(i)} &= \frac{-1}{\lambda^{(k+1)}} \sum_i (-\lambda^{(k+1)}) D^{(i)} \\
&= \frac{-1}{\lambda^{(k+1)}} \sum_i (\lambda^{(i)} - \lambda^{(k+1)}) D^{(i)} \le \frac{-1}{\lambda^{(k+1)}} \sum_{i=1}^{k} (\lambda^{(i)} - \lambda^{(k+1)}) D^{(i)} \\
&\le \frac{-1}{\lambda^{(k+1)}} \sum_{i=1}^{k} (\lambda^{(i)} - \lambda^{(k+1)}) m^{(i)} = \sum_{i=1}^{k} m^{(i)} + \frac{\sum_{i=1}^{k} \lambda^{(i)} m^{(i)}}{|\lambda^{(k+1)}|} .
\end{aligned}$$

It is easily seen that the right-hand side is just the Liapunov dimension $\dim_\Lambda \rho$, and Eq. (4.17) follows.

(d) Suppose that we have equalities in the proof above, i.e., that the Kaplan-Yorke conjecture holds for ρ. Then we must have $D^{(i)} = m^{(i)}$ for $i = 1, \ldots, k$ and $D^{(i)} = 0$ for $i = k+2, \ldots, r$ (conversely these properties imply $\sum D^{(i)} = \dim_\Lambda \rho$). In particular, *if the Kaplan-Yorke conjecture holds for ρ then ρ is an SRB measure.*

Remarks.

(a) If $\sum \lambda^{(i)} m^{(i)} > 0$ we have $\dim_\Lambda \rho = m$ (the dimension of the manifold), which provides a trivial bound for $\dim_H \rho$. However, if one replaces f by f^{-1}, changing the sign of the $\lambda^{(i)}$, one gets a new Liapunov dimension, which is $< m$ and provides a nontrivial bound on the dimension of ρ.

(b) If there are only two *distinct* characteristic exponents, then $D^{(1)}$ and $D^{(2)}$ can be computed from Eq. (4.14).

(c) Let ρ be an SRB measure with r characteristic exponents such that $\lambda^{(1)} > \cdots > \lambda^{(r-1)} > 0 > \lambda^{(r)}$ and $\sum_1^r \lambda^{(i)} m^{(i)} < 0$. Then what we have said shows that $\sum D^{(i)} = \dim_\Lambda \rho$.

E. Escape from almost attractors

Before asymptotic behavior is reached by a dynamical system, transients of considerable duration are often observed experimentally. This is the case, for instance, for the Lorenz system [Sec. II.B, example (b)] as observed by Kaplan and Yorke (1979): for some values of the parameters *preturbulence* occurs in the form of long chaotic transients, even though the system does not yet have a strange attractor. One may say that the system has an *almost attractor* and try to estimate the escape rate from this set. More generally, one would like to have a precise description of *transient chaos* (see Grebogi, Ott, and Yorke, 1983b).

The situation, as usual, is best understood for the Axiom-A systems, where the *basic sets* (see Sec. III.F) are the natural candidates to describe almost attractors. Let Ω_i be a basic set, U a small neighborhood of Ω_i, and μ a measure with positive continuous density with respect to Lebesgue measure on U. Let

$$p(T) = \mu \left[\bigcap_{0 \le t \le T} f^t U \right]$$

be the amount of mass that has not left U by time T. One finds that $p(T) \approx e^{Pt}$, where

$$P = \max \left\{ h(\rho) - \sum \text{positive } \lambda_i(\rho): \right.$$

$$\left. \rho \text{ ergodic with support in } \Omega_i \right\} . \quad (4.18)$$

Note that P vanishes, as it should, if Ω_i is an attractor; the maximum is given in that case by the SRB measure. If Ω_i is not an attractor, then $P<0$; there is again a unique measure ρ_i realizing the maximum of Eq. (4.18), but it is no longer SRB (see Bowen and Ruelle, 1975).

If our dynamical system is not necessarily Axiom A, the following is a natural guess.

Conjecture. Write

$$P = h(\rho) - \sum \text{positive } \lambda_i(\rho) . \quad (4.19)$$

Then $|P|$ is the rate of escape from the support K of ρ, provided

$$P \geq h(\sigma) - \sum \text{positive } \lambda_i(\sigma)$$

for all ergodic σ with support in K. If $P > h(\sigma) - \sum$ positive $\lambda_i(\sigma)$ when $\sigma \neq \rho$ then ρ describes the time averages over transients near K.

A heuristic argument following the Axiom-A case makes this plausible, but it is unknown how generally the conjecture holds. Some satisfactory experimental verifications have been given by Kantz and Grassberger (1984). They write Eq. (4.19) as follows in terms of the *partial dimensions* $D^{(i)}$ discussed in Sec. IV.D:

$$|P| = -P = \sum_{i:\lambda^{(i)}>0} \lambda^{(i)}(m^{(i)} - D^{(i)}) .$$

F. Topological entropy*

The measure-theoretic entropy of Sec. IV.A gave the rate of information creation with respect to an ergodic measure. A related concept, involving the *topology* rather than a measure, will be discussed here.

Let K be a compact set and $f:K \to K$ a continuous map. If $\mathscr{A} = (\mathscr{A}_1, \ldots, \mathscr{A}_\alpha)$ is a finite open cover of K (i.e., $\cup_i \mathscr{A}_i \supset K$), we write

$$f^{-k}\mathscr{A} = (f^{-k}\mathscr{A}_1, \ldots, f^{-k}\mathscr{A}_k) ,$$

$$\mathscr{A}^{(n)} = \mathscr{A} \vee f^{-1}\mathscr{A} \vee \cdots \vee f^{-n+1}\mathscr{A}$$

$$= (\mathscr{A}_{i_1} \cap f^{-1}\mathscr{A}_{i_2} \cap \cdots \cap f^{-n+1}\mathscr{A}_{i_n}) .$$

Now let $N(\mathscr{A}, n)$ be the smallest number of sets in $\mathscr{A}^{(n)}$ that still covers K. The following limit is asserted to exist:

$$h_{\text{top}}(K, \mathscr{A}) = \lim_{n\to\infty} \frac{1}{n} \log N(\mathscr{A}, n) ,$$

and one defines the *topological entropy* of K by

$$h_{\text{top}}(K) = \sup_{\mathscr{A}} h_{\text{top}}(K, \mathscr{A}) .$$

If we have a metric on K we may write more conveniently

$$h_{\text{top}}(K) = \lim_{\text{diam}\mathscr{A} \to 0} h_{\text{top}}(K, \mathscr{A}) .$$

The following important theorem relates the topological entropy and the measure-theoretic entropies.

Theorem. If K is compact and $f:K \to L$ continuous, then $h_{\text{top}}(K) = \sup\{h(\rho): \rho$ is an ergodic measure with respect to $f\}$.

[This was conjectured by Adler, Konheim, and McAndrew (1965), and proved by Goodwyn, Dinaburg, and Goodman.] For references and more details on topological entropy we must refer the reader to Walters (1975) and Denker, Grillenberger, and Sigmund (1976).

G. Dimension of attractors*

The estimates of $\dim_H \rho$ in Sec. IV.C can be completed by estimates of the dimension of compact invariant sets (like the support of ρ, or attractors and attracting sets).

Theorem. Let A be a compact invariant set for a differentiable map f. Then

$$\dim_H A \leq \sup\{\dim_\Lambda \rho : \rho \text{ is ergodic with support in } A\} . \quad (4.20)$$

This result is due to Ledrappier (1981a), based on Douady and Oesterlé (1980); it is not known whether one can write $\dim_K A$ instead of $\dim_H A$ in Eq. (4.20). Note that, contrary to what Eq. (4.20) might suggest, there are cases where $\dim_H A > \sup\{\dim_H \rho : \rho$ is ergodic with support in $A\}$ (see McCluskey and Manning, 1983).

Lower bounds on $\dim_K A$ are also known. For instance, if a dynamical system has an attracting set A and a fixed point P with unstable dimension $m_+(P)$, then $\dim_K A \geq m_+(P)$ (see the corollary in Sec. III.E). For better estimates see Young (1981).

H. Attractors and small stochastic perturbations*

In this section we discuss how physical measures and attractors are selected by their stability under small stochastic perturbations.

1. Small stochastic perturbations

In Sec. II.F we discussed how the introduction of a small amount of noise in a deterministic system could select a particular invariant measure, the Kolmogorov measure. We can now be more precise. Consider first a discrete-time dynamical system generated by the map $f:M \to M$, where M has finite dimension m. Let $\varepsilon > 0$, and for each $x \in M$, let μ_x^ε be a probability measure with support in the ball $\bar{B}_x(\varepsilon) = \{y : d(x,y) \leq \varepsilon\}$.

More specifically, we assume that $\mu_x^\varepsilon(dy)$

$=\varepsilon^{-m}\varphi[x,\varepsilon^{-1}(y-x)]dy$, where φ is continuous, $\varphi\geq 0$, $\varphi(x,0)>0$, $\varphi(x,x-y)=0$ if $d(x,y)\geq 1$, and dy denotes the Lebesgue volume element if $M=\mathbb{R}^m$ (if M is not $\mathbb{R}^m$ this prescription is modified by using a Riemann metric on M; see Kifer, 1974). A *stochastic dynamical system* is a time evolution defined not on M, but at the level of probability measures on M. In our case we replace $f:M\to M$ by the stochastic perturbation

$$\nu\to\int\mu^{\varepsilon}_{fx}\nu(dx)$$
$$=\left[\int\varepsilon^{-m}\varphi[fx,\varepsilon^{-1}(y-fx)]\nu(dx)\right]dy\ . \qquad (4.21)$$

The limit of a small stochastic perturbation corresponds to taking $\varepsilon\to 0$.

In the continuous-time case, the dynamical system defined on $\mathbb{R}^m$ by the equation

$$\frac{dx_i}{dt}=F_i(x)$$

is replaced by an evolution equation for the density Φ of ν. We write $\nu(dx)=\Phi(x)dx$, and

$$\frac{d\Phi(x)}{dt}=\sum_i F_i\partial_i\Phi(x)+\varepsilon\Delta\Phi(x)\ . \qquad (4.22)$$

If we have a Riemann manifold, the Laplacian Δ should be replaced by the Laplace-Beltrami operator. The limit of a small stochastic perturbation corresponds again to taking $\varepsilon\to 0$.

Theorem. Let an Axiom-A dynamical system on the compact manifold M be defined by a twice differentiable diffeomorphism f or a twice differentiable vector field F. Let $A_1,\ldots,A_r$ be the attractors, and $\rho_1,\ldots,\rho_r$ the corresponding SRB measures.

(a) In the discrete-time case, for ε small enough, let ρ_i^{ε} be a stationary measure for the process (4.21) with support near A_i. Then $\rho_i^{\varepsilon}\to\rho_i$ when $\varepsilon\to 0$.

(b) In the continuous-time case, there is a unique stationary measure ρ^{ε} for the process (4.22), and any limit of ρ^{ε} when $\varepsilon\to 0$ is a convex combination $\sum\alpha_i\rho_i$ where $\alpha_i\geq 0$, $\sum\alpha_i=1$.

These results have been established by Kifer (1974), following Sinai's work on Anosov systems (1972). Another proof has been announced by Young. The idea behind the theorem is as follows. The noisiness of the stochastic time evolution yields measures which have continuous densities on M. The deterministic part of the time evolution will improve this continuity in the unstable directions by stretching, and roughen it in other directions due to contraction. In the limit one gets measures that are continuous along unstable directions, i.e., SRB measures.

Note the difference between the discrete-time and the continuous-time cases, which is due to the fact that in the discrete-time case, for small ε, a point near one attractor cannot jump out of its basin of attraction.

If we have a general dynamical system (not Axiom A), the stationary states for small stochastic perturbations will again tend to be continuous along unstable directions, but the limit when $\varepsilon\to 0$ need not be SRB (see the counterexample of Sec. IV.C). Moreover, there need not be an open basin of attraction associated with each SRB measure, so that, even in the discrete-time case, the stochastic perturbation may switch from one measure to another, and the limit may be a convex combination of many SRB measures (in particular be nonergodic). That basins of attraction may indeed be a mess is shown by the following result.

Theorem (Newhouse, 1974,1979). There is an open set $S\neq\emptyset$ in the space of twice differentiable diffeomorphisms of a compact two-dimensional manifold, and a dense subset R of S such that each $f\in R$ has an infinite number of attracting periodic orbits.

A variation of this result implies that for the Hénon map [see Sec. II.D, example (a)] the presence of infinitely many attracting periodic orbits is assured for some b and a dense set of values of a in some interval (a_0,a_1). The basins of such attracting periodic orbits are mostly very small and interlock in a ghastly manner.

The study of stochastic perturbations of differentiable dynamical systems is at present quite active; see in particular Carverhill (1984a,1984b) and Kifer (1984).

2. A mathematical definition of attractors

We have defined attractors operationally in Sec. II.C. Here, finally, we discuss a mathematical definition.

If $a,b\in M$, let us write $a\to b$ (a goes to b) provided for arbitrarily small $\varepsilon>0$ there is a chain $a=x_0,x_1,\ldots,x_n=b$ such that $d(x_k,f^{\Theta_k}x_{k-1})<\varepsilon$ with $\Theta_k\geq 1$ for $k=1,\ldots,n$. We accept $a\to a$ (corresponding to a chain of length 0), and it is clear that $a\to b,b\to a$ imply $a\to c$. If for every $\varepsilon>0$ there is a chain $a\to a$ of length ≥ 1 we say that a is *chain recurrent.* If a is chain recurrent, we define its *basic class* $[a]=\{b:a\to b\to a\}$. If $[a]$ consists only of a, then a is a fixed point. Otherwise if $b\in[a]$ then b is chain recurrent and $[b]=[a]$.

We shall say that a basic class $[a]$ is an *attractor* if $a\to x$ implies $x\to a$ (i.e., $x\in[a]$). This definition ensures that $[a]$ is attracting, but in a weaker sense than the definition of attracting sets in Sec. II.B. Here, however, we have irreducibility: an attractor cannot be decomposed into two distinct smaller attractors. (More generally, the set of chain-recurrent points decomposes in a unique way into the union of basic classes.)

It can be shown that any limit when $\varepsilon\to 0$ of a measure stable under small stochastic perturbations of a discrete-time dynamical system is "carried by attractors," at least in a weak sense. More precisely, this can be stated as the following theorem.

Theorem. Let Λ be a compact attracting set for a discrete-time dynamical system, m a probability measure with support close to Λ, and ε sufficiently small. Let also m_{ε}^{k} be obtained at time k from the stochastic evolution (4.22). If $[a]$ is not an attractor, then

$$\lim_{k\to\infty}m_{\varepsilon}^{k}[B_a(\delta)]=0\ ,$$

when δ is sufficiently small. [$B_a(\delta)$ denotes the ball of radius δ centered at a.] In particular, if m_ε^∞ is a limit of m_ε^k when $k\to\infty$, then a does not belong to the support of m_ε^∞. If ρ is a limit when $\varepsilon\to 0$ of m_ε^∞, and if m is ergodic, then its support is contained in an attractor.

Proof. See Ruelle (1981).

The topological definition of an attractor given in this section follows the ideas of Conley (1978) and Ruelle (1981). A rather different definition has been proposed recently by Milnor (1985), based on the privileged role which the Lebesgue measure should play for physical dynamical systems.

I. Systems with singularities and systems depending on time*

The theory of differentiable dynamical systems may to some extent be generalized to differentiable dynamical system *with singularities*. This is of interest, for instance, in Hamiltonian systems with collisions (billiards, hard-sphere problems). On this problem we refer the reader to the considerable work by Katok and Strelcyn (1985).

Another conceptually important extension of differentiable dynamical systems is to systems with *time-dependent forces*. One does not allow here for arbitrary non-autonomous systems, but assumes that

$$x(t+1)=f(x(t),\omega(t)) \quad\text{or}\quad \frac{dx}{dt}=F(x,\omega(t)) ,$$

where ω has a stationary distribution. (For instance, ω is defined by a continuous dynamical system.) It is surprising how many results extend to this more general situation; the extension is without pain, but the formalism more cumbersome. Here again we can only refer to the literature. See Ruelle (1984) for a general discussion, and Carverhill (1984a,1984b) and Kifer (1984) for problems involving stochastic differential equations and random diffeomorphisms.

V. EXPERIMENTAL ASPECTS

Now that we have developed a theoretical background and a language in which to formulate our questions, it is time to discuss their experimental aspects. A basic conceptual problem is that of confronting the limited information that can be obtained in a real experiment with the various limits encountered in the mathematical theory. A similar situation occurs, for instance, in the application of statistical mechanics to the study of phase transitions. Other important problems in the relation between theory and experiment concern numerical efficiency and accuracy. The present section will address those problems.

We shall describe two different fields of experimentation—computer experiments and experiments with real physical systems. There is a quantitative difference between the two fields, since one can study dynamical evolution equations with *fixed* experimental conditions more accurately on a computer than in reality. However, there is also a more important qualitative difference: Since the evolution equations are explicitly known in a computer experiment, it is generally easy to compute directly the "tangent map" Df^t. In a physical experiment, by contrast, only points on a trajectory are directly measurable, and the derivatives (tangents) have to be obtained by a delicate interpolation, to be discussed below.

It must be understood that the information currently being extracted from experiments goes a long way beyond the solid mathematical foundations that we have described in the previous sections. It is a challenge for the mathematical physicist to clarify the relations between the various quantities measured on dynamical systems. Most of them seem indeed very interesting, and very promising, but a lot of work is still necessary to prove the existence of these quantities and establish their relations. Our selection below reflects to some extent our personal taste for measurements based on sound ideas and for which a mathematical foundation can be expected.

A. Dimension

The measurement of dimensions is discussed first because it is most straightforward. We concentrate on the determination of the *information dimension*, using the method advocated by Young (1982), and Grassberger and Procaccia (1983b). The idea is described in the first theorem (by Young) in Sec. IV.C. The method, developed independently by Grassberger and Procaccia, has gained wide acceptance through their work.

We start with an experimental time series $u(1),u(2),\ldots$, corresponding to measurements *regularly spaced in time*. We assume that $u(i)\in\mathbb{R}^\nu$, where $\nu=1$ in the (usual) case of scalar measurements. From the $u(i)$, a sequence of points $x(1),\ldots$, in $\mathbb{R}^{m\nu}$ is obtained by taking $x(i)=[u(i+1),\ldots,u(i+m-1)]$. This construction associates with points $X(i)$ in the phase space of the system (which is, in general, infinite dimensional) their projections $x(i)=\pi_m X(i)$ in $\mathbb{R}^{m\nu}$. In fact, if ρ is the physical measure describing our system (ρ is carried by an attractor in phase space), then the points $x(i)$ are equidistributed with respect to the projected measure $\pi_m\rho$ in $\mathbb{R}^{m\nu}$. [Actually this is not always true: if the time spacing Δt between consecutive measurements $u(i),u(i+1)$ is a "natural period" of the system—for instance, when the system is quasiperiodic—one does not have equidistribution. This exception is easily recognized and handled.] We wish to deduce $\dim_H\rho$ from this information (with the possibility of varying m in the above construction). Before seeing how this is done, a general word of caution is in order. In any given experiment we have only a finite time series, and therefore there are natural limits on what can be extracted from it: some questions are too detailed (or the statistical fluctuations too large) for a reasonable answer to come out. See, for instance, Guckenheimer (1982) for a discussion of such matters.

A serious difficulty seems to arise here from the fact that $\dim_H \pi_m \rho$ need not be equal to the desired $\dim_H \rho$. We remove this objection with the observation that, if $\dim_H \rho \leq M$, then, for most M-dimensional projections p, $\dim_H p\rho = \dim_H \rho$. More precisely, we have the following result.

Theorem. Let $0<M<n$. If E is a Suslin set in $\mathbf{R}^n$, and $\dim E \leq M$, then there is a Borel set G in the space of orthogonal projections $p:\mathbf{R}^n \to \mathbf{R}^M$ such that its complement has measure zero with respect to the natural rotation-invariant measure on projections, and $\dim pE = \dim E$ for all p in G.

Proof. See Lemma 5.3 in Mattila (1975). We are indebted to C. McCullen for this reference. It is interesting to compare this result with that of Mañé in Sec. II.G, where one obtains (with stronger restrictions) the injectivity of p.

Of course we have not proved that our projection π_m belongs to the good set G of the above theorem (with $M=m\nu$), but this appears to be a reasonable guess, and we shall proceed with the assumption that $\dim H_H \pi_m \rho = \dim_H \rho$ for large enough m.

We now use our sequence $x(1), x(2), \ldots, x(N)$ in $\mathbf{R}^{m\nu}$ to construct functions C_i^m and C^m as follows:

$$C_i^m(r) = N^{-1}\{\text{number of } x(j) \text{ such that } d[x(i),x(j)] \leq r\}\ , \tag{5.1}$$

$$C^m(r) = N^{-1}\sum_i C_i^m(r)$$
$$= N^{-2}\{\text{number of ordered pairs } [x(i),x(j)] \text{ such that } d[x(i),x(j)] \leq r\}\ . \tag{5.2}$$

[C_i^m is obtained by sorting the $x(j)$ according to their distances to $x(i)$; C^m is obtained more efficiently directly by sorting pairs than as an average of the C_i^m.] We may use $d(x,x')$ = Euclidean norm of $x'-x$, or any other norm, such as

$$|x'-x| = \max_\alpha |u'(\alpha) - u(\alpha)|\ ,$$

where the $u(\alpha)$ are the m components of x, and $|u'(\alpha)-u(\alpha)|$ is for instance the Euclidean norm in $\mathbf{R}^\nu$ (this will be used in Sec. V.B). Note that when $N\to\infty$, we have

$$\lim C_i^m = (\pi_m \rho)[B_{x(i)}(r)] \tag{5.3}$$

(except perhaps at discontinuity points of the right-hand side). Suppose now that

$$\lim_{r\to 0}\lim_{N\to\infty} \frac{\log C_i^m(r)}{\log r} = \lim_{r\to 0} \frac{\log(\pi_m\rho)[B_{x(i)}(r)]}{\log r} = \alpha_m\ . \tag{5.4}$$

Then $\dim_H \pi_m \rho = \alpha_m$ (first theorem of Sec. IV.C). Provided the projection π_m is in the "good set" G, we have thus $\alpha_m = m\nu$ if $\dim_H \rho \geq m\nu$ and $\alpha_m = \dim_H \rho$ if $\dim_H \rho \leq m\nu$. Experimentally, α_m may be obtained by plotting $\log C_i^m(r)$ vs $\log r$ and determining the slope of the curve (see below). With a little bit of luck [existence of the limit (5.4), and π_m in the good set] we may thus obtain $\dim_H \rho$ experimentally: *we choose m such that $\alpha_m < m\nu$; then we have* $\dim_H \rho = \alpha_m$. Although we cannot completely verify that π_m is in the good set, we can in principle check (within experimental accuracy) the existence of the limits α_m, and the fact that α_m becomes independent of m when m increases beyond a value such that $\alpha_m \leq m\nu$. Note that α_m should also be independent of the index i in Eq. (5.4).

The information dimension $\dim_H \rho$ may also be obtained by a modification of Eq. (5.4). We describe the method of Grassberger and Procaccia, which has been tested experimentally in a number of cases. This consists in writing

$$\lim_{r\to 0}\lim_{N\to\infty} \frac{\log C^m(r)}{\log r} = \beta_m\ , \tag{5.5}$$

and asserting that for m sufficiently large, β_m is the information dimension. The only relation that can easily be established rigorously between Eqs. (5.4) and (5.5) is that if both limits exist, then $\alpha_m \geq \beta_m$. However, it seems quite reasonable to assume that in general $\alpha_m = \beta_m$ (i.e., if the C_i^m behave like r^α, then their linear superposition C^m also behaves like r^α). The β_m obtained experimentally do become independent of m for m large enough, as expected. See Fig. 19.

To summarize: the method of Grassberger and Procaccia is a highly successful way of determining the information dimension experimentally. Values between 3 and 10 are obtained reproducibly. The method is not entirely justified mathematically, but nevertheless quite sound. The study of the limit (5.4) is also desirable, even though the statistics there is poorer.

1. Remarks on physical interpretation

a. *The meaningful range for $C^m(r)$*

Suppose we plot $\log C(r)/\log r$ as a function of $\log r$ (we suppress the superscript m and possibly the subscript i of C). First, for small r, we have a large scatter of points due to poor statistics; then there is a range (r_0, r_1) of near constancy (the constant is the information dimension if m is suitably large). For r larger than r_1 we have deviation from constancy due to nonlinear effects. The "meaningful range" (r_0, r_1) is that in which the distribution of distances between pairs of points is statistically useful.

b. *Curves with "knees"*

It is not uncommon that the $\log C(r)$ vs $\log r$ plot shows a "knee" (see Fig. 20), so that it has slope α in the range

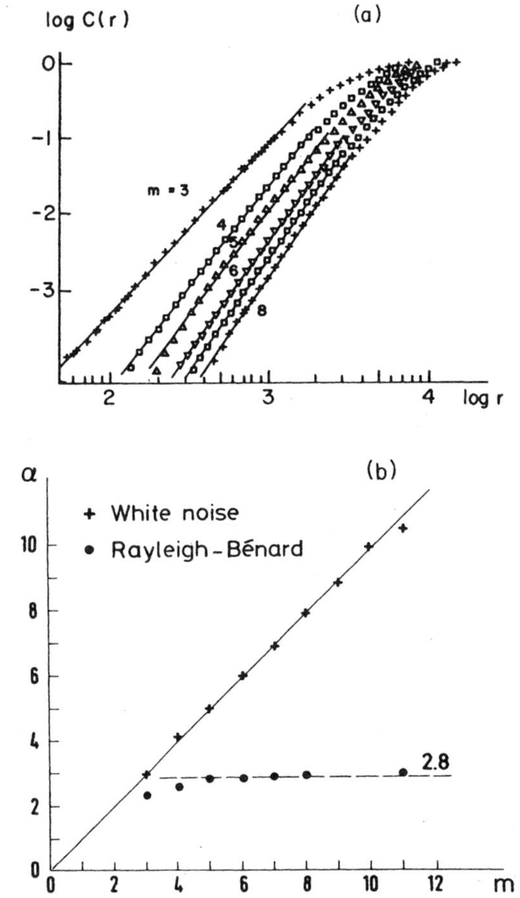

FIG. 19. Experimental results from Malraison *et al.* (1983) and Atten *et al.* (1984); see also Dubois (1982): (a) The plots show $\log C$ vs $\log r$ for different values of the embedding dimension, for the Rayleigh-Bénard experiment. (b) The measured dimension α as a function of the embedding dimension m, both for the Rayleigh-Bénard experiment and for numerical white noise. Note that α becomes nearly constant (but not quite) at $m=3$. The α for white noise is nearly equal to m (but not quite).

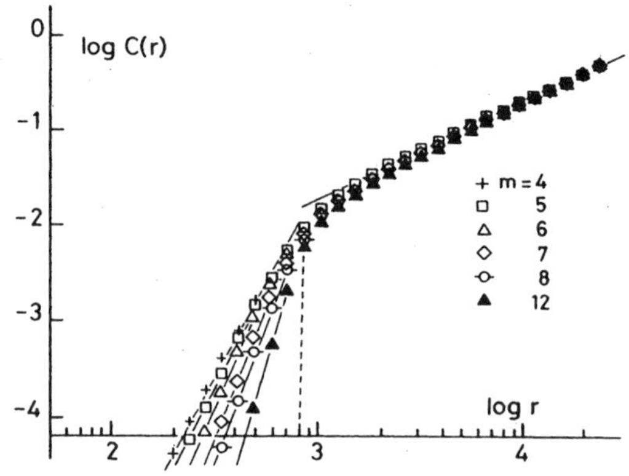

FIG. 20. A $\log C(r)$ vs $\log r$ plot may show a "knee," so that a dimension α appears in the range $(\log r_0, \log r_1)$, and a dimension α' in the range $(\log r_1, \log r_2)$. Here the scalar signal of a deterministic system with information dimension α' is perturbed by the addition of random noise, which yields a dimension α equal to the embedding dimension m (see Atten *et al.*, 1984).

$(\log r_0, \log r_1)$ and a smaller slope α' in the range $(\log r_1, \log r_2)$. The dimension, or "number of degrees of freedom" is thus different for r above and below r_1 (see, for instance, Riste and co-workers, 1985). To see how this situation can arise, let us consider a *product* dynamical system $\mathrm{I}\times\mathrm{II}$ formed of two noninteracting subsystems I and II. Take an observable $u=u_{\mathrm{I}}+u_{\mathrm{II}}$, where u_{I} and u_{II} depend only on the subsystems I and II, respectively, and let the amplitude r_1 of the signal u_{I} be much smaller than that of u_{II}. In the range $r<r_1$ we have statistical information on the complete system $\mathrm{I}\times\mathrm{II}$, giving an information dimension α. In the range $r\gg r_1$ we have statistical information only on the subsystem II, giving an information dimension α'. More generally, suppose that system II evolves independently of I, but that I has a time evolution that may depend on II; then the same conclusions persist for this "semidirect product." (The small-amplitude modes of I are driven here by system II, an example of Haken's "slaving principle.") The above argument makes clear, for instance, how the information dimension found by analysis of a turbulent hydrodynamic system does not take into account small ripples of amplitude less than the discrimination level r_0 of the analysis. (We thank P. C. Martin for useful discussions on this point.) A knee will also appear if the signal from a deterministic chaotic system is perturbed by adding random noise of smaller amplitude (see Fig. 20).

c. Spatially localized degrees of freedom

We have just discussed dynamical systems that have a product structure $\mathrm{I}\times\mathrm{II}$, or where a subsystem II evolves independently and drives other degrees of freedom. Strictly speaking, such decoupling does not seem to occur in realistic situations like that of a turbulent viscous fluid (except for the trivial case where the fluid is in two different uncoupled containers). Normally, in a nonlinear system one may say that "every mode is coupled with all other modes," and *exact* factorization is impossible. An apparent exception is constituted by quasiperiodic motions where factorization is present, but the independent frequencies do not correspond to independent physical subsystems. In other words, if a physical variable $u(t)$ of the system is monitored (for instance a component of the velocity of a viscous fluid at one point), the whole dynamics of the system (on the appropriate attractor) can in principle be reconstructed from the time series $[u(t):t$ varying from 0 to $\infty]$. In particular, the information dimension of the system can be obtained indifferently from monitoring u or any other physical variable.

As noted in Secs. V.A.1.a and V.A.1.b, the experimental uncertainties change this situation. At the level of accuracy of an experiment, some degrees of freedom may effectively be driven by others and, having small amplitude, pass unnoticed. (A case in point would be that of eddies of small size in three-dimensional turbulence.) Another frequent and important case occurs when some "oscillators" (possibly complex oscillators) are strongly localized in some region of space.[3] Consider, for instance, the flow between coaxial rotating cylinders in a regime where there is some turbulence superposed with Taylor cells. Some features of the flow are global (like the very existence of the Taylor cells), others seem to be restricted to one Taylor cell, having very little interaction with neighboring cells.

The information dimension d_c, obtained from moderate-precision measurements of one cell, is then likely to be different from the global information dimension d_ν of a column of ν cells (and one expects formulas like $d_c=a+b$, $d_\nu=a+\nu b$). Note that d_ν could be obtained by monitoring a vector signal with ν components, each corresponding to a scalar signal from one cell.

2. Other dimension measurements

The most straightforward way to find the "fractal" dimension of a set A is to cover it with a grid of size r, to count the number $N(r)$ of occupied cells, and to compute

$$\lim_{r\to 0}\frac{\log N(r)}{|\log r|} .$$

This box-counting is computationally ineffective (Farmer, 1984). It gives access to the dimension of attractors rather than to the information dimension (the latter seems for the moment to have greater theoretical interest). Another problem with box-counting is that usually the population of boxes is very uneven, so that it may take a considerable amount of time before some "occupied" boxes really become occupied. For all these reasons, the box-counting approach is not used currently.

B. Entropy

The entropy (or Kolmogorov-Sinai invariant) $h(\rho)$ of a physical measure ρ is an important quantity, as we have seen in Sec. IV. Early attempts to measure $h(\rho)$ were based directly on the definitions and used a partition $\mathscr{A}$ (see Sec. IV.A). These attempts (Shimada, 1979; Curry, 1981; Crutchfield, 1981) were interesting but not entirely successful. We describe here another approach due to Grassberger and Procaccia (1983a); see also Cohen and Procaccia (1984). [Similar ideas were developed independently by Takens (1983).] This approach has far greater potential for implementation in experimental situations.

The idea of Grassberger and Procaccia is to exploit the m dependence of the functions $C_i^m(r)$ and $C^m(r)$ defined in Eqs. (5.1) and (5.2). As before, they use $C^m(r)$, which has better statistics, but it is easier to argue with the $C_i^m(r)$, which satisfy

$$C_i^m(r)\approx(\pi_m\rho)[B_{x(i)}(r)] \tag{5.6}$$

for large N [see Eq. (5.3)]. In view of Eq. (5.6), $C_i^m(r)$ is the probability that $x(j)$ satisfies $d[x(j),x(i)]\le r$. Grassberger and Procaccia use the Euclidean norm, but we prefer to follow Takens and to take

$$d[x(j),x(i)]=\max\{|u(j)-u(i)|,\ldots,|u(j+m-1)-u(i+m-1)|\} .$$

Usually $\nu=1$ (scalar signal), but the general case is not harder to handle [with $|u(i)-u(j)|$ being the Euclidean norm of $u(i)-u(j)$ in $\mathbf{R}^\nu$]. We may thus interpret $C_i^m(r)$ as the probability that the signal $u(j+k)$ remains in the ball $B^\nu_{u(i+k)}(r)$ for m consecutive units of time [$B^\nu_{u(i)}(r)$ is the ball of radius r in $\mathbf{R}^\nu$ centered at $u(i)$].

With this interpretation, and the fact that $C^m(r)$ is the average of the $C_i^m(r)$, it can be argued that

$$\lim_{r\to 0}\lim_{m\to\infty}\frac{1}{m}\lim_{N\to\infty}[-\log C^m(r)]=\Delta t K_2(\rho) , \tag{5.7}$$

where K_2 has been defined at the end of Sec. IV.A, and Δt is the spacing between measurements of the signal u. Since $K_2(\rho)$ is a lower bound to $h(\rho)$, we see that if one obtains $K_2(\rho)>0$ from Eq. (5.7) then one can conclude that $h(\rho)>0$, i.e., that the system is chaotic.

It is, however, also possible to obtain $h(\rho)$ directly as follows. Define

$$\Phi^m(r)=\frac{1}{N}\sum_i \log C_i^m(r) .$$

Then

$$\Phi^{m+1}(r)-\Phi^m(r)=\text{average over } i \text{ of } \log[\text{probability that } u(j+m)\in B^\nu_{u(i+m)}(r)$$
$$\text{given that } u(j+k)\in B^\nu_{u(i+k)}(r) \text{ for } k=0,\ldots,m-1] .$$

Therefore,

[3]For some discussion of the difficult problem of localization in hydrodynamics, see Ruelle (1982b).

$$\lim_{r\to 0}\lim_{m\to\infty}\lim_{N\to\infty}[\Phi^{m+1}(r)-\Phi^m(r)]=\Delta th(\rho)\ . \quad (5.8)$$

Remarks.

(a) Like the expressions of Sec. V.A, the identities (5.7) and (5.8) hold "if all goes well." Basically, the condition is that the monitored signal should reveal enough of what is going on in the system.

(b) While the information dimension could be obtained from $C^m(r)$ for one single m (sufficiently large), we have a limit $m\to\infty$ in Eqs. (5.7) and (5.8). In this respect, (5.7) is not optimal (it will contain errors of order $1/m$); it is better to write

$$\lim_{r\to 0}\lim_{m\to\infty}\lim_{N\to\infty}\log\frac{C^m(r)}{C^{m+1}(r)}=\Delta tK_2(\rho)\ ,$$

or to use Eq. (5.8).

C. Characteristic exponents: computer experiments

We recall that the characteristic exponents measure the exponential separation of trajectories in time and are computed from the *derivative* D_xf^t. In computer experiments, the derivative is often directly calculable, whereas in physical experiments it has to be obtained indirectly from the experimental signal. Therefore the methods for evaluating characteristic exponents are somewhat different in the two cases and will be treated separately. In this section, we discuss computed experiments, which have served and still serve an important purpose in the exploration of dynamical systems.

Let us mention here some interesting open problems. What is the distribution of characteristic exponents for a large or a highly excited system? Can one define a density of exponents per unit volume for a spatially extended system? What is the behavior near zero exponent? For a theoretical study in the case of turbulence, see Ruelle (1982b,1984). For an experimental study in the case of the Kuramoto-Sivashinsky model, see Manneville (1985).

In the case of a discrete-time dynamical system defined by a map $f:\mathbf{R}^m\to\mathbf{R}^m$, let

$$T(x)=D_xf\ .$$

This is the matrix of partial derivatives of the m components of $f(x)$ with respect to the m components of x. Write

$$T_x^n=T(f^{n-1}x)\cdots T(fx)T(x) \quad (5.9)$$

(matrix multiplication on the right-hand side). Then the largest characteristic exponent is given by

$$\lambda_1=\lim_{n\to\infty}\frac{1}{n}\log\|T_x^n u\| \quad (5.10)$$

for almost any vector u, and this is a very efficient way to obtain λ_1. The other characteristic exponents can in principle be obtained by diagonalizing the positive matrices $(T_x^n)^*T_x^n$ and using the fact that their eigenvalues behave like $e^{2n\lambda_1},e^{2n\lambda_2},\ldots$. Obviously, for large n, the different eigenvalues have very different orders of magnitude, and this creates a problem if T_x^n is computed without precaution. When $(T_x^n)^*T_x^n$ is diagonalized, the small relative errors on the large eigenvalues might indeed contaminate the smaller ones, causing intolerable inaccuracy. We shall see below how to avoid this difficulty.

Consider next a continuous-time dynamical system defined by a differential equation

$$\frac{dx(t)}{dt}=F(x(t))\ , \quad (5.11)$$

in $\mathbf{R}^m$. An early proposal to estimate λ_1 (Benettin, Galgani, and Strelcyn, 1976) used solutions $x(t),x'(t),x''(t),\ldots$, chosen as follows. The initial condition $x'(0)$ is chosen very close to $x(0)$, and $x(t)$ remains close to $x'(t)$ up to some time T_1; one then replaces the solution x' by a solution x'' such that $x''(T_1)-x(T_1)=\alpha[x'(T_1)-x(T_1)]$ with α small. Thus $x''(T_1)$ is again very close to $x(T_1)$, and $x''(t)$ remains close to $x(t)$ up to some time $T_2>T_1$, and so on. The rate of deviation of nearby trajectories from $x(\ )$ can thus be determined, yielding λ_1. This simple method has also been applied to physical experiments (Wolf *et al.*, 1984); we shall return to this topic in Sec. V.D.

In the case of (5.11) one can, however, do much better. Namely, one differentiates to obtain

$$\frac{d}{dt}u(t)=(D_{x(t)}F)[u(t)]\ , \quad (5.12)$$

which is *linear* in u, but with nonconstant coefficients. The solution of (5.11) yields $x(t)=f^t(x(0))$, and the solution of (5.12) yields

$$u(t)=(D_{x(0)}f^t)u(0)\ .$$

Therefore one can readily compute the matrices $T_x^t=D_xf^t$ by integrating Eq. (5.12) with m different initial vectors u. Better yet, one can use the matrix differential equation

$$\frac{d}{dt}T_{x(0)}^t=(D_{x(t)}F)T_{x(0)}^t\ ,$$

with $T_{x(0)}^0$ the identity matrix.

As in the discrete case, it is not advisable to compute T_x^t for large t. We choose a reasonable unit of time τ: not too large, so that the $e^{\lambda_i\tau}$ do not differ too much in their orders of magnitude, but not too small either, because we have to multiply a number of matrices proportional to τ^{-1}. Having chosen τ, we discretize the time (setting $\tilde{f}=f^\tau$) and proceed as in the discrete-time case. If the characteristic exponents for $\tilde{f}$ are $\tilde{\lambda}_i$, then the characteristic exponents for the continuous-time system are $\lambda_i=\tau^{-1}\tilde{\lambda}_i$.

Before discussing the accurate calculation of the λ_i for $i>1$, let us mention that the knowledge of λ_1 [obtained from Eq. (5.10)] is sometimes sufficient to determine *all* characteristic exponents. This is certainly the case for one-dimensional systems, as well as for the Hénon map and the Lorenz equation, as we have seen in the examples of Sec. III.D.1. It is also possible to estimate successively

λ_1 by (5.10), then $\lambda_1+\lambda_2$ as the rate of growth of surface elements, $\lambda_1+\lambda_2+\lambda_3$ as the rate of growth of three-volume elements, etc. This approach was first proposed by Benettin *et al.* (1978). In what follows, we discuss a somewhat different method.

The algorithm we propose for the calculation of the λ_i is very close to the method presented by Johnson *et al.* (1984) for proving the multiplicative ergodic theorem of Oseledec. Remember that we are interested in the product (5.9):

$$T_x^n = T(f^{n-1}x)\cdots T(fx)T(x) \ .$$

To start the procedure, we write $T(x)$ as

$$T(x)=Q_1R_1 \ , \tag{5.13}$$

where Q_1 is an orthogonal matrix and R_1 is upper triangular with non-negative diagonal elements. [If $T(x)$ is invertible, this decomposition is unique.] Then for $k=2,3,\ldots$, we successively define

$$T_k' = T(f^{k-1}x)Q_{k-1}$$

and decompose

$$T_k' = Q_k R_k \ ,$$

where Q_k is orthogonal and R_k upper triangular with non-negative diagonal elements. Clearly, we find

$$T_x^n = Q_n R_n \cdots R_1 \ .$$

To exploit this decomposition, we shall make use of the results of Johnson *et al.*, but note that those are only proved in the "invertible case" of a dynamical system defined for negative as well as positive times. In the paper referred to, an orthogonal matrix Q is chosen at random (i.e., Q is equidistributed with respect to the *Haar* measure on the orthogonal group), and the initial $T(x)$ is replaced by $T(x)Q$ in Eq. (5.13), the matrices $T(f^{k-1}x)$ for $k>1$ being left unchanged. It is then shown that the diagonal elements $\lambda_{ii}^{(n)}$ of the upper triangular matrix product $R_n \cdots R_1$ obtained from this modified algorithm satisfy

$$\lim_{n\to\infty} \frac{1}{n}\log\lambda_{ii}^{(n)} = \lambda_i \tag{5.14}$$

almost surely with respect to the product of the invariant measure ρ and the Haar measure (corresponding to the choice of Q). *On the right-hand side of Eq. (5.14) we have the characteristic exponents arranged in decreasing order.* For practical purposes, it is clearly legitimate to take $Q=$identity.

In the case of constant $T(f^kx)$, i.e., $T(f^kx)=A$ for all k, the above algorithm is known as the "Analog of the treppen-iteration using orthogonalization."[4] See Wilkinson (1965, Sec. 9.38, p. 607). The multiplicative ergodic theorem can thus be viewed as the generalization of this algorithm to the case when the $T(f^kx)$ are randomly chosen.

Let us again call $\lambda^{(1)},\ldots,\lambda^{(r)}$ the *distinct* characteristic exponents, and $m^{(1)},\ldots,m^{(r)}$ their multiplicities. The space $E^{(i)}$ associated with the characteristic exponents $\leq\lambda^{(i)}$ (see Sec. III.A) is obtained as follows. Consider the last $m_-^{(i)}=m^{(i)}+\cdots+m^{(r)}$ columns of the matrix

$$R_1^{-1}\cdots R_n^{-1}\Delta = (R_n\cdots R_1)^{-1}\Delta \ ,$$

where Δ is the diagonal matrix equal to the diagonal part of $R_n\cdots R_1$. Let $E^{(i)}(n)$ be the space generated by these $m_-^{(i)}$ column vectors. Then $E^{(i)}=\lim_{n\to\infty}E^{(i)}(n)$.

Note that if we are only interested in the largest s characteristic exponents, $\lambda_1\geq\cdots\geq\lambda_s$, then it suffices to do the decomposition to triangular form only in the upper left $s\times s$ submatrix, leaving the matrices untouched in the lower right $(m-s)\times(m-s)$ corner.

The practical task of decomposing a matrix T_k' as Q_kR_k, as discussed above, is abundantly treated in the literature, and library routines exist for it. According to Wilkinson (1965, Secs. 4.47—4.56), the Householder triangularization is preferable to Schmidt orthogonalization, since it leads to more precisely orthogonal matrices. This algorithm is available in Wilkinson and Reinsch (1971, Algorithm I/8, procedure "decompose"). It exists as part of the packages EISPACK and NAG. This algorithm is numerically very stable, and in fact the size of the eigenvalues should not matter.

D. Characteristic exponents: physical experiments

By contrast with computer experiments, experiments in the laboratory do not normally give direct access to the derivatives D_xf^t. These derivatives must thus be estimated by a detailed analysis of the data. Once the derivatives D_xf^t are known, the problem is analogous to that encountered in computer experiments. The same algorithms can be applied to obtain either the largest characteristic exponent λ_1 or other exponents. Only the positive exponents will be determined, however, or part of them. We have seen above how to restrict the computation to the largest s characteristic exponents, and we shall see below why one can only hope to determine the *positive* λ_i in general.

As in Sec. V.A we start with a time series $u(1),u(2),\ldots$, in $\mathbf{R}^\nu$, and from this we construct a sequence $x(1),x(2),\ldots$, in $\mathbf{R}^{m\nu}$, with $x(i)=[u(i),\ldots,u(i+m-1)]$. We shall discuss in remark (c) below how large m should be taken. We shall now try to estimate the derivatives $T_{x(i)}^\tau = D_{x(i)}f_\tau$. As in Sec. V.C, τ should be such that the $e^{\lambda_k\tau}$ are not too large (we are only interested in positive λ_k); this means that τ should not be larger (and rather smaller) than the "characteristic time" of the system. Also, τ should not be too small, since we have to multiply later a number of matrices $T_{x(i)}^\tau$ proportional to τ^{-1}. Of course, τ will be a multiple $p\Delta t$ of the time interval Δt between measurements, so that

[4]We thank G. Wanner for helpful discussions in relation to this problem.

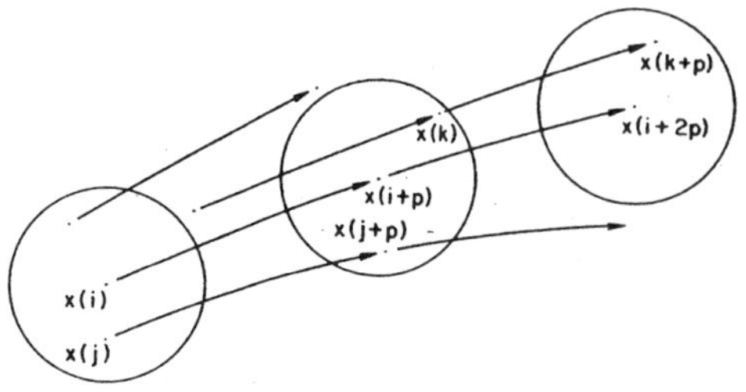

FIG. 21. The balls of radius $\tilde{r}$ centered at $x(i)$, $x(i+p)$, and $x(i+2p)$.

$f_\tau x(i)=x(i+p)$.

The derivatives $T^\tau_{x(i)}$ will be obtained by a best linear fit of the map which, for $x(j)$ close to $x(i)$, sends $x(j)-x(i)$ to $f^\tau(x(j))-f^\tau(x(i))=x(j+p)-x(i+p)$ (see Fig. 21). Close here means that the map should be approximately linear. This will be ensured by choosing $\tilde{r}$ sufficiently small and taking only those $x(j)$ for which

$$d[x(i),x(j)]\le\tilde{r} \text{ and } d[x(i+p),x(j+p)]\le\tilde{r}$$

(these conditions imply that $x(j+k)$ is close to $x(i+k)$ for $0<k<p$; one could also require $d[x(i+k), x(j+k)]\le\tilde{r}$ for each k separately). The choice of $\tilde{r}$ should probably be made by trial and error, monitoring how good a linear fit is obtained. Certainly $\tilde{r}$ should be less than the upper bound r_1 of the "meaningful range for $C^m(r)$," discussed in remark (a) of Sec. V.A. Having chosen $\tilde{r}$, we have to assume that the length N of the original time series is sufficiently long so that there is a fair number of points $x(j)$ in the ball of radius $\tilde{r}$ around $x(i)$, and such that $x(j+p)$ is also in the ball of radius $\tilde{r}$ around $x(i+p)$. In principle, m points are enough to determine a linear map, but we want many points: (a) to overcome the statistical scatter of the $x(j)$, (b) because a symmetric distribution of the $x(j)$ will yield a linear best fit from which the quadratic nonlinear terms have been eliminated. We repeat how $T^\tau_{x(i)}=D_{x(i)}f^\tau$ is obtained, with $\tau=p\Delta t$. Take all $x(j)$ such that $d[x(i),x(j)]\le\tilde{r}$ and $d[x(i+p),x(j+p)]\le\tilde{r}$, and determine the $m\nu\times m\nu$ matrix $T^\tau_{x(i)}$ by a least-squares fit[5] such that

$$T^\tau_{x(i)}[x(j)-x(i)]\approx x(j+p)-x(i+p)\ . \qquad (5.15)$$

Note that when we estimate $T^\tau_{x(i+p)}$ we have to start looking again for *all* $x(k)$ such that $d[x(i+p),x(k)]\le\tilde{r}$ and $d[x(i+2p),x(k+p)]\le\tilde{r}$, and not just for the $x(k)$ of the form $x(j+p)$!

In general, the points $x(j)$ will not be uniformly distributed in all directions from $x(i)$. In other words, the vectors $x(j)-x(i)$ may not span $\mathbf{R}^{m\nu}$, and therefore the matrix $T^\tau_{x(i)}$ may not be well defined by our prescription. Even if $T^\tau_{x(i)}$ is defined, there will in general be directions in which there are many fewer points than in others, so that the uncertainty on the elements of $T^\tau_{x(i)}$ corresponding to those directions is large. In fact, we can only expect with confidence that the vectors $x(j)-x(i)$ span the *expanding directions* around $x(i)$, i.e., the linear space tangent to the unstable manifold at $x(i)$ (because an SRB measure is absolutely continuous along the unstable manifold or because an attracting set contains the unstable manifolds of the points on it). The fact that the matrix $T^\tau_{x(i)}$ is only known with confidence in the unstable directions need not distress us: It means that we can determine with confidence only the *positive characteristic exponents.* This is done with the method of Sec. V.C, constructing triangular matrices from the T^τ_x for $x=x(i_0)$, $x(i_0+p), x(i_0+2p),\ldots$, computing the characteristic exponents, and discarding those which are ≤ 0. The latter will usually (although not necessarily) be meaningless.

Remarks.

(a) The detailed method presented above for deriving characteristic exponents from experiments seems new. Up to now, attention has been concentrated on obtaining the largest exponent λ_1, using basically the method discussed in Sec. V.C after Eq. (5.11). For a different approach, see Wolf *et al.* (1984).

(b) The example of a time series $u(i)=$const, corresponding to an attracting fixed point, shows that it is not possible in general to obtain the negative characteristic exponents from the long-term behavior of a dynamical system. It is conceivable that our method works up to that k after which the sum of the largest k characteristic exponents becomes negative. If one has access to transients, then negative characteristic exponents are in principle accessible.

(c) We have discussed in this section the determination of the characteristic exponents of a measure ρ from its projection $\pi_m\rho$. How large should one choose m to be? For the determination of the information dimension, it was sufficient to take $m\nu\ge\dim_H\rho$. Here, however, this will usually be insufficient, because we have to reconstruct the *dynamics* in the support of ρ from its projection in $\mathbf{R}^{m\nu}$. We want π_m therefore to be *injective* on the support of ρ. According to Mañé's theorem in Sec. II.G. this may require $m\nu>2\dim_K(\text{support }\rho)+1$. Probably the best evidence that π_m is a "good" projection for the present purposes would be a reasonably good linear fit for Eq. (5.15).

E. Spectrum, rotation numbers

The ergodic quantities that we have discussed in this paper—characteristic exponents, entropy, information dimension—are those which appear at this moment most important and most easily accessible. They are, however, not the only quantities one might consider. For a quasiperiodic system, the *generating frequencies* are of course important. More generally, Frisch and Morf (1981) have drawn attention to the *complex singularities* of the signal

[5]The most convenient algorithms are given in Wilkinson and Reinsch (1971), contribution I/8. (All algorithms in this book are available in the large libraries such as EISPACK and NAG.)

$u(t)$ and their relation to the high-frequency behavior of the power spectrum. One may also look for poles at complex values of the frequency in the power spectrum (*resonances*). Finally one should mention *rotation numbers*, which are not always defined but are interesting quantities when they make sense (see Ruelle, 1985).

VI. OUTLOOK

Our review has led from the definitions of dynamical systems theory to a discussion of those quantities accessible today through the *statistical* analysis of time series for deterministic nonlinear systems. Together with the more geometrical aspects of bifurcation theory, this represents the main body of theoretically and experimentally successful ideas concerning nonlinear dynamics at this time. The purpose of this review is to make this knowledge accessible to a large number of scientists. The results presented here are the combined achievement of many investigators, only incompletely cited. We believe that the next step in the study of dynamical systems should lead to a better understanding of space-time patterns, for which only timid beginnings are now seen. We hope that the present review serves as an encouragement for the undertaking of this difficult problem.

Note added in proof. Another useful reprint collection to be added to the list of Sec. I is Hao Bai-Lin, 1984, *Chaos* (World Scientific, Singapore).

ACKNOWLEDGMENTS

We wish to thank E. Lieb, A. S. Wightman, and especially P. Bergé and F. Ledrappier for their critical reading of the manuscript. This work was partially supported by the Fonds National Suisse.

REFERENCES

Abraham, N. B., J. P. Gollub, and H. L. Swinney, 1984, "Testing nonlinear dynamics," Physica D **11**, 252.

Adler, R. L., A. G. Konheim, and M. H. McAndrew, 1965, "Topological entropy," Trans. Am. Math. Soc. **114**, 309.

Alexander, J. C., and J. A. Yorke, 1984, "Fat baker's transformations," Ergod. Theory Dynam. Syst. **4**, 1.

Anosov, D. V., 1967, "Geodesic flows on a compact Riemann manifold of negative curvature," Trudy Mat. Inst. Steklov **90**, [Proc. Steklov Math. Inst. **90** (1967)].

Arnold, V. I., and A. Avez, 1967, *Problèmes Ergodiques de la Mécanique Classique* (Gauthier-Villars, Paris) [*Ergodic Problems of Classical Mechanics* (Benjamin, New York, 1968)].

Atten, P., J. G. Caputo, B. Malraison, and Y. Gagne, 1984, "Détermination de dimension d'attracteurs pour différents écoulements," preprint.

Benedicks, M., and L. Carleson, 1984, "On iterations of $1-ax^2$ on $(-1,1)$," Ann. Math., in press.

Benettin, G., L. Galgani, J. M. Strelcyn, 1976, "Kolmogorov entropy and numerical experiments," Phys. Rev. A **14**, 2338.

Benettin, G., L. Galgani, A. Giorgilli, and J. M. Strelcyn, 1978, "Tous les nombres de Liapounov sont effectivement calculables," C. R. Acad. Sci. Paris **286A**, 431.

Bergé, P., Y. Pomeau, and C. Vidal, 1984, *L'ordre dans le Chaos* (Herman, Paris).

Billingsley, P., 1965, *Ergodic Theory and Information* (Wiley, New York).

Bowen, R., 1975, *Equilibrium States and the Ergodic Theory of Anosov Diffeomorphisms,* Lecture Notes in Mathematics **470** (Springer, Berlin).

Bowen, R., 1978, *On Axiom A Diffeomorphisms,* Regional conference series in mathematics, No. 35 (Amer. Math. Soc., Providence, R.I.)

Bowen, R., and D. Ruelle, 1975, "The ergodic theory of Axiom *A* flows," Inventiones Math. **29**, 181.

Brin, H., and Z. Nitecki, 1985, private communication.

Caffarelli, R., R. Kohn, and L. Nirenberg, 1982, "Partial regularity of suitable weak solutions of the Navier-Stokes equations," Commun. Pure Appl. Math. **35**, 771.

Campanino, M., 1980, "Two remarks on the computer study of differentiable dynamical systems," Commun. Math. Phys. **74**, 15.

Campbell, D., and H. Rose, 1983, Eds., *Order in Chaos: Proceedings of the International Conference . . . Los Alamos, . . . 1982* (Physica D **7**, 1).

Carverhill, A., 1984a, "Flows of stochastic systems: ergodic theory," Stochastics, in press.

Carverhill, A., 1984b, "A 'Markovian' approach to the multiplicative ergodic (Oseledec) theorem for nonlinear dynamical systems," preprint.

Cohen, A., and I. Procaccia, 1984, "On computing the Kolmogorov entropy from the time signals of dissipative and conservative dynamical systems," Phys. Rev. A **31**, 1872.

Collet, P., J.-P. Eckmann, and O. E. Lanford, 1980, "Universal properties of maps of an interval," Commun. Math. Phys. **76**, 211.

Collet, P., and J.-P. Eckmann, 1980a, "On the abundance of aperiodic behavior for maps on the interval," Commun. Math. Phys. **73**, 115.

Collet, P., and J.-P. Eckmann, 1980b, *Iterated Maps on the Interval as Dynamical Systems* (Birkhauser, Cambridge, MA).

Collet, P., and J.-P. Eckmann, 1983, "Positive Liapunov exponents and absolute continuity for maps of the interval," Ergod. Theory Dynam. Syst. **3**, 13.

Conley, C., 1978, *Isolated Invariant Sets and the Morse Index,* Regional conference series in mathematics, No. 38 (Amer. Math. Soc., Providence, R.I.)

Croquette, V., 1982, "Déterminisme et chaos," Pour la Science **62**, 62.

Crutchfield, J., 1981, private communication.

Curry, J. H., 1979, "On the Hénon transformation," Commun. Math. Phys. **68**, 129.

Curry, J. H., 1981, "On computing the entropy of the Hénon attractor," J. Stat. Phys. **26**, 683.

Cvitanović, P., 1984, Ed., *Universality in Chaos* (Adam Hilger, Bristol).

Denker, M., C. Grillenberger, and K. Sigmund, 1976, *Ergodic Theory on Compact Spaces,* Lecture Notes in Mathematics 527 (Springer, Berlin).

Douady, A., and J. Oesterlé, 1980, "Dimension de Hausdorff des attracteurs," C. R. Acad. Sci. Paris **290A**, 1135.

Dubois, M., 1982, "Experimental aspects of the transition to turbulence in Rayleigh-Bénard convection," in *Stability of Thermodynamic Systems,* Lecture Notes in Physics 164, (Springer, Berlin), pp. 177—191.

Dunford, M., and J. T. Schwartz, 1958, *Linear Operators* (Inter-

science, New York).
Eckmann, J.-P., 1981, "Roads to turbulence in dissipative dynamical systems," Rev. Mod. Phys. **53**, 643.
Farmer, D., 1984, private communication.
Fathi, A., M. R. Herman, and J.-C. Yoccoz, 1983, "A proof of Pesin's stable manifold theorem," in *Geometric Dynamics*, Lecture Notes in Mathematics 1007 (Springer, Berlin), pp. 177—215.
Fauve, S., and A. Libchaber, 1982, "Rayleigh-Bénard experiment in a low Prandtl number fluid, mercury," in *Chaos and Order in Nature: Proceedings of the International Symposium on Synergetics at Schloss Elmau, 1981*, edited by H. Haken (Springer, Berlin), pp. 25—35.
Feigenbaum, M. J., 1978, "Quantitative universality for a class of nonlinear transformations," J. Stat. Phys. **19**, 25.
Feigenbaum, M. J., 1979, "The universal metric properties of nonlinear transformations," J. Stat. Phys. **21**, 669.
Feigenbaum, M. J., 1980, "The transition to aperiodic behavior in turbulent systems," Commun. Math. Phys. **77**, 65.
Feit, S. D., 1978, "Characteristic exponents and strange attractors," Commun. Math. Phys. **61**, 249.
Feller, W., 1966, *An Introduction to Probability Theory and its Applications*, Vol. II (Wiley, New York).
Foias, C., and R. Temam, 1979, "Some analytic and geometric properties of the solutions of the evolution Navier-Stokes equations," J. Math. Pures Appl. **58**, 339.
Frederickson, P., J. L. Kaplan, E. D. Yorke, and J. A. Yorke, 1983, "The Lyapunov dimension of strange attractors," J. Diff. Equ. **49**, 185.
Frisch, U., and R. Morf, 1981, "Intermittency in nonlinear dynamics and singularities at complex times," Phys. Rev. A **23**, 2673.
Ghil, M., R. Benzi, and G. Parisi, 1985, *Turbulence and Predictability in Geophysical Fluid Dynamics and Climate Dynamics* (North-Holland, Amsterdam).
Grassberger, P., 1981, "On the Hausdorff dimension of fractal attractors," J. Stat. Phys. **26**, 173.
Grassberger, P., 1984, "Information aspects of strange attractors," preprint, Wuppertal.
Grassberger, P., and I. Procaccia, 1983a, "Estimating the Kolmogorov entropy from a chaotic signal," Phys. Rev. A **28**, 2591.
Grassberger, P., and I. Procaccia, 1983b, "Measuring the strangeness of strange attractors," Physica D **9**, 189.
Grebogi, C., E. Ott, and J. A. Yorke, 1983a, "Are three-frequency quasiperiodic orbits to be expected in typical nonlinear dynamical systems?" Phys. Rev. Lett. **53**, 339.
Grebogi, C., E. Ott, and J. A. Yorke, 1983b, "Crises, sudden changes in chaotic attractors and transient chaos," Physica **7D**, 181.
Guckenheimer, J., 1982, "Noise in chaotic systems," Nature **298**, 358.
Guckenheimer, J., and P. Holmes, 1983, *Nonlinear Oscillations, Dynamical Systems, and Bifurcations of Vector Fields* (Springer, New York).
Gurel, O., and O. E. Rössler, 1979, Eds., *Bifurcation Theory and Applications in Scientific Disciplines* (Ann. N. Y. Acad. Sci. **316**).
Haken, H., 1983, *Advanced Synergetics* (Springer, Berlin).
Helleman, R. H. G., 1980, Ed., *International Conference on Nonlinear Dynamics* (Ann. N. Y. Acad. Sci. **357**).
Hénon, M., 1976, "A two-dimensional mapping with a strange attractor," Commun. Math. Phys. **50**, 69.
Hénon, M., and C. Heiles, 1964, "The applicability of the third integral of motion, some numerical experiments," Astron. J. **69**, 73.
Hurewicz, W., and H. Wallman, 1948, *Dimension Theory* (Princeton University, Princeton).
Ilyashenko, Yu. S., 1983, "On the dimension of attractors of k-contracting systems in an infinite dimensional space," Vestn. Mosk. Univ. Ser. 1 Mat. Mekh. 1983 No. 3, 52—58.
Iooss, G., R. Helleman, and R. Stora, 1983, Eds., *Chaotic Behavior of Deterministic Systems* (North-Holland, Amsterdam).
Jakobson, M., 1981, "Absolutely continuous invariant measures for one-parameter families of one-dimensional maps," Commun. Math. Phys. **81**, 39.
Johnson, R. A., K. J. Palmer, and G. Sell, 1984, "Ergodic properties of linear dynamical systems," preprint, Minneapolis.
Kantz, H., and P. Grassberger, 1984, "Repellers, semi-attractors, and long-lived chaotic transients," preprint, Wuppertal.
Kaplan, J. L., and J. A. Yorke, 1979, "Preturbulence: a regime observed in a fluid flow model of Lorenz," Commun. Math. Phys. **67**, 93.
Katok, A., 1980, "Liapunov exponents, entropy and periodic orbits for diffeomorphisms," Publ. Math. IHES **51**, 137.
Katok, A., and J.-M. Strelcyn, 1985, "Smooth maps with singularities, invariant manifolds, entropy and billiards" (in preparation).
Kifer, Yu. I., 1974, "On small random perturbations of some smooth dynamical systems," Izv. Akad. Nauk SSSR Ser. Mat. **38**, No. 5, 1091 [Math. USSR Izv. **8**, 1083 (1974)].
Kifer, Yu. I., 1984, "A multiplicative ergodic theorem for random transformations," preprint.
Ladyzhenskaya, O. A., 1969, *The Mathematical Theory of Viscous Incompressible Flow*, 2nd ed. (Nauka, Moscow, 1970). [2nd English edition Gordon and Breach, New York (1969)].
Lanford, O. E., 1977, in Turbulence Seminar, Proceedings 1976/77, edited by P. Bernard and T. Ratiu, Lecture Notes in Mathematics 615 (Springer, Berlin), pp. 113—116.
Ledrappier, F., 1981a, "Some relations between dimension and Liapunov exponents," Commun. Math. Phys. **81**, 229.
Ledrappier, F., 1981b, "Some properties of absolutely continuous invariant measures on an interval," Ergod. Theory Dynam. Syst. **1**, 77.
Ledrappier, F., and M. Misiurewicz, 1984, "Dimension of invariant measures," preprint.
Ledrappier, F., and L.-S. Young, 1984, "The metric entropy of diffeomorphisms. Part I. Characterization of measures satisfying Pesin's formula. Part II. Relations between entropy, exponents and dimension," preprints, University of California, Berkeley, 1984. "The metric entropy of diffeomorphisms," Bull. Am. Math. Soc. (New Ser.) **11**, 343.
Li, T., and J. A. Yorke, 1975, "Period three implies chaos," Am. Math. Monthly **82**, 985.
Libchaber, A., and J. Maurer, 1979, "Une expérience de Rayleigh-Bénard de géométrie réduite; multiplication, accrochage, et démultiplication de fréquences," J. Phys. (Paris) Colloq. **41**, C3-51.
Lieb, E. H., 1984, "On characteristic exponents in turbulence," Commun. Math. Phys. **92**, 473.
Lorenz, E. N., 1963, "Deterministic nonperiodic flow," J. Atmos. Sci. **20**, 130.
Lundqvist, Stig, 1985, Ed., *The Physics of Chaos and Related Problems: Proceedings of the 59th Nobel Symposium, 1984* (Phys. Scr. **T9**, 1).
Mallet-Paret. J., 1976, "Negatively invariant sets of compact

maps and an extension of a theorem of Cartwright," J. Diff. Equ. 22, 331.
Malraison, B., P. Atten, P. Bergé, and M. Dubois, 1983, "Dimension of strange attractors: an experimental determination for the chaotic regime of two convective systems," J. Phys. (Paris) Lett. 44, L-897.
Mandelbrot, B., 1982, *The Fractal Geometry of Nature* (Freeman, San Francisco).
Mañé, R., 1981, "On the dimension of the compact invariant sets of certain nonlinear maps," in *Dynamical Systems and Turbulence, Warwick 1980*, Lecture Notes in Mathematics 898 (Springer, New York), pp. 230—242.
Mañé, R., 1983, "Liapunov exponents and stable manifolds for compact transformations," in *Geometrical Dynamics*, Lecture Notes in Mathematics 1007 (Springer, Berlin), pp. 522—577.
Manneville, P., 1985, "Liapunov exponents for the Kuramoto-Sivashinsky model," preprint.
Manning, A., 1984, "The dimension of the maximal measure for a polynomial map," Ann. Math. 119, 425.
Mattila, P., 1975, "Hausdorff dimension, orthogonal projections and intersections with planes," Ann. Acad. Sci. Fenn. Ser. A1, Math., 1, 227.
McCluskey, H., and A. Manning, 1983, "Hausdorff dimension for horseshoes," Ergod. Theory Dynam. Syst. 3, 251.
Milnor, J., 1985, "On the concept of attractor," Commun. Math. Phys., in press.
Misiurewicz, M., 1981, "Structure of mappings of an interval with zero entropy," Publ. Math. IHES 53, 5.
Newhouse, S., 1974, "Diffeomorphisms with infinitely many sinks," Topology 13, 9.
Newhouse, S., 1979, "The abundance of wild hyperbolic sets and non-smooth stable sets for diffeomorphisms," Publ. Math. IHES 50, 102.
Newhouse, S., D. Ruelle, and F. Takens, 1978, "Occurrence of strange axiom A attractors near quasiperiodic flows on T^m, $m \geq 3$," Commun. Math. Phys. 64, 35.
Nobel Symposium, 1984, see Lundqvist, 1985.
Oseledec, V. I., 1968, "A multiplicative ergodic theorem. Lyapunov characteristic numbers for dynamical systems," Trudy Mosk. Mat. Obsc. 19, 179 [Moscow Math. Soc. 19, 197 (1968)].
Pesin, Ya. B., 1976, "Invariant manifold families which correspond to nonvanishing characteristic exponents," Izv. Akad. Nauk SSSR Ser. Mat. 40, No. 6, 1332 [Math. USSR Izv. 10, No. 6, 1261 (1976)].
Pesin, Ya. B., 1977, "Lyapunov characteristic exponents and smooth ergodic theory," Usp. Mat. Nauk 32, No. 4 (196), 55 [Russian Math. Survey 32, No. 4, 55 (1977)].
Procaccia, I., 1984, *The static and dynamic invariants that characterize chaos and the relations between them in theory and experiments*, proceedings of the 59th Nobel Symposium (Phys. Scr. T9, 40).
Pugh, C. C., and M. Shub, 1984, "Ergodic attractors" (in preparation).
Raghunathan, M. S., 1979, "A proof of Oseledec's multiplicative ergodic theorem," Israel J. Math. 32, 356.
Riste, T., and K. Otnes, 1984, *Neutron scattering from a convecting nematic: multicriticality, multistability and chaos*, proceedings of the 59th Nobel Symposium (Phys. Scr. T9, 76).
Rössler, O. E., 1976, "An equation for continuous chaos," Phys. Lett. A 57, 397.
Roux, J.-C., R. H. Simoyi, and H. L. Swinney, 1983, "Observation of a strange attractor," Physica D 8, 257.
Roux, J.-C., and H. L. Swinney, 1981, "Topology of chaos in a chemical reaction," in *Nonlinear Phenomena in Chemical Dynamics*, edited by C. Vidal and A. Pacault (Springer, Berlin), pp. 38—43.
Ruelle, D., 1976, "A measure associated with Axiom *A* attractors," Am. J. Math. 98, 619.
Ruelle, D., 1978, "An inequality for the entropy of differentiable maps," Bol. Soc. Bras. Mat. 9, 83.
Ruelle, D., 1979, "Ergodic theory of differentiable dynamical systems," Phys. Math. IHES 50, 275.
Ruelle, D., 1980, "Measures describing a turbulent flow," Ann. N. Y. Acad. Sci. 357, 1.
Ruelle, D., 1981, "Small random perturbations of dynamical systems and the definition of attractors," Commun. Math. Phys. 82, 137.
Ruelle, D., 1982a, "Characteristic exponents and invariant manifolds in Hilbert space," Ann. Math. 115, 243.
Ruelle, D., 1982b, "Large volume limit of the distribution of characteristic exponents in turbulence," Commun. Math. Phys. 87, 287.
Ruelle, D., 1984, "Characteristic exponents for a viscous fluid subjected to time dependent forces," Commun. Math. Phys. 93, 285.
Ruelle, D., 1985, "Rotation numbers for diffeomorphisms and flows," Ann. Inst. Henri Poincaré, 42, 109.
Ruelle, D., and F. Takens, 1971, "On the nature of turbulence," Commun. Math. Phys. 20, 167. Note concerning our paper "On the nature of turbulence," Commun. Math. Phys. 21, 21.
Shaw, R. S., 1981, "Strange attractors, chaotic behavior and information flow," Z. Naturforsch. 36a, 80.
Shimada I., 1979, "Gibbsian distribution on the Lorenz attractor," Prog. Theor. Phys. 62, 61.
Sinai, Ya. G., 1972, "Gibbs measures in ergodic theory," Usp. Mat. Nauk 27, No. 4, 21 [Russian Math. Surveys 27, No. 4, 21 (1972)].
Smale, S., 1967, "Differentiable dynamical systems," Bull. Am. Math. Soc. 73, 747.
Takens, F., 1981, "Detecting strange attractors in turbulence," in *Dynamical Systems and Turbulence, Warwick 1980*, Lecture Notes in Mathematics 898 (Springer, Berlin), pp. 366—381.
Takens, F., 1983, "Invariants related to dimension and entropy," in *Atas do 13. Col. brasiliero de Matematicas, Rio de Janerio, 1983*.
Temam, R., 1979, *Navier-Stokes Equations*, 2nd ed. (North-Holland, Amsterdam).
Thieullen, P., 1985, "Exposant de Liapounov des fibrés dynamiques pseudocompacts," in preparation.
Ushiki, S., M. Yamaguti, and H. Matano, 1980, "Discrete population models and chaos," Lecture Notes Num. Appl. Anal. 2, 1.
Vidal, C., and A. Pacault, 1981, Eds., *Nonlinear Phenomena in Chemical Dynamics: Proceedings* (Springer, Berlin).
Vul, E. B., Ya. G. Sinai, and K. M. Khanin, 1984, "Universality of Feigenbaum and thermodynamic formalism," Usp. Mat. Nauk 39, No. 3 (237), 3.
Walden, R. W., P. Kolodner, A. Passner, and C. M. Surko, 1984, "Nonchaotic Rayleigh-Bénard convection with four and five incommensurate frequencies," Phys. Rev. Lett. 53, 242.
Walters, P., 1975, *Ergodic Theory—Introductory Lectures*, Lecture Notes in Mathematics 458 (Springer, Berlin).
Wilkinson, J. H., 1965, *The Algebraic Eigenvalue Problem* (Clarendon, Oxford).
Wilkinson, J. H., and C. Reinsch, 1971, *Linear Algebra* (Springer, Berlin).
Wolf, A., J. B. Swift, H. L. Swinney, and J. A. Vastano, 1984,

"Determining Liapunov exponents from a time series," Physica D, in press.
Young, L.-S., 1981, "Capacity of attractors," Ergod. Theory Dynam. Syst. **1**, 381.
Young, L.-S., 1982, "Dimension, entropy and Liapunov exponents," Ergod. Theory Dynam. Syst. **2**, 109.
Young, L.-S., 1984, "Dimension, entropy and Liapunov exponents in differentiable dynamical systems," Physica A **124**, 639.

Another Proof of Jakobson's Theorem and Related Results

Marek Ryszard Rychlik

Department of Mathematics
University of Arizona
Tucson, AZ 85721 USA
phone: (602) 621-6864, FAX: (602) 621-8322
rychlik@math.arizona.edu **

Summary. The author shows that any family C^2-close to $f_\alpha(x) = 1 - ax^2$ $(2 - \epsilon \leq a \leq 2)$ satisfies Jakobson's theorem: For a positive measure set of α the transformation f_α has an absolutely continuous invariant measure. He also indicates some generalizations.

Introduction

In recent years there has been major interest in the following theorem of Jakobson:

Theorem. *Let $f_r(x) = rx(1-x)$, $0 \leq r \leq 4$, be a one parameter family of mappings of the unit interval. There is a positive measure set of those r for which f_r has an absolutely continuous invariant measure (abbreviation: a.c.i.m.).*

The author of the current paper uses his earlier ideas from [5] to give another proof of this theorem. It seems to be less technical than other existing proofs (see [2], [1]) and therefore it yields some interesting generalizations (Section 6). In particular, any family C^2-close to the one above satisfies Jakobson's theorem. Also, the families that contain $f_4(x) = 4x(1-x)$ and do not satisfy Jakobson's theorem form a 'set of codimension ∞' in the set of C^2-families that contain f_4 (or any mapping C^2-close to f_4 with the property that f^2(critical point) = fixed point). In particular, any analytic family of this type satisfies Jakobson's theorem.

** This article was originally published in journal form in *Ergod. The. & Dynam. Sys.* (1988), **8**, 93–109. With kind permission from the publishers it is included in this volume. The author's address at the time of writing of this article was: University of Washington, Department of Mathematics, GN-50, Seattle, Washington 98195 USA and Institute of Mathematics, University of Warsaw, Poland.

One reason to understand Jakobson's result is a possible generalization to higher dimensions. Similar phenomena seem to accompany every period-doubling bifurcation, when we pass the critical value of the parameter (and the 'chaos' is born!). So far M. Rees has found an analogue for rational mappings of the Riemann sphere [4]. (Probably our proof can be modified to work in that case also.)

Let us say a few things about our notation. The Lebesgue measure of a set A is denoted $|A|$. Variables as subscripts mean differentiation. Occasionally we use prime for the derivative over x or when the parameter is the only variable. Sometimes we do not say explicitly that an object depends on the parameter. We also call a set an interval, where obviously the set has two components. For technical reasons we work with the family $f(x, a) = 1 - ax^2$ and interval $[-1, 1]$.

The author would like to apologize if some ideas are being used without references. It is difficult, though, to write a proof disjoint with the existing work.

We would like to express our thanks to the referee, whose work allowed us to eliminate many mistakes and other deficiencies of the first version.

The author would like to thank Laura Plaut for her careful typesetting of this paper.[3]

This work was partially supported by NSF Grant DMS 8320356.

1 Sketch of the proof

For $a = 2$ the point $f^2(0) = -1$ is a repelling fixed point.

Let us define $I_n = [-3^{-n}, 3^{-n}]$ for $n = 0, 1, \ldots$. Let us fix $\rho \in (0, 1)$ very close to 1. Let $V_n = f(\rho^{-1} I_n)$. It is easy to show (see Appendix) that $f^{n+1} | \rho^{-1} I_n \backslash \rho I_{n+1}$ is expanding with constants $\Lambda_0 = \frac{4}{3}\rho$. It is also easy to show that the set $V_{n+1} \cup f(V_{n+1}) \cup \cdots \cup f^n(V_{n+1}) \subset \{x : |f'(x)| \geq 3\}$. Let $W_n = f^n(V_{n+1})$. We will show that $|W_n| \leq \rho^{-1} \left(\frac{4}{9}\right)^{n-1}$, so $|W_n| \to 0$, as $n \to \infty$.

Let us define $T : I \to I$ by

$$T(x) = f^n(x), \quad \text{as } x \in M_n \stackrel{\text{def}}{=} I_{n-1} \backslash I_n \tag{1}$$

for $n = 1, 2, \ldots$. This transformation is piecewise expanding. From [5] we know that T has an a.c.i.m. ν_0 with bounded density. We can easily verify that the measure

$$\nu = \sum_{n=1}^{\infty} \sum_{k=0}^{n-1} f_*^k (\nu_0 | M_n) \tag{2}$$

[3] The current version was typeset by the author, and all typographical errors are due to him, as well as several corrections of the typographical errors of the original article.

is an a.c.i.m. for f. It is finite, since

$$\nu(I) = \sum_{n=1}^{\infty} n\nu(M_n) \leq \text{const.} \sum_{n=1}^{\infty} n\,|M_n| < \infty. \tag{3}$$

We would like to point out that this measure is well known and has density const. $\left(1 - x^2\right)^{-1/2}$. The construction we have just presented is a starting point to our construction of a.c.i.m. for $\alpha \neq 2$. It also produces a.c.i.m. for families like $f(x) = 1 - 2x^2 + \xi x^2\left(1 - x^2\right)$, where ξ is a small parameter. The author does not know any explicit formula for the density of the a.c.i.m. in this case.

Let us start with the observation that given arbitrarily large $N \in \mathbb{Z}^+$ there is $\alpha_N < 2$ such that $f^{n+1}|\rho^{-1}I_n \backslash \rho I_{n+1}$ is an expanding for $n \leq N$ and $\alpha \in [\alpha_N, 2]$. For $\alpha = 2$ the set W_n contains -1 for every n. For $\alpha \neq 2$ the set W_n approaches the critical point. The extreme case is when for some n we have $f^n(0) = 0$. In this situation 0 attracts a.e. orbit of f and no a.c.i.m. exists. Our hope is that by varying α we can push W_n away from 0. Actually, we will put W_n into a set $\rho^{-1}I_k \backslash \rho I_{k+1}$ for some $k < n$.

Let us fix N sufficiently large and let $\alpha_N < 2$ be such that for all $\alpha \in [\alpha_N, 2]$ and $n = 0, 1, \ldots, N$ we have $W_n \subset \{x : |f'(x)| \geq 3\}$. Let $T_{(n)} = f^{n+1}|\rho^{-1}I_n \backslash \rho I_{n+1}$ for $n = 0, 1, \ldots, N-1$ and let $S_{(n)} = f^n|V_n$ for $n = 1, 2, \ldots, N$. We have

$$T_{(n)} = S_{(n)} \circ f \quad \text{and} \quad S_{(n+1)} = T_{(0)} \circ S_{(n)} \tag{4}$$

for $n = 0, 1, \ldots, N-1$.

We will try to extend the definition of expandings $T_{(n)}$ and $S_{(n)}$ on $n > N$ (at least for some parameters). We are going to use induction.

Suppose that I_m, $S_{(m)}$ have been defined for $m \leq n$ $(n \geq N)$ and let $T_{(m)} = S_{(m)} \circ f$ for $m < n$. Let us fix $\tau \in \left(\frac{1}{2}, 1\right)$ to the solution of the equation $4^\tau = 3$ and let $a = \frac{1}{3}$.

We define

$$\begin{aligned} I_{n+1} &= \left[-a\left|S'_{(n)}(1)\right|^{-\tau}, \ a\left|S'_{(n)}(1)\right|^{-\tau}\right], \\ V_{n+1} &= f\left(\rho^{-1}I_{n+1}\right). \end{aligned} \tag{5}$$

Now we can define $T_{(n)} = S_{(n)} \circ f|\rho^{-1}I_n \backslash \rho I_{n+1}$ and $W_n = S_{(n)}(V_{n+1})$.

Suppose that $W_n \subset \rho^{-1}I_k \backslash \rho I_{k+1}$ for some $k = k(n) < n$. In this case we set

$$S_{(n+1)} = T_{(k)} \circ S_{(n)}|V_{n+1} \quad (k = k(n)). \tag{6}$$

As we have already mentioned, for some parameters there is no k with the property $W_n \subset \rho^{-1}I_k \backslash I_{k+1}$. We need to discard those parameters to proceed with the next step of our induction. Let us describe in detail how we do it.

Let $(s_m)_{m=0}^{n}$ be a sequence of integers such that $T_{(m)} = f^{s_m}$ (on $\rho^{-1}I_m \backslash \rho^{-1}I_{m+1}$). We impose an extra condition in our construction.

We fix a sufficiently small $\beta > 0$ and require

$$s_{k(m)} \leq \max(\beta m, 1), \quad (m = 0, 1, \ldots, n). \tag{7}$$

Along with I_n and $S_{(n)}$ we construct families of intervals $\mathcal{A}_n$ in the space of parameters. This is the corresponding inductive definition:

(i) $\mathcal{A}_N$ consists of a single interval $[\alpha_N, 2]$.

(ii) If $J \in \mathcal{A}_n$ then for $\alpha \in J$ all n steps of our construction work and yield the same sequence $(s_m)_{m=0}^{n-1}$. Let us consider intervals

$$J_k = \left\{\alpha \in J : \; S_{(n)}(1, \alpha) \in I_k \backslash I_{k+1}\right\}. \tag{8}$$

The family $\mathcal{A}_{n+1}$ consists of all intervals J_k for all $J \in \mathcal{A}_n$ and k such that $s_k \leq \beta n$.

We will show that if $\alpha \in J_k$ then $W_n \subset \rho^{-1}I_k \backslash \rho I_{k+1}$. This is possible because the length of $|W_n|$ decays much faster than the length of $I_{k(n)}$.

The set of parameters $J \backslash \bigcup J_k$ is in general nonempty, but we will see that there is $\lambda \in (0, 1)$ such that

$$J \backslash \bigcup J_k \subset \left\{\alpha \in J : \; \mathrm{dist}\left(S_{(n)}(1, \alpha), 0\right) \leq a\lambda^{\sqrt{n}}\right\}. \tag{9}$$

Let $A_n = \bigcup \mathcal{A}_n$ and let B_n be the union of the sets on the right-hand side of (9). The crucial part of the proof is to show that $|B_n| \leq \text{const.}\ \lambda^{\sqrt{n}}\, |A_N|$. Once we have obtained this estimate, we can write (in view of $A_n \backslash A_{n+1} \subset B_n$):

$$\begin{aligned} |A_n| &\geq \left| A_N \backslash \bigcup_{m=N}^{n-1} B_m \right| \geq |A_N| - \sum_{m=N}^{n-1} |B_m| \\ &\geq |A_N| \left(1 - \text{const.} \sum_{m=N}^{n-1} \lambda^{\sqrt{m}}\right). \end{aligned} \tag{10}$$

Let $A_\infty = \bigcap_{n=N}^{\infty} |A_n|$. Letting $n \to \infty$ in (10) we get

$$|A_\infty| \geq |A_N| \left(1 - \text{const.} \sum_{m=N}^{\infty} \lambda^{\sqrt{m}}\right). \tag{11}$$

By fixing N sufficiently large we can make the ratio $|A_\infty| / |A_N|$ arbitrarily close to 1, in particular $|A_\infty| > 0$.

From our estimates it easily follows that if $\alpha \in A_\infty$ then f has an a.c.i.m. First we construct a piecewise expanding $T(x) = f^{s_n}(x)$, as $x \in I_n \backslash I_{n+1}$ ($n = 0, 1, \ldots$) which has an a.c.i.m. ν_0 with bounded density. Formulas analogous to (2) and (3) yield an a.c.i.m. ν for f.

2 Certain consequences of the Chain Rule

Let u and v be functions of x and α, where α is a parameter. For example, $w = v \circ u$ means that $w(x,\alpha) = v(u(x,\alpha),\alpha)$. We can easily verify:

Lemma 1. *The following formulas hold:*

(i) $w_x = (v_x \circ u)\, u_x$,

(ii) $\dfrac{w_\alpha}{w_x} = \left(\dfrac{v_\alpha}{v_x} \circ u\right) \dfrac{1}{u_x} + \dfrac{u_\alpha}{u_x}$,

(iii) $\dfrac{w_{xx}}{w_x^2} = \dfrac{v_{xx}}{v_x^2} \circ u + \left(\dfrac{1}{v_x} \circ u\right) \dfrac{u_{xx}}{u_x^2}$,

(iv) $\dfrac{w_{\alpha x}}{w_x^2} = \left(\dfrac{v_{\alpha x}}{v_x^2} \circ u\right) \dfrac{1}{u_x} + \left(\dfrac{v_{xx}}{v_x^2} \circ u\right) \dfrac{u_\alpha}{u_x} + \left(\dfrac{1}{v_x} \circ u\right) \dfrac{u_{\alpha x}}{u_x^2}$,

(v) $\dfrac{w_{\alpha\alpha}}{w_x^2} = \left(\dfrac{v_{\alpha\alpha}}{v_x^2} \circ u\right) \dfrac{1}{u_x^2} + 2\left(\dfrac{v_{\alpha x}}{v_x^2} \circ u\right) \dfrac{u_\alpha}{u_x} \dfrac{1}{u_x} + \left(\dfrac{v_{xx}}{v_x^2} \circ u\right) \left(\dfrac{u_\alpha}{u_x}\right)^2 + \left(\dfrac{1}{v_x} \circ u\right) \dfrac{u_{\alpha\alpha}}{u_x^2}$.

Let us introduce the following notations:

$$\begin{aligned}
\Delta_{xx}(u) &= \frac{|u_{xx}|}{|u_x|^2}, \quad \Delta_{\alpha x}(u) = \frac{|u_{\alpha x}|}{|u_x|^2}, \quad \Delta_{\alpha\alpha}(u) = \frac{|u_{\alpha\alpha}|}{|u_x|^2}, \\
\Delta(u) &= \max\left(\Delta_{xx}(u), \Delta_{\alpha x}(u), \Delta_{\alpha\alpha}(u)\right), \\
\delta(u) &= \frac{u_\alpha}{u_x}, \\
R_{xx}(u) &= 1, \\
R_{\alpha x}(u) &= \left|\frac{1}{u_x}\right| + \left|\frac{u_\alpha}{u_x}\right|, \\
R_{\alpha\alpha}(u) &= \frac{1}{|u_x|^2} + 2\left|\frac{u_\alpha}{u_x}\right| \frac{1}{|u_x|} + \left|\frac{u_\alpha}{u_x}\right|^2, \\
R(u) &= \max\left(R_{xx}(u), R_{\alpha x}(u), R_{\alpha\alpha}(u)\right).
\end{aligned} \tag{12}$$

Lemma 2. *Let $w = v \circ u$. Then*

$$\Delta(w) \le \left(\frac{1}{|v_x|} \circ u\right) \Delta(u) + R(u)(\Delta(v) \circ u). \tag{13}$$

Moreover, this formula holds with $\Delta(w)$, $\Delta(u)$ and $R(u)$ (but not $\Delta(v)$) subscripted with 'xx', 'αx' or '$\alpha\alpha$'.

Proof. By inspection of formulas (iii)-(v) of Lemma 1.

Remark 1. If $|u_x| \ge 1$ then $R(u) \le 4 \max\left(1, |\delta(u)|^2\right)$.

3 Basic concepts

Let f be a C^2-function of x and α. As a function of x, f is a transformation $I \to I$, where $I = [-1, 1]$. We will deal with transformations $T = f^d : \mathcal{D}(T) \to \mathcal{R}(T)$, where $\mathcal{D}(T)$ and $\mathcal{R}(T)$ are subintervals of I. The integer $d \geq 1$ is called the degree of T and denoted by $\deg(T)$.

Let $\Lambda_0 > 1$. We will call T Λ_0-expanding, if $|T'| \geq \Lambda_0$.

Definition 1. *A sequence of expanding maps $(T_i)_{i=1}^n$ is called β-homogeneous ($\beta \geq 0$) if $\deg(T_i) \leq max(\beta i, 1)$ for $i = 1, 2, \ldots, n$.*

This notion proves useful because of the following:

Lemma 3. *Let $(T_i)_{i=1}^n$ be a β-homogeneous sequence of Λ_0-expandings and let $S_i = T_i \circ T_{i-1} \circ \cdots \circ T_1$ for $i = 1, 2, \ldots, n$. Suppose that $\beta \leq 1$. Then*

$$|T_i'| \circ S_{i-1} \leq R_0 \left|S_{i-1}'\right|^\epsilon , \tag{14}$$

where R_0 is a constant such that $\sup |f'| \leq R_0$ *and* $\epsilon = \beta \log R_0 / \log \Lambda_0$.

Proof. $|T_i'| \leq R_0^{\deg(T_i)} \leq R^{\max(\beta i, 1)} \leq \max\left(R_0, R_0^{\beta i}\right)$. Also, $R_0 \left|S_{i-1}'\right|^\epsilon \geq R_0 \left(\Lambda_0^{i-1}\right)^\epsilon = R_0 R_0^{\beta(i-1)} \geq R_0^{\beta i}$.

Remark 2. (14) holds with T_i replaced by any T verifying $\deg(T) \leq \max(\beta i, 1)$.

Definition 2. *A sequence of functions $(\phi_i)_{i=1}^n$ is called (C, q)-stable ($C \geq 0$, $q \in (0, 1)$) if*

$$|\phi_i - \phi_j| \leq C q^{\min(i,j)} \tag{15}$$

for $i, j = 1, 2, \ldots, n$.

Definition 3. (i) *We say that T has rank (μ, A) if*

$$|\delta(T)| \leq A \left|T_x\right|^\mu . \tag{16}$$

(ii) *We say that T has type (σ, B) if*

$$\Delta(T) \leq B \left|T_x\right|^\sigma . \tag{17}$$

Theorem 1. *Suppose $(T_i)_{i=1}^n$ is a β-homogeneous sequence of Λ_0-expandings and T_i has rank (μ, A) for $i = 1, 2, \ldots, n$. Let $S_i = T_i \circ \cdots \circ T_1$ for $i \leq n$ and let $S = S_n$, $S_0 = id$. There are constants C_1 and $q \in (0, 1)$ independent of A, μ or n such that for sufficiently small β we have:*

(i) *$\delta(S) \leq C_1$ (equivalently, S has rank $(0, C_1)$).*

(ii) *The sequence of functions $(\delta(S_i))_{i=0}^n$ is (C_1, q)-stable.*

Proof. From Lemma 1 we derive by induction the following formula ($i < j$):

$$\delta(S_j) = \sum_{l=i+1}^{j} (\delta(T_l) \circ S_{l-1}) \left(S'_{l-1}\right)^{-1} + \delta(S_i). \tag{18}$$

This gives

$$|\delta(S_j) - \delta(S_i)| \leq \sum_{l=i+1}^{j} (|\delta(T_l)| \circ S_{l-1}) \left|S'_{l-1}\right|^{-1}. \tag{19}$$

For $l \leq [\beta^{-1}]$ we have $|\delta(T_l)| \leq R_1$ where R_1 depends on f only (note: $\deg(T_l) = 1$). For $l > [\beta^{-1}]$ we have

$$|\delta(T_l)| \circ S_{l-1} \leq A \left|T'_l\right|^{\mu} \circ S_{l-1} \leq A R_0^{\mu} \left|S'_{l-1}\right|^{\epsilon\mu}$$

by Lemma 3. We always have $\left|S'_{l-1}\right| \geq \Lambda_0^{l-1}$. Therefore,

$$|\delta(S_j) - \delta(S_i)| \leq \sum_{1+i \leq l \leq [\beta^{-1}]} R_1 \Lambda_0^{-(l-1)} + \sum_{[\beta^{-1}] < l < j} R_0^{\mu} A \Lambda_0^{-(1-\epsilon\mu)(l-1)}. \tag{20}$$

Suppose $\epsilon\mu < 1/3$ (i.e. $\beta < \log \Lambda_0 / 3\mu \log R_0$). Let $q = \Lambda_0^{-\frac{1}{3}}$ and let β be such that $R_0^{\mu} A \leq R_1 q^{-[\beta^{-1}]}$. The right-hand side of (20) does not exceed

$$\sum_{l=i+1}^{j} R_1 q^{l-1} \leq R_1 (1-q)^{-1} q^i$$

(the terms of the first sum are obviously $\leq R_1 q^{l-1}$ and the second sum $\leq R_1 q^{2(l-1)-[\beta^{-1}]} \leq R_1 q^{l-1}$, since $l - 1 \geq [\beta^{-1}]$). Therefore, the Lemma holds with $C_1 = R_1(1-q)^{-1}$. We notice that (i) can be easily obtained from (ii), since $\delta(S_0) = 0$.

Theorem 2. *Let ν be an arbitrary positive number. Suppose that the assumptions of Theorem 1 are satisfied and, in addition, T_i has type (σ, B) for $i = 1, 2, \ldots, n$. There is a constant C_2 independent of A, B, μ, ν, σ or n such that if β is sufficiently small then S has type (ν, C_2).*

Proof. Lemma 2, Remark 1 and Theorem 1(i) give

$$\Delta(S_n) \leq \Lambda_0^{-1} \Delta(S_{n-1}) + C_3 (\Delta(T_n) \circ S_{n-1}), \tag{21}$$

where $C_3 = 4 \max\left(1, C_1^2\right)$. Therefore, by induction we get

$$\Delta(S_n) \leq C_3 \sum_{i=1}^{n} \Lambda_0^{-(n-1)} \Delta(T_i) \circ S_{i-1} \leq C_3 \left(1 - \Lambda_0^{-1}\right)^{-1} \max_{1 \leq i \leq n} \Delta(T_i) \circ S_{i-1}. \tag{22}$$

We can assume that for $i \leq [\beta^{-1}]$ we have $\Delta(T_i) \leq R_2$ and for $i > [\beta^{-1}]$ we have $\Delta(T_i) \circ S_{i-1} \leq B\,|T_i'|^{\sigma} \circ S_{i-1} \leq R_0^{\sigma} B\,|S'|^{\epsilon\sigma}$. Therefore

$$\Delta(S) \leq \frac{C_3}{1-\Lambda_0^{-1}} \max\left(R_2, R_0^{\sigma} B\,|S'|^{\epsilon\sigma}\right). \tag{23}$$

Suppose that $\epsilon\sigma < \nu/2$ (i.e. $\beta < \nu \log \Lambda_0 / 2\sigma \log R_0$) and β is such that $R_2 \Lambda_0^{[\beta^{-1}]\nu/2} \geq R_0^{\sigma} B$. Then

$$R_0^{\sigma} B\,|S'|^{\epsilon\sigma} \leq R_0^{\sigma} B\,|S'|^{\nu/2} \leq \left(R_0^{\sigma} B / |S'|^{\nu/2}\right) |S'|^{\nu}.$$

We also have

$$R_0^{\sigma} B / |S'|^{\nu/2} \leq R_0^{\sigma} B \Lambda_0^{-n\nu/2} \leq R_0^{\sigma} B \Lambda_0^{-[\beta^{-1}]\nu/2} \leq R_2.$$

Therefore

$$\Delta(S) \leq \frac{C_3}{1-\Lambda_0^{-1}} R_2\,|S'|^{\nu}$$

and we can set C_2 to $R_2 C_3 / \left(1-\Lambda_0^{-1}\right)$.

Definition 4. *Let $V \subset \mathcal{D}(T)$. The number $\sup_V |T'| / \inf_V |T'|$ is called the* distortion *of T on V. We omit V, if $V = I$.*

The next lemma is well known in the theory of expandings.

Lemma 4. *Let $S = T_n \circ \cdots \circ T_1$, where T_i is a Λ_0-expanding for $i = 1, 2, \ldots, n$ and $\deg(T_i) = 1$. The distortion of S does not exceed $C_4 = \exp\left(2R_2\left(1-\Lambda_0^{-1}\right)^{-1}\right)$.*

Proof. In a similar way as in the proof of Theorem 2 we get $\left|S''/(S')^2\right| \leq C_5 = R_2\left(1-\Lambda_0^{-1}\right)^{-1}$. Therefore, for $y, z \in \mathcal{D}(S)$ we have

$$\left|\ln \frac{S'(y)}{S'(z)}\right| \leq \int_{[y,z]} \left|\frac{S''}{S'}\right| \leq C_5 \int_{[y,z]} |S'| = C_5\,|S(y)-S(z)| \leq 2C_5. \tag{24}$$

This gives $S'(y)/S'(z) \leq \exp(2C_5) = C_4$ (note: $S'(y)$, $S'(z)$ have the same sign).

Possessing type (σ, B) does not imply bounded distortion on the whole of I, though the distortion is bounded on sufficiently small intervals.

Theorem 3. *Let S be an expanding of type (ν, C_2). For every $\theta_0 > 1$ there is $\eta_0 > 0$ s.t. if for some interval $V \subset \mathcal{D}(S)$ and $y_0 \in V$ we have $|V| \leq \eta_0\,|S'(y_0)|^{-(1+\nu)}$ then the distortion of S on V is bounded by θ_0.*

Proof. Let $y \in V$. We can assume that $S' \geq 0$. We have

$$\left|\frac{1}{S'(y)^{1+\nu}} - \frac{1}{S'(y_0)^{1+\nu}}\right| \leq (1+\nu)\int_{[y,y_0]}\left|\frac{S_{xx}}{S_x^{2+\nu}}\right|$$
$$\leq (1+\nu)C_2|V| \leq (1+\nu)\eta_0 C_2 |S'(y_0)|^{-(1+\nu)} \quad (25)$$

Multiplying (25) by $|S'(y_0)|^{1+\nu}$ yields

$$\left|\left(\frac{S'(y_0)}{S'(y)}\right)^{1+\nu} - 1\right| \leq (1+\nu)\eta_0 C_2. \quad (26)$$

Now it's clear that the distortion is arbitrarily close to 1, if η_0 is sufficiently small.

Now let us go back to the construction of Section 1. We will apply our results to the β-homogeneous sequence $T_i = T_{(k(i))}$ $(i = 1, 2, \ldots, n)$ and $\alpha \in J \in \mathcal{A}_n$. Obviously, $S_i = T_i \circ \cdots \circ T_1 = S_{(i)}$. From now on we will often write S_n instead of $S_{(n)}$, which is consistent with the formula and simplifies our notation.

Theorem 4. *Suppose that $S = S_n$ has rank $(0, C_1)$, type (ν, C_2) and distortion $\leq C_4$. Let $T = T_{(n)} = S \circ f|\rho^{-1}I_n \backslash \rho I_{n+1}$. There exist constants C_7, A, B, μ, σ independent of n such that T has rank (μ, A) and type (σ, B). Moreover, $|T'| \geq C_7 |S'(1)|^{1-\tau}$.*

Proof. Let us fix C_0 such that

$$\begin{aligned} C_0^{-1}|x| \leq |f_x| &\leq C_0|x|, \\ \max(R(f), \Delta(f)) &\leq C_0|x|^{-2}, \\ |\delta(f)| &\leq C_0. \end{aligned} \quad (27)$$

We have $|T'| = (|S'| \circ f)|f'| \geq C_4^{-1}|S'(1)| C_0^{-1}|x| \geq C_4^{-1}C_0^{-1}a|S'(1)|^{1-\tau}$ (see (5)). We set $C_7 = C_4^{-1}C_0^{-1}a$. We notice that

$$|S'(1)| \leq C_8 |T'|^{1/(1-\tau)}, \quad (28)$$

where $C_8 = C_7^{-1/(1-\tau)}$.

Lemma 2 gives

$$\begin{aligned} \Delta(T) &\leq (|S'|^{-1} \circ f)\Delta(f) + (\Delta(S) \circ f) \cdot R(f) \\ &\leq \max(\Delta(f), R(f)) \times (1 + \Delta(S) \circ f) \\ &\leq C_0|x|^{-2}\left(1 + C_2(C_4|S'(1)|)^{\nu}\right) \\ &\leq C_0(1 + C_2C_4^{\nu})a^{-2}|S'(1)|^{2\tau+\nu} \\ &\leq C_0(1 + C_2C_4^{\nu})a^{-2}C_8^{2\tau+\nu}|T'|^{(2\tau+\nu)/(1-\tau)}. \end{aligned} \quad (29)$$

This yields $\sigma = (2\tau + \nu)/(1 - \tau)$ and $B = C_0 \left(1 + C_2 C_4^{\nu}\right) a^{-2} C_8^{2\tau+\nu}$. We have

$$\begin{aligned} |\delta(T)| &= \left|(\delta(S) \circ f) f_x^{-1} + \delta(f)\right| \\ &\le C_1 \cdot C_0 \left|x\right|^{-1} + C_0 \le a^{-1} C_1 C_0 \left|S'(1)\right|^{\tau} + C_0 \\ &\le \left(1 + a^{-1} C_1\right) C_0 C_8^{\tau} \left|T'\right|^{\tau/(1-\tau)} . \end{aligned} \tag{30}$$

This yields $A = \left(1 + a^{-1} C_1\right) C_0 C_8^{\tau}$, $\mu = \tau/(1 - \tau)$.

Remark 3. We can apply the method of the proof to $\Delta_{xx}(T)$, $\Delta_{\alpha x}(T)$, $\Delta_{\alpha\alpha}$ separately. We obtain useful estimates

$$\begin{aligned} \Delta_{xx}(T) &\le C_9 \left|S'(1)\right|^{\max(\nu, 2\tau-1)} \\ \Delta_{\alpha x}(T) &\le C_9 \left|S'(1)\right|^{\nu+\tau} \\ \Delta_{\alpha\alpha}(T) &\le C_9 \left|S'(1)\right|^{\nu+2\tau} . \end{aligned} \tag{31}$$

From now on we assume $\nu < 2\tau - 1$, which reduces the first exponent to $2\tau - 1$.

Remark 4. There exists $\theta_0 > 1$ such that for sufficiently large N, $\alpha \in [\alpha_N, 2]$ and all $n \ge 1$ we have $|S_n'(1)| \ge \theta_0^n$ and $\left|T_{(n)}'\right| \ge \max\left(\theta_0, C_7 \theta_0^{n(1-\tau)}\right)$. In particular, $\inf \left|T_{(n)}'\right|$ grows exponentially.

Proof. Let us pick $\theta_0 \in \left(1, \frac{4}{3}\rho\right)$ arbitrarily and let N be large enough, so that for all $\alpha \in \mathcal{A}_N = [\alpha_N, 2]$ we have $\left|T_{(n)}'\right| \ge \theta_0$ for all n satisfying $(*)$ $C_7 \theta_0^{n(1-\tau)} < \theta_0$. The number of such n is bounded, so this is possible by the Appendix and C^1-continuity of our family (since $\alpha_N \to 2$, as $N \to \infty$).

We will show by induction that

(i) $\left|S_{(n)}'(1)\right| \ge \theta_0^n$,

(ii) $\left|T_{(k)}'\right| \ge \theta_0$ for all $k < n$.

This is obvious for $n = 1$. Using the induction hypothesis and Theorem 4, we get $\left|T_{(n)}'\right| \ge C_7 \theta_0^{n(1-\tau)}$, which is $\ge \theta_0$, if n does not satisfy $(*)$. If n does satisfy $(*)$ then (ii) is obvious for $k = n$. Now we apply the definition $S_{(n+1)} = T_{(k)} \circ S_{(n)}$ $(k < n)$ and get $\left|S_{n+1}'(1)\right| \ge \theta_0^{n+1}$, which completes the proof.

Corollary 1. *There exist constants C_1, C_2, C_4, A, B, Λ_0, ν, μ (all positive and $\Lambda_0 > 1$) s.t. for N sufficiently large and β sufficiently small and for every $n \ge 1$ and $\alpha \in A_{\max(n,N)}$ the mapping $S_{(n)}$ has rank $(0, C_1)$, type (ν, C_2), distortion $\le C_4$ and is Λ_0^n-expanding, and $T_{(n)}$ has rank (μ, A), type (σ, B) and is Λ_0-expanding.*

Proof. Induction. Suppose that $T_{(k)}$ has rank (μ, A) and type (σ, B) for all $k < n$. Theorems 1, 2 (applied to the sequence $T_i = T_{(k(i))}$) and Remark 4 imply that $S_{(n)}$ has rank $(0, C_1)$ and type (ν, C_2). If $n \le N$ then Lemma 4 implies that the distortion of $S_{(n)}$ does not exceed C_4.

Let us notice that $\mathcal{D}(S_{(n)}) = V_n$ and

$$|V_n| \le \frac{1}{2} C_0 \rho^{-2} |I_n|^2 \le \frac{1}{2} C_0 \rho^{-2} (2a)^2 \left| S'_{(n)} \right|^{-2\tau}.$$

This means that Theorem 3 applies to $S_{(n)}$ for $n \ge N$ with $\theta_0 = C_4$, if N is large enough (note: $2\tau > 1 + \nu$).

Eventually from Theorem 4 we get that $T_{(n)}$ has rank (μ, A) and type (σ, B). This ends the proof.

4 Families with a prerepelling critical point

Here we consider families a little more general than $f(x, \alpha) = 1 - \alpha x^2$.

Suppose that f has a critical point $c(\alpha)$ for α close to α_0 and suppose that for $\alpha = \alpha_0$ and some $m \in \mathbb{Z}^+$ the point $f^m(c(\alpha_0), \alpha_0) = x_0$ is a repelling point of period κ. We consider the following nondegeneracy condition (cf. [4]). Let $x(\alpha) = f^m(c(\alpha))$ be a differentiable function and let

$$\left. \frac{d}{d\alpha} \right|_{\alpha=\alpha_0} (f^\kappa(x(\alpha)) - x(\alpha)) \neq 0. \tag{32}$$

Let us define a sequence of functions

$$\chi_n(\alpha) = \left(\frac{d}{d\alpha} f^n(x(\alpha), \alpha) \right) \Big/ (f^n)_x (x(\alpha), \alpha). \tag{33}$$

Proposition 1. *The limit* $\chi = \lim_{n \to \infty} \chi_n(\alpha_0)$ *exists. It is* $\neq 0$ *iff* (32) *is satisfied.*

Proof. Let $T = f^\kappa$. Let $n = l\kappa + l_1$, where $0 \le l_1 \le \kappa - 1$. We have by Lemma 1(ii):

$$\delta(f^n) = \sum_{i=1}^{l} \left(\delta(T) \circ T^{i-1} \right) \frac{1}{(T^{i-1})'} + \frac{1}{(T^l)'} \left(\sum_{j=1}^{l_1} (\delta(f) \circ f^{j-1}) \cdot \frac{1}{(f^{j-1})'} \right) \circ T^l. \tag{34}$$

Substituting (x_0, α_0) and letting $l \to \infty$ yields:

$$\lim_{n \to \infty} \delta(f^n)(x_0, \alpha_0) = \delta(T)(x_0, \alpha_0) \left(1 - T'(x_0)^{-1} \right)^{-1} = \bar{\chi}. \tag{35}$$

Condition (32) is equivalent to

$$(T_x(x_0, \alpha_0) - 1) x'(\alpha_0) \neq -T_\alpha(x_0, \alpha_0) \tag{36}$$

or $x'(\alpha_0) \neq -\bar{\chi}$. On the other hand,

$$\left.\frac{d}{d\alpha}\right|_{\alpha=\alpha_0} f^n(x(\alpha),\alpha)\Big/ (f^n)_x(x_0,\alpha_0)$$
$$= x'(\alpha_0) + \delta(f^n)(x_0,\alpha_0) \to x'(\alpha_0) + \bar{\chi} = \chi, \quad \text{as } n \to \infty. \tag{37}$$

This completes the proof.

Theorem 5. *Let $\chi \neq 0$. For every $\eta > 0$ there is $n_0 \in \mathbb{Z}^+$ and $\delta > 0$ such that for every $n \geq n_0$ and $\alpha \in (\alpha_0 - \delta,\ \alpha_0 + \delta)$: if for some interval $U \subset I$, $f(\cdot,\alpha)|U$ is Λ_0-expanding and $\{x(\alpha), f(x(\alpha),\alpha), \dots, f^n(x(\alpha),\alpha)\} \subset U$ then*

$$|\chi_n(\alpha)/\chi - 1| < \eta. \tag{38}$$

Proof. First, we choose n_0 large enough, so that

$$\begin{gathered} |\chi_{n_0}(\alpha_0) - \chi| < \frac{1}{4}\eta\,|\chi| \\ |\delta(f^{n_0})(x,\alpha) - \delta(f^n)(x,\alpha)| < C_1 q^{n_0} < \frac{1}{4}\eta\,|\chi| \end{gathered} \tag{39}$$

(we applied Theorem 1).

We pick δ such that if $|\alpha - \alpha_0| < \delta$ then

$$\begin{gathered} |\delta(f^{n_0})(\chi,\alpha) - \delta(f^{n_0})(x_0,\alpha_0)| < \frac{1}{4}\eta\,|\chi| \\ |x'(\alpha) - x'(\alpha_0)| < \frac{1}{4}\eta\,|\chi|\,. \end{gathered} \tag{40}$$

We get Theorem 5 by the triangle inequality.

Remark 5. We will not dwell on the general case, leaving the details to the reader. We notice that in the case of $f(\alpha,x) = 1 - \alpha x^2$ we have $\chi_n(\alpha) = -2\alpha\delta(S_{(n+1)})(1,\alpha)$. It is easy to see that condition (32) holds for this family.

Corollary 2. *Let $D = |\chi/(2\alpha_0)|$. For every $\theta > 1$ we can choose sufficiently large n_1 such that for every $n \geq n_1$ and every α, α_1, $\alpha_2 \in A_n$*

$$\begin{gathered} h \leq \left|\delta(S_{(n)})(1,\alpha)\right| \leq H, \\ \theta^{-1} \leq \frac{\delta(S_{(n)})(1,\alpha_1)}{\delta(S_{(n)})(1,\alpha_2)} \leq \theta, \end{gathered} \tag{41}$$

where $h = \theta^{-1}D$, $H = \theta D$ (see Remark 5 and Theorem 1(ii)).

Let us finish this section with one more distortion estimate, this time over α.

Lemma 5. *For every $\theta_1 > 1$ there is $\eta_1 > 0$ such that if $|\alpha_2 - \alpha_1| \leq \eta_1 \left|(S_n)_\alpha(1,\alpha_1)\right|^{-(1+\nu)}$, $\alpha_1, \alpha_2 \in A_n$, then*

$$\theta_1^{-1} \leq \frac{(S_n)_\alpha(1,\alpha_2)}{(S_n)_\alpha(1,\alpha_1)} \leq \theta_1.$$

Proof. We notice that

$$\left|\frac{(S_n)_{\alpha\alpha}}{(S_n)_\alpha^{2+\nu}}\right| = |\delta(S_n)|^{-(2+\nu)}\left|\frac{(S_n)_{\alpha\alpha}}{(S_n)_\alpha^{2+\nu}}\right| \leq h^{-(2+\nu)}C_2. \tag{42}$$

Integrating over α we get

$$\begin{aligned}\left|\frac{1}{(S_n)_\alpha^{1+\nu}(\alpha_2)} - \frac{1}{(S_n)_\alpha^{1+\nu}(\alpha_1)}\right| &\leq (1+\nu)h^{-(2+\nu)}C_2\,|\alpha_2-\alpha_1| \\ &\leq (1+\nu)h^{-(2+\nu)}C_2\eta_1\,|(S_n)_\alpha(1,\alpha_1)|^{-(1+\nu)}.\end{aligned}$$

Multiplying by $|(S_n)_\alpha(1,\alpha_1)|^{1+\nu}$ leads to:

$$\left|\left(\frac{(S_n)_\alpha(1,\alpha_1)}{(S_n)_\alpha(1,\alpha_2)}\right)^{1+\nu} - 1\right| \leq (1+\nu)h^{-(2+\nu)}C_2\eta_1. \tag{43}$$

If η_1 is sufficiently small, (43) yields the lemma.

Corollary 3. *Using the assumptions and notation of Lemma* 5 *we have*

$$\theta_1^{-1}\theta^{-1} \leq \frac{(S_n)_x(1,\alpha_2)}{(S_n)_x(1,\alpha_1)} \leq \theta_1\theta. \tag{44}$$

Remark 6. The sets J_k have not more components than $I_k\backslash I_{k+1}$, since by simple differentiation we can see that the absolute value of the derivative of $S_{(n)}(1,\alpha)$ over α is $>$ absolute value of the derivative of the ends of $I_k\backslash I_{k+1}$.

Proof. Indeed, we ask if

$$\begin{aligned} a\left|\frac{d}{d\alpha}|S_k'(1,\alpha)|^{-\tau}\right| &= a\tau\,|(S_k)_{\alpha x}(1,\alpha)|\cdot|S_k'(1,\alpha)|^{-(\tau+1)} \\ &< |(S_n)_\alpha(1,\alpha)|. \end{aligned} \tag{45}$$

This is equivalent to

$$a\Delta_{\alpha x}(S_k)\cdot\tau\,|S_k'|^{-(\tau-1)} < |\delta(S_n)|\,|S_n'| \tag{46}$$

and because of $\Delta_{\alpha x}(S_k) \leq C_2\,|S_k'|^\nu$ we need

$$a\tau C_2\,|S_k'|^{\nu-\tau+1} < |S_n'|\,|\delta(S_n)|. \tag{47}$$

We have $|S_k'| \leq |S_n'|$. Also $|\delta(S_n)| \geq h > 0$ for large n. Therefore (47) holds for large n, since $\nu < \tau$ (note: $\nu < 2\tau - 1 < \tau$).

5 The measure of A_∞

We use the notation of Section 1.

Proposition 2. (i) *Suppose that* $N \geq 2\beta^{-1}$. *The numbers* s_m, $m = N, N+1, \ldots, n$ *satisfy the inequality*

$$s_m \leq \frac{\beta}{2} m^2 + 1. \tag{48}$$

(ii) *If* $\alpha \in J_k \in \mathcal{A}_{n+1}$ *then* $W_n \subset \rho^{-1} I_k \backslash \rho I_{k+1}$.
(iii) *There is* $\lambda \in (0,1)$ *such that* (9) *is satisfied.*

Proof. (i) Induction. We notice that $s_N = N+1 \leq (\beta/2)N^2+1$, if $N \geq 2/\beta$. The induction step:

$$s_{n+1} = s_n + s_{k(n)} \leq \frac{\beta}{2} n^2 + 1 + \beta n \leq \frac{\beta}{2}(n+1)^2 + 1.$$

(ii) We notice that since $S_{(n)}(1) \in I_k \backslash I_{k+1}$:

$$\begin{aligned} \operatorname{dist}\left(S_{(n)}(1), \rho I_{k+1}\right) &\geq \frac{1}{2}(1-\rho)\left|I_{k+1}\right| \\ \operatorname{dist}\left(S_{(n)}(1), \rho^{-1} I_k\right) &\geq \frac{1}{2}\left(\rho^{-1}-1\right)\left|I_k\right|. \end{aligned} \tag{49}$$

It suffices to show that

$$\begin{aligned} 2\left|W_n\right| &\leq \min\left((1-\rho)\left|I_{k+1}\right|, \left(\rho^{-1}-1\right)\left|I_k\right|\right) \\ &= (1-\rho)\left|I_{k+1}\right|. \end{aligned} \tag{50}$$

Obviously, we have

$$\begin{aligned} \left|W_n\right| &\leq \sup\left|S'_{(n)}(1)\right| \cdot \left|V_{n+1}\right| \leq C_4 \left|S'_{(n)}\right| \cdot \frac{1}{2}\rho^{-2} C_0 \left|I_{n+1}\right|^2 \\ &= \frac{1}{2} C_0 \rho^{-2} C_4 a^{-2} \left|S'_{(n)}(1)\right|^{1-2\tau} \leq \frac{1}{2} C_0 \rho^{-2} C_4 a^{-2} \Lambda_0^{n(1-2\tau)}. \end{aligned} \tag{51}$$

On the other hand, $\left|I_{k+1}\right| = 2a\left|S'_{(k)}(1)\right|^{-\tau} \geq 2aR_0^{-\tau s_k}$, since $\deg(S_{(k)}) = s_k - 1 \leq s_k$. Because $s_k \leq \beta n$ for $n \geq N$, we get $R_0^{-\tau s_k} \geq R_0^{-\tau\beta n} = \Lambda_0^{-\epsilon\tau n}$. Suppose that $\epsilon\tau < 2\tau - 1$. For sufficiently large n we have

$$2a(1-\rho)\Lambda_0^{-\epsilon n\tau} \geq C_0 \rho^{-2} C_4 a^{-2} \Lambda^{-n(2\tau-1)} \tag{52}$$

and (ii) holds if N is sufficiently large.
(iii) Suppose that $S_{(n)}(1,\alpha) \in I_k$ and $\deg(T_{(k)}) > \beta n$. Then

$$\operatorname{dist}\left(S_n(1,\alpha), 0\right) \leq \frac{1}{2}|I_k| = a\left|S'_{(k)}(1)\right|^{-\tau} \leq a\Lambda_0^{-k\tau}.$$

From (i) we now have that $\deg(T_{(k)} = s_k \leq (\beta/2)k^2 + 1$. Hence, $\beta n < (\beta/2)k^2 + 1$ or $k > \sqrt{2n - 2/\beta} \geq \sqrt{n}$, since $n \geq N \geq 2/\beta$. This yields (iii) with $\lambda = \Lambda_0^{-\tau}$.

Let us consider the transformation ψ_n defined on every $J \in \mathcal{A}_n$ by $\psi_n(\alpha) = S_n(1, \alpha)$. We assume $N \geq 2\beta^{-1}$.

Theorem 6. *There is a constant C_{10} such that for every $n \geq N$*

$$\sum_{J \in \mathcal{A}_n} \sup_J \frac{1}{|\psi_n'|} \leq C_{10} |A_N| . \tag{53}$$

Proof. Let π be the partition of I into intervals of equal length d ($d^{-1} \in \mathbb{Z}$). Let $\mathcal{A}_n'$ be the family of intervals defined in a similar way to $\mathcal{A}_n$, except we replace the definition of J_k with

$$J_k(P) = \{\alpha \in J : \ S_n(1, \alpha) \in (I_k \backslash I_{k+1}) \cap P\} , \tag{54}$$

where $P \in \pi$. Since every $J \in \mathcal{A}_n$ is a union of element of $\mathcal{A}_n'$, it is sufficient to prove (53) for $\mathcal{A}_n'$ instead of $\mathcal{A}_n$. We will be able to do it, if d is sufficiently small.

Let us introduce two sequences

$$\begin{aligned} \gamma_n &= \sum_{J \in \mathcal{A}_n'} \sup_J \frac{1}{|\psi_n'|} \\ \eta_n &= \sum_{J \in \mathcal{A}_n'} \sup_J \left| \frac{\psi_n''}{(\psi_n')^2} \right| |J| . \end{aligned} \tag{55}$$

We are going to derive estimates of $(\gamma_{n+1}, \eta_{n+1})$ in terms of (γ_n, η_n).

We fix an arbitrary $\theta > 1$. First, by Corollary 2 we have for n large enough

$$\begin{aligned} \sup_{J_k(P)} \frac{1}{|\psi_{n+1}'|} &\leq \frac{1}{h} \sup_{J_k(P)} \frac{1}{|S_{n+1}'(1)|} \\ &\leq \frac{1}{h} \sup \frac{1}{\left|T_{(k)}'\right|} \sup_{J_k(P)} \frac{1}{|S_n'(1)|} \\ &\leq \frac{1}{h} \sup \frac{1}{\left|T_{(k)}'\right|} H \sup_{J_k(P)} \frac{1}{|\psi_n'|} \\ &= \theta^2 \sup \left|T_{(k)}'\right|^{-1} \sup_{J_k(P)} |\psi_n'|^{-1} . \end{aligned} \tag{56}$$

The first supremum is over $\alpha \in J_k(P)$ and $x \in P \cap (I_k \backslash I_{k+1})$ (note: I_k depends on α). There is $k_1 \in \mathbb{Z}^+$ such that

$$\sum_{k \geq k_1} \sup \left|T_{(k)}'\right| \leq \Lambda_0^{-1} \tag{57}$$

(see Remark 4).

Let us choose d small enough, so that for every $P \in \pi$ one of the two alternatives holds: either (1°) $P \subset I_{k_1}$ or (2°) P intersects not more than two of the sets $I_k \backslash I_{k+1}$.

We get:

$$\sum_k \sup_{J_k(P)} \frac{1}{|\psi'_{n+1}|} \leq 2\theta^2 \Lambda_0^{-1} \sup_{J(P)} \frac{1}{|\psi'_n|}, \tag{58}$$

where $J(P) = J \cap \psi_n^{-1}(P)$.

We notice that $\psi_n(J(P)) \neq P$ for at most two $P \in \pi$, namely those for which (*) $P \cap \partial\psi_n(J) \neq \emptyset$. Summing up over these P, we get

$$\sum_{k,P}^{*} \sup_{J_k(P)} \frac{1}{|\psi'_{n+1}|} \leq 4\theta^2 \Lambda_0^{-1} \sup_{\alpha \in J} \frac{1}{|\psi'_n|}. \tag{59}$$

For those P that (**) $\psi_n(J(P)) = P$ we have

$$\sup_{J(P)} \frac{1}{|\psi'_n|} \leq \frac{1}{|\psi_n(J(P))|} \int_{\psi_n(J(P))} \frac{1}{|\psi'_n|} \circ \psi_n^{-1} + \sup_{\alpha \in J(P)} \left| \frac{\psi''_n}{(\psi'_n)^2} \right| |J(P)|, \tag{60}$$

according to the rule 'maximum $\leq$ average+max. of derivative$\times$length of the interval' applied to $(1/|\psi'_n|) \circ \psi_n^{-1}$ on $P = \psi_n(J(P))$; we notice that the average is just $d^{-1}|J(P)|$. Totalling over P satisfying (**) we get

$$\sum_{P,k}^{**} \sup_{J_k(P)} \frac{1}{|\psi'_{n+1}|} \leq 2\theta^2 \Lambda_0^{-1} \left(\frac{1}{d} |J| + \sup_J \left| \frac{\psi''_n}{(\psi'_n)^2} \right| \cdot |J| \right). \tag{61}$$

Inequalities (59) and (61) and summation over J yield

$$\gamma_{n+1} \leq 4\theta^2 \Lambda_0^{-1} \gamma_n + 2\theta^2 \Lambda_0^{-1} \eta_n + \frac{2\theta^2 \Lambda_0^{-1}}{d} |A_N|. \tag{62}$$

Now we intend to estimate η_{n+1}. We have $\psi_{n+1}(\alpha) = T_{(k)}(\psi_n(\alpha), \alpha)$ and this yields

$$\begin{aligned} \psi''_{n+1}(\alpha) = &\left(T_{(k)}\right)_{\alpha\alpha} (\psi_n(\alpha), \alpha) \\ &+ 2\left(T_{(k)}\right)_{\alpha x} (\psi_n(\alpha), \alpha)\, \psi'_n(\alpha) \\ &+ \left(T_{(k)}\right)_{xx} (\psi_n(\alpha), \alpha)\, (\psi'_n(\alpha))^2 \\ &+ \left(T_{(k)}\right)_x (\psi_n(\alpha), \alpha)\, \psi''_n(\alpha). \end{aligned} \tag{63}$$

We also have $\psi'_n(\alpha) = (S_n)_\alpha (1, \alpha)$. Therefore

$$\begin{aligned} \frac{|\psi''_{n+1}|}{(\psi'_{n+1})^2} \leq \delta(S_{n+1})^{-2} \Bigg[&\Delta_{xx}(T_{(k)}) \delta(S_n)^2 + \frac{2}{|S'_n|} \Delta_{\alpha x}\left(T_{(k)}\right) |\delta(S_n)| \\ &+ \frac{1}{(S'_n)^2} \Delta_{\alpha\alpha}(T_{(k)}) \Bigg] + \frac{1}{\left|\left(T_{(k)}\right)_x\right|} \frac{\delta(S_n)^2}{\delta(S_{n+1})^2} \frac{|\psi''_n|}{(\psi'_n)^2}, \end{aligned} \tag{64}$$

where we omitted the arguments for simplicity. This yields

$$\left|\frac{\psi''_{n+1}}{\left(\psi'_{n+1}\right)^2}\right| \leq \Phi_n\left(T_{(k)}\right) + \Lambda_0^{-1}\theta^2 \left|\frac{\psi''_n}{\left(\psi'_n\right)^2}\right| \tag{65}$$

where θ^2 comes from Corollary 2 and

$$\Phi_n(T_{(k)}) = \delta(S_{n+1})^{-2}\left[\Delta_{xx}(T_{(k)})\delta(S_n)^2 + \frac{2}{|S'_n|}\Delta_{\alpha x}(T_{(k)})\delta(S_n) + \frac{1}{\left(S'_n\right)^2}\Delta_{\alpha\alpha}(T_{(k)})\right]. \tag{66}$$

Therefore, we can write

$$\eta_{n+1} = \sum_{J,k,P} \sup_{J_k(P)} \left|\frac{\psi''_{n+1}}{\left(\psi'_{n+1}\right)^2}\right| |J_k(P)| \leq \sum_{J,k,P} \Phi_n(T_{(k)})\,|J_k(P)| + \Lambda_0^{-1}\theta^2\eta_n. \tag{67}$$

Because of the obvious inequality

$$|J_k(P)| \leq \sup_{J_k(P)} \frac{1}{|\psi'_n|} \cdot \left|P \cap \bigcup_{\alpha\in J} (I_k \backslash I_{k+1})\right| \tag{68}$$

we get

$$\eta_{n+1} \leq r\gamma_n + \Lambda_0^{-1}\theta^2\eta_n, \tag{69}$$

where

$$r = \sup_n \max_{P\in\pi} \sum_k \sup_{\alpha,x} \Phi_n(T_{(k)}) \left|P \cap \bigcup_{\alpha} (I_k \backslash I_{k+1})\right|. \tag{70}$$

We shall soon see that r is finite and even arbitrarily small. Inequalities (62) and (69) imply the theorem, if the eigenvalues of the matrix

$$\mathbf{P} = \begin{bmatrix} 4\theta^2\Lambda_0^{-1} & 2\theta^2\Lambda_0^{-1} \\ r & \theta^2\Lambda_0^{-1} \end{bmatrix} \tag{71}$$

have modulus < 1. As a matter of fact, we can write (62) and (69) as a single matrix inequality

$$\boldsymbol{\xi}_{n+1} \leq \mathbf{P}\boldsymbol{\xi}_n + \mathbf{c} \quad (n \leq N), \tag{72}$$

where $\boldsymbol{\xi}_n = (\gamma_n, \eta_n)$. Here $\mathbf{c}$ is a constant vector and $\|\mathbf{c}\| \leq \text{const.}\,|A_N|$. Here and in what follows, const. means an unspecified constant independent of N. Also, we can easily see that $\|\boldsymbol{\xi}_N\| \leq \text{const.}\,|A_N|$. In fact, it is sufficient to show that $\eta_N \leq \text{const.}\,|A_N|$. The inequality $\gamma_N \leq \text{const.}\,|A_N|$ will follow by the same principle we used to prove (60). First, we notice that

$$\eta_N \leq \sup_{A_N} \left|\frac{\psi''_N}{\left(\psi'_N\right)^2}\right| |A_N|$$

(by definition of Section 1, $\mathcal{A}_N$ has only one element). Let us fix n_1 such that (41) hold for $n \geq n_1$. We can apply (65) for $\alpha \in A_N$ and $n = n_1, n_1+1, \ldots, N$. For these n we have $k = 1$ and $\Phi_n(T_{(1)}) \leq h^{-1}\left(R_2 H^2 + 2R_2 H + R_2\right) = C_{11}$. If $\Lambda_0^{-1}\theta^2 < 1$, (65) yields

$$\sup_{\mathcal{A}_n}\left|\frac{\psi_N''}{(\psi_N')^2}\right| \leq \frac{C_{11}}{1-\Lambda_0\theta^2}\sup_{A_N}\left|\frac{\psi_{n_1}''}{(\psi_{n_1}')^2}\right|. \tag{73}$$

Therefore (72) yields the Theorem if the spectral radius of $\mathbf{P}$ is < 1. We recall that θ is arbitrarily close to 1, as β is sufficiently small. Also, we will see that r is arbitrarily small, as d is sufficiently small.

By Remark 3 we can check that $\Phi_n(T_{(k)}) \leq \text{const.}\,|S_k'(1)|^{2\tau-1}$ (note: $\nu < 2\tau - 1$). We also have $|I_k| \leq \text{const.}\left|S_{k-1}'(1)\right|^{-\tau}$, where α on both sides may be different (see Corollary 3). Also, const. $\cdot\left|S_{k-1}'(1)\right| \geq |S_k'(1)|^{1/(1+\epsilon)}$ by Lemma 3 (indeed, $S_k = T_{(k)} \circ S_{k-1}$ and $|S_k'| = \left(\left|T_{(k)}'\right| \circ S_{k-1}\right)\left|S_{k-1}'\right| \leq R_0\left|S_{k-1}'\right|^{\epsilon}\left|S_{k-1}'\right| = R_0\left|S_{k-1}'\right|^{1+\epsilon}$). So, $|I_k| \leq \text{const.}\,|S_k'|^{-\tau/(1+\epsilon)}$. Hence

$$\Phi_n(T_{(k)})\left|P \cap \bigcup_{\alpha\in J}(I_k\backslash I_{k+1})\right| \leq \text{const.}\,|S_k'(1)|^{-\zeta}, \tag{74}$$

where $\zeta = \tau/(1+\epsilon)+1-2\tau$, if β is sufficiently small. We also have $|S_k'(1)|^{-\zeta} \leq \Lambda_0^{-k\zeta}$. So, the sum in (70) has uniformly exponentially decreasing terms. By fixing sufficiently small d we can make it arbitrarily close to 0.

This proves the theorem, if $4\Lambda_0^{-1} < 1$. When this condition is not satisfied, we pick $p \in \mathbb{Z}^+$ such that $\Lambda_0^p > 4$ and examine the relation between $(\gamma_{n+p}, \eta_{n+p})$ and (γ_n, η_n). The corresponding matrix, as $r \to 0$ looks like

$$\begin{bmatrix} 2M\Lambda_0^{-p} & * \\ 0 & \Lambda_0^{-p} \end{bmatrix}, \tag{75}$$

where M is an integer which is the maximal number of $J' \in \mathcal{A}_{n+p}'$ contained in a fixed $J \in \mathcal{A}_n'$ and such that there is $i \in \{n, n+1, \ldots, n+p-1\}$ with the property: there is $k \leq k_1$ such that $\psi_i(J') \subset I_k\backslash I_{k+1}$. This number can be made $= 2$, if we slightly perturb intervals $I_1, \ldots, I_{k_1}$.

Corollary 4. *We have* $|B_n| \leq 2C_{10}a\lambda^{\sqrt{n}}\,|A_N|$.

Proof. Clearly, $B_n \subset \psi_n^{-1}\left(\left[-a\lambda^{\sqrt{n}}, a\lambda^{\sqrt{n}}\right]\right)$. Theorem 6 gives

$$\left|\psi_n^{-1}(E)\right| \leq \left(\sum_{J\in\mathcal{A}_n}\sup_J\frac{1}{|\psi_n'|}\right)|E| \leq C_{10}\,|E|\,|A_N| \tag{76}$$

for an arbitrary measurable set $E \subset I$.

Remark 7. The existence of an a.c.i.m. for T follows directly from [5], since the function

$$g(x) = \begin{cases} \dfrac{1}{|T'(x)|}, & \text{where } T \text{ is continuous,} \\ 0 & \text{on discontinuities} \end{cases} \tag{77}$$

verifies $\mathrm{Var} g < +\infty$ and $\sup g < 1$ (a sufficient condition for the existence of an a.c.i.m.). One can obtain another proof (and make this paper self-contained) by imitating the proof of Theorem 6 for the sequence of mappings $\psi_n(x) = T^n(x)$ ($\mathcal{A}_n$ becomes the partition into the pieces of continuity of ψ_n). The new functions satisfy the recursive formula $\psi_{n+1}(x) = T_{(k)}(\psi_n(x))$ (if $\psi_n(x) \in I_k \backslash I_{k+1}$). This formula is much easier to differentiate, since x is the only variable.

Proof. We have

$$\mathrm{Var} g < 2 \sum_k \sup \left| T'_{(k)} \right|^{-1} + \sum_k \Delta_{xx}(T_{(k)}) \, |I_k| \, . \tag{78}$$

The first sum converges by Remark 4. The second sum converges by an argument similar to the proof of (74).

In view of Section 1, Proposition 2(iii), Corollary 4 and Remark 7 the main theorem is proved.

6 Some generalizations and final remarks

Our method applies without any changes to families $f(x) = 1 - a\,|x|^{\xi}$, where ξ is sufficiently close to 2. For $\xi < 2$ we get an example of a family which is not C^2. It is easy to get examples of families with singularities like $|x|^{\xi}$ with any $\xi > 1$.

Another class of examples would be families with a finite or infinite number of singularities like

$$f(x) = \begin{cases} 1 - ax^2 & \text{for } -1 \le x \le 0 \\ [10/x] - 1 & \text{for } 1 \ge x \ge 0. \end{cases}$$

Let us discuss a little different result now. Let $f_0(x) = 1 - 2x^2$ and let for every $\eta > 0$ $X(\eta)$ be the η-neighborhood of f_0 in the C^2-topology. If η is sufficiently small then there is a C^1 function $c : X(\eta) \to I$ such that $c(f)$ is the only critical point of $f \in X(\eta)$. Let $M \subset X(\eta)$ be a submanifold of codimension 1 defined as follows:

$$M = \left\{ f \in X(\eta) : \; f^3(c(f)) = f^2(c(f)) \right\} . \tag{79}$$

For $f \in M$, $x(f) \overset{\text{def}}{=} f^2(c(f))$ is a hyperbolic fixed point.

Theorem 7. *Suppose that a path $\alpha \to f_\alpha$ intersects M transversally for some α_0. The α_0 is a one-sided Lebesgue density point of the set of those α that f_α has an a.c.i.m.*

Corollary 5. *There is a open set of C^2-families satisfying Jakobson's Theorem.*

One can consider a transversality condition in higher jets

$$\left.\frac{d^k}{d\alpha^k}\right|_{\alpha=\alpha_0} (f_\alpha(x(f_\alpha)) - x(f_\alpha)) \neq 0, \tag{80}$$

where $k \geq 1$ is arbitrary.

Theorem 8. *Theorem 7 remains true if transversality to M is replaced with condition* (80).

The idea of the proof. Let $G : X(\eta) \to \mathbb{R}$ be defined by

$$G(f) = f(x(f)) - x(f). \tag{81}$$

It is easy to check that $dG(f) \neq 0$ for $f \in X(\eta)$, if η is sufficiently small. Therefore, there is a mapping $p : X(\eta) \to M$ such that $X \ni f \mapsto (p(f), G(f)) \in M \times \mathbb{R}$ is a diffeomorphism (one can write down suitable formulas explicitly).

For every $f \in M$ we have a standard family f_γ corresponding to the fiber $\{f\} \times \mathbb{R}$, so that $p(f_\gamma) = f$ and $G(f_\gamma) = \gamma$ for $|\gamma| < \gamma_0$. We repeat our proof simultaneously for all families f_γ. We obtain a family $\mathcal{F}$ of submanifolds $F \subset X(\eta)$ such that if $g \in F \in \mathcal{F}$ then g has an a.c.i.m. Moreover, there is a constant K with the following properties:

(i) Every $F \in \mathcal{F}$ corresponds to a graph of a Lipschitz function $h_F : M \to \mathbb{R}$ with a Lipschitz constant $\leq K$ ($M \in \mathcal{F}$; $h_M \equiv 0$).

(ii) Let $A_\infty(f) = \{\gamma : f_\gamma \in F \text{ for some } F \in \mathcal{F}\}$. The family $\mathcal{F}$ defines a measurable mapping $A_\infty(f_1) \to A_\infty(f_2)$. This mapping has a measure-theoretic Jacobian $\in [K^{-1}, K]$ a.e.

From our estimates it follows easily that for every $f \in M$ and $h > 0$

$$\frac{|A_\infty(f) \cap [0, h]|}{h} \geq 1 - c_1 \exp\left(-c_2 \left(\log \frac{1}{h}\right)^{1/2}\right), \tag{82}$$

where c_1, $c_2 > 0$.

These estimates are sufficient to show that every family g_α such that $G(g_{\alpha_0}) = 0$ and $\left.(d^k/d\alpha^k)\, G(g_\alpha)\right|_{\alpha=\alpha_0} \neq 0$ for some k intersects the leaves of $\mathcal{F}$ for a positive measure set of parameters α.

Corollary 6. *Any analytic family which contains f_0, satisfies Jakobson's theorem.*

Let us list two more facts concerning our construction.

(i) The density of an a.c.i.m. we constructed $\in L^p(I)$ for every $p \in [1,2)$. This result is analogous to one of Carleson's results.

(ii) If α_0 is sufficiently close to 2 and $\inf_{n\geq 1} |f^n_{\alpha_0}(0)| > 0$ then α_0 is also a density point of the set of α such that f_α has an a.c.i.m. The condition that α_0 is close to 2 can be replaced with the condition given in [3].

Appendix

In this Appendix $f(x) = 1 - 2x^2$.

Theorem 9. *Let $I_n = [-3^{-n}, 3^{-n}]$ for $n \in \mathbb{Z}^+$. Let $V_n = f(I_n)$. Then for every n and $\rho \in (0,1)$:*

(i) $\left| (f^n)' \,|\, |V_n| \right| \geq \left(3\frac{1}{9}\right)^n$ *and* $\sim 4^n$ *for large* n,

(ii) $\left| (f^{n+1})' \, |I_n \backslash \rho I_{n+1}| \right| \geq \Lambda = 4\rho/3$ *and* $\geq$ const. $\rho \left(\frac{4}{3}\right)^n$,

(iii) $|W_n| = |f^n(V_{n+1})| \leq \left(\frac{4}{9}\right)^{n-1} |V_1|$.

Proof. (i) We have $V_n = f(I_n) = [1 - 2 \cdot 9^{-n}, 1]$. We have $|f'| \leq 4$. By the Mean Value Theorem we have for $k < n$

$$|f^k(V_n)| \leq 2 \cdot 9^{-n} \cdot 4^k = 4^{-(n-k-1)} \cdot \left(\frac{4}{9}\right)^{n-1} |V_1| \leq \left(\frac{4}{9}\right)^{n-1} |V_1| . \tag{83}$$

This implies $f^k(V_n) \subset \pm V_1$ for $n \geq 2$. We also have $\left| f' \right| \pm V_1 \Big| \geq 4\,(1 - |V_1|) = \frac{28}{9} = 3\frac{1}{9}$.

We can write

$$\left| (f^n)' \, \Big| V_n \right| \geq \left| \prod_{k=0}^{n-1} f' |f^k(V_n)| \right| . \tag{84}$$

Clearly, it is $\geq \left(3\frac{1}{9}\right)^n$ and also $\sim 4^n$ for large n. Indeed, the product is $\geq$

$$\prod_{k=0}^{n-1} 4\,(1 - |f^k(V_n)|) \geq 4^n \prod_{k=0}^{n-1} \left(1 - \left(\frac{4}{9}\right)^{n-1} |V_1|\, 4^{-(n-k-1)} \right) . \tag{85}$$

Hence,

$$\left| (f^{n+1})' \, |I_n \backslash \rho I_{n+1}| \right| \geq \left| (f^n)' \, \Big| V_n \right| \cdot 4 \cdot \frac{1}{2} \, |\rho I_{n+1}| \geq 3^n 4 \rho 3^{-(n+1)} = \frac{4}{3}\rho = \Lambda. \tag{86}$$

Using $\left| (f^n)' \, \Big| V_n \right| \geq$ const.$\cdot 4^n$ we also get $\left| (f^{n+1})' \, |I_n \backslash \rho I_{n+1}| \right| \geq$ const. $\rho \left(\frac{4}{3}\right)^n$.

If $\rho \geq 27/28$ one gets the same inequalities on $\rho^{-1} I_n \backslash \rho I_n$ with $3\frac{1}{9}$ replaced by 3.

References

1. M. Benedicks & L. Carleson. On iteration of $1 - ax^2$ on $(-1, 1)$. *Ann. of Math.* (2) **122** (1985), no. 1, 1–25.
2. M. V. Jakobson. Absolutely continuous Invariant Measures for one-parameter families of one-dimensional maps. *Comm. in Math. Phys.* **81** (1981), 39–88.
3. M. Misiurewicz. Absolutely continuous measures for certain maps of an interval. *I.H.E.S. Publications Mathématiques*, #**53** (1981), 17–52.
4. M. Rees. Positive measure sets of ergodic rational maps. Preprint. [4]
5. M. Rychlik. Bounded variation and invariant measures. *Studia Math.* t. **76**, (1983), 69-80.

[4] This paper has been published as: M. Rees. Positive measure sets of ergodic rational maps. *Ann. Scient. Ec. Norm. Sup.*, (4) 19 (1986), 383-407.

Unstable periodic orbits and the dimensions of multifractal chaotic attractors

Celso Grebogi
Laboratory for Plasma and Fusion Energy Studies, University of Maryland, College Park, Maryland 20742

Edward Ott
Laboratory for Plasma and Fusion Energy Studies, University of Maryland, College Park, Maryland 20742
and Department of Electrical Engineering and Department of Physics, University of Maryland, College Park, Maryland 20742

James A. Yorke
Institute for Physical Science and Technology and Department of Mathematics, University of Maryland, College Park, Maryland 20742

(Received 28 September 1987)

The probability measure generated by typical chaotic orbits of a dynamical system can have an arbitrarily fine-scaled interwoven structure of points with different singularity scalings. Recent work has characterized such measures via a spectrum of fractal dimension values. In this paper we pursue the idea that the infinite number of unstable periodic orbits embedded in the support of the measure provides the key to an understanding of the structure of the subsets with different singularity scalings. In particular, a formulation relating the spectrum of dimensions to unstable periodic orbits is presented for hyperbolic maps of arbitrary dimensionality. Both chaotic attractors and chaotic repellers are considered.

I. INTRODUCTION

The long time distribution generated by a typical orbit of a chaotic nonconservative dynamical system is generally highly singular. The subset of phase space to which the orbit asymptotes with time, the attractor, can be geometrically fractal. Furthermore, the distribution of orbit points on the attractor can have an arbitrarily fine-scaled interwoven structure of hot and cold spots. Sets with such distributions have been called *multifractals. By hot and cold spots we mean points on* the attractor for which the frequency of close approach of typical orbits is either much greater than typical (a hot spot) or much less than typical (a cold spot). Recently there has been much work developing ways of quantitatively characterizing how such chaotic orbits distribute themselves on attractors.[1-4] In particular, the spectrum of fractal dimensions introduced in Refs. 2–4 are sensitive to the characteristics of the structure of hot and cold spots on the attractor. In this paper we present results which show that, for a large class of chaotic attractors, the infinite number of unstable periodic orbits embedded in the attractor provide the key to an understanding of such issues. (A brief preliminary report of some of this work appears in Grebogi, Ott, and Yorke.[5])

The importance of unstable periodic orbits in determining ergodic properties of chaotic systems has long been recognized in the mathematical literature (e.g., Bowen[6] and Katok[7]). For some more recent work see Refs. 8 and 9 which also illustrate the important point that information about unstable periodic orbits is readily accessible from numerical computation (and perhaps experimentally[9,10]) and can be used for determining ergodic properties. In addition, in the theory of quantum chaos, the distribution of energy levels can be related to unstable periodic orbits of the classical Hamiltonian.[11-13] Another case where unstable periodic orbits appear[10] is in determining the behavior near parameter values where sudden changes in chaotic attractors occur [the argument in connection with our Fig. 1 is similar to that for Eq. (2) of Ref. 10].

The organization of this paper is as follows. Section II presents a discussion of the pointwise dimension for attractors and shows that hot and cold spots occur on the unstable manifolds of saddle periodic orbits in the attractor. Numerical experiments illustrating this are also presented. Section III reviews recent work on the dimensions of attractors including the partition function formalism.[4] Section IV presents our results relating the distribution of typical chaotic orbits on attractors and the associated fractal dimensions to the unstable periodic orbits. Section V illustrates the material of Sec. IV with examples. Arguments yielding the results stated in Sec. IV are presented in Sec. VI for the case of hyperbolic attractors. Section VII treats the case of chaotic sets which are repelling rather than attracting.

The dynamical systems to be discussed throughout this paper are d-dimensional maps of the form $\underline{x}_{n+1}=\underline{F}(\underline{x}_n)$, where $\underline{x}$ is a vector in the d-dimensional phase space of the system. An *attractor* A for such a system is a closed set, invariant under $\underline{F}$, which is the limit set as time goes to $+\infty$ for almost every initial condition in some neighborhood of A. (By "almost every" we mean that the set of initial conditions in the neighborhood that do not approach A can be covered by a set of d-dimensional cubes of arbitrarily small total volume.) The *basin of attraction* for the attractor is the closure of the set of points which asymptote to the attractor as time goes to $+\infty$. In the case of continuous time systems (flows), we can think of $\underline{F}(\underline{x})$ as arising from a Poincaré surface of section.

II. POINTWISE DIMENSION

For most purposes we may think of the *natural measure* of an attractor as follows: For a subset S of the phase space and an initial condition $\underline{x}$ in the basin of attraction of the attractor, we define $\mu(\underline{x},S)$ as the fraction of time the trajectory originating at $\underline{x}$ spends in S in the limit that the length of the trajectory goes to infinity. If $\mu(\underline{x},S)$ is the same for almost every $\underline{x}$ in the basin of attraction, then we denote this value $\mu(S)$ and say that μ is the natural measure of the attractor (cf. Appendix). Henceforth, we assume that the attractor has a natural measure. In particular, this means that the attractor is ergodic (i.e., it cannot be split into two disjoint pieces that each have positive natural measure and are invariant under application of $\underline{F}$).

Let $B(l,\underline{x})$ denote a d-dimensional ball of radius l centered at a point $\underline{x}$ on an attractor embedded in the d-dimensional phase space of the dynamical system being considered. Then the pointwise dimension (at the point $\underline{x}$) of the attractor is defined as

$$D_p(\underline{x})=\lim_{l\to 0}\frac{\log\mu(B(l,\underline{x}))}{\log l} \tag{2.1}$$

or $\mu(B(l,\underline{x}))\sim l^{D_p(\underline{x})}$. For *almost every* point with respect to the natural measure on the attractor, $D_p(\underline{x})$ takes on a common value and is equal to the information dimension (defined in Sec. III). That is, the set of points on the chaotic attractor for which $D_p(\underline{x})$ is not this common value may be covered with a set of d-dimensional cubes of varying sizes which together contain an arbitrarily small amount of the natural measure of the attractor. [Points $\underline{x}$ where $D_p(\underline{x})$ is greater than (less than) the common value it assumes at almost every point with respect to the natural measure are the hot (cold) spots referred to in Sec. I.] For example, a chaotic attractor typically has a dense set of unstable periodic orbits embedded within it, and, as we shall see, $D_p(\underline{x})$ with $\underline{x}$ on one of these periodic orbits does not take on the typical values. The periodic points, however, are countable and so have zero measure. Nevertheless, it is a main point of this paper that this zero measure set is important and leads to interesting properties of the attractor.

To see why $D_p(\underline{x})$ is the same for almost every $\underline{x}$ with respect to the natural measure, assume that it is not. Then we can pick some D_{p0} such that there is a nonzero amount for the natural measure of the attractor for which $D_p(\underline{x})>D_{p0}$ and another nonzero amount for which $D_p(\underline{x})\le D_{p0}$. Thus the attractor is divided into two disjoint sets, $A_>$ and $A_<$, $\mu(A_>)+\mu(A_<)=1$, $\mu(A_{\gtrless})>0$. From the definition (2.1), one can show that $D_p(\underline{x})$ is invariant to smooth changes of coordinates. In particular, for a smooth map $\underline{x}_{n+1}=\underline{F}(\underline{x}_n)$, we can take $\underline{y}=\underline{F}(\underline{x})$ as the change of coordinates. It follows that $D_p(\underline{x})=D_p(\underline{F}(\underline{x}))$. Thus the sets $A_>$ and $A_<$ are invariant under $\underline{F}$. This implies that every orbit on the attractor is confined either to $A_>$ or $A_<$. Hence, contrary to our assumption that the attractor has a natural measure, the attractor has been decomposed into two disjoint $\underline{F}$ invariant sets. We conclude that $D_p(\underline{x})$ must be the same for almost every $\underline{x}$ with respect to the natural measure on the attractor.

Returning now to consideration of the zero measure set of points $\underline{x}$ for which $D_p(\underline{x})$ is not typical (i.e., is not the common value assumed at almost every $\underline{x}$ on the attractor) and taking the map to be two dimensional ($d=2$), we will obtain the following result. Let j be an index labeling the fixed points of the n times iterated map $\underline{F}^n$. (The components of a period n orbit are fixed points of $\underline{F}^n$.) We assume that the Jacobian matrix of $\underline{F}^n$ at fixed point j has one unstable direction and one stable direction. Then for any point $\underline{x}$ *on the unstable manifold* of fixed point j of $\underline{F}^n$,

$$D_p(\underline{x})=1-\frac{\log\lambda_{1j}}{\log\lambda_{2j}}, \tag{2.2}$$

where $\lambda_{1j}>1$ and $\lambda_{2j}<1$ are the magnitudes of the unstable and stable eigenvalues of the Jacobian matrix of $\underline{F}^n$. Since points on different periodic orbits typically have different eigenvalues, $D_p(\underline{x})$ will clearly be different for different periodic orbits and hence will not be the typical $D_p(\underline{x})$.

To obtain (2.2) consider a point $\underline{x}_0$ on the unstable manifold of a saddle periodic point and two small circular disks centered at $\underline{x}_0$ with radii l_1 and l_2, where $l_1/l_2=\lambda_{2j}^{-1}$. We iterate the two disks backward a large integral number of periods so that the two disks are now similar ellipses close to the saddle and with their major axes parallel to the stable manifold of the saddle (cf. Fig. 1). We now iterate the l_2 ellipse backward one more period. Since it is close to the saddle, its backward iteration by one period is governed by the linearized map at the saddle (i.e., by the eigenvalues λ_{1j} and λ_{2j}). Thus, since we choose $l_1/l_2=\lambda_{2j}^{-1}$, the major diameter of the l_2 ellipse is now the same as that for the l_1 ellipse, while its minor diameter is smaller than that for the l_1 ellipse by the factor $\lambda_{2j}/\lambda_{1j}$. The inverse images of the disks contain the same natural measure as the original disks. Thus, treating the attractor measure as if it were smooth along the unstable direction, we have $\mu(B(l_2,\underline{x}_0))/\mu(B(l_1,\underline{x}_0))=\lambda_{2j}/\lambda_{1j}$. Setting $\mu(B(l,\underline{x}_0))\sim l^{D_p}$ and $l_2=l_1\lambda_{2j}$, this yields Eq. (2.2), the desired result.

We have tested Eq. (2.2) numerically using the Henon map given by

$$\begin{aligned} x_{n+1}&=1.42-x_n^2+0.3y_n\,,\\ y_{n+1}&=x_{n+1}\,. \end{aligned} \tag{2.3}$$

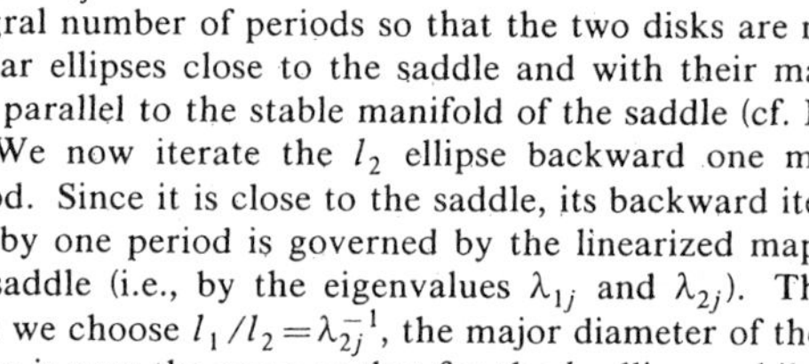

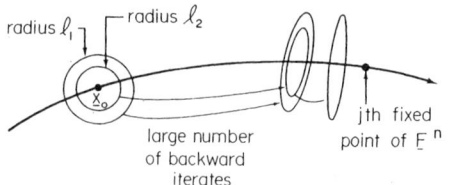

FIG. 1. Schematic for the derivation of Eq. (2.2).

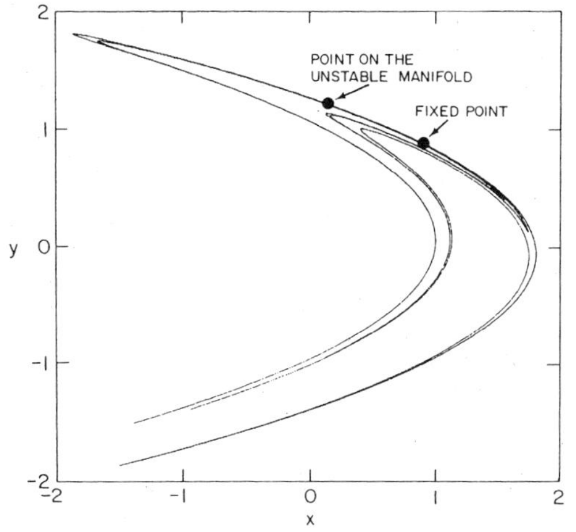

FIG. 2. Plot of the iterates of the Henon map (2.3).

In Fig. 2, we plot iterates of the map (2.3) to show the attractor. In this figure we also indicate the location of a point which we have accurately determined to be on the unstable manifold of an unstable fixed point (i.e., a period one unstable orbit) of the map (2.3). In Fig. 3, we display the result of a calculation of the pointwise dimension $D_p(\underline{x})$ for $\underline{x}$ at this point. In this figure, we plot $\mu(B(l,\underline{x}))$ versus l on a log-log scale. Here $\mu(B(l,\underline{x}))$ is obtained by iterating a randomly chosen initial condition in the basin of the attractor 10^6 times (so that the orbit is essentially on the attractor) and then determining the fraction of subsequent orbit points which fall in $B(l,\underline{x})$ on further iteration. The numerically determined pointwise dimension is the slope of the straight line through the points in Fig. 3; we obtain $D_p(\underline{x}) \cong 1.36$. The magnitudes of the eigenvalues at the fixed point are $1.94 \cdots$ and $0.155 \cdots$, which, when inserted in Eq. (2.2), yield $D_p(\underline{x}) = 1.36$, in agreement with the data in Fig. 3.

Next we obtain a typical point $\underline{x}$ on the attractor which is within a small distance (1.5×10^{-4} in this case) of the previously chosen point on the unstable manifold of the fixed point. We do this by randomly picking an initial condition, preiterating 2×10^6 times, and then iterating the map until the orbit falls within a circle of radius 2×10^{-4} centered at the previously obtained point on the unstable manifold. Results for $\mu(B(l,\underline{x}))$ versus l for this point are shown in Fig. 4. For $l > 1.5 \times 10^{-4}$ the slope of the straight line fitted to the numerical data is 1.36, which is the same value as obtained in Fig. 3 for the point on the unstable manifold. This agreement is as expected, since for $l \gg 10^{-4}$, the two points are essentially indistinguishable. However, for data points in the range $6 \times 10^{-6} < l < 1.5 \times 10^{-4}$, we find that the slope of a fitted line is 1.21, which is significantly different from the 1.36 slope for $l > 1.5 \times 10^{-4}$.

There are an infinite number of unstable periodic orbits on the attractor. Thus, although it is true that all typical $\underline{x}$ must have a common value for $D_p(\underline{x})$, one might suspect that there will be significant fluctuations in numerically determined values of D_p, since such calculations are necessarily restricted to a finite range of l. This seems to be the case: We have numerically determined $D_p(\underline{x})$ for the map (2.3) using a range $10^{-1} \leq l \leq 10^{-5}$ at 20 different typical points, and we find considerable variation in the resulting numerically determined pointwise dimensions. A list of the 20 values obtained appears in Table I. In obtaining these values, a least square fit was used in $10^{-1} \leq l \leq 1.22 \times 10^{-5}$, and the root-mean-square deviation of the least square fit is also shown in Table I. This root-mean-square deviation of the fit for $D_p(\underline{x})$ at individual $\underline{x}$ values is small compared with the standard deviation (0.10) about the mean (1.27) obtained by using the twenty $D_p(\underline{x})$ values (given in the first column of Table I). (The 20 typical $\underline{x}$ values used were obtained by choosing a random initial condition in the basin of the attractor, iterating it 20×10^6 times, and selecting every millionth iterate.)

The Kaplan-Yorke formula, discussed in Sec. IV, predicts the typical value of the pointwise dimension in terms of the Lyapunov numbers. For the case considered here, the predicted value is 1.26, which is in good agreement with our mean of the twenty numerically obtained D_p values but is far from the value obtained in Fig. 3.

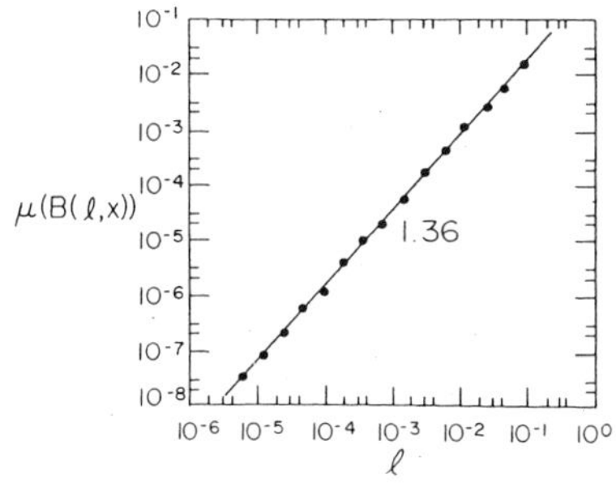

FIG. 3. Log-log plot of $\mu(B(l,x))$ vs l for a point on the unstable manifold of the fixed point.

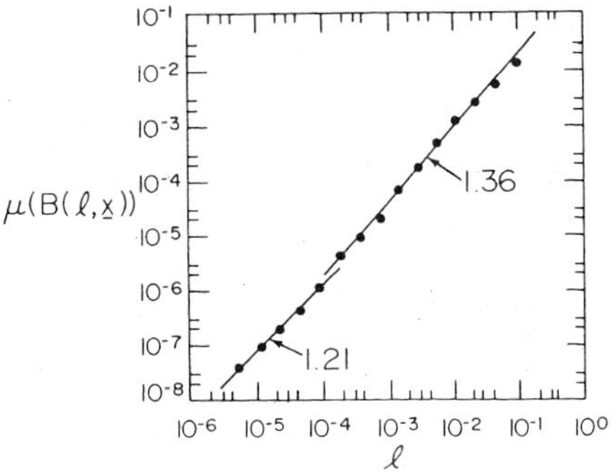

FIG. 4. Log-log plot of $\mu(B(l,\underline{x}))$ vs l for a typical point on the attractor which is within 1.5×10^{-4} of the point on the unstable manifold of the fixed point.

TABLE I. Result of pointwise dimension calculations for 20 typical points. Mean equals 1.27±0.10

$D_p(\underline{x})$	rms deviation of fit
1.29	0.03
1.31	0.03
1.12	0.02
1.34	0.03
1.21	0.02
1.25	0.02
1.21	0.04
1.34	0.03
1.31	0.04
1.38	0.02
1.25	0.04
1.28	0.03
1.26	0.03
1.34	0.03
1.33	0.03
1.15	0.02
1.02	0.01
1.32	0.03
1.41	0.02
1.37	0.02

III. DIMENSIONS OF MULTIFRACTAL CHAOTIC ATTRACTORS

In this section we review past work on the dimensions characterizing multifractal chaotic attractors.[2–4] References 2 and 3 consider the quantity $\tilde{D}_q$ defined by

$$\tilde{D}_q = \frac{1}{q-1} \lim_{l \to 0} \frac{\log \sum_{i=1}^{N(l)} p_i^q}{\log l} , \tag{3.1}$$

where the attractor is covered with $N(l)$ d-dimensional cubes from a grid of unit length l, and p_i is the natural measure of the attractor in the ith cube. Taking the limit $q \to 1$, Eq. (3.1) yields[2,3]

$$\tilde{D}_1 = \lim_{l \to 0} \left[\sum_i p_i \log p_i \right] \Big/ \log l , \tag{3.2}$$

which is called the information dimension of the attractor. It is $\tilde{D}_1$ which is the common value assumed by $D_p(\underline{x})$ for almost all $\underline{x}$ with respect to the natural measure on the attractor.[1] We may think of the information dimension as the capacity dimension (cf. below) of the smallest set which contains most of the natural measure of the attractor.[1]

As q is increased past 1, the contribution of the sum $\sum_i p_i^q$ from a relatively few boxes with very little of the total attractor measure but with larger p_i than typical (i.e., hot spots) becomes relatively more important. Similarly, as q is decreased from one, the contribution from low probability boxes begins to be more important. For example, for $q=0$, Eq. (3.1) yields

$$\tilde{D}_0 = -\lim_{l \to 0} \log N(l) / \log l , \tag{3.3}$$

which is known as the capacity or box-counting dimension of the attractor. Note that all boxes on the attractor contribute democratically to (3.3) no matter what their natural measure p_i is. For typical chaotic attractors, it is to be expected that $\tilde{D}_0 > \tilde{D}_1$, since low-probability boxes (cold regions) containing very little of the total natural measure on the attractor may be vastly more numerous than those required to cover most of the natural measure on the attractor.[1]

At this point it is appropriate to discuss another definition of dimension, the Hausdorff dimension,[14] which is, in fact, an older concept than either the capacity[15] or information dimensions.[16] To define the Hausdorff dimension, we cover the attractor with d-dimensional cubes of variable edge length l_i, all of which we restrict to be no bigger than some value l ($l_i \le l$). We then form the quantity

$$\Gamma_H(D, l, \{l_i\}) = \sum_i l_i^D . \tag{3.4}$$

Next the covering of cubes is optimized so as to make the sum $\sum l_i^D$, minimum,

$$\Gamma_H(D, l) = \inf_{\{l_i\}} \sum_i l_i^D , \tag{3.5}$$

where the infimum is taken over all possible collections of cubes that cover the attractor subject to the constraint $l_i \le l$. Finally, the limit $l \to 0$ is taken,

$$\Gamma_H(D) = \lim_{l \to 0} \Gamma_H(D, l) . \tag{3.6}$$

The quantity $\Gamma_H(D)$ can be shown to be either zero or infinity except at a critical value of D [cf. Fig. 5(a)]. This critical value defines the Hausdorff dimension which we denote D_H.

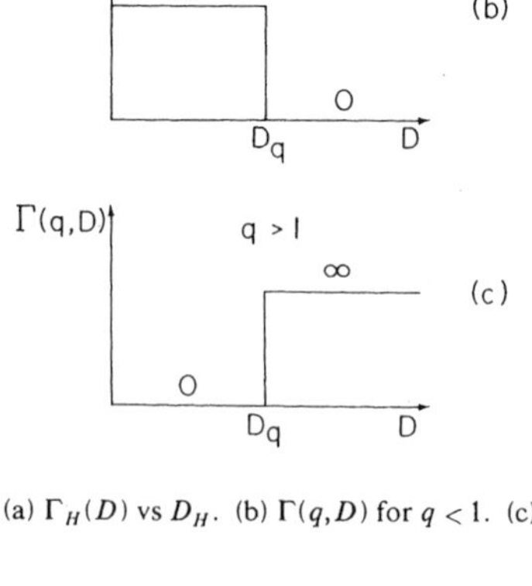

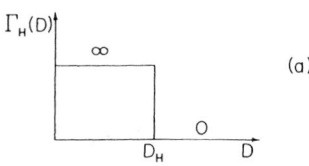

FIG. 5. (a) $\Gamma_H(D)$ vs D_H. (b) $\Gamma(q,D)$ for $q<1$. (c) $\Gamma(q,D)$ for $q>1$.

The capacity dimension $\tilde{D}_0$ and the Hausdorff dimension are closely related. In particular, if we do not optimize over the cube covering, but instead use a cubic grid covering of fixed length l, then all the l_i in (3.4) are equal to l, and we have $\Gamma_H^{(\text{grid})}(D,l)=N(l)l^D$. Equation (3.3) is satisfied by $N(l)\sim l^{-\tilde{D}_0}$, which yields $\Gamma_H^{(\text{grid})}(D,l) \sim l^{D-\tilde{D}_0}$. In the limit $l\to 0$, $\Gamma_H^{(\text{grid})}$ is zero or infinity [as in Fig. 5(a)] with the transition at $D=\tilde{D}_0$. Since the covering using a cubic grid may not be optimal, $\Gamma_H^{(\text{grid})}(D,l)\geq\Gamma_H(D,l)$ and

$$\tilde{D}_0\geq D_H \ .$$

For examples of ergodic chaotic attractors, where $\tilde{D}_0$ and D_H are analytically calculable, it is found[1] that $\tilde{D}_0=D_H$, and it has been conjectured that this is to be expected, in general, for typical ergodic chaotic attractors (although it is easy to construct sets[1] which are not attractors which have $\tilde{D}_0>D_H$).

Reference 4 gives a formalism which essentially generalizes the Hausdorff procedure and its relation to $\tilde{D}_0$ so that $\tilde{D}_q$ for arbitrary q is treated. In particular, again say we cover the attractor with cubes of variable edge length l_i, but now consider the following quantity, which in analogy with statistical mechanics, has been called the partition function,

$$\Gamma(q,D,l,\{l_i\})=\sum p_i^q/l_i^\tau, \quad \tau=(q-1)D \tag{3.7}$$

where $D\geq 0$. Optimize over covering $\{l_i\}$,

$$\Gamma(q,D,l)=\begin{cases}\sup[\Gamma(q,D,l,\{l_i\})] & \text{for } q>1 \ ,\\ \inf[\Gamma(q,D,l,\{l_i\})] & \text{for } q<1 \ ,\end{cases} \tag{3.8}$$

and take the limit $l\to 0$,

$$\Gamma(q,D)=\lim_{l\to 0}\Gamma(q,D,l) \ . \tag{3.9}$$

Again $\Gamma(q,D)$ is zero or infinity. The transition point for $\Gamma(q,D)$ is denoted D_q. [$\Gamma(q,D)=0$ for $(q-1)D<(q-1)D_q$, and $\Gamma(q,D)=\infty$ for $(q-1)D>(q-1)D_q$ (cf. Figs. 5(b) and 5(c)).]

Again say that we do not optimize over coverings, but instead use a cubic grid, of basic length l, then $l_i=l$ and we have $\sum_i p_i^q/l_i^\tau=l^{-\tau}\sum_i p_i^q$. From Eq. (3.1), $\sum_i p_i^q\sim l^{(q-1)\tilde{D}_q}$ and hence $\sum_i p_i^q/l_i^\tau\sim l^{(q-1)(\tilde{D}_q-D)}$, which, in the limit $l\to 0$, pass from 0 to ∞ at $D=\tilde{D}_q$. Since in this the optimization over coverings, prescribed in Eq. (3.8), is not done, the quantity D_q defined by Eqs. (3.7)–(3.9) is necessarily less than or equal to $\tilde{D}_q$,

$$D_q\leq\tilde{D}_q \ . \tag{3.10}$$

However, as for the Hausdorff and capacity dimensions, it is to be expected that, in practice, the equality in (3.10) typically holds for chaotic attractors.

A central result of Halsey *et al.*[4] was the demonstration of the way in which D_q is connected to the hot and cold points on the attractor (i.e., those points $\underline{x}$ on the attractor for which the pointwise dimension $(D_p(\underline{x})\neq\tilde{D}_1)$. In particular, they consider the set of $\underline{x}$ values such that $D_p(\underline{x})=\alpha$, and they denote the Hausdorff dimension of this set by $f(\alpha)$. They show that the D_q can be explicitly obtained from the dimensions $f(\alpha)$ via the formulas,

$$df(\alpha)/d\alpha=q \ , \tag{3.11a}$$

$$(q-1)D_q=[q\alpha(q)-f(\alpha(q))] \ . \tag{3.11b}$$

IV. UNSTABLE PERIODIC ORBITS

In this section we state and discuss results obtained in Sec. VI on the relation of unstable periodic orbits on chaotic attractors to the ergodic properties of these attractors. The results to be quoted are for the case of chaotic attractors that are *mixing* and *hyperbolic.* By mixing we mean that for any two sets S_a and S_b in the phase space of the system, we have

$$\lim_{n\to\infty}\mu[S_a\cap\underline{F}^n(S_b)]=\mu(S_a)\mu(S_b) \ ,$$

where μ is the natural measure of the attractor. A chaotic attractor would not be mixing in the commonly encountered situation where the attractor consists of a finite number h of disjoint pieces, and the orbit cycles from piece to piece. In this case, instead of the map $\underline{F}$ one can consider the map $\underline{F}^h$. For $\underline{F}^h$ each of the h pieces of the attractor for $\underline{F}$ is a separate attractor in its own right and is typically mixing. Our result would then apply to the attractors of $\underline{F}^h$. Henceforth, we assume that the attractor is mixing. Hyperbolic attractors are defined in Sec. VI.

We consider a d-dimensional twice differentiable map, $\underline{x}_{m+1}=\underline{F}(\underline{x}_m)$. The magnitudes of the eigenvalues of the Jacobian matrix of the n times iterated map $\underline{F}^n$ at the jth fixed point of $\underline{F}^n$ are denoted $\lambda_{1j},\lambda_{2j},\ldots,\lambda_{uj}$, $\lambda_{(u+1)j},\ldots,\lambda_{dj}$, where we order the eigenvalues as follows: $\lambda_{1j}\geq\lambda_{2j}\geq\cdots\geq\lambda_{uj}>1\geq\lambda_{(u+1)j}\geq\cdots\geq\lambda_{dj}$. Thus in this notation, the number of unstable eigenvalues is u. Let L_j be the product of the unstable eigenvalues at the jth fixed point of $\underline{F}^n$,

$$L_j=\lambda_{1j}\lambda_{2j}\cdots\lambda_{uj} \ . \tag{4.1}$$

Then, as shown in Sec. VI for mixing hyperbolic (axiom A) attractors, the natural probability measure of the attractor contained in some closed subset S of the d-dimensional phase space is the limit as $n\to\infty$ of the sum of the L_j^{-1} over all the fixed points j of $\underline{F}^n$ which lie in S,

$$\mu(S)=\lim_{n\to\infty}\left[\sum_{\substack{\text{fixed points}\\ \text{in } S}} L_j^{-1}\right] \ . \tag{4.2}$$

Thus (4.2) is essentially a representation of the natural measure in terms of the unstable periodic orbits on the attractor. In particular, in the special case where S covers the entire attractor, we have $\mu(S)=1$, and, hence, we obtain a relation amongst the unstable eigenvalues,

$$1=\lim_{n\to\infty}\sum_j L_j^{-1} \ , \tag{4.3}$$

where the sum is over all fixed points of $\underline{F}^n$ on the attractor. Equation (4.3) has been conjectured to apply in general for Hamiltonian chaotic systems[12] and has been used to derive an interesting correspondence between the eigenvalues of a random matrix and the statistics of the semiclassical limit of the energy levels of a bound, time-independent, quantum system whose classical limit is chaotic.[13]

We now give a partition function formulation of the multifractal properties of chaotic attractors. This formulation is in terms of the eigenvalues of the dense set of periodic saddles on the attractor [rather than in terms of the measure of coverings of the attractor, i.e., the p_i in Eqs. (3.7)]. Let

$$D \equiv \Delta + \delta ,$$

where Δ is the integer part of D, and $\delta \equiv D - \Delta = (D$ modulo 1) is the fractional part of D. In addition, let

$$S_j(D) \equiv \lambda_{ij}\lambda_{2j} \cdots \lambda_{\Delta j}(\lambda_{(\Delta+1)j})^{\delta} . \tag{4.4}$$

[Note that $S_j(D)$ is a continuous function of D.] In terms of L_j and $S_j(D)$, the result obtained in Sec. VI for the partition function is

$$\hat{\Gamma}(q,D,n) = \sum_j L_j^{-1}[S_j(D)]^{-(q-1)} . \tag{4.5}$$

In the two-dimensional case with $\lambda_{1j} > 1 > \lambda_{2j}$ this reduces to

$$\hat{\Gamma}(q,D,n) = \sum_j \lambda_{1j}^{-q} \lambda_{2j}^{-(D-1)(q-1)} , \tag{4.6}$$

which appears in Grebogi *et al.*[5] and Morita *et al.*[5] and, for the case of the Hausdorff dimension ($q=0$), in Ref. 9. Taking the limit $n \to \infty$ is analogous to taking the limit $l \to 0$ in Eq. (3.9),

$$\hat{\Gamma}(q,D) = \lim_{n\to\infty} \hat{\Gamma}(q,D,n) . \tag{4.7}$$

The quantity $\hat{\Gamma}(q,D)$ is zero or infinity in analogy with the quantity $\Gamma(q,D)$ in Eq. (3.9) [cf. Figs. 5(b) and 5(c)]. We denote the value of D at the transition of $\hat{\Gamma}(q,D)$ from zero to infinity by $\hat{D}_q$ and call it the periodic point dimension.

$$\hat{\Gamma}(q,D) = \begin{cases} 0 & \text{for } (q-1)D > (q-1)\hat{D}_q \\ +\infty & \text{for } (q-1)D < (q-1)\hat{D}_q . \end{cases}$$

In Sec. VI we show that

$$\hat{D}_q \geq D_q . \tag{4.8}$$

We conjecture that for typical chaotic attractors of two-dimensional maps with $\lambda_{1j} > 1 > \lambda_{2j}$

$$\hat{D}_q = D_q . \tag{4.9}$$

Indeed for the two-dimensional case with $\lambda_{1j} > 1 > \lambda_{2j}$, examples are worked out in Sec. V verifying that Eq. (4.9) holds. (For further discussion see Sec. VI.)

Setting $q=1$ and comparing (4.5) with (4.3) we have $\hat{\Gamma}(1,D)=1$. Formally expanding (4.5) around $q=1$ we obtain

$$\hat{\Gamma}(q,D,n) = 1 - (q-1)\sum_j \frac{\log S_j(D)}{L_j} + O[(q-1)^2] .$$

Letting $n \to \infty$ the coefficient of the $(q-1)$ term may be expressed using (4.4) as

$$\lim_{n\to\infty} \left[\left(\sum_j L_j^{-1} \log(\lambda_{1j}\lambda_{2j} \cdots \lambda_{\Delta j}) \right) + \delta \left[\sum_j L_j^{-1} \log \lambda_{(\Delta+1)j} \right] \right] .$$

As will become evident from a correspondence with Lyapunov numbers to be discussed subsequently, the first term in large parentheses becomes positive infinite and the second term negative infinite (both sums behave like n for large n). Thus this formal expansion in $(q-1)$ indicates that the transition value $\hat{D}_1 = \Delta_1 + \delta_1$ occurs at

$$\hat{D}_1 = \Delta_1 - \lim_{n\to\infty} \frac{\sum_j L_j^{-1} \log(\lambda_{1j} \cdots \lambda_{\Delta_1 j})}{\sum_j L_j^{-1} \log(\lambda_{(\Delta_1+1)j})} , \tag{4.10}$$

where (to satisfy $0 \leq \delta_1 \leq 1$) we define Δ_1 as the largest integer such that $\sum_j L_j^{-1} \log(\lambda_{1j} \cdots \lambda_{\Delta_1 j})$ is positive. We now compare (4.10) with the Kaplan-Yorke conjecture.[1,17] The Kaplan-Yorke conjecture gives a formula for $\tilde{D}_1$ [defined in (3.2)] in terms of the Lyapunov numbers of typical orbits (i.e., orbits obtained for almost every choice of initial condition in the basin of the attractor). Denoting these (typical) Lyapunov numbers

$$\lambda_1 \geq \lambda_2 \geq \cdots \geq \lambda_d ,$$

the Kaplan-Yorke formula states that for typical systems,

$$\tilde{D}_1 = \Delta_1 - \frac{\ln(\lambda_1\lambda_2 \cdots \lambda_{\Delta_1})}{\ln(\lambda_{\Delta_1+1})} , \tag{4.11}$$

where here Δ_1 is the largest integer such that $\lambda_1\lambda_2 \cdots \lambda_{\Delta_1} > 1$. Comparing (4.11) and (4.10), see that these equations are the same if we interpret the sums over periodic orbits in (4.10) in terms of Lyapunov numbers,

$$\log \lambda_p = \lim_{n\to\infty} \frac{1}{n} \sum_j L_j^{-1} \log \lambda_{pj} . \tag{4.12}$$

Equation (4.12) is reasonable since our construction, to be discussed in Sec. IV A, shows that, during n iterates of the map, each orbit on the attractor stays close to some orbit of period n, and, furthermore, the natural measure of the orbits which stay close to a given period n orbit is L_j^{-1}. [This latter statement is related to Eq. (4.2).] Thus $\tilde{D}_1 = \hat{D}_1$.

The Kaplan-Yorke conjecture Eq. (4.11) has been shown to apply in a variety of examples and numerical experiments[1] (although it has not yet been proven in general). The correspondence of the $q \to 1$ limit of Eq. (4.5) [i.e., Eq. (4.10)] with the Kaplan-Yorke formula (4.11) supports our conjecture, Eq. (4.9).

Note that the two-dimensional map result for the pointwise dimension on the unstable manifolds of periodic orbits, Eq. (2.2), is the same as (4.11) with $d=2$ and the Lyapunov numbers of a typical orbit replaced by the magnitudes of the eigenvalues of the periodic orbit, $\lambda_{1j} > 1 > \lambda_{2j}$. A generalization of Eq. (2.2) to d-dimensional maps $d > 2$ is the statement that all points on the unstable manifold of the jth fixed point of $\underline{F}^n$ have the same pointwise dimension which is given by (4.11) with the Lyapunov numbers λ_p replaced by the eigenvalues λ_{pj}.

The Hausdorff dimension $f(\alpha)$ of the set of points on the attractor with $D_p(\underline{x})=\alpha$ can also be obtained directly from this formulation. In particular, define $\hat{\Gamma}_\alpha$ to be the same as the expression in (4.6) with $q=0$ (for the Hausdorff dimension) and the sum restricted only to those fixed points j which satisfy $\alpha+\Delta\alpha \geq D_p(\underline{x}) \geq \alpha$,

$$\hat{\Gamma}_\alpha(D,\Delta\alpha,n)=\sum_{\substack{j \\ \alpha+\Delta\alpha\geq D_p(\underline{x})\geq\alpha}} \lambda_{2j}^{(D-1)} .$$

We then take limits,

$$\hat{\Gamma}_\alpha(D)=\lim_{\Delta\alpha\to 0}\lim_{n\to\infty}\hat{\Gamma}_\alpha(D,\Delta\alpha,n) ,$$

and we obtain $f(\alpha)$ as the transition value of D for which $\hat{\Gamma}_\alpha(D)$ goes from infinity to zero as D increases.

The results quoted in this section are for hyperbolic attractors (Sec. VI). For the two-dimensional hyperbolic case with $\lambda_{1j}>1>\lambda_{2j}$, we have $D_p(\underline{x})\geq 1$ for every point on the attractor. For the nonhyperbolic case, $D_p(\underline{x})$ can be less than 1, and we conjecture that (4.6) and (4.9) also hold in the nonhyperbolic case but only for q values corresponding to $\alpha(q)>1$ [cf. (3.11)].

Another result concerning ergodic properties of a map is that for the topological entropy S in terms of N_n, the number of fixed points of the n times iterated map $\underline{F}^n$,

$$S=\lim_{n\to\infty}\frac{1}{n}\log N_n . \tag{4.13}$$

For this result for the case of axiom A attractors see the papers by Bowen[6] and by Katok.[7]

V. EXAMPLES

We now illustrate the results on periodic orbits with two analytically tractable two-dimensional, hyperbolic map examples.

A. Example 1: The generalized baker's map

The generalized baker's map was introduced in Ref. 1 as a model for dimension studies which is amenable to analysis yet also has nonconstant stretching and contraction. We divide the square $0\leq(x,y)\leq 1$ into a bottom part, $0\leq y<a$, and a top part, $a<y\leq 1$. This is illustrated in Fig. 6(a) ($b=1-a$ in the figure). We compress the bottom (top) part by a factor λ_a (λ_b) along x, and stretch it in y by a factor a^{-1} (b^{-1}). We then have two rectangles, both of vertical height unity, one of width λ_a and the other of width λ_b [Fig. 6(b)]. We then move the λ_b width strip so that its lower left corner is at $x=\frac{1}{2}$, $y=0$ [Fig. 6(c)]. Thus we have a map of the unit square into itself: $x_{n+1}=\lambda(y_n)x_n+(\frac{1}{2})u(y_n-a)$; $y_{n+1}=\gamma(y_n)[y_n-au(y_n-a)]$; where $\lambda(y)=(\lambda_a,\lambda_b)$ for $y\gtrless a$, $\gamma(y)=(a^{-1},b^{-1})$ for $y\gtrless a$, and $u(y)$ is the unit step function.

Using similarity arguments[1] it can be shown directly from the map that the following transcendental equation determines D_q (cf. Refs. 2 and 3),

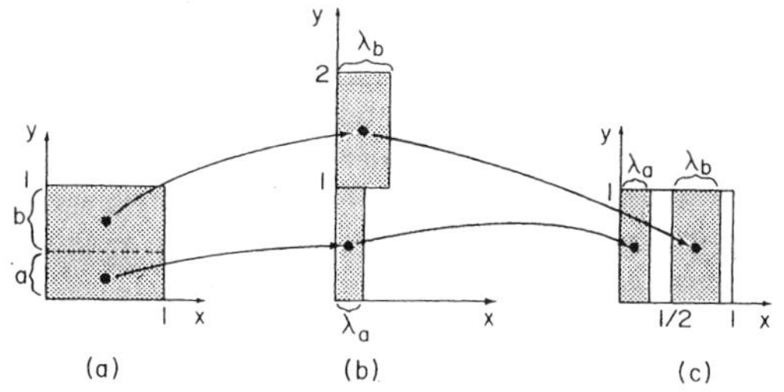

FIG. 6. Schematic of generalized baker's map.

$$1=\overline{A}+\overline{B} , \tag{5.1}$$

where $\overline{A}=\lambda_a^{-(q-1)(D_q-1)}a^q$, $\overline{B}=\lambda_b^{-(q-1)(D_q-1)}b^q$.

We now show that the equation determining the periodic point dimension $\hat{D}_q$ is the same as (5.1). We can specify an orbit for the generalized baker's transformation by its symbolic itinerary which specifies whether the orbit's location on successive iterates is in the top (symbolized by a 1) or in the bottom (symbolized by a 0). Thus a periodic orbit of period n which spends $k\leq n$ of its n iterates in $y>a$ is represented by a string of n symbols with k ones and $n-k$ zeros. The eigenvalues associated with such an orbit are $\lambda_2=\lambda_a^{n-k}\lambda_b^k$, $\lambda_1=a^{-(n-k)}b^{-k}$. Equation (4.6) yields

$$\hat{\Gamma}(q,D,n)=\sum_{k=0}^{n}N_{nk}A^{(n-k)}B^k , \tag{5.2}$$

where $A=a^q\lambda_a^{-(q-1)(D-1)}$, $B=b^q\lambda_b^{-(q-1)(D-1)}$, and N_{nk} is the number of fixed points of the n times iterated map which belong to periodic orbits which spend k iterates in the top ($y>a$). It can be shown that N_{nk} is the number of ways of arranging k zeros and $n-k$ ones,

$$N_{nk}=\binom{n}{k} . \tag{5.3}$$

Hence Eq. (5.2) is just a binomial expansion, $\hat{\Gamma}=(A+B)^n$. Letting $n\to\infty$, we see that the transition of $\hat{\Gamma}(q,D)$ from zero to infinity occurs at $D=D_q$ with D_q given by (5.1). Thus $D_q=\hat{D}_q$ for this example. [This calculation is algebraically equivalent to one in Halsey *et al.*,[4] although the basis for their calculation is Eq. (3.7), while the basis here is the periodic orbit formula, Eq. (4.6).]

To find $f(\alpha)$, we use (2.2) with λ_{1j} set equal to $a^{-(n-k)}b^{-k}$ and λ_{2j} set equal to $\lambda_a^{n-k}\lambda_b^k$ to obtain an equation relating α and $k/n\equiv\kappa$. This gives

$$\alpha=1+\frac{(1-\kappa)\log a+\kappa\log b}{(1-\kappa)\log\lambda_a+\kappa\log\lambda_b} .$$

Setting $q=0$ and using only the terms in the sum (5.2) with k/n values near that required by the specification of α and letting $n\to\infty$, we have (again using Stirling's approximation).

$$f(\alpha)=1+\frac{(1-\kappa)\log(1-\kappa)+\kappa\log\kappa}{(1-\kappa)\log\lambda_a+\kappa\log\lambda_b} , \tag{5.4}$$

with

$$\kappa=[\log a-(\alpha-1)\log\lambda_a]/[(\alpha-1)\log(\lambda_b/\lambda_a)+\log(a/b)] .$$

To obtain the topological entropy we note that $N_n=\sum_{k=0}^{n}N_{nk}$. Since N_{nk} is the binomial coefficient (5.3), this yields $N_n=2^n$, a well-known result for the baker's map. Equation (4.13) then gives $S=\log 2$, also well known.

It is also straightforward to verify Eq. (4.3). In particular, the right-hand side of (4.3) is $\sum \binom{n}{k} a^{n-k}b^k=(a+b)^n=1$ (recall $a+b=1$).

B. Example 2: The baker's apprentice's map

Example 1 has the property that setting $\hat{\Gamma}(q,D,n)=1$ gives precisely the desired result, Eq. (5.1), for all n, rather than only in the $n\to\infty$ limit. The generalized baker's map is exceptional in this regard, and this is due to the exact self-similarity of the attractor. A more typical example, which is still analytically tractable, is illustrated in Figs. 7. Again, we divide the unit square into top $(y>a)$ and bottom $(y<a)$ parts. We again horizontally compress the two parts by λ_a and λ_b. The bottom part is vertically stretched by a^{-1}, as before. The difference is that we now vertically compress the top part by a/b. The parts are then reassembled in the square as shown in Fig. 7(b). [We call this the baker's apprentice's map because the baker squashes and stretches the "dough" just right, so that it is exactly twice its original length, while the less experienced apprentice misses by not stretching enough to make the length double.]

Again we can specify orbits by a string of ones (tops) and zeros (bottoms). In this case, however, an orbit point in the top is always mapped to the bottom. Thus a one is *always* followed by a zero. Replacing B by $\tilde{B}=(b/a)^q\lambda_b^{-(q-1)(D-1)}$ (to account for the compression by a/b as opposed to the stretching by $1/b$ in example 1), we see that Eq. (5.2) still applies. Equation (5.3) for N_{nk}, however, does not.

To find N_{nk} we first note that the number of fixed points of the n times iterated map is the number of possible sequences of length n which contain k ones and $n-k$ zeros, subject to the constraint that a zero always follows a one (except when the last symbol is a one). We consider two cases: (a) the last symbol is a zero, and (b) the last symbol is a one. In case (a), to find the contribution to N_{nk} from such sequences, we regard the sequence (1,0) as a single symbol denoted by a 2. Thus a period n orbit which is located in the top k times is represented by a string of $(n-k)$ symbols of which k are twos and $n-2k$ are zeros (clearly $k\le n/2$). There are $\binom{n-k}{k}$ such symbol

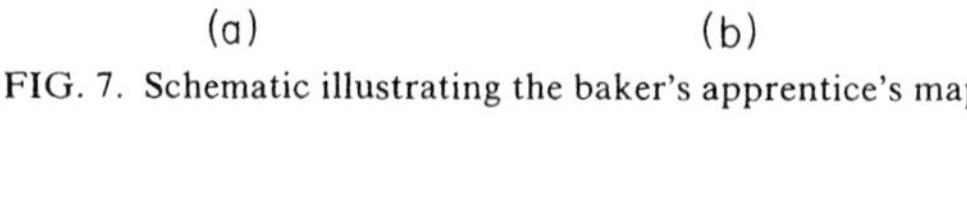

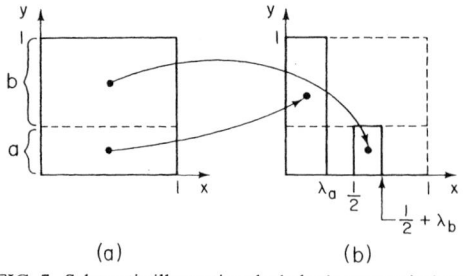

FIG. 7. Schematic illustrating the baker's apprentice's map.

sequences. Sequences ending in the top [case (b)], on the other hand, end in a one. Since the sequence represents a periodic orbit it must also start with a zero. All the rest of the symbols can be thought of as zeros and twos. For this case the zero-two sequence has $n-k-1$ symbols of which $k-1$ are twos. There are $\binom{n-k-1}{k-1}$ sequences of this type. Thus we have

$$N_{nk}=\binom{n-k}{k}+\binom{n-k-1}{k-1}, \tag{5.5}$$

and $N_{nk}=0$ for $k>n/2$.

Using Stirling's approximation to expand $Z(\kappa)\equiv(1/n)\log(N_{nk}A^{(n-k)}\tilde{B}^{k})$ for large n, we have

$$Z(\kappa)\cong(1-\kappa)\log(1-\kappa)-\kappa\log\kappa-(1-2\kappa)\log(1-2\kappa)+\kappa\log\tilde{B}+(1-\kappa)\log A\ , \tag{5.6}$$

where $\kappa=k/n$ and $\frac{1}{2}>\kappa>0$. The quantity Z is concave down $(d^2Z/d\kappa^2<0)$ and has one maximum in $\frac{1}{2}>\kappa>0$. The location of this maximum is given by $\kappa_0(1-\kappa_0)=(1-2\kappa_0)^2\tilde{B}/A$. Since the summand in Eq. (5.2) is $\exp[nZ(\kappa)]$, if $n\to\infty$ and $Z(\kappa_0)<0$, then $\hat{\Gamma}(q,D)\to 0$. On the other hand, if $Z(\kappa_0)>0$, then $\hat{\Gamma}(q,D)\to\infty$. Thus at the transition we have the condition $Z(\kappa_0)=0$. This gives a transcendental equation for $\hat{D}_q$, $A+A\tilde{B}=1$, or

$$1=a^q\lambda_a^{(1-q)(\hat{D}_q-1)}+b^q(\lambda_a\lambda_b)^{(1-q)(\hat{D}_q-1)}\ . \tag{5.7}$$

We now show that D_q also satisfies Eq. (5.7). We employ the similarity technique[1-3] [used, for example, in Refs. 2 and 3 to obtain Eq. (5.1)]. We write the sum in the partition function, $\sum_i p_i^q/l_i^\tau$, as a sum over the top region plus a sum over the bottom region, $\Gamma(l)=\Gamma_T(l)+\Gamma_B(l)$. Similarly we write Γ_B as a sum over the bottom left $(x<\frac{1}{2})$, region plus a sum over the bottom right region,

$$\Gamma_B(l)=\Gamma_{BL}(l)+\Gamma_{BR}(l)\ . \tag{5.8}$$

Applying the map to one of the coverings of size l_i in the bottom, we see that it is compressed by λ_a and elongated by $1/a$. Thus this l_i covering can be covered by $(a\lambda_a)^{-1}$ coverings of size $(l_i\lambda_a)$. Each of these new coverings has a probability $(p_i a\lambda_a)$. Inserting this information in the partition function Eq. (3.7) we have that

$$\Gamma_T(l\lambda_a)=\frac{b}{a\lambda_a}\left[\frac{(a\lambda_a)^q}{\lambda_a^\tau}\Gamma_B(l)\right]=\frac{b}{a}\frac{a^q}{\lambda_a^{(q-1)(D_q-1)}}\Gamma_B(l)\ .$$

Thus,

$$\Gamma_T(l)=\frac{b}{a}\frac{a^q}{\lambda_a^{(q-1)(D_q-1)}}\Gamma_B(l/\lambda_a)\ .$$

Similarly,

$$\Gamma_{BR}(l)=\frac{(b/a)^{q-1}}{\lambda_b^{(q-1)(D_q-1)}}\Gamma_T(l/\lambda_b)\ .$$

Also, $\Gamma_{BL}=(a/b)\Gamma_T$. Combining these in (5.8) and letting $l \to 0$, we obtain (5.7), and hence $D_q=\hat{D}_q$ for this example as well.

To find $f(\alpha)$ for the apprentice's map, we again use $\alpha=D_p(\underline{x})$ with $D_p(\underline{x})$ given by (2.2). This gives

$$\alpha=1+\frac{(1-\kappa)\log a+\kappa\log(b/a)}{(1-\kappa)\log\lambda_a+\kappa\log\lambda_b} . \tag{5.9}$$

Including only values of k/n in the sum for $\hat{\Gamma}$ which nearly satisfies this relation and letting $n\to\infty$, we see that the Hausdorff dimension $f(\alpha)$ of the set where $\alpha=D_p(\underline{x})$ is given by setting $Z(\kappa)=0$ in Eq. (5.6) (with $q=0$),

$$f(\alpha)=1+\frac{(1-\kappa)\log(1-\kappa)-\kappa\log\kappa-(1-2\kappa)\log(1-2\kappa)}{(1-\kappa)\log\lambda_a+\kappa\log\lambda_b} \tag{5.10}$$

where from (5.9)

$$\kappa(\alpha)=\frac{\log a-(\alpha-1)\log\lambda_a}{(\alpha-1)\log(\lambda_b/\lambda_a)+\log(a^2/b)} .$$

To find the topological entropy we perform the sum

$$N_n=\sum_{k=0}^{n/2} N_{nk} ,$$

again by using Stirling's approximation. We obtain $N_n \sim G^n$, where $G=(1+\sqrt{5})/2$ is the golden mean. Thus from (4.13) the topological entropy is $S=\log G$.

Again one may verify Eq. (4.3) by direct calculation. The quantity to be obtained is

$$\lim_{n\to\infty}\sum_{k=0}^{n/2} N_{nk}(b/a)^k a^{n-k} .$$

The computation is somewhat tedious, but straightforward (use Stirling's approximation yet again), and yields 1, as it should.

To conclude this section, we emphasize that these two-dimensional map examples have both been shown to satisfy the conjectured equality of D_q with the periodic point dimension $\hat{D}_q$ [Eq. (4.9)].

VI. DEMONSTRATION OF THE RELATION OF UNSTABLE PERIODIC ORBITS TO THE ERGODIC PROPERTIES OF ATTRACTORS

In this section we obtain the results stated in Sec. IV and illustrated in Sec. V concerning the relationship of periodic orbits to ergodic properties of chaotic attractors. We shall do this only for the case of hyperbolic attractors which have a dense set of periodic orbits (i.e., axiom A attractors). Although our arguments are strictly only for the hyperbolic case, we believe that the results may be valid much more generally. For any point $\underline{x}$ in the phase space let $W^s(\underline{x})$ and $W^u(\underline{x})$ denote the stable and unstable manifolds of $\underline{x}$. The stable manifold of $\underline{x}$ is the set of points $\underline{y}$ such that $\|\underline{F}^n(\underline{x})-\underline{F}^n(\underline{y})\|\to 0$ as $n\to+\infty$ (Fig. 8); the unstable manifold is the set of points $\underline{z}$ such that $\|\underline{F}^{-n}(\underline{x})-\underline{F}^{-n}(\underline{z})\|\to 0$ as $n\to+\infty$ (we assume here that

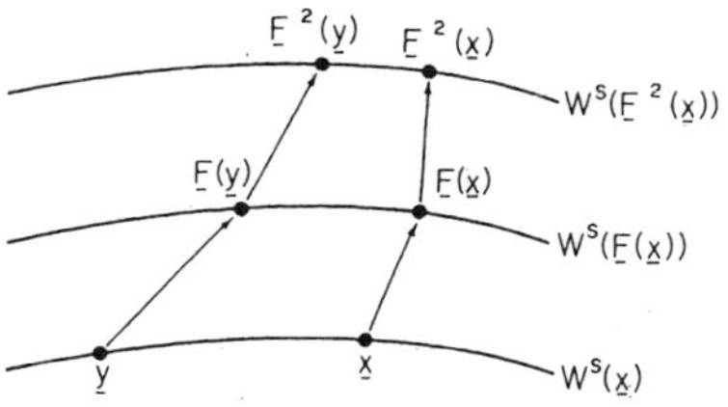

FIG. 8. Schematic illustrating the stable manifold of a point $\underline{x}$.

$\underline{F}$ is invertible). A hyperbolic attractor is one for which the following two conditions are satisfied.

(a) There exist stable and unstable manifolds $W^s(\underline{x})$ and $W^u(\underline{x})$ at each point $\underline{x}$ on the attractor that are not tangent and whose dimensions, d_s and d_u are the same for all $\underline{x}$ on the attractor, with $d_s+d_u=d$, where d is the dimension of the space. [Here $W^s(\underline{x})$ and $W^u(\underline{x})$ are smooth surfaces and d_s and d_u denote their Euclidean dimensions.]

(b) There exists a constant $K>1$ such that for all $\underline{x}$ on the attractor, if a vector $\underline{v}$ is chosen tangent to the unstable manifold, then

$$\|D\underline{F}(\underline{x})\underline{v}\| \geq K\|\underline{v}\| ,$$

and if $\underline{v}$ is chosen tangent to the stable manifold, then

$$\|D\underline{F}(\underline{x})\underline{v}\| \leq \|\underline{v}\|/K .$$

[Here $D\underline{F}(\underline{x})$ denotes the Jacobian matrix of partial derivatives of $\underline{F}(\underline{x})$ evaluated at $\underline{x}$.]

From condition (b) nearby points on the same stable (unstable) manifold approach (separate from) each other with time at least as fast as $\exp(-\kappa n)$ [$\exp(\kappa n)$]. For example, the magnitudes of the eigenvalues of $\underline{F}^n$ at a periodic point of a period n orbit must satisfy $\lambda_{pj}\geq K^n$ for $p\leq u$ and $\lambda_{pj}\leq K^{-n}$ for $p>u$. In particular, there can be no zero eigenvalues. In addition, it is very common for chaotic attractors encountered in practice to not be hyperbolic because they have points where $W^s(\underline{x})$ and $W^u(\underline{x})$ are tangent [in violation of condition (a)]. For example, the Henon attractor is of this type.

We first deal with two-dimensional maps (Secs. VI A–VI B) with $d_u=d_s=1$ and then indicate how the results can be extended to higher dimensions (Sec. VI C).

A. Measure

Imagine that we partition the space into cells C_i, where each cell has as its boundaries stable and unstable manifolds [Fig. 9(a)]. If the cells are very small, the curvature of the boundaries will be slight, and we can regard them as parallelograms [Fig. 9(b)]. Say we consider a given cell C_k and a large number of initial conditions sprinkled within the cell according to the natural probability measure on the attractor. Imagine that we iterate each of these initial conditions n times. After n iterates, a small fraction of the initial conditions may return to the small cell C_k. Since we assume the attractor to be ergodic and

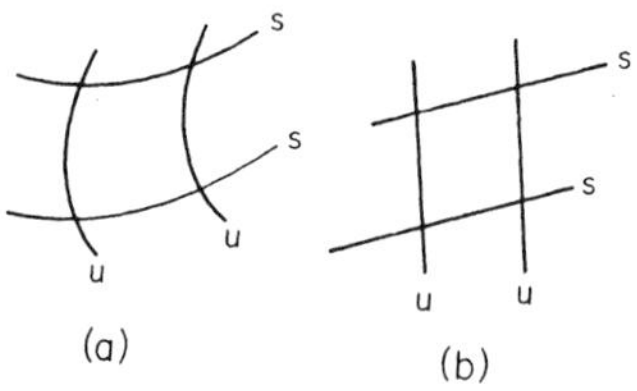

FIG. 9. Cells in a partition of the phase space. The letters s and u label stable and unstable manifold segments bounding the cell.

mixing this fraction is asymptotically (i.e., in the limit $n \to \infty$) equal to the natural measure of the attractor in the cell, $\mu(C_k)$. Let $\underline{x}_0$ be an initial condition that returns and $\underline{x}_n$ its nth iterate. This is illustrated in Fig. 10(a), where we take the stable direction as horizontal and the unstable direction as vertical. The line $ab(c'd')$ through $\underline{x}_0$ ($\underline{x}_n$) is a stable (unstable) manifold segment traversing the cell. Now take the nth forward iterate of ab and the nth backward iterate of $c'd'$. These map to $a'b'$ and cd as shown in Fig. 10(b). Now consider a rectangle constructed by passing unstable manifold segments $e'f'$ through a' and $g'h'$ through b'. By the construction, the nth preimages of these segments are the stable manifold segments ef and gh shown in Fig. 10(c). Thus we have constructed a rectangle $efgh$ in C_k such that all the points in $efgh$ return to C_k in n iterates. That is, $efgh$ maps to $e'f'g'h'$ in n iterates. The intersection of these two rectangles must contain a single saddle fixed point of the n times iterated map [cf. Fig. 10(d)]. Conversely, given a saddle fixed point, we can construct a rectangle of initial conditions $efgh$ which returns to C_k by closely following the periodic orbit which goes through the given fixed point j of $\underline{F}^n$ [the construction is the same as in Figs. 10(a)–10(c) except that $\underline{x}_0 = \underline{x}_n$]. Thus all initial conditions which return after n iterates lie in some

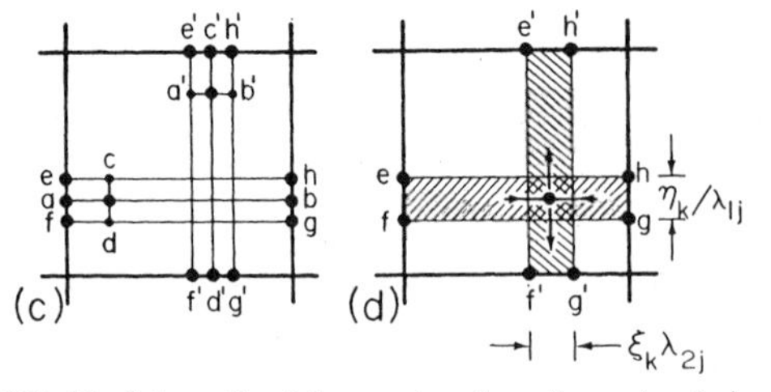

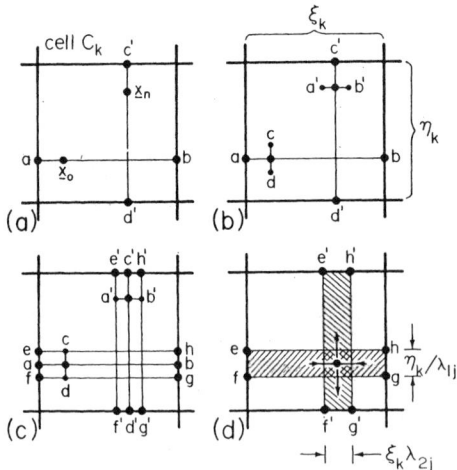

FIG. 10. Schematic of the construction of a rectangle in cell C_k which returns to C_k.

long thin horizontal strip (like $efgh$) which contains a fixed point of the n times iterated map. We label this fixed point j and the magnitudes of its unstable and stable eigenvalues λ_{1j} and λ_{2j}. Denoting the horizontal and vertical lengths of the sides of the cell C_k by ξ_k and η_k [cf. Fig. 10(b)], we see that the initial strip $efgh$ has dimensions ξ_k by (η_k/λ_{1j}) and the final strip has dimensions $\xi_k\lambda_{2j}$ by η_k [cf. Fig. 10(d)]. Since the dynamics is expanding in the vertical direction, the attractor measure varies smoothly in this direction.[18] Since the cell is assumed small, we can treat the attractor as if it were essentially uniform along the vertical direction. Thus the fraction of the measure of C_k occupied by the strip $efgh$ is $1/\lambda_{1j}$. Since, for $n \to \infty$, the fraction of initial conditions starting in C_k which return to it is $\mu(C_k)$, we have

$$\mu(C_k) = \lim_{n \to \infty} \left[\sum_{\substack{\text{fixed points} \\ \text{in } C_k}} \lambda_{1j}^{-1} \right] . \tag{6.1}$$

(Also note that, as n gets larger, λ_{1j}^{-1} and λ_{2j} get exponentially smaller and the number of fixed points in C_k grows exponentially.) Since we imagine that we can make the partition into cells as small as we wish, we can approximate any subset S of the phase space (with reasonably smooth boundaries) by a covering of cells. Thus we obtain the result, Eq. (4.2), of Sec. IV (with $L_j = \lambda_{1j}$ for the case treated here; i.e., $d=2$ with saddle periodic orbits).

In the construction which we used in arriving at Eq. (6.1), we have made two implicit assumptions. Namely, we have assumed that the segment ab maps to a segment $a'b'$ which lies entirely within C_k [i.e., we assume that the situation in Fig. 11(a) does not occur], and we have assumed that the preimage of $c'd'$ is entirely within C_k [i.e., Fig. 11(b) does not occur]. These situations might conceivably occur if $\underline{x}_n$ is too close ($\sim \xi_k \lambda_{2j}$) to a stable boundary or if $\underline{x}_0$ is too close ($\sim \eta_k/\lambda_{1j}$) to an unstable boundary. The point we wish to make here is that, for hyperbolic systems, the partition into cells can be chosen in such a way that the situations depicted in Fig. 11 do not occur. Such partitions are called Markov partitions.[18]

A Markov partition[18] satisfies the following condition. Say we have a Markov partition into cells C_i. Then, if $\underline{x}$ is in cell C_k and $\underline{F}(\underline{x})$ is in cell C_l, we have that

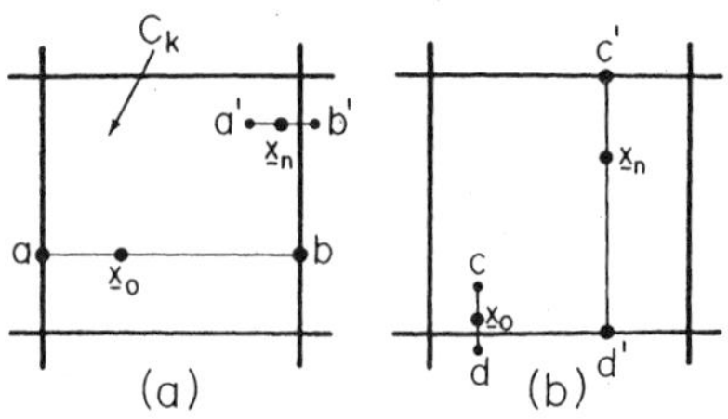

FIG. 11. Schematic depicting what can happen if $\underline{x}_0$ or $\underline{x}_n$ is too close to the boundary of C_k and a Markov partition is not used.

$$\underline{F}(W^u(\underline{x},C_k)) \supset W^u(\underline{F}(\underline{x}),C_l)\ , \tag{6.2a}$$

$$\underline{F}(W^s(\underline{x},C_k)) \subset W^s(\underline{F}(\underline{x}),C_l)\ , \tag{6.2b}$$

where $W^u(\underline{x},C_k)$ denotes a segment of the unstable manifold of $\underline{x}$ which traverses C_k passing through $\underline{x}$, and similarly for $W^s(\underline{x},C_k)$. Equations (6.2) are illustrated in Fig. 12. [Note that, to satisfy (6.2a), the end points of $\underline{F}(W^u(\underline{x},C_k))$ map to cell boundaries.] As an example, Fig. 13 shows a succession of finer Markov partitions for the baker's apprentice's map. The unit square, Fig. 13(a), is mapped into itself and bounded by segments of stable (horizontal) and unstable (vertical) manifolds. The boundaries of the partitions are obtained by taking forward iterates of the vertical (unstable) boundaries and backward iterates of the horizontal (stable) boundaries (as is explained further in the caption to Fig. 13).

Remark. The arguments as presented are not rigorous, but can be made rigorous. In particular, our arguments assume that we can approximate the map on each cell C_k as if it were linear. Also we assume that n is very large, giving the mixing sufficient time to make the probability of returning to C_k nearly equal to the measure $\mu(C_k)$. Detailed calculations based on the second derivative of the map show that our linear estimate (6.1) can be in error by at most a factor of $1\pm\epsilon$, where ϵ depends on the size of C_k and the size of the second-order partial derivatives of the map. Since we can make the size of C_k uniformly small, we can make ϵ uniformly small. Hence Eq. (4.2) holds exactly.

B. Dimension

In Fig. 10 the attractor measure contained in $e'f'g'h'$ is $1/\lambda_{1j}$. Say we cover $e'f'g'h'$ with boxes of edge length ξ_k/λ_{2j} corresponding to the narrow width of $e'f'g'h'$. There are $m_j=\eta_k/(\xi_k\lambda_{2j})$ such boxes. Since the attractor measure in the cell is essentially uniform along the vertical direction, the measure contained within one of the small boxes of edge length $\xi_k\lambda_{2j}$ is $1/(\lambda_{1j}m_j)$. Thus the contribution from $e'f'h'g'$ to the sum $\sum_i p_i^q/l_i^\tau$ in Eq. (3.7) is $m_j(\lambda_{1j}m_j)^{-q}(\xi_k\lambda_{2j})^{-\tau}$, or

$$\chi_k \frac{1}{\lambda_{1j}^q\lambda_{2j}^{(q-1)(D-1)}}\ ,$$

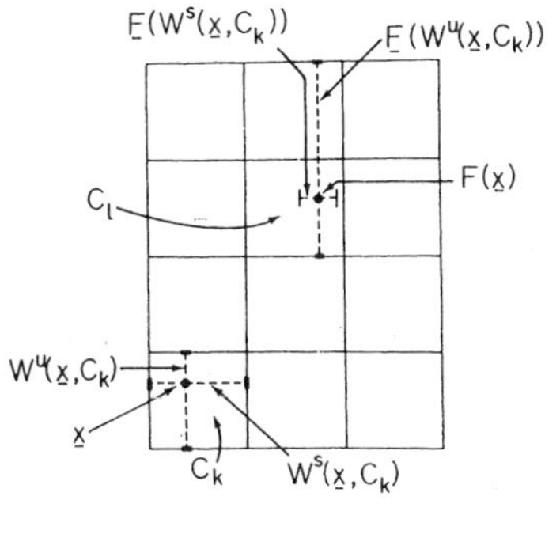

FIG. 12. Schematic illustrating Eqs. (6.2).

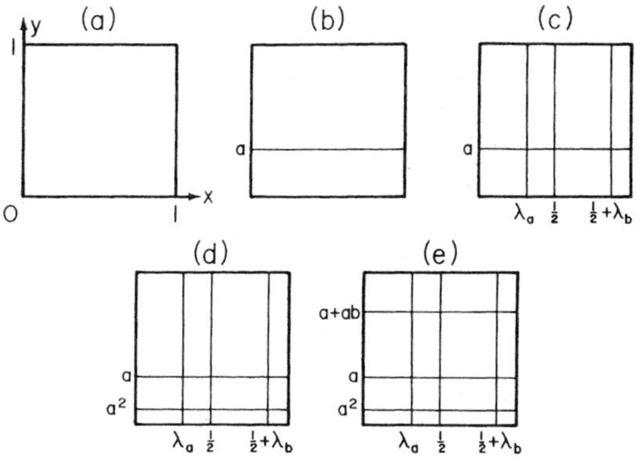

FIG. 13. Several successively finer Markov partitions for the baker's apprentice's map. Starting with (a), (b) is obtained by taking a preimage of the horizontal line $y=1$ (which is a stable manifold segment). (c) is obtained by taking a forward iterate of the vertical (unstable manifold) segments, $x=0$ and $x=1$. (d) is obtained by taking a preiterate of $y=a$ in (b), and (e) is obtained from (d) by preiterating $y=a^2$, etc.

where

$$\chi_k=(\eta_k\xi_k^{D-1})^{-(q-1)}\ .$$

Note that χ_k depends only on the partition and not on n. For a given partition there will be some cell $k=k_M$ where χ_k is the largest and some cell $k=k_m$ where χ_k is the smallest

$$\chi_{k_M}\ge\chi_k\ge\chi_{k_m}\ . \tag{6.3}$$

Now summing over all boxes we have

$$\sum_i p_i^q/l_i^\tau=\sum_j \chi_k\frac{1}{\lambda_{1j}^q\lambda_{2j}^{(q-1)(D-1)}}\ . \tag{6.4}$$

Using the bound (6.3), we have for the quantity in (6.4)

$$\chi_{k_M}\hat{\Gamma}(q,D,n)\ge\sum_j\chi_k\frac{1}{\lambda_{1j}^q\lambda_{2j}^{(q-1)(D-1)}}\ge\chi_{k_m}\hat{\Gamma}(q,D,n)\ , \tag{6.5}$$

where $\hat{\Gamma}(q,D,n)$ is given by (4.6). Letting $l\to 0$ in (3.9) is analogous to letting $n\to\infty$, since the box edge length used for the cover of $e'f'g'h'$ is edge $\xi_k\lambda_{2j}$, and λ_{2j} decreases exponentially with n. Letting n go to infinity, we see that the right-hand side of (6.4) goes through a transition from zero to infinity at the same value of D (denoted $\hat{D}_q$) as does $\sum_i p_i^q/l_i^\tau$. Furthermore, by (6.5) this transition also occurs at $D=\hat{D}_q$ for $\hat{\Gamma}(q,D)$ given by (4.6). This is the desired result for the two-dimensional map case.

In the above we have used a particular covering in obtaining the result (6.4) for $\sum_i p_i^q/l_i^\tau$. In particular, we have used the covering suggested by the dynamics and the Markov partition [cf. Fig. 10(d)]. However, since we have not optimized over all coverings, $\hat{D}_q$ the transition value of $\lim_{l\to 0}\sum_i p_i^q/l_i^\tau$ might overestimate D_q. That is

$\hat{D}_q$ is an upper bound to D_q, Eq. (4.8). It appears, however, that the covering of the attractor that we use is a rather efficient one, and we therefore believe that $D_q = \hat{D}_q$ for typical chaotic attractors. This is confirmed for the examples in Sec. V. [Furthermore, note that our result has turned out to be independent of the particular choice of the Markov partition (i.e., $\hat{\Gamma}$ does not involves $\mathcal{X}_k$).]

C. Systems of higher dimension

Again we consider a Markov partition, this time in a d-dimensional ($d > 2$) space. A fixed point j in a small cell C_k has associated with it a d-dimensional parallelepiped of initial conditions extending across the cell in the stable directions and thin in the unstable directions. The nth iterate of this parallelepiped is a d-dimensional "slab" which extends across the cell in the unstable directions and is thin in the stable directions. This is illustrated for the case $d = 3$ and $\lambda_{1j} > \lambda_{2j} > 1 > \lambda_{3j}$ (i.e., two unstable and one stable direction) in Fig. 14. Using this construction, it is readily seen that the derivation given in Sec. VI A for Eq. (6.1) extends to the higher dimensional case, for which we obtain Eq. (4.2). Now we turn to a consideration of the partition function for $d > 2$. As shown in Eqs. (6.3)–(6.5) of Sec. VI B, the dimensions of the cells do not affect the final result; thus, for simplicity we set the d edge lengths of the cell C_k equal to 1 (e.g., $\xi_k = \eta_k = \beta_k = 1$ in Fig. 14). Now we cover the slab (cf. Fig. 14) with small d-dimensional cubes. We choose the edge length of the small cubes to be $\lambda_{(k+1)j}$, where we leave k unspecified for the moment, except to say that k is large enough so that $\lambda_{(k+1)j} < 1$ (or $k \geq u$ where u denotes the number of unstable eigenvalues). The number of such small cubes necessary to cover the slab is

$$m_j = \frac{\lambda_{1j}\lambda_{2j} \cdots \lambda_{kj}}{(\lambda_{(k+1)j})^k} \frac{1}{L_j} .$$

The probability measure of the attractor in the slab is L_j^{-1}. Thus, if the measures in each cube used to cover the slab are equal, then

$$p_i = \frac{1}{m_j L_j} \equiv \bar{p} . \tag{6.6}$$

The assumption of equal probabilities was justified in the two-dimensional case (Sec. VI B) by the fact that the measure varies smoothly along unstable manifolds. In the case considered here, however, k can be greater than u. The directions corresponding to subscripts $u+1, u+2, \ldots, k$ are not stretching. Thus the assumption of equal probabilities is not as well founded as it was in Sec. VI B. Nevertheless, it is still useful, as we now show. The contribution to $\sum_i p_i^q / l_i^\tau$ from the slab is $\lambda_{(k+1)j}^{-\tau} \sum_{\text{slab}} p_i^q$. The sum $\sum_{\text{slab}} p_i^q$ subject to the constraint $\sum_{\text{slab}} p_i = L_j^{-1}$ is bounded by the value it assumes when all of the p_i in the slab are equal,

$$\sum_{\text{slab}} p_i^q \leq m_j \bar{p}^{\,q} \quad \text{for} \quad q < 1 ,$$

$$\sum_{\text{slab}} p_i^q \geq m_j \bar{p}^{\,q} \quad \text{for} \quad q > 1 .$$

Thus using (6.6) for $q < 1$ can only increase $\hat{\Gamma}(q,D)$. From Fig. 5(b) we see that this can possibly increase the transition value of D but not decrease it. Similarly, using (6.6) for $q > 1$ can only decrease $\hat{\Gamma}(q,D)$, again leading to a possible increase in the transition value of D [cf. Fig. 5(c)]. Thus, we conclude that using (6.6) to obtain $\hat{\Gamma}$ will give a transition value of D for $\hat{\Gamma}$ which is an upper bound on D_q [Eq. (4.8)]. Using (6.6) the contribution to $\sum_i p_i^q / l_i^\tau$ from the slab is

$$m_j \left[\frac{1}{m_j L_j} \right]^q \frac{1}{(\lambda_{(k+1)j})^{(q-1)D}} = L_j^{-1} [S_{kj}(D)]^{-(q-1)} ,$$

where

$$S_{kj}(D) = m_j L_j (\lambda_{(k+1)j})^D$$
$$= \lambda_{1j}\lambda_{2j} \cdots \lambda_{kj} (\lambda_{(k+1)j})^{(D-k)} .$$

In accord with the supremum for $q > 1$ and the infimum for $q < 1$ in Eq. (3.8), we choose k to make $S_{kj}(D)$ as small as possible. Increasing k decreases $S_{kj}(D)$ so long as $D - k > 0$ (recall that $\lambda_{(k+1)j} < 1$). Thus $k = \Delta$, where Δ is the integer part of D, and the minimum $S_{kj}(D)$ is $S_j(D)$ given in Eq. (4.4). Now summing over all fixed points and all cells, we obtain the desired result, Eq. (4.5). We note that due to the possible nonuniformity of p_i within the slabs, our conjecture (4.9) is not on as firm ground for $d > 2$ as it is for the case $d = 2$ treated in Sec. VI B.

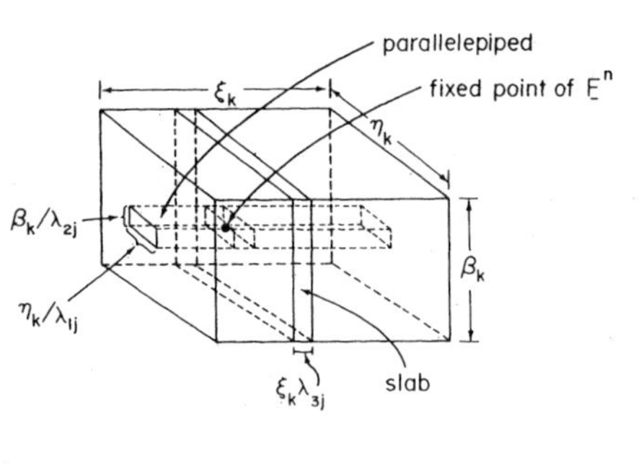

FIG. 14. Schematic illustrating a cell in the partition for $d = 3$ with two unstable and one stable direction.

VII. TRANSIENT CHAOS

In many situations there are sets with chaotic dynamics which are not attractors. While such sets do not lead to long-term chaotic behavior of typical initial conditions, they do have an important effect on the gross dynamics.[10,19–21] In particular, they manifest themselves by the presence of chaotic transients. It is the purpose of this section to extend some of the results of the previous section to the case of chaotic sets which are not attracting.

Imagine a region of space Λ which encloses what we shall call a strange saddle. This is illustrated for $d = 2$ in Fig. 15. This figure shows the saddle as the intersection of its stable and unstable manifolds, both of which may

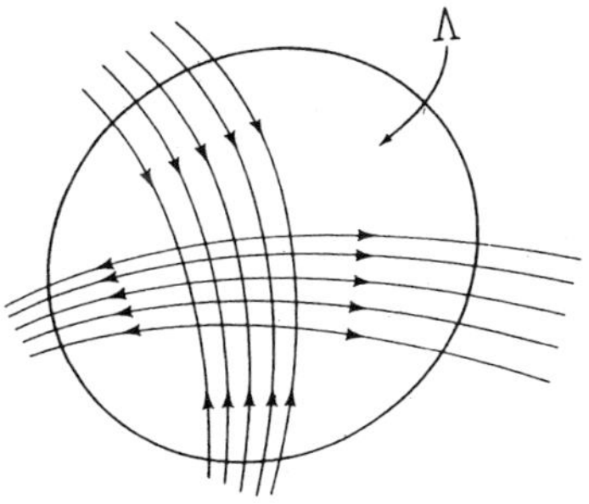

FIG. 15. Strange saddle and its stable and unstable manifolds, and a region Λ containing the strange saddle.

be thought of as a Cantor set of roughly parallel lines. If we sprinkle N_0 initial conditions uniformly in Λ, on subsequent iterates, the orbits leave Λ never to return. Let N_n be the number of orbits that have not left Λ after n iterates. For large n this number will decay exponentially with time,

$$(N_n/N_0) \sim \exp-(n/\tau) , \tag{7.1}$$

where we call τ the lifetime of the chaotic transient. We can define a natural measure for this process as follows. Let W be a subset of Λ. The natural measure of W is

$$\mu(W) = \lim_{n\to\infty} \lim_{N_0\to\infty} N_n(W)/N_n , \tag{7.2}$$

where $N_n(W)$ is the number of orbit points which fall in W at time n. Equations (7.1) and (7.2) imply that if initial conditions are distributed in accord with the natural measure and evolved in time, then the distribution will decay exponentially at the rate $1/\tau$. (This is not an invariant measure.) Points which leave Λ after a long time do so by being attracted along the stable manifold of the saddle, bouncing around on the saddle in a, perhaps, chaotic way, and then exiting along the unstable manifold.[10,19] The natural measure (7.2) is thus concentrated along the unstable manifold of the strange saddle.

Proceeding as in Sec. VI A, one can show that [compare with (4.3)]

$$\lim_{n\to\infty} e^{n/\tau} \sum_j L_j^{-1} = 1 , \tag{7.3}$$

from which the transient lifetime can be determined,

$$\frac{1}{\tau} = \lim_{n\to\infty} \left[\frac{1}{n} \log \left[\sum_j L_j^{-1} \right]^{-1} \right] . \tag{7.4}$$

This equation has been previously conjectured in Ref. 8 where numerical experiments verifying it are also reported. The quantity on the right-hand side of Eq. (7.4) is called the pressure and plays an important role in the ergodic theory of dynamical systems.

One can also ask what the dimensions D_q and $f(\alpha)$ are for the natural measure,[20,21] Eq. (7.2). Proceeding as in Secs. VI B and VI C, we find that the relevant partition function is

$$\hat{\Gamma}(q,D,n) = e^{qn/\tau} \sum_j \frac{1}{L_j} [S_j(D)]^{-(q-1)} . \tag{7.5}$$

In (7.5) the sum is over all fixed points of $\underline{F}^n$ which lie on the strange saddle. [The factors $e^{n/\tau}$ and $e^{qn/\tau}$ that appear in Eqs. (7.3) and (7.5) do so essentially in order to compensate for the exponential time decay of orbits in Λ which start with initial conditions distributed in accord with the natural measure (7.2).]

Formally expanding Eq. (7.5) in powers of $q-1$ [as was done in Sec. IV to obtain Eq. (4.10)], we find that the information dimension of the natural measure is

$$D_1 = \Delta + \frac{\log(\bar{\lambda}_1 \bar{\lambda}_2 \cdots \bar{\lambda}_\Delta) - 1/\tau}{\log(1/\bar{\lambda}_{\Delta+1})} , \tag{7.6}$$

where the Lyapunov numbers $\bar{\lambda}_p$ in this equation are defined in terms of the periodic orbits as follows:

$$\log\bar{\lambda}_p = \lim_{n\to\infty} \frac{e^{n/\tau}}{n} \sum_j \frac{1}{L_j} \log\lambda_{pj} . \tag{7.7}$$

Equation (7.6) has been given in Ref. 20 for the two-dimensional case, with the Lyapunov numbers defined in terms of typical orbits rather than periodic orbits [Eq. (7.7)], and we conjecture that the two definitions yield the same Lyapunov numbers. Note that for attractors $\tau\to\infty$ and (7.6) recovers the Kaplan-Yorke formula, Eq. (4.11).

VIII. CONCLUSION

The main result of this paper is a partition function formalism for determining the spectrum of fractal dimensions in terms of unstable periodic orbits. Beyond its conceptual appeal and potential use in further theoretical developments, the utility of this formalism will depend on how easy or difficult it is to extract information on periodic orbits from numerical computations and experimental data.[8-10]

ACKNOWLEDGMENTS

This work was supported by the U.S. Department of Energy (Basic Energy Sciences) and by the U.S. Office of Naval Research.

APPENDIX: NATURAL MEASURE

At the beginning of Sec. II we have given a definition of the natural measure which is adequate for "most purposes." A difficulty with the definition as stated can occur in special cases. For example, if the attractor has zero phase-space volume and we let the set S be the attractor itself, then, for almost every $\underline{x}$ in the basin of attraction, $\mu(\underline{x},S)$ as defined is zero (for finite length trajectories, the orbit approaches S but is not on S). A proper definition should give $\mu(S)=1$ for this S. To correct this, one can define $\mu(\underline{x},S)$ in a slightly different way. Let $\mu_\epsilon(\underline{x},S)$ be the fraction of time the trajectory originating at $\underline{x}$ spends in the ϵ neighborhood of S, and define

$$\mu(\underline{x},S) = \lim_{\epsilon\to 0^+} \mu_\epsilon(\underline{x},S) .$$

If this $\mu(\underline{x},S)$ is the same for almost every $\underline{x}$ in the attractor basin, then we denote this value $\mu(S)$ and call it the natural measure of the attractor.

[1]J. D. Farmer, E. Ott, and J. A. Yorke, Physica D **7**, 153 (1983).
[2]P. Grassberger, Phys. Lett. **97A**, 227 (1983).
[3]H. G. E. Hentschel and I. Procaccia, Physica D **8**, 435 (1983).
[4]P. Grassberger, Phys. Lett. **107A**, 101 (1985); T. C. Halsey, M. J. Jensen, L. P. Kadanoff, I. Procaccia, and B. I. Shraiman, Phys. Rev. A **33**, 1141 (1986).
[5]C. Grebogi, E. Ott, and J. A. Yorke, Phys. Rev. A **36**, 3522 (1987). Recent related works are the following: T. Morita, H. Hata, H. Mori, T. Horita, nd K. Tomita, Progr. Theor. Phys. **78**, 511 (1987); G. Gunaratne and I. Procaccia, Phys. Rev. Lett. **59**, 1377 (1987); M. H. Jensen (unpublished).
[6]R. Bowen, Trans. Am. Math. Soc. **154**, 377 (1971).
[7]A. B. Katok, Publ. Math. IHES **51**, 137 (1980).
[8]L. P. Kadanoff and C. Tang, Proc. Natl. Acad. Sci. USA **81**, 1276 (1984).
[9]D. Auerbach, P. Cvitanovic, J.-P. Eckmann, G. Gunaratne, and I. Procaccia, Phys. Rev. Lett. **58**, 2387 (1987).
[10]C. Grebogi, E. Ott, and J. A. Yorke, Phys. Rev. Lett. **57**, 1284 (1986); Phys. Rev. A **36**, 5365 (1987).
[11]M. V. Berry, in *Chaotic Behavior of Deterministic Systems*, edited by G. Iooss, R. H. G. Helleman, and R. Stora (North-Holland, Amsterdam, 1983), p. 171–271.
[12]J. H. Hannay and A. M. Ozorio de Almeida, J. Phys. A **17**, 3429 (1984).
[13]M. V. Berry, Proc. R. Soc. London A **400**, 229 (1985).
[14]F. Hausdorff, Math. Annalen **79**, 157 (1918).
[15]A. N. Kolmogorov, Dokl. Akad. Nauk SSSR **119**, 861 (1958).
[16]A. Renyi, Acta Mathematica (Hungary) **10**, 193 (1959).
[17]J. Kaplan and J. Yorke, in *Functional Differential Equations and the Approximation of Fixed Points*, Vol. 730 of *Lecture Notes in Mathematics*, edited by H. O. Peitgen and H. O. Walther (Springer, Berlin, 1978), p. 228.
[18]R. Bowen, *On Axiom A Diffeomorphisms*, CBMS Regional Conference Series in Mathematics (American Mathematical Society, Providence, 1978), Vol. 35.
[19]C. Grebogi, E. Ott, and J. A. Yorke, Physica D **7**, 181 (1983).
[20]H. Kantz and P. Grassberger, Physica D **17**, 75 (1985); G. Hsu, E. Ott, and C. Grebogi, Phys. Lett. A (to be published).
[21]P. Szépfalusy and T. Tél, Phys. Rev. A **34**, 2520 (1986).

ISRAEL JOURNAL OF MATHEMATICS, Vol. 67, No. 3, 1989

ABSOLUTELY CONTINUOUS INVARIANT MEASURES FOR PIECEWISE EXPANDING C^2 TRANSFORMATIONS IN R^N

BY
P. GÓRA[a] AND A. BOYARSKY[b,†]
[a]*Department of Mathematics, Warsaw University, Warsaw, Poland; and*
[b]*Department of Mathematics, Concordia University, 7141 Sherbrooke St. West, Montreal, Canada H4B 1R6*

ABSTRACT

Let S be a bounded region in R^N and let $\mathcal{P} = \{S_i\}_{i=1}^m$ be a partition of S into a finite number of subsets having piecewise C^2 boundaries. We assume that where C^2 segments of the boundaries meet, the angle subtended by tangents to these segments at the point of contact is bounded away from 0. Let $\tau : S \to S$ be piecewise C^2 on $\mathcal{P}$ and expanding in the sense that there exists $0 < \sigma < 1$ such that for any $i = 1, 2, \ldots, m$, $\| D\tau_i^{-1} \| < \sigma$, where $D\tau_i^{-1}$ is the derivative matrix of τ_i^{-1} and $\| \quad \|$ is the euclidean matrix norm. The main result provides an upper bound on σ which guarantees the existence of an absolutely continuous invariant measure for τ.

1. Introduction

In 1973 Lasota and Yorke [14] proved a general sufficient condition for the existence of an absolutely continuous invariant measure (a.c.i.m.) for expanding, piecewise C^2 transformations on the interval. In spite of the suggestion at the end of [14] that the "bounded variation" techniques of [14] can be easily used to obtain analogous results in higher dimensions, the generalization of the main result of [14] has taken much longer than expected. This was partly due to the difficulty in finding the right definition of variation in higher dimensions. For smooth maps on boundaryless domains, general results for the existence of a.c.i.m. were known as early as 1969 [12]. For piecewise C^2 maps in R^N, the first major attempt to prove an existence result came in 1979 [11].

† The research of the second author was supported by NSERC and FCAR grants.
Received March 7, 1989

The authors do not use a bounded variation argument but the proof, based on a one-dimensional version [10], is flawed. The first correct, but partial result, appeared in [8]. There, the author considers expanding, piecewise analytic transformations on the unit square partitioned by smooth boundaries. A complicated definition of bounded variation is used and the method cannot be extended beyond dimension 2. For boundaries which are not analytic, the sufficient condition that arises is rather complicated [9].

Working on rectangular partitions and with expanding, piecewise C^2 transformations which are very restrictive (the ith component of the transformation depends only on the ith variable), Jabłoński [7] proved the existence of an a.c.i.m. using the Tonnelli definition of bounded variation. The technique in this special setting is exactly analogous to that in [14].

In [17] a necessary and sufficient condition for the existence of a.c.i.m. is presented, but in most cases it cannot be applied.

With the publication of [3], a major new tool became available. The definition of variation of a function in R^N as the integral of its generalized derivative [3] led to the following partial result [1]: piecewise C^2 transformations on a rectangular partition satisfying a strong expansiveness condition (which depends on the dimension N of the space) have an a.c.i.m. Another partial result was obtained in [6].

In this note we follow through the approach of [1] in a more general setting. With only C^2 restrictions on the boundaries of the partition and with a mild restriction on how these boundaries meet, we prove the existence of an absolutely continuous invariant measure for τ if the slope of τ is sufficiently large.

In this setting, we can invoke the powerful Ionescu Tulcea and Marinescu Theorem [5] to obtain a useful spectral decomposition for the Perron–Frobenius operator of τ and, as a consequence, prove strong ergodic properties of the transformation itself.

Applications of ergodic theory for higher dimensional transformations can be found in [16, 18].

2. Main result

Let S be a bounded region in R^N and let τ be a transformation from S into S. We assume that τ is piecewise C^2 and expanding, i.e.,

(a) there exists a partition $\mathscr{P} = \{S_i\}_{i=1}^m$ of S, where m is a positive integer, and each S_i is a bounded closed domain having a piecewise C^2 boundary of finite $(N-1)$-dimensional measure;

(b) $\tau_i = \tau_{|S_i}$ is a C^2, 1–1 transformation from $\text{int}(S_i)$ onto its image and can be extended as a C^2 transformation onto S_i, $i = 1, 2, \ldots, m$;

(c) there exists $0 < \sigma < 1$ such that for any $i = 1, 2, \ldots, m$,

$$\| D\tau_i^{-1} \| < \sigma, \tag{1}$$

where $D\tau_i^{-1}$ is the derivative matrix of τ_i^{-1} and $\| \quad \|$ is the euclidean matrix norm.

We remark that condition (1) implies, for $\tau_i^{-1}(x)$, $\tau_i^{-1}(y)$ close enough,

$$\rho(\tau_i^{-1}(x), \tau_i^{-1}(y)) < \sigma\rho(x, y),$$

where $x, y \in R_i \equiv \tau(\text{int}(S_i))$ and ρ is the euclidean metric in R^N.

Condition (1) is implied by any of the following equivalent conditions:

(c1) all the eigenvalues of $D\tau_i^{-1}$ are smaller than 1;

(c2) all the eigenvalues of $D\tau_i$ are larger than 1.

If $|\partial\tau_{ij}^{-n}/\partial x_k| < 1$ for some n, where τ_{ij}^{-1} is the jth component of τ_i^{-1}, for $i = 1, \ldots, m$, and $1 \leqq j, k \leqq N$, then condition (c) is true for some iterate τ^l.

Let $Z = \bigcup_{i=1}^m \text{int}(S_i)$. We will consider τ as a transformation from Z into S. Our assumptions imply it is nonsingular, i.e., $\lambda_N(\tau^{-1}(\quad))$ is absolutely continuous with respect to Lebesgue measure λ_N on S. This is enough for τ to induce the Perron–Frobenius operator

$$P_\tau : L_1(S) \to L_1(S),$$

defined by

$$P_\tau f(x) = \sum_{i=1}^m \frac{f(\tau_i^{-1}(x))}{\mathcal{J}(\tau_i^{-1}(x))} \chi_{R_i}(x),$$

where χ_R is the characteristic function of the set R and $\mathcal{J}(\beta)$ is the absolute value of the Jacobian of β. The properties of P_τ are described in [13], for example. It is well known that f is a τ-invariant density if and only if $P_\tau f = f$.

The main tool of the paper is the multidimensional notion of variation defined using derivatives in the distributional sense [3]:

$$V(f) = \int_{R^N} \| Df \| = \sup\left\{ \int_{R^N} f \operatorname{div}(g) d\lambda_N : g = (g_1, \ldots, g_N) \in C_0^1(R^N, R^N) \right\},$$

where $f \in L_1(R^N)$ has bounded support, Df denotes the gradient of f in the distributional sense, and $C_0^1(R^N, R^N)$ is the space of continuously differentiable functions from R^N into R^N having compact support. We will use the following property of variation which is easily derived from [3, Remark 2.14]:

If $f = 0$ outside a closed domain A whose boundary is Lipschitz continuous, $f_{|A}$ is continuous, $f_{|\mathrm{int}(A)}$ is C^1, then

$$V(f) = \int_{\mathrm{int}(A)} \| Df \| \, d\lambda_N + \int_{\partial A} |f| \, d\lambda_{N-1},$$

where λ_{N-1} is the $(N-1)$-dimensional measure on the boundary of A.

In the sequel we shall consider the Banach space [3, Remark 1.12],

$$BV(S) = \{ f \in L_1(S) : V(f) < +\infty \},$$

with the norm $\| f \|_{BV} = \| f \|_{L_1} + V(f)$.

Before stating the main theorem, we shall need a number of lemmas.

Consider an element $S_i \in \mathscr{P}$. Let x be a point in ∂S_i and $y = \tau(x)$ a point in $\partial(\tau(S_i))$. Let $\mathscr{J}$ be the Jacobian of $\tau_{|S_i}$ at x and $\mathscr{J}_0$ the Jacobian of $\tau_{|\partial(S_i)}$ at x.

LEMMA 1. $\mathscr{J}_0 / \mathscr{J} \leqq \sigma$.

PROOF. Let C_n be a neighbourhood of y in $\tau(S_i)$ and $B_n = C_n \cap \partial(\tau(S_i))$, $n = 1, 2, \ldots$. Let γ be a curve perpendicular to $\partial(\tau(S_i))$ at y extending into C_n, and let $\gamma_n = \gamma \cap C_n$. We foliate C_n into hypersurfaces $B_n(t)$, $t \in \gamma_n$, each $B_n(t)$ being perpendicular to γ_n, and thus approximately parallel to B_n. We assume that for any n, $\lambda_{N-1}(B_n(t)) = \lambda_{N-1}(B_n)$ for all $t \in \gamma_n$. Then, if C_n is small enough, we have:

$$\lambda_N(C_n) = (1 + \varepsilon_n) \int_{\gamma_n} \lambda_{N-1}(B_n(t)) d\gamma_n(t) = (1 + \varepsilon_n) \lambda_{N-1}(B_n) \lambda_1(\gamma_n),$$

where $\varepsilon_n \to 0$ as $\mathrm{diam}(C_n) \to 0$. On the other hand, we have

$$1/\mathscr{J} = \lim_{\mathrm{diam}(C_n) \to 0} \lambda_N(\tau^{-1}(C_n)) / \lambda_N(C_n).$$

To estimate $\lambda_N(\tau^{-1}(C_n))$, let $\eta_n = \tau^{-1}(\gamma_n)$, $D_n(t) = \tau^{-1}(B_n(t))$, $t \in \gamma_n$. Let $\mathscr{J}_0(t, \zeta)$ be the Jacobian of $\tau_{|D_n(t)}$, where $t \in \gamma_n$, $\zeta \in D_n(t)$. Since $\tau_{|S_i}$ is a C^2-diffeomorphism, $1/\mathscr{J}_0(t, \zeta)$ is a C^1-function. Thus

$$\begin{aligned} \lambda_{N-1}(D_n(t)) &= \int_{B_n(t)} (1/\mathscr{J}_0(t, \zeta)) d\lambda_{N-1}(\zeta) \\ &\leqq \int_{B_n(t)} (1/\mathscr{J}_0 + K \, \mathrm{diam}(C_n)) d\lambda_{N-1}(\zeta) \\ &= (1/\mathscr{J}_0 + K \, \mathrm{diam}(C_n)) \lambda_{N-1}(B_n), \end{aligned}$$

for a constant $K > 0$.

We have

$$\lambda_N(\tau^{-1}(C_n)) \leqq \int_{\eta_n} \lambda_{N-1}(D_n(t))d\eta_n(t)$$

$$= (1/\mathcal{J}_0 + K\,\mathrm{diam}(C_n))\lambda_{N-1}(B_n)\lambda_1(\eta_n)$$

$$= \sigma(1/\mathcal{J}_0 + K\,\mathrm{diam}(C_n))\lambda_{N-1}(B_n)\lambda_1(\gamma_n).$$

Thus

$$\lambda_N(\tau^{-1}(C_n))/\lambda_N(C_n) \leqq \sigma(1/\mathcal{J}_0 + K\,\mathrm{diam}(C_n))(1+\varepsilon_n).$$

Taking the limit as $\mathrm{diam}(C_n) \to 0$, we get $1/\mathcal{J} \leqq \sigma/\mathcal{J}_0$. ∎

We note that this result is a considerable improvement over the condition $\mathcal{J}_0/\mathcal{J} \leqq N\sigma$ derived in [1] for transformations on rectangular partitions.

Let S be a closed domain in R^N with $W = \partial S$, which is piecewise C^2 and of finite $(N-1)$-dimensional measure. Let D denote the set of singular points of W, and let $\nu(x)$ denote the normalized outward normal vector at x (if $x \in D$, there are several possible outward normals). For any $x \in W$, let W_x be a small neighbourhood of x in W, contained completely in one face of W. If $x \in D$, we use a half-neighbourhood which is in one face.

For any $x \in W$, we define an R^N-neighbourhood of x, $\mathcal{U}(\delta, \alpha, x)$, $\delta > 0$, $\pi/2 < \alpha < \pi$, as follows: let $H(y)$, $y \in W_x$ be a C^1 normalized vector field, such that $\angle(H(x), \nu(x)) = \alpha$ ($\angle$ denotes angle). For any point $y \in W_x$, let $L_y = [y, y + \delta H(y)]$, the segment joining y and $y + \delta H(y)$. Now, let $\mathcal{U}(\delta, \alpha, x) = \bigcup_{y \in W_x} L_y$. If W_x and δ are small enough $\mathcal{U}(\delta, \alpha, x)$ lies completely on one side of W_x.

LEMMA 2. *For any $x \in W$, and for any $\varepsilon > 0$ sufficiently small, we can choose W_x so that:*

$$\text{(2)} \qquad \frac{(1+\varepsilon)^2}{|\cos\alpha| - \varepsilon} \int_{\mathcal{U}(\delta,\alpha,x)} f\,d\lambda_N \geqq \int_{W_x} \left(\int_{L_y} f(\xi)d\xi \right) d\lambda_{N-1}(y)$$

for any $f \in C^1(R^N)$.

PROOF. The inequality (2) obviously holds if W_x is a piece of an $(N-1)$-dimensional hyperplane. The idea of the proof is to convert our situation to that simple case.

By an orthogonal change of variables to the variables $(z_1, \ldots, z_N)$ in R^N, we can ensure that the hyperplane T_x tangent to W_x at x is given by $z_N = 0$, and the angle α between $\nu(x)$ and L_x is contained in the plane $z_1 = z_2 = \cdots = z_{N-2} = 0$. We choose W_x so small that it can be described by the equation: $z_N =$

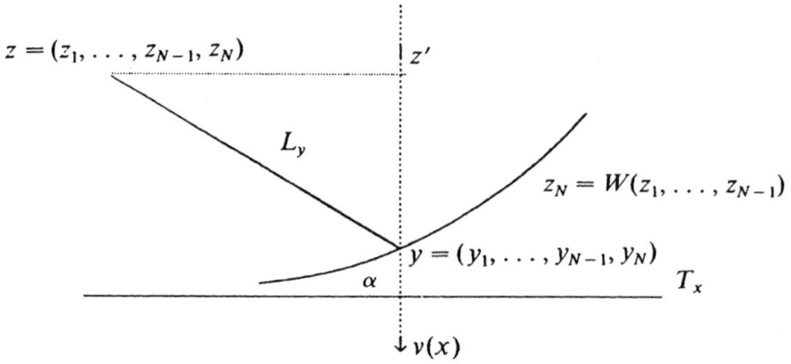

Fig. 1.

$W(z_1, \ldots, z_N)$; see Fig. 1. First, we straighten out all the segments L_y. We shall do this by keeping the same point y and shifting the point z to z', as shown in Fig. 1. If $z = (z_1, \ldots, z_N)$ is any point on the segment L_y, $y = (y_1, \ldots, y_N)$, then:

$$z = (y_1, \ldots, y_N) + |\operatorname{tg} \alpha| (z_N - y_N)[h_1, \ldots, h_N]$$

and $z' = (y_1, \ldots, y_{N-1}, z_N)$, where $[h_1, \ldots, h_N] = H(y)$, $\alpha = \alpha(y_1, \ldots, y_N) = \angle(H(y), v(x))$. Thus the straightening out is accomplished by the transformation φ_1 defined by:

$$\varphi_1(z_1, \ldots, z_N) = (z_1, \ldots, z_N) - |\operatorname{tg} \alpha| (z_N - y_N)[h_1, \ldots, h_{N-1}, 0],$$

where α, $h_1, \ldots, h_{N-1}$ are C^1 functions of $y_1, \ldots, y_N$, $y_N = W(y_1, \ldots, y_{N-1})$ and $y_1, \ldots, y_{N-1}$ are C^2 functions of $z_1, \ldots, z_N$.

Let

$$\mathcal{J}(\varphi_1) = \left| \det \left(\frac{\partial \varphi_{1,i}}{\partial z_j} \right)_{i,j=1}^N \right|$$

be the Jacobian of φ_1.

Notice that for $z = x = y$ we have $h_1 = h_2 = \cdots = h_{N-2} = \partial W / \partial z_1 = \partial W / \partial z_2 = \partial W / \partial z_{N-1} = z_N - y_N = 0$ and that the derivatives of all involved functions are continuous and bounded. This implies that $\mathcal{J}(\varphi_1)(x) = 1$ and that choosing W_x and δ small enough we can ensure $\mathcal{J}(\varphi_1) \leqq 1 + \varepsilon$ on $\mathcal{U}(\delta, \alpha, x)$.

We now straighten out the surface W_x. This is done by the transformation

$$\varphi_2(z_1, \ldots, z_N) = (z_1, \ldots, z_{N-1}, z_N - W(z_1, \ldots, z_{N-1})).$$

The Jacobian of φ_2, $\mathcal{J}(\varphi_2)$, is equal to 1. Let $\mathcal{U} = \varphi_2 \circ \varphi_1(\mathcal{U}(\delta, \alpha, x))$ and $W'_x = \varphi_2\varphi_1(W_x)$. Then, we have:

$$\int_{\mathcal{U}(\delta,\alpha,x)} f\, d\lambda_N = \int_{\mathcal{U}} f \frac{1}{J(\varphi_2)} \frac{1}{J(\varphi_1)}\, d\lambda_N \geqq \frac{1}{(1+\varepsilon)} \int_{\mathcal{U}} f\, d\lambda_N$$

$$= \frac{1}{(1+\varepsilon)} \int_{W'_x} \left(\int_{\varphi_2(\varphi_1(L_y))} f(\eta)\, d\eta \right) d\lambda_{N-1}(w),$$

where $\eta = \varphi_2 \circ \varphi_1(\xi)$ and $w = \varphi_2(y)$. To obtain (2), we need $d\eta/d\xi$ and $J(\bar{\varphi}_2)$, where $\bar{\varphi}_2$ *is* φ_2 treated as a transformation from W_x into W'_x. From Fig. 1, it is easy to see that $d\eta/d\xi = |\cos\alpha|$, where $\alpha = \alpha(z_1, \ldots, z_N)$. If W_x is small enough $|\cos\alpha(z_1, \ldots, z_N)| \geqq |\cos\alpha| - \varepsilon$. Also,

$$J(\bar{\varphi}_2)(y) = |\mathrm{Det}(\langle g_i, g_j \rangle)_{i,j=1}^{N-1}|^{-1},$$

where $g(z_1, \ldots, z_{N-1}) = (z_1, \ldots, z_{N-1}, W(z_1, \ldots, z_{N-1}))$ and $g_i = \partial g/\partial z_i$, $i = 1, \ldots, N-1$. Since this determinant is 1 at the point x and all derivatives involved are at least continuous, we can choose W_x so small that

$$J(\bar{\varphi}_2)(y) \geqq \frac{1}{1+\varepsilon}, \qquad \text{for } y \in W_x.$$

We obtain

$$\int_{\mathcal{U}(\delta,\alpha,x)} f\, d\lambda_N \geqq \frac{1}{(1+\varepsilon)} \int_{W_x} \left(\int_{L_y} f(\xi) |\cos\alpha|\, d\xi \right) J(\bar{\varphi}_2)(y)\, d\lambda_{N-1}(y)$$

$$\geqq \frac{|\cos\alpha| - \varepsilon}{(1+\varepsilon)^2} \int_{W_x} \left(\int_{L_y} f(\xi)\, d\xi \right) d\lambda_{N-1}(y). \qquad \blacksquare$$

By a regular cone in R^N we mean a cone whose base is a $(N-1)$-dimensional disk B and such that the central ray L joining the vertex to the center of the disk B is perpendicular to the disk. We define the angle subtended at the vertex of a regular cone to be the angle between L and any line joining the vertex to a point on the boundary of B.

Let S, like before, be a closed domain in R^N having piecewise C^2 boundary of finite $(N-1)$-dimensional measure.

Let us now construct at any singular point $x \in D$, the largest possible regular cone having its vertex at x and which lies completely in S. Let $\theta(x)$ denote the angle subtended at the vertex of this cone. Then define

$$\beta(S) = \min_{x \in D} \theta(x).$$

Since the faces of W meet at angles bounded away from 0, $\beta(S) > 0$. Let $\alpha(S) = \pi/2 + \beta(S)$ and

$$a(S) = |\cos \alpha(S)|.$$

Now we will construct a C^1 field of segments L_y, $y \in W = \partial S$, every L_y being a central ray of a regular cone contained in S, with angle subtended at the vertex y greater than or equal to $\beta(S)$.

We start at the points $y \in D$, where the minimal angle $\beta(S)$ is attained, defining L_y to be central rays of the largest regular cones contained in S. Then we extend this field of segments to the C^1 field we want, making L_y short enough to avoid overlapping. Let $\delta(y)$ be the length of L_y, $y \in W$. By compactness of W we have

$$\delta(S) = \inf_{y \in W} \delta(y) > 0.$$

Now we shorten L_y of our field, making them all of the length $\delta(S)$.

LEMMA 3. *If some S is a closed domain with piecewise C^2 boundary of finite $(N-1)$-dimensional measure, whose smooth faces meet at angles bounded away from zero, and f is a C^1 function on S, then*

$$\int_{\partial S} f(y) d\lambda_{N-1}(y) \leqq \frac{1}{a(S)} \left(\frac{1}{\delta(S)} \int_S f \, d\lambda_N + V(f, \operatorname{int}(S)) \right).$$

PROOF. Fix $\varepsilon > 0$, sufficiently small. We partition $W = \partial S$ into sets W_{x_i}, $i = 1, \ldots, M$, for which inequality (2) holds and define sets $\mathcal{U}(\sigma(S), \alpha(S), x_i)$, $i = 1, \ldots, M$, using the field of segments L_y, $y \in W$, constructed above. Now, for any $y \in W \setminus D$, we have:

$$f(y) \leqq \min\{f(x) : x \in L_y\} + V_{L_y}(f),$$

and

$$f(y) \leqq \frac{1}{|L_y|} \int_{L_y} f(\xi) d\xi + V_{L_y}(f) \leqq \frac{1}{\delta(S)} \int_{L_y} f(\xi) d\xi + V_{L_y}(f),$$

where $V_L(f)$ is the variation of f along the line L_y. We also have

$$V_{L_y}(f) \leqq \int_{L_y} \| Df(x) \| \, dx.$$

Integrating over W_{x_i}, $i = 1, \ldots, M$, we get

$$\int_{W_x} f(y)d\lambda_{N-1}(y)$$

$$\leqq \frac{(1+\varepsilon)^2}{a(S)-\varepsilon}\frac{1}{\delta(S)}\int_{\mathscr{U}(\delta(S),\alpha(S),x_i)} f\,d\lambda_N + \frac{(1+\varepsilon)^2}{a(S)-\varepsilon}\int_{\mathscr{U}(\delta(S),\alpha(S),x_i)} \| Df \| \,d\lambda_N.$$

Summing up, and noting that $\mathscr{U}(\delta(S), \alpha(S), x_i)$ do not overlap, we get:

$$\int_{\partial S} f(y)d\lambda_{N-1}(y) \leqq \frac{(1+\varepsilon)^2}{a(S)-\varepsilon}\left(\frac{1}{\delta(S)}\int_S f\,d\lambda_N + \int_S \| Df \| \,d\lambda_N\right).$$

Since ε is arbitrary, Lemma 3 is proved. ■

Let $\tau : S \to S$ be a piecewise C^2 expanding transformation. We assume that the sets S_i, $i = 1, \ldots, m$, of its defining partition have piecewise C^2 boundaries of finite $(N-1)$-dimensional measure and that

$$a = \min\{a(S_i) : i = 1, \ldots, m\} > 0.$$

Let

$$\delta = \min\{\delta(S_i) : i = 1, \ldots, m\} > 0.$$

Under these assumptions, we prove the following results.

LEMMA 4. *Let $f \in L_1(S)$. If $V(f) < +\infty$, then*

$$V(P_\tau f) \leqq \sigma(1 + 1/a)V(f) + K \| f \|_{L_1},$$

for some constant $K < +\infty$.

PROOF. First we assume that $f \in C^1(S)$. Then

$$P_\tau f = \sum_{i=1}^{m} \frac{f(\tau_i^{-1})}{\mathscr{J}(\tau_i^{-1})}\chi_{R_i}.$$

Let $F_i = f(\tau_i^{-1})/\mathscr{J}(\tau_i^{-1})$, $i = 1, \ldots, m$. Then,

$$\int_{R^N} \| DP_\tau f \| \,d\lambda_N \leqq \sum_{i=1}^{m}\int_{R^N} \| D(F_i\chi_{R_i}) \| \,d\lambda_N$$

$$\leqq \sum_{i=1}^{m}\left(\int_{R^N} \| (DF_i)\chi_{R_i} \| \,d\lambda_N + \int_{R^N} \| F_i(D\chi_{R_i}) \| \,d\lambda_N\right).$$

We have

$$\begin{aligned}
&\int_{R^N} \| (DF_i)\chi_{R_i} \| \, d\lambda_N \\
&= \int_{R_i} \| (DF_i) \| \, d\lambda_N \\
&\leqq \int_{R_i} \| (Df(\tau_i^{-1}))/\mathscr{J}(\tau_i^{-1}) \| \, d\lambda_N + \int_{R_i} \| f(\tau_i^{-1}) D(1/\mathscr{J}(\tau_i^{-1})) \| \, d\lambda_N \\
&\leqq \sum_{j=1}^{N} \int_{R_i} \frac{\left| \frac{\partial f}{\partial x_j}(\tau_i^{-1}) \right| \left| \sum_{k=1}^{N} \frac{\partial \tau_{ij}^{-1}}{\partial x_k} \right|}{\mathscr{J}(\tau_i^{-1})} \, d\lambda_N + \int_{R_i} |f(\tau_i^{-1})| \frac{K_0}{\mathscr{J}(\tau_i^{-1})} \, d\lambda_N \\
&\leqq \sigma \int_{S_i} \| Df \| \, d\lambda_N + K_0 \int_{S_i} \| f \| \, d\lambda_N,
\end{aligned}$$

where τ_{ij}^{-1} is the jth component of τ_i^{-1}, $1 \leqq i \leqq m$, $1 \leqq j \leqq N$, and K_0 is an upper bound for $\| D(\mathscr{J}(\tau_i^{-1}))^{-1} \| \mathscr{J}(\tau_i^{-1})$, which exists by the C^2 assumption on τ_i, $1 \leqq i \leqq m$.

Now, using Example 1.4 of [3], we obtain:

$$\begin{aligned}
\int_{R^N} \| F_i(D\chi_{R_i}) \| \, d\lambda_N &= \int_{\partial R_i} |F_i| \, d\lambda_{N-1} \\
&= \int_{\partial R_i} |f(\tau_i^{-1})| \mathscr{J}(\tau_i^{-1})^{-1} d\lambda_{N-1} = \int_{\partial S_i} |f| (\mathscr{J}_0/\mathscr{J}) d\lambda_{N-1},
\end{aligned}$$

where $\mathscr{J}_0$ is the Jacobian of $\tau_i : \partial S_i \to \partial R_i$.

By Lemma 1, $\mathscr{J}_0/\mathscr{J} \leqq \sigma$. Therefore, using Lemma 3, we obtain:

$$\begin{aligned}
\int_{R^N} \| F_i(D\chi_{R_i}) \| \, d\lambda_N &\leqq \sigma \int_{\partial S_i} |f| \, d\lambda_{N-1} \\
&\leqq \frac{\sigma}{a} V(f, S_i) + \frac{\sigma}{a\delta} \int_{S_i} |f| \, d\lambda_N.
\end{aligned}$$

Summing up, we get:

$$V(P_\tau f) \leqq \sigma(1 + 1/a) V(f) + (K_0 + \sigma/a\delta) \| f \|_{L_1}.$$

Let $K = K_0 + \sigma/a\delta$.

In general, let $f \in BV(S)$. There exists a sequence of C^1 functions f_n, $n = 1, 2, \ldots$, which approximates f in $BV(S)$. We have:

$$\begin{aligned} V(P_\tau f) &= \sup_h \int_{R^N} (P_\tau f)\operatorname{div}(h)d\lambda_N \\ &= \sup_h \lim_{n\to+\infty} \int_{R^N} (P_\tau f_n)\operatorname{div}(h)d\lambda_N \\ &= \sup_h \limsup_{n\to+\infty} \int_{R^N} (P_\tau f_n)\operatorname{div}(h)d\lambda_N \\ &\leqq \limsup_{n\to+\infty} V(P_\tau f_n) \\ &\leqq \lim_{n\to+\infty} (\sigma(1+1/a)V(f_n) + K\,\| f_n \|_{L_1}) \\ &= \sigma(1+1/a)V(f) + K\,\| f \|_{L_1}, \end{aligned}$$

where $h \in C^1(R^N, R^N)$, $\| h \| \leqq 1$. ■

Lemma 5. *For any* $f \in BV(S)$

(3) $$\| P_\tau f \|_{BV} \leqq \sigma(1+1/a)\,\| f \|_{BV} + (K+1)\,\| f \|_{L_1}.$$

Proof. Follows directly from Lemma 4 and the definition of $\| \ \|_{BV}$. ■

Remark. For $N = 1$, the inequality (3) yields the same slope condition as in the original Lasota–Yorke Theorem [14].

We can now state the main result of this paper.

Theorem 1. *Let* $\tau : S \to S$, $S \subset R^N$, *be a piecewise* C^2, *expanding transformation. If* $\sigma(1+1/a) < 1$, *then* τ *admits an absolutely continuous invariant measure.*

Proof. From inequality (3) it follows that the set $\{ \| P_\tau^i(1) \|_{BV} \}_{i \geqq 1}$ is uniformly bounded. Hence the set $\{P_\tau^i(1)\}_{i \geqq 1}$ is weakly compact in L_1 (actually it is strongly compact), and it follows from the Kakutani–Yoshida Theorem that P_τ has a nontrivial fixed point f^* which is the density of an a.c.i.m. ■

Corollary 1. *Let* $\tau : S \to S$, $S \subset R^N$, *be piecewise* C^2 *and such that some iterate* τ^k *satisfies* $\sigma(1+1/a) < 1$ (σ *and* a *corresponds to* τ^k). *then* τ *admits an a.c.i.m.*

Proof. Straightforward. ■

Example. For a rectangular partition of a rectangular domain in R^N, we have $1/a = \sqrt{N}$, which gives the expansion condition $\sigma(1+\sqrt{N}) < 1$.

REMARK. The expansion condition $\sigma < (1 + 1/a)^{-1}$ depends only on the domain S and not on the transformation τ. Under certain conditions on τ, we can obtain an improved expansion condition such as is done in [9] in dimension 2. These conditions are usually very complex and require accurate knowledge of the transformation.

3. Spectral decomposition

In this section we will use the Ionescu Tulcea and Marinescu Theorem [5]. First we have to check that the assumptions of the theorem are satisfied:

We consider the space BV, $\| \quad \|_{BV}$ as included in L_1, $\| \quad \|_{L_1}$.

(1) By the semicontinuity property of variation (Theorem 1.9 of [3]), if $\{f_n\} \in BV$, $\| f_n \|_{BV} \leqq D$, for $n = 1, 2, \ldots$ and $f_n \to f$ in L_1, then $f \in BV$ and $\| f \|_{BV} \leqq D$.

(2) The operator norm of the Perron–Frobenius operator P_τ is 1.

(3) There exist constants $R > 0$, $0 < r < 1$ such that

$$\| P_\tau f \|_{BV} \leqq r \| f \|_{BV} + R \| f \|_{L_1}, \qquad \text{for } f \in BV.$$

This follows by Lemma 5 if $\sigma(1 + 1/a) < 1$.

(4) The image of any bounded subset of BV under the Perron–Frobenius operator is relatively compact in L_1. This follows from the compactness Theorem 1.19 of [3].

The Ionescu Tulcea and Marinescu Theorem implies the following result:

THEOREM 2. *Let $\tau : s \to S$, $S \subset R^N$, $N \geqq 1$, be a piecewise C^2 and expanding transformation with $\sigma(1 + 1/a) < 1$ and let $P = P_\tau$ be its Perron–Frobenius operator. Then:*

(a) *P (as an operator from BV into BV) has a finite number of eigenvalues of modulus 1: $\alpha_1, \ldots, \alpha_t$. They are roots of unity and*

$$P = \sum_{i=1}^{t} \alpha_i P_i + T,$$

where $P_i : BV \to BV$ are linear projections with finite dimensional range, and $T : BV \to BV$ is a continuous linear operator;

(b) $P_i^2 = P_i$, $P_i P_j = 0$ $(i \neq j)$, $P_i T = T P_i = 0$, $1 \leqq i, j \leqq t$;

(c) $\| T^n \|_{BV} \leqq M/(1 + h)^n$, $n = 1, 2, \ldots,$ *for some* M, $h > 0$.

REMARK (see [15]). Operators P_i, $i = 1, \ldots, t$ and T have unique extensions onto L_1. Moreover $P_i(L_1) \subset BV$, $\| P_i \|_{L_1} \leqq 1$ and $\sup_n \| T^n \| < +\infty$. For any $f \in L_1$, $T^n f \to 0$ in L_1, as $n \to +\infty$.

The following theorem and corollaries are consequences of the representation of the Perron–Frobenius operator obtained in Theorem 2.

THEOREM 3 (see [15]). *Assume that* 1 *is the only eigenvalue of* P *with modulus* 1 (*we can consider* P^k, *where* k *is the smallest common multiplier of orders of* $\alpha_1, \ldots, \alpha_t$). *Let* $Ug = g \circ \tau$, *for* $g \in L_\infty$. *Then there exist nonnegative functions* $\phi_1, \ldots, \phi_s \in BV$ *and* $\psi_1, \ldots, \psi_s \in L_\infty$ *such that*:

(a) *For any* $f \in L_1$

$$P_1 f = \sum_{i=1}^{s} \left(\int_{R^N} f \psi_i d\lambda_N \right) \phi_i,$$

(b) $P\phi_i = \phi_i$, $U\psi_i = \psi_i$, $i = 1, \ldots, s$.

(c) $\int_{R^N} \phi_i \psi_j d\lambda_N = \delta_{ij}$, $\inf\{\phi_i, \phi_j\} = 0 = \inf\{\psi_i, \psi_j\}$ *as* $i \neq j$, *and* $\int_{R^N} \phi_i d\lambda_N = 1$, $1 \leqq i, j \leqq s$,

(d) *There exist measurable sets* $C_1, \ldots, C_s \subset S$ *such that* $\psi_i = \chi_{C_i}$ *a.e., for* $i = 1, \ldots, s$ *and* $S = \bigcup_{i=1}^{s} C_i$ *a.e.*

(e) $\bigcap_{n=1}^{\infty} U^n(L_1) = \bigcap_{n=1}^{\infty} U^n(L_\infty) = \mathrm{Span}\{\psi_1, \ldots, \psi_s\}$.

(f) *For any* $f \in L_1$, $U^n f \to P_1^* f$ *in* $\sigma(L_1, BV)$-*topology*; *for any* $f \in L_\infty$, $U^n f \to P_1^* f$ *in* $\sigma(L_\infty, L_1)$-*topology*;

$$P_1^* f = \sum_{i=1}^{s} \left(\int_{R^N} f \phi_i d\lambda_N \right) \psi_i.$$

COROLLARY 2 (see [15]). *For any* $1 \leqq i \leqq s$, $\tau^k_{|C_i}$ *is an exact transformation.*

COROLLARY 3 (see [15]). *If we assume that* τ *is mixing* (*or weakly mixing, which is equivalent in this situation*), *and* μ *is its unique a.c.i.m., then* τ *has the property of exponential decay of correlation*: *Let* $f \in BV$, $g \in L_\infty$ *and* $\mu(f) = \int_{R^N} f\, d\mu$, $\mu(g) = \int_{R^N} g\, d\mu$. *Then*

$$\int_{R^N} (fg(\tau^n) - \mu(f)\mu(g)) d\mu \leqq \mu(f) V(f) \| g \|_{L_\infty} r^n, \qquad n = 1, 2, \ldots,$$

where $0 < r < 1$ *is the constant of condition* (3).

COROLLARY 4 (see [15]). *If we assume that* τ *is mixing, then the defining partition* $\{S_i\}_{i=1}^{m}$ *is weakly Bernoulli for* τ, *whch implies that the natural extension of the dynamical system* (τ, μ) *is isomorphic to a Bernoulli shift* (μ *is the* τ-*invariant absolutely continuous measure*).

COROLLARY 5 (see [4]). *Assume that* τ, μ *is weakly mixing (it is equivalent to being mixing or exact in our situation). Let* $f \in BV$ *and* $\mu(f) = \int_{R^N} f\,d\mu = 0$. *Define*

$$S(t) = \sum_{i=0}^{t-1} f \circ \tau^i,$$

which is a stochastic process on (S, μ). *Then the series* (σ below has nothing to do with the contraction constant of formula (1), we use it here only for historical reasons)

$$\sigma^2 = \int_S f^2 d\mu + 2 \sum_{k=1}^{\infty} \int_S f(f \circ \tau^k) d\mu$$

converges absolutely, $\int_S S(t)^2 d\mu = t\sigma^2 + O(1)$ *and, if* $\sigma^2 \neq 0$, *the following holds:*

(i) $\sup_{z \in R} |\mu((\sigma^2 t)^{-1/2} S(t) \leqq z) - (2\pi)^{-1/2} \int_{-\infty}^{z} \exp(-x^2/2)dx| = O(t^{-\nu})$, *for some* $\nu > 0$.

(ii) *Without changing its distribution, one can redefine the process* $(S(t))_{t \geqq 0}$ *on a richer probability space together with the standard Brownian motion* $(B(t))_{t \geqq 0}$ *such that*

$$|\sigma^{-1} S(t) - B(t)| = O(t^{(1/2)-\varepsilon}), \qquad \mu\text{-}a.e.,$$

for some $0 < \varepsilon < 1/2$.

(iii) *The process* $(S(t))_{t \geqq 0}$ *satisfies the iterated log law and other properties of Brownian motion.*

ACKNOWLEDGEMENT

We are grateful to H. Proppe who found errors in earlier drafts of this paper and made many helpful suggestions.

REFERENCES

1. D. Candeloro, *Misure invariante per transformazioni in piú dimensioni*, Atti Sem. Mat. Fis. Univ. Modena **XXXV** (1987), 33–42.

2. H. Federer, *Geometric Measure Theory*, Springer-Verlag, New York, 1969.

3. E. Giusti, *Minimal Surfaces and Functions of Bounded Variation*, Birkhauser, 1984.

4. F. Hofbauer and G. Keller, *Ergodic properties of invariant measures for piecewise monotonic transformations*, Math. Z. **180** (1982), 119–140.

5. C. T. Ionescu Tulcea and G. Marinescu, *Théorie ergodique pour des classes d'opérations non completement continues*, Ann. of Math. **52** (1950), 140–147.

6. V. V. Ivanov and A. G. Kachurowski, *Absolutely continuous invariant measures for locally*

expanding transformations, Preprint No. 27, Institute of Mathematics AN USSR, Siberian Section (in Russian).

7. M. Jabłoński, *On invariant measures for piecewise C^2-transformations of the n-dimensional cube*, Ann. Polon. Math. **XLIII** (1983), 185–195.

8. G. Keller, *Ergodicité et mesures invariantes pour les transformations dilatantes par morcaux d'une région bornée du plan*, C.R. Acad. Sci. Paris **289**, Série A (1979), 625–627.

9. G. Keller, *Proprietés ergodiques des endomorphismes dilatants, C^2 par morceaux, des régions bornées du plan*, Thesis, Université de Rennes, 1979.

10. A. A. Kosyakin and E. A. Sandler, *Ergodic properties of a class of piecewise-smooth transformations of an interval*, Izv. VUZ Matematika (3)[118] (1972), 32–40. (English translation from the British Library, Translation Service.)

11. A. A. Kosyakin and E. A. Sandler, *Stochasticity of a certain class of discrete system*, translated from Automatika and Telemekhanika No. 9 (1972), 87–94.

12. K. Krzyżewski and W. Szlenk, *On invariant measures for expanding differentiable mappings*, Studia Math. **33** (1969), 82–92.

13. A. Lasota and M. Mackey, *Probabilistic Properties of Deterministic Systems*, Cambridge University Press, 1985.

14. A. Lasota and J. A. Yorke, *On the existence of invariant measures for piecewise monotonic transformations*, Trans. Am. Math. Soc. **186** (1973), 481–488.

15. M. Rychlik, *Bounded variation and invariant measures*, Studia Math. **LXXVI** (1983), 69–80.

16. F. Schweiger, *Invariant measures and ergodic properties of numbertheoretical endomorphisms*, Banach Center Publications, to appear.

17. E. Straube, *On the existence of invariant absolutely continuous measures*, Commun. Math. Phys. **81** (1981), 27–30.

18. M. Yuri, *On a Bernouilli property for multi-dimensional mappings with finite range structure*, in *Dynamical Systems and Nonlinear Oscillations*, Vol. 1 (Giko Ikegami, ed.), World Scientific, Singapore, 1986.

Invent. math. 112, 541–576 (1993)

Inventiones mathematicae

Sinai–Bowen–Ruelle measures for certain Hénon maps

Michael Benedicks[1] **and Lai-Sang Young**[2,*]

[1] Department of Mathematics, Royal Institute of Technology, S-100 44 Stockholm, Sweden
[2] Department of Mathematics, UCLA, Los Angeles, CA 90024, USA

Oblatum 20-III-1992

0 Introduction

We study maps $T_{a,b}: \mathbb{R}^2 \to \mathbb{R}^2$ defined by

$$T_{a,b}(x, y) = (1 - ax^2 + y, bx), \quad 0 < a < 2, b > 0.$$

In [BC2] it was proved that for a positive measure set of parameters (a, b), $T_{a,b}$ has a topologically transitive attractor $\Lambda = \Lambda_{a,b}$ on which there is some hyperbolic behavior. The aim of this paper is to study the *statistical* properties of these attractors. Using the machinery developed in [BC2] we prove the following

Theorem. *There is a set* $\Delta \subset \mathbb{R}^2$ *with* $\text{Leb}(\Delta) > 0$ *such that for all* $(a, b) \in \Delta$, $T = T_{a,b}$ *admits a unique* SBR *measure* λ^*. *Moreover,* $\text{supp}(\lambda^*) = \Lambda$ *and* (T, λ^*) *is Bernoulli.*

A T-invariant Borel probability measure μ is called a Sinai–Bowen–Ruelle (SBR) measure if there is a positive Lyaponov exponent μ-a.e. and the conditional measures of μ on unstable manifolds are absolutely continuous with respect to the Riemannian measure induced on these manifolds. (A more precise definition is given in Sect. 3.4.1.) This notion is due to Sinai [S1, S2]. See also [LS]. The significance of these measures is evident in the following corollary.

Corollary. *For* $(a, b) \in \Delta$, $T = T_{a,b}$ *has the following property: Let* U *be a neighborhood of* Λ. *Then there is a set* $\tilde{U} \subset U$ *with positive Lebesgue measure such that for all continuous functions* $\varphi: U \to \mathbb{R}$,

$$\frac{1}{n} \sum_{i=0}^{n-1} \varphi \circ T^i(x) \to \int \varphi \, d\lambda^*$$

for every $x \in \tilde{U}$.

---

* Benedicks is partially supported by the Swedish Natural Science Research Council (NFR) and the Swedish Board of Technical Development (STU). Young is partially supported by NSF

This corollary follows from our theorem and general nonuniform hyperbolic theory (see [PS, Theorem 1] or Sect. 4.1.2). The property expressed in this corollary is often assumed to be true in physics; it is also taken for granted in numerical experiments. In the case of Axiom A attractors, mathematical justification of this property is provided by the theory of Sinai, Bowen and Ruelle (see e.g. [B], [Ru1] and [S2]). It has been conjectured that many other attractors admit SBR measures, but as far as we know, the Hénon family and similar examples (see next paragraph) are the only nonuniformly hyperbolic attractors for which SBR measures have been constructed.

Mora and Viana [MV] proved recently that in homoclinic bifurcations of surface diffeomorphisms, very small attractors with high periods appear for a positive measure set of parameters. These attractors have the same qualitative estimates as those in [BC2]. Since our proofs rely only on these qualitative estimates, our results apply there also.

This paper is organized as follows. We assume throughout that $T = T_{a,b}$, where (a, b) is a pair of "good" parameters, meaning the ones selected in [BC2]. We do not concern ourselves with how these parameters are selected. In Sect. 1 we give a summary of the results from [BC2] that are relevant to this work. The primary focus of [BC2] is on a certain set of points called "critical points". These points live on a curve W that is the unstable manifold of a fixed point. In Sect. 2 we shift our attention from the critical points to the action of T on all of W. Our main proposition establishes a near complete correspondence between the dynamics of $T|W$ and that of certain interval maps. An SBR measure is constructed in Sect. 3, where Lebesgue measure on a small piece of W is pushed forward. The ergodic properties of this invariant measure are studied in Sect. 4.

1 Dynamics of certain Hénon maps: results from [BC2]

In this section we try to give a self-contained summary of what is known about the "good" Hénon maps. For conceptual simplicity, we will isolate for separate discussion several aspects of their dynamics. The reader should be aware that many of these properties are related, in the sense that their proofs are very delicately intertwined. Some of the ideas in [BC2] will be expressed here in slightly different language. For instance, we will speak of orbits and vectors as being "controlled" (see Sects. 1.4 and 1.5). We will also introduce the notion of "generalized tangential position" (Sect. 1.6).

1.1 The one-dimensional model (Sect. 2 of [BC2])

The b-values considered in [BC2] are very small and the entire analysis there is modeled on that of a certain class of 1-d maps. These are maps of the form $f = f_a : [-1, 1] \circlearrowleft$, where $f_a x = 1 - ax^2$ and a is very near 2. In addition the following conditions are imposed of f:

(1) There is $c > 0$ ($c \approx \log 2$) such that $|Df^n(f0)| \geqq e^{cn}$ for all $n \geqq 0$.

(2) There is a small real number $\alpha > 0$ (e.g. $\alpha = 10^{-6}$) such that $|f^n 0| \geqq e^{-\alpha n}$ for all $n \geqq 1$.

It is proved in [BC2] that the set of parameter values $\mathscr{A} = \{a \mid f_a$ satisfies (1) and (2)$\}$ has positive Lebesgue measure.

To study the growth of $(f^n)'x$ for $x \in [-1, 1]$, we have three types of derivative estimates given in Sects. 1.1.1–1.1.3. First some notation: Let $\delta > 0$ be a small real number that is nevertheless $\gg 2 - a$. We assume that $\delta = e^{-\mu_0}$ for some $\mu_0 \in \mathbb{Z}^+$. For bookkeeping purposes write

$$(-\delta, \delta) = \dot{\bigcup_{|\mu| \geqq \mu_0}} I_\mu ,$$

where $I_\mu = (e^{-(\mu+1)}, e^{-\mu})$ for $\mu > 0$ and $I_\mu = -I_{-\mu}$ for $\mu < 0$. Each I_μ is further subdivided into μ^2 intervals $\{I_{\mu,j}\}$ of equal length.

For $a \in \mathscr{A}$, $f = f_a$ has the following properties:

1.1.1 Derivative estimates away from the critical point

There is $c_0 > 0$ and $M_0 \in \mathbb{Z}^+$ such that

(i) if $x, \ldots, f^{j-1}x \notin (-\delta, \delta)$ and $j \geqq M_0$, then $|Df^j(x)| \geqq e^{c_0 j}$;

(ii) if $x, fx, \ldots, f^{k-1}x \notin (-\delta, \delta)$ and $f^k x \in (-\delta, \delta)$, any $k \in \mathbb{Z}^+$, then $|Df^k(x)| \geqq e^{c_0 k}$.

(iii) if $x, fx, \ldots, f^{k-1}x \notin (-\delta, \delta)$, then $|Df^k(x)| \geqq \delta e^{c_0 k}$.

1.1.2 Derivative estimates when bound to the critical orbit

Let $\beta = 14\alpha$. For $x \in (-\delta, \delta)$ define $p(x)$ to be the largest integer p such that

$$|f^j x - f^j 0| < e^{-\beta j} \quad \forall j < p.$$

Then

(i) $\frac{1}{2}|\mu| \leqq p(x) \leqq 5|\mu| \;\; \forall x \in I_\mu$;

(ii) $|Df^p(x)| \geqq e^{c' p}$ for some $c' > 0$.

The orbit of x is said to be *bound* to the critical orbit during the period $j < p$. We may assume that p is constant on each $I_{\mu,j}$.

1.1.3 Distortion of Df^n

Let $\mathscr{P}$ be the partition of $[-1, 1]$ into $[-1, -\delta] \cup [\delta, 1] \cup \bigcup_{\mu,j} I_{\mu,j}$. For $J \in \mathscr{P}$, let J^+ denote the interval consisting of J and its two adjacent intervals in $\mathscr{P}$. Then the following holds: Assume that $\omega \subset [-1, 1]$ is such that for all $k < N$, $f^k\omega \subset J^+$ for some $J \in \mathscr{P}$. Then there is a constant C, independent of N such that

$$\frac{|Df^N(x)|}{|Df^N(y)|} \leqq C \quad \forall x, y \in \omega.$$

1.2 Some geometric properties of Hénon maps

The material in this subsection is quite elementary and is valid for $T = T_{a,b}$ where (a, b) belongs to a set Δ' of the form $\{(a, b): a_1 < a < a_2 < 2, 0 < b < b_0\}$.

1.2.1 The unstable manifold W

T has a unique fixed point $\hat{z}$ in the first quadrant. This fixed point is hyperbolic. Its expanding direction has slope of order $-b/2$ and eigenvalue ≈ -2. Its contractive direction has slope ≈ 2 and eigenvalue $\approx b/2$. The global unstable manifold at $\hat{z}$ is called W. It will play the rôle of the interval $[-1, 1]$ in our 1-d model.

1.2.2 Attracting sets

To guarantee that T has a compact attracting set, we first choose $a_0 < a_1 < 2$ with a_0 sufficiently near 2. Then there exists $b_0 > 0$ sufficiently small compared to $2 - a_1$ such that for all $(a, b) \in \Delta' := [a_0, a_1] \times (0, b_0]$, W stays in a bounded region. Moreover, if $\Lambda = \Lambda_{a,b}$ is the closure of W, then there is an open neighborhood $U = U_{a,b}$ of Λ such that for all $z \in U$, dist $(T^n z, \Lambda) \to 0$ as $n \to \infty$ ([BC2, Theorem 4]; see also [BM]).

1.2.3 Hyperbolicity and resemblance to 1-d behavior outside of $(-\delta, \delta) \times \mathbb{R}$

Let $\delta > 0$ be at least as small as needed in our one-dimensional analysis and assume that $b_0 \ll 2 - a_0 \ll \delta$. Let $s(v)$ denote the absolute value of slope of the vector v. A simple calculation shows that if $z = (x, y) \notin (-\delta, \delta) \times \mathbb{R}$ and $s(v) \leqq \delta$, then

$$s(DT_z v) \leqq \frac{b}{|x|} \leqq \frac{b}{\delta},$$

which we assume to be $\ll \delta$. ([BC2, Lemma 4.5].) This defines invariant cones outside of $(-\delta, \delta) \times \mathbb{R}$.

For $z = (x, y) \notin (-\delta, \delta) \times \mathbb{R}$ and a unit vector v with $s(v) \leqq \delta$, we have essentially the same estimates as in 1-dimension. That is, there is $c_0 > 0$ and $M_0 \in \mathbb{Z}^+$ such that

(i) if $z, \ldots, T^{j-1} z \notin (-\delta, \delta) \times \mathbb{R}$ and $j \geqq M_0$, then $|DT_z^j v| \geqq e^{c_0 j}$;
(ii) if $z, Tz, \ldots, T^{k-1} z \notin (-\delta, \delta) \times \mathbb{R}$ and $T^k z \in (-\delta, \delta) \times \mathbb{R}$, any $k \in \mathbb{Z}^+$, then $|DT_z^k v| \geqq e^{c_0 k}$;
(iii) if $z, Tz, \ldots, T^{k-1} z \notin (-\delta, \delta) \times \mathbb{R}$ then $|DT^k v| \geqq \delta e^{c_0 k}$.

1.2.4 Curvature estimates

Tangencies between stable and unstable manifolds of T are inevitable. In order to show that some of these tangencies are quadratic, we will need to control the curvature of W. In [BC2], a curve γ in $\mathbb{R}^2$ is called a $C^2(b)$ *curve* if $s(\gamma') \leqq 10b$ and its curvature $\kappa \leqq 10b$. The following lemma is used to verify the $C^2(b)$ property.

Lemma (proved in Section 7.6 of [BC2]). *Let* $t \mapsto \gamma(t)$ *be a smooth curve. Write* $\gamma_n = T^n\gamma$ *and let* κ_n *denote the curvature of* γ_n. *Assume there is a* $c > 0$ *such that for all* t

(i) $\kappa_0(t) \leqq 1$;
(ii) $|\gamma'_n(t)|/|\gamma'_{n-j}(t)| \geqq e^{cj}$ $\forall j \leqq n$;
(iii) $\gamma_{n-j}(t) \notin (-\delta, \delta) \times \mathbb{R}$ and $s(\gamma'_{n-j}(t)) < \delta$ *for* $j = 1, \ldots, 4$.

Then $\kappa_n(t) \leqq 10b$ *for all* t.

The discussion from here on applies only to $T = T_{a,b}$, *where* (a, b) *belongs in an inductively constructed set* $\Delta \subset \Delta'$. The set Δ has the property that for each sufficiently small $b > 0$, $\Delta_b := \{a : (a, b) \in \Delta\}$ is a Cantor set and has positive Lebesgue measure.

1.3 The critical set (mostly Sects. 5 and 6 of [BC2])

We use the following notation. For $A \in \mathrm{GL}(2, \mathbb{R})$, if $v \mapsto |Av|/|v|$ is not constant, let $e(A)$ denote the unit vector that is contracted the most by A. Write $e_n(z) = e(DT^n_z)$ if it makes sense . For $z \in W$, let $\tau \in T_z\mathbb{R}^2$ denote a unit vector, tangent to W.

A certain subset $\mathscr{C}$ of W called the critical set is singled out in [BC2] to play the rôle of 0 in 1-d. This set is constructed according to the rule in 1.3.1. Each $z_0 \in \mathscr{C}$ has the property that $\tau(z_0)$ is the most contracted direction, i.e. $\lim_{n \to \infty} e_n(z_0) = \tau(z_0)$. This can be thought of as the moral equivalent to $f'(0) = 0$.

1.3.1 Location and rule of construction of $\mathscr{C}$

$\mathscr{C}$ is located in $W \cap ((-10b, 10b) \times \mathbb{R})$. It seems likely that it does not lie on any smooth curve. To give a more precise description of where points of $\mathscr{C}$ are located, we divide W into segments of different "generations": First, there is a unique point $z_0 \in \mathscr{C}$ on the roughly horizontal segment of W containing the fixed point $\hat{z}$. In [BC2], the segment of W between $T^2 z_0$ and $T z_0$ is called G_1, the *leaf of generation* 1. Leaves of generation $g \geqq 2$ are defined by

$$G_g := T^{g-1} G_1 - G_{g-1} .$$

We assume (a, b) is sufficiently near (2, 0) so that $\bigcup_{g \leqq 27} G_g$ consists of 2^{26} roughly horizontal segments linked by sharp turns near $x = \pm 1, y = 0$, and that $(\bigcup_{g \leqq 27} G_g) \cap ((-\delta, \delta) \times \mathbb{R})$ consists of 2^{26} $C^2(b)$ curves (see Sect. 1.2.4). If $\mathscr{C}_g := \mathscr{C} \cap \bigcup_{i \leqq g} G_i$, then $\mathscr{C}_{27}$ contains 2^{26} points, one on each of these $C^2(b)$ curves.

For $g > 27$, the following rule is used. Let ρ be a number with $b \ll \rho \ll e^{-72}$, and assume $\mathscr{C}_{g-1}$ is already defined. Consider a maximal piece of $C^2(b)$ curve $\gamma \subset G_g$. If γ contains a segment of length $2\rho^g$ centered at $z'_0 = (x'_0, y'_0)$, and there is a critical point $z_0 = (x_0, y_0) \in \mathscr{C}_{g-1}$ with $x'_0 = x_0$, and $|y'_0 - y_0| \leqq b^{g\sigma}$, $\sigma = \frac{1}{540}$, then there is a unique point $\hat{z}_0 \in \mathscr{C}_g \cap \gamma$. Moreover, $|\hat{z}_0 - z'_0| \leqq |y'_0 - y_0|^{1/2}$. These are the only points of $\mathscr{C}_g$.

1.3.2 Remarks on the construction of $\mathscr{C}$

Roughly speaking, certain points z on W with $e_n(z) = \tau$ are picked out as approximate critical points. Once a point is designated an approximate critical point, parameters are excluded to ensure that a point in $\mathscr{C}$ is eventually constructed nearby. The construction of $\mathscr{C}$ itself involves an inductive procedure that is guaranteed to work only for the "good" parameters.

As to when to initiate the construction of a critical point, the rules in Sect. 1.3.1 are obviously quite arbitrary. There are two guiding principles. The first is that each "distinguishable" critical orbit of length n causes a certain measure set of parameters to be discarded, so we cannot afford to have too many critical points. On the other hand, as in 1-d, when a critical orbit returns to $(-\delta, \delta) \times \mathbb{R}$, we want it to be "bound" to a suitable critical point of earlier generation. So we must be sure there are enough of these "suitable" critical points. These issues are discussed in Sects. 3 and 6 of [BC2].

1.3.3 Dynamical properties of $\mathscr{C}$

The parameter selection procedure is also designed to guarantee that every $z_0 \in \mathscr{C}$ has the following properties:

(1) For all $n \geqq 0$,

$$|DT^n_{z_0}\binom{0}{1}| \geqq e^{cn} \text{ for some } c \approx \log 2,$$

$$|DT^n_{z_0}\tau| \leqq (Cb)^n \text{ for some } C \text{ independent of } b;$$

(2) there is a small real number α, say $\alpha = 10^{-6}$, such that $T^n z_0$ stays a distance of $\geqq e^{-\alpha n}$ from certain points of $\mathscr{C}$.

The uniform hyperbolicity expressed in (1) is analogous to condition (1) in Sect. 1.1. We will elaborate on (2) and other dynamical properties of $\mathscr{C}$ later on.

1.3.4 Contracting fields in neighborhoods of a critical point

Let $z_0 \in \mathscr{C}$. When an orbit comes near z_0, the course of its interaction with z_0 in the next n iterates is determined to a great extent by the contracting field e_n and expanding field $e_n^\perp$ near z_0. The following facts are proved in Sect. 5 of [BC2][1]:

(i) e_1 is defined everywhere and has slope $= 2ax + \mathcal{O}(b)$.

(ii) There is λ with $b \ll \lambda < 1$ such that for all $z_0 \in \mathscr{C}$, e_n is defined on $B_n(z_0)$:= the disk of radius $(\lambda/5)^n$ about z_0. Moreover,

$$|DT^n_z e_n| \leqq (Cb)^n, \quad \forall z \in B_n .$$

(iii) There is a constant $C > 0$ such that for all $z_1, z_2 \in B_n$

$$|e_n(z_1) - e_n(z_2)| \leqq C|z_1 - z_2| .$$

[1] The slight difference between the definition of e_n here and that in [BC2] is unimportant

(iv) For $(x_1, y_1), (x_2, y_2) \in B_n$ with $|y_1 - y_2| \leqq |x_1 - x_2|$

$$|e_n(x_1, y_1) - e_n(x_2, y_2)| = (2a + \mathcal{O}(b))|x_1 - x_2| .$$

(v) For $m < n$, $|e_n - e_m| \leqq \mathcal{O}(b^m)$ on $B_n(z_0)$.

Suppose z_0 lies on a $C^2(b)$ curve γ. Then it follows from (i)–(v) that there is a unique z_0^n in $\gamma \cap B_n(z_0)$ at which $e_n = \tau$, and that $|z_0^n - z_0| = \mathcal{O}(b^n)$.

1.4 Controlled orbits

1.4.1 Definitions

(i) Let $z_0 = (x_0, y_0)$ be a critical point lying on a piece of $C^2(b)$ curve $\gamma \subset W$, and let $\zeta_0 = (\xi_0, \eta_0)$ be an arbitrary point in $(-\delta, \delta) \times \mathbb{R}$. We say that ζ_0 is in *tangential position* with respect to z_0 if there is $\hat{z}_0 = (\hat{x}_0, \hat{y}_0) \in \gamma$ with $\hat{x}_0 = \xi_0$ and $|\hat{y}_0 - \eta_0| < |\hat{x}_0 - x_0|^4$.

(ii) Let $\zeta_0 \in W$, and let $\zeta_n = T^n \zeta_0$. We say that *the orbit of* ζ_0 *is controlled on the time interval* $[0, N)$, $N \leqq \infty$, if the following holds. If $\zeta_0 \notin \mathscr{C}$, let $n_1 < n_2 < \ldots$ be the times when ζ_n is in $(-\delta, \delta) \times \mathbb{R}$. If $\zeta_0 \in \mathscr{C}$, take $n_1 > 0$ to be the first time ζ_n returns to $(-\delta, \delta) \times \mathbb{R}$. Then for every $n_i \in [0, N)$, ζ_{n_i} is attached to a critical point with respect to which it is in tangential position. This critical point is denoted by $z(\zeta_{n_i})$ and is called the *binding point* of ζ_{n_i}. See 1.4.3 for a further requirement on the assignment $\zeta_{n_i} \mapsto z(\zeta_{n_i})$.

1.4.2 Control of critical orbits

A key idea in [BC2] is that through parameter exclusion, it is arranged so that for every $z_0 \in \mathscr{C}$

(i) the orbit of z_0 is controlled on the time interval $[0, \infty)$;
(ii) $|z_{n_i} - z(z_{n_i})| \geqq e^{-\alpha n_i}$ $\forall i \geqq 1$;
(iii) $\text{gen}(z(z_{n_i})) < \theta \cdot n_i$ for some small $\theta > 0$.

(ii) is the precise statement of (2) in 1.3.3. To explain the availability of binding points, consider first the following ideal situation. Suppose that z_n is a free return, and that $\bigcup_{g < \theta n} G_g$ contains $2^{[\theta n] - 1}$ $C^2(b)$ curves stretched across $(-\delta, \delta) \times \mathbb{R}$, each containing a critical point. In this case one needs only to consider the curve nearest to z_n. Most locations of z_n are in tangential position relative to the critical point on this curve; bad locations of z_n correspond to deleted parameters. In general, some of the $2^{[\theta n] - 1}$ pieces of W are missing or partially missing. What is true is that at free returns, z_n is always surrounded by a "fairly regular" collection of $C^2(b)$ segments $\{\gamma_j\}$ of W. "Fairly regular" here means that $\text{gen}(\gamma_j) \sim 3^j$, $\text{length}(\gamma_j) \sim \rho^{3^j}$ and $\text{dist}(z_n, \gamma_j) \sim b^{3^j}$ $(3^j < \theta n)$. It is then shown that such a family contains enough critical points for our purposes. (See Sects. 6 and 7 of [BC2].)

1.4.3 Bound periods

Let $z_0 \in \mathscr{C}$, and let ζ_0 be in tangential position with respect to z_0. As a first approximation we define the bound period $\tilde{p}(\zeta_0, z_0)$ to be the largest k such that

$$|T^j\zeta_0 - T^j z_0| < e^{-\beta j} \quad \forall j < k .$$

(Recall that $\beta = 14\alpha$.)

Now consider ζ_0 whose orbit is controlled on $[0, \infty)$, and let $\tilde{p}_i = \tilde{p}(\zeta_{n_i}, z(\zeta_{n_i}))$. Note that it is entirely possible for ζ_n with $n_i < n < n_i + \tilde{p}_i$ to return to $(-\delta, \delta) \times \mathbb{R}$, so that bound periods can be initiated in the middle of bound periods. It is shown in Sect. 6.2 of [BC2] that by modifying slightly the definition of $\tilde{p}$, bound periods can be made "nested". More precisely, one can choose $\{p_i\}_{i=1}^{\infty}$ in such a way that for all i,

(i) $p_i \leqq \tilde{p}_i$ and $|T^{p_i}\zeta_{n_i} - T^{p_i}z(\zeta_{n_i})| \geqq e^{-\beta^* p_i}$ for some $\beta^* \approx \beta$,
(ii) if n_j is such that $n_i < n_j < n_i + p_i$, then $n_j + p_j \leqq n_i + p_i$.

The bound period between ζ_{n_i} and $z(\zeta_{n_i})$ is then defind to be p_i. It is further required that if the bound relation between z_{n_i} and $z(z_{n_i})$ is still in effect at time $n_j > n_i$, then we must have $z(z_{n_j}) = z(T^{n_j - n_i} z(z_{n_i}))$.

1.4.4 Bound of free states

Conceptually, it is convenient to divide a controlled orbit into "free" and "bound" states. Let ζ_0 be as above.

- For $n \leqq n_1$, ζ_n is *free.*
- For $n_1 < n < n_1 + p_1$, ζ_n is said to be in *bound* state. (During this period ζ_n may make multiple returns to $(-\delta, \delta) \times \mathbb{R}$, at which times new bindings are formed. But by our definition of p_i, these "secondary bound periods" expire at or before $n_1 + p_1$.)
- Let n_i be the first return to $(-\delta, \delta) \times \mathbb{R}$ after $n_1 + p_1$. Then ζ_n is free for $n_1 + p_1 \leqq n \leqq n_i$, and ζ_{n_i} is called a *free return.*
- ζ_n is in bound state again during the period $n_i < n < n_i + p_i$, and so on.

1.5 Keeping track of derivatives

Throughout this subsection we assume that the orbit ζ_0 is controlled on $[0, \infty)$, with return times $n_1 < n_2 < \ldots$ and binding points $\zeta_{n_i} \leftrightarrow z(\zeta_{n_i})$ as defined in 1.4.1.

1.5.1 The fold period

For each n_i, the fold period of ζ_{n_i} with respect to $z(\zeta_{n_i})$ is defined to be a number l_i with $2m \leqq l_i \leqq 3m$, where

$$(5b)^m \leqq |z(\zeta_{n_i}) - \zeta_{n_i}| \leqq (5b)^{m-1} .$$

For convenience we will choose l_i such that $n_i + l_i \neq n_j, n_j + 1$ for any j. (The precise definition is given in [BC2, Sect. 6.3] and does not particularly concern us.) We mention three facts about fold periods:

(i) The fold period initiated at a return to $(-\delta, \delta) \times \mathbb{R}$ is very short compared to the bound period initiated at that time. In fact, $l/p \leqq \text{const}/\log(1/b)$, which tends to 0 as $b \to 0$.

(ii) It follows from (i) above and Sect. 1.4.2 that every $z_0 \in \mathscr{C}$ has the following property: for every $m \in \mathbb{Z}^+$, there are integers m_1 and m_2, with $m_1 \leqq m \leqq m_2$ and $m_2 - m_1 \leqq \text{const.}\, m/\log(1/b)$, such that z_{m_1} and z_{m_2} are outside all fold periods. (See Lemma 6.5 in [BC2].)

(iii) If $n_i < n_j \leqq n_i + l_i$ then $n_j + l_j \leqq n_i + l_i$, i.e. a fold period initiated within another fold period does not extend beyond that fold period.

The role of the fold period will become clear in 1.5.4.

1.5.2 The splitting algorithm (Sect. 7.1 of [BC2])

Let $v \in T_{\zeta_0}\mathbb{R}^2$ be a tangent vector. The following algorithm is devised in [BC2] to keep track of $w_n(\zeta_0, v) := DT^n_{\zeta_0} v$. For each n, we decompose $w_n(\zeta_0, v)$ into

$$w_n(\zeta_0, v) = w_n^*(\zeta_0, v) + E_n(\zeta_0, v)\,,$$

where w_n^* and E_n are defined according to the following rules. (We will omit all references to (ζ_0, v) from now on.)

- For $n < n_1$, let $w_n^* = w_n$.
- At time n_1, let $e = e_{l_1}$ be the contractive vector field around $z(\zeta_{n_1})$. Write

$$w_{n_1} = A_{n_1} e + B_{n_1}\binom{0}{1}$$

  and let

$$w_{n_1}^* = B_{n_1}\binom{0}{1}, \qquad E_{n_1} = A_{n_1} e\,.$$

- For $n > n_1$, if $n \neq$ any n_i, define

$$E_n = \sum_{\substack{n_k < n \\ \text{s.t. } n_k + l_k > n}} DT^{n-n_k}_{\zeta_{n_k}}(A_{n_k} e_{l_k})\,.$$

  This of course defines w_n^* as well.
- If $n = n_i$ for some $i > 1$, we let $e = e_{l_i}$, write

$$DT_{\zeta_{n-1}} w_{n-1}^* = A_n e + B_n\binom{0}{1}$$

  and let

$$w_n^* = B_n\binom{0}{1}\,.$$

This algorithm has no geometric meaning for arbitrary (ζ_0, v). See 1.5.4 for a discussion of the situation for which it is designed.

1.5.3 Controlled derivatives along controlled orbits

Definition. Let $\zeta_0 \in W$ and let $v \in T_{\zeta_0}\mathbb{R}^2$. We say that the pair (ζ_0, v) is *controlled* on the time interval $[0, N)$ if the orbit of ζ_0 is controlled on $[0, N)$ and whenever $\zeta_n \in (-\delta, \delta) \times \mathbb{R}$, the splitting algorithm in Sect. 1.5 gives

$$(*) \qquad 3|\zeta_n - z(\zeta_n)| \leqq |\bar{B}_n| \leqq 5|\zeta_n - z(\zeta_n)|, \quad 0 \leqq n < N,$$

where

$$\bar{B}_n = \frac{B_n}{|DT_{\zeta_{n-1}} w_{n-1}^*|}.$$

If $(*)$ holds, we say that the vector $v_n = DT_{\zeta_0}^n v$ *splits correctly*.

One of the most important properties of T proved in [BC2] is that

for every $z_0 \in \mathscr{C}$, $(z_0, \binom{0}{1})$ *is controlled during the time interval* $[0, \infty)$.

It is also proved that if ζ_0 is bound to $z_0 \in \mathscr{C}$ then $(\zeta_0, \binom{0}{1})$ is also controlled during the bound period $[0, p)$.

1.5.4 1-dimensional behavior in 2-dimensions

We now give some idea of how the splitting algorithm is used to study $D_{\zeta_0}^n v$ where (ζ_0, v) is controlled on $[0, \infty)$. For definiteness consider $\zeta_0 \in \mathscr{C}$ and $v = \binom{0}{1}$. We claim that $\{w_n^*\}_{n=0}^{\infty}$ is essentially 1-d in character, and has properties similar to $(f^n)'(f0)$ for $f(x) = 1 - ax^2$.

As before, let $n_1 < n_2 < \ldots$ be the return times of ζ_0 to $(-\delta, \delta) \times \mathbb{R}$. First $w_1^* = \binom{1}{0}$. For $0 < n < n_1$, $\{|w_n^*|\}$ behaves qualitatively like $\{|(f^n)'(f0)|\}$; see 1.2.3. Let us consider $n = n_1$, and let $w_n = A_n e + B_n\binom{0}{1}$ be the decomposition given by the splitting algorithm.

Ignoring the A-term for now, and using the fact that $(\zeta_0, \binom{0}{1})$ is controlled, we see that

$$|w_{n+1}^*| = |w_n^*| = |\bar{B}_n| \cdot |DT_{\zeta_{n-1}} w_{n-1}^*|$$
$$\sim 2a|\zeta_n - z(\zeta_n)| \cdot |DT_{\zeta_1}^{n-1} w_1^*|,$$

which has an obvious resemblance to $|(f^n)'(f0)|$. Moreover, since $w_{n+1}^* = B_n\binom{1}{0}$, the action of DT^j on w_{n+1}^* for $j < n_2 - n_1$ is again essentially 1-d. The vector $DT^{n_2 - n_1} w_{n_1}^*$ is split at time n_2, and the new B-term is treated similarly.

The general philosophy is that the A- terms can, in some sense, be neglected. Let us consider a simple situation, where the fold period l at time $n = n_1$ expires before the next return to $(-\delta, \delta) \times \mathbb{R}$. We claim that $DT_{\zeta_n}^l A_n e$ is extremely short compared to $DT_{\zeta_n}^l B_n\binom{0}{1}$, so that the effect of adding this term at time $n + l$ is negligible. This is because

$$|DT_{\zeta_n}^l A_n e| \leqq |DT_{\zeta_{n-1}} w_{n-1}^*| \cdot (5b)^l,$$

whereas

$$|DT_{\zeta_n}^l B_n\tbinom{0}{1}| \geqq |DT_{\zeta_{n-1}} w_{n-1}^*| \cdot |\zeta_n - z(\zeta_n)| \cdot |DT_{\zeta_n}^l\tbinom{0}{1}|,$$

and we know that $|\zeta_n - z(\zeta_n)| \geqq (5b)^{l/2}$ by the definition of l, and also that $|DT^{l-1}_{\zeta_{n+1}}\binom{1}{0}| \geqq \delta e^{c(l-1)}$ by 1.2.3(iii).

In general, the computation is more complicated, but if (ζ_0, v) is controlled one has $|\bar{B}_j| \geqq |\zeta_j - z(\zeta_j)|$ at all returns, and the rejoining of the A-terms of w_n^* at the end of fold periods has negligible effects. (See Sect. 7.3 of [BC2].)

We have made a point of comparing $w_n^*(z_0, \binom{0}{1})$, $z_0 \in \mathscr{C}$, to $(f^n)'(f0)$. This parallel is not complete unless we could guarantee, through parameter exclusion, that for all $z_0 \in \mathscr{C}$, $|w_n^*(z_0, \binom{0}{1})| \geqq e^{cn}$ for some $c > 0$. This is not exactly what we stated in 1.3.3, but it is in fact true.

1.6 Dynamics near the "turns"

1.6.1 Bound state estimates

Assume that ζ_0 is in tangential position with respect to $z_0 \in \mathscr{C}$, and let p and l be its bound and fold periods. We now record some estimates from [BC2] pertaining to the bound states $\{\zeta_j\}_{j=0}^{p}$. These estimates rely on the fact that critical orbits have the properties in Sects. 1.3.1 and 1.5.3.

(i) $|T^j\zeta_0 - T^j z_0| \sim |\zeta_0 - z_0|^2 |DT^j_{z_0}\binom{0}{1}|$ for $l \leqq j \leqq p$.
(For a precise statement, see Sect. 7.5, in particular Lemma 7.4, of [BC2].)

(ii) If $|\zeta_0 - z_0| \approx e^{-\mu}$, $\mu > 0$, then $\frac{1}{2}\mu \leqq p \leqq 5\mu$.
(This follows essentially from (i) and Sect. 1.3.1.)

(iii) $|DT^p_{\zeta_0}\binom{0}{1}| \cdot |\zeta_0 - z_0| \geqq e^{c'p}$ for some $c' > 0$.
(See Lemma 7.5 of [BC2].)

Two points are to be noted here. The first is the striking resemblance between these estimates and those in 1-d. Assertion (i), for instance, says that ζ_0 experiences a quadratic contraction from its interaction with z_0. The second observation we wish to make is that unlike the situation in 1-d, not all the points near z_0 have this "quadratic" behavior. For instance, ζ_0 may lie on a piece of stable manifold through z_0, and be attracted to z_0 forever. Or, if ζ_0 is directly above z_0, then we will have $|T^j\zeta_0 - T^j z_0| \approx |\zeta_0 - z_0| \cdot |DT^j_{z_0}\binom{0}{1}|$ instead of (i) for the first few iterates.

1.6.2 Generalized tangential positions

We wish to say more precisely for which region the estimates in 1.6.1 hold. For $z \in W$, we say that (x', y') is the natural coordinate system at z if $(0, 0)$ is at z, the x'-axis lines up with $\tau(z)$, and the y'-axis with $\tau(z)^\perp$. The following definition is *not* contained in [BC2] but will be useful for us in the next section.

Definition. Let $c > 0$ be a small number $\ll 2a$, say $c = \frac{1}{100}$, and let $z_0 \in \mathscr{C}$. A point ζ_0 near z_0 is said to be in *generalized tangential position* with respect to z_0 if in the natural coordinate system at z_0, $\zeta_0 = (\xi', \eta')$ with $|\eta'| \leqq c\xi'^2$.

Clearly, if ζ_0 is in tangential position wrt z_0, then it is in generalized tangential position, because the segment of W containing z_0 is a $C^2(b)$ curve tangent to the x'-axis at $(0, 0)$. While it is not explicitly stated this way, the proofs in [BC2] show in fact that the bound estimates in 1.6.1 hold for all ζ_0 in generalized tangential position wrt $z_0 \in \mathscr{C}$. Let us recall briefly why this is so:

Let p be the bound period between ζ_0 and z_0. From 1.3.4 we know that there is a contractive direction field defined on a small ball about z_0. The integral curves of e_p are roughly parabolas of the form $y' = \text{const} + a(x' - x'_p)^2$, and have a unique tangency with the x'-axis at $z_0^{(p)} := (x'_p, 0)$. Since $|z_0^{(p)} - z_0| = O(b^p) \ll |z_0 - \zeta_0|$, for our purposes we may as well confuse $z_0^{(p)}$ with z_0. With this simplification we can think of z_0 and ζ_0 as lying on the graph of a function $\varphi: x'\text{-axis} \to y'\text{-axis}$ with $|\varphi'(x')| \leqq 2c|x'|$. The estimates in Sects. 7.4 and 7.5 of [BC2] apply to points on such a curve. What really matters is that all the tangent vectors to graph(φ) split correctly wrt e_p, which a priori is defined only in a small ball about z_0 but is shown in a local induction in [BC2] to be defined on the entire segment of graph(φ) between z_0 and ζ_0. These estimates imply those in 1.6.1.

1.7 Distortion estimates during bound periods

In the course of proving some of the estimates in the last few subsections, the following estimates are used.

(i) For all $z_0 \in \mathscr{C}$, if ζ_0 is bound to z_0 with bound period p then

$$\frac{1}{2} \leqq \frac{|w_j^*(\zeta_0, \binom{0}{1})|}{|w_j^*(z_0, \binom{0}{1})|} \leqq 2 \quad \forall j < p\,.$$

This is Assertion 4(a) in [BC2]. It is proved in Lemmas 7.8, 7.9 and on p. 151.

(ii) There is a constant $C_0 > 0$ such that $\forall z_0 \in \mathscr{C}$, if ζ_0 and ζ'_0 are bound to z_0 during $[0, v]$, then

(a) $$\frac{|w_v^*(\zeta_0, \binom{0}{1})|}{|w_v^*(\zeta'_0, \binom{0}{1})|} \leqq \exp\left\{ C_0 \sum_{j=0}^{v-1} \frac{\Delta_j(\zeta_0, \zeta'_0)}{d_{\mathscr{C}}(z_j)} \right\},$$

where $\Delta_j(\zeta_0, \zeta'_0) = \max_{0 \leqq i \leqq j} |\zeta_i - \zeta'_i|$ and

$$d_{\mathscr{C}}(z_j) := \begin{cases} |x_j| & \text{if } z_j = (x_j, y_j) \text{ and } |x_j| \geqq \delta, \\ |z_j - z(z_j)| & \text{if } |x_j| \leqq \delta \text{ and } z(z_j) \text{ is the binding point.} \end{cases}$$

(b) $\measuredangle(w_v^*(\zeta_0, \binom{0}{1}), w_v^*(\zeta'_0, \binom{0}{1})) < 2b^{\frac{1}{4}} \Delta_v$.

These estimates are proved the same time (i) is proved.

(iii) There is a constant $C'_0 > 0$ such that if z_0, ζ_0 and ζ'_0 are as in (ii) and in addition ζ_0 and ζ'_0 lie on a curve all the tangent vectors of which split correctly with respect to z_0 (compare 1.5.3) then

$$\sum_{j=0}^{p} \frac{\Delta_j(\zeta_0, \zeta'_0)}{d_{\mathscr{C}}(z_j)} \leqq C'_0 \frac{|\zeta_0 - \zeta'_0|}{|\zeta_0 - z_0|}\,.$$

This is proved on pp. 151 and 163 of [BC2].

The constant C_0 and C'_0 above remain uniformly bounded as $\delta \to 0$.

2 Dynamics on the unstable manifold W

2.1 Statements of results

Let $T = T_{a,b}$ be as in Sect. 1. The purpose of this section is to use the dynamics of points in $\mathscr{C}$ to help us understand the dynamics of all points on W. We will prove that the derivative estimates for $DT^n_\zeta \tau$, $\zeta \in W$, are completely parallel to those for $(f^n)'x$ where $f: [-1, 1] \circlearrowleft$ satisfies the conditions in Sect. 1.1.

As usual, $\{\zeta_n\}_{n=-\infty}^{\infty}$ denotes the orbit of ζ_0. Before we begin we need to modify or extend slightly a couple of our definitions from Sect. 1. First we relax our definition of "control" for the orbit of ζ_0 to requiring only that whenever $\zeta_n \in (-\delta, \delta) \times \mathbb{R}$, there is a critical point $z(\zeta_n)$ wrt which ζ_n is in *generalized* tangential position. (See 1.6.2.) Second, we say that the pair (ζ_0, v) is controlled on the time interval $[j, N)$, $-\infty < j < N$, if $(\zeta_j, DT^j_{\zeta_0} v)$ is controlled on $[0, N-j)$, and that (ζ_0, v) is controlled on $(-\infty, N)$ if it is controlled on $[j, N)$ for all $j < N$. Note that for all $\zeta_0 \in W$, ζ_j tends to the unique hyperbolic fixed point $\hat{z}$ as $j \to -\infty$.

2.1.1 Main Proposition and Main Corollary

Proposition 1 (Main proposition) *For all $\zeta_0 \in W$, if $\zeta_i \notin \mathscr{C}$ for all $i < N$, then the pair $(\zeta_0, \tau(\zeta_0))$ is controlled on the time interval $(-\infty, N)$.*

Proposition 1 enables us to describe each ζ_i, $i < N$, as being in a "free state" or "bound state" as discussed in 1.4.4. (Technically one needs to specify a starting point for this to make sense, but since $\zeta_i \to \hat{z}$ as $i \to -\infty$, we can think of our trajectory as starting from ζ_j for some negatively very large j.

Corollary 1 (Main corollary) *Let $\zeta_0 \in W$, and assume that $\zeta_i \notin \mathscr{C}$ for all i. Then ζ_i is in a "free state" infinitely often, and the following holds for $\tau_i := DT^i_{\zeta_0}\tau$:*

I. Expansion outside of $(-\delta, \delta) \times \mathbb{R}$
There is $c_0 > 0$ and $M_0 \in \mathbb{Z}^+$ such that

(i) *if ζ_i is free and $\zeta_i, \ldots, \zeta_{i+M_0} \notin (-\delta, \delta) \times \mathbb{R}$, then*

$$\frac{|\tau_{i+M_0}|}{|\tau_i|} \geqq e^{cM_0};$$

(ii) *if $\zeta_i \notin (-\delta, \delta) \times \mathbb{R}$ is free, and $k > i$ is the first time $\zeta_k \in (-\delta, \delta) \times \mathbb{R}$ then*

$$\frac{|\tau_k|}{|\tau_i|} \geqq e^{c_0(k-i)}.$$

II. Bound period estimates
There is $c_1 \approx \log 2$ such that the following holds: if $\zeta_i \in (-\delta, \delta) \times \mathbb{R}$ is free and becomes bound at this time to $z(\zeta_i) \in \mathscr{C}$ with bound period p, then

(i) *if $e^{-\mu-1} \leqq |\zeta_i - z(\zeta_i)| \leqq e^{-\mu}$, then $\frac{1}{2}\mu \leqq p \leqq 5\mu$;*
(ii) $|\tau_{i+j}|/|\tau_i| \geqq 3|\zeta_i - z(\zeta_i)| \cdot |DT^j_{z(\zeta_i)}\binom{0}{1}| \geqq |\zeta_i - z(\zeta_i)| e^{c_1 j}$, $0 < j < p$;
(iii) $|\tau_{i+p}|/|\tau_i| \geqq e^{c_1 p/3}$.

2.1.2. Global distortion estimates

We will need the following distortion estimate for $DT^n\tau$. Let $\mathscr{P} = \{I_{\mu j}\}$ be the partition of $(-\delta, \delta)$ as described in Sect. 1.1, i.e.

$$(-\delta, \delta) = \dot{\bigcup_{|\mu| \geqq \mu_0}} I_\mu ,$$

where $I_\mu = (e^{-(\mu+1)}, e^{-\mu})$ for $\mu > 0$, $I_\mu = -I_{-\mu}$ for $\mu < 0$, and each I_μ is further subdivided into μ^2 intervals $\{I_{\mu j}\}$ of equal length.

For x_0 with $|x_0| \ll \delta$, we let $\mathscr{P}_{[x_0]}$ denotes a copy of $\mathscr{P}$ with 0 "moved" to x_0. More precisely, let $h:(-\delta, \delta) \to (-\delta, \delta)$ be the piecewise linear homeomorphism taking the points $-\delta, -e^{-(\mu_0+1)}, x_0, e^{-(\mu_0+1)}, \delta$ to $-\delta, -e^{-(\mu_0+1)}, 0, e^{-(\mu_0+1)}, \delta$ respectively and let $\mathscr{P}_{[x_0]} = h^{-1}\mathscr{P}$.

Furthermore, if γ is a roughly horizontal curve and $z_0 = (x_0, y_0)$ is such that $|x_0| \ll \delta$, then $\mathscr{P}_{[z_0]} = \mathscr{P}_{[x_0]}$ is the obvious partition on $\gamma \cap ((-\delta, \delta) \times \mathbb{R})$. Once γ and z_0 are specified, we will speak about $I_{\mu j}$ as though it was a subsegment of γ. Also, for $J = I_{\mu j}$, let J^+ denote the union of J with its adjacent intervals in $\mathscr{P}_{[\cdot]}$

We will use the following notation: if γ_0 is a curve segment then $\gamma_j = T^j\gamma_0$; ζ_j denotes $T^j\zeta_0$ and $\tau_j(\zeta_0) = DT^j_{\zeta_0}\tau$ etc.

Proposition 2 *Let $\gamma_0 \subset W \cap ((-\delta, \delta) \times \mathbb{R})$ be a curve segment, and let $0 = t_0 < t_1 < \ldots < t_q = N$ be its free return times. More precisely, for all $k < q$,*

(1) *all points in γ_{t_k} have a common binding point, which we denote by $z^{(k)}$, and $\gamma_{t_k} \subset J^+$ for some $J \in \mathscr{P}_{z^{(k)}}$;*

(2) *If p_k is the bound period between γ_{t_k} and $z^{(k)}$, then t_{k+1} is the smallest $j \geqq t_k + p_k$ such that $\gamma_j \cap ((-\delta, \delta) \times \mathbb{R}) \neq \emptyset$.*
Then for all $\zeta_0, \zeta_0' \in \gamma_0$,

$$\frac{|\tau_N(\zeta_0)|}{|\tau_N(\zeta_0')|} \leqq C_1$$

for some C_1 independent of γ_0 or N.

2.1.3 Remarks on the critical set

Recall that in [BC2], the critical set $\mathscr{C}$ is constructed according to some seemingly arbitrary rules. We wish to point out here that $\mathscr{C}$ in fact admits certain intrinsic characterizations. For instance, Corollary 1 gives the following dynamical characterization of $\mathscr{C}$: Let $z_0 \in W$. Then

$$z_0 \text{ lies on a critical orbit} \Leftrightarrow \limsup_{n\to\infty} |DT^n_{z_0}\tau| < \infty \Leftrightarrow \lim_{n\to\infty} |DT^n_{z_0}\tau| = 0 .$$

In fact, $z_0 \in \mathscr{C}$ iff

$$|DT^j_{z_0}\tau| \leqq (5b)^j \ \forall j > 0 \quad \text{and} \quad |DT_{z_{-1}}\tau| > 1 .$$

We mention also that $\mathscr{C}$ admits geometric characterizations. For instance, it is straight-forward to verify using the curvature computations in Sect. 7.6 of [BC2] that $z_0 \in \mathscr{C}$ iff

$$\kappa(z_0) \ll 1 \quad \text{and} \quad \kappa(z_n) > b^{-n} \quad \forall n > 1 .$$

2.2 Some lemmas

The lemmas in Sect. 2.2 are not explicitly stated or proved in [BC2], but their proofs resemble those in [BC2]. We will repeat the shorter arguments and refer the reader to [BC2] for the longer ones.

We will write $w_j(z) := w_j(z, \binom{0}{1})$ and $\tau_j(z) := w_j(z, \tau)$, and let $w_j^*(z)$ and $\tau_j^*(z)$ have the obvious meanings. When we say that a segment $\gamma \subset W$ is "free", it will be assumed implicitly that $\forall z \in \gamma$, (z, τ) is controlled during the time interval $(-\infty, 0)$. The absolute value of the slope of a vector is denoted $s(v)$.

2.2.1 Slope, curvature and derivative estimates

The purpose of this subsection is to prove the following three lemmas:

Lemma 1 *Let γ be a free segment of W. Then*

(i) $\forall z \in \gamma$, $s(\tau(z)) < 2b/\delta$;
(ii) $\forall z \in \gamma \cap ((-\delta, \delta) \times \mathbb{R})$, $s(\tau(z)) < 10b$.

Lemma 2. *Let γ be a free segment of W in $(-\delta, \delta) \times \mathbb{R}$. Then $\kappa(\gamma) < 10b$.*

It follows immediately from these two lemmas that free segments of W in $(-\delta, \delta) \times \mathbb{R}$ are $C^2(b)$ curves. The next lemma is needed in Sect. 3.

Lemma 3 *There exists $c_1 > 0$ such that if $z \in W \cap ((-\delta, \delta) \times \mathbb{R})$ is in a free state, then*

$$|DT_z^{-j}\tau| \leqq e^{-c_1 j} \quad \forall j \geqq 0 .$$

We begin with the following sublemma:

Sublemma 1 *Let z_0 be a point in G_1 near the fixed point $\hat{z}$, and assume that (z_0, τ) is controlled on $[0, j)$. If $z_j \notin (-\delta, \delta) \times \mathbb{R}$ then, $s(\tau_j^*(z_0)) < 2b/\delta$.*

Proof. Proceed by induction exactly as is done in the proof of LI(v')(a) in Sect. 7.3 of [BC2]. □

Proof of Lemma 1 (i) follows immediately from Sublemma 1. To see (ii), note that $T^{-1}z$ is outside of all fold periods. Apply Sublemma 1 and Sect. 1.2.3. □

Sublemma 2 *Let ζ_0 be in generalized tangential position with respect to $z_0 \in \mathscr{C}$. Let p be the bound period between ζ_0 and z_0, and let n be the first free return of ζ_0 to $(-\delta, \delta) \times \mathbb{R}$. Then there is $c > 0$ such that*

(i) $|w_j(\zeta_0)| \geqq e^{cj} \quad \forall j < p$;

(ii) $\dfrac{|w_n(\zeta_0)|}{|w_j(\zeta_0)|} \geqq e^{c(n-j)} \quad \forall j < n$.

Proof. (i) If z_j is outside of all fold periods, then

$$|w_j(\zeta_0)| = |w_j^*(\zeta_0)| \gtrsim |w_j^*(z_0)| \geqq e^{c^* j} .$$

(See 1.7 for "$\gtrsim$" and 1.5.4 for the inequality.) If not, choose $k > j$ such that z_k is outside of all fold periods and $k - j < (C/\log(1/b))j$. (See 1.5.1.) We then have

$$|w_j(\zeta_0)| \geqq 5^{-(C/\log(1/b))j} e^{c^*k} \geqq e^{cj} .$$

(ii) If $j \geqq p$ the estimate follows from 1.2.3(ii) so we only need to consider $j < p$.

Case 1 z_j is outside of all fold periods. We have

$$\frac{|w_n(\zeta_0)|}{|w_j(\zeta_0)|} \gtrsim \frac{|w_n(\zeta_0)|}{|w_p(\zeta_0)|} \cdot \frac{|w_p^*(z_0)|}{|w_j^*(z_0)|} .$$

By 1.2.3 (ii) the first factor is $\geqq e^{c(n-p)}$. To see that the second factor $\geqq e^{c(p-j)}$, let $n_1 < n_2 < \ldots < n_k$ be the times between j and p when z_j returns freely to $(-\delta, \delta) \times \mathbb{R}$. Write $\bar{w}_i(z_0) = DT_{z_{i-1}} w_{i-1}^*(z_0)$. We know from 1.2.3, 1.5.4 and 1.6.1(iii) that every factor in

$$\frac{|w_p^*(z_0)|}{|\bar{w}_{n_k}(z_0)} \cdot \frac{|\bar{w}_{n_k}(z_0)|}{|\bar{w}_{n_{k-1}}(z_0)|} \cdot \ldots \cdot \frac{|\bar{w}_{n_1}(z_0)|}{|w_j^*(z_0)|}$$

is exponential.

Case 2 z_j is in a fold period. Let k be the last time it was outside all fold periods. Because of the rule for construction of the fold periods (see 1.5.1(iii)) ζ_j must still be in fold relation to $z(\zeta_k)$. Let p' and l' be the length of the bound period and fold period resp., initiated at time k. Then $p' < n - k$ since z_n is free and

$$j - k \leqq l' \leqq \frac{Cp'}{\log(1/b)} \leqq \frac{C(n-k)}{\log(1/b)}$$
$$\leqq \frac{C(n-j)}{\log(1/b)} + \frac{C(j-k)}{\log(1/b)} .$$

We move the second term to the left and conclude that

$$j - k \leqq \frac{C'(n-j)}{\log(1/b)} .$$

The same argument as before finishes the proof. □

Proof of Lemma 3 Let $z \in \gamma$. If $T^{-i}z \notin (-\delta, \delta) \times \mathbb{R}\ \forall i > 0$ our exponential estimate follows directly from (1.2.3)(ii). Otherwise suppose $\zeta_0 = T^{-n}z$ is the previous free return. It is enough to verify

$$\frac{|\tau_n(\zeta_0)|}{|\tau_j(\zeta_0)|} \geqq e^{c_1(n-j)}, \quad 0 \leqq j \leqq n .$$

(If there is more then one free return between $T^{-j}z$ and z, repeat the argument and use the chain rule.)

From Sublemma 2 (i) we know that $e_n(\zeta_0)$ is well-defined. We write

$$\tau(\zeta_0) = Ae_n(\zeta_0) + B\binom{0}{1}$$

and let l be the fold period between ζ_0 and $z_0 = z(\zeta_0)$.

Case 1 $j \geqq l$. Since τ splits correctly at ζ_0, we have

$$|DT^k_{\zeta_0}B\binom{0}{1}| \geqq |B|e^{ck} \gg |DT^k_{\zeta_0}Ae_n|$$

$\forall k \geqq l$, and so

$$\frac{|\tau_n(\zeta_0)|}{|\tau_j(\zeta_0)|} \gtrsim \frac{|B||w_n(\zeta_0)|}{|B||w_j(\zeta_0)|} \geqq e^{c(n-j)}$$

by Sublemma 2(ii).

Case 2 $j < l$. At most we have $|\tau_j(\zeta_0)| \leqq 5^l$. Since $l \leqq (C/\log(1/b))n$, we are still guaranteed that

$$\frac{|\tau_n(\zeta_0)|}{|\tau_j(\zeta_0)|} \geqq \frac{e^{cn}}{5^{(C/\log(1/b))n}} \geqq e^{c_1(n-j)} .$$

This completes the proof. □

Proof of Lemma 2 Choose n so that $\zeta_0 = T^{-n}z$ is on the first generation G_1 of W close to the fixed point. Then $\kappa(\zeta_0) \leqq 1$ and the curvature estimate follows immediately from Lemma 3 and 1.2.4. □

2.2.2 Abundance of $C^2(b)$ segments of W near free returns

Lemma 4 *Let $\zeta_0 \in W \cap ((-\delta, \delta) \times \mathbb{R})$ be free. Then for every $m \in \mathbb{Z}^+$ with $3m < \text{gen}(\zeta_0)$, $\exists m'$ with $m < m' \leqq 3m$ and two $C^2(b)$ curves γ and γ' in W with the following properties*:

(i) ζ_0 *is sandwiched between* γ *and* γ' *extending* $\geqq 3\rho^{m'}$ *to each side of* ζ_0;
(ii) $\text{dist}(\zeta_0, \gamma)$, $\text{dist}(\zeta_0, \gamma') \leqq (Cb)^{m'}$, $C = 5e^{72}$;
(iii) *if* η *and* η' *are the two points in* γ *and* γ' *resp. with the same x-coordinate as* ζ_0, *then* $|\tau(\eta) - \tau(\zeta_0)|$, $|\tau(\eta') - \tau(\zeta_0)| \leqq (Cb)^{m'/6}$;
(iv) $\text{gen}(\gamma) = m' + 1$, $\text{gen}(\gamma') \leqq m'$.

The proof of Lemma 4 is virtually identical to that of the corresponding result in [BC2] for z_n (instead of ζ_0), where $z_0 \in \mathscr{C}$ and z_n is a free return to $(-\delta, \delta) \times \mathbb{R}$. (See Sect. 1.4.2.) We sketch here the main steps of the proof, referring to [BC2] for more details:

Outline of proof of Lemma 4

Step 1 Show that $\exists m' \in [m, 3m]$ such that $\zeta_{-m'}$ is in a "favorable" position relative to ζ_0. This means that $\zeta_{-m'}$ is outside of all folding periods, and that $\forall k < m'$, $|\zeta_{-m'+k} - z(\zeta_{-m'+k})| \geqq e^{-36k}$ whenever $\zeta_{-m'+k} \in (-\delta, \delta) \times \mathbb{R}$. (See Lemma 6.6 in [BC2] for a proof.)

Step 2 Show that $|DT^k_{\zeta_{-m'}}\tau| \geqq e^{-36k}$ $\forall k < m'$. Idea of proof: Let $t_1 < \ldots < t_j$ be the free return times of $\zeta_{-m'}$. First note that $s(\tau(\zeta_{-m'})) < 2b/\delta$ (Sublemma 1), so that $|DT^{t_1}_{\zeta_{-m'}}\tau| \geqq e^{ct_1}$. Between t_i and t_{i+1}, use the fact that τ splits correctly, and apply Sublemma 2(i) and 1.6.1(iii). (This is also proved in Lemma 7.2 of [BC2].)

Step 3 Show that there is a contractive vector field $e = e_{m'}$ defined on a strip containing $\zeta_{-m'}$ such that

(i) $|DT^j e| \leqq (Cb)^j \quad \forall j < m, C = 5e^{72}$;
(ii) there is an integral curve joining $\zeta_{-m'}$ to two points η_0^1 and η_0^2 on G_1 and G_2 respectively;
(iii) About $\eta_0^r, r = 1, 2$, there is a segment $\gamma^r \subset W$ extending $\geqq 3\rho^{m'}$ to each side of η_0^r such that each point on γ^1 (resp. γ^2) is joined to a point on G_2 (resp. G_1) by an integral curve of e.

This is proved in Sect. 5.3 of [BC2].

Step 4 Show $T^{m'}\gamma^1$ and $T^{m'}\gamma^2$ contains $C^2(b)$ curves with the properties as claimed. Idea of proof: Again consider the free return times $t_1 < \ldots < t_j$ of $\zeta_{-m'}$. Assuming that γ^r is not so near the "tips", we verify using the criteria in 1.2.4 that $T^{t_1}\gamma^r$ $(r = 1, 2)$ is a $C^2(b)$ curve. Assume now that $T^{t_i}\gamma^r$ contains a $C^2(b)$ curve $\gamma^{r,i}$, trimmed so that it extends by $3\rho^{m'}$ to each side of $\eta_{t_i}^r$. We need to know that τ splits correctly at every point on $\gamma^{r,i}$. Assuming that, we may view $\gamma^{r,i}$ as being tied to $\zeta_{-m'+t_i}$ for the next $t_{i+1} - t_i$ iterates, and the same argument as in Lemma 2 will tell us that $\gamma^{r,i+1}$ is again $C^2(b)$. To see that τ splits correctly on $\gamma^{r,i}$, use the following facts: (i) τ splits correctly at $\zeta_{-m'+t_i}$; (ii) $|\tau(\eta_{t_i}^r) - \tau(\zeta_{-m'+t_i})| < (Cb)^{t_i/6}$ (see Lemma 7.3 in [BC2]), and $(Cb)^{t_i/6} \ll e^{-36t_i} < |\zeta_{-m'+t_i} - z(\zeta_{-m'+t_i})|$; (iii) $\gamma^{r,i}$ is $C^2(b)$ and has length $\ll |\zeta_{-m'+t_i} - z(\zeta_{-m'+t_i})|$. □

Let ζ_0 be as in Lemma 4. Then for every j with $3^{j+1} < \text{gen}(\zeta_0)$, there is m_j with $3^j < m_j \leqq 3^{j+1}$ and $C^2(b)$ curves γ_j and γ'_j with the properties in Lemma 4. In the future we will refer to $\{\gamma_j\}$ and $\{\gamma'_j\}$ as "stacks captured by ζ_0".

2.3 Geometry of the critical set

The purpose of this subsection is to establish some regularity in the structure of the fractal set $\mathscr{C}$.

2.3.1 A basic lemma

Lemma 5 *Let $z_0 \in \mathscr{C}$ be located in a $C^2(b)$ curve $\gamma \subset W$. Assume that γ extends to $\geqq 2d$ to each side of z_0 and let $\zeta_0 \in \gamma$ be such that $|\zeta_0 - z_0| = d$. Then there are no critical points in the disk $B_{d^2}(\zeta_0) := \{z : |z - \zeta_0| \leqq d^2\}$.*

Proof. We will assume that there is a critical point $\tilde{z}_0 \in B_{d^2}(\zeta_0)$ and try to obtain a contradiction.

Let us first assume that $\tilde{z}_0$ lies on a $C^2(b)$ curve $\tilde{\gamma} \subset W$ that extends $\geqq d$ to each side of $\tilde{z}_0$. Using the notation in Sect. 1.3, we choose m with $b^m \ll d^2$ and $2d < (\lambda/5)^m$, and estimate $|\tau(\zeta_0) - e_m(\zeta_0)|$ in 2 different ways:

First note that e_m is defined on all of $B_{2d}(z_0)$, and that for purpose of comparing $e_m(z_0)$ with $e_m(\zeta_0)$, we may assume $e_m(z_0) = \tau(z_0)$ (see 1.3.4). We therefore have

$$|\tau(\zeta_0) - e_m(\zeta_0)| = (2a + \mathcal{O}(b))d \,.$$

Next, we let $\tilde{\zeta}_0$ be the point in $\tilde{\gamma}$ with the same x-coordinate as ζ_0 and write

$$|e_m(\zeta_0) - \tau(\zeta_0)| \leqq |e_m(\zeta_0) - e_m(\tilde{z}_0)| + |e_m(\tilde{z}_0) - \tau(\tilde{\zeta}_0)| + |\tau(\tilde{\zeta}_0) - \tau(\zeta_0)| .$$

The first two terms are $\leqq 5d^2$ (again see 1.3.4). As for the third term, since γ and $\tilde{\gamma}$ are nonintersecting $C^2(b)$ curves of length $\geqq d$ and $|\zeta_0 - \tilde{\zeta}_0| \leqq d^2$, an easy computation gives $|\tau(\zeta_0) - \tau(\tilde{\zeta}_0)| \leqq 2d$. These three terms add up to $2d + 10d^2$, which is $< (2a + \mathcal{O}(b)) \cdot d$, and we obtain a contradiction.

Now the $C^2(b)$ curve containing $\tilde{z}_0$ may not be long enough. In that case we use the rules of construction of $\mathscr{C}$ (see 1.3.1) to successively obtain critical points $\tilde{z}_0^{(1)}, \tilde{z}_0^{(2)}, \ldots, \tilde{z}_0^{(k)}$ of lower and lower generation. Let G be such that $\rho^{2G} \leqq d \leqq \rho^G$, and let i be the smallest integer with $\text{gen}(\tilde{z}_0^{(i)}) < G$. Then

$$|\tilde{z}_0^{(i)} - \tilde{z}_0| \leqq \sum_{j = \text{gen}(\tilde{z}_0^{(i-1)})}^{\infty} (b^{\sigma j})^{\frac{1}{2}} \leqq 2b^{\sigma G/2} ,$$

which we may assume is $< d^2$ and $\tilde{z}_0^{(i)}$ lies on a $C^2(b)$ curve extending $\geqq 3\rho^G > d$ to each side on $\tilde{z}_0^{(i)}$. The previous argument can now be repeated with $\tilde{z}_0^{(i)}$ in place of $\tilde{z}_0$. □

2.3.2 View of $\mathscr{C}$ from a point in the free state

Lemma 6 *Let $\zeta \in G_n \cap (-\delta, \delta) \times \mathbb{R}$ be free, and let z_0 and z'_0 be two critical points wrt which ζ is in tangential position. Assume that $|\zeta - z_0|, |\zeta - z'_0| \geqq b^{\frac{n}{100}}$. Then the x-coordinate of ζ cannot lie between those of z_0 and z'_0.*

Proof. We assume this scenario occurs. Let $d = |\zeta - z_0|$, $d' = |\zeta - z'_0|$, and suppose that $d \leqq d'$. Let η be the point on $W^u_{\text{loc}}(z'_0)$ with the same x-coordinate as z_0, and let $D = |\eta - z_0|$. We will show that $D \ll d^2 + d'^2 < (d + d')^2$, which will contradict Lemma 5.

The difficulty in comparing $W^u_{\text{loc}}(z_0)$ and $W^u_{\text{loc}}(z'_0)$ directly is that the two $C^2(b)$ curves may be too far apart and one of them may not be long enough. (We had a similar problem in the proof of Lemma 5.) So again we rely on curves captured by ζ.

First we use ζ to capture a segment W_0 of W of length $> d$ and with $\text{dist}(\zeta, W_0) < d^4$. (This is clearly possible, even with $d \approx b^{\frac{n}{100}}$.) We further require that W_0 be on the opposite side of ζ as $W^u_{\text{loc}}(z_0)$. (See Lemma 4.) Let ζ_{z_0} and ζ_{W_0} be the points on $W^u_{\text{loc}}(z_0)$ and W_0 respectively with the same x-coordinate as ζ. Then $|\zeta_{z_0} - \zeta_{W_0}| < 2d^4$; and $|\tau(\zeta_{z_0}) - \tau(\zeta_{W_0})| < 2d^2$ because $W^u_{\text{loc}}(z_0)$ and W_0 are nonintersecting $C^2(b)$ curves that extend $> d$ to each side of ζ. Putting this together we have

$$D_0 := \text{dist}(z_0, W_0) < 2d^4 + 2d^2 \cdot d + 20b \cdot d^2 \ll d^2 .$$

Similarly we let W'_0 be a segment captured by ζ on the opposite side of $W^u_{\text{loc}}(z'_0)$. We require that W'_0 has length $> d'$ and $\text{dist}(\zeta, W'_0) < d'^4$, so that

$$D'_0 := \text{dist}(\eta, W'_0) < 2d'^4 + 2d'^2 \cdot d + 20b \cdot d^2 \ll d'^2 .$$

Now D is easily estimated as follows: If $W^u_{\text{loc}}(z_0)$ and $W^u_{\text{loc}}(z'_0)$ are on the same side of ζ, then $D \leqq \max(D_0, D'_0)$. If $W^u_{\text{loc}}(z_0)$ and $W^u_{\text{loc}}(z'_0)$ are on opposite sides of ζ, then $D \leqq D_0 + D'_0$. □

Consider a point $\zeta \in (-\delta, \delta) \times \mathbb{R}$ in the free state. Let $\{\gamma_j\}_{j=1}^{k}$ be a stack captured by ζ as discussed in 2.2.2. We may assume that there is a critical point on γ_1. Let $j(\zeta)$ be such that $\gamma_{j(\zeta)}$ is the γ_j of highest generation that contains a critical point, and call this critical point $\hat{z}(\zeta)$. We now fix some terminology to describe the location of $\mathscr{C}$ relative to ζ: We say that $\mathscr{C}$ is "*in the middle*" if there exists a stack $\{\gamma_j\}_{j=1}^{k}$ captured by ζ with the property that $j(\zeta) = k$ and $|\zeta - \hat{z}(\zeta)| < b^{\mathrm{gen}(\zeta)/100}$. If this is not the case, then we say that $\mathscr{C}$ is "*on the left*" (or "*on the right*") if for every stack captured by ζ, $\hat{z}(\zeta)$ lies on the left half (resp. right half) of $\gamma_{j(\zeta)}$.

We remark that if $\mathscr{C}$ is not "in the middle", then it has to be either "on the left" or "on the right" of ζ, because ζ is always in tangential position wrt $\hat{z}(\zeta)$, and Lemma 6 applies.

2.4 Proof of main proposition

2.4.1 General strategy

Let G_n be the leaf of generation n (see 1.3.1 for definition). We assume it has been proved that for all $\zeta_0 \in G_n$ the pair (ζ_0, τ) is controlled on the time interval $(-\infty, 0)$, and will show that (ζ_0, τ) is controlled on $(-\infty, 0]$. Clearly, we need only to consider $\zeta_0 \in (-\delta, \delta) \times \mathbb{R}$ and may assume that $\zeta_0 \notin T^i \mathscr{C}$ for any $i \geqq 0$.

If ζ_0 is in a bound state, let ζ_{-i} be the last time it was free and let $\tilde{z}_0 = z(\zeta_{-i})$. Then ζ_0 is in tangential position wrt $z(\tilde{z}_i)$. Moreover, since $(\zeta_{-i}, \binom{0}{1})$ is controlled on the time interval $[0, i]$ (see Sect. 1.5.3) and $\tau(\zeta_{-i})$ splits correctly into $A_{-i}e + B_{-i}\binom{0}{1}$ by our induction hypothesis, the rejoining of the A_{-i}-term (if it has already taken place) has negligible effect. This proves that $\tau(\zeta_0)$ again splits correctly and we are done.

The case where ζ_0 is free is handled in the following lemma:

Lemma 7 *Let γ be a maximal free segment of G_n. If $\gamma \cap ((-\delta, \delta) \times \mathbb{R}) \neq \emptyset$, then there is a critical point z_0 with respect to which every $\zeta \in \gamma \cap ((-\delta, \delta) \times \mathbb{R})$ is in generalized tangential position. Moreover, if $\tau(\zeta) = A(\zeta)e + B(\zeta)\binom{0}{1}$ is the spilitting in* 1.5.2 *with respect to our binding point $z_0 := z(\gamma)$, then*

$$3|\zeta - z_0| \leqq |B(\zeta)| \leqq 5|\zeta - z_0| \, .$$

Let γ_- and γ_+ denote the left and right endpoints of γ respectively. (This makes sense since free segments are roughly horizontal.) If $\gamma_+ \in (-\delta, \delta) \times \mathbb{R}$, then it is attached to some $\tilde{z}_+ \in \mathscr{C}$ because it is also in a bound state. Similarly, let $\tilde{z}_- = z(\gamma_-)$ if $\gamma_- \in (-\delta, \delta) \times \mathbb{R}$. The following are all the possible geometric configurations:

Case 1 γ is stretched across $(-\delta, \delta) \times \mathbb{R}$, i.e. neither γ_+ nor γ_- is in $(-\delta, \delta) \times \mathbb{R}$. We will show that z_0 lies on γ.

Case 2 γ_+ is in $(-\delta, \delta) \times \mathbb{R}$ and $\tilde{z}_+$ lies to the left of γ_+; $\gamma_- \notin (-\delta, \delta) \times \mathbb{R}$. As in Case 1, we will show that $z_0 \in \gamma$.

Case 3 Both γ_- and γ_+ are in $(-\delta, \delta) \times \mathbb{R}$; $\tilde{z}_-$ lies to the right of γ_- and $\tilde{z}_+$ to the left of γ_+. We will also show that $z_0 \in \gamma$.

Case 4 $\gamma_+ \in (-\delta, \delta) \times \mathbb{R}$ and $\tilde{z}_+$ is to the right of γ_+. In this case we will show that $\tilde{z}_+$ is a viable candidate for z_0.

2.4.2 *Proof of Lemma 7*

We continue to use the terminology introduced in Sect. 2.3.2. For $n \in \mathbb{Z}^+$, let $\Delta_n > 0$ be sufficiently small that the following holds: Let γ be a free segment in $G_n \cap (-\delta, \delta) \times \mathbb{R}$, and let $\zeta = (\xi, \eta)$ and $\zeta' = (\xi', \eta')$ be two points on γ with $\xi - \Delta_n \leqq \xi' \leqq \xi$. If $\mathscr{C}$ is "on the left" of ζ, then it is either "on the left" or "in the middle" for ζ'. We assume also the analogous statement if $\mathscr{C}$ is "on the right" of ζ'. The existence of Δ_n follows from the proof of Lemma 6. In the following $\hat{z}(\zeta)$ is the "captured critical point" as defined in Sect. 2.3.2.

We now deal with the four cases discussed in Sect. 2.4.1.

Case 1 Since $\gamma \cap (-\delta, \delta) \times \mathbb{R}$ is a $C^2(b)$ curve (Corollary to Lemmas 1 and 2), once we produce a critical point $z_0 \in \gamma$, the rest of the assertion will follow. To product z_0 we start with $\zeta^{(0)}$ on γ with x-coordinate $= \frac{1}{2}\delta$. Then $\hat{z}(\zeta^{(0)})$ must be on the left. We move left along γ by steps of Δ_n to obtain successively $\zeta^{(1)}, \zeta^{(2)}, \ldots$. If for some k, $\mathscr{C}$ is "in the middle", then we are done, because some point on γ "sees" $\hat{z}(\zeta^{(k)})$ (see Sublemma 3 below) and the rules of construction of $\mathscr{C}$ says that a critical point must have been constructed on γ. Clearly we cannot move left indefinitely and continue to have $\mathscr{C}$ "on our left."

Sublemma 3 *Let $\zeta = \zeta^{(k)}$ and γ are as above. Then vert* $\text{dist}(\hat{z}(\zeta), \gamma) < b^{\frac{1}{540}n}$.

Proof. Let η be the point on γ with the same x-coordinate as $\hat{z}(\zeta)$, and let $\hat{\zeta}$ be the point on $\gamma_{j(\zeta)}$ with the same x-coordinate as ζ. Then

$$|\hat{z}(\zeta) - \eta| \leqq |\zeta - \hat{\zeta}| + 20b|\hat{z}(\zeta) - \hat{\zeta}|$$

$$< (Cb)^{\frac{n}{9}} + 20b \cdot b^{\frac{n}{100}} \ll b^{\frac{n}{540}}. \qquad \square$$

Case 2 Let $d_+ = |\gamma_+ - \tilde{z}_+|$. We need the following simple estimate:

Sublemma 4 *Let γ be as in Case 2, and let ζ be a point on γ with $|\zeta - \gamma_+| < d_+$. Then*

$$\text{dist}(\zeta, W^u_{\text{loc}}(\tilde{z}_+)) \ll d_+^2 .$$

Proof. Use γ_+ to capture an unstable leaf γ_j on the opposite side of $W^u_{\text{loc}}(\tilde{z}_+)$ with length$(\gamma_j) \approx d_+$. Let η_j and $\hat{\eta}$ be the points on γ_j and $W^u_{\text{loc}}(\tilde{z}_+)$ resp with the same x-coordinate as γ_+. Then $|\hat{\eta} - \gamma_+| < d_+^4$ because γ_+ is in tangential position wrt $\tilde{z}_+$, and $|\eta_j - \gamma_+| < (Cb)^{m_j} \ll d_+^4$. Also, γ_j and $W^u_{\text{loc}}(\tilde{z}_+)$ are long enough to guarantee that $|\tau(\eta_j) - \tau(\hat{\eta})| < d_+^2$. So

$$\text{dist}(\zeta, W^u_{\text{loc}}(\tilde{z}_+)) < 2d_+^4 + 2d_+^2 \cdot d_+ + (20b)d_+^2 \ll d_+^2 . \qquad \square$$

Let $R_+ = \{z \in \gamma : |z - \gamma_+| < \frac{1}{2}d_+\}$. We note that if $\zeta \in R_+$ then $\mathscr{C}$ cannot be "in the middle". This is because $d_+ > e^{-\alpha n}$, which is $\gg b^{\frac{n}{100}}$, and if $|\hat{z}(\zeta) - \zeta| < b^{\frac{n}{100}}$ for any stack, then we will have

$$\text{horiz dist}(\hat{z}(\zeta), \tilde{z}_+) > \tfrac{1}{3}d_+ ,$$

and

$$\text{dist}(\hat{z}(\zeta), W^u_{\text{loc}}(\tilde{z}_+)) \ll d_+^2,$$

which contradicts the geometry of the critical set.

We start with $\zeta^{(0)} = \gamma_+$. First we claim that $\mathscr{C}$ is on the left: "middle" has been ruled out in the last paragraph, and "right" is not compatible with the position of $\tilde{z}_+$ (Lemma 6). We move left by steps of Δ_n as before until we reach $\zeta^{(k)}$ with $\mathscr{C}$ in the middle. Now $\zeta^{(k)} \notin R_+$, which guarantees that γ extends $> \frac{1}{2}\, e^{-\alpha n} \gg 2\rho^n$ to each side of $\zeta^{(k)}$, and so a critical point z_0 on γ is assured. The correct splitting of τ wrt z_0 is automatic as before.

Case 3 Wc begin as in Case 2, taking $\zeta^{(0)} = \gamma_+$ and moving left in small steps. We must reach some $\zeta^{(k)}$ with $\hat{z}(\zeta^{(k)})$ in the middle before arriving at γ_-, otherwise $\hat{z}(\gamma_-)$ would be on the left, contradicting the fact that $\tilde{z}_-$ is to the right of γ_-. Now $\zeta^{(k)} \notin R_+ \cup R_-$, where R_- has the obvious definition. So again a critical point on γ is guaranteed.

Case 4 Let $d = |\gamma_+ - \tilde{z}_+|$, and let η be the point on $W^u_{\mathrm{loc}}(\tilde{z}_+)$ with the same x-coordinate as γ_+. Then $|\eta - \gamma_+| < 2d^4$ and $|\tau(\eta) - \tau(\tau_+)| < d^2$ (an exercise: cf. the proof of Sublemma 4). We need to show that every point in γ is in a generalized tangential position with respect to $\tilde{z}_+$. Let (x', y') be the natural coordinate system at $\tilde{z}_+$ (see 1.6.2) and let φ be the function whose graph is the curve γ. Then

$$|\varphi(-d)| < 2d^4 + (10b)d^2 \ll \tfrac{1}{100}d^2\,,$$

$$|\varphi'(-d)| < d^2 + (10b)d \ll \tfrac{2}{100}d\,,$$

and $|\varphi''| \leqq 10b$. This proves that $|\varphi(x')| < \frac{1}{100}\, x'^2$ for all the relevant x'.

We also need to know that τ splits "correctly" at every point on γ. This is true because $\tau(\gamma_+)$ splits correctly (it is in a bound state) and γ is a $C^2(b)$ curve.

This completes the proof of Lemma 7. □

We remark that an alternate proof of Lemma 7 is to try to carry out the "capture" argument in Lemma 4 simultaneously for all points on γ.

2.4 Proof of Corollary 1

Part I of Corollary 1 follows from Lemma 1 and Sect. 1.2. Part II follows from the correct splitting guaranteed in Proposition 1, the properties of $\mathscr{C}$ as discussed in 1.3.3, the distortion estimates in 1.7.1, and the bound period estimates in 1.6.

2.5 Proof of Proposition 2

A proof of Proposition 2 is given in Sects. 2 and 8 of [BC2]; see Lemma 8.9 in particular. Because the arguments there are a bit sketchy, and this estimate is of central importance in the construction of SBR measures, we are going to fill in some details. We will assume the distortion estimate for $DT^j_{(\cdot)}\binom{0}{1}$ during bound periods (see 1.7.1), and try to explain how these estimates can be used to control the distortion of τ_j over arbitrarily long periods of time. The fact that (ζ, τ) is controlled for all ζ in question is used implicitly throughout.

We write

$$\log \frac{|\tau_N(\zeta_0)|}{|\tau_N(\zeta'_0)|} = \sum_{k<q} S'_k + \sum_{k<q} S''_k,$$

where

$$S'_k = \log \frac{|\tau_{p_k}(\zeta_{t_k})|}{|\tau_{p_k}(\zeta'_{t_k})|} \quad \text{and} \quad S''_k = \log \frac{|\tau_{t_{k+1}-p_k}(\zeta_{t_k+p_k})|}{|\tau_{t_{k+1}-p_k}(\zeta'_{t_k+p_k})|}.$$

First we show that $\sum S''_k <$ some C''_1. Consider j with $t_k + p_k \leqq j < t_{k+1}$. Since γ_j is free and free segments have uniformly bounded slopes and curvatures, it follows that

$$|DT_{\zeta_j}\tau - DT_{\zeta'_j}\tau| \leqq \text{const}\,|\gamma_j|.$$

So

$$S''_k \lesssim \sum_{j=t_k+p_k}^{t_{k+1}-1} \frac{|DT_{\zeta_j}\tau - DT_{\zeta'_j}\tau|}{|DT_{\zeta'_j}\tau|} \leqq \frac{1}{\delta}\cdot \text{const} \sum_j |\gamma_j|,$$

which is $< \text{const}\,|\gamma_{t_{k+1}}|$ because Corollary 1 (part I) tells us that

$$|\gamma_{t_{k+1}}| \geqq e^{c_0(t_{k+1}-j)}|\gamma_j|.$$

Using Corollary 1 again (both parts I and II) we conclude that

$$|\gamma_{t_{k+1}}| \geqq |\gamma_{t_k+p_k}| \geqq 2|\gamma_{t_k}|.$$

Hence

$$\sum S''_k < \text{const}\,|\gamma_N| < \text{some } C''_1.$$

To estimate S'_k we first prove the following

Sublemma 5 *Let $\eta_0, \eta'_0 \in W$ lie in the same $I_{\mu j}$ with respect to some $\tilde{z}_0 \in \mathscr{C}$, and let p be their common bound period. Then there is $C > 0$ not depending on η_0, η'_0 or z_0 such that*

$$\log \frac{|\tau_p(\eta_0)|}{|\tau_p(\eta'_0)|} \leqq C \frac{|\eta_0 - \eta'_0|}{e^{-\mu}}.$$

Throughout the proof we will use c to denote a generic constant that is positive and very small.

Proof of Sublemma 5. Split

$$\tau(\eta_0) = Ae + B\binom{0}{1},$$

$$\tau(\eta'_0) = A'e' + B'\binom{0}{1},$$

where e and e' are the contractive (unit) vectors for the period $[0, p]$. We will use the notation

$$e_j := DT^j_{\eta_0}e, \qquad e'_j := DT^j_{\eta'_0}e',$$

$$w_j := DT^j_{\eta_0}\binom{0}{1}, \qquad w'_j := DT^j_{\eta'_0}\binom{0}{1}.$$

Let R_θ be rotation by θ, and let θ be chosen such that R_θ carries w_p to a positive multiple of w'_p. We will write

$$\frac{|\tau_p(\eta_0)|}{|\tau_p(\eta'_0)|} \leqq \text{(I)} + \text{(II)} + \text{(III)} + \text{(IV)} + \text{(V)},$$

where (I)–(V) are defined and estimated as follows:

$$(\mathrm{I}) = \frac{\left| R_\theta B w_p + \dfrac{|Bw_p|}{|B' w'_p|} A' e'_p \right|}{|B' w'_p + A' e'_p|} = \frac{|Bw_p|}{|B' w'_p|},$$

where

$$\frac{|B|}{|B'|} \leqq 1 + 5 \frac{|\eta_0 - \eta'_0|}{e^{-\mu}}$$

because of the way τ and e change with z, and

$$\frac{|w_p|}{|w'_p|} \leqq 1 + 2C_0 C'_0 \frac{|\eta_0 - \eta'_0|}{e^{-\mu}}$$

by 1.7.1(ii), (iii). (Note that the constants C_0 and C'_0 are purely numerical so

$$C_0 C'_0 \frac{|\eta_0 - \eta'_0|}{e^{-\mu}}$$

is small if μ_0 is chosen sufficiently large.)

$$(\mathrm{II}) = \frac{1}{|\tau_p(\eta'_0)|} \cdot \left| \frac{|Bw_p|}{|B' w'_p|} A' e_p - \frac{|Bw_p|}{|B' w'_p|} A' e'_p \right| \leqq |e_p - e'_p|,$$

which is $< c|\eta_0 - \eta'_0|$ by Lemma 5.5 of [BC2].

$$(\mathrm{III}) = \frac{1}{|\tau_p(\eta'_0)|} \cdot \left| A' e_p - \frac{|Bw_p|}{|B' w'_p|} A' e_p \right| \leqq \left| 1 - \frac{|Bw_p|}{|B' w'_p|} \right|,$$

which is $< \mathrm{const}|\eta_0 - \eta'_0| e^{\mu}$ (same as (I)).

$$(\mathrm{IV}) = \frac{1}{|\tau_p(\eta'_0)|} \cdot |R_\theta A' e_p - A' e_p| \leqq |\theta|,$$

which is $< c\Delta_p \leqq cC'_0 |\eta_0 - \eta'_0| e^{\mu}$ by 1.7. (ii) and (iii).

$$(\mathrm{V}) = \frac{1}{|\tau_p(\eta'_0)|} \cdot |R_\theta A e_p - R_\theta A' e_p| \leqq c|A' - A| \leqq c|\eta_0 - \eta'_0|.$$

Together these estimates prove the sublemma. □

Applying this sublemma to each S'_k, we obtain

$$\sum_{k=0}^{q-1} S'_k \leqq C \sum_{k=0}^{q-1} \frac{|\gamma_{t_k}|}{e^{-\mu_k}},$$

where $\gamma_{t_k} \subset I_{\mu_k j_k}$. To estimate this sum, we let $m(\mu) = \max\{t_k : \mu_k = \mu\}$ for each μ, and use the fact that $|\gamma_{t_{k+1}}| \geqq 2|\gamma_{t_k}|$ to conclude that

$$\sum_{k<q} \leqq \frac{|\gamma_{t_k}|}{e^{-\mu_k}} \leqq \mathrm{const} \sum_{\mu} \frac{|\gamma_{m(\mu)}|}{e^{-\mu}} \leqq \mathrm{const} \sum_{\mu} \frac{1}{\mu^2}.$$

This completes the proof of Proposition 2. □

3 Construction of SBR-measures

We continue to assume that $T = T_{a,b}$, where (a, b) is one of the "good" parameters. Our strategy is as follows. Put Lebesgue measure m on a piece of W. Transport m forward by T, and take the ergodic averages of these measures. We will show that any limit point of these ergoidic averages contains at least one component that has absolutely continuous conditional measures on unstable manifolds. This construction is standard for Axiom A attractors. The piecewise uniformly hyperbolic case is dealt within e.g. [Y2], which contains a simple version of what is done here.

3.1 Bookkeeping on the unstable manifold

We showed in Lemma 7, Sect. 2, that every maximal free segment γ in $W \cap ((-\delta, \delta) \times \mathbb{R})$ is assigned to a critical point with respect to which it is in generalized tangential position. This critical point will be denoted by $z(\gamma)$. On γ, it is natural to consider the partition $\mathscr{P}_{[z(\gamma)]}$, where $\mathscr{P}_{[\cdot]}$ is the partition defined in Sect. 2.1.2. In the construction below we will often speak of $I_{\mu j}$ on γ without explicit mention of $\mathscr{P}_{[z(\gamma)]}$.

We select a piece of W on which to begin our construction. Let, for instance, z_0 be the critical point on G_1, and let $\Delta \subset G_1 \cap ((-\delta, \delta) \times \mathbb{R})$ correspond to $(e^{-(\mu_0+1)}, e^{-\mu_0})$. Let $\mathscr{P}_0 = \mathscr{P}_{[z_0]}|\Delta$. We will describe in the next few paragraphs a sequence of partitions $\mathscr{P}_0 \prec \mathscr{P}_1 \prec \mathscr{P}_2 \ldots$, such that each $\mathscr{P}_n$ divides Δ into a countable number of intervals – or curve segments rather. Points in $\mathscr{P}_n(z)$ can be regarded as having trajectories "indistinguishable" from that of z up to time n.

We consider one element ω of $\mathscr{P}_0$ at a time. Regarding ω as bound to $z(\omega) = z_0$, we let $j_1 > 0$ be the first time when $T^j\omega$ is free and intersects $(-\delta, \delta)$. (We may assume that all points in $T^j\omega$ become free simultaneously.) If $T^{j_1}\omega$ contains some $I_{\mu j}$ then we let $k_1 = j_1$ and go to the next paragraph. If not, then we consider $T^{j_1}\omega$ as bound to $z(T^{j_1}\omega)$ and wait for it to return again in a free state, say at time j_2. From Corollary 1 in Sect. 2 it is clear that $|T^{j_2}\omega| \geqq 2|T^{j_1}\omega|$, so repeating this process for at most a finite number of times, there will be a free return at time j_l, when $T^{j_l}\omega \supseteq$ some $I_{\mu j}$. Set $k_1 = j_l$.

We now define $\mathscr{P}_n|\omega$ for $n \leqq k_1$. Let $\mathscr{P}_n = \mathscr{P}_0$ for $n < k_1$. For $n = k_1$, we first let $\mathscr{P}'_{k_1}|\omega = T^{-k_1}(\mathscr{P}_{[z(T^{k_1}\omega)]} \cup \{(-1, -\delta), (\delta, 1)\})$, and obtain $\mathscr{P}_{k_1}$ from $\mathscr{P}'_{k_1}$ by adjoining each of the two end intervals on $\mathscr{P}'_{k_1}|\omega$ to its neighbor unless the T^{k_1}-image of this end interval lies outside of $(-\delta, \delta)$ and has length $\geqq |I_{\mu_0 j}|$.

We then repeat the argument in the last two paragraphs for each element ω' of $\mathscr{P}_{k_1}$. That is, if k_2 is the first time after k_1 when part of $T^{k_2}\omega'$ returns freely to $(-\delta, \delta)$ and $T^{k_2}\omega' \supset$ some $I_{\mu j}$ then we cut up ω' again at this time according to the locations of $T^{k_2}z$.

Next we introduce a sequence of stopping times $t_0 < t_1 \ldots$ on Δ. Let Δ^+ and Δ^- be the rightmost and leftmost intervals in the partition $\mathscr{P}$ of $(-\delta, \delta)$. (We may assume that Δ^+ and Δ^- are fixed intervals not depending on the location of critical points.) Let $t_0 \equiv 0$. For $z \in \Delta$, we define $t_1(z)$ to be the smallest $k > 0$ such that $T^k(\mathscr{P}_{k-1}(z))$ contains either Δ^+ or Δ^-, $t_2(z)$ to be the smallest $k > t_1(z)$ when $T^k(\mathscr{P}_{k-1}(z))$ contains either Δ^+ or Δ^-, and so on. Note that $t_n(z)$ could take on the value ∞, since it is possible for a point to keep returning to the shorter intervals which get cut before they get a chance to grow long. We will prove in Sect. 3.3, however, that this is an extremely improbable event.

Our construction of $\mathscr{P}_n$ is virtually identical to that of a similar partition in Sect. 1 of [BC2] – except of course that our construction takes place on W whereas the one in [BC2] is carried out in parameter space. Our t_n's correspond essentially (though not exactly) to the "escape times" in [BC2].

3.2 Derivative estimates

Let m denote Lebesgue measure on W, i.e. if $\gamma \subset W$ is a curve segment, then $m(\gamma)$ is equal to the arc length of γ. Let $T^j_*(m|\Delta)$ denote the measure with $T^j_*(m|\Delta)(E) = m(T^{-j}E \cap \Delta)$. Clearly the density of $T^j_*(m|\Delta)$ on $T^j\Delta$ is given by

$$\frac{dT^j_*(m|\Delta)}{dm}(z) = |DT_z^{-j}\tau| \,.$$

Our first task is to study how $T^j_*(m|\Delta)$ is distributed along W for $j = 1, 2, \ldots$ For this we use the derivative estimates in Corollary 1. The distortion estimate in Proposition 2 is crucial for controlling local fluctuations in densities along certain segments of W. This will be important in our construction. Lemma 3 will also be used.

3.3 Frequency of returns

The aim of this section is to show that a positive measure set of points in Δ return with positive frequency to $\Delta^+ \cup \Delta^-$.

3.3.1 First escape time estimates

Consider first the 1-d map $f:[-1, 1]\circlearrowright$ satisfying the conditions in Sect. 1.1. Let γ be an interval in $[-1, 1]$. We assume that γ is either $\approx I_{rj}$ for some r, j, or $\gamma \cap (-\delta, \delta) = \emptyset$. (The notation $\gamma \approx I_{rj}$ means that $I_{rj} \subset \gamma \subset I^+_{rj}$, where I^+_{rj} is defined to be the union of I_{rj} and its two adjacent intervals.) We define the *first escape time function* $t|_\gamma$ exactly as t_1 is defined in Section 3.1 – except that we start from γ. This definition is related to our earlier definition of stopping times $t_1 < t_2 < \ldots$ on Δ (had we defined them for our interval map) by

$$t_{i+1}(x) - t_i(x) = (t|_{f^{t_i}(\mathscr{P}_{t_i}(x))})(x) \,.$$

Note that if $\gamma = f^{t_i}(\mathscr{P}_{t_i}(x))$ for some x and does not intersect $(-\delta, \delta)$, then it has δ or $-\delta$ as one of its end points and has length $\geqq |\Delta^+|$. We claim that for such an interval γ, $t|_\gamma$ is constant on γ and is $\leqq M$ for some M independent of γ. To see this let k_1 be the first time $f^{k_1}\gamma \supset$ some $I_{\mu j}$. Using bound estimates similar to those for $\Delta^\pm$ it is easy to verify that $k_1 \leqq M = C\log(1/\delta)$ and that $f^j\gamma \geqq C/\log(1/\delta) \gg 2\delta$. Therefore $f^{k_1}\gamma$ must contain either Δ^+ or Δ^- and $t|_\gamma = k_1$.

When $\gamma \approx I_{rj}$, the following large deviation estimate for $t|_\gamma$ is proved in Sect. 2.2 of [BC2].

Lemma 8 [BC2] *For $\gamma \approx I_{rj}$ and $n \geqq 6r$*

$$m\{x \in \gamma : t|_\gamma(x) \geqq n\} \leqq e^{-n/20} m(\gamma) \,.$$

Because of the close correspondence between the derivative estimates in 1 and 2 dimensions (see Corollary 1), these first escape time estimates apply without change to free segments $\gamma \subset T^k \Delta$. We now derive from these estimates the main lemma of Sect. 3.3.

3.3.2 *A lemma*

Lemma 9 *There is a constant C^* such that for all $i \geqq 0$,*

$$\int_{\Delta} (t_{i+1} - t_i)\, dm \leqq C^* .$$

Proof. For $i \geqq 1$, let $\hat{\mathscr{P}}_i$ be the partition of Δ into "distinguishable orbits" up to time $t_i - 1$, i.e. $\hat{\mathscr{P}}_i$ refines the partition of Δ by values of t_i and

$$\hat{\mathscr{P}}_i|_{\{t_i = k\}} = \mathscr{P}_{k-1}|_{\{t_i = k\}} .$$

This means in particular that for each $\omega \in \hat{\mathscr{P}}_i$, $T^{t_i}\omega \supset \Delta^+$ or Δ^-. Also, since $m\{t_i = \infty\} = 0$, $\hat{\mathscr{P}}_i$ is a genuine partition of Δ up to a set of measure 0. It suffices to give a uniform upper bound for

$$\frac{1}{m(\omega)} \int_{\omega} (t_{i+1} - t_i)\, dm, \quad \omega \in \hat{\mathscr{P}}_i .$$

We fix $\omega \in \hat{\mathscr{P}}_i$ and let $\gamma = T^{t_i}\omega$ Then

$$\int_{\gamma} (t_{i+1} - t_i) \circ T^{-t_i}\, dm \leqq \sum_{r,j} \int_{I_{r,j}} t|_{I_{r,j}} + \int_{\gamma_-} t|_{\gamma_-} + \int_{\gamma_+} t|_{\gamma_+} ,$$

where γ_- and γ_+ are those parts of γ in $(-1, -\delta) \times \mathbb{R}$ and $(\delta, 1) \times \mathbb{R}$ resp. From Lemma 8, we see that

$$\begin{aligned}\sum_{r,j} \int_{I_{r,j}} t|_{I_{r,j}}\, dm &\geqq \sum_{r,j} \sum_{n=0}^{\infty} m\{z \in I_{r,j} : t|_{I_{r,j}}(z) \geqq n\} \\ &\leqq \sum_{n=0}^{\infty} \left\{ m\left(\bigcup_{|r| \geqq n/6} I_{r,j} \right) + e^{-\frac{n}{20}} m\left(\bigcup_{|r| \leqq n/6} I_{r,j} \right) \right\},\end{aligned}$$

so that

$$\begin{aligned}\frac{1}{m(\gamma)} \int_{\gamma} (t_{i+1} - t_i) \circ T^{-t_i}\, dm &\leqq \frac{1}{|\Delta^+|} \left\{ \sum_{n=0}^{\infty} (2e^{-\frac{n}{6}} + e^{-\frac{n}{20}}(2\delta)) + 2M \right\} \\ &\leqq \text{some } C .\end{aligned}$$

Our desired estimate then is given by

$$\begin{aligned}\frac{1}{m(\omega)} \int_{\omega} (t_{i+1} - t_i)\, dm &= \frac{1}{m(\omega)} \int_{\gamma} (t_{i+1} - t_i) \circ T^{-t_i} d(T^{t_i}(m|\omega)) , \\ &\leqq C_1 C ,\end{aligned}$$

where C_1 is the distortion constant in Proposition 2. The case $i = 0$ is simple. □

3.3.3 A lower bound on the frequency of returns to $\Delta^{\pm}$

Let $\mathscr{P}_{n,\Delta^{\pm}} = \bigcup\{\omega\in\mathscr{P}_n : T^n\omega = \Delta^+ \text{ or } \Delta^-\}$.

Lemma 10 *There exists a constant $\alpha^* > 0$ such that*

$$\varliminf_{N\to\infty} \frac{1}{N}\sum_{n=0}^{N} m\mathscr{P}_{n,\Delta^{\pm}} \geqq \alpha^* > 0 .$$

Proof. From Lemma 9 we know that

$$\int_{\Delta} t_n\, dm \leqq C^* n$$

and so

$$m\{z\in\Delta : t_n(z) \leqq 2nC^*\} \geqq \tfrac{1}{2}m(\Delta) .$$

This means that

$$\sum_{k=1}^{2nC^*} m\{z\in\Delta : k = t_i(z) \text{ some } i\} \geqq n\cdot\tfrac{1}{2}m(\Delta) .$$

Now for each $\omega\in\mathscr{P}_{k-1}$ with $k = t_i(\omega)$, there is $\omega'\in\mathscr{P}_k$, $\omega'\subset\omega$, such that $T^k\omega \approx \Delta^+$ or Δ^-. Moreover

$$m(\omega') \geqq m(\omega)\cdot\frac{|\Delta^+|}{2}\cdot\frac{1}{C_1} .$$

These estimates together give the desired result. □

3.4 SBR measures as limits of Lebesgue measure on W

3.4.1 Definition of SBR measures

In this subsection we define precisely what we mean by SBR-measures. Let $F: M ⟲$ be an arbitrary C^2 diffeomorphism of a finite dimensional manifold and let μ be an F-invariant Borel probability measure on M with compact support. We will assume throughout that at μ-a.e. point, there is a strictly positive Lyaponov exponent. Under these conditions, the unstable manifold theorem of Pesin [P1] or Ruelle [Ru2] tells us that passing through μ-a.e. x there is an *unstable manifold* which we denote by $W^u(x)$.

A measurable partition $\mathscr{Q}$ of M is said to be *subordinate to W* (with respect to the measure μ) if at μ-a.e. x, $\mathscr{Q}(x)$ is contained in $W^u(x)$ and contains an open neighborhood of x in $W^u(x)$. On each $\mathscr{Q}(x)$, there are two measures that are of interest to us. One is the restriction to $\mathscr{Q}(x)$ of the Riemann measure induced on $W^u(x)$; let us call this $m_x^{\mathscr{Q}}$. The other is $\mu_x^{\mathscr{Q}}$, where $\{\mu_x^{\mathscr{Q}}\}$ is a canonical family of conditional measures of μ with respect to the partition $\mathscr{Q}$. (For a reference see e.g. Rohlin [Ro].)

Definition. Let $F:(M,\mu)⟲$ be as above. We say that μ has *absolutely continuous conditional measures on unstable manifolds* if for every measurable partition $\mathscr{Q}$ subordinate to W^u, $\mu_x^{\mathscr{Q}}$ is absolutely continuous with respect to $m_x^{\mathscr{Q}}$ for μ-a.e. x.

For ease of reference, we will in this paper refer to invariant probability measures with absolutely continuous conditional measures on unstable manifolds as *Sinai–Bowen–Ruelle measures* or simply *SBR-measures*[2].

3.4.2 Pushing forward Lebesgue measure on W

Let

$$m_0 = m|_\Delta$$

and

$$m_n = \frac{1}{n} \sum_{k=0}^{n-1} T_*^k m_0 \,.$$

We define $\hat{m}_n^+$ to be the restriction to $\Delta^+ \times \mathbb{R}$ of

$$\frac{1}{n} \sum_{\substack{\omega \in \hat{\mathscr{P}}_i \\ t_i(\omega) < n}} T_*^{t_i}(m_0 | \omega) \,,$$

and let $\hat{m}_n^-$ be defined similarly. From Lemma 10 we know that for either $\{\hat{m}_n^+\}$ or $\{\hat{m}_n^-\}$ – let us say $\{\hat{m}_n^+\}$ – there is a sequence $N_1 < N_2 < \ldots$ such that

$$\hat{m}_{N_i}^+(\mathbb{R}^2) \geqq \frac{\alpha^*}{3} \quad \text{for all } i \,.$$

Passing to a subsequence if necessary, we may assume that

$$\hat{m}_{N_i}^+ \xrightarrow{\text{weakly}} \text{some } \hat{\lambda}$$

and

$$m_{N_i} \xrightarrow{\text{weakly}} \text{some } \lambda \,.$$

The following are immediate:

(1) λ is a T-invariant Borel measure, whose support is contained in the attractor $\Lambda = \bar{W}$;
(2) the total mass of $\hat{\lambda}$ is $\geqq \alpha^*/3$;
(3) λ is the sum of $\hat{\lambda}$ and another Borel measure.

Our plan is to use the geometric properties of $\hat{\lambda}$ to show that λ has at least one component with absolutely continuous conditional measures on unstable manifolds.

3.4.3 Geometric properties of $\hat{\lambda}$

Let c_1 be the constant in Lemma 3, and let

$$\Gamma = \{z \in \Lambda \cap (\Delta^+ \times \mathbb{R}) : \exists v \neq 0 \in T_z\mathbb{R}^2 \text{ with } |DT_z^{-j} v| \leqq e^{-c_1 j} \ \forall j \geqq 0\} \,.$$

[2] SBR measures are sometimes defined differently. All the definitions are equivalent for Axiom *A* attractors, but not all of them have been shown to be equivalent in the nonuniformly hyperbolic setting

Since $|\det DT^{-1}| = b^{-1}$ and the contraction above is uniform, it follows easily that Γ is compact, and that for $z \in \Gamma$, if v_z is the direction contracted by DT^{-j} in the definition of Γ, then $z \mapsto v_z$ is continuous. Moreover, by Lemma 3 we know that $\operatorname{supp} \hat{m}_n^+ \subset \Gamma$ for all n and that $\tau(z) = v_z$.

Sublemma 6 *Let z be an accumulation point of* $\bigcup_n \operatorname{supp}(\hat{m}_n^+)$. *Then there is a* C^1-*curve* $\gamma(z)$ *passing through z such that*

(1) $\gamma(z) = \operatorname{graph}(\varphi)$ *for some* $\varphi : \Delta^+ \to \mathbb{R}$;
(2) $\gamma(z) \subset \Gamma$ *and its tangent vector at* z' *is* $v_{z'}$.

Proof. Let $z_i \in \bigcup_n \operatorname{supp}(\hat{m}_n^+)$ be such that $z_i \to z$. For each z_i, let $\varphi_i : \Delta^+ \to \mathbb{R}$ be the function whose graph is the component of $W \cap (\Delta^+ \times \mathbb{R})$ containing z_i. By the $C^2(b)$-property of free segments the sequence of second derivatives $\{\varphi_i''\}$ is uniformly bounded. Hence a subsequence $\{\varphi_{i_k}'\}$ of $\{\varphi_i'\}$ converges uniformly. It follows immediately that φ_{i_k} converges in the C^1 sense to some φ with the properties in Sublemma 6. □

From Oseledec's theorem we know that Lyapunov exponents are well defined $\hat{\lambda}$-a.e. Sublemma 6 tells us that one of the exponents is positive. We also know from general theory (see [P1] or [Ru2]) that for a.e. point, its local unstable manifold is the unique curve passing through that point that contracts exponentially in backward time. Thus for a typical z, $\gamma(z)$ in Sublemma 6 must be the component of $W^u(z) \cap (\Delta^+ \times \mathbb{R})$ containing z. Let us call it $W^u_{\Delta^+}(z)$.

Let $X \subset \Delta^+ \times \mathbb{R}$ be a measurable set with the following properties: (i) $\hat{\lambda}(\mathbb{R}^2 - X) = 0$, and (ii) X is the disjoint union of $W^u_{\Delta^+}$-curves. Let $\mathcal{Q}$ be the partition of X into $W^u_{\Delta^+}$-leaves, and let $\{\hat{\lambda}_z^{\mathcal{Q}}\}$ be a canonical family of conditional measures of $\hat{\lambda}$. (See Sect. 3.4.1 for notations.)

Sublemma 7 *The measures* $\hat{\lambda}_z^{\mathcal{Q}}$ *and* $m_z^{\mathcal{Q}}$ *are equivalent for* $\hat{\lambda}$-a.e. z.

Proof. We claim that there is $C > 0$ such that for all intervals $J \subset \Delta^+$, one has

$$\frac{1}{C}|J| \leqq \hat{\lambda}_z^{\mathcal{Q}}(J \times \mathbb{R}) \leqq C|J|$$

for a.e. z. To see this let $\mathcal{Q}_n$ be a sequence of finite partitions of X such that $\forall z \in X$, $\mathcal{Q}_n(z) \supset W^u_{\Delta^+}(z)$ for all n, and $\bigcap_n \mathcal{Q}_n(z) = W^u_{\Delta^+}(z)$. If some $W^u_{\Delta^+}$-curve γ is contained in the support of $\hat{m}_n^+$, let ρ_n denote the density of $\hat{m}_n^+ | \gamma$. Proposition 2 tells us that for all $z_1, z_2 \in \gamma$,

$$\frac{\rho_n(z_1)}{\rho_n(z_2)} \leqq C_1 .$$

Integrating over $W^u_{\Delta^+}$-curves in each element of $\mathcal{Q}_n$, we obtain

$$\frac{1}{C}|J| \leqq E_{\hat{\lambda}}((J \times \mathbb{R})|\mathcal{Q}_n) \leqq C|J|$$

for some C independent of J. Our assertion follows from the martingale convergence theorem. □

3.4.4 Completing the construction

First we observe that $(\lambda \mid X)^{\mathcal{Q}}_z$ is equivalent to $m^{\mathcal{Q}}_z$ for $\hat{\lambda}$-a.e. z. This is true because the σ-algebra of T-invariant measurable sets is contained in the σ-algebra of measurable set made up of entire W^u-leaves, and that for every ergodic measure v, the conditional measures of v on local W^u-manifolds are either equivalent to m on a.e. leaf, or they are singular to m on a.e. leaf.

Let

$$X' = \left\{ z : \frac{d\hat{\lambda}}{d\lambda} > 0 \right\}.$$

and let λ' be the saturation of $\lambda|_{X'}$ under T. That is, let $r: X' \to \mathbb{Z}^+$ be the first return time to X' under T and let

$$\lambda' = \sum_{n=0}^{\infty} T^n_*(\lambda \mid (X' \cap \{r > n\})) .$$

We noted above that $\lambda \mid X'$ has absolutely continuous conditional measures on W^u-leaves. Hence the same is true for λ'. Moreover, an invariant measure with absolutely continuous conditional measures on W^u-leaves is the sum of at most a countable number of ergodic measures with the same property (see e.g. [L]). We may therefore assume for the rest of this paper that λ^* is one of these ergodic components, normalized to give $\lambda^*(\Lambda) = 1$. This is our SBR-measure.

4 Properties of SBR measures on Λ

Let λ^* be the T-invariant ergodic probability measure we constructed in Sect. 3. The purpose of this section is to prove

(1) the support of λ^* is the entire attractor Λ (Sect. 4.2);
(2) T does not admit any other SBR measures (Sect. 4.3);
(3) (T, λ^*) is Bernoulli (Sect. 4.4);

and to indicate how the existence of λ^* implies the corollary in the introduction (see Sect. 4.1.2).

4.1 Some known facts from the general theory of nonuniformly hyperbolic systems

The material in this subsection is not particular to the Hénon maps. We consider an arbitrary C^2 diffeomorphism $F: M \circlearrowleft$ of a finite dimensional Riemannian manifold preserving an ergodic Borel probability measure μ with compact support. It will be assumed throughout that (F, μ) has no zero Lyapunov exponents.

4.1.1 Stable and unstable manifolds

In [P1] Pesin proved the existence of stable and unstable manifolds in this nonuniform setting and studied their properties. (Pesin assumed that μ is equivalent to Lebesgue.) The case of arbitrary invariant measures is considered by Ruelle

[Ru2].) We recall here a couple of their results, giving precise statements only of what we will use and leaving out much more that is proved.

We write $T_xM = E^u(x) \oplus E^s(x)$ wherever it makes sense, and for $\delta > 0$, we let $B^u_\delta(x)$ and $B^s_\delta(x)$ denote the balls of radius δ about 0 in $E^u(x)$ and $E^s(x)$ respectively. Let $B_\delta(x) = B^u_\delta(x) \times B^s_\delta(x)$. It is sometimes convenient geometrically to introduce a new inner product $\langle \cdot , \cdot \rangle'_x$ on T_xM: under $\langle \cdot , \cdot \rangle'_x$, $E^u(x)$ and $E^s(x)$ are perpendicular; whereas restricted to $E^u(x)$ and $E^s(x)$, $\langle \cdot , \cdot \rangle'_x$ agrees with the given Riemannian metric. Let $\| \cdot \|'_x$ denote the corresponding norm. The following is true:

There exist Borel subset $\Gamma_1 \subset \Gamma_2 \subset \ldots$ of M with $\mu(\bigcup \Gamma_i) = 1$ and sequences of positive numbers δ_n, ε_n and θ_n, possibly $\downarrow 0$ as $n \uparrow \infty$, such that (1) and (2) below hold for every $x \in \Gamma_n$. (Think of the Γ_n's as uniformly hyperbolic sets that are not necessarily invariant, with the strength of hyperbolicity deteriorating as $n \to \infty$.)

(1) Let $\Gamma_n(x) = \{y \in \Gamma_n : d(x, y) < \varepsilon_n\}$. For $y \in \Gamma_n(x)$, let $W^u_x(y)$ denote the connected component of $(\exp_x^{-1} W^u(y)) \cap B_{\delta_n}(x)$ that contains $\exp_x^{-1} y$. Then for all $y \in \Gamma_n(x)$, $W^u_x(y)$ is the graph of a function $\varphi : B^u_{\delta_n}(x) \to B^s_{\delta_n}(x)$ with $\|D\varphi\|'_x \leqq \frac{1}{100}$. Moreover, as a C^1 embedded disk, $W^u_x(y)$ varies continuously with y. An analogous statement holds for $W^s_x(y)$.

(2) For $i = 1, 2$, let Σ_i be either $W^u_x(y)$ for some $y \in \Gamma_n(x)$ or a plane in $B_{\delta_n}(x)$ parallel to $E^u(x)$. Let $\Sigma'_1 = \Sigma_1 \cap \bigcup_{y \in \Gamma_n(x)} W^s_x(y)$, and let π be the map that takes $z \in \Sigma'_1$ to Σ_2 by sliding along $W^s_z(\cdot)$. Then for every Borel set $A \subset \Sigma'_1$.

$$\mathrm{Leb}(\pi A) \geqq \theta_n \mathrm{Leb}(A) .$$

Property (2) is the precise statement of what is called the "absolute continuity of the W^s-foliation". It is proved for "dissipative" systems in [PS].

4.1.2 Generic points for SBR measures

Let $F : (M, \mu) \circlearrowleft$ be as above. A point $x \in M$ is said to be *future-generic with respect to* μ or simply μ-generic if for every continuous function $\varphi : M \to \mathbb{R}$, $n^{-1} \sum_{i=0}^{n-1} \varphi \circ F^i(x) \to \int \varphi \, d\mu$ as $n \to \infty$. In particular, if μ is ergodic, then μ-a.e. x is μ-generic, and if x is μ-generic, then every $y \in W^s(x)$ is μ-generic as well.

The corollary stated in the introduction is an immediate consequence of the following general fact.

Proposition 3 *Let $F : (M, \mu) \circlearrowleft$ be as above. If μ is an ergodic* SBR *measure with no zero Lyapunov exponents, then there is a Borel subset $Y \subset M$ with positive Riemannian measure such that every $y \in Y$ is μ-generic.*

Proof. Since μ is an SBR measure, there is a piece of unstable manifold γ and a set $A \subset \gamma$ with $mA > 0$ such that every $x \in A$ is μ-generic. (As usual, m denotes the induced Riemannian measure on γ.) Using the absolute continuity of the W^s-foliation discussed in the last subsection, we see that

$$Y := \bigcup_{x \in A} W^s(x)$$

has the desired properties. $\square$

4.1.3 Ergodic properties of SBR measures

The following is proved by Pesin [P2] when μ is equivalent to Riemannian volume and generalized by Ledrappier [L] to the situation where μ has absolutely continuous conditional measures on W^u:

Let μ be as above and (F, μ) be ergodic. Then there are pairwise disjoint Borel sets $A_1, \ldots, A_n \in M$, such that

(1) $F(A_i) = A_{i+1}$ for $i < n$, $F(A_n) = A_1$;

and

(2) $(F^n | A_i, \mu | A_i)$ is Bernoulli for all i.

4.2 The support of λ^*

Recall the sequence of choices leading to the selection of "good" parameters in [BC2]: First $\delta > 0$ is fixed. Then $a_0 < a_1 < 2$ are chosen with a_0 very near 2. Next, b is chosen sufficiently small depending on a_0 and a_1; and finally, for fixed b, "good" maps $T_{a,b}$ are selected by varying $a \in (a_0, a_1)$.

Consider first the 1-d situation. Assume that $\delta > 0$ is fixed, so that $\Delta^{\pm}$, the outermost intervals on the partition of $(-\delta, \delta)$ are determined. (See Subsect. 3.1 for definitions.) Let $\hat{x}_a$ be the unique fixed point of $f_a : [-1, 1] \circlearrowleft$. For $a = 2$ since f_2 is topologically conjugate to its piecewise linear model, $\bigcup_{n \geq 0} f_2^{-n} \hat{x}_2$ is dense in $[-1, 1]$. So $\exists N_0 \in \mathbb{Z}^+$, intervals $\tilde{\Delta}^{\pm} \subset \Delta^{\pm}$, and a neighborhood V of $\hat{x}_2$, such that $f_2^{N_0} \tilde{\Delta}^{\pm} \supset V$. Moreover, one can choose $N_0, \tilde{\Delta}^{\pm}$ and V such that for all $x \in \tilde{\Delta}^{\pm}, f_2^{N_0} x$ is "free". Chose a_0 such that for all $a \in (a_0, 2)$, this picture persists with the same $N_0, \tilde{\Delta}^{\pm}$ and V.

Returning now to our Hénon maps $T_{a,b}$, we assume b is sufficiently small that the fixed point $\hat{z}_{a,b}$ lies in $V \times \mathbb{R}$, and that if γ is a curve with small Hausdorff distance from either $\Delta^+ \times \{0\}$ or $\Delta^- \times \{0\}$, then $T_{a,b}^{N_0} \gamma$ contains a curve with small Hausdorff distance from $V \times \{0\}$. Moreover, it is clear from our derivative estimates is Sect. 2 that if, in addition, $\gamma \subset W$ and is a free segment, then $T^{N_0} \gamma$ is a $C^2(b)$ curve, which must then intersect $W^s_{\text{loc}}(\hat{z})$ with an angle $\geqq \pi/4$ in at least one point.

From our construction of λ^* in Sect. 3, it follows that the support of λ^* contains a curve γ near $\Delta^+ \times \{0\}$ or $\Delta^- \times \{0\}$ that is the C^1-limit of free segments in W. This guarantees that γ intersects $W^s(\hat{z})$ transversally, which in turn implies that $W^u_{\text{loc}}(\hat{z}) \subset \operatorname{supp} \lambda^*$. Since $\Lambda = \bar{W}$, it follows that $\Lambda \subset \operatorname{supp} \lambda^*$. The reverse inclusion is obvious.

4.3 Uniqueness of SBR measures

Suppose that λ^* is not unique, so that there is another SBR measure μ on Λ. Without loss of generality we may assume that μ is ergodic. Fix $N_1 \in \mathbb{Z}^+$ with $\mu \Gamma_{N_1} > 0$. (See 4.1.1 for definition.) We claim that $\exists z_1 \in \Gamma_{N_1}$ such that in a neighborhood of z_1 we have the following picture: (For notational simplicity let us confuse $\Lambda \subset B_{\delta_{N_1}}(z_1)$ with $\exp_{z_1} \Lambda$ in the next few paragraphs.)

In $B_{\delta_{N_1}}(z_1)$, there is a "rectangle" two of whose sides, γ_1 and γ_2, are $W^u_{z_1}$-manifolds. Let us assume that $\gamma_1 = W^u_{z_1}(z_1)$. The other two "sides" of this "rectangle" are sets of the form $W^s_{z_1}(A_i)$, $i = 1, 2$, where A_i is a subset of γ_1 and $W^s_{z_1}(A_i) := \bigcup_{z \in A_i} W^s_{z_1}(z)$. The sets A_1 and A_2 are to have the following properties:

(1) $m(A_1), m(A_2) > 0$ $(m = \text{Leb on } \gamma_1)$;
(2) every $z \in A_1 \cup A_2$ is *generic* with respect to μ.

Moreover, there is at least one point $w \in \Lambda$ in the "interior" of our "rectangle". (See Fig. 1.)

Let us assume this picture for now and complete the proof. We claim that there is a W^u-leaf γ with the following properties:

(3) m-a.e. $z \in \gamma$ is generic with respect to λ^*;
(4) γ connects the "outside" of our "rectangle" to the "inside".

To see that this claim is valid, recall the geometric properties of $\hat{\lambda}$ in Sect. 3 and the argument in Sect. 4.2 showing that $W^u_{\text{loc}}(\hat{z})$ is the C^1 limit of curves with property (3). Iterating forward, we see that every compact segment of W is the C^1 limit of curves with property (3). Now $w \in \Lambda = \bar{W}$ is "inside" our "rectangle", whereas we may assume that $\hat{z}$ is "outside". (3) and (4) should now be obvious.

Since γ clearly cannot intersect γ_1 or γ_2, it must intersect $W^s_{z_1}(A_1) \cup W^s_{z_2}(A_2)$. If we show that this intersection has positive m-measures in γ, then we will have proved that a positive m-measure set in γ is generic with respect to both λ^* and μ, forcing $\mu = \lambda^*$.

Two points need to be justified. First, the "rectangle" in our picture. Let $D = \{z \in \mathbb{R}^2 : z \text{ is generic with respect to } \mu\}$. Chose $z_1 \in \Gamma_{N_1}$ such that

(i) $\mu\{z \in \Gamma_{N_i} : d(z, z_1) < \varepsilon\} > 0$ for all $\varepsilon > 0$,
(ii) $m\{z \in W^u_{\text{loc}}(z_1) : z \in \Gamma_{N_1} \cap D$ and $d^u(z, z_1) < \varepsilon\} > 0$ for all $\varepsilon > 0$. (Here $d^u :=$ dist along $W^u(z_1)$.)

Let $\gamma_1 = W^u_{z_1}(z_1)$. Then A_1 and A_2 can be chosen as subsets of $\gamma_1 \cap \Gamma_{N_1} \cap D$. To complete our "rectangle" and to guarantee that some point of Λ lies inside, it suffices to argue that arbitrarily near γ_1, there are infinitely many $W^u_{z_1}$-curves. If this was not the case, then by (i) above we must have $\mu\gamma_1 > 0$. A standard argument in nonuniform hyperbolic theory then tells us that for some $n > 0$, $T^n\gamma \supset \gamma$ and $T^n|_{T^{-n}\gamma}$ is, in suitable coordinates, expanding. (See e.g. [K] for more details.) This implies that $\mu\{z_1\} > 0$, contradicting our assumption that μ is SBR.

The other point which perhaps needs some justification is our assertion that

$$m(\gamma \cap (W^s_{z_1}(A_1) \cup W^s_{z_1}(A_2))) > 0 .$$

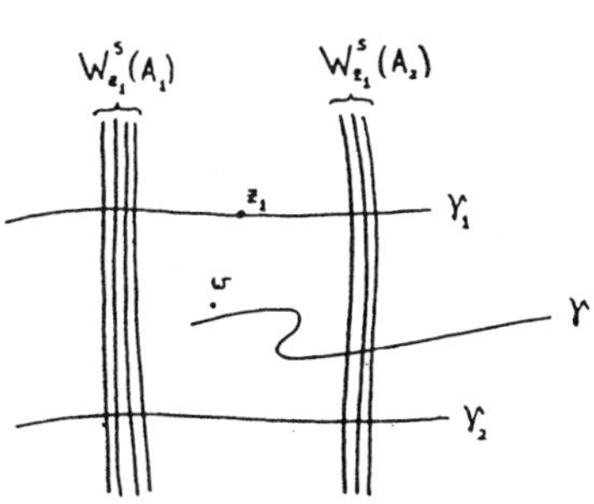

Fig. 1

We may assume that A_1 is a Cantor set, and is $= \bigcap_{n=1}^{\infty} E_n$ where each E_n is the disjoint union of a finite number of curve segments $\{E_{n,i}: 1 \leqq i \leqq i_n\}$ in γ_1. For each (n, i) let $y_{n,i}^l$ and $y_{n,i}^r$ be the left and right endpoint of $E_{n,i}$, and let $S_{n,i}$ be the strip in $B_{\delta_{N_1}}(z_1)$ bounded by $W_{z_1}^s(y_{n,i}^l)$, $W_{z_1}^s(y_{n,i}^r)$ and $\partial B_{\delta_{N_1}}(z_1)$. Then clearly $\bigcap_n \bigcup_i S_{n,i} = W_{z_1}^s(A_1)$. Let us assume that γ connects the two side of $S_{1,1}$ and let τ_1 be a subsegment of γ joining $W_{z_1}^s(y_{1,1}^l)$ to $W_{z_1}^s(y_{1,1}^r)$. For each (n, i) with $n > 1$, choose inductively $\tau_{n,i} \subset$ some $\tau_{n-1,j}$ such that $\tau_{n,i}$ joins $W_{z_1}^s(y_{n,i}^l)$ to $W_{z_1}^s(y_{n,i}^r)$. Then for every (n, i), we have

$$m(\tau_{n,i}) \geqq \text{min dist between } W_{z_1}^s(y_{n,i}^l) \text{ and } W_{z_1}^s(y_{n,i}^r)$$

$$\geqq c \cdot \text{dist}(L \cap W_{z_1}^s(y_{n,i}^l), L \cap W_{z_1}^s(y_{n,i}^r)),$$

where L is any line in $B_{\delta_{N_1}}(z_1)$ parallel to $E^u(z_1)$ and $c > 0$ is a constant depending only on the angle between $E^u(z_1)$ and $E^s(z_1)$. Using the absolute continuity of the W^s-foliation (see Sect. 4.1.1) we may then conclude that

$$m(\tau_{n,i}) \geqq c \cdot \theta_{N_1} \cdot m(A_1 \cap E_{n,i})$$

and hence

$$m(\gamma \cap W_{z_1}^s(A_1)) \geqq m\left(\bigcap_n \bigcup_i \tau_{n,i}\right) \geqq c \cdot \theta_{N_1} \cdot mA_1 .$$

4.4 *Bernoulliness of* (T, λ^*)

In light of Sect. 4.1.3 it suffices to show that (T^n, λ^*) is ergodic for all $n \geqq 1$. Let us fix $n_0 \in \mathbb{Z}^+$ and let μ_0 be one of the ergodic components of (T^{n_0}, λ^*) with absolutely continuous conditional measures on W^u-manifolds. From our construction of λ^* it follows that for some $k \in \mathbb{Z}^+$, $\mu_0(T^k X) > 0$, where X is the set in Sect. 3.4.3. Using again the fact that points in the same W^u-leaf belong in the same ergodic component of T^{n_0}, we see that $W_{\text{loc}}^u(\hat{z}) \subset \text{supp}\, \mu_0$, from which it follows that $\text{supp}\, \mu_0 = \Lambda$. Let $\bar{\mu}_0$ be μ_0 normalized. Repeating the argument in Sect. 4.3 with $(T^{n_0}, \bar{\mu}_0)$ in place of (T, λ^*), we see that $\bar{\mu}_0$ is the unique SBR measure for T^{n_0}. Thus (T^{n_0}, λ^*) has only one ergodic component and our proof is complete.

4.5 *Further properties of* (T, λ^*) *and an open problem*

We mention a couple of facts about (T, λ^*) that follow from general nonuniform hyperbolic theory. Let $\chi_1 > 0 > \chi_2$ be the Lyaponov exponents of (T, λ^*). We have the entropy formula

$$h_{\lambda^*}(T) = \chi_1 ,$$

where the quantity on the left is metric entropy. (See [P2, LS].) Using [Y1] and the entropy formula above, we obtain the following formula for the dimension of λ^*:

$$\text{HD}(\lambda^*) = h_{\lambda^*}(T) \cdot \left(\frac{1}{\chi_1} - \frac{1}{\chi_2} \right) = 1 - \frac{\chi_1}{\chi_2} = 1 + \frac{\chi_1}{\chi_1 - \log b} .$$

We finish by mentioning a problem the resolution of which would give a more complete geometric picture of these "good" Hénon maps. Let

$$B = \{z \in \mathbb{R}^2 : d(T^n, \Lambda) \to 0 \text{ as } n \to \infty\} .$$

The set B is called the *basin of attraction* of Λ and is known to contain an open neighborhood of Λ. Proposition 3 in Sect. 4.1 tells us that a positive Lebesgue measure subset of B consists of points that are λ^*-generic. That is to say, the statistics of these orbits are completely governed by the invariant measure λ^*. It would be nice to know if this property holds not just on a large set but almost everywhere in B.

References

[B] Bowen, R.: Equilibrium states and the ergodic theory of Anosov diffeomorphisms. (Lect. Notes Math., vol. 470) Berlin Heidelberg New York Springer: 1975

[BC1] Benedicks, M., Carleson, L.: On iterates of $x \mapsto 1 - ax^2$ on $(-1, 1)$. Ann. Math. **122**, 1–25 (1985)

[BC2] Benedicks, M., Carleson, L.: The dynamics of the Hénon map. Ann. Math. **133**, (1991) 73–169

[BM] Benedicks, M., Moeckel, R.: An attractor for the Hénon map. Zürich: ETH (Preprint)

[BY] Benedicks, M., Young, L.S.: Random perturbations and invariant measures for certain one-dimensional maps. Ergodic Theory Dyn. Syst. (to appear)

[L] Ledrappier, F.: Propriétés ergodiques des mesures de Sinai. Publ. Math., Inst. Hautes Etud. Sci. **59**, 163–188 (1984)

[LS] Ledrappier, F., Strelcyn, J.-M.: A proof of the estimation from below in Pesin's entropy formula. Ergodic Theory Dyn. Sys **2**, 203–219 (1982)

[LY] Ledrappier, F., Young, L.-S.: The metric entropy of diffemorphisms. Part I. Characterization of measures satisfying Pesin's formula. Part II. Relations between entropy, exponents and dimension. Ann. Math. **122**, 509–539, 540–574 (1985)

[MV] Mora, L., Viana, M.: Abundance of strange attractors. IMPA reprint (1991)

[K] Katok, A.: Lyapunov exponents, entropy and periodic orbits for diffeomorphisms. Publ. Math., Inst. Hautes Étud. Sci. **51**, 137–174 (1980)

[P1] Pesin, Ja.G.: Families of invariant manifolds corresponding to non-zero characteristic exponents. Math. USSR, Izv. **10**, 1261–1305 (1978)

[P2] Pesin, Ja.G.: Characteristic Lyaponov exponents and smooth ergodic theory. Russ. Math. Surv. **32.4**, 55–114 (1977)

[PS] Pugh, C., Shub, M.: Ergodic Attractors. Trans. Am. Math. Soc. **312**, 1–54 (1989)

[Ro] Rohlin, V.A.: On the fundamental ideas of measure theory. Transl., Am. Math. Soc. **10**, 1–52 (1962)

[Ru1] Ruelle, D.: A measure associated with Axiom A attractors. Am. J. Math. **98**, 619–654 (1976)

[Ru2] Ruelle, D.: Ergodic theory of differentiable dynamical systems. Publ. Math., Inst. Hautes Étud. Sci. **50**, 27–58 (1979)

[S1] Sinai, Ya. G.: Markov partitions and *C*-diffeomorphisms. Func. Anal. Appl. **2**, 64–89 (1968)

[S2] Sinai, Ya. G.: Gibbs measures in ergodic theory. Russ. Math. Surv. **27:4**, 21–69 (1972)

[Y1] Young, L.-S.: Dimension, entropy and Lyapunov exponents. Ergodic Theory Dyn. Syst. **2**, 163–188 (1982)

[Y2] Young, L.-S.: A Bowen-Ruelle measure for certain piecewise hyperbolic maps. Trans. Am. Math. Soc. **287**, 41–48 (1985)

SIAM J. NUMER. ANAL.
Vol. 36, No. 2, pp. 491–515

ON THE APPROXIMATION OF COMPLICATED DYNAMICAL BEHAVIOR*

MICHAEL DELLNITZ† AND OLIVER JUNGE†

Abstract. We present efficient techniques for the numerical approximation of complicated dynamical behavior. In particular, we develop numerical methods which allow us to approximate Sinai–Ruelle–Bowen (SRB)-measures as well as (almost) cyclic behavior of a dynamical system. The methods are based on an appropriate discretization of the Frobenius–Perron operator, and two essentially different mathematical concepts are used: our idea is to combine classical convergence results for finite dimensional approximations of compact operators with results from ergodic theory concerning the approximation of SRB-measures by invariant measures of stochastically perturbed systems. The efficiency of the methods is illustrated by several numerical examples.

Key words. computation of invariant measures, approximation of the Frobenius–Perron operator, computation of SRB-measures, almost invariant set, cyclic behavior

AMS subject classifications. 58F11, 65L60, 58F12

PII. S0036142996313002

1. Introduction.

The approximation of the behavior of a dynamical system is typically done by direct simulation. This method is particularly useful in the situation where a specific trajectory has to be approximated for a finite period of time. However, if one is interested in the long term behavior and if the underlying system exhibits complicated dynamics, then the information derived from a single trajectory is not always satisfying. Rather in this case it seems more appropriate to determine a statistical description of the dynamical behavior, and this information is encoded in an underlying (natural) invariant measure.

In this paper we describe a numerical method for the approximation of such invariant measures based on a discretization of the *Frobenius–Perron operator*. Using the fact that invariant measures are fixed points of this operator, we first approximate it by a Galerkin projection and then compute eigenvectors of the discretized operator corresponding to the eigenvalue 1. This allows us to identify regions in state space where trajectories are likely to be observed or, on the other hand, hardly observed. In addition to this information we show how to use other parts of the spectrum of the Frobenius–Perron operator to determine the dynamical behavior of the system. First we describe how to decompose an invariant set into components which are *cyclically permuted* by the dynamics. Second we develop techniques for the approximation of *almost invariant sets*, that is, regions in state space which are visited for a "long" period of time before the dynamical process leads to different areas. More generally the same techniques allow us to detect *almost cyclic behavior*, that is, to identify components of invariant sets which are "frequently" cyclically permuted by the dynamical process. Moreover, we can quantify the probability by which the cycle occurs depending on the absolute value of a corresponding eigenvalue of the Frobenius–Perron operator. Roughly speaking, we construct an approximation of the *essential dynamical behavior*,

*Received by the editors November 27, 1996; accepted for publication (in revised form) February 6, 1998; published electronically February 19, 1999. This research was partially supported by the Deutsche Forschungsgemeinschaft under grant De 448/5-2.

http://www.siam.org/journals/sinum/36-2/31300.html

†Mathematisches Institut, Universität Bayreuth, D-95440 Bayreuth (dellnitz@uni-bayreuth.de, http://www.uni-bayreuth.de/departments/math/~mdellnitz; junge@uni-bayreuth.de, http://www.uni-bayreuth.de/departments/math/~ojunge).

that is, the dynamics modulo complex (unpredictable) behavior which is due to the presence of chaos.

From a practical point of view the most important invariant measures are the so-called *SRB-measures*. The reason is that for these measures the spatial and temporal averages of observables are identical for a set of initial conditions which has positive Lebesgue measure. The introduction of the underlying concept goes back to Y. Sinai (see [27]), and the existence of these measures has been shown for Axiom A systems by D. Ruelle and R. Bowen [26, 3]. In this article we suggest a numerical method for the approximation of SRB-measures, and in this context their stochastic stability is particularly important: first we use this fact as an analytical tool in our main convergence result in section 4, and second it is of practical importance if we view the numerical approximation as a small random perturbation. Indeed, stochastic stability of SRB-measures is guaranteed for Axiom A systems [17, 18].

More precisely there are two essential mathematical ingredients which allow us to develop a numerical method for the approximation of SRB-measures. We use a result of Yu. Kifer on the convergence of invariant measures in stochastic perturbations of the underlying dynamical system to the SRB-measure (see [17]) and combine this with results on the convergence of eigenspaces of discretized compact operators (see, e.g., [25]). The same technique is used for the approximation of the subsets in state space which are (almost) cyclically permuted by the dynamical process. With respect to the approximation of SRB-measures a similar result has previously been obtained by F. Hunt (see [15]). However, our methods are quite different from the ones used in that work. In particular, the results stated here cover the important situation where the random perturbations have a probability distribution with local support. In fact, this is the relevant case having in mind that the round off error in the numerical approximation can be interpreted as such a local perturbation. Another approach for the computation of SRB-measures—avoiding the approximation of the Frobenius–Perron operator—has recently been suggested by Yu. Kifer [19].

As mentioned above, in addition to the approximation of SRB-measures the main development in this paper is a numerical method which allows us to identify (almost) cyclic behavior. To accomplish this we use a Galerkin method to discretize the Frobenius–Perron operator in such a way that the discretization has the same cyclic properties as the operator itself. More precisely, if the underlying dynamical system has a cycle of order r, then the rth roots of unity are eigenvalues of the Frobenius–Perron operator, and we will show that the corresponding eigenmeasures $\nu_0, \ldots, \nu_{r-1}$ yield the desired information on the cyclic components: these components can be identified as supports of probability measures obtained by specific linear combinations of $\nu_0, \ldots, \nu_{r-1}$. Our Galerkin approximation respects the cyclic behavior in the sense that the rth roots of unity are also eigenvalues of the discretized operator and that the corresponding eigenvectors converge to the eigenmeasures $\nu_0, \ldots, \nu_{r-1}$ with increasing dimension of the approximating space. We will illustrate how to use these results to determine the subsets in state space which are almost cyclically permuted.

Finally, let us remark that the results on the approximation of the essential dynamical behavior obtained in this article can also be used to compute other statistical quantities such as the *entropy*, *dimensions* (depending on the particular invariant measure), or *Lyapunov exponents*. In fact, the efficient numerical use of invariant measures for the computation of Lyapunov exponents is currently under investigation.

An outline of the paper is as follows. In section 2 we begin with a brief review of the results on Markov processes which will be needed later on. The Frobenius–Perron operator is introduced in section 3. In that section we also describe the Galerkin projection that we use in the numerical approximation. In section 4 we use Kifer's result on small random perturbations of diffeomorphisms to prove convergence of the approximations to an SRB-measure in the hyperbolic case (Theorem 4.4). In the main section of this article, section 5, we show how to extract numerically the information on the (almost) cyclic components from the spectrum of the Frobenius–Perron operator. In particular, we present a method to identify regions in phase space where, on average, trajectories stay for a long period of time. Finally, in section 6 we illustrate by several examples the usefulness of our methods as tools in the numerical analysis of dynamical behavior.

2. Stochastic transition functions. For our main theoretical results we are using the concept of *small random perturbations* of dynamical systems. Since we assume that the typical reader is not familiar with this concept we begin by recalling some basic notions and results on Markov processes that will be needed later on. For a detailed introduction the reader is referred to [11].

Invariant measures. Our aim is to approximate the dynamical behavior of discrete dynamical systems of the form

$$x_{i+1} = f(x_i), \quad i = 0, 1, 2, \ldots,$$

where $f : X \to X$ is a diffeomorphism on a compact subset $X \subset \mathbb{R}^n$. We denote by $\mathcal{B}$ the Borel σ-algebra on X and by m the Lebesgue measure on $\mathcal{B}$. Moreover, let $\mathcal{M}$ be the space of probability measures on $\mathcal{B}$. Recall that a measure $\mu \in \mathcal{M}$ is *invariant* if

$$\mu(B) = \mu(f^{-1}(B)) \quad \text{for all } B \in \mathcal{B}.$$

On the other hand, a set $A \in \mathcal{B}$ is *invariant* if

$$f(A) \subset A.$$

We now turn our attention to the more general stochastic framework.

Definition 2.1. *A function* $p : X \times \mathcal{B} \to [0, 1]$ *is a* stochastic transition function *if*

(i) $p(x, \cdot)$ *is a probability measure for every* $x \in X$,
(ii) $p(\cdot, A)$ *is Lebesgue-measurable for every* $A \in \mathcal{B}$.

Let δ_y denote the Dirac measure supported on the point $y \in X$. Then $p(x, A) = \delta_{h(x)}(A)$ is a stochastic transition function for every m-measurable function h. We will see below that the specific choice $h = f$ represents the deterministic situation in this more general setup.

We set $p^{(1)}(x, A) = p(x, A)$ and define recursively the *i-step stochastic transition function* $p^{(i)} : X \times \mathcal{B} \to \mathbb{R}$ by

$$p^{(i+1)}(x, A) = \int p^{(i)}(y, A)\, p(x, dy), \quad i = 1, 2, \ldots,$$

where $p(x, dy)$ indicates that the integration is done with respect to the measure $p(x, \cdot)$. It is easy to see that $p^{(i)}$ is indeed a stochastic transition function. In particular, for the case where $p(x, A) = \delta_{f(x)}(A)$ we obtain for $i \geq 1$

$$p^{(i)}(x, A) = \delta_{f^i(x)}(A).$$

We now define the notion of an invariant measure in the stochastic setting.

DEFINITION 2.2. *Let p be a stochastic transition function. If $\mu \in \mathcal{M}$ satisfies*

$$\mu(A) = \int p(x, A)\, d\mu(x)$$

for all $A \in \mathcal{B}$, then μ is an invariant measure *of p.*

Remark 2.3.

(a) In the literature on Markov processes (e.g., [11]) an invariant measure is typically referred to as a *stationary absolute probability measure*. However, having the situation in mind that we consider stochastically perturbed dynamical systems we prefer the notion of an invariant measure (see also [17, 24]).

(b) If μ is an invariant measure of p, then it follows that

$$\mu(A) = \int p^{(i)}(x, A)\, d\mu(x)$$

for all $i = 1, 2, \ldots$.

The following example illustrates the previous remark that we recover the deterministic situation in the case where $p(x, \cdot) = \delta_{f(x)}$.

EXAMPLE 2.4. *Suppose that $p(x, \cdot) = \delta_{f(x)}$ and let μ be an invariant measure of p. Then we compute for $A \in \mathcal{B}$*

$$\mu(A) = \int p(x, A)\, d\mu(x) = \int \delta_{f(x)}(A)\, d\mu(x) = \int \chi_A(f(x))\, d\mu(x) = \mu(f^{-1}(A)),$$

where we denote by χ_A the characteristic function of A. Hence μ is an invariant measure for the diffeomorphism f.

DEFINITION 2.5. *A set $A \in \mathcal{B}$ is called a* consequent set *of x if $p^{(i)}(x, A) = 1$ for all $i \geq 1$. The set A is* invariant *if it is the consequent set of all of its points. Furthermore if $C \in \mathcal{B}$ is a set for which*

$$\lim_{i \to \infty} p^{(i)}(x, C) = 0 \quad \textit{for all } x \in X,$$

then C is called a transient set.

Considering our guiding example let $p(x, \cdot) = \delta_{f(x)}$ and let A be an invariant set for p. Then we have for $y \in A$

$$1 = p(y, A) = \delta_{f(y)}(A).$$

Hence $f(A) \subset A$ and A is an invariant set for the diffeomorphism f.

Absolutely continuous stochastic transition functions. Now we assume that for every $x \in X$ the probability measure $p(x, \cdot)$ is absolutely continuous with respect to the Lebesgue measure m. Hence we may write $p(x, \cdot)$ as

$$p(x, A) = \int_A k(x, y)\, dm(y) \quad \text{for all } A \in \mathcal{B},$$

with an appropriate *transition density function* $k : X \times X \to \mathbb{R}$. Obviously,

$$k(x, \cdot) \in L^1(X, m) \quad \text{and} \quad k(x, y) \geq 0.$$

In this case we also call the stochastic transition function p *absolutely continuous.* Note that

$$\int k(x,y)\,dm(y) = p(x,X) = 1 \quad \text{for all } x \in X.$$

We let $k^{(1)}(x,y) = k(x,y)$ and define the *i-step transition density function* as

$$k^{(i+1)}(x,y) = \int k(x,\xi)k^{(i)}(\xi,y)\,dm(\xi), \quad i = 1,2,\ldots.$$

With this definition we obtain for $A \in \mathcal{B}$

$$p^{(i)}(x,A) = \int_A k^{(i)}(x,y)\,dm(y),$$

that is, the i-step transition density function $k^{(i)}$ is the stochastic transition density function for $p^{(i)}$.

The following theorem provides a characterization of all invariant measures of a certain class of stochastic transition functions.

THEOREM 2.6. *Let p be an absolutely continuous stochastic transition function with density function k. Suppose that $k(x,y) \leq M$ for some positive constant $M > 0$ and all $x, y \in X$.*

Then X can be decomposed into finitely many disjoint invariant sets $E_1, E_2, \ldots, E_e$ and a transient set $F = X - \cup_{j=1}^{e} E_j$ such that for each E_j there is a unique probability measure $\pi_j \in \mathcal{M}$ with $\pi_j(E_j) = 1$ and

$$\lim_{N\to\infty} \frac{1}{N} \sum_{i=1}^{N} p^{(i)}(x,A) = \pi_j(A) \quad \textit{for all } A \in \mathcal{B} \textit{ and all } x \in E_j. \tag{2.1}$$

Furthermore the left-hand side in (2.1) *exists uniformly in x and defines for every fixed $x \in X$ an invariant measure. Finally, every invariant measure of p is a convex combination of the π_j's.*

A proof of this theorem can be found in [11].

Remark 2.7.

(a) The E_j's are called the *ergodic sets* of p.

(b) One can show that the invariant measures π_j are absolutely continuous with density functions $\kappa_j \in L^1$, that is, we have

$$\pi_j(A) = \int_A \kappa_j(y)\,dm(y), \quad j = 1,\ldots,e.$$

It follows that for every fixed $x \in X$ the limit in (2.1) is also absolutely continuous with a density function $\ell(x,\cdot) \in L^1$.

3. Approximation of the Frobenius–Perron operator. The main purpose of this section is to describe an appropriate Galerkin method for the approximation of the Frobenius–Perron operator. But first we introduce this operator and derive certain spectral properties.

The Frobenius–Perron operator.

DEFINITION 3.1. *Let p be a stochastic transition function. Then the* Frobenius–Perron operator *$P : \mathcal{M}_\mathbb{C} \to \mathcal{M}_\mathbb{C}$ is defined by*

$$P\mu(A) = \int p(x,A)\,d\mu(x),$$

where $\mathcal{M}_{\mathbb{C}}$ is the space of bounded complex valued measures on $\mathcal{B}$. If p is absolutely continuous with density function k, then we may define the Frobenius–Perron operator P on L^1 by

$$Pg(y) = \int k(x,y)g(x)\,dm(x) \quad \textit{for all } g \in L^1.$$

Remark 3.2.

(a) By definition a measure $\mu \in \mathcal{M}$ is invariant if and only if it is a fixed point of P. In other words, invariant measures correspond to eigenmeasures of P for the eigenvalue 1.
Moreover, let $\lambda \in \mathbb{C}$ be an eigenvalue of P with corresponding eigenmeasure ν, that is, $P\nu = \lambda\nu$. Then in particular

$$\lambda\nu(X) = P\nu(X) = \int p(x,X)\,d\nu(x) = \nu(X)$$

since $p(x,X) = 1$ for all $x \in X$. It follows that $\nu(X) = 0$ if $\lambda \neq 1$.

(b) Observe that in the deterministic situation where $p(x,\cdot) = \delta_{f(x)}$ we obtain

$$P\mu(A) = \int p(x,A)\,d\mu(x) = \mu(f^{-1}(A))$$

(cf. Example 2.4). This is indeed the standard definition of the Frobenius–Perron operator in the deterministic setting (see, e.g., [21]).

(c) Note that in the case where p is absolutely continuous we have $P : L^1 \to L^1$ since for each $g \in L^1$

$$\begin{aligned}\int Pg(y)\,dm(y) &= \iint k(x,y)g(x)\,dm(x)\,dm(y)\\ &= \int g(x)\int k(x,y)\,dm(y)\,dm(x)\\ &= \int g(x)\,dm(x) < +\infty.\end{aligned}$$

Correspondingly, a nonnegative fixed point $g \in L^1$ of P with $\|g\|_1 = 1$ is the density of an invariant probability measure and, conversely, the density of every absolutely continuous invariant probability measure is a fixed point of P.

We are particularly interested in approximating cyclic dynamical behavior of the underlying dynamical system. In the stochastic setting this corresponds to the situation where there are disjoint compact subsets $X_j \subset X$, $j = 0,\ldots,r-1$, such that

$$X = \bigcup_{j=0}^{r-1} X_j,$$

and for which the stochastic transition function p satisfies

$$p(x, X_{j+1 \bmod r}) = \begin{cases} 1 & \text{if } x \in X_j, \\ 0 & \text{otherwise.} \end{cases} \tag{3.1}$$

We now relate the cyclic dynamical behavior described by (3.1) to spectral properties of the corresponding Frobenius–Perron operator P.

PROPOSITION 3.3. *If the stochastic transition function p satisfies* (3.1), *then we have the following for the corresponding Frobenius–Perron operator* P*:*

(a) *The* r*th power* P^r *has an eigenvalue* 1 *of multiplicity at least* r*. Moreover, there are* r *corresponding invariant measures* $\mu_k \in \mathcal{M}$, $k = 0, 1, \ldots, r-1$, *with support on* X_k*, that is,* $\mathrm{supp}(\mu_k) \subset X_k$*. These measures can be chosen to satisfy*

$$\mu_k = P^k \mu_0, \quad k = 0, 1, \ldots, r-1.$$

(b) *The* r*th roots of unity* ω_r^k, $k = 0, 1, \ldots, r-1$, *where* $\omega_r = e^{2\pi i/r}$, *are eigenvalues of* P.

Proof.

(a) Observing that for each $\mu \in \mathcal{M}_{\mathbb{C}}$

$$P^j \mu(A) = \int p^{(j)}(x, A)\, d\mu(x),$$

where $p^{(j)}$ is the j-step stochastic transition function, the existence of the measures μ_k, $k = 0, 1, \ldots, r-1$, follows from standard results on Markov processes (see, e.g., [11, Chap. V]). Moreover, these measures can be chosen so that

$$\mu_{k+1 \bmod r} = P\mu_k, \quad k = 0, 1, \ldots, r-1.$$

Simply note that if μ_k is invariant for P^r, then

$$P^r(P\mu_k) = P(P^r\mu_k) = P\mu_k$$

and hence $P\mu_k$ is an invariant measure with support on X_{k+1}.

(b) Let μ be one of the probability measures which exist by part (a). We show that for $k \in \{0, 1, \ldots, r-1\}$

$$\nu_k = \sum_{j=0}^{r-1} \omega_r^{-kj} P^j \mu \in \mathcal{M}_{\mathbb{C}} \tag{3.2}$$

is an eigenmeasure of P for the eigenvalue ω_r^k. Indeed, using the fact that $P^r\mu = \mu$ we compute

$$\begin{aligned} P\nu_k &= P\mu + \omega_r^{-k} P^2 \mu + \cdots + \omega_r^{-k(r-2)} P^{r-1} \mu + \omega_r^{-k(r-1)} \mu \\ &= \omega_r^k \left(\mu + \omega_r^{-k} P\mu + \cdots + \omega_r^{-k(r-1)} P^{r-1} \mu \right) \\ &= \omega_r^k \nu_k. \end{aligned}$$

Finally, $\nu_k \neq 0$ since $\nu_k(X_j) \neq 0$ for $j = 0, 1, \ldots, r-1$. □

Approximation by a Galerkin method. We begin with the following observation which immediately follows from standard results on integral operators (see, e.g., [28, p. 277]).

LEMMA 3.4. *Suppose that the transition density function k satisfies*

$$\iint |k(x, y)|^2\, dm(x) dm(y) < \infty. \tag{3.3}$$

Then the Frobenius–Perron operator $P : L^2 \to L^2$ *is compact.*

From now on we consider the case where P is given by a dynamical process with a transition density function k satisfying the condition (3.3). The aim is to use a Galerkin method for the approximation of such a Frobenius–Perron operator together with its spectrum. More precisely, let V_d, $d \geq 1$, be a sequence of d-dimensional subspaces of L^2 and let $Q_d : L^2 \to V_d$ be a projection such that Q_d converges pointwise to the identity on L^2. If we define the approximating operators by $P_d = Q_d P$, then we have

$$\|P_d - P\|_2 \to 0 \quad \text{as } d \to \infty.$$

Denote by $\sigma(P)$ and $\rho(P)$ the spectrum and resolvent set of P, respectively, and by $R_z = (zI - P)^{-1}$, $z \in \rho(P)$, the resolvent operator. Let $\lambda \neq 0 \in \sigma(P)$ be a nonzero eigenvalue of P and let $\Gamma \subset \mathbb{C}$ be a circle in $\rho(P)$ with center λ such that no other point of $\sigma(P)$ is inside Γ. Then the operator defined by

$$E = E(\lambda) = \frac{1}{2\pi i} \int_\Gamma R_z(P)\, dz$$

is a projection onto the space of generalized eigenvectors associated with λ and P. The following theorem—which is a specific application of the main result of [25] on compact operators—allows us to approximate eigenvectors of P by eigenvectors of P_d.

THEOREM 3.5 (see [25]). *Let λ_d be an eigenvalue of P_d such that $\lambda_d \to \lambda$ for $d \to \infty$, and let g_d be a corresponding eigenvector of unit length. Then there is a vector $h_d \in R(E)$ and a constant $C > 0$ such that $(\lambda I - P)h_d = 0$ and*

$$\|h_d - g_d\|_2 \leq C\|(P - P_d)|_{R(E)}\|_2,$$

where $R(E)$ denotes the range of E.

Next we use Theorem 3.5 to approximate the eigenvalues of P which are lying on the unit circle. For this we construct a Galerkin projection which possesses the same cyclic behavior in the approximation. Suppose that (3.1) holds and let $\{\varphi_i^j\}$, $j = 0, 1, \dots, r-1$, $i = 1, 2, \dots, d_j$, be a basis of V_d with the following properties:

$$\text{(3.4)} \qquad \begin{aligned} &\text{(i)} \quad \operatorname{supp}(\varphi_i^j) \subset X_j \quad (j = 0, 1, \dots, r-1, \quad i = 1, 2, \dots, d_j), \\ &\text{(ii)} \quad \sum_{i=1}^{d_j} \varphi_i^j(x) = 1 \quad \text{for all } x \in X_j,\ j = 0, 1, \dots, r-1. \end{aligned}$$

Remark 3.6.

(a) In section 5 we will see how to generate a basis satisfying (3.4). In that case, V_d consists of functions which are locally constant.

(b) Observe that by construction

$$\sum_{i,j} \varphi_i^j(x) = 1 \quad \text{for all } x \in X.$$

The Galerkin projection $Q_d g$ of $g \in L^2$ is defined by

$$(Q_d g, \varphi_i^j) = (g, \varphi_i^j) \quad \text{for all } i, j,$$

where $(\cdot, \cdot)$ is the usual inner product in L^2. The following result is a generalization of Lemma 8 in [10], where just the fixed point of P is considered. Recall that $\omega_r = e^{2\pi i/r}$.

PROPOSITION 3.7. *Suppose that the Galerkin projection satisfies* (3.4). *Then the approximating operators $P_d = Q_d P$ possess the eigenvalues ω_r^k, $k = 0, 1, \dots, r-1$.*

Proof. Suppose that λ is an eigenvalue of P_d with corresponding eigenvector $\psi(x) = \sum_{i,j} \beta_i^j \varphi_i^j(x)$. Then $P_d\psi = \lambda\psi$ is equivalent to

$$\sum_{i_1,k_1} \beta_{i_1}^{k_1}(P\varphi_{i_1}^{k_1}, \varphi_{i_2}^{k_2}) = \lambda \sum_{i_1,k_1} \beta_{i_1}^{k_1}(\varphi_{i_1}^{k_1}, \varphi_{i_2}^{k_2}) \quad \text{for all } i_2, k_2.$$

Introducing the coefficient vector $\beta = (\beta_i^j)$ we may write this equation in matrix form as

$$M_1\beta - \lambda M_2\beta = 0, \tag{3.5}$$

where both M_1 and M_2 have nonnegative entries. Moreover, noting that

$$\int Q_d g \, dm = \int P_d g \, dm = \int g \, dm \quad \text{for every } g \in L^2,$$

and using the fact that $\sum_{i,j} \varphi_i^j(x) = 1$ we can proceed in the same way as in [10, Lemma 8], to see that $(1, 1, \ldots, 1)$ is a left eigenvector with eigenvalue 1 for the generalized eigenvalue problem (3.5). The fact that M_2 is invertible—since $\{\varphi_i^j\}$ is a basis of V_d—now implies that there is an eigenvector α with

$$M_2^{-1}M_1\alpha = \alpha.$$

We claim that $M_2^{-1}M_1$ has a cyclic structure so that $(M_2^{-1}M_1)^r$ is of block diagonal form where the blocks have the dimensions d_j, $j = 0, 1, \ldots, r-1$. Decomposing α with respect to this block structure we may proceed as in the proof of part (b) of Proposition 3.3 to show that ω_r^k, $k = 0, 1, \ldots, r-1$, are eigenvalues of (3.5) and hence of P_d as desired.

We now prove the claim. Since the matrix M_2 in (3.5) already has the desired block diagonal form (by (i) in (3.4)), it remains to show that the basis functions are cyclically permuted by P_d respecting the block structure of M_2. More precisely, we will show that $(P\varphi_{i_1}^{k_1}, \varphi_{i_2}^{k_2}) = 0$ if $k_2 \neq (k_1 + 1) \bmod r$.

By (3.1) we have

$$\int_{X_{j+1 \bmod r}} k(x, y) \, dm(y) = \begin{cases} 1 & \text{if } x \in X_j \\ 0 & \text{otherwise.} \end{cases}$$

It follows that

$$\int_{X_{k_2}} \int_{X_{k_1}} k(x, y) \, dm(x) dm(y) = \int_{X_{k_1}} \int_{X_{k_2}} k(x, y) \, dm(y) dm(x) = 0$$

if $k_2 \neq (k_1 + 1) \bmod r$, and therefore

$$(P\varphi_{i_1}^{k_1}, \varphi_{i_2}^{k_2}) = \int_{X_{k_2}} \int_{X_{k_1}} k(x, y)\varphi_{i_1}^{k_1}(x)\varphi_{i_2}^{k_2}(y) \, dm(x) dm(y) = 0$$

if $k_2 \neq (k_1 + 1) \bmod r$ as desired. □

Now we may combine Theorem 3.5 and Proposition 3.7 to obtain a convergence result for eigenvectors corresponding to eigenvalues of P of modulus one.

COROLLARY 3.8. *Suppose that P and its approximation P_d satisfy the hypotheses stated above. Then each simple eigenvalue $e^{2\pi ik/r}$ of P on the unit circle is an eigenvalue of P_d and there are corresponding eigenvectors g_d of P_d converging to an eigenfunction h of P. More precisely, there is a constant $C > 0$ such that for all $d \geq 1$*

$$\|h - g_d\|_2 \leq C\|P_d - P\|_2.$$

4. The computation of SRB-measures.

SRB-measures. Let us briefly recall the notion of an SRB-measure. In the existing literature several different definitions can be found which are all equivalent in the case where the underlying dynamical behavior is Axiom A. This is precisely the situation we will consider, and hence we can, without loss of generality, work with just one of them.

DEFINITION 4.1. *An ergodic measure μ is an* SRB-measure *if there exists a subset $U \subset X$ with $m(U) > 0$ and such that for each continuous function ψ*

$$\lim_{N\to\infty} \frac{1}{N} \sum_{j=0}^{N-1} \psi(f^j(x)) = \int \psi \, d\mu \tag{4.1}$$

for all $x \in U$.

Remark 4.2.

(a) Recall that (4.1) always holds for μ-a.e. $x \in X$ by the Birkhoff ergodic theorem. The crucial difference for an SRB-measure is that the temporal average equals the spatial average for a set of initial points $x \in X$ which has positive Lebesgue-measure. This is the reason why this measure is also referred to as the *natural* or the *physically relevant* invariant measure.

(b) The concept of SRB-measures in the context of Anosov systems had been introduced by Y. Sinai in the 1960s (e.g., [27]). Later the existence of SRB-measures was shown for Axiom A systems by R. Bowen and D. Ruelle (see [26, 3]). More recently, M. Benedicks and L.-S. Young have shown that the Hénon-map has an SRB-measure for a "large" set of parameter values [1]. However, it is still one of the major problems in ergodic theory to establish the existence of SRB-measures for a more general class of dynamical systems.

Small random perturbations. We specify concretely the stochastic transition function p underlying the numerical realization. Recall that the purpose is to approximate the Frobenius–Perron operator of a *deterministic* dynamical system represented by a diffeomorphism f. Hence the *stochastic* system that we consider should be a small perturbation of this original deterministic system.

For $\varepsilon > 0$ we set

$$k_\varepsilon(x, y) = \frac{1}{\varepsilon^n m(B)} \chi_B \left(\frac{1}{\varepsilon} \left(y - x \right) \right), \quad x, y \in X. \tag{4.2}$$

Here $B = B_0(1)$ denotes the open ball in $\mathbb{R}^n$ of radius one and χ_B is the characteristic function of B. Obviously $k_\varepsilon(f(x), y)$ is a transition density function and we may define a stochastic transition function p_ε by

$$p_\varepsilon(x, A) = \int_A k_\varepsilon(f(x), y) \, dm(y). \tag{4.3}$$

Remark 4.3. Note that $p_\varepsilon(x, \cdot) \to \delta_{f(x)}$ for $\varepsilon \to 0$ uniformly in x in a weak*-sense. Hence the Markov process defined by any initial probability measure μ and the transition function p_ε is a *small random perturbation* of the deterministic system f in the sense of Yu. Kifer [17].

Observe that we can apply the results from section 3 since

$$\iint |k_\varepsilon(f(x), y)|^2 \, dm(x) dm(y) \le \left(\frac{m(X)}{\varepsilon^n m(B)} \right)^2 < \infty,$$

and therefore the Frobenius–Perron operator $P_\varepsilon : L^2 \to L^2$ is compact (see Lemma 3.4).

Approximation of SRB-measures. We now combine Corollary 3.8 with a result of Yu. Kifer [17] to show that the approximations of the invariant measures converge to an SRB-measure with decreasing magnitude of the random perturbations.

Let us be more precise. Suppose that the diffeomorphism f possesses a hyperbolic attractor Λ with an SRB-measure μ_{SRB}, and let p_ε be a small random perturbation of f. Then, under certain hypotheses on p_ε, it is shown in [17] that the invariant measures of p_ε converge in a weak*-sense to μ_{SRB} as $\varepsilon \to 0$. On the other hand we can approximate the relevant eigenmeasures of P_ε by Corollary 3.8 and this leads to the desired result.

THEOREM 4.4. *Suppose that the diffeomorphism f has a hyperbolic attractor Λ, and that there exists an open set $U_\Lambda \supset \Lambda$ such that*

$$k_\varepsilon(x,y) = 0 \quad \text{if } x \in \overline{f(U_\Lambda)} \text{ and } y \notin U_\Lambda.$$

Then the transition function p_ε in (4.3) *has a unique invariant measure π_ε with support on Λ, and the approximating measures*

$$\mu_d^\varepsilon(A) = \int_A g_d^\varepsilon \, dm$$

converge in a weak-sense to the SRB-measure μ_{SRB} of f as $\varepsilon \to 0$ and $d \to \infty$,*

$$\lim_{\varepsilon\to 0} \lim_{d\to\infty} \mu_d^\varepsilon = \mu_{SRB}. \tag{4.4}$$

Proof. It is straightforward to check that the conditions of Theorem 1 in [17] are satisfied for the densities

$$q_x^\varepsilon(y) = k_\varepsilon(x,y),$$

provided $\varepsilon < 1$. Hence—denoting the unique invariant measure of the transition function p_ε with support in $\overline{U_\Lambda}$ by π_ε—this theorem implies that

$$\pi_\varepsilon \xrightarrow{\text{weak*}} \mu_{SRB} \quad \text{for } \varepsilon \to 0. \tag{4.5}$$

By Remark 2.7 we know that π_ε is absolutely continuous, and we denote its density function by κ_ε. Then Corollary 3.8 guarantees that the fixed points g_d^ε of P_d^ε converge to κ_ε as $d \to \infty$. Therefore

$$\begin{aligned}\left|\int h\, d\mu_d^\varepsilon - \int h\, d\pi_\varepsilon\right| &= \left|\int h(g_d^\varepsilon - \kappa_\varepsilon)\, dm\right| \\ &\le \|h\|_2 \|g_d^\varepsilon - \kappa_\varepsilon\|_2 \to 0\end{aligned}$$

as $d \to \infty$ for every $h \in L^2$ and, in particular,

$$\mu_d^\varepsilon \xrightarrow{\text{weak*}} \pi_\varepsilon \quad \text{as } d \to \infty.$$

Combining this with (4.5) leads to (4.4), as desired. □

5. Extracting dynamical behavior. In the previous section we have seen that we can approximate the physically relevant invariant measure—the SRB-measure—of our original deterministic system by the computation of the invariant measure of a randomly perturbed system. However, dynamically the nonstationary behavior is also interesting, and we will now describe how to detect numerically components in state space which are (almost) cyclically permuted.

Extraction of cyclic behavior. Suppose that the stochastic transition function of the randomly perturbed dynamical system satisfies the cycle condition (3.1). Then the purpose is to identify the components X_j. By Proposition 3.7 we know that the approximating operator P_d^ε has the eigenvalues ω_r^k, $k = 0, 1, \ldots, r-1$, and we now show that the cyclic components can be approximated by certain linear combinations of the corresponding eigenvectors.

It is instructive to consider the simplest case where $r = 2$ before we turn our attention to the general situation. Suppose that we have two components X_0 and X_1 which are cyclically permuted by our process. Then the aim is to find approximations of eigenmeasures μ_0 and $\mu_1 = P_\varepsilon\mu_0$ of P_ε^2 with support on X_0 and X_1, respectively; see Proposition 3.3. By the same proposition we know that $\omega^0 = 1$ and $\omega^1 = -1$ are eigenvalues of P_ε. Let ν_0 and ν_1 be corresponding (real) eigenmeasures. Then, by (3.2), there are $\alpha_0, \alpha_1 \in \mathbb{R}$ such that

$$\nu_0 = \alpha_0(\mu_0 + P_\varepsilon\mu_0) \quad \text{and} \quad \nu_1 = \alpha_1(\mu_0 - P_\varepsilon\mu_0).$$

Rescaling ν_0 and ν_1 so that $\nu_0(X_0) = \nu_1(X_0) = 1$ we can compute μ_0 and μ_1 by

$$\mu_0 = \frac{1}{2}(\nu_0 + \nu_1) \quad \text{and} \quad \mu_1 = \frac{1}{2}(\nu_0 - \nu_1).$$

This process also shows how to construct eigenvectors of the Galerkin approximation for which linear combinations are appropriate approximations of the probability measures μ_0 and μ_1.

We now consider the general case. For $\ell = 0, 1, \ldots, r-1$ denote by $\mu_\ell = P_\varepsilon^\ell\mu_0$ the invariant measure of P_ε^r with support on X_ℓ (see Proposition 3.3).

LEMMA 5.1. *For $s \in \{0, 1, \ldots, r-1\}$ let*

$$\nu_k^s = \sum_{j=0}^{r-1} \omega_r^{-kj} P_\varepsilon^j \mu_s \tag{5.1}$$

be a specific choice for the eigenmeasures of P_ε corresponding to the eigenvalues ω_r^k, $k = 0, 1, \ldots, r-1$ (see (3.2)). Then

$$\frac{1}{r}\sum_{k=0}^{r-1} \omega_r^{\ell k} \nu_k^s = \mu_{\ell+s \bmod r}.$$

Proof. We compute

$$\frac{1}{r}\sum_{k=0}^{r-1} \omega_r^{\ell k}\nu_k^s = \sum_{j=0}^{r-1}\left(\frac{1}{r}\sum_{k=0}^{r-1}\omega_r^{(\ell-j)k}\right) P_\varepsilon^j\mu_s = P_\varepsilon^\ell\mu_s = \mu_{\ell+s \bmod r}.$$

Here we have used the identity

$$\frac{1}{r}\sum_{k=0}^{r-1}\omega_r^{(\ell-j)k} = \delta_{\ell j},$$

where $\delta_{\ell j}$ is the Kronecker symbol. □

The previous lemma indicates how to approximate the cyclic components of X: we have to find eigenvectors $v_0^s, \ldots, v_{r-1}^s$ of the matrix $M_2^{-1}M_1$ (see (3.5)) which are approximations of the eigenmeasures ν_k^s in (5.1) for an $s \in \{0, 1, \ldots, r-1\}$. Then we can compute

$$u_{\ell+s \bmod r} = \frac{1}{r} \sum_{k=0}^{r-1} \omega_r^{\ell k} v_k^s$$

for $\ell = 0, 1, \ldots, r-1$, and the positive components of u_j provide the desired information about the support of μ_j on X_j ($j = 0, 1, \ldots, r-1$).

Hence it remains to describe how to construct eigenvectors $v_0^s, \ldots, v_{r-1}^s$ approximating the ν_k^s in (5.1) for an $s \in \{0, 1, \ldots, r-1\}$. We do this for the case where the eigenvalues ω_r^k are simple—the case of several coexisting cycles will be treated in the following section.

Suppose that we have a set of eigenmeasures ρ_k corresponding to the eigenvalues ω_r^k, $k = 0, 1, \ldots, r-1$. Since the eigenvalues are simple we know that for each $s \in \{0, 1, \ldots, r-1\}$ there is a constant $\alpha_k^s \in \mathbb{C}$ such that ρ_k can be written as

$$\rho_k = \alpha_k^s \nu_k^s.$$

Hence the task is to rescale ρ_k so that $\alpha_k^s = 1$ for all k. By (3.2) it is easy to see that for each $s \in \{0, 1, \ldots, r-1\}$

$$\rho_k(X_s) \neq 0.$$

We choose a particular s and rescale the ρ_k's by (complex) factors so that

$$\rho_k(X_s) = 1 \quad \text{for all } k = 0, 1, \ldots, r-1.$$

With this choice it follows that $\rho_k = \nu_k^s$.

In the realization of the approximation of the ν_k^s's we proceed with the eigenvectors of $M_2^{-1} M_1$ in an analogous way: by our choice of the Galerkin approximation we just need to find an index such that the corresponding components of the eigenvectors do not vanish. The scaling as described above can then be done in a similar way: find complex multiples of the eigenvectors so that they possess (real) positive components which add up to one for each vector. We will illustrate the method by examples in section 6.

Identification of several coexisting cycles or invariant sets. In applications it may occur that there exist several different cycles in the dynamical system under consideration. We now show how to identify these sets numerically. Replacing P_ε by an appropriate power P_ε^r if necessary we can, without loss of generality, restrict our attention to the case where there are different invariant sets.

Again we begin with the simplest case and assume that P_ε has two linearly independent invariant probability measures ν_1 and ν_2. Then, by Theorem 2.6, there are constants α_j^i ($i, j = 1, 2$) with

$$\alpha_1^1 \nu_1 + \alpha_2^1 \nu_2 = \pi_1 \quad \text{and} \quad \alpha_1^2 \nu_1 + \alpha_2^2 \nu_2 = \pi_2,$$

where π_1 and π_2 are probability measures with $\pi_i(E_i) = 1$ for invariant sets E_1 and E_2. Let $B_i \subset E_i$ be subsets with $\nu_i(B_i) \neq 0$. Then the coefficients α_j^i can be found as the solutions of the equations

$$\alpha_1^1 \nu_1(B_2) + \alpha_2^1 \nu_2(B_2) = 0,$$
$$\alpha_1^2 \nu_1(B_1) + \alpha_2^2 \nu_2(B_1) = 0,$$

with the additional requirement that $\pi_i = \alpha_1^i \nu_1 + \alpha_2^i \nu_2$ are probability measures. Indeed, let β_j^i be constants such that

$$\alpha_1^1 \nu_1 + \alpha_2^1 \nu_2 = \beta_1^1 \pi_1 + \beta_2^1 \pi_2,$$
$$\alpha_1^2 \nu_1 + \alpha_2^2 \nu_2 = \beta_1^2 \pi_1 + \beta_2^2 \pi_2.$$

Then, in particular,

$$0 = \alpha_1^1 \nu_1(B_2) + \alpha_2^1 \nu_2(B_2) = \beta_1^1 \pi_1(B_2) + \beta_2^1 \pi_2(B_2) = \beta_2^1 \pi_2(B_2),$$
$$0 = \alpha_1^2 \nu_1(B_1) + \alpha_2^2 \nu_2(B_1) = \beta_1^2 \pi_1(B_1) + \beta_2^2 \pi_2(B_1) = \beta_1^2 \pi_1(B_1).$$

Since $\nu_i(B_i) \neq 0$, it follows that $\pi_i(B_i) \neq 0$ and therefore $\beta_2^1 = \beta_1^2 = 0$ as desired.

We now generalize this observation. Suppose that the eigenspace of the Frobenius–Perron operator corresponding to the eigenvalue 1 is e-dimensional. Then, by Theorem 2.6, there are e distinct invariant sets $E_1, E_2, \ldots, E_e$ and invariant measures $\pi_1, \pi_2, \ldots, \pi_e$ such that $\pi_j(E_j) = 1$.

LEMMA 5.2. *Let ν_j, $j = 1, \ldots, e$, be invariant measures spanning the e-dimensional eigenspace of the Frobenius–Perron operator corresponding to the eigenvalue 1. Let $B_i \subset E_{k_i}$ be subsets such that $\nu_i(B_i) \neq 0$, $i = 1, 2, \ldots, e$. We have that*

(a) *the matrix $M = (\nu_j(B_i))_{i,j=1,\ldots,e}$ has full rank if and only if*

$$\{k_1, k_2, \ldots, k_e\} = \{1, 2, \ldots, e\}; \tag{5.2}$$

(b) *if $\operatorname{rank}(M) = e$, then the invariant measures π_j, $j = 1, \ldots, e$, are given by*

$$\pi_{k_\ell} = \sum_{j=1}^{e} \alpha_j^\ell \nu_j,$$

where $\alpha^\ell = (\alpha_j^\ell)$ is the (rescaled) null vector of $M_\ell = (\nu_j(B_i))_{i,j=1,\ldots,e,\ i \neq \ell}$.

Proof. Suppose that (5.2) holds. Let $\alpha \in \mathbb{R}^e$ be an element of the kernel of the matrix $M = (\nu_j(B_i))$, that is,

$$\sum_{j=1}^{e} \alpha_j \nu_j(B_i) = 0, \quad i = 1, 2, \ldots, e.$$

Using the fact that there are constants β_j, $j = 1, \ldots, e$, such that

$$\sum_{j=1}^{e} \alpha_j \nu_j = \sum_{j=1}^{e} \beta_j \pi_j,$$

we obtain

$$0 = \sum_{j=1}^{e} \beta_j \pi_j(B_i) = \beta_{k_i} \pi_{k_i}(B_i), \quad i = 1, 2, \ldots, e.$$

Since $0 \neq \nu_i(B_i) = \gamma_i \pi_{k_i}(B_i)$ for appropriate constants γ_i, it follows that $\beta_i = 0$ for $i = 1, 2, \ldots, e$ (here we have used (5.2)). Hence $\sum_{j=1}^{e} \alpha_j \nu_j = 0$, and since the ν_j's span the e-dimensional eigenspace of the Frobenius–Perron operator corresponding to the eigenvalue 1, we may conclude that $\alpha = 0$.

To complete the proof of part (a) it remains to show that $M = (\nu_j(B_i))$ is singular if (5.2) is not satisfied, that is, if there is a $k \in \{1, 2, \dots, e\}$ which is not in $\{k_1, k_2, \dots, k_e\}$. Writing π_k as

$$\pi_k = \sum_{j=1}^{e} \alpha_j \nu_j$$

and using $B_i \not\subset E_k$ we obtain for $i = 1, \dots, e$

$$0 = \pi_k(B_i) = \sum_{j=1}^{e} \alpha_j \nu_j(B_i).$$

Hence $\alpha = (\alpha_j)$ is a nontrivial null vector of M. The statement in part (b) is an immediate consequence. □

For the numerical identification of the sets $E_1, \dots, E_e$ we choose nonvanishing components of the e approximations $v^1, \dots, v^e$ of eigenmeasures $\nu_1, \dots, \nu_e$ in such a way that the matrix $M = (v_i^j)_{i,j=1,\dots,e}$ is nonsingular. Then we identify the distinct invariant components as the support of the eigenmeasures approximated by the scaled null vectors of the matrices $M_\ell = (v_i^j)_{i,j=1,\dots,e,\ i \neq \ell}$.

Extraction of almost cyclic behavior. We distinguish two different scenarios by which an *almost cyclic* behavior may occur in a dynamical system:

(i) The first scenario we have in mind is that cyclic components $X_0, X_1, \dots, X_{r-1}$ merge while a control parameter is varied in the system. If this has happened, then the cyclic behavior can frequently be observed although it is strictly no longer present.

(ii) Second, it may happen that two *different* cycles merge while a control parameter is varied. For instance, if two invariant sets "collide," then immediately after the collision there are still two subsets in state space which are *almost invariant.*

In section 6 we will illustrate both scenarios by numerical examples.

We know that the rth roots of unity are eigenvalues of P_ε if there are r cyclic components. If these components merge, then these eigenvalues leave the unit circle. The main purpose of this section is to relate the modulus of these eigenvalues to the probability that the cyclic behavior is still observed. As in the previous subsection, we may just consider *almost invariant* sets by replacing P_ε by P_ε^r if necessary. In this case a cluster of eigenvalues moves away from one along the real line while several (precisely) invariant sets disappear.

DEFINITION 5.3. *A subset $A \subset X$ is* δ-almost invariant with respect to $\rho \in \mathcal{M}$ *if $\rho(A) \neq 0$ and*

$$\int_A p_\varepsilon(x, A)\, d\rho(x) = \delta \rho(A).$$

Remark 5.4.

(a) Using the definition of p_ε we compute for a subset $A \subset X$

$$p_\varepsilon(x, A) = \frac{m(A \cap B_{f(x)}(\varepsilon))}{m(B_0(\varepsilon))}.$$

Hence

$$\delta = \frac{1}{\rho(A)} \int_A \frac{m(A \cap B_{f(x)}(\varepsilon))}{m(B_0(\varepsilon))}\, d\rho(x).$$

(b) We have seen that $p_\varepsilon(x,\cdot) \to \delta_{f(x)}$ for $\varepsilon \to 0$. Thus, we obtain in the deterministic limit

$$\int_A p_0(x,A)\, d\rho(x) = \int_A \delta_{f(x)}(A)\, d\rho(x) = \rho(f^{-1}(A) \cap A).$$

Therefore in this case δ is the relative ρ-measure of the subset of points in A which are mapped into A.

According to the classification of the occurrence of almost cyclic behavior given at the beginning of this section we identify almost cyclic behavior in the numerical realization as follows:

(i) If cyclic components $X_0, X_1, \ldots, X_{r-1}$ merge while a control parameter is varied in the dynamical system, then we use the same linear combinations as in the unperturbed case to identify the components of the "almost-cycle." The overlap of the different components indicates the subset of points which no longer follow the cyclic behavior.

(ii) If two invariant sets merge, then we use the results obtained for the identification of coexisting invariant sets. However, observe that in the perturbed case we can no longer expect that an approximation of the matrix M in Lemma 5.2 will be singular although the chosen subsets do not properly represent the (almost) cyclic behavior. In the numerical realization we take this into account and use as a criterion the condition number of these matrices: in the construction of the matrix M we choose the components of the approximating vectors in such a way that the condition number of M is as small as possible.

From now on we assume that $\lambda \neq 1$ is an eigenvalue of P_ε with corresponding real valued eigenmeasure $\nu \in \mathcal{M}_\mathbb{C}$, that is,

$$P_\varepsilon \nu = \lambda \nu.$$

Recall that in this case $\nu(X) = 0$ (see Remark 3.2 (a)). The aim is to relate the value of this eigenvalue to the probability δ in Definition 5.3. We begin with the following elementary observation.

LEMMA 5.5. *Suppose that ν is scaled so that $|\nu| \in \mathcal{M}$, and let $A \subset X$ be a set with $\nu(A) = \frac{1}{2}$. Then $\nu = |\nu|$ on A.*

Proof. Obviously, $\nu(B) \leq |\nu|(B)$ for all measurable B. For contradiction let $B \subset A$ be a measurable set with $\nu(B) < |\nu|(B)$. It follows that there is a $C \subset A$ with $\nu(C) < 0$. Hence we have for $E = A - C$

$$\nu(E) > \nu(A) = \frac{1}{2}.$$

Using $\nu(X) = 0$ and $|\nu|(X) = 1$, this leads to a contradiction

$$1 = |\nu|(E) + |\nu|(X - E) \geq |\nu(E)| + |\nu(X - E)| > \frac{1}{2} + \frac{1}{2} = 1. \qquad \square$$

Remark 5.6. Observe that by the *Hahn decomposition* (see, e.g., [28]) the existence of a set A with $\nu(A) = \frac{1}{2}$ is guaranteed.

PROPOSITION 5.7. *Suppose that ν is scaled so that $|\nu| \in \mathcal{M}$, and let $A \subset X$ be a set with $\nu(A) = \frac{1}{2}$. Then*

$$\delta + \sigma = \lambda + 1, \tag{5.3}$$

if A is δ-almost invariant and $X - A$ is σ-almost invariant with respect to $|\nu|$.

Proof. By Lemma 5.5 we have

$$\int_A p_\varepsilon(x, A)\, d\nu(x) = \int_A p_\varepsilon(x, A)\, d|\nu|(x) = \delta\nu(A)$$

since $\nu(A) = \frac{1}{2}$. Similarly,

$$\begin{aligned}\int_{X-A} p_\varepsilon(x, A)\, d\nu(x) &= \int_{X-A} 1 - p_\varepsilon(x, X - A)\, d\nu(x)\\ &= \nu(X - A) - \sigma\nu(X - A)\\ &= (\sigma - 1)\nu(A),\end{aligned}$$

since $\nu(X - A) = -\nu(A)$. Finally, using the fact that ν is an eigenmeasure we compute

$$\begin{aligned}\lambda\nu(A) &= \int_A p_\varepsilon(x, A)\, d\nu(x) + \int_{X-A} p_\varepsilon(x, A)\, d\nu(x)\\ &= \delta\nu(A) + (\sigma - 1)\nu(A) = (\delta + \sigma - 1)\nu(A),\end{aligned}$$

yielding (5.3). □

Remark 5.8.

(a) Observe that in the case where λ is close to one we may assume that the probability measure $|\nu|$ is close to the invariant measure μ of the system. In this sense we have derived the desired relation between the eigenvalue λ and the probability that the system is still behaving in a cyclic way.

(b) In our numerical computations we will work with the unperturbed equations rather than introduce noise artificially. Thus, it would be important to know whether the eigenvalues of P_0 and P_ε are close to each other for small ε. First results concerning the stochastic stability of the spectrum of the Frobenius–Perron operator are obtained in [2].

In (5.3) both δ and σ occur, and in general there will be no relation between these constants. However, if the underlying system possesses an additional symmetry, then we can express one of them in terms of the other one.

To illustrate this fact let us consider the simplest case where we have a symmetry transformation κ in the problem with $\kappa^2 = id$. In that case

$$p_\varepsilon(x, B) = p_\varepsilon(\kappa x, \kappa B) \quad \text{for all measurable } B \subset X, \tag{5.4}$$

which implies that for any $\rho \in \mathcal{M}$

$$\int_B p_\varepsilon(x, B)\, d\rho = \int_{\kappa B} p(x, \kappa B)\, d\kappa^* \rho.$$

Hence we have the following corollary.

COROLLARY 5.9. *Suppose in addition to the assumptions in Proposition* 5.7 *that*

(i) *p_ε is symmetric, that is,* (5.4) *holds,*

(ii) *the set A satisfies $\kappa A = X - A$, and*

(iii) *the measure $|\nu|$ is κ-symmetric, that is, $\kappa^*|\nu| = |\nu|$.*

Then $X - A$ is δ-almost invariant with respect to $|\nu|$ if and only if A is δ-almost invariant, and in particular

$$\delta = \frac{\lambda + 1}{2}. \tag{5.5}$$

Numerical solution of the eigenvalue problems. Since in this article we were mainly interested in the description of how to extract numerically information on the dynamical behavior from the spectrum of P_ε, we just outline the algorithmic steps which are necessary for the numerical approximation and solution of the eigenvalue problem (3.5). For the details concerning the implementation the reader is asked to consult the references listed below.

All the algorithms described in the following are integrated into the software package

GAIO (**G**lobal **A**nalysis of **I**nvariant **O**bjects),

which can be obtained online via http. (See the first page of this paper for the authors' homepages.)

(i) *Construction of a box-covering.* We begin with the construction of a box-covering of the interesting (randomly perturbed) dynamics in state space. This can be done either by a subdivision technique (see [12, 6, 5, 7]) or by a cell-mapping approach (see, e.g., [20, 14]). This way we obtain a collection of boxes B_k, $k = 1, 2, \ldots, N$, such that the part of state space containing the interesting dynamics is covered by their union.

Remark 5.10. To simplify the description we assume that all the boxes have the same volume. However, it turned out that a box-covering can be constructed in an even more efficient way if the size of the boxes is chosen in an adaptive way in each step of the subdivision procedure. This fact is explored in [9].

(ii) *Galerkin approximation.* The basis functions we have chosen are the characteristic functions of the B_k's,

$$\varphi_k = \chi_{B_k}.$$

We assume that the Hausdorff distance between X and the covering $\cup B_k$ is small enough so that the assumption (3.4) on the φ_k's is satisfied.

Remark 5.11. In practice we use the subdivision technique from [6] to construct a box-covering for which (3.4) holds. In fact, we simply neglect those boxes which have measure zero once a certain number of steps in the subdivision algorithm has been performed.

(iii) *Approximation and solution of the eigenvalue problem.* We approximate the coefficients of the matrices M_1 and M_2 in (3.5) by a numerical evaluation of the integrals which are involved. Observe that by our specific choice of the boxes and the φ_k's the matrix M_2 is a multiple of the identity. Hence the main numerical effort lies in the computation of the inner products to approximate the operator P_ε. This is done either by a Monte Carlo method or by an exhaustion technique as described in [13].
For the computation of eigenvalues and corresponding eigenvectors we use ARPACK which provides an iterative eigenvalue solver for sparse matrices (see [22]). Alternatively we use an approach based on *bordered matrices* for finding specific single eigenvectors. For the solution of the corresponding system of linear equations we use an iterative method taking into account the fact that the matrix M_1 is extremely sparse.

6. Examples. In this section we illustrate our numerical methods by three examples in which the transition from a true cyclic behavior to an almost cycling one becomes apparent. First we use the well-known Hénon map as an example to analyze a 2-cycle and an almost 2-cycle. The second example is a $\mathbb{Z}_3$-equivariant mapping in the complex plane which shows a cycling behavior of period six. Finally we investi-

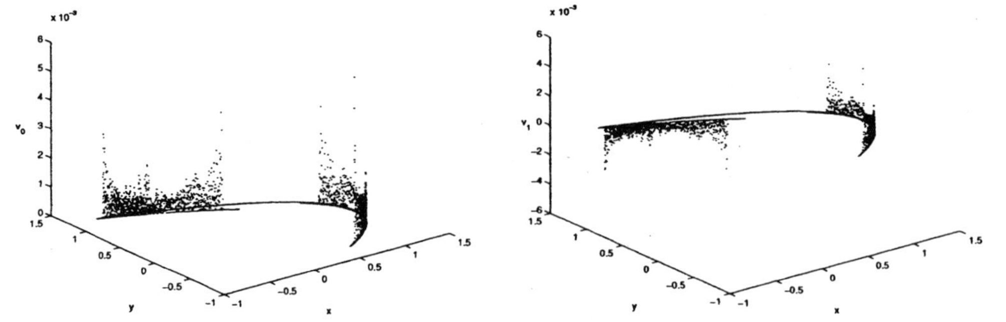

(a) v_0 (b) v_1

FIG. 6.1. *Eigenvectors of the approximation of the Frobenius–Perron operator for the Hénon map* ($a = 1.2$, $b = 0.2$).

gate numerically the *Chua circuit.* Since this is an ordinary differential equation, no nontrivial cycling is expected to be seen. However, in this case we will identify two sets in phase space which are almost invariant with respect to the flow.

All the computations were done without an artificial introduction of noise. Rather it turned out to be sufficient for our purposes to interpret the round off error as a small random perturbation.

A 2-cycle in the Hénon map. We consider a scaled version of the well-known *Hénon map,*

$$f(x, y) = (1 - ax^2 + y/5,\ 5bx),$$

where we fix $b = 0.2$ and vary a. For $a = 1.2$ this map possesses a 2-cycle, and we have used the approximation procedure described in section 5 to identify the two components X_0 and X_1. In Figure 6.1 we show the approximations v_0 and v_1 of the two eigenmeasures of the Frobenius–Perron operator corresponding to the eigenvalues $\lambda_0 = 1$ and $\lambda_1 = -1$.

By Lemma 5.1

$$u_0 = \frac{1}{2}(v_0 + v_1) \quad \text{and} \quad u_1 = \frac{1}{2}(v_0 - v_1)$$

are approximations of probability measures μ_0 and μ_1 which have support on X_0 and X_1, respectively. These are shown in Figure 6.2.

Remark 6.1. In the computation the box-covering was obtained by the continuation algorithm described in [5]. The boxes were of size $1/2^{10}$ in each coordinate direction and the continuation was restricted to the square $Q = [-2, 2]^2 \subset \mathbb{R}^2$. This way we have produced a covering of the closure of the one-dimensional unstable manifold of the hyperbolic fixed point in the first quadrant by 2525 boxes.

Next we set $a = 1.272$. For this parameter value the 2-cycle has disappeared, but in simulations the cycling behavior can still be observed for most iterates. Correspondingly we find that $\lambda_1 = -0.9944$ is an eigenvalue of the approximation of the Frobenius–Perron operator. Using the same notation as before we show in Figures 6.3

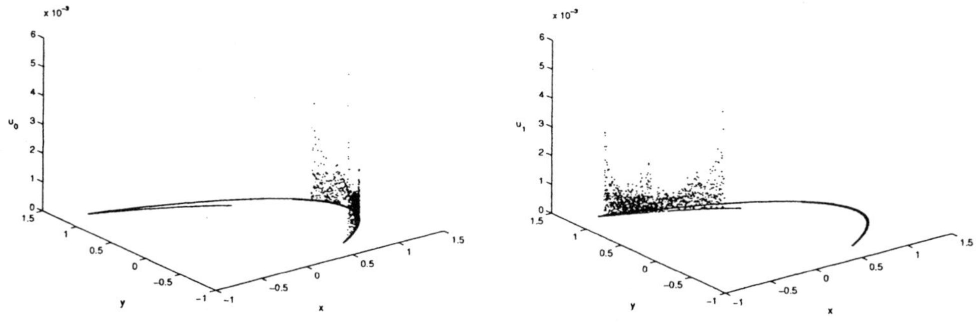

(a) u_0 (b) u_1

FIG. 6.2. *Approximations of probability measures with support on the two components of the 2-cycle* ($a = 1.2$, $b = 0.2$).

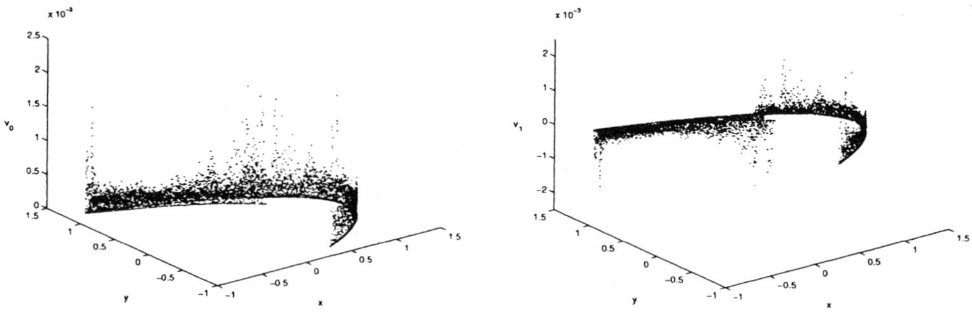

(a) v_0 (b) v_1

FIG. 6.3. *Eigenvectors of the approximation of the Frobenius–Perron operator for the Hénon map* ($a = 1.272$, $b = 0.2$).

and 6.4 the approximations of the eigenmeasures. In this case the box-covering has 3101 elements. Note that the supports of u_0 and u_1 have a nonempty intersection. This fact is illustrated in Figure 6.5 where we have marked by black circles all boxes on which $u_0 > 10$ and $u_1 > 10$.

A period six cycle. As the second example we slightly modify a mapping from [4] and consider the dynamical system $f : \mathbb{C} \to \mathbb{C}$,

$$f(z) = e^{-\frac{2\pi i}{3}} \left((|z|^2 + \alpha)z + \frac{1}{2}\bar{z}^2 \right),$$

for the parameter value $\alpha = -1.7$. For the computation of the box-covering we have used the subdivision algorithm described in [6]. Starting with the square $Q = [-1.5, 1.5]^2$ we have subdivided Q seven times by bisection in each coordinate direction

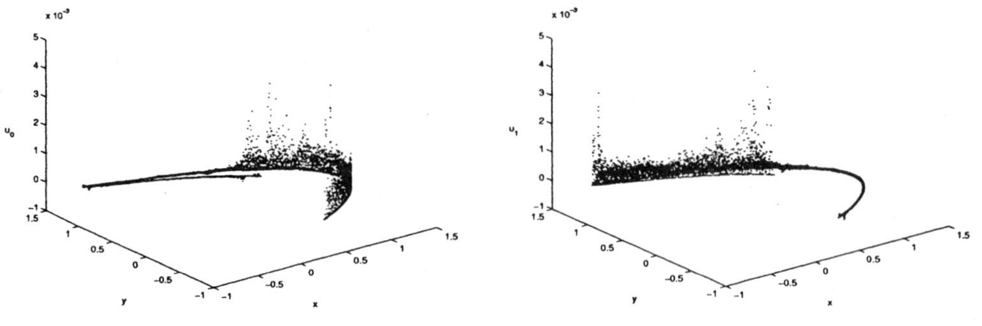

(a) u_0 (b) u_1

FIG. 6.4. *Approximations of probability measures which correspond to the two components of the almost 2-cycle* ($a = 1.272$, $b = 0.2$).

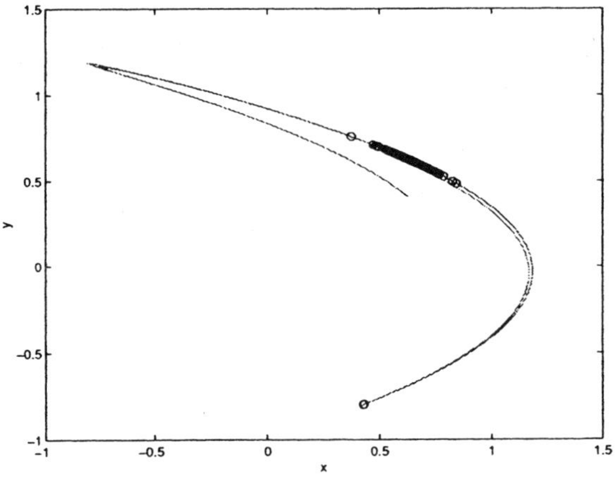

FIG. 6.5. *Approximation of a subset which is in the supports of both probability measures corresponding to the almost 2-cycle.*

which leads to a box-covering by 3606 boxes. In Figure 6.6 we show the approximation of the invariant measure, that is, the eigenvector v_0 corresponding to the eigenvalue $\lambda_0 = 1$ of the discretized Frobenius–Perron operator. In this case this operator additionally has the eigenvalues ω_6^k, $k = 1, \dots, 5$, and hence we may use Lemma 5.1 to compute approximations $v_0, \dots, v_5$ of the probability measures with support on the cyclic components $X_0, \dots, X_5$. These supports are shown in Figure 6.7.

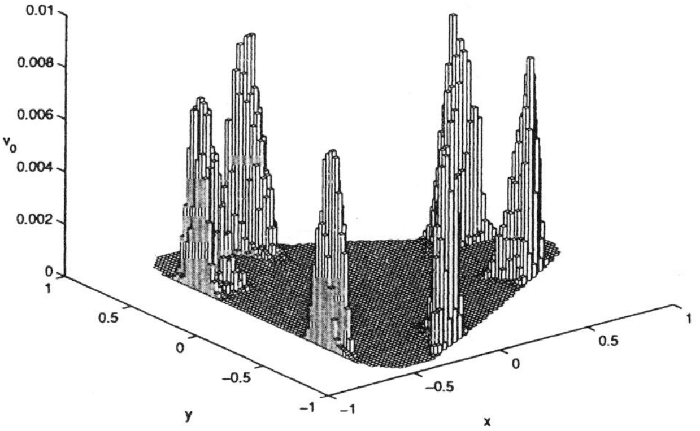

FIG. 6.6. *Approximation of the invariant measure for* $\alpha = -1.7$.

TABLE 1

j	λ_j	$\lvert\lambda_j\rvert$
0	1	1
1,5	$0.4918 \pm 0.8534i$	0.985
2,4	$-0.4880 \pm 0.8437i$	0.9747
3	-0.9709	0.9709

Now we vary the parameter and set $\alpha = -1.8$. For this value of α the strict cyclic behavior disappears and there is an almost 6-cycle. In Figure 6.8(a) we show the "essential" supports of the approximations $v_0, \dots, v_5$ of the six almost cyclic components. More precisely we show all boxes B_ℓ for which $(v_i)_\ell > 0.1$, $i = 0, \dots, 5$ (by $(v_i)_\ell$ we denote the ℓth component of the vector v_i). In Figure 6.8(b) we demonstrate that the intersection of these supports is nonempty. We have shown all boxes B_ℓ for which there are at least two indices $i, j \in \{0, \dots, 5\}$, $i \neq j$, such that $(v_i)_\ell > 0.1$ and $(v_j)_\ell > 0.1$.

In Table 1 we list the corresponding eigenvalues together with their absolute values. We remark that for this parameter value the subdivision algorithm leads to a covering by 4364 boxes.

Two almost invariant sets in the Chua circuit. Finally we present a system of three first order ordinary differential equations in which two almost invariant sets can be identified numerically. The system that we are considering is the *Chua circuit*,

$$\begin{aligned}\dot{x} &= \alpha\left(y - m_0 x - \frac{1}{3} m_1 x^3\right),\\ \dot{y} &= x - y + z,\\ \dot{z} &= -\beta y,\end{aligned}$$

where we have chosen the parameter values $\alpha = 18$, $\beta = 33$, $m_0 = -0.2$, and $m_1 = 0.01$.

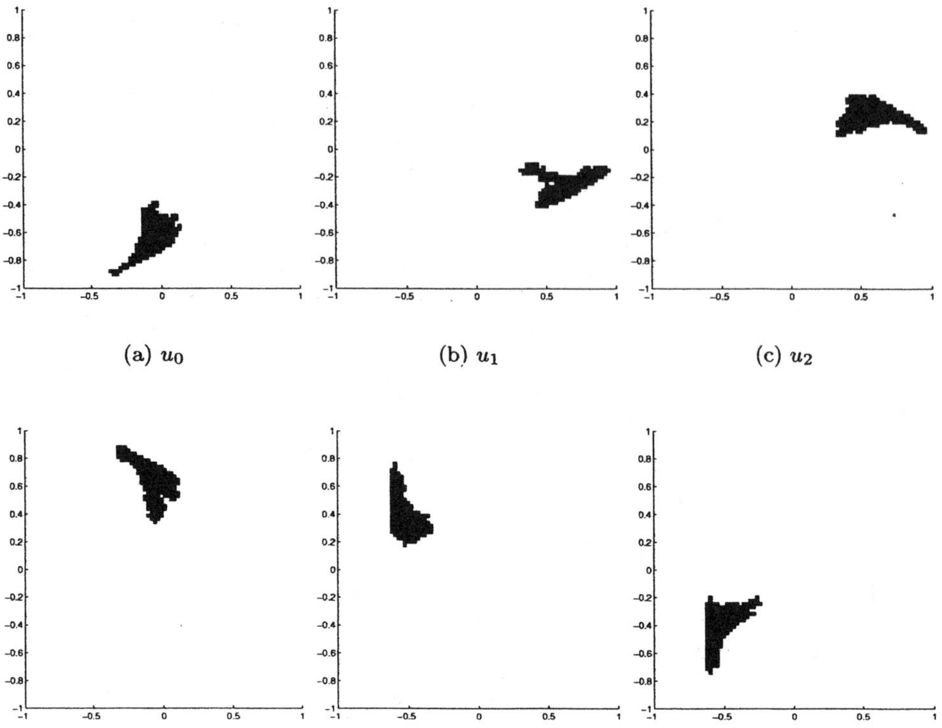

(a) u_0 (b) u_1 (c) u_2

(d) u_3 (e) u_4 (f) u_5

FIG. 6.7. *Approximation of the cyclic components $X_0, \ldots, X_5$ for $\alpha = -1.7$.*

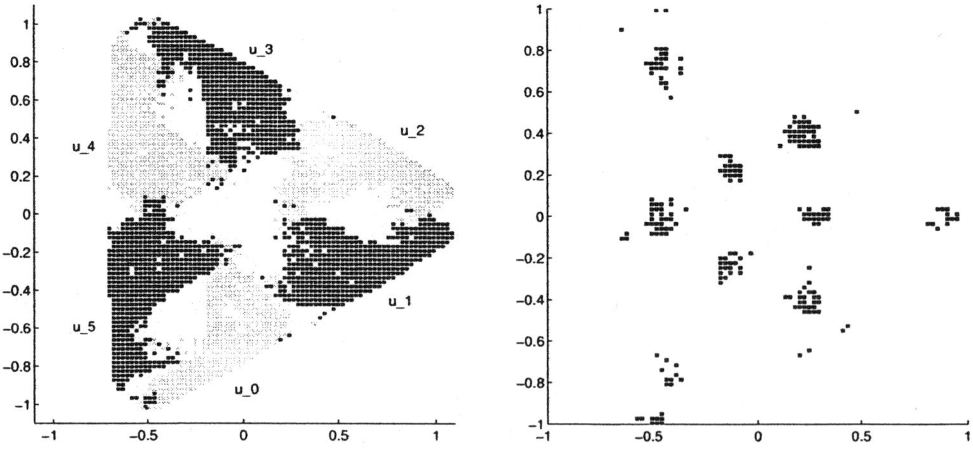

(a) "Essential" supports (b) Intersection of the supports

FIG. 6.8. *Approximation of the almost cyclic components for $\alpha = -1.8$.*

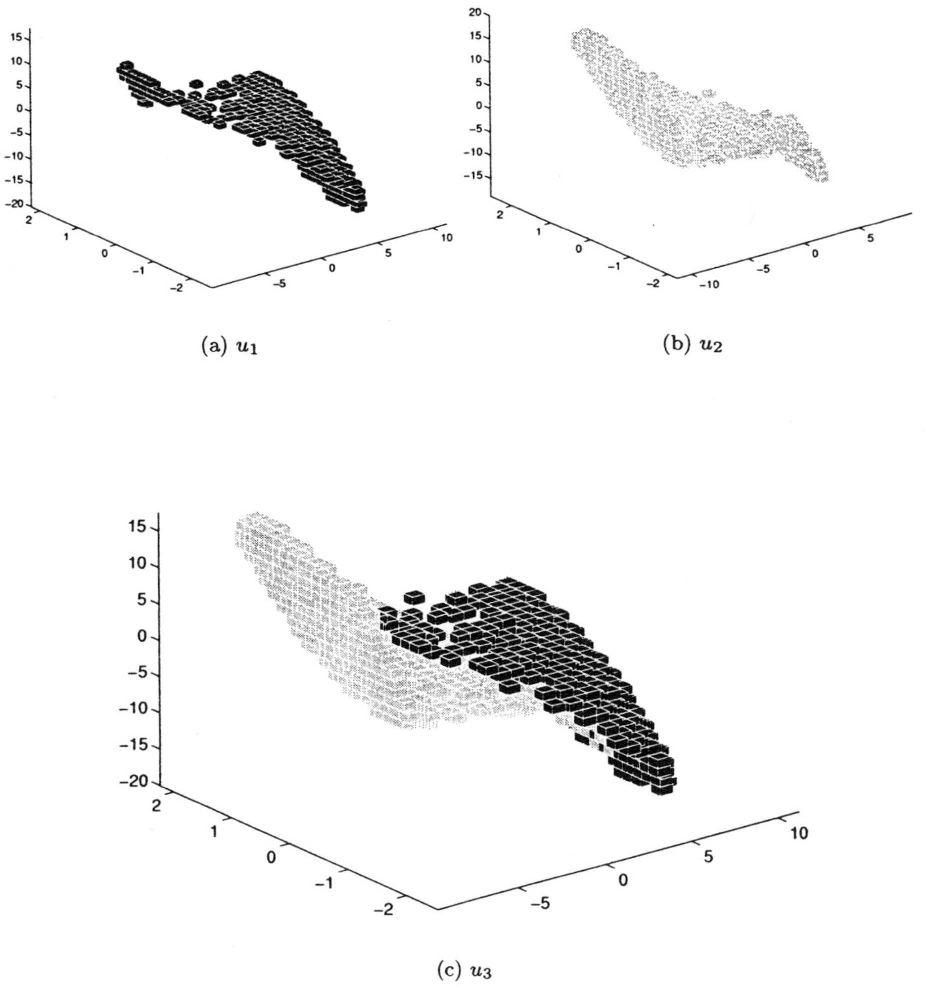

FIG. 6.9. *Illustration of the existence of two almost invariant sets in the Chua circuit.* (a) *Boxes corresponding to components of the approximating densities with value bigger than* 10^{-4}; (b) *boxes corresponding to components of the approximating densities with value less than* -10^{-4}; (c) *superposition of the two almost invariant sets.*

A detailed discussion of the dynamical behavior of this system can be found in [16]; see also [23]. We consider the time-0.1 map and—using the continuation method described in [6]—cover the unstable manifold of the origin by 10372 boxes. In addition to the eigenvalue 1 the discretized Frobenius–Perron operator does also possess the eigenvalue $\lambda_1 = 0.9272$. We may conclude from this result that there are two almost invariant sets. Indeed, a numerical approximation of the corresponding regions in phase space leads to the result shown in Figure 6.9. A detailed numerical study of this particular example can be found in [8].

Acknowledgments. We are grateful to Bernold Fiedler, who inspired our work on the detection of cyclic behavior during a provocative dinner conversation, and we thank Gerhard Keller for his patience throughout several discussions concerning our questions related to the theoretical background in ergodic theory.

REFERENCES

[1] M. Benedicks and L.-S. Young, *Sinai-Bowen-Ruelle measures for certain Hénon maps*, Invent. Math., 112 (1993), pp. 541–576.

[2] M. Blank and G. Keller, *Random perturbations of chaotic dynamical systems: Stability of the spectrum*, Nonlinearity, 11 (1998), pp. 1351–1364.

[3] R. Bowen and D. Ruelle, *The ergodic theory of axiom a flows*, Invent. Math., 29 (1975), pp. 181–202.

[4] P. Chossat and M. Golubitsky, *Symmetry-increasing bifurcation of chaotic attractors*, Phys. D, 32 (1988), pp. 423–436.

[5] M. Dellnitz and A. Hohmann, *The computation of unstable manifolds using subdivision and continuation*, in Nonlinear Dynamical Systems and Chaos, Progr. Nonlinear Differential Equations Appl. 19, H. W. Broer, S. A. van Gils, I. Hoveijn, and F. Takens, eds., Birkhäuser, Basel, 1996, pp. 449–459.

[6] M. Dellnitz and A. Hohmann, *A subdivision algorithm for the computation of unstable manifolds and global attractors*, Numer. Math., 75 (1997), pp. 293–317.

[7] M. Dellnitz, A. Hohmann, O. Junge, and M. Rumpf, *Exploring invariant sets and invariant measures*, Chaos, 7 (1997), p. 221.

[8] M. Dellnitz and O. Junge, *Almost invariant sets in Chua's circuit*, Internat. J. Bifur. Chaos Appl. Sci. Engrg., 7 (1997), pp. 2475–2485.

[9] M. Dellnitz and O. Junge, *An adaptive subdivision technique for the approximation of attractors and invariant measures*, Comput. Visual. Sci., 1 (1998), pp. 63–68.

[10] J. Ding, Q. Du, and T. Y. Li, *High order approximation of the Frobenius-Perron operator*, Appl. Math. Comput., 53 (1993), pp. 151–171.

[11] J. L. Doob, *Stochastic Processes*, John Wiley, New York, 1960.

[12] M. Eidenschink, *Exploring Global Dynamics: A Numerical Algorithm Based on the Conley Index Theory*, Ph.D. thesis, Georgia Institute of Technology, Atlanta, GA, 1995.

[13] R. Guder, M. Dellnitz, and E. Kreuzer, *An adaptive method for the approximation of the generalized cell mapping*, Chaos Solitons Fractals, 8 (1997), pp. 525–534.

[14] H. Hsu, *Global analysis by cell mapping*, Internat. J. Bifur. Chaos Appl. Sci. Engrg., 2 (1992), pp. 727–771.

[15] F. Y. Hunt, *Approximating the invariant measures of randomly perturbed dissipative maps*, J. Math. Anal. Appl., 198 (1996), pp. 534–551.

[16] A. Khibnik, D. Roose, and L. O. Chua, *On periodic orbits and homoclinic bifurcations in Chua's circuit with a smooth nonlinearity*, Internat. J. Bifur. Chaos Appl. Sci. Engrg., 3 (1993), pp. 363–384.

[17] Yu. Kifer, *General random perturbations of hyperbolic and expanding transformations*, J. Anal. Math., 47 (1986), pp. 111–150.

[18] Yu. Kifer, *Random Perturbations of Dynamical Systems*, Birkhäuser, Basel, 1988.

[19] Yu. Kifer, *Computations in dynamical systems via random perturbations*, Discrete Continuous Dynam. Systems, 3 (1997), pp. 457–476.

[20] E. Kreuzer, *Numerische Untersuchung nichtlinearer dynamischer Systeme*, Springer-Verlag, New York, 1987.

[21] A. Lasota and M. C. Mackey, *Chaos, Fractals and Noise*, Springer-Verlag, New York, 1994.

[22] R. B. Lehoucq, *Analysis and Implementation of an Implicitly Restarted Arnoldi Iteration*, Rice University Technical Report TR95-13, Houston, TX, 1995.

[23] T. Matsumoto, L. O. Chua, and M. Komuro, *The double scroll*, IEEE Trans. Circuits Systems, 32 (1985), pp. 797–818.

[24] S. P. Meyn and R. L. Tweedie, *Markov Chains and Stochastic Stability*, Springer-Verlag, New York, 1993.

[25] J. E. Osborn, *Spectral approximation for compact operators*, Math. Comp., 29 (1975), pp. 712–725.

[26] D. Ruelle, *A measure associated with axiom a attractors*, Amer. J. Math., 98 (1976), pp. 619–654.

[27] Y. G. Sinai, *Gibbs measures in ergodic theory*, Russian Math. Surveys, 166 (1972), pp. 21–69.

[28] K. Yosida, *Functional Analysis*, Springer-Verlag, New York, 1980.

Commun. Math. Phys. 208, 605 – 622 (2000)

Communications in
**Mathematical
Physics**

Absolutely Continuous Invariant Measures for Piecewise Real-Analytic Expanding Maps on the Plane

Masato Tsujii

Department of Mathematics, Hokkaido University, Sapporo 060-0810, Japan.
E-mail: tsujii@math.sci.hokudai.ac.jp

Received: 5 June 1998 / Accepted: 11 May 1999

Abstract: We prove the existence of absolutely continuous invariant measures for piecewise real-analytic expanding maps on bounded regions in the plane.

1. Introduction

Expanding properties of dynamical systems give rise to chaotic behavior of the orbits. On the other hand, they often lead to good ergodic properties such as the existence of absolutely continuous invariant measures. One typical example is the fork-lore theorem [12] that shows the existence of a smooth ergodic measure for every expanding C^2 self-map on a closed manifold. Hence, one interest in the study of chaotic dynamical systems is the relations between expanding properties and the ergodic properties they produce.

Lasota and Yorke showed, in their famous work [11], the existence of absolutely continuous invariant measures for piecewise C^2 expanding maps on intervals. They made use of the Perron–Frobenius operator and functions of bounded variation, and their idea has been used extensively in the study of one dimensional dynamical systems. This paper concerns the generalization of their result towards higher dimension. Though it is natural to expect similar results, it has turned out that things are not simple in higher dimension. In fact, at present, we do not know whether piecewise C^2 expanding maps on bounded regions in higher dimensional Euclidean space always have absolutely continuous invariant measures. The main difficulty in higher dimension is the fact that the partition of the domain into the regions where an iteration of the map is smooth can be very complicated.

Gerhard Keller treated piecewise C^2 expanding maps on bounded regions in the plane in his thesis [7,8] and gave some criterion for the existence of absolutely continuous invariant measure. The most effective result we have so far is that of Góra and Boyarski [4], which gives a lower bound for the minimum expansion rate that assures the existence of absolutely continuous invariant measures. Their result is valid for arbitrary dimension.

But their lower bound depends on the minimal angle on the boundaries of the regions in the partition associated to the map. See [1] for a modification of their result.

In this article, we consider the problem for piecewise real- analytic maps on bounded regions in the plane. (We will give the definition of piecewise real-analytic maps in the next section.) The real-analytic property somewhat relaxes the difficulty we mentioned above. In fact, we can prove the following theorem as the main result of this paper.

Theorem 1. *An absolutely continuous invariant finite measure exists for every piecewise real-analytic expanding map on a bounded region in the plane.*

This result improves a theorem of Keller in his thesis [7,8], which gives the same conclusion under one additional assumption that the map is piecewise conformal.

Actually we will prove the so-called Lasota–Yorke type inequality for some iterations of piecewise real-analytic expanding maps. It is known that we can derive many other properties of the maps from that kind of inequality. We will mention some of them in the appendix.

The author learned from Gerhard Keller that Jérôme Buzzi (C.N.R.S., Institut de Mathématiques de Luminy) obtained a similar result [2] when he was preparing the manuscript of this paper.

2. Piecewise Real-Analytic Map

We call a map $c : [a, b] \to \mathbb{R}^2$ a real-analytic curve if it is a restriction of a real-analytic map defined on a neighborhood of $[a, b]$ and satisfies $c'(t) \neq 0$ for $t \in [a, b]$. In what follows, we will assume

$$\|c'(t)\| \equiv 1 \quad \text{for } t \in [a, b] \tag{1}$$

for real-analytic curves, by real-analytic change of the variable t. Also we will denote the image of a real-analytic curve $c : [a, b] \to \mathbb{R}^2$ by the same symbol c, as an abuse of the symbol.

A continuous map $c : [a, b] \to \mathbb{R}^2$ is called a piecewise real-analytic curve if there is a sequence $a = \xi_0 < \xi_1 < \xi_2 < \cdots < \xi_n = b$ such that the restrictions $c|_{[\xi_i, \xi_{i+1}]}$, $0 \leq i < n$, are real-analytic curves.

Let D be a region on the plane $\mathbb{R}^2$ whose boundary consists of finite simple closed piecewise real-analytic curves. We consider a finite (quasi)-partition $\xi = \{D_i\}_{i=1}^k$ of the domain D such that

- $D_i \subset D$ is a region whose boundary is a finite union of simple closed piecewise real-analytic curves,
- $D_i \cap D_j = \emptyset$ if $i \neq j$, and
- $\cup_{i=1}^k \overline{D}_i = \overline{D}$, where $\overline{D}$ and $\overline{D}_i$ denote the closures of D and D_i respectively.

We call such a partition a real-analytic partition of D. We denote $E = \cup_{i=1}^k \partial D_i = \overline{D} - \cup_{i=1}^k D_i$.

For a real-analytic partition $\xi = \{D_i\}_{i=1}^k$, we can choose a finite set of real-analytic curves $\{\gamma_i : [a_i, b_i] \to E \subset \mathbb{R}^2\}_{i=1}^m$ in E with the following properties

- each γ_i is a simple curve, that is, has no multiple point,
- the boundary of each region D_j, $1 \leq j \leq k$, is a union of the image of some γ_i's, and

- the images of curves γ_i, $1 \leq j \leq k$, except their end points are mutually disjoint, that is,

$$\gamma_i(t) \neq \gamma_j(s) \quad \text{if } i \neq j, t \in (a_i, b_i) \text{ and } s \in (a_j, b_j).$$

We call these curves the dividing curves of the partition ξ. Remark that there are only finitely many points that belong to more than two dividing curves.

A map $f : D \to D$ is called a piecewise real-analytic map on D if there is a real-analytic partition $\xi = \{D_i\}_{i=1}^k$ of D as above such that each restriction $f|_{D_i}$ of f to D_i, $1 \leq i \leq k$, can be extended to a neighborhood of $\overline{D}_i$ as a real-analytic map. We will denote, by f_{D_i}, the real- analytic extension of $f|_{D_i}$ to a neighborhood of $\overline{D}_i$.

For a tangent vector v at $x \in D - E$, we define its expansion rate $\rho(v, f)$ by

$$\rho(v, f) = \frac{\|Df(v)\|}{\|v\|}.$$

The minimum expansion rate $\rho(f)$ of the map f is the infimum of the expansion rate over all non-zero vectors at all points in $D - E$. If $\rho(f) > 1$ for a piecewise real-analytic map f, the map f is called a piecewise real-analytic expanding map. We will fix a piecewise real-analytic expanding map f, the partitions $\xi = \{D_i\}_{i=1}^k$ and the dividing curves $\{\gamma_i : [a_i, b_i] \to \mathbb{R}^2\}_{i=1}^m$ throughout this paper.

An important consequence from the above definitions is the fact that the iterations f^n of a piecewise real-analytic map f are also piecewise real-analytic maps. We will use this fact repeatedly in the proof of Theorem 1.

Remark 2. Iterations of piecewise C^r maps with $r \leq \infty$ are not necessarily piecewise C^r maps since the partition associated to them may have infinitely many connected components. We refer to [14] for examples of piecewise C^r expanding maps with $r < \infty$ that have singular ergodic properties.

3. Germs of Real-Analytic Curves

Let p be a point on the plane $\mathbb{R}^2$, and let $c_i : [0, \epsilon_i] \to \mathbb{R}^2$, $i = 1, 2$, be two real-analytic curves satisfying $c_i(0) = p$ and (1). We say that these two curves give the same germ at p if $c_1(t) = c_2(t)$ for $0 \leq t < \min\{\epsilon_1, \epsilon_2\}$. This is an equivalence relation between real-analytic curves c satisfying $c(0) = p$ and (1). The equivalence classes are called germs of real-analytic curves at p.

We say that two open subsets U_1 and U_2 on the plane $\mathbb{R}^2$ give the same germ at p if there exists $\delta > 0$ such that $U_1 \cap B(p, \delta) = U_2 \cap B(p, \delta)$, where $B(p, \delta) = \{x \in \mathbb{R}^2; \|x - p\| < \delta\}$. This is also an equivalence relation. We call the equivalence classes germs of open subsets at p.

Let β_1 and β_2 be distinct germs of real-analytic curves at p, and let $b_i : [0, \epsilon_i] \to \mathbb{R}^2$, $i = 1, 2$ be simple real-analytic curves that represent β_i respectively. If $\delta > 0$ is sufficiently small, the open set $B(p, \delta)\backslash(b_1 \cup b_2)$ consists of two connected components. The germs of an open subset represented by the connected component of $B(p, \delta)\backslash(b_1 \cup b_2)$ that is located in the counterclockwise direction of the curve b_1 is called the region between β_1 and β_2. From this definition, the region U between β_1 and β_2 and that between β_2 and β_1 are complementary. The germs of real-analytic curves β_1 and β_2 are called the boundary curves of the region U.

Let $\text{Angle}_1(\beta_1, \beta_2) \in [0, 2\pi]$ be the angle that is formed by the region between β_1 and β_2 at p. So we have $\text{Angle}_1(\beta_1, \beta_2) = 2\pi - \text{Angle}_1(\beta_2, \beta_1)$. If $\text{Angle}_1(\beta_1, \beta_2) \neq 0$

we define $\mathrm{Ord}(\beta_1, \beta_2) = 1$. On the other hand, if $\mathrm{Angle}_1(\beta_1, \beta_2) = 0$, we define $\mathrm{Ord}(\beta_1, \beta_2)$ as the contact order of the two germs of real-analytic curves β_1 and β_2 at p, that is,

$$\begin{aligned}\mathrm{Ord}(\beta_1, \beta_2) &= \lim_{t\to+0} \frac{\log \min\{\|b_1(t) - b_2(s)\| \mid s \in [0, \epsilon_2)\}}{\log t}\\ &= \lim_{t\to+0} \frac{\log \min\{\|b_2(t) - b_1(s)\| \mid s \in [0, \epsilon_1)\}}{\log t}.\end{aligned}$$

When $\mathrm{Ord}(\beta_1, \beta_2) = d > 1$, we define $\mathrm{Angle}_d(\beta_1, \beta_2)$ by

$$\begin{aligned}\mathrm{Angle}_d(\beta_1, \beta_2) &= \lim_{t\to+0} \frac{\min\{\|b_1(t) - b_2(s)\| \mid s \in [0, \epsilon_2)\}}{t^d}\\ &= \lim_{t\to+0} \frac{\min\{\|b_2(t) - b_1(s)\| \mid s \in [0, \epsilon_1)\}}{t^d}.\end{aligned}$$

For the region U between β_1 and β_2, we define $\mathrm{Ord}(U) = \mathrm{Ord}(\beta_1, \beta_2)$ and $\mathrm{Angle}_d(U) = \mathrm{Angle}_d(\beta_1, \beta_2)$. We will need the following elementary lemmas.

Lemma 3. *Let $H : W \to \mathbb{R}^2$ be a real-analytic map defined on a neighborhood W of a point $p \in \mathbb{R}^2$. Assume that $\|DH(p)w\|/\|w\| \geq 1$ for all tangent vectors $w \neq 0$ at p. Let β_1 and β_2 be germs of real-analytic curves at p, and let v_i, $i = 1, 2$, be the unit tangent vectors of them at the point p respectively. Let U be the region between β_1 and β_2.*

(a) *If* $0 < \mathrm{Angle}_1(\beta_1, \beta_2) < 2\pi$, *we have*

$$|\det DH(p)| = \left| \frac{\sin(\mathrm{Angle}_1(H(U)))}{\sin(\mathrm{Angle}_1(U))} \right| \cdot \rho(v_1, H)\rho(v_2, H). \tag{2}$$

(b) *If* $\mathrm{Ord}(U) = d > 1$, *we have*

$$|\det DH(p)| = \frac{\mathrm{Angle}_d(H(U))}{\mathrm{Angle}_d(U)} \cdot \rho(v_1, H)^{d+1}. \tag{3}$$

Remark 4. In the claim (b), $v_1 = v_2$ and $\rho(v_1, H) = \rho(v_2, H)$.

Proof. In the case $\mathrm{Ord}(U) = 1$, the formula (2) says nothing but the fact that the absolute value of the determinant of H at p is the ratio between the area of the infinitesimal parallelogram at p spanned by the vectors v_1 and v_2 and that of its image under $DH(p)$. Let us consider the case $\mathrm{Ord}(U) = d > 1$. Since two curves b_i, $i = 1, 2$, representing the germs β_i are almost parallel in small neighborhoods of the point p, the minimum $\min\{\|b_1(t) - b_2(s)\| | s \in [0, \epsilon_2)\}$ is attained when $b_1(t) - b_2(s)$ is almost orthogonal to the vector $v_1 = v_2$ when t is small. Hence we can see, by elementary geometric argument,

$$\begin{aligned}\lim_{t\to+0} \frac{\min\{\|H \circ b_1(t) - H \circ b_2(s)\| \mid s \in [0, \epsilon_2)\}}{\min\{\|b_1(t) - b_2(s)\| \mid s \in [0, \epsilon_2)\}} &= \langle DH(v_1^{\perp}), DH(v_1)^{\perp}\rangle\\ &= \frac{|\det DH(p)|}{\rho(v_1, H)},\end{aligned}$$

where $v_1^\perp$ and $DH(v_1)^\perp$ are the unit vectors that are orthogonal to the vectors v_1 and $DH(v_1)$ respectively and $\langle\cdot,\cdot\rangle$ is the inner product in the ordinary sense. Taking into account the change of the variable t so that the map $H \circ b_1(t)$ satisfies the condition (1), we obtain the claim (b). □

Under the assumption of Lemma 3, we define $\rho(U, H)$ as the maximum of the expansion rate $\rho(v, H) = \|DH(v)\|/\|v\|$ over all tangent vectors v at p that is in between v_1 and v_2 or, in other words, that is contained in the closure of the angle formed by U at p.

Lemma 5. *Under the same assumption as in the last lemma, we have*

$$\frac{\mathrm{Angle}_{\mathrm{Ord}(U)}(H(U))}{\mathrm{Angle}_{\mathrm{Ord}(U)}(U)} \le 2\pi \cdot \frac{|\det DH(p)|}{\rho(U, H)\rho_0}, \tag{4}$$

where $\rho_0 \ge 1$ is the minimum of the expansion rate $\rho(w, H)$ over all tangent vectors $w \ne 0$ at p.

Proof. If $\mathrm{Ord}(U) > 1$, (4) is obvious from the last lemma. Let us consider the case $\mathrm{Ord}(U) = 1$. We first prove, for $K = 2\pi$,

$$\frac{\mathrm{Angle}_1(H(U))}{\mathrm{Angle}_1(U)} \le K \cdot \frac{|\det DH(p)|}{\rho(v_1, H)\rho_0}. \tag{5}$$

If $\mathrm{Angle}_1(U) > \pi/2$, we get (5) for $K = 4$ because the left-hand side is smaller than $2\pi/(\pi/2) = 4$, while the right-hand side except K is not smaller than 1. If $\mathrm{Angle}_1(U) \le \pi/2$ and $\mathrm{Angle}_1(H(U)) \le \pi/2$, we get (5) for $K = \pi/2$ from the last proposition because

$$2/\pi \le \sin(x)/x \le 1 \quad \text{for } 0 < x \le \pi/2.$$

Finally, we consider the case when $\mathrm{Angle}_1(U) \le \pi/2$ and $\mathrm{Angle}_1(H(U)) > \pi/2$. In this case, we can take a unit tangent vector v between v_1 and v_2 such that $DH(p)v$ is orthogonal to $DH(p)v_1$. Let γ be a germ of real-analytic curve passing through U that is tangent to the vector v at p, and let V be the region between β_1 and γ. Then we can apply the above argument to V and get (5) with U replaced by V for $K = \pi/2$. Since

$$\frac{\mathrm{Angle}_1(H(U))}{\mathrm{Angle}_1(U)} \le \frac{2\pi}{\mathrm{Angle}_1(V)} = 4\frac{\mathrm{Angle}_1(H(V))}{\mathrm{Angle}_1(V)},$$

we get (5) for $K = 2\pi$. Therefore we have (5) for $K = 2\pi$ in any case.

Remark that we can replace v_1 by v_2 in (5) by symmetry. Now let us take a vector v at p that is in between v_1 and v_2 and satisfy $\rho(v, H) = \rho(U, f)$. If $v = v_1$ or $v = v_2$, the conclusion of the lemma is nothing but (5) or that with v_1 replaced by v_2. Otherwise, we consider a germ γ of a real-analytic curve that is tangent to v at p. The germ of curve γ divide U into two regions U_1 and U_2. Applying (5) to these regions, we get

$$\begin{aligned}\frac{\mathrm{Angle}_1(H(U))}{\mathrm{Angle}_1(U)} &\le \max\left\{\frac{\mathrm{Angle}_1(H(U_1))}{\mathrm{Angle}_1(U_1)}, \frac{\mathrm{Angle}_1(H(U_2))}{\mathrm{Angle}_1(U_2)}\right\} \\ &\le 2\pi \cdot \frac{|\det DH(p)|}{\rho(v, H)\rho_0}.\end{aligned}$$

The lemma is proved. □

4. Weighted Multiplicity

In this section we introduce what we call weighted multiplicity that count the multiplicity of the intersection of dividing curves $\{\gamma_i\}_{i=1}^m$ with appropriate weight. Let p be a point in $E = \cup_{i=1}^k \partial D_i$. Let $\gamma_i : [a_i, b_i] \to \mathbb{R}^2$ be a dividing curve. If the curve γ_i passes through the point p, it gives germs of real-analytic curves at p in the following manner: If $\gamma_i(t) = p$ for $t \in (a_i, b_i)$, the curve γ_i gives two germs of real-analytic curve at p represented by the curves $s \mapsto \gamma_i(t+s)$ and $s \mapsto \gamma_i(t-s)$. If $\gamma_i(a_i) = p$ (resp. $\gamma_i(b_i) = p$), the curve γ_i gives one germ represented by a curve $s \mapsto \gamma_i(a_i + s)$ (resp. $s \mapsto \gamma_i(b_i - s)$).

Let $\mathcal{B}(p) = \{\beta_i(p)\}_{i=1}^{m(p)}$ be the collection of the distinct germs of real-analytic curves given in such a way by all dividing curves. Remark that $m(p) = 2$ for all points $p \in E$ except for finite points. These germs of real-analytic curves are called the germs of curves at p given by the dividing curves. We always assume that the germs of curves $\beta_i(p), i = 1, 2, \cdots, m(p)$, are arranged in counterclockwise order around the point p.

Let $U_i, 1 \le i < m(p)$, be the region between $\beta_i(p)$ and $\beta_{i+1}(p)$, and let $U_{m(p)}$ be that between $\beta_{m(p)}(p)$ and $\beta_1(p)$. We denote the set of these regions by $\mathcal{U}(p) = \{U_i(p)\}_{i=1}^{m(p)}$. For $U \in \mathcal{U}(p)$, let f_U be the germ of a real-analytic map at p that is obtained as the real-analytic extension of the restriction of f to a representative of U.

We define the weight $W(U_i(p))$ of the region $U_i(p) \in \mathcal{U}(p)$ by

$$W(U_i(p)) = \frac{\|Df_{U_i(p)}(p)v_1\|/\|v_1\| + \|Df_{U_i(p)}(p)v_2\|/\|v_2\|}{|\det Df_{U_i(p)}(p)|}$$

for $1 \le i \le m(p)$, where v_1 and v_2 are the tangent vectors of the boundary curves of $U_i(p)$ at p. The weighted multiplicity $M(p, f)$ at a point $p \in E$ is defined by

$$M(p, f) = \sum_{i=1}^{m(p)} W(U_i(p)).$$

The weighted multiplicity $M(f)$ of a piecewise real-analytic expanding map f is the supremum of $M(f, p)$ over all points $p \in E$. Remark again that $M(f, p) \le 4\rho(f)^{-1}$ for all $p \in E$ except for finite points. Weighted multiplicity $M(f)$ is the quantity that we are most concerned with in the argument below.

5. Functions of Bounded Variation

We use the theory of bounded variation functions in higher dimensional space, which is developed in the book [3]. We recall some definitions and properties of functions of bounded variation from [3].

Let U be an open subset of the plane $\mathbb{R}^2$. Let $C^r(U, \mathbb{R}^2)$ be a set of bounded vector-valued C^r functions $g = (g_1, g_2) : U \to \mathbb{R}^2$ and let $C_0^r(U, \mathbb{R}^2)$ be the subset of $C^r(U, \mathbb{R}^2)$ that consists of functions with compact support. Similarly, let $\Omega^r(U)$ be the set of 1-forms $\Psi = \Psi_1 dx + \Psi_2 dy$ of class C^r on U and let Ω_0^r be the subset of $\Omega^r(U)$ that consists of 1-forms with compact support. We denote the d-dimensional Hausdorff measure by μ_d. We define the variation $\mathrm{Var}(\varphi, U)$ of the function $\varphi \in L^1(U)$ as the supremum of

$$\int_U \varphi(z)\mathrm{Div}g(z)d\mu_2(z) \tag{6}$$

over all $g = (g_1, g_2) \in C_0^1(U, \mathbb{R}^2)$ satisfying $\|g(z)\| \le 1$ for $z \in U$, where

$$\mathrm{Div} g(x, y) = \frac{\partial}{\partial x} g_1(x, y) + \frac{\partial}{\partial y} g_2(x, y).$$

A function $\varphi \in L^1(U)$ is said to be of bounded variation if $\mathrm{Var}(f, \mathbb{R}^2) < \infty$. We denote, by $\mathcal{BV}(U)$, the set of functions $\varphi \in L^1(U)$ of bounded variation. Sometimes it is convenient to write the variation $\mathrm{Var}(\varphi, U)$ as

$$\mathrm{Var}(\varphi, U) = \sup \left\{ \int_U \varphi d\Psi \ ; \ \Psi \in \Omega_0^1(U) \text{ and } \|\Psi(z)\| \le 1 \text{ for } z \in U. \right\}, \tag{7}$$

where $\|\Psi(z)\| = \sqrt{\Psi_1^2(z) + \Psi_2^2(z)}$ for $\Psi = \Psi_1 dx + \Psi_2 dy$. We can obtain this formula from the correspondence between $g = (g_1, g_2) \in C_0^1(U, \mathbb{R}^2)$ and $\Psi = -g_2 dx + g_1 dy \in \Omega_0^1(U)$.

Remark that, for $\varphi \in \mathcal{BV}(U)$, the functional

$$\Phi_\varphi : g \in C_0^1(U, \mathbb{R}^2) \mapsto \int_U \varphi \mathrm{Div} g d\mu_2 \in \mathbb{R}$$

satisfies

$$|\Phi(g_1) - \Phi(g_2)| \le \left(\sup_{z \in U} \|g_1(z) - g_2(z)\| \right) \cdot \mathrm{Var}(\varphi, U).$$

Hence Φ can be extended uniquely as a continuous linear functional on $C_0^0(U, \mathbb{R}^2)$. We can consider this extension as a vector-valued Radon measure on U with total variation $\mathrm{Var}(\varphi, U)$. (See [10, Ch.6] for example.) We denote this vector-valued Radon measure by $D\varphi$. Let $\int_U \langle g, D\varphi \rangle$ be the integration of a vector-valued function $g \in C^0(U, \mathbb{R}^2)$ with respect to the vector-valued measure $D\varphi$. Let $|D\varphi|$ be the measure that is obtained as the total variation of $D\varphi$. Obviously we have $\int_U \langle g, D\varphi \rangle \le \int \|g\| \cdot |D\varphi|$ for $g \in C^0(U, \mathbb{R}^2)$, where $\|g\|$ denotes the function $\|g\|(x) = \|g(x)\|$ on U.

The bounded variation norm of the function $g \in \mathcal{BV}(U)$ is defined as

$$\|\varphi\|_{BV(U)} = \mathrm{Var}(\varphi, U) + \int_U |\varphi| d\mu_2.$$

This norm makes $\mathcal{BV}(U)$ a Banach space. See [3, Remark 1.12]. We make use of the following fact when we prove the existence of an absolutely continuous invariant measure in Sect. 6.

Proposition 6. *Let $U \subset \mathbb{R}^2$ be a bounded open set with C^1 boundary. Then sets of functions in $\mathcal{BV}(U)$ that are uniformly bounded in the bounded variation norm $\|\cdot\|_{BV(U)}$ are relatively compact in $L^1(U)$.*

Another important property of functions of bounded variation is that they give traces on the boundary. Let $U \subset \mathbb{R}^2$ be a bounded region whose boundary is a finite union of real-analytic simple closed curves. We denote by $L^1(\partial U)$ the set of functions that is integrable with respect to the one dimensional Hausdorff measure μ_1. We put $B(x, r) = \{y \in \mathbb{R}^2 \mid \|x - y\| < r\}$. Then we have

Proposition 7. *For $\varphi \in \mathcal{BV}(U)$, there is a unique function $\varphi^- \in L^1(\partial U)$ such that*

$$\lim_{r\to 0} \mu_2(B(x,r))^{-1} \int_{B(x,r)\cap U} |\varphi(z) - \varphi^-(x)| d\mu_2(z) = 0$$

for μ_1-almost all $x \in \partial U$. Moreover,

(a) *for $\zeta \in C_0^1(\mathbb{R}^2, \mathbb{R}^2)$, we have*

$$\int_{\partial U} \varphi^-(z)\langle \zeta(z), \nu(z)\rangle d\mu_1(z) = \int_U \varphi(z)\mathrm{Div}\zeta(z)d\mu_2(z) + \int_U \langle \zeta, D\varphi\rangle,$$

where $\nu(z)$ is the unit outer normal vector for the boundary ∂U at z,

(b) *if we define $\varphi(z) = 0$ for $z \notin U$, we have*

$$\mathrm{Var}(\varphi, \mathbb{R}^2) = \mathrm{Var}(\varphi, U) + \int_{\partial U} |\varphi^-(x)| d\mu_1(x).$$

The function $\varphi^- \in L^1(\partial U)$ in the above proposition is called the trace of φ on the boundary ∂U. We refer to Theorem 1.19 of [3] for Proposition 6, and Theorem 2.10 and Remark 2.14 of [3] for Proposition 7.

Remark 8. In Theorem 2.10 and Remark 2.14 of [3], the boundary of the region U is assumed to be Lipschitz. Hence Proposition 7 is not a direct consequence of that theorem when there are cusps on the boundary of U. But, with slight modification in the proof, we can derive Proposition 7, because the proposition is essentially a local one.

6. An Existence Theorem for Absolutely Continuous Invariant Measures

Let $f : D \to D$ be a piecewise real-analytic expanding map and let $\xi = \{D_i\}_{i=1}^k$ be the partition of the domain D associated to it as in Sect. 2. In this section we prove

Theorem 9. *If f satisfies*

(a) *$M(f) + \rho(f)^{-1} < 1$ and,*

(b) *the continuous extension of each restriction $f|_{D_i}$, $1 \le i \le k$, to the closure $\overline{D_i}$ is injective,*

then there exists an absolutely continuous invariant finite measure for f.

Theorem 9 above is a modification of the result of Góra and Boyarski in [4], and the essential part of the proof below is a repetition of the argument in [4]. We define the Perron–Frobenius operator $P_f : L^1(D) \to L^1(D)$ by

$$P_f(\varphi)(x) = \sum_{f(y)=x} \frac{\varphi(y)}{|\det Df(y)|},$$

where the sum is taken over all $y \in \cup_{i=1}^k D_i$ such that $f(y) = x$. Remark that if there exists a non-negative valued function $h \neq 0$ in $L^1(D)$ such that $P_f(\varphi) = \varphi$, the measure $h \cdot \mu_2$ is an absolutely continuous invariant finite measure for f. From the definition of Perron–Frobenius operator, we have

$$\int_D P_f g d\mu_2 = \int_D g d\mu_2 \tag{8}$$

for a non-negative valued function $g \in L^1(D)$. In what follows, we consider each element $\varphi \in L^1(D)$ as an element of $L^1(\mathbb{R}^2)$ by defining $\varphi(x) = 0$ for $x \notin D$. The following is the key in the proof of Theorem 9.

Proposition 10. *For any $\epsilon > 0$, there exists a constant $C > 0$ such that*

$$\begin{aligned}\mathrm{Var}(P_f\varphi, \mathbb{R}^2) &\leq \sum_{j=1}^{k} \mathrm{Var}(P_f(\varphi \cdot \chi_{D_j}), \mathbb{R}^2) \\ &\leq (M(f) + \rho^{-1}(f) + \epsilon)\mathrm{Var}(\varphi, \mathbb{R}^2) + C\|\varphi\|_{L^1} \qquad (9)\end{aligned}$$

for $\varphi \in \mathcal{BV}(D)$.

This kind of inequality appeared in the original work of Lasota and Yorke and can be seen in the papers of Keller [7,8] and Góra&Boyarski [4].

First, we prove that Proposition 10 implies Theorem 9. Take a small number $\epsilon > 0$ such that $M(f) + \rho(f)^{-1} + \epsilon < 1$. Then take the constant C in Proposition 10 for that ϵ. Let $\varphi \in \mathcal{BV}(D)$ be a non-negative valued function such that $\int_D \varphi d\mu_2 = 1$. From (9) and (8), we obtain

$$\mathrm{Var}(P_f^n\varphi, \mathbb{R}^2) \leq (1 - M(f) - \rho(f)^{-1} - \epsilon)^{-1}C + \mathrm{Var}(\varphi, \mathbb{R}^2)$$

by induction. Hence the set of functions $\{\zeta_n = n^{-1}\sum_{i=0}^{n-1} P_f^i\varphi\}_{n=1}^{\infty}$ are contained in $\mathcal{BV}(D)$ and uniformly bounded in the bounded variation norm $\|\cdot\|_{BV(\mathbb{R}^2)}$ on $\mathbb{R}^2$. Applying Theorem 6 to a bounded open subset $U \supset D$ with C^1 boundary, we can find a subsequence ζ_{n_k} that converges to a function $\varphi_\infty \in L^1(D)$ in the L^1 norm. Obviously φ_∞ is non-negative valued. We have $\int_D \varphi_\infty d\mu_2 = 1$ from (8). Moreover, φ_∞ is a fixed point of the Perron–Frobenius operator P_f because

$$\begin{aligned}\|P_f\varphi_\infty - \varphi_\infty\|_{L^1} &= \lim_{k\to\infty} \left\| \frac{1}{n_k}\sum_{i=0}^{n_k-1} P_f^{i+1}\varphi - \frac{1}{n_k}\sum_{i=0}^{n_k-1} P_f^i\varphi \right\|_{L^1} \\ &\leq \lim_{k\to\infty} \frac{1}{n_k}\|P_f^{n_k}\varphi - \varphi\|_{L^1} \leq \lim_{k\to\infty} \frac{2}{n_k} = 0.\end{aligned}$$

Therefore $\varphi_\infty \cdot \mu_2$ is an absolutely continuous invariant finite measure for f. Theorem 9 is proved.

Now let us go into the proof of Proposition 10. We first study the situation that the image of a dividing curve $\gamma_i : [a_i, b_i] \to \mathbb{R}^2$ is contained in the boundary of some region $D_j \in \xi$. From Proposition 7, a function $\varphi \in \mathcal{BV}(D)$ viewed as a function on D_j gives the trace φ_j^- on the curve γ_i. We consider one side of the tubular neighborhood of the curve γ_i,

$$\Gamma_{ij} : [a_i, b_i] \times [0, \delta] \to \mathbb{R}^2, \quad (t, s) \mapsto \gamma_i(t) + s \cdot \nu(t),$$

where $\nu(t)$ is the unit inner normal vector for the boundary ∂D_j at $\gamma_i(t)$ and $\delta > 0$ is a small constant that we will specify in the argument below. We first take $\delta > 0$ so small that Γ_{ij} is a diffeomorphism. We will denote the image of Γ_{ij} by the same symbol Γ_{ij}.

Let $V_i(x)$ be the unit tangent vector of the curve γ_i at $x \in \gamma_i$. We define a real-analytic function $h_{ij} : \gamma_i \to \mathbb{R}$ by

$$h_{ij}(x) = \frac{\rho(V_i(x), f_{D_j})}{|\det Df_{D_j}(x)|}.$$

Let $\pi : [a_i, b_i] \times [0, \delta] \to [a_i, b_i]$ be the projection. We define a function

$$\tilde{h}_{ij} = (0, h_{ij} \circ \Gamma_{ij} \circ \pi) : [a_i, b_i] \times [0, \delta] \to \mathbb{R}^2.$$

Put $\|\tilde{h}_{ij}\|(x) = \|\tilde{h}_{ij}(x)\|$. Then we have

Lemma 11. *If $\varphi \in \mathcal{BV}(D)$, the composition $\varphi \circ \Gamma_{ij}$ viewed as a function on the open rectangle $(a_i, b_i) \times (0, \delta)$ is of bounded variation. We have*

$$\int_{\gamma_i} h_{ij} \cdot \varphi_j^- d\mu_1 \le \frac{1}{\delta} \int_{[a_i,b_i]\times[0,\delta]} \|\tilde{h}_{ij}\| \cdot |\varphi \circ \Gamma_{ij}| d\mu_2 + \int_{(a_i,b_i)\times(0,\delta)} \|\tilde{h}_{ij}\| \cdot |D(\varphi \circ \Gamma_{ij})|.$$

Proof. We can get the first claim easily from formula (7). For $y \in (0, \delta)$, let us consider the rectangle $R_y = [a_i, b_i] \times [0, y]$. The function $\varphi \circ \Gamma_{ij}$, viewed as a function on the interior of R_y, gives the trace G_y^- on the boundary ∂R_y of the rectangle. Remark that the restriction of G_y^- on the edge $[a_i, b_i] \times \{0\}$ does not depend on y and equals the function $\varphi_i^- \circ \Gamma_{ij}$ on $[a_i, b_i] \times \{0\}$ from the definition of the trace. Obviously, we have

$$\int_{[a_i,b_i]\times\{0\}} G_y^- \cdot h_{ij} \circ \Gamma_{i,j} \circ \pi d\mu_1 = \int_{\gamma_i} h_{ij} \cdot \varphi_j^- d\mu_1.$$

Let us define a function $B : (a_i, b_i) \times (0, \delta) \to \mathbb{R}$ by $B(x, y) = G_y^-(x, y)$. Then, from Lebesgue's theorem [15, Theorem 1.3.8], $B(x, y) = \varphi \circ \Gamma_{ij}(x, y)$ for almost every $(x, y) \in (a_i, b_i) \times (0, \delta)$. Applying Proposition 7 to $\varphi \circ \Gamma_{ij}(x, y)$ and $\tilde{h}_{ij}$ on R_y, we obtain

$$\left| \int_{[a_i,b_i]\times\{y\}} B \cdot h_{ij} \circ \Gamma_{ij} \circ \pi d\mu_1 - \int_{\gamma_i} h_{ij} \cdot \varphi_j^- d\mu_1 \right| \le \int_{\mathrm{int} R_y} \|\tilde{h}_{ij}\| \cdot |D(\varphi \circ \Gamma_{ij})| \tag{10}$$

because $\mathrm{Div}\, \tilde{h}_{ij} \equiv 0$. Since $\|\tilde{h}_{ij}\| = h_{ij} \circ \Gamma_{ij} \circ \pi$, we get

$$\delta \left(\int_{\gamma_i} h_{ij} \cdot \varphi_j^- d\mu_1 - \int_{(a_i,b_i)\times(0,\delta)} \|\tilde{h}_{ij}\| \cdot |D(\varphi \circ \Gamma_{ij})| \right) \le \int_{[a_i,b_i]\times[0,\delta]} \|\tilde{h}_{ij}\| \cdot |\varphi \circ \Gamma_{ij}| d\mu_2$$

from Fubini's theorem. This implies the lemma. □

Let us take a small number $\eta > 0$ such that $(1+\eta)^2 M(f) < M(f) + \epsilon$. Remark that, if $\delta > 0$ is sufficiently small, the map Γ_{ij} is almost an isometry on $[a_i, b_i] \times [0, \delta]$. Hence we can take $\delta > 0$ so small that

$$(1+\eta)^{-1}\|v\| < \|D\Gamma_{ij}(v)\| \leq (1+\eta)\|v\| \tag{11}$$

for all tangent vectors v at all points in $[a_i, b_i] \times [0, \delta]$. Let us define a function $\hat{h}_{ij} : \mathbb{R}^2 \to \mathbb{R}$ by

$$\hat{h}_{ij}(z) = \begin{cases} h_{ij} \circ \Gamma_{ij} \circ \pi \circ \Gamma_{ij}^{-1} & \text{for } x \in \Gamma_{ij}; \\ 0 & \text{for } x \notin \Gamma_{ij}. \end{cases}$$

Since $\hat{h}_{ij} \circ \Gamma_{ij} = \|\tilde{h}_{ij}\|$ for $x \in \Gamma_{ij}$, we obtain, from (11),

$$\int_{(a_i,b_i)\times(0,\delta)} \|\tilde{h}_{ij}\| \cdot |D(\varphi \circ \Gamma_{ij})| \leq (1+\eta) \int_{int\Gamma_{ij}} \hat{h}_{ij}|D\varphi|.$$

Therefore we get, from the last lemma,

Proposition 12. *We have*

$$\int_{\gamma_i} h_{ij} \cdot \varphi_j^- d\mu_1$$
$$\leq (1+\eta)^2 \left(\frac{1}{\delta} \int_{\Gamma_{ij}} \hat{h}_{ij}|\varphi| d\mu_2 + \int_{\mathbb{R}^2} \hat{h}_{ij}|D\varphi| \right).$$

Next we prove the following proposition.

Proposition 13. *Let U be one of the regions in the partition $\xi = \{D_j\}_{j=1}^k$. Let $V(x)$ be the unit tangent vector for the boundary ∂U at $x \in \partial U$. Let $\varphi \in \mathcal{BV}(U)$ be non-negative valued. Let φ^- be the trace of φ on the boundary ∂U. Then we have*

$$\begin{aligned} \mathrm{Var}(P_f\varphi, \mathbb{R}^2) &\leq \rho(f)^{-1}\mathrm{Var}(\varphi, U) + C(f, U)\|\varphi\|_{L^1} \\ &\quad + \int_{\partial U} \frac{\rho(V(x), f_U)}{|\det Df_U(x)|} \varphi^-(x) d\mu_1(x), \end{aligned}$$

where $C(f, U)$ is a constant depending only on the restriction $f|_U$ of f (defined in (12) below).

Proof. We have

$$\begin{aligned} \int_{\mathbb{R}^2} P_f \varphi d\Psi &= \int_{\mathbb{R}^2} \frac{\varphi}{|\det Df|} df^*\Psi \\ &= \int_{\mathbb{R}^2} \varphi d\left(\frac{f^*\Psi}{|\det Df|}\right) - \int_{\mathbb{R}^2} \varphi d\left(\frac{1}{|\det Df|}\right) \wedge f^*\Psi \end{aligned}$$

for $\Psi \in \Omega_0^1(f(U))$. Hence we get, from formula (7),

$$\mathrm{Var}(P_f\varphi, f(U)) \leq \rho(f)^{-1}\mathrm{Var}(\varphi, U) + C(f, U)\|\varphi\|_{L^1}$$

if we put $\sigma(f, U) = \sup\{\|Df(v)\|/\|v\| \mid 0 \neq v \in T_x\mathbb{R}^2, x \in U\}$ and

$$C(f, U) = \sigma(f, U) \sup\{\|D((\det Df)^{-1})(x)\| \mid x \in U\}. \tag{12}$$

On the other hand, we have

$$\int_{\partial f(U)} (P_f\varphi)^- d\mu_1 = \int_{\partial U} \frac{\rho(V(x), f_U)}{|\det Df_U(x)|}\varphi^-(x) d\mu_1(x).$$

From these and Proposition 7(b), we obtain the conclusion. □

Now we complete the proof of Proposition 10. From Proposition 13,

$$\begin{aligned}\operatorname{Var}(P_f\varphi, \mathbb{R}^2) &\le \sum_{j=1}^{k} \operatorname{Var}(P_f(\varphi \cdot \chi_{D_j}), \mathbb{R}^2)\\ &\le \sum_{j=1}^{k} \left(\frac{\operatorname{Var}(\varphi, D_j)}{\rho(f)} + C(f, D_j)\|\varphi\|_{L^1} + \sum_{i\sim j} \int_{\gamma_i} h_{ij}\varphi_j^- d\mu_1 \right),\end{aligned}$$

where $\sum_{i\sim j}$ is the sum over i satisfying $\gamma_i \subset \partial D_j$. We have

$$\sum_{j=1}^{k} \operatorname{Var}(\varphi, D_j) \le \operatorname{Var}(\varphi, \mathbb{R}^2).$$

Hence, in order to prove Proposition 10, we show

$$\sum_{j=1}^{k}\sum_{i\sim j} \int_{\gamma_i} h_{ij}\varphi_j^- d\mu_1 \le (M(f) + \epsilon)\operatorname{Var}(\varphi, \mathbb{R}^2) + K\|\varphi\|_{L^1}$$

for some constant $K > 0$. But, from Proposition 12, it is sufficient to show

$$(1+\eta)^2 \sum_{j=1}^{k}\sum_{i\sim j} \int_{\mathbb{R}^2} \hat{h}_{ij}|D\varphi| \le (M(f) + \epsilon)\operatorname{Var}(\varphi, \mathbb{R}^2)$$

or, more simply,

$$(1+\eta)^2 \sum_{j=1}^{k}\sum_{i\sim j} \hat{h}_{ij}(x) \le M(f) + \epsilon \quad \text{for } x \in D. \tag{13}$$

Notice that $\hat{h}_{ij}(x) = h_{ij}(x)$ on the dividing curves γ_i. From the definition of weighted multiplicity $M(p, f)$, we have

$$\sum_{j=1}^{k}\sum_{i\sim j} \hat{h}_{ij}(x) \le M(x, f)$$

for $x \in E$, if δ is sufficiently small. Let F be the set of points that is contained in more than two dividing curves. From the choice of η, we can take a small open neighborhood W of the finite set F in $\overline{D}$ such that the left-hand side of (13) is not larger than $(M(f)+\epsilon)$ on W. If δ is sufficiently small, the intersections of two distinct subsets in $\{\Gamma_{ij} \mid \gamma_i \subset D_j\}$ are contained in W. By continuity, we easily see that the left-hand side of (13) is smaller than $M(f) + \epsilon$ for $x \in \overline{D} - W$ if δ is sufficiently small. Therefore (13) holds for sufficiently small δ. Proposition 10 is proved. □

7. Estimates of the Weighted Multiplicity for the Iterations

In this last section, we complete the proof of Theorem 1 by considering the iterations of a piecewise real-analytic expanding map f. First, remark that a map has an absolutely continuous finite invariant measure if and only if an iteration of it does. Since f^n is a piecewise real-analytic map with minimum expansion rate $\rho(f^n) \geq \rho(f)^n$, we can assume

$$\rho(f) > 64\pi$$

without loss of generality. We prove the following theorem.

Theorem 14. $M(f^n) \to 0$.

Subdividing the partition ξ by real-analytic curves artificially, we can assume that f satisfies condition (b) in Theorem 9. It means that all iterations of f also satisfy that condition. Thus, from Theorem 14 above, iterations f^n of f satisfy assumptions (a) and (b) of Theorem 9 if n is sufficiently large. Therefore we can get Theorem 1 from Theorem 14.

Let us prepare some notations in order to prove Theorem 14. Let $\xi_n = \{D_i^{(n)}\}_{i=1}^{k(n)}$ be the real-analytic partition associated to the piecewise real-analytic map f^n. Let $E(n) = \overline{D} - \cup_{i=1}^{k(n)} D_i^{(n)} = \cup_{i=1}^{k(n)} \partial D_i^{(n)}$. For $p \in E(n)$, let $\{\beta_i(p,n)\}_{i=1}^{m(p,n)}$ be the germs of real-analytic curves at $p \in E(n)$ given by the dividing curves of the partition ξ_n. We assume that these germs of curves are arranged in counterclockwise order around p as before. Let $U_i(p,n)$ be the region between $\beta_i(p,n)$ and $\beta_{i+1}(p,n)$ for $1 \leq i \leq m(p,n)$. Let us denote $\mathcal{U}(p,n) = \{U_i(p,n)\}_{i=1}^{m(p,n)}$. For $U \in \mathcal{U}(p,n)$, let f_U^n be the germ of the real-analytic map at p obtained as a real-analytic continuation of the restriction of f^n to a representative of U.

Let Δ be the maximum of $\mathrm{Ord}(U_i(p,1))$ over all $1 \leq i \leq m(p,1)$ and all $p \in E(1)$. Let θ and Θ be the minimum and maximum of

$$\mathrm{Angle}_{\mathrm{Ord}(U_i(p,1))}(f(U_i(p,1)))$$

over all $1 \leq i \leq m(p,1)$ and all $p \in E(1)$ respectively. Let μ be the maximum of $m(p,1)$ over all $p \in E(1)$. Thus Δ, θ, Θ and μ depend only on the single map f.

Let us consider a point $p \in E(n)$. Each $V \in \mathcal{U}(p,n+1)$ is contained in some $U \in \mathcal{U}(p,n)$ as a germ. Remark that, if $V \subset U$ and $V \neq U$, the image $f_U^n(p)$ is contained in $E(1)$ and a dividing curve of the partition $\xi = \xi_1$ passing through $f_U^n(p)$ divides $f^n(U)$ into more than two regions.

We say that $V \in \mathcal{U}(p,n+1)$ is a kid of $U \in \mathcal{U}(p,n)$ if $V \subset U$. If V is a kid of U and if V and U have at least one germ of real-analytic curve as a boundary curve in common additionally, we say V is a daughter of U. Especially, if $V = U$, V is a daughter of U. Obviously, each $U \in \mathcal{U}(p,n)$ has at most two daughters. If $\mathrm{Ord}(V) > \mathrm{Ord}(U)$, we say that V is a small kid of U.

The reason why we distinguish daughters is the following. Let $V \in \mathcal{U}(n+1,p)$ be a kid of $U \in \mathcal{U}(n,p)$ and assume that $U \neq V$. If V is not a daughter, $f^n(V)$ should coincide with an element of $\mathcal{U}(f_U^n(p),1)$. So $\mathrm{Angle}_{\mathrm{Ord}(V)}(f^{n+1}(V)) \leq \Theta$. On the other hand, if V is a daughter and it is small, we can not expect such anestimate on $\mathrm{Angle}_{\mathrm{Ord}(V)}(f^{n+1}(V))$. For the same reason, we put the following definition. An element U of $\mathcal{U}(p,n)$ is called special if $\mathrm{Ord}(U) > \Delta$ or if there is a chain $U_i \in \mathcal{U}(p, n-\ell+i)$, $i = 0, 1, 2, \cdots, \ell$, of regions with length $\ell + 1 \geq 2$ such that

- $U_\ell = U$,

- U_1 is a small kid and daughter of U_0, and
- U_{i+1} is a daughter of U_i for $1 \le i < \ell$.

In order to estimate $M(f^n)$, we introduce what we call the modified weight $\mathcal{W}(U)$ of $U \in \mathcal{U}(p,n)$ in the following manner. We fix a small number $0 < \eta < 1$ that will be specified later in the condition (16). We define the level $\ell(U)$ of $U \in \mathcal{U}(p,n)$, $p \in E(n)$, by

$$\ell(U) = \begin{cases} 2\min\{\mathrm{Ord}(U), \Delta+1\} & \text{if } U \text{ is not special;} \\ 2\min\{\mathrm{Ord}(U), \Delta+1\} - 1 & \text{if } U \text{ is special.} \end{cases}$$

Remark that we always have $\ell(V) \ge \ell(U)$ if V is a kid of U. If $U \in \mathcal{U}(p,n)$ is special, we define

$$\mathcal{W}(U) = \frac{\eta^{\ell(U)} \rho(U, f_U^n)}{|\det Df_U^n(p)|}.$$

(For the definition of $\rho(U, f_U^n)$, see Sect. 3.) On the other hand, if U is not special, we define

$$\mathcal{W}(U) = \frac{\eta^{\ell(U)} \rho(U, f_U^n)}{|\det Df_U^n(p)|} \left[\frac{\mathrm{Angle}_{\mathrm{Ord}(U)}(f^n(U))}{\theta} + 1 \right],$$

where $[\cdot]$ is Gauss' symbol. We put

$$\mathcal{M}(p, f^n) = \sum_{U \in \mathcal{U}(p,n)} \mathcal{W}(U, f^n) \quad \text{and} \quad \mathcal{M}(f^n) = \sup_{p \in E(n)} \mathcal{M}(p, f^n).$$

Clearly, we have $M(f^n) \le 2\eta^{-2\Delta-2}\mathcal{M}(f^n)$. In order to prove Theorem 14, it is enough to show the following proposition.

Proposition 15. *If we take $\eta > 0$ sufficiently small, we have*

$$\sum_{V: a \text{ kid of } U} \mathcal{W}(V) \le \mathcal{W}(U)/2 \tag{14}$$

for all $U \in \mathcal{U}(p,n)$, $p \in E(n)$, $n \ge 1$.

In fact, if this is true, we have

$$\mathcal{M}(p, f^{n+1}) \le (1/2)\mathcal{M}(p, f^n) \le (1/2)\mathcal{M}(f^n)$$

for $p \in E(n)$. On the other hand, we have

$$\mathcal{M}(p, f^{n+1}) \le \rho(f)^{-n}\mathcal{M}(f^n(p), f) \le (1/2)^n \mathcal{M}(f)$$

for $p \in E(n+1) - E(n)$. These show $\mathcal{M}(f^n) \le (1/2)^n \mathcal{M}(f)$ inductively. Therefore $M(f^n) \le (1/2)^{n-1}\eta^{-2\Delta-2}\mathcal{M}(f) \to 0$ as $n \to \infty$.

Proof. We prove Proposition 15. Let us consider a region $U \in \mathcal{U}(p,n)$ and its kids. We assume that $\mathrm{Ord}U \le \Delta$ until the end of this proof where we treat the case $\mathrm{Ord}U > \Delta$. We classify the kids V of U into the following four classes:

1. V is a daughter of U, and V is a small kid of U,
2. V is not a daughter of U, and V is a small kid of U,
3. V is a daughter of U, and V is not a small kid of U, and
4. V is not a daughter of U, and V is not a small kid of U.

We estimate the sums of modified weights over the kids V in each class.

First we consider the kids in class 1. Kids V in this class is special. Hence we have

$$\mathcal{W}(V) \leq \frac{\rho(f^n(V), f_{f^n(U)})}{|\det Df_{f^n(U)}(f_U^n(p))|}\mathcal{W}(U) \leq \rho(f)^{-1}\mathcal{W}(U).$$

Since the number of kids in this class is at most 2, the sum of $\mathcal{W}(V)$ over the kids V in this class is not larger than $2\rho(f)^{-1}\mathcal{W}(U) < (1/8)\mathcal{W}(U)$.

We consider class 2. Note that $\ell(V) \geq \ell(U)+1$ in this case. The number of the kids in this class is not larger than μ. For each kid V in this class, we have

$$\text{Angle}_{\text{Ord}(V)}(f^{n+1}(V)) \leq \Theta.$$

Thus it holds

$$\begin{aligned}\mathcal{W}(V) &\leq \eta(\Theta/\theta+1)\frac{\rho(f^n(V), f_{f^n(V)})}{|\det Df_{f^n(V)}(f_V^n(p))|}\mathcal{W}(U)\\ &\leq \eta(\Theta/\theta+1)\mathcal{W}(U).\end{aligned}$$

Hence the sum of $\mathcal{W}(V)$ over the kids V in this class is not larger than $\mu\eta(\Theta/\theta+1)\mathcal{W}(U)$.

We consider case 3. Remark that U is special if and only if so is V. In the case U is special, we easily see that

$$\mathcal{W}(V) \leq \frac{\rho(f^n(V), f)}{|\det Df_U(p)|}\mathcal{W}(U) \leq \rho(f)^{-1}\mathcal{W}(U).$$

And the sum is not larger than $2\rho(f)^{-1}\mathcal{W}(U) < (1/8)\mathcal{W}(U)$. In case U is not special, we have

$$\begin{aligned}\mathcal{W}(V) &\leq \frac{[\text{Angle}_{\text{Ord}(V)}(f^{n+1}(V))/\theta]+1}{[\text{Angle}_{\text{Ord}(U)}(f^n(U))/\theta]+1}\,\frac{\rho(f^n(V), f_{f^n(V)})}{|\det Df_{f^n(V)}(f_V^n(p))|}\mathcal{W}(U)\\ &\leq \left\{1+\frac{\text{Angle}_{\text{Ord}(V)}(f^{n+1}(V))}{\text{Angle}_{\text{Ord}(V)}(f^n(V))}\right\}\frac{\rho(f^n(V), f_{f^n(V)})}{|\det Df_{f^n(V)}(f_V^n(p))|}\mathcal{W}(U).\end{aligned}$$

Here we used the fact $\text{Angle}_{\text{Ord}(U)}(f^n(U)) \geq \text{Angle}_{\text{Ord}(V)}(f^n(V))$ and an inequality

$$\frac{[y+1]}{[x+1]} \leq \frac{y+1}{\max\{x, 1\}} \leq \frac{y}{x}+1 \quad \text{for } x>0 \text{ and } y>0.$$

Using Lemma 5 in case $\frac{\text{Angle}(f^{n+1}(V))}{\text{Angle}(f^n(V))} > 1$, we get

$$\mathcal{W}(V) \leq 4\pi\rho(f)^{-1}\mathcal{W}(U).$$

Since the number of kids in this class is at most 2, the sum is not larger than $8\pi\rho(f)^{-1}\mathcal{W}(U) < (1/8)\mathcal{W}(U)$.

We see class 4. In this case, let V_i, $i=1,2,\cdots,\ell$, be the kids of this class and let $d=\text{Ord}(U)$. Remark that V_i, $i=1,2,\cdots,\ell$, are not special. In case U is special, we have $\ell(V_i) \geq \ell(U)+1$ for all i. Hence, in this case, we can see that the sum $\sum_{i=1}^{\ell}\mathcal{W}(V_i)$

is not larger than $\mu\eta(\Theta/\theta + 1)\mathcal{W}(U)$ by just the same argument as above for class 2. Let us consider the case that U is not special. Obviously we have

$$\sum_{i=1}^{\ell} \mathrm{Angle}_d(f^n(V_i)) \leq \mathrm{Angle}_d(f^n(U)). \tag{15}$$

Since $\mathrm{Angle}_d(f^{n+1}(V_i)) \geq \theta$ for $i = 1, 2, \cdots, \ell$, we have

$$[\mathrm{Angle}_d(f^{n+1}(V_i))/\theta + 1] \leq 2\mathrm{Angle}_d(f^{n+1}(V_i))/\theta.$$

By using Lemma 5, we obtain

$$\begin{aligned}\frac{\mathcal{W}(V_i)}{\mathcal{W}(U)} &\leq \frac{[\mathrm{Angle}_d(f^{n+1}(V_i))/\theta + 1]}{[\mathrm{Angle}_d(f^n(U)/\theta + 1]} \cdot \frac{\rho(f^n(V_i), f)}{|\det Df_{f^n(V_i)}(f^n_{V_i}(p))|}\\ &\leq \frac{[\mathrm{Angle}_d(f^{n+1}(V_i))/\theta + 1]}{[\mathrm{Angle}_d(f^n(U))/\theta + 1]} \cdot \frac{2\pi\,\mathrm{Angle}_d(f^n(V_i))}{\rho(f)\mathrm{Angle}_d(f^{n+1}(V_i))}\\ &\leq \frac{4\pi\,\mathrm{Angle}_d(f^n(V_i))}{\rho(f)\mathrm{Angle}_d(f^n(U))}.\end{aligned}$$

Thus we have, from (15),

$$\sum_{i=1}^{\ell} \mathcal{W}(V_i) \leq 4\pi\rho(f)^{-1}\mathcal{W}(U)$$

in this case.

Summing up all the above arguments for the four classes, we obtain (14) for the case $\mathrm{Ord}(U) \leq \Delta$, if we take $\eta > 0$ so small that

$$\mu\eta(\Theta/\theta + 1) < 1/8 \tag{16}$$

because the sums of modified weights over each of four classes are smaller than $(1/8)\mathcal{W}(U)$.

Finally, let us consider the case $\mathrm{Ord}(U) > \Delta$. In this case every kid of U should be a daughter. So the number of the kids is at most 2. Since U and its kid V are special, we have

$$\mathcal{W}(V) \leq \frac{\rho(f^n(V), f)}{|\det Df_{f^n(V)}(f^n_V(p))|}\mathcal{W}(U) \leq \rho(f)^{-1}\mathcal{W}(U).$$

Therefore the left-hand side of (14) is smaller than $2\rho(f)^{-1}\mathcal{W}(U) < (1/2)\mathcal{W}(U)$. We completed the proof of Proposition 15. □

Appendix: Other Ergodic Properties of Piecewise Real-Analytic Expanding Maps

Proposition 10 and Theorem 9 imply that, if n is sufficiently large, the iteration f^n satisfies the inequality (9) with f replaced by f^n and the coefficient $(M(f^n)+\rho^{-1}(f^n)+\epsilon)$ is smaller than 1. As is pointed out in the papers of Keller [7] and Góra & Boyarski [4], once we get such an inequality, we can derive many properties of Perron–Frobenius operator P_f and those of the absolutely continuous invariant measures for f. In fact, we can apply an ergodic theorem of Ionescu-Tulcea and Marinescu [6] and show that, if we put $\mathcal{BV}(U, \mathbb{C}) = \{\xi + \sqrt{-1}\eta \mid \xi, \eta \in \mathcal{BV}(U)\}$ and $\|\xi + \sqrt{-1}\eta\|_{\mathcal{BV}(U,\mathbb{C})} = \|\xi\|_{\mathcal{BV}(U)} + \|\eta\|_{\mathcal{BV}(U)}$,

- the operator $P_f : \mathcal{BV}(U) \to \mathcal{BV}(U)$ has finitely many eigenvalues $\lambda_1, \lambda_2, \cdots, \lambda_r$ of modulus 1, and the corresponding eigenspaces $E_i = \{\varphi \in \mathcal{BV}(U, \mathbb{C}) \mid P_f\varphi = \lambda_i\varphi\}$ are of finite dimension,
- the natural extension of P_f to $\mathcal{BV}(U, \mathbb{C})$ is written as

$$P_f = \sum_{i=1}^{r} \lambda_i \Psi_i + Q,$$

where Ψ_i are projections to E_i and
  - $\limsup_{n\to\infty} \sqrt[n]{\|Q^n\|_{\mathcal{BV}(U,\mathbb{C})}} < 1$,
  - $\Psi_i \circ \Psi_j = 0$ for $i \neq j$ and $\Psi_i \circ Q = Q \circ \Psi_i = 0$ for $1 \leq i \leq r$.

Also one of the eigenvalues λ_i must be 1. See [7] for details. Let us assume $\lambda_1 = 1$. Since P_f does not increase L^1-norm of functions and since $\mathcal{BV}(U)$ is dense in $L^1(U)$, we can see, by approximation argument,

- For any $\varphi \in L^1(U)$, the sequence $1/n \sum_{k=0}^{n-1} P_f^k(\varphi)$ converges in L^1 to an element in the eigenspace E_1 for $\lambda_1 = 1$.

Especially the density function of each absolutely continuous invariant measure is contained in the finite dimensional space E_1. So we have

- there exists only finitely many absolutely continuous ergodic measures μ_k, $1 \leq k \leq q$, and all other absolutely continuous invariant measures are convex combinations of them.

Furthermore, using the argument in [5, 13], we can derive the following ergodic properties of the measure μ_i's.

- For each μ_k, there exist a positive integer p and Borel measurable mutually disjoint subsets C_i, $1 \leq i \leq p$, such that $\mu_k(C_i) = 1/p$, $f(C_i) \subset C_{i+1}$ for $1 \leq i \leq p-1$, $f(C_p) \subset C_1$ and $f^p|_{C_i}$ are exact.

We refer to [4, Sect. 3], [9] and the references given there for further results.

References

1. Adl-Zarabi, K.: Absolutely continuous invariant measure for piecewise expanding C^2 transformations in $\mathbb{R}^n$ on domains with cusps on the boundaries. Ergod. Th.& Dynam. Sys. **16**, 1–18 (1996)
2. Buzzi, J.: A.C.I.M.'S for arbitrary expanding piecewise $\mathbb{R}$-analytic mappings of the plane. Ergod. Th.& Dynam. Sys. (to appear)
3. Giusti, E.: *Minimal Surfaces and Functions of Bounded Variation*. Monographs in Mathematics, Vol. **80**, Boston: Birkhauser, 1984
4. Góra, P., Boyarski, A.: Absolutely continuous invariant measures for piecewise expanding transformations in $\mathbb{R}^N$. Israel J. Math. **67**, 272–276 (1989)
5. Hofbauer, F. and Keller, G.: Ergodic properties of invariant measures for piecewise monotonic transformations. Math. Z. **180**, 119–140 (1982)
6. Ionescu-Tulcea, C., and Marinescu, G.: Theorie ergodique pour des classes d'operations non completement continues. Ann. Math.(2) **52**, 140–147 (1950)
7. Keller, G.: *Propriété ergodique des endomorphismes dilatants, C^2 par morceaux, des rgions bornées du plan*. Thesis, Universite de Rennes, 1979
8. Keller, G.: Ergodicité et mesures invariantes pour les transformations dilatantes par morceaux d'une région bornée du plan. C.R.Acad. Sci. Paris **289** Serie A, 625–627 (1979)
9. Keller, G.: Generalized bounded variation and applications to piecewise monotonic transformations. Z. Wahr. verw. Geb. **69**, 461–478 (1985)

10. Lang, S.: *Real and Functional Analysis*. Graduate Text in Math. **142**, Berlin–Heidelberg–New York: Springer, 1993
11. Lasota, A., Yorke, J.: On the existence of invariant measure for piecewise monotonic transformations. Trans. A.M.S. **186**, 481–488 (1973)
12. Renyi,A.: Representation of real numbers and their ergodic properties. Acta. Math. Akad. Sc. Hungar. **8**, 477–493 (1957)
13. Rychlik, M.: Bounded variation and invariant measures. Studia math. **76**, 69–80 (1983)
14. Tsujii, M.: Piecewise expanding maps on the plane with singular ergodic properties. Ergod. Th.& Dynam. Sys. (to appear)
15. Ziemer, W.: *Weakly Differentiable Functions*. Graduate Text in Math. 120, Berlin–Heidelberg–New York: Springer, 1989

Communicated by Ya. G. Sinai

Invent. math. 140, 351–398 (2000)
Digital Object Identifier (DOI) 10.1007/s002220000057

Inventiones mathematicae

SRB measures for partially hyperbolic systems whose central direction is mostly expanding*

José F. Alves[1], **Christian Bonatti**[2], **Marcelo Viana**[3]

[1] Dep. Matemática Pura, Faculdade de Ciências do Porto, P-4099-002 Porto, Portugal (e-mail: jfalves@fc.up.pt)
[2] Lab. Topologie, CNRS – UMR 5584, Univ. Bourgogne, B.P. 47870, 21078 Dijon Cedex, France (e-mail: bonatti@satie.u-bourgogne.fr)
[3] IMPA, Estrada D. Castorina 110, Jardim Botânico, 22460-320 Rio de Janeiro RJ, Brazil (e-mail: viana@impa.br)

Oblatum 16-IV-1999 & 29-X-1999
Published online: 21 February 2000 –

Abstract. We construct Sinai-Ruelle-Bowen (SRB) measures supported on partially hyperbolic sets of diffeomorphisms – the tangent bundle splits into two invariant subbundles, one of which is uniformly contracting – under the assumption that the complementary subbundle is *non-uniformly* expanding. If the rate of expansion (Lyapunov exponents) is bounded away from zero, then there are only finitely many SRB measures. Our techniques extend to other situations, including certain maps with singularities or critical points, as well as diffeomorphisms having only a dominated splitting (and no uniformly hyperbolic subbundle).

1. Introduction

The following approach has been most effective in studying the dynamics of complicated systems: one tries to describe the average time spent by typical orbits in different regions of the phase space. According to the ergodic theorem of Birkhoff, such times are well defined for almost all point, with respect to any invariant probability measure. However, the notion of typical orbit is usually meant in the sense of volume (Lebesgue measure), which is not always captured by invariant measures. Indeed, it is a fundamental open problem to understand under which conditions the behaviour of typical points is well defined, from this statistical point of view.

* Work carried out at the Laboratoire de Topologie, CNRS-UMR 5584, Dijon, and IMPA, Rio de Janeiro. Partially supported by IMPA and PRONEX-Dynamical Systems, Brazil, Université de Bourgogne, France, and Praxis XXI-Física Matemática and Centro de Matemática da Universidade do Porto, Portugal.

This problem can be precisely formulated by means of the following notion, introduced by Sinai, Ruelle, and Bowen. Let us consider discrete time systems, namely maps $f : M \to M$ on a manifold M. Given an f-invariant Borelian probability μ in M, we call *basin* of μ the set $B(\mu)$ of the points $x \in M$ such that the averages of Dirac measures along the orbit of x converge to μ in the weak* sense:

$$\lim_{n\to+\infty} \frac{1}{n} \sum_{j=0}^{n-1} \varphi(f^j(x)) = \int \varphi \, d\mu \quad \text{for any continuous } \varphi : M \to \mathbb{R}. \tag{1}$$

Then we say that μ is a *physical* or *Sinai-Ruelle-Bowen (SRB) measure* for f if the basin $B(\mu)$ has positive Lebesgue measure in M. Then, one would like to know whether, for most systems, the basins of all SRB measures cover a full Lebesgue measure subset of the whole manifold.

This question has an affirmative answer in the context of uniformly hyperbolic systems, after [Sin72,BR75,Bow75,Rue76]. A detailed picture is also available for maps of the circle or the interval, see [Jak81,Lyu]. However, in higher dimensions the problem is mostly open, outside the uniformly hyperbolic setting, despite substantial progress in the study of certain classes of maps and flows with some properties of non-uniform hyperbolicity, including the Lorenz-like attractors, see e.g. [Pes92,Sat92,Tuc99], and the Hénon-like attractors, see [BC91,BY92,BV].

In this work we deal with diffeomorphisms admitting *partially hyperbolic* invariant sets: the tangent bundle over the set has an invariant dominated splitting into two subbundles, one of which is uniformly hyperbolic (contracting or expanding). Precise definitions are given in the next subsection. This property yields a fair amount of geometric information, e.g. invariant foliations, so that it is natural to try to recover for these systems as much as possible of the geometric and ergodic properties of the hyperbolic ones.

Indeed, some knowledge of such properties is already available, through works of several authors. For the foundations concerning invariant foliations see [BP74,HPS77]. Gibbs u-states were constructed by [PS82], and used by [You90,Car93,BV99] to construct SRB measures for some types of partially hyperbolic systems. SRB measures and decay of correlations were also studied by [Alv97,Cas98,Dol]. Moreover, [Shu71,Mañ78,GPS94,Kan94,BD96,Via97,SW] constructed examples where partial hyperbolicity is used to get robustness (stability) of topological or ergodic properties, which is also a main topic in [BP74]. Conversely, [Mañ82,DPU99,BDP], showed that partial hyperbolicity (or, in high dimensions, existence of a dominated splitting) is in fact a necessary condition for robust topological transitivity.

The interest in this class of systems was further stressed by these recent results, that suggest that partial hyperbolicity (or, at least, existence of a dominated splitting) should be a crucial ingredient in a global theory of Dynamics. A program towards such a theory has been proposed a few years ago by [Pal99], the cornerstone of which is a conjecture containing the following statement: every system can be approximated by one having finitely many SRB measures, whose basins cover a full Lebesgue measure subset of the phase space.

In the present paper we obtain results of existence and finiteness of SRB measures for a large class of partially hyperbolic maps. Previous constructions depended on the existence of a uniformly expanding (strong-unstable) invariant subbundle. In fact, except for the situation of [Via97,Alv97], the SRB measures coincide with the Gibbs u-states (invariant measures absolutely continuous in the strong-unstable direction) constructed by [PS82]. As shown in [BV99], this happens whenever the central subbundle is (at least) non-uniformly contracting.

One main novelty here is that we do not assume existence of a strong-unstable direction. Using only a condition of non-uniform expansion in an invariant (centre-unstable) subbundle, we are able to construct invariant measures absolutely continuous in this centre-unstable direction. Actually, this does not even require the full strength of partial hyperbolicity, it suffices to have a dominated splitting. In some cases, these *Gibbs cu-states* are SRB measures: in particular, this is always the case if the complementary subbundle is uniformly contracting on the support of the measure.

1.1. Partially hyperbolic diffeomorphisms

Let $f : M \to M$ be a C^1 diffeomorphism on a manifold M. Here we say that a compact set $K \subset M$ is *partially hyperbolic* for f if it is positively invariant, i.e. $f(K) \subset K$, and there exists a continuous Df-invariant splitting $T_K M = E^{ss} \oplus E^{cu}$ of the tangent bundle restricted to K and a constant $\lambda < 1$ satisfying (for some choice of a Riemannian metric on M)

1. E^{ss} is uniformly contracting: $\|Df \mid E^{ss}_x\| \leq \lambda$ for all $x \in K$;
2. E^{cu} is dominated by E^{ss}: $\|Df \mid E^{ss}_x\| \cdot \|Df^{-1} \mid E^{cu}_{f(x)}\| \leq \lambda$ for all $x \in K$.

We call E^{ss} strong-stable subbundle and E^{cu} centre-unstable subbundle.

Theorem A. *Let $f : M \to M$ be a C^2 diffeomorphism having a partially hyperbolic set K. Assume that f is non-uniformly expanding along the centre-unstable direction, meaning that*

$$\limsup_{n\to+\infty} \frac{1}{n} \sum_{j=1}^{n} \log \left\|Df^{-1} \mid E^{cu}_{f^j(x)}\right\| < 0 \tag{2}$$

for all x in a positive Lebesgue measure set $H \subset K$. Then f has some ergodic SRB measure with support contained in $\cap_{j=0}^{\infty} f^j(K)$. In fact, Lebesgue almost every point in H belongs in the basin of some such SRB measure.

As we shall explain below, for the proof it is enough to have condition (2) on a positive Lebesgue measure subset of some disk transverse to the strong-stable direction.

The SRB measures produced by this theorem have dim E^{cu} positive Lyapunov exponents and dim E^{ss} negative Lyapunov exponents. Moreover, they have absolutely continuous conditional measures along corresponding Pesin unstable manifolds (which are tangent to the centre-unstable subbundle E^{cu} at almost every point).

We ignore whether the conclusion of Theorem A remains true if the non-uniform expansion condition is replaced by

$$\limsup_{n\to+\infty} \frac{1}{n} \log \left\| Df^{-n} \mid E^{cu}_{f^n(x)} \right\| < 0 \tag{3}$$

(i.e. positivity of all the Lyapunov exponents in the centre-unstable direction). Clearly, the two formulations (2) and (3) are equivalent when the non-uniformly expanding direction is 1-dimensional, that is, when E^{cu} can be split as $E^c \oplus E^{uu}$ with E^c having dimension 1 and E^{uu} being uniformly expanding. Let us also note that for a set of points with full probability (full measure with respect to any invariant probability measure) condition (3) implies condition (2) for some f^k, $k \geq 1$. Of course, the theorem is not affected when f is replaced by any positive iterate of it.

As a consequence of the proof of the theorem we also get

Corollary B. *Under the assumptions of Theorem A, if the limit in (2) is bounded away from zero, then the set H is contained in the union of the basins of finitely many SRB measures, up to a zero Lebesgue measure subset.*

These results have the following curious consequence: in the setting of Theorem A the set of values of the limit in (2) over a full Lebesgue measure subset of H is discrete, with zero as the unique possible accumulation value.

The assumption of partial hyperbolicity in Theorem A can be somewhat relaxed, as we explain in Sect. 6. Assuming only the existence of a dominated splitting $E^{cs} \oplus E^{cu}$ with E^{cu} non-uniformly expanding, we prove that the diffeomorphism admits invariant probability measures with absolutely continuous conditional measures along the centre-unstable direction. This may be thought of as a non-uniform version of the results of [PS82] on existence of Gibbs u-states. We also show that in some cases where E^{cs} is non-uniformly contracting these invariant measures are SRB measures for the corresponding system.

1.2. Maps with singular sets

Similar methods allow us to construct SRB measures for certain maps with singularities and/or critical points. Apart from a condition of non-uniform expansion on a positive Lebesgue measure subset H, similar to (2), we also need the points of H not to spend most of the time too close to the singular set. This is properly expressed in (6) below. Before that, let us explain what we mean by singular set of a map.

Let M be a compact manifold, $\mathcal{S} \subset M$ a compact subset, and $f : M \setminus \mathcal{S} \to M$ a C^2 map on $M \setminus \mathcal{S}$. We assume that f behaves like a power of the distance to $\mathcal{S}$ close to the *singular set* $\mathcal{S}$, in the following sense: there exist constants $B > 1$ and $\beta > 0$ such that

$$\text{(S1)} \quad \frac{1}{B}\operatorname{dist}(x,\mathcal{S})^{\beta} \leq \frac{\|Df(x)v\|}{\|v\|} \leq B\operatorname{dist}(x,\mathcal{S})^{-\beta};$$

$$\text{(S2)} \quad \left|\log\|Df(x)^{-1}\| - \log\|Df(y)^{-1}\|\right| \leq B\frac{\operatorname{dist}(x,y)}{\operatorname{dist}(x,\mathcal{S})^{\beta}};$$

$$\text{(S3)} \quad \left|\log|\det Df(x)| - \log|\det Df(y)|\right| \leq B\frac{\operatorname{dist}(x,y)}{\operatorname{dist}(x,\mathcal{S})^{\beta}};$$

for every $v \in T_xM$ and $x, y \in M \setminus \mathcal{S}$ with $\operatorname{dist}(x,y) < \operatorname{dist}(x,\mathcal{S})/2$.

Given $\delta > 0$ and $x \in M \setminus \mathcal{S}$ we define the *δ-truncated distance* from x to $\mathcal{S}$

$$\operatorname{dist}_{\delta}(x,\mathcal{S}) = \begin{cases} 1 & \text{if } \operatorname{dist}(x,\mathcal{S}) \geq \delta \\ \operatorname{dist}(x,\mathcal{S}) & \text{otherwise.} \end{cases} \tag{4}$$

Note that this is not really a distance function: $\operatorname{dist}(x,y) + \operatorname{dist}_{\delta}(y,\mathcal{S})$ may be smaller than $\operatorname{dist}_{\delta}(x,\mathcal{S})$. We also denote $\mathcal{S}_{\infty} = \cup_{n=0}^{\infty} f^n(\mathcal{S})$.

Theorem C. *Assume that f satisfies (S1), (S2), (S3) and is non-uniformly expanding, in the sense that*

$$\limsup_{n\to+\infty} \frac{1}{n}\sum_{j=0}^{n-1} \log\|Df(f^j(x))^{-1}\| < 0 \tag{5}$$

for all x in a positive Lebesgue measure set $H \subset M \setminus \mathcal{S}_{\infty}$. Assume moreover that, given any $\varepsilon > 0$ there exists $\delta > 0$ such that for every $x \in H$

$$\limsup_{n\to+\infty} \frac{1}{n}\sum_{j=0}^{n-1} -\log\operatorname{dist}_{\delta}(f^j(x),\mathcal{S}) \leq \varepsilon. \tag{6}$$

Then Lebesgue almost every point in H belongs in the basin of some ergodic absolutely continuous invariant measure.

As a by-product, corresponding to the case when the singular set is empty,

Corollary D. *Let $f : M \to M$ be a C^2 covering map (local diffeomorphism) on a compact manifold M, which is non-uniformly expanding: (5) holds for all x in a set $H \subset M$ with positive Lebesgue measure. Then Lebesgue almost every point $x \in H$ belongs in the basin of some ergodic absolutely continuous invariant measure.*

In the settings of these last two results, we also have an analogue of Corollary B: if the limit in (5) is bounded away from zero then H is covered by the basins of finitely many ergodic absolutely continuous invariant measures. See also the comments at the end of Sect. 5 concerning similar results for partially hyperbolic maps with singularities.

For unimodal maps of the interval with negative Schwarzian derivative, results in a similar spirit had been obtained by [Led81] and [Kel90]. In particular, [Kel90] contains a strong version of Corollary D for such maps.

This paper is organized as follows. To construct the SRB measures in Theorem A we fix any C^2 disk transverse to the strong-stable direction and intersecting H on a positive Lebesgue measure subset, and we consider the sequence of averages of forward iterates of Lebesgue measure restricted to such a disk. We prove, in Sect. 3, that a definite fraction of each average corresponds to a measure that is absolutely continuous with respect to Lebesgue measure along the iterates of the disk, with uniformly bounded densities. This uses distortion bounds that we obtain in Sect. 2, together with the key notion of *hyperbolic times*, first introduced in [Alv97]. The construction of the SRB measures is completed in Sect. 4, where we show that the absolute continuity property passes to the limit. In Sect. 4 we also obtain enough information about the basins of these measures to prove the finiteness statement in Corollary B.

In Sect. 5 we explain how condition (6) allows us to bypass the difficulty caused by the presence of singularities, and obtain Theorem C and Corollary D, as well as the results on partially hyperbolic maps with singularities that we mentioned before. Sect. 6 contains related results for systems with a dominated splitting $E^{cs} \oplus E^{cu}$ where neither of the factors needs have any uniform subbundle, that we also mentioned above. In the Appendix we give a few simple criteria allowing to check the assumptions of our results in specific situations.

2. Curvature and distortion

A first main step in the proof of Theorem A is to prove a bounded distortion property for iterates of f over disks whose tangent space is contained in a centre-unstable cone at each point. This section is devoted to the precise statement and proof of this property.

Remark 2.1. Throughout this section, as well as in Sect. 3 and Subsect. 4.1, we do not need the full strength of partial hyperbolicity. In Subsect. 2.1 we only use existence of a dominated splitting $E^{cs} \oplus E^{cu}$. Subsections 2.2 and 2.3, Sect. 3, and Subsect. 4.1 depend also on the condition of non-uniform expansion along the centre-unstable direction (2). Existence of a strong-stable subbundle E^{ss} is used for the first time in Subsect. 4.2.

We fix continuous extensions of the two subbundles E^{cs} and E^{cu} to some neighbourhood V_0 of K, that we denote $\tilde{E}^{cs}$ and $\tilde{E}^{cu}$. It should be noted that we do not require these extensions to be invariant under Df. Then, given $0 < a < 1$, we define the *centre-unstable cone field* $C_a^{cu} = \left(C_a^{cu}(x)\right)_{x\in V_0}$ *of width* a by

$$C_a^{cu}(x) = \left\{v_1 + v_2 \in \tilde{E}_x^{cs} \oplus \tilde{E}_x^{cu} \text{ such that } \|v_1\| \leq a\|v_2\|\right\}. \tag{7}$$

We define the *centre-stable cone field* $C_a^{cs} = \left(C_a^{cs}(x)\right)_{x\in V_0}$ *of width* a in a similar way, just reversing the roles of the subbundles in (7).

We fix $a > 0$ and V_0 small enough so that, up to slightly increasing $\lambda < 1$, the domination condition 2. in Subsect. 1.1 remains valid for any pair of vectors in the two cone fields:

$$\|Df(x)v^{cs}\| \cdot \|Df^{-1}(f(x))v^{cu}\| \leq \lambda\|v^{cs}\|\,\|v^{cu}\|$$

for every $v^{cs} \in C_a^{cs}(x)$, $v^{cu} \in C_a^{cu}(f(x))$, and any point $x \in V_0 \cap f^{-1}(V_0)$. Note that the centre-unstable cone field is positively invariant: $Df(x)C_a^{cu}(x) \subset C_a^{cu}(f(x))$ whenever x and $f(x)$ are in V_0. Indeed, the domination property together with the invariance of $E^{cu} = (\tilde{E}^{cu} \mid K)$ imply

$$Df(x)C_a^{cu}(x) \subset C_{\lambda a}^{cu}(f(x)) \subset C_a^{cu}(f(x)),$$

for every $x \in K$, and this extends to any $x \in V_0 \cap f^{-1}(V_0)$ by continuity.

Wherever we presume E^{cs} to be uniformly contracting (as already mentioned, this will happen not happen before Subsect. 4.2), we denote it by E^{ss} instead, and represent by $\tilde{E}^{ss}$ its extension to V_0. Moreover, in that case the cone field C_a^{cs} is denoted C_a^{ss}, and called *strong-stable*.

2.1. Hölder control of the tangent direction

We say that an embedded C^1 submanifold $N \subset V_0$ is *tangent to the centre-unstable cone field* C_a^{cu} if the tangent subspace to N at each point $x \in N$ is contained in the corresponding cone $C_a^{cu}(x)$. Then $f(N)$ is also tangent to the centre-unstable cone field, if it is contained in V_0.

The tangent bundle of N is said to be Hölder continuous if $x \mapsto T_xN$ defines a Hölder continuous section from N to the corresponding Grassman bundle of M. In this subsection we show that the tangent bundle of the iterates of a C^2 submanifold are Hölder continuous (as long as they do not leave V_0), with uniform Hölder constants.

The basic idea is contained in the following observation, which the reader may easily check. Let E_1, E_2 be two Euclidean spaces and L be a linear isomorphism on $E_1 \oplus E_2$ leaving both factors invariant. Assume that we have the domination property

$$\|L \mid E_1\| \cdot \|L^{-1} \mid E_2\| < 1.$$

Then there exist $C > 0$ and $0 < \zeta \leq 1$ such that if $\Gamma \subset E_1 \oplus E_2$ is the graph of a $C^{1+\zeta}$ map $\phi : E_1 \to E_2$, with Hölder constant C, then the same is true for $L(\Gamma)$. In fact, it suffices to take any ζ such that $\|L \mid E_1\| \cdot \|L^{-1} \mid E_2\|^{1+\zeta} < 1$.

In order to apply similar arguments in our situation, it is useful to express the notion of Hölder variation of the tangent bundle in local coordinates, as follows.

We choose $\delta_0 > 0$ small enough so that the inverse of the exponential map $\exp_x$ is defined on the δ_0 neighbourhood of every point x in V_0. From now on we identify this neighbourhood of x with the corresponding neighbourhood U_x of the origin in $T_x N$, through the local chart defined by $\exp_x^{-1}$. Accordingly, we identify x with $0 \in T_x N$. Reducing δ_0, if necessary, we may suppose that $\tilde{E}^{cs}_x$ is contained in the centre-stable cone $C^{cs}_a(y)$ of every $y \in U_x$. In particular, the intersection of $C^{cu}_a(y)$ with $\tilde{E}^{cs}_x$ reduces to the zero vector. Then, the tangent space to N at y is parallel to the graph of a unique linear map $A_x(y) : T_x N \to \tilde{E}^{cs}_x$. Given constants $C > 0$ and $0 < \zeta \le 1$, we say that *the tangent bundle to N is (C, ζ)-Hölder* if

$$\|A_x(y)\| \le C d_x(y)^\zeta \quad \text{for every } y \in N \cap U_x \text{ and } x \in V_0 . \tag{8}$$

Here, $d_x(y)$ denotes the distance from x to y along $N \cap U_x$, defined as the length of the shortest curve connecting x to y inside $N \cap U_x$.

Recall that we have chosen the neighbourhood V_0 and the cone width a sufficiently small so that the domination property remains valid for vectors in the cones $C^{cs}_a(z)$, $C^{cu}_a(z)$, and for any point z in V_0. Then, there exist $\lambda_1 \in (\lambda, 1)$ and $\zeta \in (0, 1]$ such that

$$\|Df(z)v^{cs}\| \cdot \|Df^{-1}(f(z))v^{cu}\|^{1+\zeta} \le \lambda_1 < 1 \tag{9}$$

for every norm 1 vectors $v^{cs} \in C^{cs}_a(z)$ and $v^{cu} \in C^{cu}_a(z)$, at any $z \in V_0$. Then, up to reducing $\delta_0 > 0$ and slightly increasing $\lambda_1 < 1$, (9) remains true if we replace z by any $y \in U_x$, $x \in V_0$ (taking $\| \cdot \|$ to mean the Riemannian metric in the corresponding local chart).

We fix ζ and λ_1 as above in all that follows. Then, given a C^1 submanifold $N \subset V_0$, we denote

$$\kappa(N) = \inf\{C > 0 : \text{the tangent bundle of } N \text{ is } (C, \zeta)\text{-Hölder}\}. \tag{10}$$

Proposition 2.2. *There exist $\lambda_0 < 1$ and $C_0 > 0$ so that if $N \subset V_0 \cap f^{-1}(V_0)$ is any C^1 submanifold tangent to the centre-unstable cone field then*

$$\kappa(f(N)) \le \lambda_0 \, \kappa(N) + C_0 .$$

Proof. Of course, we only need to consider the case when $\kappa(N)$ is finite, that is, the tangent bundle of N is (C, ζ)-Hölder for some $C > 0$. Let $x \in N$ be fixed. We use $(u, s) \in T_x N \oplus \tilde{E}^{cs}_x$ and $(u_1, s_1) \in T_{f(x)} f(N) \oplus \tilde{E}^{cs}_{f(x)}$, respectively, to represent the local coordinates in U_x and $U_{f(x)}$ introduced above. We write the expression of our map in these local coordinates as

$f(u, s) = (u_1(u, s), s_1(u, s))$. Observe that if $x \in K$ then the partial derivatives of u_1 and s_1 at the origin $0 \in T_x N$ are

$$\partial_u u_1(0) = Df|T_x N, \quad \partial_s u_1(0) = 0, \quad \partial_u s_1(0) = 0, \quad \partial_s s_1(0) = Df|\tilde{E}^{cs}_x .$$

This is because $E^{cs}_x = \tilde{E}^{cs}_x$ is mapped to $E^{cs}_{f(x)} = \tilde{E}^{cs}_{f(x)}$ under $Df(x)$ and, similarly, $T_x N$ is mapped to $T_{f(x)} N$. Then, given any small $\varepsilon_0 > 0$ we have that

$$\|\partial_u u_1(y) - Df|T_x N\|, \ \|\partial_s u_1(y)\|, \ \|\partial_u s_1(y)\|, \ \|\partial_s s_1(y) - Df|\tilde{E}^{cs}_x\|, \tag{11}$$

are all less than ε_0 for every $x \in V_0$ and $y \in U_x$, as long as δ_0 and V_0 are small. Taking the cone width a also small, we get

$$\begin{aligned} \left\|Df|T_y N - Df|\tilde{E}^{cu}_x\right\| &\leq \varepsilon_0 \quad \text{and} \\ \left\|Df^{-1}|T_{f(y)} f(N) - Df^{-1}|\tilde{E}^{cu}_{f(x)}\right\| &\leq \varepsilon_0 , \end{aligned} \tag{12}$$

for every $x \in V_0$ and $y \in U_x$. Since f is C^2, there is also some constant $K_2 > 0$ such that

$$\|\partial_s u_1(y)\| \leq K_2 d_x(y) \quad \text{and} \quad \|\partial_u s_1(y)\| \leq K_2 d_x(y). \tag{13}$$

For y_1 in $U_{f(x)}$, let $A_{f(x)}(y_1)$ be the linear map from $T_{f(x)} f(N)$ to $\tilde{E}^{cs}_{f(x)}$ whose graph is parallel to $T_{y_1} f(N)$. We are going to prove that, fixing ε_0 sufficiently small, then $A_{f(x)}(y_1)$ satisfies (8) for any $C > \lambda_0 \kappa(N) + C_0$, with convenient λ_0 and C_0. Let us begin by noting that $\|A_{f(x)}(y_1)\|$ is bounded by some uniform constant $K_1 > 0$, since $f(N)$ is tangent to the centre-unstable cone field. We will choose the constant $C_0 \geq K_1/(\delta_0/\|Df^{-1}\|)^\zeta$, so that (8) is immediate when $d_{f(x)}(y_1) \geq \delta_0/\|Df^{-1}\|$:

$$\|A_{f(x)}(y_1)\| \leq K_1 \leq C_0(\delta_0/\|Df^{-1}\|)^\zeta \leq C_0 d_{f(x)}(y_1)^\zeta .$$

Here $\|Df^{-1}\|$ is the supremum of all $\|Df^{-1}(z)\|$ with $z \in U_w$, $w \in V_0$, where the norms are taken with respect to the Riemannian metrics in the local charts. This permits us to restrict to the case when $d_{f(x)}(y_1) < \delta_0/\|Df^{-1}\|$ in all that follows. Let Γ_1 be any curve on $f(N) \cap U_{f(x)}$ joining $f(x)$ to y_1 and whose length approximates $d_{f(x)}(y_1)$. Then $\Gamma = f^{-1}(\Gamma_1)$ is a curve in $N \cap U_x$ joining x to $y = f^{-1}(y_1)$, with length less than δ_0. In fact, cf. (12),

$$d_x(y) \leq \text{length}(\Gamma) \leq \left(\|Df^{-1}|\tilde{E}^{cu}_{f(x)}\| + \varepsilon_0\right) \text{length}(\Gamma_1).$$

This shows that $d_x(y) \leq (\|Df^{-1}|\tilde{E}^{cu}_{f(x)}\| + \varepsilon_0) d_{f(x)}(y_1)$.

Now we observe that

$$A_{f(x)}(y_1) = \left[\partial_u s_1(y) + \partial_s s_1(y) \cdot A_x(y)\right] \cdot \left[\partial_u u_1(y) + \partial_s u_1(y) \cdot A_x(y)\right]^{-1}.$$

On the one hand, by (11) and (13),

$$\begin{aligned}\|\partial_u s_1(y) + \partial_s s_1(y) \cdot A_x(y)\| &\le K_2 d_x(y) + \big(\|Df|\tilde{E}^{cs}_x\| + \varepsilon_0\big)\kappa(N) d_x(y)^\zeta \\ &\le \big(K_2 + \big(\|Df|\tilde{E}^{cs}_x\| + \varepsilon_0\big)\kappa(N)\big) d_x(y)^\zeta .\end{aligned}$$

On the other hand, $\|\partial_s u_1(y) \cdot A_x(y)\| \le \epsilon_0 K_1$, which can be made much smaller than $1/\|(\partial_u u_1(y)^{-1}\|$. As a consequence, recall (12) and (13),

$$\big\|\big[\partial_u u_1(y) + \partial_s s_1(y) \cdot A_x(y)\big]^{-1}\big\| \le \big\|Df^{-1}\big|\tilde{E}^{cu}_{f(x)}\big\| + \varepsilon_1 ,$$

where ε_1 can be made arbitrarily small by reducing ε_0. Putting these bounds together, we conclude that $\|A_{f(x)}(y_1)\|\, d_{f(x)}(y_1)^{-\zeta}$ is less than

$$\frac{\big(\|Df|\tilde{E}^{cs}_x\| + \varepsilon_0\big)\big(\|Df^{-1}|\tilde{E}^{cu}_{f(x)}\| + \varepsilon_1\big)}{\big(\|Df^{-1}|\tilde{E}^{cu}_{f(x)}\| + \varepsilon_0\big)^{-\zeta}}\kappa(N) + \frac{K_2\big(\|Df^{-1}|\tilde{E}^{cu}_{f(x)}\| + \varepsilon_1\big)}{\big(\|Df^{-1}|\tilde{E}^{cu}_{f(x)}\| + \varepsilon_0\big)^{-\zeta}}.$$

So, choosing δ_0, V_0, a sufficiently small, we can make ε_0, ε_1, sufficiently close to zero so that the factor multiplying $\kappa(N)$ is less than some $\lambda_0 \in (\lambda_1, 1)$; recall (9). Moreover, the second term in the expression above is bounded by some constant that depends only on f. We take C_0 larger than this constant. □

Remark 2.3. The proof remains valid if the diffeomorphism f is only of class $C^{1+\zeta}$: it suffices to replace (13) by $\|\partial_s u_1(y)\|$, $\|\partial_u s_1(y)\| \le K_2 d_x(y)^\zeta$, which is sufficient for the rest of the argument.

Corollary 2.4. *There exists $C_1 > 0$ such that, given any C^1 submanifold $N \subset V_0$ tangent to the centre-unstable cone field,*

a) there exists $n_0 \ge 1$ such that $\kappa(f^n(N)) \le C_1$ for every $n \ge n_0$ such that $f^k(N) \subset V_0$ for all $0 \le k \le n$;
b) if $\kappa(N) \le C_1$, then the same is true for every iterate $f^n(N)$, $n \ge 1$, such that $f^k(N) \subset V_0$ for all $0 \le k \le n$;
c) in particular, if N and n are as in b), then the functions

$$J_k : f^k(N) \ni x \mapsto \log\big|\det\big(Df \mid T_x f^k(N)\big)\big|, \quad 0 \le k \le n,$$

are (L_1, ζ)-Hölder continuous with $L_1 > 0$ depending only on C_1 and f.

Proof. It suffices to choose any $C_1 \ge C_0/(1 - \lambda_0)$. □

Remark 2.5. Suppose that we have the following stronger form of domination

$$\|Df|E^{cs}_x\| \cdot \|Df^{-1}|E^{cu}_{f(x)}\|^i \le \lambda \quad \text{for } i = 1, 2,$$

and any $x \in K$. Then, assuming that f is a C^2 diffeomorphism, we may take $\zeta = 1$ in the previous arguments. In that case, $\kappa(N)$ yields a bound on the curvature tensor of N. So, Corollary 2.4 asserts that if N is C^2 then the curvature of all its iterates $f^n(N)$, $n \ge 1$, is bounded by some constant that depends only on the curvature of N.

2.2. *Hyperbolic times and distortion bounds*

The following notion will allow us to derive *uniform behaviour* (expansion, distortion) from the hypothesis of non-uniform expansion in (2).

Definition 2.6. *Given* $\sigma < 1$, *we say that* n *is a* σ-hyperbolic time *for a point* $x \in K$ *if*

$$\prod_{j=n-k+1}^{n} \left\| Df^{-1} \mid E^{cu}_{f^j(x)} \right\| \le \sigma^k \qquad \textit{for all } 1 \le k \le n.$$

In particular, if n is a σ-hyperbolic time for x then $Df^{-k} \mid E^{cu}_{f^n(x)}$ is a contraction for every $1 \le k \le n$:

$$\left\| Df^{-k} | E^{cu}_{f^n(x)} \right\| \le \prod_{j=n-k+1}^{n} \left\| Df^{-1} | E^{cu}_{f^j(x)} \right\| \le \sigma^k.$$

Moreover, if a is taken sufficiently small in the definition of our cone fields, and we choose $\delta_1 > 0$ also small (in particular, the δ_1-neighbourhood of K should be contained in V_0), then, by continuity,

$$\|Df^{-1}(f(y))v\| \le \frac{1}{\sqrt{\sigma}} \|Df^{-1}|E^{cu}_{f(x)}\| \, \|v\| \tag{14}$$

whenever $x \in K$, $\operatorname{dist}(x, y) \le \delta_1$, and $v \in C^{cu}(y)$.

Let D be any C^1 disk contained in V_0 and tangent to the centre-unstable cone field. We use $\operatorname{dist}_D(\cdot\,,\cdot)$ to denote distance between two points in the disk, measured along D. The distance from a point $x \in D$ to the boundary of D is $\operatorname{dist}_D(x, \partial D) = \inf_{y \in \partial D} \operatorname{dist}_D(x, y)$.

Lemma 2.7. *Given any* C^1 *disk* $D \subset V_0$ *tangent to the centre-unstable cone field,* $x \in D \cap K$, *and* $n \ge 1$ *a* σ*-hyperbolic time for* x,

$$\operatorname{dist}_{f^{n-k}(D)}(f^{n-k}(y), f^{n-k}(x)) \le \sigma^{k/2} \operatorname{dist}_{f^n(D)}(f^n(y), f^n(x)).$$

for any point $y \in D$ *with* $\operatorname{dist}(f^n(x), f^n(y)) \le \delta_1$.

Proof. Let η_0 be a curve of minimal length in $f^n(D)$ connecting $f^n(x)$ to $f^n(y)$. For $1 \le k \le n$ write $\eta_k = f^{n-k}(\eta_0)$. We prove the lemma by induction. Let $1 \le k \le n$ and assume that

$$\operatorname{length}(\eta_j) \le \delta_1 \quad \text{for } 0 \le j \le k-1.$$

Denote by $\dot\eta_0(z)$ the tangent vector to the curve η_0 at the point z. Then, in view of the choice of δ_1 in (14) and the definition of σ-hyperbolic times,

$$\|Df^{-k}(z)\dot\eta_0(z)\| \le \sigma^{-k/2} \|\dot\eta_0(z)\| \prod_{j=n-k+1}^{n} \|Df^{-1}|E^{cu}_{f^j(x)}\| \le \sigma^{k/2} \|\dot\eta_0(z)\|.$$

As a consequence,

$$\text{length}(\eta_k) \leq \sigma^{k/2}\,\text{length}(\eta_0) = \sigma^{k/2}\,\text{dist}_{f^{n-k}(D)}(f^{n-k}(y), f^{n-k}(x)) \leq \delta_1 .$$

This completes our induction, thus proving the lemma. □

Proposition 2.8. *There exists $C_2 > 1$ such that, given any C^1 disk D tangent to the centre-unstable cone field with $\kappa(D) \leq C_1$, and given any $x \in D \cap K$ and $n \geq 1$ a σ-hyperbolic time for x, then*

$$\frac{1}{C_2} \leq \frac{|\det Df^n \mid T_y D|}{|\det Df^n \mid T_x D|} \leq C_2$$

for every $y \in D$ such that $\text{dist}(f^n(y), f^n(x)) \leq \delta_1$.

Proof. For $0 \leq i < n$ and $y \in D$, we denote $J_i(y) = |\det Df \mid T_{f^i(y)} f^i(D)|$. Then,

$$\log \frac{|\det Df^n \mid T_y D|}{|\det Df^n \mid T_x D|} = \sum_{i=0}^{n-1} \big(\log J_i(y) - \log J_i(x)\big)$$

By Corollary 2.4, $\log J_i$ is (L_1, ζ)-Hölder continuous, for some uniform $L_1 > 0$. Moreover, by Lemma 2.7, the sum of all $\text{dist}_D(f^j(x), f^j(y))^\zeta$ over $0 \leq j \leq n$ is bounded by $\delta_1/(1-\sigma^{\zeta/2})$. Now it suffices to take $C_2 = \exp(L_1\delta_1/(1-\sigma^{\zeta/2}))$. □

2.3. Curvature at hyperbolic times

It is possible to obtain control of the curvature at hyperbolic times, without having to assume the stronger form of domination in Remark 2.5. As before, we assume that f is a C^2 diffeomorphism with a partially hyperbolic set K. Let $\sigma < 1$ be fixed, and $\delta_1 > 0$ be chosen as in (14).

Proposition 2.9. *Let D be a C^2 disk tangent to the centre-unstable cone field, $x \in D \cap K$, and $n \geq 1$ be a σ-hyperbolic time for x. Then, the curvature of the δ_1-neighbourhood of $f^n(x)$ in $f^n(D)$ is bounded by a constant $K_0 > 0$ that depends only on f, σ, and the curvature of D. In fact, if n is sufficiently large then K_0 may be taken depending only on f and σ.*

The main idea in the proof of this proposition is to show that, up to conformal changes of the Riemannian metric, we may suppose that $Df \mid E^{cu}$ is uniformly expanding at every point $f^j(x), 0 \leq j < n$. As a consequence, the domination condition 2. in Subsect. 1.1 implies the condition in Remark 2.5 (with respect to the modified metrics). In doing this, it is important that all metric changes can be done by dilation, which is due to the hyperbolic time condition.

Lemma 2.10. *Let $n \geq 1$ and $a_1, \dots, a_n$, c_0, be real numbers such that*

$$\sum_{j=k+1}^{n} a_j \geq (n-k)c_0 \quad \text{for all } 0 \leq k < n. \tag{15}$$

Then, there exist $b_1, \dots, b_n$ such that

1. $|b_j| \leq \sup_{1\leq j\leq n} |c_0 - a_j|$ *for all* $1 \leq j \leq n$;
2. $a_j + b_j \geq c_0$ *for all* $1 \leq j \leq n$;
3. $\sum_{j=1}^{k} b_j \geq 0$ *for* $1 \leq k < n$ *and* $\sum_{j=1}^{n} b_j = 0$.

Proof. Define b_j by recurrence, through

$$b_1 = \max\{0, c_0 - a_1\} \quad \text{and} \quad b_j = \max\{-\sum_{i=1}^{j-1} b_j, c_0 - a_j\}$$
$$\text{for } j = 2, \dots, n.$$

The first condition in the statement is clear, in the case when $b_j = c_0 - a_j$. Otherwise, $b_j = -\sum_{i=1}^{j-1} b_i$ which, by construction, is always non-positive. So, in this second case we must have $0 \geq b_j \geq c_0 - a_j$, so that the bound in 1. remains valid. The second condition follows immediately from the construction, and the same is true for the first statement in 3. To obtain the last claim, we begin by proving the following fact, by induction on j:

$$\sum_{i=1}^{j} b_i \leq \sum_{i=j+1}^{n} (a_i - c_0) \quad \text{for } j = 1, \dots, n-1. \tag{16}$$

In view of (15),

$$\sum_{i=2}^{n}(a_i - c_0) \geq 0 \quad \text{and} \quad \sum_{i=2}^{n}(a_i - c_0) \geq \sum_{i=1}^{n}(a_i - c_0) + (c_0 - a_1) \geq (c_0 - a_1).$$

This gives $\sum_{i=2}^{n}(a_i - c_0) \geq \max\{0, c_0 - a_1\} = b_1$, corresponding to case $j = 1$. Now we suppose that, by recurrence, $\sum_{i=1}^{j-1} b_i \leq \sum_{i=j}^{n}(a_i - c_0)$. Then, either $b_j = c_0 - a_j$ in which case, adding b_j to both sides of the previous inequality immediately gives the conclusion. Or else, $b_j = -\sum_{i=1}^{j-1} b_i$ and then

$$\sum_{i=1}^{j} b_i = 0 \leq \sum_{i=j+1}^{n} (a_i - c_0),$$

due to our assumption (15). This completes the proof of (16). In particular, taking $j = n-1$, we get that $-\sum_{i=1}^{n-1} b_i \geq c_0 - a_n$, and so $b_n = -\sum_{j=1}^{n-1} b_i$. □

Now we use this lemma to prove Proposition 2.9.

Proof. Let D_n be the neighbourhood of radius δ_1 around $f^n(x)$ in $f^n(D)$, and $D_j = f^{j-n}(D_n)$ for $0 \leq j < n$. Take $c_0 = -\log\sigma$ and $a_j = -\log \|Df^{-1}|E^{cu}_{f^j(x)}\|$, $1 \leq j \leq n$, in Lemma 2.10. Let b_j, $1 \leq j \leq n$, be the corresponding sequence, and denote $t_j = \exp(\sum_{i=1}^{j} b_i)$ for $1 \leq j \leq n$, and $t_0 = 1$. Conclusion 3. in the lemma implies that

$$t_j \geq 1 \text{ for every } j \quad \text{and} \quad t_n = 1.$$

In all that follows $\|\cdot\|_j$ denotes the metric obtained by multiplying the initial Riemannian metric of M by t_j, $0 \leq j \leq n$. Accordingly, we denote $\|Df\|_{j-1,j}$ and $\|D^2 f\|_{j-1,j}$ the norms of the derivatives of f from $(M, \|\cdot\|_{j-1})$ to $(M, \|\cdot\|_j)$. We use similar notations for the restrictions of Df, Df^{-1} to the subbundles $\tilde{E}^{cu}$ and $\tilde{E}^{cs}$. Observe that

$$\|Df\|_{j-1,j} = \exp b_j \|Df\| \quad \text{and} \quad \|D^2 f\|_{j-1,j} = \frac{\exp b_j}{t_{j-1}} \|D^2 f\|.$$

Since the b_j are bounded, and the t_j are larger than 1, $\|Df\|_{j-1,j}$ and $\|D^2 f\|_{j-1,j}$ are bounded by constants that depend only on f and σ. Note also that the domination property is not affected by this conformal change of metrics:

$$\left\|Df|\tilde{E}^{cs}_y\right\|_{j-1,j} \cdot \left\|Df^{-1}|\tilde{E}^{cu}_{f(y)}\right\|_{j,j-1} = \left\|Df|\tilde{E}^{cs}_y\right\| \cdot \left\|Df^{-1}|\tilde{E}^{cu}_{f(y)}\right\| \leq \lambda, \tag{17}$$

at every point y where these subbundles are defined.

Conclusion 2. in the lemma now means that

$$\left\|Df^{-1}|E^{cu}_{f^j(x)}\right\|_{j,j-1} = \exp(-b_j - a_j) \leq \sigma.$$

So, by (14) and Lemma 2.7, $\|Df^{-1}|\tilde{E}^{cu}_{f^j(y)}\|_{j,j-1} \leq \sqrt{\sigma} < 1$ for every y in D_0 and $1 \leq j \leq n$. Together with (17), this gives

$$\left\|Df|\tilde{E}^{cs}_{f^{j-1}(y)}\right\|_{j-1,j} \cdot \left\|Df^{-1}|\tilde{E}^{cu}_{f^j(y)}\right\|^2_{j,j-1} \leq \lambda\sqrt{\sigma} \leq \lambda$$

for every $y \in D_0$ and $1 \leq j \leq n$. This means that a strong domination property as in Remark 2.5 is valid, with respect to the relevant modified metrics, at every point of $D_0 \cup f(D_0) \cup \cdots \cup f^{n-1}(D_0)$. Since we already checked that the first and second derivatives have uniformly bounded norms relative to these modified metrics, the arguments in Proposition 2.2 carry on completely to the present context to prove Proposition 2.9. □

Closing this section we observe that these arguments could also be used to give an alternative proof of the distortion bounds we obtained in the previous section.

3. Lebesgue measure at hyperbolic times

The following lemma, due to Pliss [Pli72], will permit us to prove that a point x satisfying assumption (2) has many (positive density at infinity) hyperbolic times.

Lemma 3.1. *Given $A \geq c_2 > c_1 > 0$, let $\theta_0 = (c_2 - c_1)/(A - c_1)$. Then, given any real numbers $a_1, \ldots, a_N$ such that*

$$\sum_{j=1}^{N} a_j \geq c_2 N \qquad \text{and} \qquad a_j \leq A \text{ for every } 1 \leq j \leq N,$$

there are $l > \theta_0 N$ and $1 < n_1 < \cdots < n_l \leq N$ so that

$$\sum_{j=n+1}^{n_i} a_j \geq c_1(n_i - n) \quad \text{for every } 0 \leq n < n_i \text{ and } i = 1, \ldots, l.$$

Proof. (cf. [Mañ87, Section 2]) Define $S(n) = \sum_{j=1}^{n}(a_j - c_1)$, for each $1 \leq n \leq N$, and also $S(0) = 0$. Then define $1 < n_1 < \cdots < n_l \leq N$ to be the maximal sequence such that $S(n_i) \geq S(n)$ for every $0 \leq n < n_i$ and $i = 1, \ldots, l$. Note that l can not be zero, since $S(N) > S(0)$. Moreover, the definition means that

$$\sum_{j=n+1}^{n_i} a_j \geq c_1(n_i - n) \quad \text{for } 0 \leq n < n_i \text{ and } i = 1, \ldots, l.$$

So, we only have to check that $l > \theta_0 N$. Observe that, by definition,

$$S(n_i - 1) < S(n_{i-1}) \quad \text{and so} \quad S(n_i) < S(n_{i-1}) + (A - c_1)$$

for every $1 < i \leq l$. Moreover, $S(n_1) \leq (A - c_1)$ and $S(n_l) \geq S(N) \geq N(c_2 - c_1)$. This gives,

$$N(c_2 - c_1) \leq S(n_l) = \sum_{i=2}^{l} \big(S(n_i) - S(n_{i-1})\big) + S(n_1) < l(A - c_1),$$

which completes the proof. □

Clearly, the set H in the statement of Theorem A may be taken positively invariant under f. Given any $\sigma < 1$, let $H(\sigma)$ be the set of points in H for which the limit in (2) is smaller than $3 \log \sigma$. Then $H(\sigma)$ is positively invariant and, since we are assuming that H has positive Lebesgue measure, $H(\sigma)$ must also have positive Lebesgue measure if σ is close enough to 1. Then, there exists some small C^2 disk D transverse to the centre-stable subbundle, and intersecting $H(\sigma)$ in a set with positive Lebesgue measure inside D. Up to replacing it by some small disk contained in $f^l(D)$ for some large enough $l \geq 1$, we may suppose that D is tangent to the centre-unstable cone field and $\kappa(D) \leq C_1$. Here $\kappa(\cdot)$ is the Hölder constant defined by (10), recall Corollary 2.4. We fix such a disk D, once and for all.

Corollary 3.2. *There is $\theta > 0$, depending only on σ and f, such that, given any $x \in D \cap H(\sigma)$ and any sufficiently large $N \geq 1$, there exist σ-hyperbolic times $1 \leq n_1 < \cdots < n_l \leq N$ for x, with $l \geq \theta N$.*

Proof. Since $x \in H(\sigma)$, we have

$$\sum_{j=1}^{N} \log \left\| Df^{-1} | E^{cu}_{f^j(x)} \right\| \leq 2N \log \sigma,$$

for all $N \geq 1$ sufficiently large. Now it suffices to take $c_1 = |\log \sigma|$, $c_2 = 2c_1$, $A = \sup \left| \log \|Df^{-1}|E^{cu}\| \right|$, and $a_j = -\log \|Df^{-1}|E^{cu}_{f^j(x)}\|$ in the previous lemma. □

Let $\delta_1 > 0$ be the small number introduced prior to Lemma 2.7. In particular, we requested that the δ_1-neighbourhood of the set K be contained in the domain V_0 of the invariant cone fields C^{cs} and C^{cu}. Reducing δ_1 if necessary, we may suppose that the subset $A = A(D, \sigma, \delta_1)$ of points $x \in D \cap H(\sigma)$ such that $\operatorname{dist}_D(x, \partial D) \geq \delta_1$ still has positive Lebesgue measure in D.

We consider the sequence

$$\mu_n = \frac{1}{n} \sum_{j=0}^{n-1} f^j_* \operatorname{Leb}_D \tag{18}$$

of averages of forward iterates of Lebesgue measure on D. A main idea is to decompose μ_n as a sum of two measures, to be denoted ν_n and η_n, such that ν_n is uniformly absolutely continuous on iterates of the disk D (uniformly bounded density with respect to Lebesgue measure) and has total mass uniformly bounded away from zero for all large n. This is done as follows.

Given integers $n \geq 1$, define the following subset of $D \cap H(\sigma)$

$$H_n = \{x \in A : n \text{ is a } \sigma\text{-hyperbolic time for } x\}.$$

It follows from Lemma 2.7 that if $x \in H_n$ then the distance from $f^n(x)$ to the boundary of $f^n(D)$ is larger than δ_1. For $\delta > 0$, we denote $\Delta_n(x, \delta)$ the δ-neighbourhood of $f^n(x)$ inside $f^n(D)$. Clearly, $(f^n_* \operatorname{Leb}_D) \mid \Delta_n(x, \delta_1)$ is absolutely continuous with respect to Lebesgue measure on $\Delta_n(x, \delta_1)$. Moreover, if $x \in H_n$ and one normalizes both measures, then Proposition 2.8 means that the density of the former with respect to the latter is uniformly bounded from below and above.

Proposition 3.3. *There exists a constant $\tau > 0$ such that for any n there exists a finite subset $\widehat{H}_n$ of H_n such that the balls of radius $\delta_1/4$ in $f^n(D)$ around the points $x \in f^n(\widehat{H}_n)$ are two-by-two disjoint, and their union Δ_n satisfies*

$$f^n_* \operatorname{Leb}_D(\Delta_n \cap H(\sigma)) \geq f^n_* \operatorname{Leb}_D(\Delta_n \cap f^n(H_n)) \geq \tau \operatorname{Leb}_D(H_n)\,.$$

This is obtained taking $N = f^n(D) \cap V_0$, $\omega = f^n_* \mathrm{Leb}_D$, $r = \delta_1$, $\Omega = f^n(H_n)$, and $\widehat{H}_n = I$ in the following abstract result.

Lemma 3.4. *There exist $\tau > 0$ and $r_0 > 0$ such that the following holds. Let $N \subset V_0$ be a C^1 embedded submanifold of M tangent to the centre-unstable cone field, and ω be a finite Borelian measure in N. Let $0 < r \le r_0$ and $\Omega \subset N$ be a measurable subset with compact closure, whose distance to the boundary of N is larger than $r > 0$.*

Then there exists a finite subset $I \subset \Omega$ such that the balls $\Delta(x, r/4)$ in N around the points of I are two-by-two disjoint, and their union Δ satisfies

$$\omega(\Delta \cap \Omega) \ge \tau\omega(\Omega) .$$

Proof. By the continuity of the centre-unstable cone field, we may fix $r_0 > 0$ small enough so that the connected component of the intersection of N with the ball of radius r_0 around each point $z \in N$ that contains z coincides with the graph of a map g_z from a neighbourhood of 0 in E^{cu}_z to E^{cs}_z (or, to be more precise, with the image of such a graph under the exponential map $\exp_z$). Moreover, as long as r_0 is small enough, then g_z is a Lipschitz continuous map, with Lipschitz constant depending only on the constant a in the definition of our cone fields. As a consequence, there exists a constant $R > 0$ such that, given $0 < r_1 < r_2 \le r_0$, any ball of radius r_2 in N can be covered by at most $(r_2/r_1)^d R$ balls of radius r_1, with $d = \dim E^{cu}$. We assume that $\omega(\Omega) > 0$, since otherwise there is nothing to prove.

Let $0 < r \le r_0$ and $z_1 \in N$ be such that $\omega(\Omega \cap \Delta(z_1, r))$ is larger than $\omega(\Omega \cap \Delta(z, r))/2$ for any other point $z \in N$. By the previous remarks, we may find a point $y_1 \in N$ such that the ball $\Delta(y_1, r/8)$ of radius $r/8$ around y_1 intersects $\Delta(z_1, r)$ and

$$\omega(\Delta(y_1, r/8) \cap \Omega) \ge \frac{1}{R8^d}\, \omega(\Omega \cap \Delta(z_1, r)).$$

In particular $\Delta(y_1, r/8)$ contains some point $x_1 \in \Omega$. We take it to be the first point in our set I. Observe that, due to the choice of z_1,

$$\begin{aligned} \omega(\Delta(x_1, r/4) \cap \Omega) &\ge \omega(\Delta(y_1, r/8) \cap \Omega) \\ &\ge \frac{1}{R8^d}\, \omega(\Omega \cap \Delta(z_1, r)) \\ &\ge \frac{1}{2R8^d}\, \omega(\Omega \cap \Delta(x_1, r)). \end{aligned} \tag{19}$$

Now we consider $\Omega_1 = \Omega \setminus \Delta(x_1, r)$. Either this set has zero ω measure, in which case we stop, or we may apply the same construction as before to determine a second point $x_2 \in \Omega$. Observe that the balls of radius $r/4$ around x_1 and x_2 are disjoint, since x_2 belongs in Ω_1. Repeating this procedure, we find a sequence x_i, $i \ge 1$, of points in Ω whose balls of radius $r/4$ are two-by-two disjoint. By compactness, this sequence is necessarily finite.

Moreover, by construction, Ω is contained in the union of the balls of radius r around these x_i. So, using the bound in (19),

$$\omega(\Omega) \leq 2R8^d \sum_i \omega(\Delta(x_i, r/4)) \cap \Omega) = 2R8^d \omega(\Delta \cap \Omega).$$

This means that we may take $\tau = 1/(2R8^d)$. □

In the sequel we shall denote $\mathcal{D}_n$ the family of balls of radius $\delta_1/4$ in $f^n(D)$ around the points $x \in f^n(\widehat{H}_n)$, that form Δ_n. Now we define

$$\nu_n = \frac{1}{n} \sum_{j=0}^{n-1} (f_*^j \operatorname{Leb}_D) \mid \Delta_j \,, \tag{20}$$

and $\eta_n = \mu_n - \nu_n$.

Proposition 3.5. *There is $\alpha > 0$ such that $\nu_n(H(\sigma)) \geq \alpha$ for all n large enough.*

Proof. Recall that we took H and $H(\sigma)$ positively invariant under f. By Proposition 3.3 we have that $\nu_n(H(\sigma))$ is bounded from below by the product of τ by $n^{-1} \sum_{i=0}^{n-1} \operatorname{Leb}_D(H_i)$. So, it suffices to prove that this last expression is larger than some positive constant, for n large.

For all $k > 0$, denote A_k the set of points $x \in A$ such that, for any $n \geq k$ the sum $\sum_{j=1}^{n} \log \|Df^{-1}|E^{cu}_{f^j(x)}\|$ is smaller than $2n \log \sigma$. As A is the increasing union of the A_n, there is $k_0 \geq 1$ such that the Lebesgue measure of A_k is nonzero for all $k \geq k_0$. Given any $k \geq k_0$, let ξ_n be the measure in $\{1, \ldots, n\}$ defined by $\xi_n(B) = \#B/n$, for each subset B. Then, using Fubini's theorem

$$\begin{aligned}\frac{1}{n}\sum_{i=0}^{n-1} \operatorname{Leb}_D(H_n) &= \int \left(\int \chi(x,i)\, d\operatorname{Leb}_D(x) \right) d\xi_n(i) \\ &= \int \left(\int \chi(x,i)\, d\xi_n(i) \right) d\operatorname{Leb}_D(x),\end{aligned}$$

where $\chi(x, i) = 1$ if $x \in H_i$ and $\chi(x, i) = 0$ otherwise. Now, Corollary 3.2 means that the integral with respect to $d\xi_n$ is larger than $\theta > 0$, as long as $k \geq k_0$. So, the expression on the right hand side is bounded from below by $\theta \operatorname{Leb}_D(D)$. □

Remark 3.6. We proved a slightly stronger fact, that will be useful in Sect. 4: $\nu_n\big(\cup_{i=0}^{n-1} f^i(D \cap H(\sigma))\big) \geq \alpha$ for every large n.

We consider some subsequence $(n_k)_k$ such that μ_{n_k} and ν_{n_k} converge to measures μ and ν, respectively. It is easy to see that μ is a probability and f-invariant. Moreover $\nu(K) \geq \limsup_k \nu_k(K) \geq \alpha > 0$. We shall prove, in the next section, that ν has a property of absolute continuity along certain (fairly large) disks contained in its support. Here are some preparatory comments.

Recall that each ν_n is supported on a finite union $\cup_{j=0}^{n-1}\Delta_j$ of disks whose size is bounded from below and from above. Then the support of ν is contained in the set

$$\Delta_\infty = \cap_{n=1}^{\infty} \text{closure}\left(\cup_{j\geq n}\Delta_j\right)$$

of accumulation points of such Δ_j. Given $y \in \Delta_\infty$ then there exist $(j_i)_i \to \infty$, disks $D_i = \Delta_{j_i}(x_i, \delta_1/4) \subset \Delta_{j_i}$, and points $y_i \in D_i$ converging to y as $i \to \infty$. Up to considering subsequences, we may suppose that the centers x_i converge to some point x and, using the theorem of Ascoli-Arzela, the D_i converge to a disk $D(x)$ of size $\delta_1/4$ around x. Then y is in the closure $\bar{D}(x)$ of $D(x)$, and $\bar{D}(x) \subset \Delta_\infty$.

We shall denote $\mathcal{D}_\infty$ the family of disks $D(x)$ obtained in this way. Observe that these points x are in $\hat{H}_\infty = \cap_{n=1}^{\infty} \text{closure}\left(\cup_{j\geq n} f^j(\hat{H}_j)\right)$. Since every $\hat{H}_j$ is contained in K, which is compact and positively invariant, $\hat{H}_\infty$ is a subset of $\cap_{n=1}^{\infty} f^n(K)$. According to the next lemma, $D(x)$ depends only on x and not on the various choices we made in the construction.

Lemma 3.7. *The subspace E^{cu}_x is uniformly expanding: $\|Df^{-k} \mid E^{cu}_x\| \leq \sigma^{k/2}$ for all $k \geq 1$. The disk $D(x)$ is contained in the corresponding strong-unstable manifold $W^{uu}(x)$, and so it is uniquely defined by x. Moreover, $D(x)$ is tangent to the centre-unstable subbundle at every point of $\cap_{n=1}^{\infty} f^n(K) \cap D(x)$.*

Proof. Let $j_i \to \infty$, $x_i \to x$, and $D_i \to D(x)$ be as in the construction of $D(x)$ above. Note that D_i is contained in the j_i-iterate of D, which was taken tangent to the centre-unstable cone field. So, the domination property implies that the angle between D_i and E^{cu} goes to zero as $i \to \infty$, uniformly on $\cap_{n=1}^{\infty} f^n(K)$. By Lemma 3.7, given any $k \geq 1$ then f^{-k} is a $\sigma^{k/2}$-contraction on D_i for every large i. Passing to the limit, we get that every f^k is a $\sigma^{k/2}$-contraction on $D(x)$, and $D(x)$ is tangent to the centre-unstable subbundle at every point in $\cap_{n=1}^{\infty} f^n(K) \cap D(x)$, including x.

In particular, we have shown that the subspace E^{cu}_x is indeed uniformly expanding for Df. The domination property means that any expansion Df may exhibit along the complementary direction E^{cs} is weaker than this. Then, see [Pes76], there exists a unique *strong-unstable* manifold $W^{uu}_{loc}(x)$ tangent to E^{cu} and which is contracted by the negative iterates of f: for every $y \in W^{uu}(x)$, $\text{dist}(f^{-k}(x), f^{-k}(y))$ decreases at least as $\|Df^{-k} \mid E^{cu}_x\| \leq \sigma^{k/2}$ when k gets large. To see that $D(x)$ is contained in $W^{uu}(x)$ is suffices to recall that it is contracted by every f^{-k}, and that all its negative iterates are tangent to centre-unstable cone field. □

4. Existence and finiteness of SRB measures

This section contains the proofs of Theorem A and Corollary B. First we show that the measure ν obtained in the previous section has an absolute continuity property on disks as in Lemma 3.7. Then, using the uniformly contracting bundle, we conclude that some ergodic component of the invariant measure μ is an SRB measure.

4.1. Absolute continuity

We write $u = \dim E^{cu}$ and $s = \dim E^{cs}$, and use B^u, B^s to represent the unit compact balls in the Euclidean space of dimension u, s, respectively. In what follows we call cylinder any diffeomorphic image of $B^u \times B^s$.

Proposition 4.1. *There exists a cylinder $\mathcal{C} \subset M$, and there exists a family $\mathcal{K}_\infty$ of disjoint disks contained in $\mathcal{C} \cap \Delta_\infty$ and which are graphs over B^u, such that*

1. *the union of all the disks in $\mathcal{K}_\infty$ has positive ν-measure; in fact, the intersection of K with this union also has positive ν-measure;*
2. *the restriction of ν to that union has absolutely continuous conditional measures along the disks in $\mathcal{K}_\infty$.*

The first step of the proof is to construct a covering of the support of ν by cylinders, one of which will be the $\mathcal{C}$ in the statement. Let us point out that these cylinders we shall obtain are not small: each one contains some ball with radius uniformly bounded away from zero, depending only on the diffeomorphism f.

As we have seen, given any $y \in \Delta_\infty$ there exists a point $x \in \hat{H}_\infty$ and a disk $D(x)$ of size $\delta_1/4$ around x such that $y \in \bar{D}(x) \subset \Delta_\infty$. For any such x and $r > 0$ small, let $C_r(x)$ be the tubular neighbourhood of $\bar{D}(x)$, defined as the union of the images under the exponential map at each point $z \in \bar{D}(x)$ of all vectors orthogonal to $\bar{D}(x)$ at z and with norm less or equal than r. We take r to be sufficiently small, so that $C_r(x)$ is a cylinder and it is endowed with a canonical projection $\pi : C_r(x) \to D(x)$. Slightly adjusting r if necessary, we may also suppose that the boundary of $C_r(x)$ has zero ν-measure.

The covering of the support of ν by cylinders that we mentioned above will be obtained decomposing each of these $C_r(x)$ into a sufficient number of domains with small diameter in the centre-unstable direction, as we now explain.

Recall that each set Δ_j, $j \geq 0$, consists of a finite union of disks of radius $\delta_1/4$ inside $f^j(D)$. For any small $\varepsilon > 0$, we denote $\Delta_{j,\varepsilon}$ the subset of Δ_j obtained by removing the ε-neighbourhood of the boundary from each one of these disks. Moreover, for $n \geq 1$, we denote $\nu_{n,\varepsilon}$ the restriction of ν_n to $\cup_{j=0}^{n-1} \Delta_{j,\varepsilon}$. Let $\alpha > 0$ be as in Proposition 3.5.

Lemma 4.2. *If $\varepsilon > 0$ is sufficiently small then $\nu_{n,\varepsilon}(K) \geq \alpha/2$ for every large n.*

Proof. This is a simple consequence of Proposition 3.5. Indeed, the proposition implies that the ν_n-measure of K is greater or equal than α, for all large n. On the other hand, if ε is small then the Lebesgue measure of the ε-neighbourhood of the boundary of each disk in Δ_j is a small fraction of the Lebesgue measure of that disk. Then, in view of the distortion bound given by Proposition 2.8, the same is true with $f_*^j \operatorname{Leb}_D$ in the place of Lebesgue measure. So, taking ε small enough, we are certain to have $(f_*^j \operatorname{Leb}_D)(\Delta_j \setminus \Delta_{j,\varepsilon}) \leq \alpha/2$ for every $j \geq 0$. Then, by the definitions of ν_n and $\nu_{n,\varepsilon}$,

$$\nu_n(K) - \nu_{n,\varepsilon}(K) \leq \frac{1}{n}\sum_{j=0}^{n-1}(f_*^j \operatorname{Leb}_D)(\Delta_j \setminus \Delta_{j,\varepsilon}) \leq \frac{\alpha}{2}.$$

This completes the proof. □

In the sequel, we fix $\varepsilon > 0$ as in the lemma. Let $x \in \hat{H}_\infty$ and $\pi : C_r(x) \to \bar{D}(x)$ be as above. We fix a covering of $\bar{D}(x)$ by finitely many domains $D_{x,l} \subset \bar{D}(x), l = 1, \ldots, N$, small enough so that the intersection of each $C_{x,l} = \pi^{-1}(D_{x,l})$ with any smooth disk γ tangent to the centre-unstable cone has diameter less than ε inside γ. We take the $D_{x,l}$ diffeomorphic to the compact ball B^u, so that every $C_{x,l}$ is a cylinder.

We say that a disk γ *crosses* $C_{x,l}$ if π maps $\gamma \cap C_{x,l}$ diffeomorphically onto $D_{x,l}$. For each $j \geq 0$, let $K_j(x,l)$ be the union of the intersections of $C_{x,l}$ with all the disks in $\mathcal{D}_j$ (the disks in the support Δ_j of $f_*^j \operatorname{Leb}_D$) that cross $C_{x,l}$. Similarly, let $K_\infty(x,l)$ be the union of the intersections of $C_{x,l}$ with all the disks $D(x)$ in $\mathcal{D}_\infty$ that cross $C_{x,l}$.

The next lemma asserts that, for at least one of the cylinders $C_{x,l}$, the part of the measure ν that is carried by the disks in $K_\infty(x,l)$ gives positive weight to the set K. Recall that $\nu = \lim \nu_{n_k}$ for some $(n_k)_k$.

Lemma 4.3. *There exist (x,l) and $\alpha_1 > 0$ such that $\nu\big(K \cap K_\infty(x,l)\big) \geq \alpha_1$, and $\nu_n\big(K \cap \cup_{j=0}^{n-1} K_j(x,l)\big) \geq \alpha_1$ for n in some subsequence of $(n_k)_k$.*

Proof. For each $n \geq 1$, let $\tilde{\nu}_{n,\varepsilon}$ be the restriction of $\nu_{n,\varepsilon}$ to K, i.e., the measure defined by $\tilde{\nu}_{n,\varepsilon}(E) = \nu_{n,\varepsilon}(K \cap E)$ for every measurable subset E of M. Up to replacing $(n_k)_k$ by some subsequence, we may suppose that $\tilde{\nu}_{n_k,\varepsilon}$ converges to some measure $\tilde{\nu}_\varepsilon$. On the one hand, Lemma 4.2 means that $\tilde{\nu}_{n,\varepsilon}(M) \geq \alpha/2$ for every large n, and so $\tilde{\nu}_\varepsilon(M) \geq \alpha/2$. On the other hand, the support of $\tilde{\nu}_\varepsilon$ is contained in $\cap_{n=1}^\infty \operatorname{closure}\big(\cup_{j\geq n} \Delta_{j,\varepsilon}\big)$, and this set is covered by the interiors of the cylinders $C_r(x)$. By compactness, the support of $\tilde{\nu}_\varepsilon$ is contained in the union of a finite number of these $C_r(x)$, and so it is also contained in the union of finitely many cylinders $C_{x,l}$.

As a consequence, there must be (x,l) such that $\tilde{\nu}_\varepsilon(C_{x,l}) > 0$. We are going to show that any such (x,l) satisfies the conclusion of the lemma, if $\alpha_1 < \tilde{\nu}_\varepsilon(C_{x,l})$.

Given any disk D_j in $\mathcal{D}_j$, $j \geq 1$, let $D_{j,\varepsilon}$ be the subset obtained by removing from D_j the ε-neighbourhood of the boundary. As a consequence of the way we chose these cylinders, we have that if $D_{j,\varepsilon}$ intersects $C_{x,l}$ then D_j must cross $C_{x,l}$. This implies that

$$\tilde{\nu}_{n,\varepsilon}\big(C_{x,l}\big) = \nu_{n,\varepsilon}\big(K \cap C_{x,l}\big) \leq \nu_n\big(K \cap \cup_{j=0}^{n-1} K_j(x,l)\big)$$

for every $n \geq 1$. Since the boundary of $C_{x,l}$ has zero measure for ν, and $\tilde{\nu}_\varepsilon \leq \nu$,

$$\lim_k \tilde{\nu}_{n_k,\varepsilon}\big(C_{x,l}\big) = \tilde{\nu}_\varepsilon(C_{x,l}) > \alpha_1 .$$

Combining this with the previous inequality, we get the second part of the lemma. To get the first part, we observe that the accumulation set of $\cup_{j=0}^{n-1} K_j(x,l)$, as $n \to \infty$, is contained in $K_\infty(x,l)$. So, since K is compact,

$$\limsup_k \nu_n\big(K \cap \cup_{j=0}^{n-1} K_j(x,l)\big) \leq \nu(K \cap K_\infty(x,l)).$$

Thus, $\nu(K \cap K_\infty(x,l)) \geq \tilde{\nu}_\varepsilon(C_{x,l}) > \alpha_1$ as we claimed. □

In what follows, we fix (x,l) as in the lemma. We take the cylinder $\mathcal{C}$ in Proposition 4.1 to be $C_{x,l}$, and we let $\mathcal{K}_\infty$ be the family of disks forming $K_\infty(x,l)$. To complete the proof of the proposition, we now show that the restriction of ν to $K_\infty(x,l)$ has absolutely continuous conditional measures along the disks in $\mathcal{K}_\infty$.

Lemma 4.4. *There exists $C_3 > 1$ and a family of conditional measures $(\nu_\gamma)_\gamma$ of $\nu \mid K_\infty(x,l)$ along the disks $\gamma \in \mathcal{K}_\infty$, such that ν_γ is absolutely continuous with respect to Lebesgue measure* Leb_γ *on γ, with* $1/C_3\,\mathrm{Leb}_\gamma(B) \leq \nu_\gamma(B) \leq C_3\,\mathrm{Leb}_\gamma(B)$ *for any Borel set $B \subset \gamma$.*

Proof. Let us introduce $\widehat{K}(x,l) = \cup_{0\leq j\leq\infty} K_j(x,l) \times \{j\}$. In this space, we consider the sequence of (finite) measures $\hat{\nu}_n$ defined by

$$\hat{\nu}_n\left(B_0 \times \{0\} \cup \cdots \cup B_{n-1} \times \{n-1\}\right) = \frac{1}{n}\sum_{j=0}^{n-1} f_*^j\,\mathrm{Leb}_D(B_j),$$

and $\hat{\nu}_n(B) = 0$ whenever B is contained in $\cup_{n\leq j\leq\infty} K_j(x,l) \times \{j\}$. We also consider a sequence of partitions $\mathcal{P}_k$ in $\widehat{K}(x,l)$ constructed as follows. Fix an arbitrary point z in $D_{x,l}$, and let V be the inverse image $\pi^{-1}(z)$ under the canonical projection. Fix also a sequence $\mathcal{V}_k$, $k \geq 1$, of increasing partitions of V with diameter going to zero. Then, by definition, two points (x,m), $(y,n) \in \widehat{K}(x,l)$ are in a same atom of the partition $\mathcal{P}_k$ if

1. the disk in $\mathcal{D}_m$ containing x, and the disk in $\mathcal{D}_n$ containing y intersect a same element of $\mathcal{V}_k$;
2. either $m \geq k$ and $n \geq k$ or $m = n < k$.

It is clear from the construction that for any point $\xi \in K_j(x,l)$, $0 \le j \le \infty$,

$$\mathcal{P}_1(\xi) \supset \cdots \supset \mathcal{P}_k(\xi) \supset \cdots$$

and $\cap_{k=1}^{\infty} \mathcal{P}_k(\xi)$ coincides with the intersection of the cylinder $C_{x,l}$ with the disk in $\mathcal{D}_j$ that contains ξ. Let $\hat{\pi} : \widehat{K}(x,l) \to D_{x,l}$ be defined by $\hat{\pi}(x,j) = \pi(x)$. We claim that there exits $C_3 > 1$ such that, given any Borel subset B of $D_{x,l}$, $k \ge 1$, and $\xi \in \widehat{K}(x,l)$,

$$\frac{1}{C_3}\nu_n(\mathcal{P}_k(\xi))\,\mathrm{Leb}(B) \le \hat{\nu}_n\left(\hat{\pi}^{-1}(B) \cap \mathcal{P}_k(\xi)\right) \le C_3\nu_n(\mathcal{P}_k(\xi))\,\mathrm{Leb}(B)\,. \tag{21}$$

Indeed, by definition each atom $\mathcal{P}_k(\xi)$ is a union of sets $\gamma \times \{j\}$, where γ is the intersection of the cylinder with a disk in $\mathcal{D}_j$. Since the projection π maps γ diffeomorphically onto $D_{x,l}$,

$$\begin{aligned}\frac{1}{C_4}\frac{\mathrm{Leb}(B)}{\mathrm{Leb}(D_{x,l})} &\le \frac{\mathrm{Leb}(\hat{\pi}^{-1}(B) \cap (\gamma \times \{j\}))}{\mathrm{Leb}(\gamma \times \{j\})}\\ &= \frac{\mathrm{Leb}(\pi^{-1}(B) \cap \gamma)}{\mathrm{Leb}(\gamma)} \le C_4\frac{\mathrm{Leb}(B)}{\mathrm{Leb}(D_{x,l})},\end{aligned}$$

for some uniform constant C_4. By Proposition 2.8, the density of $f_*^j\,\mathrm{Leb}_D$ with respect to Lebesgue measure on each disk in Δ_j is bounded from below and from above. So, the previous inequality implies

$$\frac{1}{C_2^2C_4}\frac{\mathrm{Leb}(B)}{\mathrm{Leb}(D_{x,l})} \le \frac{(f_*^j\,\mathrm{Leb}_D)(\hat{\pi}^{-1}(B) \cap (\gamma \times \{j\}))}{(f_*^j\,\mathrm{Leb}_D)(\gamma \times \{j\})} \le C_2^2C_4\frac{\mathrm{Leb}(B)}{\mathrm{Leb}(D_{x,l})},$$

Since this holds for every γ, we get (21) with $C_3 = C_2^2C_4/\,\mathrm{Leb}(D_{x,l})$.

Clearly, any accumulation measure of the sequence $\hat{\nu}_n$ must be supported in $K_\infty(x,l) \times \{\infty\}$. We have chosen a sequence n_k such that ν_{n_k} converges to some measure ν, and it is easy to see that this is just the same as saying that $\hat{\nu}_k$ converges to the measure $\hat{\nu}_\infty$ defined by $\hat{\nu}_\infty(B \times \{\infty\}) = \nu(B)$, for any Borel set $B \subset C_{x,l}$. Then, by (21) and the theorem of Radon-Nikodym, the disintegration of $\hat{\nu}$ along the disks $\cap_{k=1}^{\infty}\mathcal{P}_k(\xi)$ is absolutely continuous with respect to Lebesgue measure on those disks, with densities almost everywhere bounded from above by C_3 and from below by $1/C_3$. Since $\hat{\nu}$ is naturally identified with ν, this gives the conclusion of the lemma. □

At this point we completed the proof of Proposition 4.1.

4.2. *Ergodicity and basin of attraction*

Let us introduce some notations that are useful for the proof of the next lemma. We denote by R the set of *regular points* z of f: this means that, given any continuous function $\varphi : M \to \mathbb{R}$, both forward and backward time averages

$$\lim_{n\to+\infty} \frac{1}{n}\sum_{j=0}^{n-1} \varphi(f^j(z)) \quad \text{and} \quad \lim_{n\to+\infty} \frac{1}{n}\sum_{j=0}^{n-1} \varphi(f^{-j}(z))$$

exist and they coincide. The ergodic theorem ensures that R has full measure, with respect to any invariant probability. We say that two points $z, w \in R$ are in a same *accessibility class*, see [BP74,PS89], if there exist $N \geq 1$ and points $z = z_0, z_1, \ldots, z_{N-1}, z_N = w$ in R such that $f^{t_i}(z_i)$ is in the union $W^s(z_{i-1}) \cup W^u(z_{i-1})$ of the stable and the unstable sets of z_{i-1}, for some integer t_i and every $1 \leq i \leq N$. It follows from the definition, that accessibility classes are invariant sets. Since forward averages are constant on stable sets, and backward averages are constant on unstable sets, the restriction of any invariant probability measure to an accessibility class is an ergodic measure (possibly identically zero).

Lemma 4.5. *The invariant measure $\mu = \nu + \eta$ has some ergodic component μ_* whose Lyapunov exponents are all non-zero, and whose conditional measures along local unstable manifolds are absolutely continuous with respect to Lebesgue measure. Moreover, we may choose μ_* with support contained in $\cap_{j=0}^{\infty} f^j(K)$ and* $\mathrm{Leb}_D(B(\mu_*) \cap H) > 0$.

Proof. Since R has full μ-measure, it must also have full ν-measure. In particular, up to replacing $\mathcal{K}_\infty$ by a convenient sub-family of disks γ, whose union has full ν measure in K_∞, we may suppose that ν_γ almost every point in γ is regular, for every $\gamma \in \mathcal{K}_\infty$. In particular, Leb_γ almost every point in any disk γ is regular. Using the fact that the strong-stable foliation is absolutely continuous, cf. [BP74, Section 2], we conclude that all such regular points are in a same accessibility class. Moreover, this accessibility class $\mathcal{A}$ has positive ν-measure, and so also positive μ-measure, by Lemma 4.3.

We let μ_* be the normalized restriction of μ to $\mathcal{A}$: $\mu_*(B) = \mu(B \cap \mathcal{A})/\mu(\mathcal{A})$ for every Borel set B. Then μ_* is an invariant ergodic probability measure. It follows from Lemma 3.7 that the Lyapunov exponents of μ_* along the tangent space of the disks γ are positive. Of course, the Lyapunov exponents along the strong-stable direction are all negative. Since μ_* is ergodic, its conditional measures along local unstable manifolds in $\mathcal{K}_\infty$ are either almost everywhere singular or almost everywhere absolutely continuous (with respect to Lebesgue measure). This is a well known fact, whose proof can be sketched as follows.

Suppose there is $A \subset K_\infty(x, l)$ such that $m_\gamma(A \cap \gamma) = 0$ for all $\gamma \in \mathcal{K}_\infty$, and yet $\mu_*(A) > 0$ (hence $\mu_{*\gamma}(A \cap \gamma) > 0$ for many $\gamma \in \mathcal{K}_\infty$). Let

$B = \cup_{j=-\infty}^{+\infty} f^j(A)$. By ergodicity, $\mu_*(B) = 1$, and so $\mu_{*,\gamma}(B \cap \gamma) = 1$ for $\hat{\mu}_*$-almost all γ in $\mathcal{K}_\infty$. On the other hand, $m_\gamma(B \cap \gamma) = 0$ for every $\gamma \in \mathcal{K}_\infty$. This is because f is a diffeomorphism, unstable manifolds are an invariant family of submanifolds, and B is given by a countable union. So, in this case, $\mu_{*\gamma}$ is singular with respect to m_γ for $\hat{\mu}_*$-almost all $\gamma \in \mathcal{K}_\infty$. Now suppose that, on the contrary, every measurable set $A \subset K_\infty(x, l)$ satisfying $m_\gamma(A \cap \gamma) = 0$ for all $\gamma \in \mathcal{K}_\infty$ has zero μ_*-measure. Then, restricted to $K_\infty(x, l)$, μ_* is absolutely continuous with respect to the product measure $m_\gamma \times \hat{\mu}_*$. Consequently, in this second case, the conditional measures $\mu_{*,\gamma}$ are absolutely continuous with respect to Lebesgue measure m_γ for $\hat{\mu}_*$-almost every γ in $\mathcal{K}_\infty$.

In the setting we are dealing with, the singular case is easily excluded: the conditional measures of μ_* can be written as the sum of the conditional measures of the restrictions of η and ν to the accessibility class, and the latter are equivalent to Lebesgue measure at least on $K_\infty(x, l)$. So, the conditional measures of μ_* must be almost everywhere absolutely continuous.

Since $\nu(K \cap K_\infty(x, l))$ is positive, by Lemma 4.3, $\mu(K \cap \mathcal{A}) > 0$ and so $\mu_*(K) > 0$. As K is compact and positively invariant, the ergodicity of μ_* implies that $\mu_*(\cap_{n=0}^{\infty} f^n(K)) = 1$, and the support of μ_* is contained in $\cap_{j=0}^{\infty} f^j(K)$.

To prove the last statement in the lemma we need the following fact:

Claim: *There exists a disk D_∞ inside $K_\infty(x, l)$ such that Lebesgue almost all the points in D_∞ are in the basin of μ_*, and there exists a sequence D_k of disks in $K_{j_k}(x, l)$ accumulating on D_∞ and such that* $\mathrm{Leb}_{D_k}(D_k \cap f^{j_k}(H(\sigma) \cap D))$ *is uniformly bounded away from zero.*

We assume this for a while, and explain how to conclude the proof of the lemma from it. Since forward averages of continuous functions are constant on strong-stable leaves, the basin of μ_* contains the union of all strong-stable leaves through Lebesgue almost all points in D_∞. As this foliation is absolutely continuous, and the D_k accumulate on D_∞, such union intersects D_k in a subset whose relative Lebesgue measure inside D_k goes to 1 when k goes to infinity. In particular, $\mathrm{Leb}_{D_k}(D_k \cap f^{j_k}(H(\sigma) \cap D) \cap B(\mu_*))$ is positive for every large k. Of course, the basin is invariant by f, so we may conclude that $\mathrm{Leb}_D(H(\sigma) \cap B(\mu_*)) > 0$. □

All that is left to do is to prove the Claim above.

Proof. We use Remark 3.6:

$$\nu_n\big(\cup_{i=0}^{n-1} f^i(D \cap H(\sigma))\big) \geq \alpha.$$

It follows that if $\epsilon > 0$ is fixed sufficiently small then there exists a subset of disks in the support of ν_n with total ν_n-mass larger than $\alpha/2$ and such that a fraction larger than ϵ of any such disk corresponds to points coming from $D \cap H(\sigma)$. Then, by the same argument as at the end of the proof of Lemma 4.3, the union E of the disks in $\mathcal{D}_\infty$ that are accumulated by disks as above has ν-mass larger than $\alpha/2$. Then some of these disks must be such that Lebesgue almost all points in it are in the basin of μ_*. Indeed, since μ_* is ergodic, its basin has full μ_*-measure. Then, a full measure subset of E consists of disks where almost all points, with respect to the conditional measure of μ_* on the disk, are in $B(\mu_*)$. Since we know that the conditional measures of μ_* along the disks in $\mathcal{K}_\infty$ are bounded away from zero (because the same is true for ν, cf. Lemma 4.4), we conclude that Lebesgue almost all points in some disk in E is in $B(\mu_*)$. □

As a consequence, we also get that

$$\lim_{n\to+\infty}\frac{1}{n}\sum_{j=1}^{n-1}\log\left\|Df^{-1}|E^{cu}_{f^j(x)}\right\| = \int\log\left\|Df^{-1}|E^{cu}_y\right\|\,d\mu_*(y) < 2\log\sigma \qquad (22)$$

for μ_*-almost every $x \in M$ (and for every $x \in B(\mu_*)$ that remains in the neighbourhood V_0 where E^{cu} has a meaning). This is a simple consequence of the ergodicity of μ_*, and the fact that its basin intersects $H(\sigma)$. In particular, all the Lyapunov exponents of μ_* in the centre-unstable direction are larger than $-\log\sigma$.

Finally, we deduce the following result which completes the proof of Theorem A and also gives Corollary B.

Corollary 4.6. *For any $\sigma < 1$, a full Lebesgue measure subset of $H(\sigma)$ is contained in the union of finitely many SRB measures supported in $\cap_{j=0}^{\infty} f^j(K)$.*

Proof. First we observe that the set of points in $H(\sigma)$ which do not belong in the basin of some SRB measure as in the statement must have zero Lebesgue measure. Indeed, otherwise we could apply the previous arguments with this set in the place of $H(\sigma)$: we would get, cf. Lemma 4.5, an extra positive Lebesgue measure subset in the basin of some SRB measure, contradicting the definition.

The main point to obtain the finiteness statement is to note that the choice of δ_1 in the context of (14) depends only on σ. Using this remark, we can deduce that for any SRB measure μ_* as we constructed above there exists a disk $D(\mu_*)$ of fixed radius δ_1, tangent to the centre-unstable cone field and such that Lebesgue almost every point in $D(\mu_*)$ is in the basin of μ_*. For this, we recall that for the SRB measures we constructed above there exist disks D_0 containing some point $x \in H(\sigma)\cap B(\mu_*)$ and on which Lebesgue almost every point is in the basin $B(\mu_*)$. In view of (22), x has many σ-hyperbolic times n. We may take $D(\mu_*)$ to be the disk of radius δ_1 around $f^n(x)$ inside $f^n(D_0)$, for any such n sufficiently large.

Then, the union of all strong-stable leaves through the points in $D(\mu_*)\cap B(\mu_*)$ is contained in $B(\mu_*)$. Using the absolute continuity property of the strong-stable foliation, we may conclude that this union contains a subset of a neighbourhood of $D(\mu_*)$ with volume bounded away from zero by some constant that depends only on δ_1 and f. For this, we fix some neighbourhood of $D(\mu_*)$ with size uniformly bounded from below, as well as a smooth foliation of it by disks C^1 close to $D(\mu_*)$. For instance, $D(\mu_*)$ could be contained in one of the leaves of this foliation. Given any leaf D, the strong-stable manifolds through the points of $D(\mu_*)$ intersect D on a subdisk D' whose Lebesgue measure inside D is bounded away from zero, by a constant that depends only on the size δ_1 of D, and on the map f. This is just by continuity of the strong-stable foliation. Moreover, absolute continuity implies that Lebesgue almost every point of D' is in the strong-stable manifold of a point of $D(\mu_*) \cap B(\mu_*)$. Now, our claim that the Lebesgue measure of the union of these strong-unstable leaves is uniformly bounded away from zero follows from Fubini's theorem.

Of course, basins of different SRB measures are two-by-two disjoint. So the conclusion of the previous paragraph implies that there can only be finitely many such measures (even if we do not assume M to be compact), since small neighbourhoods of the compact set K have finite volume. □

5. Maps with singular or critical points

Here we explain how the previous arguments can be adapted to prove Theorem C and Corollary D. A main difference concerns the notion of hyperbolic times. A key point in the previous sections was that *if n is a hyperbolic time for a point x then there exists a neighbourhood of x, in the disk D, which is mapped onto a ball of fixed radius around $f^n(x)$, in $f^n(D)$, diffeomorphically and with uniformly bounded distortion*. This was a consequence of the contraction property in Definition 2.6. Now, in the presence of a singular set $\mathcal{S}$, in order to have a similar property we must also ensure that iterates $f^j(x)$ with $0 \le j < n$ are not too close to $\mathcal{S}$.

Let $B > 1$ and $\beta > 0$ be as in the hypotheses (S1), (S2), (S3). In what follows b is any fixed constant such that $0 < b < \min\{1/2, 1/(2\beta)\}$.

Definition 5.1. *Given $\sigma < 1$ and $\delta > 0$, we say that n is a* (σ, δ)-hyperbolic time *for a point $x \in M \setminus \mathcal{S}_\infty$ if, for all $1 \le k \le n$,*

$$\prod_{j=n-k}^{n-1} \|Df(f^j(x))^{-1}\| \le \sigma^k \quad \textit{and} \quad \operatorname{dist}_\delta(f^{n-k}(x), \mathcal{S}) \ge \sigma^{bk}.$$

Let us begin by proving that this notion does imply the key property above:

Lemma 5.2. *Given $\sigma < 1$ and $\delta > 0$, there exists $\delta_1 > 0$ such that if n is a (σ, δ)-hyperbolic time for a point $x \in M \setminus \mathcal{S}_\infty$, then there exists a neighbourhood V_x of x such that*

1. *f^n maps V_x diffeomorphically onto the ball of radius δ_1 around $f^n(x)$;*
2. *for every $1 \le k < n$ and $y, z \in V_x$,*

$$\operatorname{dist}(f^{n-k}(y), f^{n-k}(z)) \le \sigma^{k/2} \operatorname{dist}(f^n(y), f^n(z)).$$

Proof. We shall prove, by induction on $j \geq 1$, that if δ_1 is chosen small enough then there exists a well defined branch of f^{-j} on the ball of radius δ_1 around $f^n(x)$, mapping $f^n(x)$ to $f^{n-j}(x)$. In addition, this branch is a $\sigma^{j/2}$-contraction. The precise condition δ_1 should satisfy is given by the following statement:

Claim: *Fix $\delta_1 > 0$ so that $4\delta_1 < \delta$ and $4B\delta_1 < \delta^\beta |\log \sigma|$. Then,*

$$\|Df(y)^{-1}\| \leq \sigma^{-1/2} \|Df(f^{n-j}(x))^{-1}\| \tag{23}$$

for any $1 \leq j < n$ and any point y in the ball of radius $2\delta_1 \sigma^{j/2}$ around $f^{n-j}(x)$.

Proof. By hypothesis $\operatorname{dist}_\delta(f^{n-j}(x), \mathcal{S}) \geq \sigma^j$. According to the definition of the truncated distance, this means that

$$\operatorname{dist}(f^{n-j}(x), \mathcal{S}) = \operatorname{dist}_\delta(f^{n-j}(x), \mathcal{S}) \geq \sigma^{bj}$$
$$\text{or else} \quad \operatorname{dist}\,(f^{n-j}(x), \mathcal{S}) \geq \delta.$$

In either case, $\operatorname{dist}(y, f^{n-j}(x)) < \operatorname{dist}(f^{n-j}(x), \mathcal{S})/2$ because we chose $b < 1/2$ and $\delta_1 < \delta/4 < 1/4$. Therefore, we may use (S2) to conclude that

$$\log \frac{\|Df(y)^{-1}\|}{\|Df(f^{n-j}(x))^{-1}\|} \leq B \frac{\operatorname{dist}(y, f^{n-j}(x))}{\operatorname{dist}(f^{n-j}(x)), \mathcal{S})^\beta} \leq B \frac{2\delta_1 \sigma^{j/2}}{\min\{\sigma^{b\beta j}, \delta^\beta\}}.$$

Since δ and σ are smaller than 1, and we took $b\beta < 1/2$, the term on the right hand side is bounded by $2B\delta_1\delta^{-\beta}$. Moreover, our second condition on δ_1 means that this last expression is smaller than $\log \sigma^{-1/2}$. □

Starting the induction argument to prove Lemma 5.2, we note that for $j = 1$ the Claim gives

$$\|Df(y)^{-1}\| \leq \sigma^{-1/2} \|Df(f^{n-1}(x))^{-1}\| \leq \sigma^{1/2},$$

since n is a hyperbolic time for x. This means that f is a $\sigma^{-1/2}$-dilation in the ball of radius $2\delta_1\sigma^{1/2}$ around $f^{n-1}(x)$. As a consequence, there exists some neighbourhood $V(n-1)$ of $f^{n-1}(x)$ contained in that ball of radius $2\delta_1\sigma^{1/2}$, that is mapped diffeomorphically onto the ball of radius δ_1 around $f^n(x)$.

Now, given any $j > 1$, let us suppose that we have constructed a neighbourhood $V(n-j+1)$ of $f^{n-j+1}(x)$ such that the restriction of f^{j-1} to $V(n-j+1)$ is a diffeomorphism onto the ball of radius δ_1 around $f^n(x)$, with

$$\|Df(f^i(z))^{-1}\| \leq \sigma^{-1/2} \|Df(f^{n-j+i+1}(x))^{-1}\| \tag{24}$$

for all z in $V(n-j+1)$ and $0 \leq i < j$. Then, by the Claim and the hypothesis that n is a hyperbolic time for x,

$$\|Df^j(y)^{-1}\| \leq \prod_{i=0}^{j-1} \|Df(f^i(y))^{-1}\| \leq \prod_{i=0}^{j-1} \sigma^{-1/2} \|Df(f^{n-j+i}(x))^{-1}\| \leq \sigma^{j/2}$$

for any point y in the ball of radius $2\delta_1\sigma^{j/2}$ whose image $z = f(y)$ is in $V(n-j+1)$.

Now we can construct an inverse branch of f^j on the ball of radius δ_1 around $f^n(x)$, by lifting geodesics in the following way. Given a geodesic γ connecting $f^n(x)$ to a point in the boundary of the ball, there is a well defined lift of the restriction of γ to a small neighbourhood of $f^n(x)$, into a curve starting at $f^{n-j}(x)$. Moreover, as far as this curve does not leave the ball of radius $2\delta_1\sigma^{j/2}$, the derivative on it is a $\sigma^{-j/2}$-dilation. This means that the length of the lifted curve is less than $\delta_1\sigma^{j/2}$, and so the curve is actually contained in a smaller ball. This proves that the lift is well defined on the whole geodesic γ. Thus, we have a well defined branch of f^{-j} on the ball of radius δ_1 around $f^n(x)$ as we claimed. We call $V(n-j)$ the image of that inverse branch. By construction, $V(n-j)$ is contained in the $2\delta_1\sigma^{j/2}$-ball around $f^{n-j}(x)$ and its image under f coincides with $V(n-j+1)$. So, in view of the Claim, we also recovered the induction assumption (24) for points in $V(n-j)$ and times $0 \le i \le j$.

In this way, we construct neighbourhoods $V(n-j)$ of $f^{n-j}(x)$ as above, for all $1 \le j \le n$. The lemma follows taking $V_x = V(0)$. □

Corollary 5.3. *There exists $C_5 > 0$ such that for every $x \in M \setminus \mathcal{S}_\infty$, any n that is a (σ, δ)-hyperbolic time for x, and every $y, z \in V_x$*

$$\frac{1}{C_5} \le \frac{|\det Df^n(y)|}{|\det Df^n(z)|} \le C_5 .$$

Proof. By construction, for $0 \le k < n$, the distance from $f^k(x)$ to either $f^k(y)$ or $f^k(z)$ is less than $\delta_1\sigma^{(n-k)/2}$, which is much smaller than $\sigma^{b(n-k)} \le \operatorname{dist}(f^k(x), \mathcal{S})$. So, assumption (S3) implies

$$\log\frac{|\det Df^n(y)|}{|\det Df^n(z)|} = \sum_{k=0}^{n-1}\log\frac{|\det Df(f^k(y))|}{|\det Df(f^k(z))|} \le \sum_{k=0}^{n-1} 2B\frac{\delta_1\sigma^{(n-k)/2}}{\sigma^{b\beta(n-k)}} .$$

Now, it suffices to take $C_5 \ge \exp\left(\sum_{i=1}^{\infty} 2B\delta_1\sigma^{(1/2-b\beta)i}\right)$, recall that $b\beta < 1/2$. □

Let $\sigma < 1$ be fixed. The assumptions of Theorem C imply that, if σ is close enough to 1, then the set $H(\sigma)$ of points $x \in M \setminus \mathcal{S}_\infty$ for which the limit in (5) is less than $3\log\sigma$ has positive Lebesgue measure. The next lemma asserts that points in $H(\sigma)$ have many (σ, δ)-hyperbolic times.

Lemma 5.4. *There are $\theta > 0$ and $\delta > 0$, depending only on σ and on the map f, such that given any $x \in H(\sigma)$ and any sufficiently large $N \ge 1$ there exist (σ, δ)-hyperbolic times $1 \le n_1 < \cdots < n_l \le N$ for x, with $l \ge \theta N$.*

Proof. The strategy is to use Lemma 3.1 twice, first for the sequence given by $a_j = -\log\|Df(f^{j-1}(x))^{-1}\|$ (up to a cut off that makes it bounded from above), and then with $a_j = \log\operatorname{dist}_\delta(f^{j-1}(x), \mathcal{S})$ for a convenient $\delta > 0$.

We prove that there exist many times n_i for which the conclusion of Pliss' Lemma 3.1 holds, simultaneously, for both sequences. Then we check that any such n_i is a (σ, δ)-hyperbolic time for x.

Let $x \in H(\sigma)$. By definition of $H(\sigma)$, for every large N we have

$$\sum_{j=0}^{N-1} -\log \|Df(f^j(x))^{-1}\| \geq 2|\log \sigma| N .$$

Fix any $\rho > \beta$. Then (S1) implies that

$$\Big| \log \|Df(x)^{-1}\| \Big| \leq \rho \, |\log \operatorname{dist}(x, \mathcal{S})| \tag{25}$$

for every x in a neighbourhood V of $\mathcal{S}$. Fix $\varepsilon_1 > 0$ so that $\rho\varepsilon_1 \leq |\log \sigma|/2$, and let $r_1 > 0$ be small enough so that

$$\sum_{j=0}^{N-1} -\log \operatorname{dist}_{r_1}(f^j(x), \mathcal{S}) \leq \varepsilon_1 N . \tag{26}$$

Assumption (6) ensures that this is possible. Fix any $H_1 \geq \rho \, |\log r_1|$ large enough so that it is also an upper bound for $-\log \|Df^{-1}\|$ on the complement of V. Then let E be the subset of times $1 \leq j \leq N$ such that $-\log \|Df(f^{j-1}(x))^{-1}\| > H_1$, and define

$$a_j = \begin{cases} -\log \|Df(f^{j-1}(x))^{-1}\| & \text{if } j \notin E \\ 0 & \text{if } j \in E. \end{cases}$$

By construction, $a_j \leq H_1$ for $1 \leq j \leq N$. Note that if $j \in E$ then $f^{j-1}(x) \in V$. Moreover, for $j \in E$ we have $\operatorname{dist}(f^{j-1}(x), \mathcal{S}) < r_1$:

$$\rho \, |\log r_1| \leq H_1 < -\log \|Df(f^{j-1}(x))^{-1}\| < \rho \, |\log \operatorname{dist}(f^{j-1}(x), \mathcal{S})|.$$

In particular, $\operatorname{dist}_{r_1}(f^{j-1}(x), \mathcal{S}) = \operatorname{dist}(f^{j-1}(x), \mathcal{S}) < r_1$ for all $j \in E$. Therefore, by (25) and (26),

$$\sum_{j \in E} -\log \|Df(f^{j-1}(x))^{-1}\| \leq \rho \sum_{j \in E} |\log \operatorname{dist}(f^{j-1}(x), \mathcal{S})| \leq \rho \, \varepsilon_1 N.$$

We have chosen ε_1 in such a way that the last term is less than $|\log \sigma| N/2$. As a consequence,

$$\begin{aligned} \sum_{j=1}^{N} a_j = &\sum_{j=1}^{N} -\log \|Df(f^{j-1}(x))^{-1}\| \\ &- \sum_{j \in E} -\log \|Df(f^{j-1}(x))^{-1}\| \geq \frac{3}{2} |\log \sigma| N . \end{aligned}$$

Thus, we have checked that we may apply Lemma 3.1 to a_j, with $c_1 = |\log\sigma|$, $c_2 = 3\,|\log\sigma|/2$, and $A = H_1$. The lemma provides $\theta_1 > 0$ and $l_1 \geq \theta_1 N$ times $1 \leq p_1 < \cdots < p_{l_1} \leq N$ such that

$$\sum_{j=n+1}^{p_i} -\log\|Df(f^{j-1}(x))^{-1}\| \geq \sum_{j=n+1}^{p_i} a_j \geq (p_i - n)\,|\log\sigma| \tag{27}$$

for every $0 \leq n < p_i$ and $1 \leq i \leq l_1$.

Now fix $\varepsilon_2 > 0$ small enough so that $\varepsilon_2/(b|\log\sigma|) < \theta_1$, and let $r_2 > 0$ be such that, cf. condition (6),

$$\sum_{j=0}^{N-1} \log \operatorname{dist}_{r_2}(f^j(x), \mathcal{S}) \geq -\varepsilon_2 N\,.$$

Let $c_1 = b\log\sigma$, $c_2 = -\varepsilon_2$, $A = 0$, and

$$\theta_2 = \frac{c_2 - c_1}{A - c_1} = 1 - \frac{\varepsilon_2}{b\,|\log\sigma|}\,.$$

Applying Lemma 3.1 to the sequence $a_j = \log \operatorname{dist}_{r_2}(f^{j-1}(x), \mathcal{S})$, we conclude that there are $l_2 \geq \theta_2 N$ times $1 \leq q_1 < \cdots < q_{l_2} \leq N$ such that

$$\sum_{j=n}^{q_i - 1} \log \operatorname{dist}_{r_2}(f^j(x), \mathcal{S}) \geq b\log\sigma\,(q_i - n) \tag{28}$$

for every $0 \leq n < q_i$ and $1 \leq i \leq l_2$.

Finally, our condition on ε_2 means that $\theta_1 + \theta_2 > 1$. Let $\theta = \theta_1 + \theta_2 - 1$. Then there exist $l = (l_1 + l_2 - N) \geq \theta N$ times $1 \leq n_1 < \cdots < n_l \leq N$ at which (27) and (28) occur simultaneously:

$$\sum_{j=n}^{n_i - 1} \log\|Df(f^j(x))^{-1}\| \leq (n_i - n)\log\sigma$$

and

$$\sum_{j=n}^{n_i - 1} \log \operatorname{dist}_{r_2}(f^j(x), \mathcal{S}) \geq b\log\sigma(n_i - n),$$

for every $0 \leq n < n_i$ and $1 \leq i \leq l$. Therefore, given $1 \leq i \leq l$ and $1 \leq k \leq n_i$,

$$\prod_{j=n_i-k+1}^{n_i} \|Df^{-1}(f^j(x))\| \leq \sigma^k \quad \text{and} \quad \operatorname{dist}_{r_2}(f^{n_i-k}(x), \mathcal{S}) \geq \sigma^{bk}.$$

In other words, all those n_i are (σ, δ)-hyperbolic times for x, for $\delta = r_2$. □

Now we prove Theorem C, the same argument gives Corollary D.

Proof. Let H be the positive Lebesgue measure set in the statement, and $H(\sigma)$ be as above.

Lemma 5.5. *Suppose σ is close enough to 1 so that $H(\sigma)$ has positive Lebesgue measure. Then f admits some invariant probability measure μ_0 absolutely continuous with respect to Lebesgue measure and giving positive weight to $H(\sigma)$.*

Proof. According to Lemma 5.4, there exists $\delta > 0$ depending only on σ, such that for any point x in $H(\sigma)$ there exist many (σ, δ)-hyperbolic times. We let μ_n be the averages of the positive iterates of Lebesgue measure on M, and ν_n be part of μ_n carried on disks of radius δ_1 around points $f^j(x)$ such that $1 \leq j \leq n$ is a (σ, δ)-hyperbolic time for x. More precisely, arguing as in Lemma 3.4 and Proposition 3.3, we may find for each $j \geq 1$ a finite set of points $x_1, \ldots, x_N$ admitting j as a (σ, δ)-hyperbolic time, such that

1. $V_{x_1}, \ldots, V_{x_N}$ are two-by-two disjoint;
2. the Lebesgue measure of $W_j = V_{x_1} \cup \ldots \cup V_{x_N}$ is larger than the Lebesgue measure of the set of points in $H(\sigma)$ having j as a (σ, δ)-hyperbolic time, up to a uniform multiplicative constant $\tau > 0$.

Then we take

$$\nu_n = \frac{1}{n} \sum_{j=0}^{n-1} f_*^j(\mathrm{Leb} \,|\, W_j).$$

As before in Proposition 3.5, each of these ν_n has total mass bounded away from zero, in fact, $\nu_n(H(\sigma)) \geq \alpha$ for some uniform $\alpha > 0$. Moreover, as a consequence of the distortion Corollary 5.3, every $f_*^j(\mathrm{Leb} \,|\, W_j)$ is absolutely continuous with respect to Lebesgue measure, with density uniformly bounded from above, and so the same is true for every ν_n.

Now take $n_k \to \infty$ such that both μ_{n_k} and ν_{n_k} converge to measures μ and ν, respectively, in the weak* sense. Then μ is an invariant probability measure, $\mu = \nu + \eta$ for some measure η, ν is absolutely continuous with respect to Lebesgue measure, and $\nu(H(\sigma)) \geq \alpha > 0$. Now, if $\eta = \eta_{ac} + \eta_s$ denotes the Lebesgue decomposition of η (as the sum of an absolutely continuous and a completely singular measure, with respect to Lebesgue measure), then $\mu_{ac} = \nu + \eta_{ac}$ gives the absolutely continuous component in the corresponding decomposition of μ. By uniqueness of the Lebesgue decomposition, and the fact that the push-forward under f preserves the class of absolutely continuous measures, we may conclude that μ_{ac} is an invariant measure. Clearly, $\mu_{ac}(H(\sigma)) \geq \nu(H(\sigma)) > 0$. □

Up to replacing μ_0 by its normalized restriction to the (positively invariant) set $H(\sigma)$ we may suppose that $\mu_0(H(\sigma)) = 1$. The next lemma will allow us to show that $H(\sigma)$ is covered by the basins of finitely many ergodic absolutely continuous invariant measures.

Lemma 5.6. *For any positively invariant set $G \subset H(\sigma)$ there exists some disk Δ with radius $\delta_1/4$ such that* $\mathrm{Leb}(\Delta \setminus G) = 0$.

Proof. It suffices to prove that there exist disks of radius $\delta_1/4$ where the relative measure of G is arbitrarily close to 1.

Let ϵ be some small number, G_c be a compact subset of G, and G_o be a neighbourhood of G_c such that both $G \setminus G_c$ and $G_o \setminus G_c$ have Lebesgue measure less than $\epsilon\,\mathrm{Leb}(G)$. By Lemma 5.4 and Fubini's theorem, there exist arbitrarily large values of $j \geq 1$ such that the Lebesgue measure of the subset G_j of points in G for which j is a (σ, δ)-hyperbolic time has Lebesgue measure is at least $\theta\,\mathrm{Leb}(G)$. So, as long as ϵ is fixed sufficiently small, the Lebesgue measure of $G_c \cap G_j$ is larger than $(\theta/2)\,\mathrm{Leb}(G)$. Assume that j is large enough so that for any point x in $G_c \cap G_j$, the neighbourhood V_x is contained in G_o. Here V_x is the neighbourhood of x constructed in Lemma 5.2: it is mapped diffeomorphically onto the ball of radius δ_1 around $f^j(x)$ by f^j. Let $W_x \subset V_x$ be the pre-image of the ball of radius $\delta_1/4$ under this diffeomorphism. Let $x_1, \ldots, x_N \in G_c \cap G_j$ be such that $W_{x_1}, \ldots, W_{x_N}$ cover the compact set $G_c \cap G_j$. Up to reordering, we may suppose that $W_{x_1}, \ldots, W_{x_n}$, some $n \leq N$, is a maximal sub-family whose elements are two-by-two disjoint. Notice that the $V_{x_1}, \ldots, V_{x_n}$ cover $G_c \cap G_j$, since their union contains every W_{x_i}, $1 \leq i \leq N$. Indeed, every W_{x_i} must intersect some W_{x_k} with $k \leq n$. Then its image under f^j intersects the ball of radius $\delta_1/4$ around $f^j(x_k)$ and so it is contained in the corresponding ball of radius δ_1. This means, precisely, that W_{x_i} is contained in V_{x_k}.

By the bounded distortion property, $\mathrm{Leb}(W_x)$ is larger than the product of $\mathrm{Leb}(V_x)$ by some uniform constant $\tau > 0$ (independent of x or j). So, the Lebesgue measure of $W_{x_1} \cup \cdots \cup W_{x_n}$ is larger than $\tau\,\mathrm{Leb}(G_c \cap G_j)$. If $\xi > 0$ is such that $\mathrm{Leb}(W_{x_i} \setminus (G_c \cap G_j)) \geq \xi\,\mathrm{Leb}(W_{x_i})$ for every $1 \leq i \leq n$, then

$$\mathrm{Leb}\left(W_{x_1} \cup \cdots \cup W_{x_n}) \setminus (G_c \cap G_j)\right) \geq \xi\tau\,\mathrm{Leb}(G_c \cap G_j) \geq \xi\tau\theta\,\mathrm{Leb}(G)\ .$$

On the other hand, since the W_{x_i} are contained in G_o and $G_c \cap G_j \subset G$, this measure must be smaller than $\epsilon\,\mathrm{Leb}(G)$. This means that by reducing ϵ (which we may, by increasing j), we can force ξ to be arbitrarily small. In other words, we may find j and W_{x_i} such that the relative Lebesgue measure of $W_{x_i} \cap G_c \cap G_j$ in W_{x_i} is arbitrarily close to 1. Then, by bounded distortion, the relative Lebesgue measure of $G \supset f^j(G_c \cap G_j)$ in the ball of radius $\delta_1/4$ around $f^j(x_i)$ is also arbitrarily close to 1. So the proof of the lemma is complete. □

Finally, we may conclude the proof of Theorem C and Corollary D.

Let μ_0 be any absolutely continuous invariant measure with $\mu_0(H(\sigma)) = 1$. If μ_0 is not ergodic then we may decompose $H(\sigma)$ into two disjoint invariant sets H_1, H_2 both with positive μ_0-measure. In particular, both H_1 and H_2 have positive Lebesgue measure. Let μ_1, μ_2 be the normalized restrictions of μ_0 to H_1, H_2, respectively. Clearly, they are also absolutely continuous invariant measures. If they are not ergodic, we continue decomposing them, in the same way as we did for μ_0. On the other hand, by Lemma 5.6, each one of the invariant sets we find in this decomposition has full Lebesgue measure in some disk with fixed radius. Since these disks must be disjoint, and the ambient manifold is compact, there can only be finitely many of them. So, the decomposition must stop after a finite number of steps, giving that μ_0 can be written $\mu_0 = \sum_{i=1}^{s} \mu_0(H_i)\mu_i$ where $H_1, \dots, H_s$ is a partition of $H(\sigma)$ into invariant sets with positive measure and each $\mu_i(\cdot) = \mu_0(\cdot \cap H_i)/\mu_0(H_i)$ is an ergodic measure.

This completes the proof of Theorem C and Corollary D, and it also gives the finiteness result stated in Subsect. 1.1 right after the corollary. □

Having in mind important classes of maps with singularities, such as Poincaré return maps of singular (or generalized Lorenz) attractors, [ABS77,GW79,Rov93,MPP98,BPV97] we now propose a natural extension of the previous results for *partially hyperbolic maps with singularities.*

Let us a manifold M, a compact subset $\mathcal{S}$, and a C^2 diffeomorphism (onto its image) $f : M \setminus \mathcal{S} \to M$. We suppose f has a compact positively invariant subset K, in the sense that $f(K \setminus \mathcal{S}) \subset K$, such that the tangent bundle of M restricted to $K \setminus \mathcal{S}$ has a Df-invariant dominated splitting $T_{K\setminus\mathcal{S}}M = E^{ss} \oplus E^{cu}$ such that E^{ss} is uniformly contracting.

We assume that f behaves like a power of the distance to $\mathcal{S}$ *along the centre-unstable* direction: for every $x \in K \setminus \mathcal{S}$ and $v \in E^{cu}_x$

(R1) $\dfrac{1}{B}\operatorname{dist}(x,\mathcal{S})^{\beta} \le \dfrac{\|Df(x)v\|}{\|v\|} \le B\operatorname{dist}(x,\mathcal{S})^{-\beta}$;

(R2) $\|D(Df(x))\| \le B\operatorname{dist}^{cu}(x,\mathcal{S})^{-\beta}$ and
$\|D(Df(x)^{-1})\| \le B\operatorname{dist}^{cu}(x,\mathcal{S})^{-\beta}$

(R3) $\big|\log\|Df^{-1} \mid E^{cu}_{f(x)}\| - \log\|Df^{-1} \mid E^{cu}_{f(y)}\|\big| \le B\frac{\operatorname{dist}^{cu}(x,y)}{\operatorname{dist}^{cu}(x,\mathcal{S})^{\beta}}$, if x and y are in a same disk tangent to the centre-unstable cone field, and $\operatorname{dist}^{cu}(x,y) < \operatorname{dist}^{cu}(x,\mathcal{S})/2$.

Here dist^{cu} denotes the shortest distance measured along curves tangent to the centre-unstable cone field, and we also define the truncated version $\operatorname{dist}^{cu}_{\delta}$ of dist^{cu}, in the same way as in (4). Let $\mathcal{S}_{\infty} = \cup_{n=-\infty}^{+\infty} f^n(\mathcal{S})$.

Although we did not try to check all the details, it seems that the following statement can be obtained by combining the arguments in the proofs of Theorems A and C:

Let f be as above, and assume that it is non-uniformly expanding along the centre-unstable direction, in the sense that (2) holds for all x in a positive Lebesgue measure set $H \subset M \setminus \mathcal{S}_{\infty}$ Assume moreover that, given any $\varepsilon > 0$ there exists $\delta > 0$ such that for every $x \in H$

$$\limsup_{n\to+\infty} \frac{1}{n}\sum_{j=0}^{n-1} -\log \operatorname{dist}^{cu}_{\delta}(f^j(x),\mathcal{S}) \le \varepsilon.$$

Then Lebesgue almost every point in H is in the basin of some SRB measure.

The main technical point in giving a full proof, is to control the curvature of iterates of disks tangent to the centre-unstable cone field, similarly to what we did in Subsect. 2.1.

6. Diffeomorphisms with a dominated splitting

Let $f : M \to M$ be a $C^{1+\zeta}$ diffeomorphism on a manifold M. Here we suppose that f has a compact positively invariant set $K \subset M$ with a continuous invariant *dominated splitting* $T_K M = E^{cs} \oplus E^{cu}$: there exists a constant $\lambda < 1$ and some choice of a Riemannian metric on M such that

- $\|Df \mid E^{cs}_x\| \cdot \|Df^{-1} \mid E^{cu}_{f(x)}\| \leq \lambda$ for all $x \in K$.

We call E^{cs} centre-stable subbundle and E^{cu} centre-unstable subbundle.

As we did at the beginning of Sect. 2, we can extend the subbundles continuously to a neighbourhood V_0 of K, and then consider cone fields C^{cs}_a and C^{cu}_a with small width $a > 0$ around these extended subbundles. As before, we assume that f is non-uniformly expanding along the centre-unstable direction:

$$\limsup_{n\to+\infty} \frac{1}{n} \sum_{j=1}^{n} \log \left\| Df^{-1} \middle| E^{cu}_{f^j(x)} \right\| < 0 \tag{29}$$

for every x in a positive Lebesgue measure subset H of $D \cap K$, where D is some C^2 disk tangent to the centre-unstable cone field. We fix any $\sigma < 1$ such that $H(\sigma)$, defined as in Sect. 3, has positive Lebesgue measure in D.

Everything we did in Sect. 2 through Subsect. 4.1 applies immediately in this context. So, cf. Proposition 4.1, there exist measures μ, ν, η, a cylinder $\mathcal{C}$ and a family $\mathcal{K}_\infty$ of disjoint disks crossing $\mathcal{C}$ such that

1. μ is an invariant probability measure and $\mu = \nu + \eta$;
2. the disks in $\mathcal{K}_\infty$ are accumulated by sub-disks of radius δ_1 in $f^n(D)$, around points $f^n(x)$ such that n is a σ-hyperbolic time for $x \in H(\sigma)$;
3. the union K_∞ of all the disks in $\mathcal{K}_\infty$ intersects K in a set with positive ν-measure;
4. the restriction of ν to K_∞ has absolutely continuous conditional measures along the disks in $\mathcal{K}_\infty$.

Let us recall a few well-known notions and facts that are useful for the proof of the next lemma. Given a point x, let us denote μ_x the probability measure given by the time average along the orbit of x:

$$\int \varphi \, d\mu_x = \lim_{n\to+\infty} \frac{1}{n} \sum_{j=0}^{n-1} \varphi(f^j(x)) \quad \text{for every continuous } \varphi : M \to \mathbb{R}. \tag{30}$$

According to the ergodic decomposition theorem, see [Mañ87, Section 2.6], μ_x is well-defined and ergodic for every x in a set $\Sigma \subset M$ that has full measure with respect to any invariant measure ξ. Moreover, $x \mapsto \int g\, d\mu_x$ is measurable and

$$\int g\, d\xi = \int \left(\int g\, d\mu_x \right) d\xi(x)$$

for every measurable bounded function $g : M \to \mathbb{R}$. In fact, for any such g the integral $\int g\, d\mu_x$ coincides almost everywhere with the time average of g over the orbit of x.

Let R be the set of regular points of f, as introduced in Subsect. 4.2: $z \in R$ if and only if the forward and backward time averages of each continuous function over the orbit of z exist and coincide. R has full measure for any f-invariant probability measure ξ, as a consequence of Birkhoff's ergodic theorem. Let us point out that μ_x is constant on the intersection of R with every disk γ of $\mathcal{K}_\infty$. This is because these disks are (exponentially) contracted by negative iterates, cf. property 2. above and Lemmas 2.7 and 3.7, and so points in a same $\gamma \in \mathcal{K}_\infty$ have the same backward average (hence points in $R \cap \gamma$ also have the same forward average) for each continuous function φ.

Lemma 6.1. *There exists $z \in K_\infty \cap K \cap \Sigma \cap R$ such that $\mu_z(K_\infty \cap K) > 0$ and μ_z has absolutely continuous conditional measures along the disks in $\mathcal{K}_\infty$. In particular, the support of μ_z is contained in $\cap_{j=0}^{\infty} f^j(K)$.*

Proof. Fix B to be some measurable subset of M such that

$$m_\gamma(B \cap \gamma) = 0 \quad \text{for every} \quad \gamma \in \mathcal{K}_\infty, \tag{31}$$

and $\mu(B)$ is maximal among all measurable sets with this property. For instance, $B = \cup_n B_n$ where the B_n, $n \geq 1$, are measurable sets with property (31) such that $\mu(B_n)$ converges to the largest value compatible with that property. Observe that $\nu(B) = 0$, because ν is absolutely continuous along the leaves of $\mathcal{K}_\infty$. Let $Z_\infty = K_\infty \cap K \cap \Sigma \cap R \setminus B$. Then,

$$\mu(Z_\infty) \geq \nu(Z_\infty) = \nu(K_\infty \cap K \cap \Sigma \cap R) = \nu(K_\infty \cap K) > 0.$$

Let $(\mu \mid Z_\infty)$ be the restriction of μ to Z_∞: by definition $(\mu \mid Z_\infty)(E) = \mu(E \cap Z_\infty)$ for any measurable set E in M.

Let A be any measurable subset of Z_∞ such that $m_\gamma(A \cap \gamma) = 0$ for every $\gamma \in \mathcal{K}_\infty$. Then $\mu(A)$ must be zero, since we took $\mu(B)$ maximal. This means that $(\mu \mid Z_\infty)$ is absolutely continuous with respect to the product $m_\gamma \times \hat{\mu}$, where $\hat{\mu}$ stands for the quotient measure induced by $(\mu \mid Z_\infty)$ on $\mathcal{K}_\infty$. As a consequence, the conditional measures $\tilde{\mu}_\gamma$ of $(\mu \mid Z_\infty)$ on the disks $\gamma \in \mathcal{K}_\infty$ are absolutely continuous with respect to Lebesgue measure

m_γ for $\hat{\mu}$-almost all $\gamma \in \mathcal{K}_\infty$. On the other hand, for any measurable set $A \subset Z_\infty$,

$$\mu(A) = \int \mu_x(A)\, d\mu(x) \tag{32}$$

where the integral is taken over M or, more precisely, over the full measure subset Σ. We want to express this in terms of an integral over Z_∞. As we mentioned before,

$$\mu_x(A) = \int \mathcal{X}_A\, d\mu_x = \lim_{n\to+\infty} \frac{1}{n} \sum_{j=0}^{n-1} \mathcal{X}_A(f^j(x))$$

almost everywhere, with respect to any invariant measure. So, up to disregarding a zero μ-measure set of points, $\mu_x(A)$ can be non-zero only if x has some iterate in $A \subset Z_\infty$. Let $k(z)$ denote the first backward return time of a point $z \in Z_\infty$, that is, the smallest positive integer such that $f^{-k(z)}(z) \in Z_\infty$. This is defined μ-almost everywhere, by Poincaré's recurrence theorem. Observe also that $\mu_z = \mu_{f^j(z)}$ for any z and any integer j. Thus, we can rewrite (32) as

$$\mu(A) = \int_{Z_\infty} k(z)\mu_z(A)\, d\mu(z)$$

for any measurable subset A of Z_∞. The next lemma can be inferred from §3 of Rokhlin [Rok62]. For the reader's convenience, we state it explicitly and prove it, after completing the proof of Lemma 6.1.

Lemma 6.2. *Let λ be a finite measure on a measure space Z, with $\lambda(Z) > 0$. Let $\mathcal{K}$ be a measurable partition of Z, and $(\lambda_z)_{z\in Z}$ be a family of finite measures on Z such that*

1. *the function $z \mapsto \lambda_z(A)$ is measurable, and it is constant on each element of $\mathcal{K}$, for any measurable set $A \subset Z$*
2. *$\{w : \lambda_z = \lambda_w\}$ is a measurable set with full λ_z-measure, for every $z \in Z$.*

Assume that $\lambda(A) = \int \ell(z)\lambda_z(A)\, d\lambda$ for some measurable function $\ell : Z \to \mathbb{R}_+$ and any measurable subset A of Z. Let $\{\tilde{\lambda}_\gamma,\ \gamma \in \mathcal{K}\}$, and $\{\tilde{\lambda}_{z,\gamma},\ \gamma \in \mathcal{K}\}$, be disintegrations of λ and λ_z, respectively, into conditional probability measures along the elements of the partition $\mathcal{K}$. Then

$$\tilde{\lambda}_{z,\gamma} = \tilde{\lambda}_\gamma$$

for λ-almost every $z \in Z$ and $\hat{\lambda}_z$-almost every γ, where $\hat{\lambda}_z$ is the quotient measure induced by λ_z on $\mathcal{K}$.

We take $Z = Z_\infty$, $\lambda = (\mu \mid Z_\infty)$, $\mathcal{K} = \mathcal{K}_\infty$, $\lambda_z = (\mu_z \mid Z_\infty)$, and $\ell(z) = k(z)$, for each $z \in Z_\infty$. It is easy to check that the hypotheses of Lemma 6.1 are satisfied. The first part of assumption 1 is contained in the ergodic decomposition theorem, and the second part follows from (30), as we explained before. Let $\mathcal{D}$ be any countable dense subset of the space of continuous functions on M. Given $z, w \in Z_\infty$, then $\mu_z = \mu_w$ if and only for every $\varphi \in \mathcal{D}$ and every $p \geq 1$ there exists $q \geq 1$ such that

$$\left| \frac{1}{n} \sum_{j=0}^{n-1} \varphi(f^{-j}(z)) - \frac{1}{n} \sum_{j=0}^{n-1} \varphi(f^{-j}(w)) \right| < \frac{1}{p} \qquad \text{for any } n \geq p .$$

This gives the measurability condition in assumption 2. In this case the last part of assumption 2 is just a restatement of the fact that $\lambda_z = \mu_z$ is ergodic.

Then, according to Lemma 6.2, the conditional probability measures $\tilde{\mu}_{z,\gamma}$ of $(\mu_z \mid Z_\infty)$ along the disks $\gamma \in \mathcal{K}_\infty$ coincide almost everywhere with the corresponding conditional measures $\tilde{\mu}_\gamma$ of $(\mu \mid Z_\infty)$. Recall that we had already shown that the latter are almost everywhere absolutely continuous with respect to Lebesgue measure m_γ. We also have that

$$\int_{Z_\infty} k(z)\mu_z(Z_\infty)\, d\mu = \mu(Z_\infty) > 0.$$

It follows that there exists a positive μ-measure subset of points $z \in Z_\infty$ such that $\mu_z(K_\infty \cap K) \geq \mu_z(Z_\infty) > 0$, and the restriction of μ_z to Z_∞ has conditional measures with respect to $\mathcal{K}$ that are μ_z-almost everywhere absolutely continuous with respect to Lebesgue measure on the corresponding disk $\gamma \in \mathcal{K}$. Thus, any such z satisfies the first two claims in the statement of the lemma.

Finally, since K is compact and positively invariant, ergodicity implies that the support of μ_z is contained in $\cap_{j=0}^{\infty} f^j(K)$. □

Now we prove Lemma 6.2:

Proof. The idea is quite simple. Let $\mathcal{E}$ be the partition of the set Z into equivalence classes for the equivalence relation $z \sim w \Leftrightarrow \lambda_z = \lambda_w$. Assumption 2 ensures that the elements of $\mathcal{E}$ are measurable sets, and assumption 1 implies that every $\gamma \in \mathcal{K}$ is contained in some element of $\mathcal{E}$. Given any $e \in \mathcal{E}$ we define $\lambda_e = \lambda_z$, where z is an arbitrary point in e. We show that, up to normalization, $\{\lambda_e, e \in \mathcal{E}\}$ is a disintegration of λ with respect to the partition $\mathcal{E}$. Now, $\{\tilde{\lambda}_{e,\gamma} = \tilde{\lambda}_{z,\gamma}, \gamma \in \mathcal{K}, \gamma \subset e\}$ is a disintegration of $\lambda_e = \lambda_z$, with respect to the partition induced by $\mathcal{K}$ on each $e \in \mathcal{E}$. It follows that $\{\tilde{\lambda}_{e,\gamma}, \gamma \in \mathcal{K}, \gamma \subset e, e \in \mathcal{E}\}$ is a disintegration of λ with respect to $\mathcal{K}$ (obtained by conditioning first to $\mathcal{E}$, then to $\mathcal{K}$). By (essential) uniqueness of the disintegration into conditional probability measures, we must have $\tilde{\lambda}_{z,\gamma} = \tilde{\lambda}_\gamma$ almost everywhere.

Now we give the detailed argument. Let $\pi_\gamma : Z \to \mathcal{K}$ and $\pi_e : \mathcal{K} \to \mathcal{E}$ be the canonical projections. We represent by $\mathcal{B}(\mathcal{E})$ the σ-algebra generated by $\mathcal{E}$. Let $g : Z \to \mathbb{R}_+$ be a conditional expectation of ℓ relative to $\mathcal{E}$, that is, a Radon-Nikodym derivative, with respect to the restriction of λ to $\mathcal{B}(\mathcal{E})$, of the measure defined by

$$\mathcal{B}(\mathcal{E}) \ni E \mapsto \int_E \ell \, d\lambda.$$

In other words, g is a $\mathcal{B}(\mathcal{E})$-measurable function satisfying

$$\int_E \ell \, d\lambda = \int_E g \, d\lambda \quad \text{for every } E \in \mathcal{B}(\mathcal{E}). \tag{33}$$

$\mathcal{B}(\mathcal{E})$-measurability implies that g is constant on elements of $\mathcal{E}$. Set $g(e) = g(z)$ for any $e \in \mathcal{E}$ and $z \in e$. Let us consider the set

$$\{h : Z \to \mathbb{R} \text{ such that } \int \ell h \, d\lambda = \int g h \, d\lambda\}.$$

By (33), every characteristic function of an element of $\mathcal{B}(\mathcal{E})$ is in this set. Using linearity of the integral and the dominated convergence theorem, we conclude that the set contains any bounded $\mathcal{B}(\mathcal{E})$-measurable function. In particular, it contains $h(z) = \lambda_z(A)$, for any measurable set $A \subset Z$. Therefore,

$$\lambda(A) = \int \ell(z)\lambda_z(A) \, d\lambda(z) = \int g(z)\lambda_z(A) \, d\lambda(z) = \int g(e)\lambda_e(A) \, de, \tag{34}$$

where $de = (\pi_e \circ \pi_\gamma)_*\lambda$ is the quotient measure induced by λ on $\mathcal{E}$. Assumption 2 implies that

$$g(e)\lambda_e(Z \setminus e) = 0 \quad \text{for any } e \in \mathcal{E}. \tag{35}$$

Then,

$$g(e)\lambda_e(Z) = g(e)\lambda_e(e) = 1 \quad \text{for } de\text{-almost all } e \in \mathcal{E}. \tag{36}$$

Indeed, let $\delta > 0$ and F_δ be the set of all $e \in \mathcal{E}$ for which $g(e)\lambda_e(Z) \geq 1+\delta$. Denote $E_\delta = (\pi_e \circ \pi_\gamma)^{-1}(F_\delta)$. Then, using (34) and (35),

$$de(F_\delta) = \lambda(E_\delta) = \int g(e)\lambda_e(E_\delta) \, de = \int_{F_\delta} g(e)\lambda_e(Z) \, de \geq (1+\delta)de(F_\delta),$$

which implies $de(F_\delta) = 0$. Analogously, the set of $e \in \mathcal{E}$ for which $g(e)\lambda_e(Z)$ is less than $1 - \delta$ has zero de-measure for any $\delta > 0$. This proves (36). In this way, we have shown that $\{g(e)\lambda_e, \, e \in \mathcal{E}\}$ is a disintegration of λ with respect to the partition $\mathcal{E}$.

Now, for each $e \in \mathcal{E}$, let $\hat{\lambda}_e = (\pi_\gamma \mid e)_* \lambda_e$ be the quotient measure of λ_e on $(\mathcal{K} \mid e)$. Moreover, let $\{\tilde{\lambda}_{e,\gamma}, \gamma \in \mathcal{K}, \gamma \subset e\}$ be a disintegration of λ_e with respect to $(\mathcal{K} \mid e)$. Of course, $\tilde{\lambda}_{e,\gamma} = \tilde{\lambda}_{z,e}$ for any $z \in e$. Then

$$\lambda_e(A) = \int \tilde{\lambda}_{e,\gamma}(A)\, d\hat{\lambda}_e(\gamma) \quad \text{and so}$$

$$g(e)\lambda_e(A) = \int \tilde{\lambda}_{e,\gamma}(A)\, d(g(e)\hat{\lambda}_e)(\gamma)$$

for any measurable set $A \subset Z$. Replacing this in (34), we find

$$\lambda(A) = \int \int \tilde{\lambda}_{e,\gamma}(A)\, d(g(e)\hat{\lambda}_e)(\gamma)\, de. \tag{37}$$

Denote $d\gamma = (\pi_\gamma)_* \lambda$, the quotient measure of λ on $\mathcal{K}$. Note that $de = (\pi_e)_* d\gamma$, that is, de coincides with the quotient measure of $d\gamma$ on $\mathcal{E}$. Moreover,

$$d\gamma(\Gamma) = \lambda\big(\pi_\gamma^{-1}(\Gamma)\big) = \int g(e)\lambda_e\big(\pi_\gamma^{-1}(\Gamma)\big)\, de = \int g(e)\hat{\lambda}_e(\Gamma)\, de$$

for every measurable set $\Gamma \subset \mathcal{K}$. This means that $\{g(e)\hat{\lambda}_e, e \in \mathcal{E}\}$, is a disintegration of $d\gamma$ with respect to the partition $\mathcal{E}$. Thus (37) gives

$$\lambda(A) = \int \tilde{\lambda}_{e,\gamma}(A)\, d\gamma,$$

and so $\{\tilde{\lambda}_{e,\gamma}, \gamma \in \mathcal{K}\}$, is a disintegration of λ with respect to the partition $\mathcal{K}$. Since disintegrations into conditional probability measures, when they exist, are uniquely defined almost everywhere, it follows that $\tilde{\lambda}_{e,\gamma} = \tilde{\lambda}_\gamma$ for $d\gamma$-almost every $\gamma \in \mathcal{K}$. Equivalently, this holds for de-almost every $e \in \mathcal{E}$ and $d\hat{\lambda}_e$-almost every $\gamma \in (\mathcal{K} \mid e)$, which is just the same as the conclusion of the lemma. □

Let $z \in K_\infty \cap K$ be as in Lemma 6.1. Property 2. above implies that the measure $\mu_* = \mu_z$ has dim E^{cu} Lyapunov exponents larger than $-\log\sigma$. The domination condition implies that all the other exponents are less than $-\log\sigma + \log\lambda < -\log\sigma$. So, by Pesin theory [Pes76], μ_z-almost every point x has a local strong-unstable manifold which is an embedded disk whose backward orbits approach the backward of x at the exponential rate $\log\sigma$. Moreover, the disks $\gamma \in \mathcal{K}_\infty$ contain the local strong-unstable manifolds of points in their interior.

Combining these remarks with Lemma 6.1, we get

Theorem 6.3. *Let f be a C^2 diffeomorphism admitting a positively invariant compact set with a dominated splitting $E^{cs} \oplus E^{cu}$. Assume that f is non-uniformly expanding along the centre-unstable direction, cf. (29). Then f has some ergodic Gibbs cu-state μ_* supported in $\cap_{j=0}^{\infty} f^j(K)$: μ_* is an* *invariant probability measure whose* $\dim E^{cu}$ *larger Lyapunov exponents are positive and whose conditional measures along the corresponding local strong-unstable manifolds are almost everywhere absolutely continuous with respect to Lebesgue measure on these manifolds.*

If the remaining $\dim E^{cs}$ Lyapunov exponents of μ_* are all negative then μ_* is an SRB measure. This is a well known consequence of the absolute continuity property of μ_* and absolute continuity of the stable lamination [Pes76]: the union of the stable manifolds through the points whose time averages are given by μ_* is a positive Lebesgue measure set contained in the basin of μ_*.

Clearly, the centre-stable Lyapunov exponents are indeed negative whenever the subbundle E^{cs} is uniformly contracting, which was precisely our setting in Sects. 2 through 4. In general, if one assumes that E^{cs} is non-uniformly contracting

$$\limsup_{n\to+\infty} \frac{1}{n} \sum_{j=0}^{n-1} \log \left\| Df \mid E^{cs}_{f^j(x)} \right\| < 0 \tag{38}$$

on a positive Lebesgue measure subset of $H \subset D$, it is not clear whether this information can be passed to a limit Gibbs cu-state. There are however some cases where this can be done, and so our methods do yield SRB measures supported in K.

A sufficient condition is that there exist a positive Lebesgue measure set of points in H with many (positive density at infinity) *simultaneous* σ*-hyperbolic times* with respect to the two subbundles:

$$\prod_{j=n-k+1}^{n} \left\| Df^{-1} \mid E^{cu}_{f^j(x)} \right\| \le \sigma^k \quad \text{and} \quad \prod_{j=n-k}^{n-1} \left\| Df \mid E^{cs}_{f^j(x)} \right\| \le \sigma^k$$

for every $0 \le k \le n$, for some $\sigma < 1$.

Proposition 6.4. *In the setting of Theorem 6.3, suppose that every point H has many (positive density at infinity) simultaneous σ-hyperbolic times for some $\sigma < 1$. Then ergodic Gibbs cu-states can be constructed as in the theorem which are SRB measures, and whose basins cover a full Lebesgue measure subset of H.*

Proof. Let $\sigma < 1$ be fixed such that the subset $H(\sigma)$ of points with many simultaneous σ-hyperbolic times has positive Lebesgue measure. Cf. (14) if n is a σ-hyperbolic time (with respect to E^{cu}) then the tangent space at every point in the ball of radius δ_1 around $f^n(x)$ in $f^n(D)$ is uniformly contracted by the first n negative iterates of f. Up to reducing δ_1, we may also suppose that, whenever n is a simultaneous σ-hyperbolic time then the centre-stable subbundle is $\sigma^{j/2}$-expanded by these iterates f^{-j}, $1 \le j \le n$, at every point in that ball. We construct measures ν_n as before, except that we take into account only simultaneous hyperbolic times. As a consequence, E^{cs} is $\sigma^{j/2}$-expanded by all negative iterates f^{-j}, $j \geq 1$, at every point in the support of the limit measure ν. In particular, any ergodic Gibbs cu-state for which the support of ν has positive measure must have $\dim E^{cs}$ negative Lyapunov exponents, and so it is an SRB measure.

Moreover, as we have seen in the proof of the Claim in Lemma 4.5, a definite fraction of measures ν_n is carried by disks where points in the corresponding iterate of $H(\sigma)$ occupy a subset with relative Lebesgue measure bounded away from zero. Then it is easy to verify that the content of that Claim is valid for, at least, some Gibbs cu-state μ_* charging the support of ν: there exists a sequence D_k of disks in $f^{j_k}(D)$ in which $f^{j_k}(H(\sigma))$ has relative Lebesgue measure bounded away from zero, converging to some disk D_∞ in the support of μ_*, tangent to the centre-unstable direction and such that almost every point in D_∞ is in the basin of μ_*. By Pesin theory (absolute continuity of the stable lamination [Pes76]) the union of the stable manifolds of these points in $D_\infty \cap B(\mu_*)$ cuts D_k in a subset with relative Lebesgue measure going to 1 as D_k approaches D_∞. In particular, since these stable manifolds are contained in $B(\mu_*)$, the basin must contain a positive Lebesgue measure subset of $H(\sigma)$. This proves that Lebesgue almost every point in $H(\sigma)$ is in the basin of some SRB measure, for every $\sigma < 1$. □

Finally, we describe a simple condition on the diffeomorphism f implying existence of many simultaneous hyperbolic times. This condition is satisfied by a non-empty C^1 open set of diffeomorphisms of the 4-torus admitting an invariant set with a dominated splitting (without uniformly hyperbolic subbundles), as will be shown in the Appendix.

Proposition 6.5. *Let f be a C^2 diffeomorphism admitting a dominated splitting $E^{cs} \oplus E^{cu}$ on some positively invariant compact set K. Let*

$$A^u = \sup_{f(K)} -\log \|Df^{-1} \mid E^{cu}\| \quad \text{and} \quad A^s = \sup_{K} -\log \|Df \mid E^{cs}\|.$$

Suppose that there exist positive constants c^u and c^s such that

$$\frac{c^u}{A^u} + \frac{c^s}{A^s} > 1 \tag{39}$$

and

$$\limsup_{n\to+\infty} \frac{1}{n} \sum_{j=1}^{n} \log \left\| Df \middle| E^{cu}_{f^j(x)} \right\| \leq -c^u,$$

$$\limsup_{n\to+\infty} \frac{1}{n} \sum_{j=0}^{n-1} \log \left\| Df \middle| E^{cs}_{f^j(x)} \right\| \leq -c^s$$

for some point $x \in K$. Then there exists $\sigma < 1$ such that the simultaneous σ-hyperbolic times of x have positive density at infinity.

Proof. This is a direct consequence of the expression of the density bound θ in Lemma 3.1: given σ close to 1 then, for every large N, there exist $\theta^s N$ σ-hyperbolic times with respect to E^{cs} and $\theta^u N$ σ-hyperbolic times with respect to E^{cu} in the time interval $\{1, \ldots, N\}$, with

$$\theta^u = \frac{c^u + \log\sigma}{A^u + \log\sigma} \quad \text{and} \quad \theta^s = \frac{c^s + \log\sigma}{A^s + \log\sigma}.$$

If σ is close enough to 1 then $\theta^u + \theta^s > 1$ and the simultaneous hyperbolic times have density $\theta^u + \theta^s - 1 > 0$. □

A. Appendix: Applications

Here we present a few simple conditions implying the assumptions of Theorem A, Corollary D, and Propositions 6.4 and 6.5. They allow us to exhibit some robust (C^1 open) classes of maps to which these results apply.

Lemma A.1. *Given a real number σ_1 and integers $p, q \geq 1$ with $\sigma_1 > q$, there exists $\varepsilon_0 > 0$ such that the following holds. Let M be a manifold with finite volume, $f : M \to M$ be a C^1 map, and $\{B_1, \ldots, B_p, B_{p+1}, \ldots, B_{p+q}\}$ be a covering of M by measurable sets, such that*

1. $|\det Df(x)| \geq \sigma_1$ *for every x in $B_{p+1} \cup \cdots B_{p+q}$;*
2. $(f \mid B_i)$ *is injective for all $1 \leq i \leq p+q$.*

Then the orbit of Lebesgue almost every point $x \in M$ spends a fraction ε_0 of the time in $B_1 \cup \cdots \cup B_p$: that is, $\#\{0 \leq j < n : f^j(x) \in B_1 \cup \cdots \cup B_p\} \geq \varepsilon_0 n$ for every large n.

Proof. Let n be fixed, for the time being. Given a sequence $\underline{i} = (i_0, i_1, \ldots, i_{n-1})$ in $\{1, \ldots, p+q\}$, we denote

$$[\underline{i}] = B_{i_0} \cap f^{-1}(B_{i_1}) \cap \cdots \cap f^{-n+1}(B_{i_{n-1}}).$$

Moreover, we define $g(\underline{i})$ to be the number of values of $0 \leq j \leq n-1$ for which $i_j \leq p$. We begin by noting that, given any $\varepsilon_0 > 0$, the total number of sequences $\underline{i}$ for which $g(\underline{i}) < \varepsilon_0 n$ is bounded by

$$\sum_{k<\varepsilon_0 n} \binom{n}{k} p^k q^{n-k} \leq \sum_{k \leq \varepsilon_0 n} \binom{n}{k} p^{\varepsilon_0 n} q^n .$$

A standard application of Stirling's formula (see e.g. [BV99, Section 6.3]) gives that the last expression is bounded by $e^{\gamma_0 n} p^{\varepsilon_0 n} q^n$, where γ_0 depends only on ε_0 and goes to zero when ε_0 goes to zero. On the other hand, as a consequence of assumptions 1 and 2, $\mathrm{Leb}([\underline{i}]) \leq \mathrm{Leb}(M)\, \sigma_1^{-(1-\varepsilon_0)n}$. Then the measure of the union I_n of all the sets $[\underline{i}]$ with $g(\underline{i}) < \varepsilon_0 n$ is less than

$$\mathrm{Leb}(M) \sigma_1^{-(1-\varepsilon_0)n} e^{\gamma_0 n} p^{\varepsilon_0 n} q^n .$$

Since we supposed $q < \sigma_1$, we may fix ε_0 small so that $e^{\gamma_0} p^{\varepsilon_0} q < \sigma_1^{1-\varepsilon_0}$. This means that the Lebesgue measure of I_n goes to zero exponentially fast as $n \to \infty$. Thus, by the lemma of Borel-Cantelli, Lebesgue almost every point $x \in M$ belongs in only finitely many sets I_n. Clearly, any such point x satisfies the conclusion of the lemma. □

Proposition A.2. *Given real numbers $\sigma_1, \sigma_2 > 0$ and integers $p, q \geq 1$ such that $\sigma_1 > q \geq 1 > \sigma_2$, there exist $\delta_0 > 0$ and $c_0 > 0$ such that the following holds. Let M, $f : M \to M$, and $B_1, \dots, B_{p+q}$ be as in Lemma A.1, and assume that*

1. $\|Df(x)^{-1}\| \leq \sigma_2$ if $x \in B_i$, $1 \leq i \leq p$, and
2. $\|Df(x)^{-1}\| \leq 1 + \delta_0$ if $x \in B_i$, $p+1 \leq i \leq p+q$.

Then f is non-uniformly expanding: for Lebesgue almost every point $x \in M$

$$\limsup_{n\to+\infty} \frac{1}{n} \sum_{j=0}^{n-1} \log \|Df(f^j(x))^{-1}\| \leq -c_0 .$$

Proof. Let $\varepsilon_0 > 0$ be the constant given by Lemma A.1. Then, fix $\delta_0 > 0$ small enough so that $\sigma_2^{\varepsilon_0}(1+\delta_0) \leq e^{-c_0}$ for some $c_0 > 0$. Let x be any point satisfying the conclusion of the lemma. Then

$$\prod_{j=0}^{n-1} \|Df(f^j(x))^{-1}\| \leq \sigma_2^{\varepsilon_0 n}(1+\delta_0)^{(1-\varepsilon_0)n} \leq e^{-c_0 n}$$

for every large enough n, This means that x satisfies the conclusion of the proposition, so the proof is complete. □

Remark A.3. We also proved that, for f as in Proposition A.2, the Lebesgue measure of the set

$$\left\{x \in M : \|Df^j(x)^{-1}\| > e^{-c_0 j} \text{ for some } j \geq n\right\}$$

goes to zero exponentially fast when $n \to \infty$.

With the aid of this proposition we can exhibit an explicit construction of a C^1 open class of maps satisfying the hypotheses of Corollary D. Let M be any compact manifold supporting some uniformly expanding map f_0: there exists $\sigma_2 < 1$ such that

$$\|v\| < \sigma_2 \|Df_0(x)v\| \qquad \text{for every } x \in M \text{ and } v \in T_x M.$$

For instance, M could be the d-dimensional torus T^d. Let $V \subset M$ be some small compact domain, so that the restriction of f_0 to V is injective.

Corollary A.4. *Let f_1 be any C^1 map coinciding with f_0 outside V, and such that f_1 is volume expanding everywhere, $|\det Df_1(x)| > 1$ for every $x \in M$, and f is not too contracting on V: $\|Df_1(x)^{-1}\| \leq 1 + \delta_0$ for every $x \in V$ and some small enough $\delta_0 > 0$. Then every map f in a C^1-neighbourhood $\mathcal{N}$ of f_1 is non-uniformly expanding.*

Proof. Taking the C^1-neighbourhood sufficiently small, we may assume that there exists $\sigma_1 > 1$ such that the Jacobian of every f in it is bounded from below by σ_1. Moreover, $\|Df(x)^{-1}\| \le \sigma_2$ for every x outside V. Let $B_1, \ldots, B_p, B_{p+1} = V$ be any partition of M into domains such that f is injective on each B_j, $1 \le j \le p+1$. The claim follows from Proposition A.2, with $q = 1$. □

Remark A.5. Maps f_1 as in the statement can be obtained, e.g. through deformation of f_0 by isotopy inside V. In general, these maps are not expanding: deformation can be made in such way that f_1 have periodic saddles.

Using very similar ideas one can also construct robust classes of partially hyperbolic diffeomorphisms (or, more generally, diffeomorphisms with a dominated splitting) whose centre-unstable direction is non-uniformly expanding. We just sketch the main points.

This time we start with a linear Anosov diffeomorphism f_0 on the d-dimensional torus $M = T^d$, $d \ge 2$. We write $TM = E^u \oplus E^s$ the corresponding hyperbolic decomposition. Let V be a small closed domain in M, in the following sense: there exist unit open cubes K^0 and K^1 in $\mathbb{R}^d$ such that $V \subset \pi(K^0)$ and $f_0(V) \subset \pi(K^1)$, where $\pi : \mathbb{R}^d \to T^d$ is the canonical projection. Now, let f be a diffeomorphism on T^d such that

(a) f admits invariant cone fields C^{cu} and C^{cs}, with small width $\alpha > 0$ and containing, respectively, the unstable bundle E^u and the stable bundle E^s of the Anosov diffeomorphism f_0;
(b) there is $\sigma_1 > 1$ so that $|\det(Df \mid T_x\mathcal{D}^{cu})| > \sigma_1$ and $|\det(Df \mid T_x\mathcal{D}^{cs})| < \sigma_1^{-1}$ for any $x \in M$ and any disks $\mathcal{D}^{cu}$ and $\mathcal{D}^{cs}$ through x tangent, respectively, to the centre-unstable cone field C^{cu} and to centre-stable cone field C^{cs}.
(c) f is C^1-close to f_0 in the complement of V, so that there exists $\sigma_2 < 1$ satisfying

$$\|(Df \mid T_x\mathcal{D}^{cu})^{-1}\| < \sigma_2 \quad \text{and} \quad \|(Df \mid T_x\mathcal{D}^{cs})\| < \sigma_2$$

for $x \in (M \setminus V)$ and any disks $\mathcal{D}^{cu}$, $\mathcal{D}^{cs}$ tangent to C^{cu}, C^{cs}, respectively.
(d) there exists some small $\delta_0 > 0$ satisfying

$$\|(Df \mid T_x\mathcal{D}^{cu})^{-1}\| < (1+\delta_0) \quad \text{and} \quad \|(Df \mid T_x\mathcal{D}^{cs})\| < (1+\delta_0)$$

for any $x \in V$ and any disks $\mathcal{D}^{cu}$, $\mathcal{D}^{cs}$ tangent to C^{cu}, C^{cs}, respectively.

Closeness in (c) should be enough to ensure that $f(V)$ is also contained in the projection of a unit open cube.

For instance, if f_1 is a torus diffeomorphism satisfying (a), (b), (d), and coinciding with f_0 outside V, then any map f in a C^1 neighbourhood of f_1 satisfies all the previous conditions. The C^1 open classes of transitive non-Anosov diffeomorphisms presented in [BV99, Section 6], as well as other robust examples from [Mañ78], are constructed in this way and they fit in the present setting: both these diffeomorphisms and their inverses satisfy (a)–(d).

In what is left of this appendix, we argue that any f satisfying (a)–(d) is non-uniformly expanding along its centre-unstable direction. More precisely, condition (2) in Theorem A holds, with limit bounded away from zero, on a full Lebesgue set of points $x \in M$.

To explain this, let $B_1, \ldots, B_p, B_{p+1} = V$ be any partition of T^d into small domains, in the same sense as before: there exist open unit cubes K_i^0 and K_i^1 in $\mathbb{R}^d$ such that

$$B_i \subset \pi(K_i^0) \quad \text{and} \quad f(B_i) \subset \pi(K_i^1). \tag{40}$$

Let $\mathcal{F}_0^u$ be the unstable foliation of f_0, and $\mathcal{F}_j = f^j(\mathcal{F}_0^u)$ for every $j \geq 0$. By (a), each $\mathcal{F}_j$ is a foliation of T^d tangent to the centre-unstable cone field C^{cu}. For any subset E of a leaf of $\mathcal{F}_j$, $j \geq 0$, we denote $\text{Leb}_j(E)$ the Lebesgue measure of E inside that leaf.

Fix any small disk D_0 contained in a leaf of the foliation $\mathcal{F}_0$. Then, for any sequence $\underline{i} = (i_0, \ldots, i_{n-1})$ in $\{1, \ldots, p, p+1\}$, define

$$[\underline{i}] = \{x \in D_0 : f^j(x) \in B_{i_j} \text{ for } 0 \leq j < n\}.$$

Claim: *There exists* $C_0 > 0$ *depending only on* f *such that* $\text{Leb}_0([\underline{i}]) \leq C_0 \sigma_1^{-n}$ *for every sequence* $\underline{i}$ *as above.*

Proof. Indeed, let $\tilde{\mathcal{F}}_j$ be the lift to $\mathbb{R}^d$ of $\mathcal{F}_j$, for $j \geq 0$. Using (40) one can easily conclude, by induction on j, that $f^j([\underline{i}])$ is contained in the image $\pi(K_{j-1}^1 \cap \tilde{F}_j)$ of the intersection of K_{j-1}^1 with some leaf $\tilde{F}_j$ of $\tilde{\mathcal{F}}_j$, for every $0 \leq j \leq n$. So, using (b) and the fact that $(\pi \mid K_{n-1}^1)$ is a diffeomorphism and an isometry onto its image,

$$\text{Leb}_0([\underline{i}]) \leq \sigma_1^{-n} \text{Leb}_n(f^n([\underline{i}])) \leq \sigma_1^{-n} \text{Leb}_n(F_n \cap K_{n-1}^1). \tag{41}$$

Recall that we took f_0 linear, so that its unstable foliation $\mathcal{F}_0^u$ lifts to a foliation $\tilde{\mathcal{F}}_0^u$ of $\mathbb{R}^d$ by affine hyperplanes. The leaves of every $\tilde{\mathcal{F}}_n$ are C^1 submanifolds of $\mathbb{R}^d$ transverse to these hyperplanes, with angles uniformly bounded away from zero at every intersection point. Consequently, the intersection of a leaf of $\tilde{\mathcal{F}}_n$ with any unit cube in $\mathbb{R}^d$ has Lebesgue measure (inside the leaf) bounded by some uniform constant C_0. In particular, the last factor in (41) is bounded by C_0. □

Now, using the same arguments as in Lemma A.1, we may conclude that Leb_0-almost every point $x \in D_0$ spends a positive fraction ε_0 of the time outside the domain V. Then, using assumptions (c) and (d) above, there exists $c_0 > 0$ such that

$$\limsup_{n \to \infty} \frac{1}{n} \sum_{j=0}^{n-1} \log \left\| \left(Df \middle| E^{cu}_{f^j(x)}\right)^{-1} \right\| \leq -c_0$$

for Leb_0-almost every point $x \in D_0$. Since D_0 was an arbitrary disk inside a leaf of $\mathcal{F}^s_0$, and the latter is an absolutely continuous foliation, we conclude that f is non-uniformly expanding along E^{cu}, Lebesgue almost everywhere in $M = T^d$.

Remark A.6. These arguments also show that f is non-uniformly contracting along the centre-stable direction, if it satisfies (a)–(d): Lebesgue almost every $x \in M$ has

$$\limsup_{n\to\infty} \frac{1}{n} \sum_{j=0}^{n-1} \log \left\| Df | E^{cs}_{f^j(x)} \right\| \le -c_0 .$$

Finally, reducing δ_0 if necessary (this can be done without changing c_0), we can make $A^u = \sup - \log \|Df^{-1} \mid E^{cu}\|$ and $A^s = \sup \log \|Df \mid E^{cs}\|$ arbitrarily close to zero. In particular, $c_0/A^u + c_0/A^s > 1$, as in Proposition 6.5.

References

[ABS77] V. S. Afraimovich, V. V. Bykov, L. P. Shil'nikov. On the appearence and structure of the Lorenz attractor. Dokl. Acad. Sci. USSR, **234**:336–339, 1977

[Alv97] J. F. Alves. SRB measures for nonhyperbolic systems with multidimensional expansion. PhD thesis, IMPA, 1997. To appear Ann. Sci. École Norm. Sup.

[BC91] M. Benedicks, L. Carleson. The dynamics of the Hénon map. Annals of Math., **133**:73–169, 1991

[BD96] C. Bonatti, L. J. Díaz. Nonhyperbolic transitive diffeomorphisms. Annals of Math., **143**:357–396, 1996

[BDP] C. Bonatti, L. J. Díaz, E. Pujals. A C^1-generic dichotomy for diffeomorphisms: weak forms of hyperbolicity or infinitely many sinks or sources. Preprint, 1999

[Bow75] R. Bowen. Equilibrium states and the ergodic theory of Anosov diffeomorphisms, volume 470 of Lect. Notes in Math. Springer Verlag, 1975

[BP74] M. Brin, Ya. Pesin. Partially hyperbolic dynamical systems. Izv. Acad. Nauk. SSSR, **1**:177–212, 1974

[BPV97] C. Bonatti, A. Pumariño, M. Viana. Lorenz attractors with arbitrary expanding dimension. C. R. Acad. Sci. Paris, **325**, Série I:883–888, 1997

[BR75] R. Bowen, D. Ruelle. The ergodic theory of Axiom A flows. Invent. math., **29**:181–202, 1975

[BV] M. Benedicks, M. Viana. Solution of the basin problem for Hénon-like attractors. Preprint, 1999

[BV99] C. Bonatti, M. Viana. SRB measures for partially hyperbolic systems whose central direction is mostly contracting. Israel Journal Math., 1999. To appear

[BY92] M. Benedicks, L.-S. Young. Absolutely continuous invariant measures and random perturbations for certain one-dimensional maps. Ergod. Th. & Dynam. Sys., **12**:13–37, 1992

[Car93] M. Carvalho. Sinai-Ruelle-Bowen measures for n-dimensional derived from Anosov diffeomorphisms. Ergod. Th. & Dynam. Sys., **13**:21–44, 1993

[Cas98] A. A. Castro. Backward inducing and exponential decay of correlations for partially hyperbolic attractors with mostly contracting central direction. PhD thesis, IMPA, 1998

[Dol] D. Dolgopyat. On dynamics of mostly contracting diffeomorphisms. Preprint, 1998

[DPU99] L. J. Díaz, E. Pujals, R. Ures. Partial hyperbolicity and robust transitivity. Acta Math., 1999. To appear
[GPS94] M. Grayson, C. Pugh, M. Shub. Stably ergodic diffeomorphisms. Annals of Math., **140**:295–329, 1994
[GW79] J. Guckenheimer, R. F. Williams. Structural stability of Lorenz attractors. Publ. Math. IHES, **50**:59–72, 1979
[HPS77] M. Hirsch, C. Pugh, M. Shub. Invariant manifolds, volume 583 of Lect. Notes in Math. Springer Verlag, 1977
[Jak81] M. Jakobson. Absolutely continuous invariant measures for one-parameter families of one-dimensional maps. Comm. Math. Phys., **81**:39–88, 1981
[Kan94] I. Kan. Open sets of diffeomorphisms having two attractors, each with an everywhere dense basin. Bull. Amer. Math. Soc., **31**:68–74, 1994
[Kel90] G. Keller. Exponents, attractors and Hopf decompositions. Ergod. Th. & Dynam. Sys., **10**:717–744, 1990
[Led81] F. Ledrappier. Some properties of absolutely continuous invariant measures on an interval. Ergod. Th. & Dynam. Sys., **1**:77–93, 1981
[Lyu] M. Lyubich. Almost every real quadratic map is either regular or stochastic. Preprint, 1997
[Mañ78] R. Mañé. Contributions to the stability conjecture. Topology, **17**:383–396, 1978
[Mañ82] R. Mañé. An ergodic closing lemma. Annals of Math., **116**:503–540, 1982
[Mañ87] R. Mañé. Ergodic theory and differentiable dynamics. Springer Verlag, 1987
[MPP98] C. Morales, M. J. Pacifico, E. Pujals. On C^1 robust singular transitive sets for three-dimensional flows. C. R. Acad. Sci. Paris, **326**, Série I:81–86, 1998
[Pal99] J. Palis. A global view of Dynamics and a conjecture on the denseness of finitude of attractors. Astérisque, **261**:339–351, 1999
[Pes76] Ya. Pesin. Families of invariant manifolds corresponding to non-zero characteristic exponents. Math. USSR. Izv., **10**:1261–1302, 1976
[Pes92] Ya. Pesin. Dynamical systems with generalized hyperbolic attractors: hyperbolic, ergodic and topological properties. Ergod. Th. & Dynam. Sys., **12**:123–151, 1992
[Pli72] V. Pliss. On a conjecture due to Smale. Diff. Uravnenija, **8**:262–268, 1972
[PS82] Ya. Pesin, Ya. Sinai. Gibbs measures for partially hyperbolic attractors. Ergod. Th. & Dynam. Sys., **2**:417–438, 1982
[PS89] C. Pugh, M. Shub. Ergodic attractors. Trans. Amer. Math. Soc., **312**:1–54, 1989
[Rok62] V.A. Rokhlin. On the fundamental ideas of measure theory. A. M. S. Transl., **10**:1–52, 1962. Transl. from Mat. Sbornik **25** (1949), 107–150
[Rov93] A. Rovella. The dynamics of perturbations of the contracting Lorenz attractor. Bull. Braz. Math. Soc., **24**:233–259, 1993
[Rue76] D. Ruelle. A measure associated with Axiom A attractors. Amer. J. Math., **98**:619–654, 1976
[Sat92] E. A. Sataev. Invariant measures for hyperbolic maps with singularities. Russ. Math. Surveys, **471**:191–251, 1992
[Shu71] M. Shub. Topologically transitive diffeomorphisms on T^4. volume 206 of Lect. Notes in Math., page 39. Springer Verlag, 1971
[Sin72] Ya. Sinai. Gibbs measure in ergodic theory. Russian Math. Surveys, **27**:21–69, 1972
[SW] M. Shub, A. Wilkinson. Pathological foliations and removable zero exponents. Preprint, 1998
[Tuc99] W. Tucker. The lorenz attractor exists. C. R. Acad. Sci. Paris, **328**, Série I:1197–1202, 1999
[Via97] M. Viana. Multidimensional nonhyperbolic attractors. Publ. Math. IHES, **85**:69–96, 1997
[You90] L.-S. Young. Large deviations in dynamical systems. Trans. Amer. Math. Soc., **318**:525–543, 1990

ELSEVIER

Physica D 170 (2002) 50–71

PHYSICA D

www.elsevier.com/locate/physd

SLYRB measures: natural invariant measures for chaotic systems

Brian R. Hunt[a,b,*], Judy A. Kennedy[c], Tien-Yien Li[d], Helena E. Nusse[a,1]

[a] *Institute for Physical Science and Technology, University of Maryland, College Park, MD 20742, USA*
[b] *Department of Mathematics, University of Maryland, College Park, MD 20742, USA*
[c] *Department of Mathematics, University of Delaware, Newark, DE 19716, USA*
[d] *Department of Mathematics, Michigan State University, East Lansing, MI 48824, USA*

Received 13 August 2001; received in revised form 18 January 2002; accepted 18 March 2002
Communicated by E. Kostelich

In honor of James A. Yorke's 60th birthday

Abstract

In many applications it is useful to consider not only the set that constitutes an attractor but also (if it exists) the asymptotic distribution of a typical trajectory converging to the attractor. Indeed, in the physics literature such a distribution is often assumed to exist. When it exists, it is called a "natural invariant measure". The results by Lasota and Yorke, and by Sinai, Ruelle and Bowen represent two approaches both of which establish the existence of an invariant measure. The goal of this paper is to relate the "Lasota–Yorke measure" for chaotic attractors in one-dimensional maps and the "Sinai–Ruelle–Bowen measure" for chaotic attractors in higher-dimensional dynamical systems. We introduce the notion of "*SLYRB measure*". (We pronounce the term "SLYRB" as a single word "slurb".) The SRB concept of measure can be motivated by asking how a trajectory from a typical initial point is distributed asymptotically. Similarly the SLYRB concept of measure can be motivated by asking what the average distribution is for trajectories of a large collection of initial points in some region not necessarily restricted to a single basin. The latter is analogous to ask where all the rain drops from a rain storm go and the former asks about where a single rain drop goes, perhaps winding up distributed throughout a particular lake.

Keywords: Chaotic attractor; Natural invariant measure; SLYRB measure

1. Introduction

In this paper, we relate the topics of "chaotic attractors" and the corresponding "natural invariant measures". We start off with a simple yet challenging example. Consider the map $F\colon \mathbb{R}^2 \to \mathbb{R}^2$ defined by

$$F(x, y) = (x^2 - y^2 + ax + by,\ 2x + cx + dy), \tag{1}$$

where $a, b, c, d \in \mathbb{R}$. We select $a = 0.9$, $b = -0.6013$, $c = 2$ and $d = 0.5$. Pick any initial condition (x_0, y_0) and compute the orbit (trajectory) $\{(x_n, y_n)\}_{n=0}^{N}$ for some large N, where $(x_{n+1}, y_{n+1}) = F(x_n, y_n)$. Choosing the

* Corresponding author. Present address: Institute for Physical Science and Technology, University of Maryland, College Park, MD 20742, USA.
E-mail address: bhunt@ipst.umd.edu (B.R. Hunt).
[1] Permanent address: Department of Econometrics, University of Groningen, P.O. Box 800, NL-9700 AV Groningen, The Netherlands.

0167-2789/02/$ – see front matter © 2002 Elsevier Science B.V. All rights reserved.
PII: S0167-2789(02)00445-1

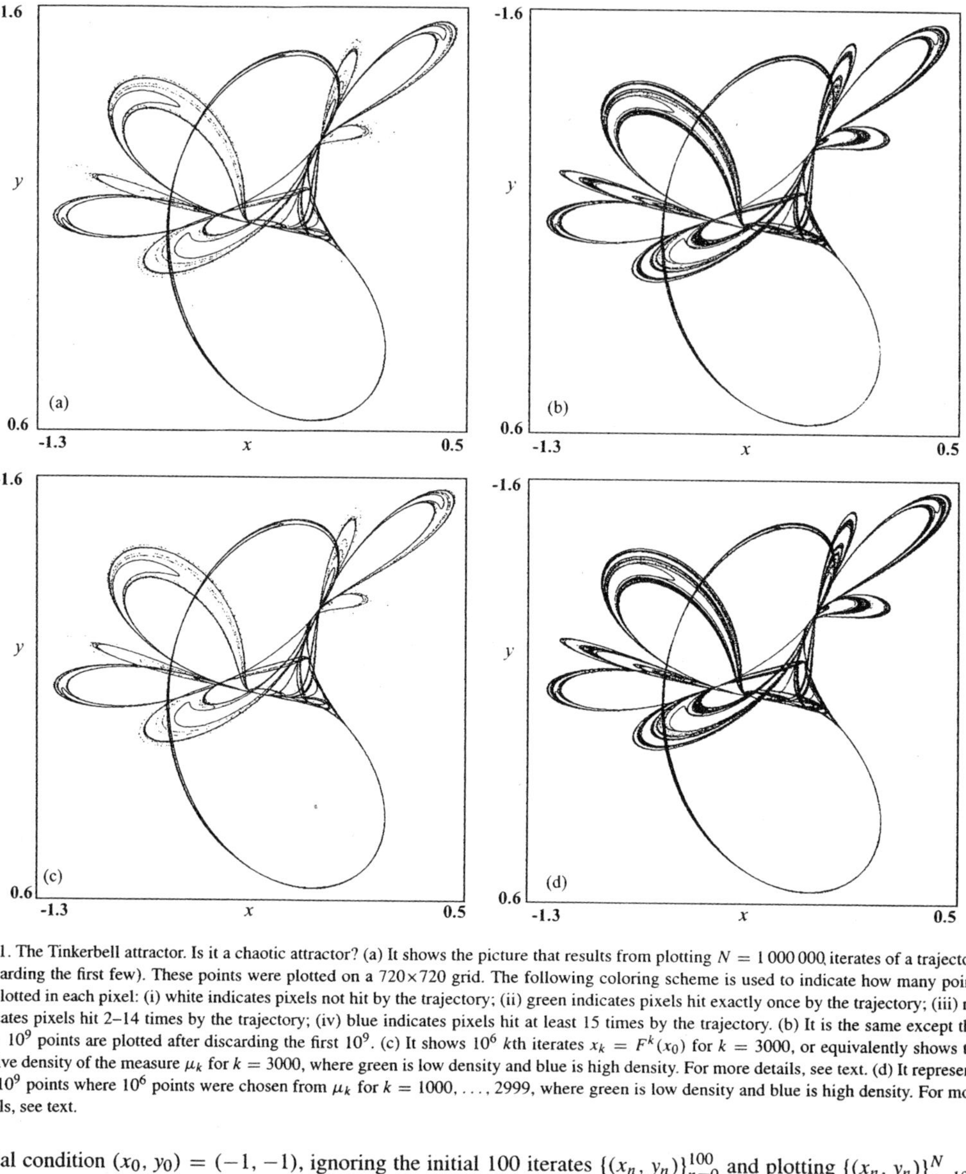

Fig. 1. The Tinkerbell attractor. Is it a chaotic attractor? (a) It shows the picture that results from plotting $N = 1\,000\,000$ iterates of a trajectory (discarding the first few). These points were plotted on a 720×720 grid. The following coloring scheme is used to indicate how many points are plotted in each pixel: (i) white indicates pixels not hit by the trajectory; (ii) green indicates pixels hit exactly once by the trajectory; (iii) red indicates pixels hit 2–14 times by the trajectory; (iv) blue indicates pixels hit at least 15 times by the trajectory. (b) It is the same except that $N = 10^9$ points are plotted after discarding the first 10^9. (c) It shows 10^6 kth iterates $x_k = F^k(x_0)$ for $k = 3000$, or equivalently shows the relative density of the measure μ_k for $k = 3000$, where green is low density and blue is high density. For more details, see text. (d) It represents 2×10^9 points where 10^6 points were chosen from μ_k for $k = 1000, \ldots, 2999$, where green is low density and blue is high density. For more details, see text.

initial condition $(x_0, y_0) = (-1, -1)$, ignoring the initial 100 iterates $\{(x_n, y_n)\}_{n=0}^{100}$ and plotting $\{(x_n, y_n)\}_{n=101}^{N}$, the resulting picture is shown in Fig. 1 with $N = 10^6$ in Fig. 1(a) and $N = 10^9$ in Fig. 1(b). The picture shows a compact set, some parts of which are heavily filled with dots and some parts of which are only sparsely filled. For every initial condition we tried in a certain region B, we got essentially the same picture. Presumably it is an example of a "Milnor attractor" for the map F. For the purpose of this paper, any continuous map $G : \mathbb{R}^m \to \mathbb{R}^m$

we call a compact set A a *Milnor attractor* for G if and only if (1) $G(A) = A$, and (2) for every open neighborhood U of A there exists a set $V \subset U$ of positive Lebesgue measure such that every initial condition $p_0 \in V$ satisfies $\text{dist}(p_n, A) \to 0$ as $n \to \infty$, where $p_n = G^n(p_0)$ is the nth iterate of p_0. We call a Milnor attractor A a *transitive Milnor attractor* if and only if A contains a trajectory that is dense in A. (In the literature there exists several (nonequivalent) definitions of the notion "attractor". For a discussion of this topic see, for example, the article by Milnor [1].) Assuming the picture in Fig. 1 represents a Milnor attractor Y^* (i.e. there is a positive probability that the trajectory of a randomly chosen point $p_0 = (x_0, y_0) \in B$ will be attracted to Y^*, so $\text{dist}(F^n(p_0), Y^*) \to 0$ as $n \to \infty$), the attractor Y^* is the Tinkerbell attractor [2], and is an example of a so-called "chaotic attractor".

In many applications it is useful to consider not only the set that constitutes an attractor but also (if it exists) the asymptotic distribution of a typical trajectory converging to the attractor. Indeed, in the physics literature such a distribution is often assumed to exist. When it exists, it is called a "natural invariant measure" [3]. Thus for the Tinkerbell attractor, we can ask: Is it true, as numerical study suggests, that for almost all (in the sense of Lebesgue measure) $p_0 \in B$ the orbit $\{p_n\}_{n=0}^{\infty}$ has a well-defined distribution and that this distribution is independent of the choice of the initial condition? We challenge the reader to come up with a right answer. Perhaps we should challenge future generations. This is a difficult question and currently we are unable to answer it. Instead, we will discuss some aspects of natural invariant measures. We want to concentrate on results by Lasota and Yorke, and by Sinai and by Ruelle and Bowen, which establish the existence of an invariant measure. *The goal is to relate the "Lasota–Yorke measure" for chaotic attractors in one-dimensional maps and the "Sinai–Ruelle–Bowen measure" for chaotic attractors in higher-dimensional dynamical systems. In Section 5 we introduce the notion of "SLYRB measure"*. (We pronounce the term "SLYRB" as a single word "slurb".) We say that a system is *SLYRB chaotic* if it admits a chaotic attractor with a SLYRB. (The ordering of the letters S-LY-RB is based on the dates of three pivotal papers of Sinai, Lasota and Yorke, and Bowen and Ruelle; see Sections 3–5 for more details.) The organization of the paper is as follows. In Section 2, we discuss briefly the two approaches and concentrate on the results by Lasota and Yorke and by Sinai and by Ruelle and Bowen which establish the existence of an invariant measure. We also present a few classical examples of maps having an invariant measure. Section 3 is devoted to the Lasota and Yorke approach to invariant measures illustrated by a simple example. In Section 4, we first discuss briefly some of the important results on invariant measures due to Bowen, Ruelle and Sinai in uniformly hyperbolic systems. Thereafter, we give a partial review on the existence of invariant measures for one-dimensional quadratic maps. Finally, we give a flavor of results on invariant measures for certain two-dimensional nonhyperbolic systems having chaotic nonhyperbolic attractors, namely the well-known Lozi maps and Hénon maps. In Section 5 we give our conclusions and introduce the notion of SLYRB measure.

2. Invariant measures

Much of this paper is devoted to a discussion of existing theorems on invariant measures. The results representing the two approaches we want to concentrate on are the results by Lasota and Yorke (Theorem LaY in Section 3), and by Sinai and by Ruelle and Bowen (Theorem SRB in Section 4), which establish the existence of an invariant measure. In this section, we discuss briefly the two approaches and also review a few classical examples of systems having an invariant measure. We start off with some definitions.

Let X be a space with Lebesgue measure m. Throughout this paper m will denote Lebesgue measure. A measure μ (on X) is said to have *density* f if $\mu(E) = \int_E f(x)\,dx$ for all measurable sets $E \subset X$, where $f : X \to [0, \infty)$ is a measurable function. A measure μ is called *absolutely continuous* if it has a density. In this paper we assume that all densities satisfy $\int_X f(x)\,dx < \infty$. (Ref. [4] requires $\int_X f(x)\,dx = 1$.) For a density f, write μ_f for the measure that has f as its density. A measure μ is a *probability measure* if $\mu(X) = 1$. Let $M : X \to X$ be a piecewise smooth map. We write $M^{-1}(E) = \{x \in X : M(x) \in E\}$ for the pre-image of E under M. A measure μ is said to be an *invariant measure* for M if $\mu(M^{-1}(E)) = \mu(E)$ for every measurable set $E \subset X$.

2.1. Invariant measures derived from Lebesgue measure: two approaches

We now briefly discuss the two approaches. Let $M : X \to X$ be a piecewise smooth map. One approach follows ideas of Sinai, who since the beginning of the 1970s, systematically developed a Markov partition method for dynamical systems. In 1972 [5], he proved the existence of invariant measures for certain systems having a Markov partition. Bowen and Ruelle [6–9] extended these techniques to Axiom A diffeomorphisms and flows in papers that were published during the period 1973–1976. The resulting measure is often called an SRB measure. To describe such an invariant measure, let $U \subset X$ be a set of positive Lebesgue measure that is mapped into itself and let $A \subset U$ be a Milnor attractor such that for every $x \in U$, $M^n(x) \to A$ as $n \to \infty$. They consider the average value along a trajectory $\{M^i(x)\}_{i=0}^{\infty}$ of any continuous bounded function φ, i.e. $\langle\varphi\rangle_x = \lim_{n\to\infty}(1/n)\sum_{i=0}^{n-1}\varphi(M^i(x))$. If for some probability measure μ, $\langle\varphi\rangle_x$ exists and equals $\int \varphi\, d\mu$ for all φ, then μ represents the asymptotic distribution of the trajectory of x. They are interested in cases in which Lebesgue almost all initial conditions $x \in U$ yield the same measure. Thus when an SRB measure μ exists, it is described by $\lim_{n\to\infty}(1/n)\sum_{i=0}^{n-1}\varphi(M^i(x)) = \int \varphi\, d\mu$ for Lebesgue almost every x in U and for every continuous bounded function φ on U (see Section 4). Hence, the significance of such an SRB measure is that it describes the asymptotic distribution of the trajectory of initial points in a set of positive Lebesgue measure in the phase space.

An alternative, but closely related, approach (based on the Perron–Frobenius techniques) was used, for example, by Lasota and Yorke [10]. Here the "evolution" of a measure μ_0 under M is studied. To explain that term, choose an initial point x_0 at random using a probability measure μ_0 and examine the trajectory defined by $x_{n+1} = M(x_n)$, $n = 0, 1, 2, \ldots$. Then x_1 has a probability distribution μ_1, where $\mu_1(E) = \mu_0(M^{-1}(E))$ for every $E \subset X$. The measure μ_n for the distribution of x_n can be defined iteratively, since the probability that $x_{n+1} \in E$ equals the probability that $x_n \in M^{-1}(E)$. Hence

$$\mu_{n+1}(E) = \mu_n(M^{-1}(E)) \quad \text{for } n = 0, 1, 2, \ldots. \tag{2}$$

In this approach an invariant measure is described as the limit of $(1/n)\sum_{i=0}^{n-1}\mu_i$ (see Section 3).

The Frobenius–Perron operator $\mathcal{P}$ for M (also denoted by $\mathcal{P}_M$) is used to describe how the corresponding densities map, i.e. if μ_0 and $\mu_1 = \mu_0(M^{-1}(\cdot))$ are as above, and if they have densities f_0 and f_1, respectively, then

$$f_1 = \mathcal{P} f_0. \tag{3}$$

When the Jacobian matrix $DM(x)$ of M at x exists and $\det DM(x) \neq 0$ for almost all x (w.r.t. the Lebesgue measure), we can equivalently define $\mathcal{P}$ $(= \mathcal{P}_M)$ by

$$\mathcal{P}_M(f)(x) = \sum_{y\in M^{-1}(x)} \frac{f(y)}{|\det DM(y)|}. \tag{4}$$

The Frobenius–Perron operator is very useful for studying the evolution of densities under iteration of M. If an initial point x is chosen using a probability distribution with density f, then $\mathcal{P}_M f$ is the density for $M(x)$. A good reference is, for example, Lasota and Mackey [4].

We now illustrate the two approaches. Let F be the Tinkerbell map defined in Eq. (1). Let the measure μ_0 denote a uniform distribution on the rectangle shown in red in Fig. 2. The rectangle is approximately 400 pixels×400 pixels. In Fig. 2 ($k = 0$) we randomly chose 10^6 points x_0 from this distribution and plotted them. The coloring scheme is as follows. A pixel is colored white if it is hit by none of these 10^6 points, green if it is hit by exactly

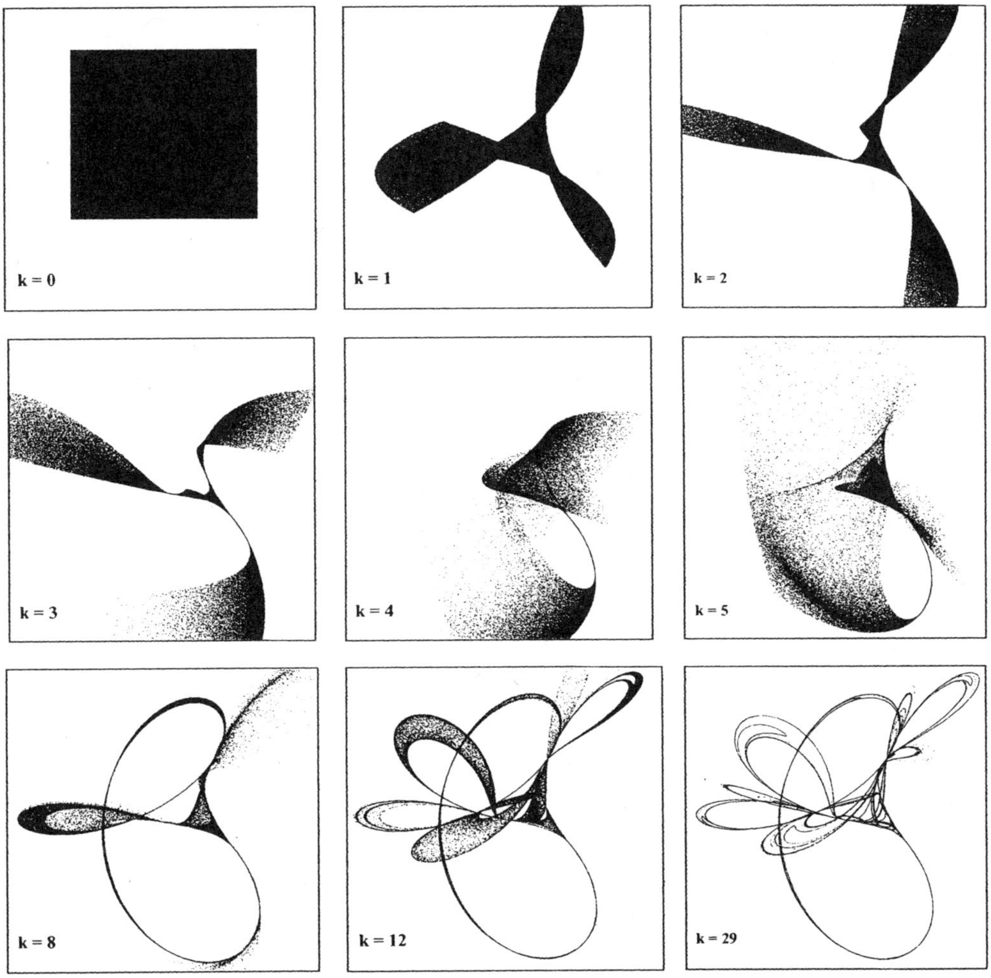

Fig. 2. This figure shows subsequently 10^6 kth iterates $x_k = F^k(x_0)$ for (a) $k = 0$, (b) $k = 1$, (c) $k = 2$, (d) $k = 3$, (e) $k = 4$, (f) $k = 5$, (g) $k = 8$, (h) $k = 12$, and (i) $k = 29$, or equivalently shows the relative densities of the measures μ_k, where green is low density and blue is high density.

one, red if it is hit 2–14 times, and blue if it is hit at least 15 times. Most pixels in the rectangle had between 2 and 14 points and were hence plotted red. Fig. 2 ($k = 1$) shows the 10^6 first iterates $x_1 = F(x_0)$, or equivalently, 10^6 points x_1 chosen at random from the probability distribution (measure) μ_1. Fig. 2 subsequently shows 10^6 kth iterates $x_k = F^k(x_0)$ for $k = 2, 3, 4, 5, 8, 12, 29$, or equivalently shows the relative densities of the measures μ_k, where green is low density and blue is high density. Fig. 1(c) shows this distribution for μ_k where $k = 3000$ and Fig. 1(d) represents 2×10^9 points where 10^6 points were chosen from μ_k for $k = 1000, \ldots, 2999$. Note that there is still a low density region in Fig. 1(d). Although Fig. 1(b) is generated by a single trajectory, Fig. 1(b) and (d) displays certain similarities. This may be of some help in answering the question posed in Section 1.

The way the two approaches describe the measures appear quite different. However, if the system has one attractor, then the two measures resulting from the two approaches are equal. The theoretical construction of a global SRB measure consists of averaging the iterates of a local measure on an arbitrary unstable manifold (see the construction for the Lozi map in Section 4). This construction resembles the Lasota and Yorke's pushing forward of the Lebesgue measure construction. Dellnitz and Junge [11] presented efficient techniques for the numerical approximation of SRB measures based on appropriate discretization of the Frobenius–Perron operator. We show how one may think of the apparent difference between the two approaches as being a change of variables.

2.2. A standard change of variables

The set of measures can be viewed as the set of linear operators on the space $\mathcal{C}$ of bounded continuous functions on the space X. Every linear operator is an integral with respect to some measure. Hence, the measure associated with $\mathcal{P}_M f$ for the map M is the operator (letting $g \in \mathcal{C}$)

$$g \mapsto \int_X (\mathcal{P}_M f)(x) g(x)\,\mathrm{d}x. \tag{5}$$

Assuming M is piecewise invertible, Eq. (5) can be rewritten as

$$g \mapsto \int_X f(y) g(M(y))\,\mathrm{d}y \tag{6}$$

as can be seen by the standard change of variables $x = M(y)$. We will see that the representation in Eq. (5) corresponds to the approach used by Lasota and Yorke [10] while the representation in Eq. (6) was utilized by Bowen and Ruelle [7]. See Section 5 for more details on the relationship between the two approaches.

2.3. Classical examples of invariant measures

We first give three examples of maps having invariant measures.

Example 1. Let T be a tent map of the interval [0, 1] into itself defined by

$$T(x) = \begin{cases} 2x & \text{if } 0 \le x \le 0.5, \\ 2(1-x) & \text{if } 0.5 < x \le 1. \end{cases}$$

For every subinterval $J \subset [0, 1]$ we have that $T^{-1}(J)$ is a union of two intervals J_1 and J_2. For example, if $J = [0.2, 0.4]$ then $T^{-1}(J) = [0.1, 0.2] \cup [0.8, 0.9]$. Every subinterval $J \subset [0, 1]$ satisfies $m(T^{-1}(J)) = m(J_1) + m(J_2) = (1/2)m(J) + (1/2)m(J) = m(J)$. Hence, m is an absolutely continuous invariant measure for T.

Example 2. Let Q be a quadratic map of [0, 1] into itself defined by

$$Q(x) = 4x(1-x).$$

In an abstract being a preliminary report, Ulam and Von Neumann [12] wrote in 1947 "... starting with almost every x_1 (in the sense of Lebesgue measure) and iterating the function $f(x) = 4x(1-x)$ one obtains a sequence of numbers on (0, 1) with a computable algebraic distribution". Although one often encounters statements suggesting that an invariant measure for Q was first constructed by Ulam and Von Neumann [12], in fact that the algebraic expression is not stated in the published text. There is a well-known simple change of variables by which one can reduce the study of the map Q to that of T in Example 1. The earliest description of this change of variables seems to be in Rechard [13], who reported in 1956 that Ulam has observed that the map Q is conjugate to T under the homeomorphism φ: $[0, 1] \to [0, 1]$ defined by $\varphi(x) = 2/\pi \arcsin\sqrt{x}$, i.e. $\varphi \circ Q \circ \varphi^{-1} = T$. (In later papers, Ulam refers to [12,13] for this conjugation φ.) To study the evolution of densities under Q, Rechard considers the Frobenius–Perron operator $\mathcal{P}_Q$ from $L^1([0, 1], m)$ to itself defined by

$$\mathcal{P}_Q f(x) = \frac{1}{4\sqrt{1-x}}\left[f\left(\frac{1}{2}(1-\sqrt{1-x})\right) + f\left(\frac{1}{2}(1+\sqrt{1-x})\right)\right].$$

Defining $\rho : (0, 1) \to \mathbb{R}$ by $\rho(x) = \varphi'(x) = 1/\pi\sqrt{x(1-x)}$ and observing that $\rho(x) = \rho(1-x)$, one verifies

$$\int_0^{4x(1-x)} \rho(t)\,dt = \varphi \circ Q(x) = T \circ \varphi(x) = \begin{cases} 2\varphi(x) = \int_0^x \rho(t)\,dt + \int_{1-x}^1 \rho(t)\,dt & (0 \le x \le \frac{1}{2}), \\ 2[1-\varphi(x)] = \int_x^1 \rho(t)\,dt + \int_1^{1-x} \rho(t)\,dt & (\frac{1}{2} \le x \le 1). \end{cases}$$

This implies that $\mathcal{P}_Q \rho = \rho$. Thus the measure m^* defined by $m^*(A) = \int_A (dx/\pi\sqrt{x(1-x)})$ is an absolutely continuous invariant measure for the map Q.

Example 3. In 1969, Krzyzewski and Szlenk [14] proved the existence of a unique absolutely continuous invariant probability measure for C^2-expanding maps on compact, connected differentiable manifolds. The following is a special case of their result. Let F be a differentiable map of the two-dimensional torus into itself defined by $F(x, y) = (ax + by + \varepsilon f(x, y), cx + dy + \varepsilon g(x, y))(\text{mod } 1 \text{ in } x \text{ and } y)$, $0 \le x, y < 1$, where (a) $a, b, c, d \in \mathbb{Z}$; (b) the eigenvalues of

$$\begin{bmatrix} a & b \\ c & d \end{bmatrix}$$

are real and their moduli are greater than 1; (c) the maps $f, g : \mathbb{R}^2 \to \mathbb{R}$ are of class C^2 and each of them is periodic with period 1 with respect to each variable; (d) ε is a positive number. If ε is sufficiently small, then the map F is expanding and so it admits an absolutely continuous invariant probability measure.

3. Lasota–Yorke measures for chaotic attractors

We first concentrate on the following example. Let S: $\mathbb{R} \to \mathbb{R}$ be the "skew tent" map defined by

$$S(x) = \begin{cases} ax + 1 & \text{for } x \le 0, \\ bx + 1 & \text{for } x > 0. \end{cases} \tag{7}$$

Yorke reports privately that the goal of [10] initially was only to handle maps like tent maps given by Eq. (7) where $1 < a < 2$ and $b = -a$, but their techniques allowed them to prove the result for more general scalar maps. The theorem of Krzyzewski and Szlenk had considered only maps for which there is in fact a C^2 invariant density. That is not the case for most tent maps. It can be shown that the absolutely continuous invariant densities for the tent maps are discontinuous at the points that are the image of $x = 0$ under τ^k for $k \ge 1$ and continuous elsewhere. For most choices of such parameter values a, the trajectory of 0 is dense in some interval, and so the support of the density f^* has a dense set of discontinuities of f^*. The set of discontinuities of f^* has Lebesgue measure zero and therefore f^* is continuous Lebesgue almost everywhere. Note that if $a = \sqrt{2}$ and $b = -a$, then the third iterate of the turning point $x = 0.5$ equals the fixed point p given by $p = 1/(1-b)$, and if $1 < a < \sqrt{2}$ and $b = -a$, the support is always disconnected. In the latter case, the attractor lies in two intervals, one lying wholly to the left of the fixed point p and the other to the right of p.

We will present the interesting case of the skew tent map (7) with values $a = 0.5$, $b = -4.4$, because there the support of f^* is disconnected. The points $p_1 = -(1+b+b^2)/(ab^2-1) < 0$, $p_2 = -(1+a+ab)/(ab^2-1) > 0$, and $p_3 = -(1+b+ab)/(ab^2-1) > 0$ constitute a period-3 orbit. Straightforward computation yields

$$\begin{aligned} S(p_1) &= -a(1+b+b^2)/(ab^2-1)+1 = -(1+a+ab)/(ab^2-1) = p_2, \\ S(p_2) &= -b(1+a+ab)/(ab^2-1)+1 = -(1+b+ab)/(ab^2-1) = p_3, \\ S(p_3) &= -b(1+b+ab)/(ab^2-1)+1 = p_1. \end{aligned}$$

Hence, the map S displays Li–Yorke chaos [15]. In particular for every $n \in \mathbb{N}$, the map S has a period-n orbit and so there are infinitely many periodic points. Plotting any trajectory (or orbit) $\{x_n\}_{n=0}^{\infty}$, it seems that it fills in three intervals. The third iterate of S, S^3, is defined by

$$S^3(x) = \begin{cases} a[a(ax+1)+1]+1 & \text{if } x \le c_1 = -6, \\ b[a(ax+1)+1]+1 & \text{if } c_1 < x \le c_2 = -2, \\ b[b(ax+1)+1]+1 & \text{if } c_2 < x \le c_3 = -2-2b^{-1}, \\ a[b(ax+1)+1]+1 & \text{if } c_3 < x \le c_4 = 0, \\ a[b(bx+1)+1]+1 & \text{if } c_4 < x \le c_5 = -b^{-1}-b^{-2}, \\ b[b(ax+1)+1]+1 & \text{if } c_5 < x \le c_6 = -b^{-1}, \\ b[a(bx+1)+1]+1 & \text{if } c_6 < x \le c_7 = -3b^{-1}, \\ a[a(bx+1)+1]+1 & \text{if } c_7 < x \end{cases} \tag{8}$$

for $a = 0.5$, $b = -4.4$. Its graph is shown in Fig. 3. It is easy to verify that for every $x \in \mathbb{R}$, there exists an $n \in \mathbb{N}$ such that $S^n(x) \in [-4, 1]$ and S maps the interval $[-4, 1]$ into itself. From now on, we consider the map S restricted to the interval $I = [-4, 1]$. One can easily verify that for every $x \in I \setminus \{c_n\}_{n=1}^{7}$, $|(S^3)'(x)| \ge |a^2b| > 1$.

Before we are able to state the fact that S has a "chaotic attractor", we need some basic notions. The trajectory (or orbit) of a point $x_0 \in I$ under the map S is the set $\{S^n(x_0)\}_{n=0}^{\infty}$, where S^n denotes the nth iterate of S. We define the Lyapunov number $L(x_0)$ of the trajectory of x_0 to be $L(x_0) = \lim_{n\to\infty}|(S^n)'(x_0)|^{1/n}$, where we assume that the limit exists. The Lyapunov exponent $\lambda(x_0)$ of the trajectory of x_0 is the logarithm of the Lyapunov number and so $\lambda(x_0) = \log L(x_0)$. Hence, $\lambda(x_0) = \lim_{n\to\infty}(1/n)\sum_{i=1}^{n}\log|S'(S^{i-1}(x_0))|$. Observe that if $\lambda(x_0) > 0$ then $|(S^n)'(x_0)|$ grows exponentially as $n \to \infty$. The trajectory of a point x is called a chaotic trajectory if (1) the trajectory of x is bounded and x is not asymptotically periodic, and (2) the trajectory of x has a positive Lyapunov exponent. Therefore, a chaotic trajectory consists of infinitely many different points. We call a compact set A a *chaotic attractor* (for S) if and only if (a) A is a Milnor attractor, and (b) A contains a chaotic trajectory that is dense in A. The following property easily follows from known results, but we provide a proof that is in the spirit of this paper.

Property 1. The map S has a chaotic attractor.

Proof. For $1 \le i \le 3$, let K_i be the interval of which one end point is the period-3 point p_i and the other point is q_i defined by $S^3(q_i) = p_i$ as shown in Fig. 3. We write $G = S^3$. One easily verifies $G(c_2) \in (q_1, p_1)$, $G(c_4) \in (q_2, p_2)$, and $G(c_7) \in (p_3, q_3)$. Therefore, G maps each of the intervals K_i into itself. Define $D = \{x \in I : G(x) \notin \cup_{i=1}^{3} \text{Int}\, K_i\}$, i.e. D is the collection of points in I that are not mapped into the interior of K_i ($1 \le i \le 3$) when the

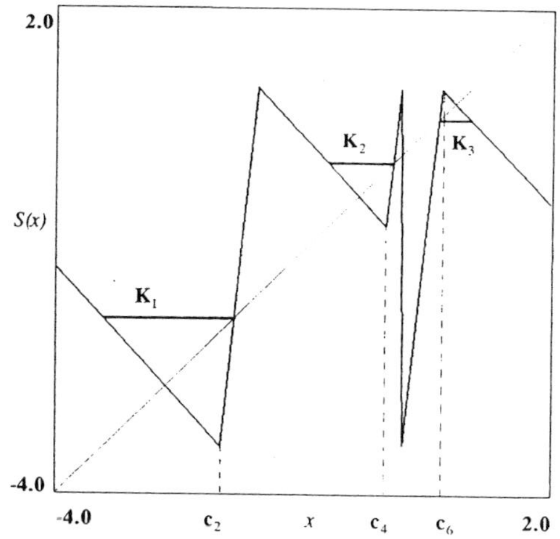

Fig. 3. The third iterate S^3 of the map S, where $S : [-4, 2] \to [-4, 2]$ is defined by $S(x) = 0.5x + 1$ if $x \le 0$ and $S(x) = -4.4x + 1$ if $x > 0$. Each interval K_i is mapped into itself when S^3 is applied ($1 \le i \le 3$).

map G is applied. Note that $c_i \notin D$ ($1 \le i \le 7$). We have that G is expanding on D, i.e. $|G'(x)| \ge |a^2b| > 1$ for all $x \in D$. The collection of points that stay in D when the map G is applied repeatedly, has Lebesgue measure zero, see [16]. (We note that there are quite a number of results in the literature which establish this fact too.) We now conclude that the trajectory of Lebesgue almost every point will enter the union of the three intervals K_i ($1 \le i \le 3$). We write $J_1 = [G(c_2), G^2(c_2)] \subset K_1$, $J_2 = [G(c_4), G^2(c_4)] \subset K_2$, and $J_3 = [G(c_7), G^2(c_7)] \subset K_3$. These compact intervals satisfy $G(J_i) = J_i$ ($1 \le i \le 3$). One easily verifies that for every $x \in \text{Int}\, K_i$, there exists an $N_x \in \mathbb{N}$ such that $G^n(x) \in J_i$ for all $n \ge N_x$ ($1 \le i \le 3$). We write $A = J_1 \cup J_2 \cup J_3$. We have $S(A) = A$ and the collection of points whose trajectories do not converge to A has Lebesgue measure zero. Therefore, A is a Milnor attractor for S.

In order that A be a chaotic attractor, it remains to be shown that there is a chaotic trajectory that is dense in A. We now concentrate on J_2. The two fixed points of G in K_2 are p_2 and $p_2^* = (ab + a + 1)/(1 - a^2b) = -1/3$. Hence, the map G has one fixed point in J_2, namely $p_2^* = -1/3$. A simple computation yields $G(c_4) = G(0) = -0.7$, $G^2(0) = 0.07$ and $G^3(0) = -0.0224$. Applying the property "the map G has a dense trajectory in $J_2 \Leftrightarrow p_2^* < G^3(c_4)$" (see, for example, [16]), we now get that the map G has a dense trajectory in J_2. Therefore, the map S has a dense trajectory in A. We now have that A is a transitive Milnor attractor for S. Let $\{x_n\}_{n=0}^{\infty} \subset A$ be a trajectory that is dense in A such that for every $n \in \mathbb{N}$, $x_n \ne 0$. Since $|S'(x_n)| \ge |a^2b| > 1$, the Lyapunov exponent $\lambda(x_0)$ of this trajectory is positive. Therefore, A has a chaotic trajectory that is dense in A. We conclude that A is a chaotic attractor. □

Property 2. The map S has a Lasota–Yorke invariant measure.

("Lasota–Yorke measure" is the invariant measure in Theorem LaY stated below.)

Proof. Consider the Frobenius–Perron operator $\mathcal{P}_S$ from $L^1([-4, 1], m)$ to itself defined by $\mathcal{P}_S f(x) = \sum_{y \in S^{-1}(x)} (f(y)/|S'(y)|)$. Let $f \in L^1([-4, 1], m)$ be a density function, so $f \ge 0$ and $\int_{-4}^{1} f(s)\, ds = 1$. Applying the theorem due to Lasota and Yorke (Theorem LaY below) yields $(1/n) \sum_{i=0}^{n} \mathcal{P}_S^i f \to f^* \in L^1([-4, 1], m)$

as $n \to \infty$, and the measure $d\mu = f^* \, dm$ is invariant under S, i.e. $\mu^*(S^{-1}(E)) = \mu(E)$ for each measurable set E. □

Conclusion. The map S has a chaotic attractor and admits a natural invariant measure, the "Lasota–Yorke invariant measure". The support of this measure is the chaotic attractor A.

3.1. The method used by Lasota and Yorke

The construction of Lasota and Yorke that we just described is an example of a method that is now commonly used for constructing natural invariant measures, namely that of "pushing forward" Lebesgue measure and Cesaro averaging. In the general case, the natural invariant measure μ_* is formed as the limit (in an appropriate topology) as $n \to \infty$ of $(1/n)\sum_{i=0}^{n-1} \mu_i$, where μ_0 is absolutely continuous with respect to Lebesgue measure and μ_i is the measure with density function $f_i = \mathcal{P}^i f_0$, where $\mathcal{P}$ is the Frobenius–Perron operator and f_0 the density function for μ_0. The initial measure μ_0 could, for example, be normalized Lebesgue measure on some compact set, representing a uniform distribution of initial points on this set. In the case of a Milnor attractor, $(1/n)\sum_{i=0}^{n-1} \mu_i$ may converge to a measure μ_* that is not absolutely continuous even though each of the μ_i is. Lasota and Yorke investigated maps g on an interval I for which $(1/n)\sum_{i=0}^{n-1} f_i$ converges (in the L^1-topology) to a function f_*, which is the density function for an absolutely continuous invariant measure μ_*. Recall that the Frobenius–Perron operator $\mathcal{P}_g\colon L^1(I, m) \to L^1(I, m)$ describes the evolution of these densities under g, namely $f_{n+1} = \mathcal{P}_g f_n$ where $\mathcal{P}_g f(x) = \sum_{y \in g^{-1}(x)} (f(y)/|g'(y)|)$. If $\lim_{n\to\infty}(1/n)\sum_{i=0}^{n-1} \mathcal{P}_g^i f = f_*$, then a formal calculation yields:

$$\mathcal{P}_g \left(\lim_{n\to\infty} \frac{1}{n} \sum_{i=0}^{n-1} \mathcal{P}_g^i f \right) = \lim_{n\to\infty} \frac{1}{n} \sum_{i=1}^{n} \mathcal{P}_g^i f,$$

$$\mathcal{P}_g f_* - f_* = \mathcal{P}_g \left(\lim_{n\to\infty} \frac{1}{n} \sum_{i=0}^{n-1} \mathcal{P}_g^i f \right) - \lim_{n\to\infty} \frac{1}{n} \sum_{i=0}^{n-1} \mathcal{P}_g^i f = \lim_{n\to\infty} \frac{1}{n} [\mathcal{P}_g^n f - f].$$

Since $[\mathcal{P}_g^n f - f]$ is the difference of two densities, its L^1-norm is at most 2, and hence $\int_X |\mathcal{P}_g f_* - f_*| \, dm = 0$. Therefore, $\mathcal{P}_g f_* = f_*$ m-almost everywhere.

3.2. The Lasota–Yorke results

We now want to present the main result due to Lasota and Yorke [10]. Let I be a compact interval $[a, b]$. Let $\tau\colon I \to I$ be a measurable nonsingular transformation, i.e. $m(\tau^{-1}(E)) = 0$ whenever $m(E) = 0$ for any measurable set E. The map $\tau\colon I \to I$ is called a piecewise C^2-map if there exist finitely many points c_i $(0 \le i \le N)$ such that $c_{i-1} < c_i$ and the restriction τ_i of τ to the open interval (c_{i-1}, c_i) is a C^2-map that can be extended to the compact interval $[c_{i-1}, c_i]$ as a C^2-map, where $1 \le i \le N$ and $I = [c_0, c_N]$. For any piecewise C^2-map $\tau\colon I \to I$, consider the Frobenius–Perron operator $\mathcal{P}_\tau$ from $L^1(I, m)$ to itself. Note that

$$\mathcal{P}_\tau f(x) = \sum_{y \in \tau^{-1}(x)} \frac{f(y)}{|\tau'(y)|} = \frac{d}{dx} \int_{\tau^{-1}([a,x])} f(s) \, ds.$$

Recall that $f \in L^1(I, m)$ is a density (function) $\Leftrightarrow$ $f \ge 0$ and $\int_a^b f(s) \, ds = 1$. Here is the first result due to Lasota and Yorke [10] published in 1973.

Theorem 1. *Let $\tau\colon I \to I$ be a piecewise C^2-map such that*

$$\inf\{|\tau'(x)| : x \in I \setminus \{c_i\}_{i=0}^{N}\} > 1.$$

Then, for every density function $f \in L^1(I, m)$, *the sequence* $(1/n)\sum_{i=0}^{n} \mathcal{P}_\tau^i f$ *converges in the* L^1*-norm to a function* $f^* \in L^1(I, m)$ *as* $n \to \infty$. *The limit function* f^* *has the following properties:*

(a) f^* *is a density function;*
(b) $\mathcal{P}_\tau f^* = f^*$ *and so* f^* *is a fixed point of the Frobenius–Perron operator* $\mathcal{P}_\tau$ *and consequently the measure* m_f^* *defined by* $\mathrm{d}m_f^* = f^*\,\mathrm{d}m$ *is invariant under* τ;
(c) *the function* f^* *is of bounded variation.*

Lasota and Yorke prove this theorem by establishing convergence for functions f of bounded variation and using the fact that these functions are dense in $L^1(I, m)$. Their original idea of considering the space of functions of bounded variation and the properties of the Frobenius–Perron operator on this space has been used many times since. We now combine a number of results in [10] into the following theorem. Note that Theorem 1 is a special case hereof.

Theorem LaY (1973). *Let* $\tau: I \to I$ *be a piecewise* C^2*-map such that for some* $K \in \mathbb{N}$, $\inf\{|(\tau^K)'(x)| : x \in I$ *at which* $(\tau^K)'(x)$ *is defined* $\} > 1$. *Then, for every density function* $f \in L^1(I, m)$, *the sequence* $(1/n)\sum_{i=0}^{n} \mathcal{P}_\tau^i f$ *converges in the* L^1*-norm to a function* $f^* \in L^1(I, m)$ *as* $n \to \infty$. *The limit function* f^* *has the following properties:*

(a) f^* *is a density function;*
(b) $\mathcal{P}_\tau f^* = f^*$ *and so* f^* *is a fixed point of the Frobenius–Perron operator* $\mathcal{P}_\tau$ *and consequently the measure* m_f^* *defined by* $\mathrm{d}m_f^* = f^*\,\mathrm{d}m$ *is invariant under* τ;
(c) *the function* f^* *is of bounded variation.*

We refer to the measure m_f^* in Theorem LaY as a *Lasota–Yorke measure*. This measure is an absolutely continuous invariant measure. It is not difficult to show that for such a map τ, each chaotic attractor has a unique absolutely continuous invariant measure and each measure m_f^* is a linear combination of these [17].

3.3. Application to tent maps

An example to which this theorem can be applied is the one-parameter family of tent maps $T_\alpha : [0, 1] \to [0, 1]$ defined by $T_\alpha(x) = \alpha x$ if $0 \le x \le 0.5$ and $T_\alpha(x) = \alpha(1 - x)$ if $0.5 < x \le 1$, where $1 < \alpha \le 2$. The above theorem guarantees that the tent map has an absolutely continuous invariant measure for every $1 < \alpha \le 2$.

Note. Lasota and Yorke also gave the following example. Consider the map $R : [0, 1] \to [0, 1]$ defined by $R(x) = x/(1 - x)$ if $0 \le x \le 0.5$ and $R(x) = 2x - 1$ if $0.5 < x \le 1$. Although $R'(0) = 1$, $R'(x) > 1$ for all $x \in (0, 1) \setminus \{0.5\}$, yet this map does not admit a (bounded) absolutely continuous invariant measure.

4. Sinai–Ruelle–Bowen invariant measures for chaotic attractors

In this section, we first discuss briefly some of the important results on invariant measures due to Bowen, Ruelle and Sinai in uniformly hyperbolic systems. Thereafter, we give a partial review on the existence of invariant measures for one-dimensional quadratic maps. Finally, we give a flavor of results on invariant measures for certain two-dimensional nonhyperbolic systems having chaotic nonhyperbolic attractors, namely the well-known Lozi maps and Hénon maps.

4.1. Sinai–Ruelle–Bowen measures for hyperbolic systems

Let $F : \mathbb{R}^n \to \mathbb{R}^n$ be a C^1-map. If F is a linear map then F is called *hyperbolic* if none of its eigenvalues lie on the unit circle. If F is nonlinear, then $p \in \mathbb{R}^n$ is a *hyperbolic fixed point* for F if and only if (1) $F(p) = p$ and (2) $DF(p)$ is a hyperbolic linear map. If $p \in \mathbb{R}^n$ is a hyperbolic fixed point for F, then one of the following possibilities occurs: (1) p is a fixed point *attractor* (all eigenvalues of $DF(p)$ are inside the unit circle), or (2) p is a fixed point *repellor* (all eigenvalues of $DF(p)$ are outside the unit circle), or (3) p is a fixed point *saddle* (at least one eigenvalue of $DF(p)$ is inside and at least one eigenvalue of $DF(p)$ is outside the unit circle). Similarly, a hyperbolic periodic point (of period m) is a hyperbolic fixed point for F^m. The notion of a "hyperbolic invariant set" is a generalization of the notion of hyperbolic fixed point. From now on, let F be a C^2-diffeomorphism (i.e. F is an invertible map and both F and F^{-1} are C^2-functions) and let $A \subset \mathbb{R}^n$ be a compact F-invariant set and so $F(A) = A$ and $F^{-1}(A) = A$. The map F is called *uniformly hyperbolic* on A if there exists a continuous splitting of the tangent bundle over A into a direct sum of two DF-invariant subbundles, written $TA = E^u \oplus E^s$, so that the following hold: there exist constants $0 \leq \lambda < 1$ and $C > 0$ such that for every $x \in A$ and $n \in \mathbb{N}$, if $v \in E^u(x)$ then $\|DF_x^{-n}v\| \leq C\lambda^n\|v\|$ and $v \in E^s(x)$ implies that $\|DF_x^n v\| \leq C\lambda^n\|v\|$. The set A is called a *uniformly hyperbolic attractor* or an *Axiom A attractor* for $F \Leftrightarrow$ (1) F is uniformly hyperbolic on A; (2) there exists a compact neighborhood U of A such that $F(U) \subset U$ and $A = \cap_{n=0}^{\infty} F^n(U)$; (3) F is topologically transitive on A, i.e. there exists a point in A whose orbit is dense in A. In a more general setting, F is defined on a Riemannian manifold M. The results in the following theorem were first proved for Anosov diffeomorphisms by Sinai, see, e.g. [5]. (Anosov diffeomorphisms are maps that are hyperbolic on the entire manifold M.) They were later generalized to Axiom A attractors by Ruelle and Bowen (see, e.g. [6–8]). Let m denote the Lebesgue measure.

Theorem SRB (1975). *Let $F : \mathbb{R}^m \to \mathbb{R}^m$ be a C^2-diffeomorphism. Let $U \subset \mathbb{R}^m$ be a compact region having nonempty interior such that $F(U) \subset \operatorname{Int}(U)$. Let $A \subset U$ be an Axiom A attractor such that for every $x \in U$, $F^n(x) \to A$ as $n \to \infty$.*

Then there exists a unique F-invariant Borel probability measure μ such that (1) $\operatorname{supp}\mu = A$ *and* (2) *for m-almost every x in U and for every continuous function φ on U, one has* $\lim_{n\to\infty}(1/n)\sum_{i=0}^{n-1}\varphi(F^i(x)) = \int \varphi\,d\mu$.

4.2. The name "SRB"

We remark that conclusion (2) in Theorem SRB is similar to the conclusion of the Birkhoff ergodic theorem, but the crucial difference is that it holds for Lebesgue almost every x in U, as opposed to μ-almost every x. This means that a Lebesgue typical initial condition chosen near the attractor A generates a trajectory that is asymptotically distributed according to μ, i.e. μ is a natural invariant measure. An invariant measure with this property has also been called a Bowen–Ruelle measure by Young [18] and a Bowen–Ruelle–Sinai (or BRS) measure by Tsujii [19]. On the other hand, Benedicks and Young [20] define a Sinai–Bowen–Ruelle (or SBR) measure as an invariant measure with a positive Lyapunov exponent and absolutely continuous conditional measures along unstable manifolds (see [5] or [20] for an explanation of the latter condition). This property also holds under the hypotheses of Theorem SRB. Above we have given just a few examples of how the names of Bowen, Ruelle, and Sinai have been applied in various orders to invariant measures with one or both of the properties we described. In the recent literature, SRB measure seems to be used more often than BRS measure or SBR measure. We now quote from Bowen [6]: "The notes as a whole constitute a version of Sinai's program [5] for applying statistical mechanics to diffeomorphisms. It was Ruelle who carried Sinai's work on Anosov diffeomorphisms over to Axiom A attractors [9] and brought in the formalism of equilibrium state [8]". Based on this statement, one could argue that SRB is a good choice for the order of the three symbols B, R and S. Another source for a definition of SRB measure is Pesin's book [21].

Note. If the attractor A in the theorem is a periodic orbit, then the SRB measure is the Dirac measure equally distributed on this orbit.

4.3. Natural invariant measures for one-dimensional quadratic maps

We now review some results for the family of quadratic maps $f_a : [0, 1] \to [0, 1]$ defined by $f_a(x) = ax(1-x)$, where $1 \leq a \leq 4$. (The proofs of these results are beyond the scope of this paper.) Note that f_a is not uniformly expanding for any a. In fact, near the critical point $c = 0.5$, f_a transforms any bounded density in one step to a density with inverse square-root singularity. We discussed the case of $a = 4$ in Section 2. One of the first general results on the existence of absolutely continuous invariant measures for quadratic maps is the following one due to Ruelle [22]. Let $3 < a \leq 4$ for which there exists $n \in \mathbb{N}$ such that $f_a^n(c) = p_0$, where p_0 is an expanding periodic point (i.e. there is an $m \geq 1$ such that $f_a^m(p_0) = p_0$ and $|(f_a^m)'(p_0)| > 1$). Then f_a admits an absolutely continuous invariant measure. In 1981, Misiurewicz [23] published a result for a class of maps including the members of the quadratic family. Let $3 < a \leq 4$ such that every iterate $f_a^n(c)$ $(n \geq 1)$ is bounded away from c (for example, if the forward orbit of c is trapped in an expanding invariant set). Then f_a admits an absolutely continuous invariant measure. Concerning the quadratic family, Misiurewicz's result can be thought of as a generalization of Ruelle's result. Related to these results is the following result due to Collet and Eckmann [24]. Let $3 < a \leq 4$ such that the orbit of $f_a(c)$ has a positive Lyapunov exponent. Then f_a admits an absolutely continuous invariant measure. An important question is: What can be said about the collection of parameters a $(3 < a \leq 4)$ for which f_a admits an absolutely continuous invariant measure? A very important result dealing with this issue is the famous theorem by Jakobson [25]. The set in the parameter space given by $\{a \in [3, 4]$: f_a admits an absolutely continuous invariant measure$\}$ has positive Lebesgue measure. For an alternative proof, see, for example [26]. The argument used in the proof of Jakobson's result is that the critical orbit does not approach the critical point too closely too quickly, and the collection of those parameters for which this approach is sufficiently controlled has positive Lebesgue measure. Therefore, these "Jakobson parameter values" include the "Misiurewicz parameter values".

In describing the distribution of the orbit $\{f_a^n(x)\}_{n=0}^{\infty}$, Tsujii [19] uses the Dirac measures δ_{x_i} concentrated at the trajectory points $x_i = f_a^i(x)$ and he examines the sequence of probability measures $\mu_n(x) = (1/n)\sum_{i=0}^{n-1}\delta_{x_i}$, $n \geq 1$. If this sequence converges to a probability measure μ as $n \to \infty$, Tsujii calls μ the *asymptotic distribution* for the orbit of x. Here the convergence is in the sense of weak topology, i.e. $\int \varphi\, d\mu_n = (1/n)\sum_{i=0}^{n-1}\varphi(f_a^i(x)) \to \int \varphi\, d\mu$ as $n \to \infty$ for every continuous function φ on the interval $[0, 1]$. Hence, the statistical properties of the orbit are given by the asymptotic distribution μ, if it exists. Tsujii calls a probability measure μ on the interval the *Bowen–Ruelle–Sinai measure* for f_a if the asymptotic distribution of the orbit exists and equals μ for almost every point in the interval with respect to the Lebesgue measure. Hence, f_a has a unique BRS measure μ if and only if $\lim_{n\to\infty}(1/n)\sum_{i=0}^{n-1}\delta_{x_i} = \mu$ for Lebesgue almost all points $x_0 \in [0, 1]$, where x_i denotes the ith iterate of x_0, so $x_i = f_a^i(x_0)$ $(i \geq 0)$. Tsujii studies how the BRS measure depends on the parameter a in the family $\{f_a\}$ and considers two cases: (1) f_a has an attracting hyperbolic periodic orbit, and (2) f_a admits an absolutely continuous invariant probability measure. In case (1) the orbit is unique and the invariant probability measure on it is the BRS measure for f_a. In case (2) the measure is unique and it is the BRS measure for f_a. Tsujii explored the continuity of BRS measures in the quadratic family. If μ_a denotes the BRS measure for f_a when it exists, then it depends continuously on a when f_a has an attracting hyperbolic periodic orbit, while μ_a does not depend continuously on a when f_a admits an absolutely continuous invariant measure. In fact, the dependence of the BRS measure μ_a on the parameter a is quite irregular when

$a \to 4$. In other recent papers on one-dimensional maps, this BRS measure is called SBR measure or SRB measure.

Another breakthrough in the analysis of the one-parameter family of quadratic maps is the following result due to Graczyk and Swiatek [27,28] and Lyubich [29]. The collection of parameters $a \in [1, 4]$ for which f_a has an attracting periodic orbit is dense in $[1, 4]$. The Lebesgue measure of this set is less than 3, since the collection of parameter values for which f_a has an absolute continuous invariant measure is positive. However, for almost every $a \in [1, 4]$, the map f_a has either an attracting periodic orbit or has an absolutely continuous invariant measure [30].

4.4. Chaotic trajectories and chaotic attractors

Let $F: \mathbb{R}^2 \to \mathbb{R}^2$ be a diffeomorphism and let $R \subset \mathbb{R}^2$ be a compact region such that F maps R into itself. Let $A \subset R$ be a compact F-invariant set which is a Milnor attractor such that for every $z \in R$, $F^n(z) \to A$ as $n \to \infty$. Recall that the collection of all $z \in \mathbb{R}^2$ whose orbits converge to A is called the *basin of attraction* of A. Notice R is contained in the basin of attraction. The trajectory $\{F^n(z_0)\}_{n=0}^{\infty}$ of any initial point z_0 has (if the relevant limits exist) two Lyapunov exponents $\lambda_1(z_0)$ and $\lambda_2(z_0)$ describing the average stretching or contracting behavior of the derivative DF along the trajectory. The trajectory of a point x is called a *chaotic trajectory* if (1) the trajectory of x is bounded and x is not asymptotic to either a fixed point or a periodic orbit, and (2) the trajectory of x has a positive Lyapunov exponent. Similar to a chaotic attractor for the skew tent map, we call a compact set C a *chaotic attractor* (for F) if and only if (a) C is a Milnor attractor, and (b) C contains a chaotic trajectory that is dense in C. Hence, the Milnor attractor A is a chaotic attractor (for F) if it contains a chaotic trajectory that is dense in A. If the trajectory of some initial condition $z_0 \in R$ is a chaotic trajectory, may we conclude that there is a chaotic attractor?

4.5. Hénon attractors and invariant measures

To illustrate the difficulty of determining if there is a chaotic attractor, we consider the Hénon map $H_{a,b} : \mathbb{R}^2 \to \mathbb{R}^2$ defined by

$$H_{a,b}(x, y) = (1 - ax^2 + by, x). \tag{9}$$

For every $a, b \in \mathbb{R}$ with $-1 < b < 1$, the map $H_{a,b}$ is a C^2-diffeomorphism. For $a = 1.4$, $b = 0.3$, Hénon [31] gave the coordinates of the four corner points of a quadrilateral having the property that it is mapped into itself when the map $H_{a,b}$ is applied once. Based on this information, we consider the quadrilateral Q defined by the points $P_1 = (-1.862, 1.96)$, $P_2 = (1.848, 0.6267)$, $P_3 = (1.743, -0.6533)$ and $P_4 = (-1.484, -2.3333)$. The superposition of the quadrilateral $Q = P_1P_2P_3P_4$ and its first iterate are shown in Fig. 4(a). The intersection of all forward iterates $H^* = \cap_{n=0}^{\infty} H_{a,b}^n(Q)$ of the quadrilateral region Q is a compact set of Lebesgue measure zero, since $|\det DH_{a,b}(z)| = 0.3 < 1$ for all $z \in \mathbb{R}^2$. Hence, for all $z \in Q$, $H_{a,b}^n(z) \to H^*$ as $n \to \infty$, so H^* is a Milnor attractor. In the literature, H^* is known as the *Hénon attractor*.

Choosing initial point $z_0 = (1, 1)$, ignoring the initial 100 iterates and plotting the next 10^6 iterates, the resulting picture is shown in Fig. 4(b). The Hénon attractor is believed to be a chaotic attractor because (according to numerical calculations) it contains a trajectory which is dense and which has a positive Lyapunov exponent. However, there exists no proof of such a fact. One has to prove that there exists a chaotic trajectory that is dense in H^*. Numerical simulations give the strong impression this is true. One may ask the following question. Does there exists a natural $H_{a,b}$-invariant Borel probability measure μ such that (1) the support of the measure μ is H^* and (2) for m-almost every x in Q and every continuous function φ on Q, $(1/n)\sum_{i=0}^{n-1} \varphi(H_{a,b}^i(x)) \to \int \varphi \, d\mu$? If there exists such a measure, then for m-almost all $z_0 \in Q$ the orbit $\{z_n\}_{n=0}^{\infty}$ has a well-defined distribution and that this distribution is independent of the choice of the initial condition. Unfortunately, the answer is still unknown.

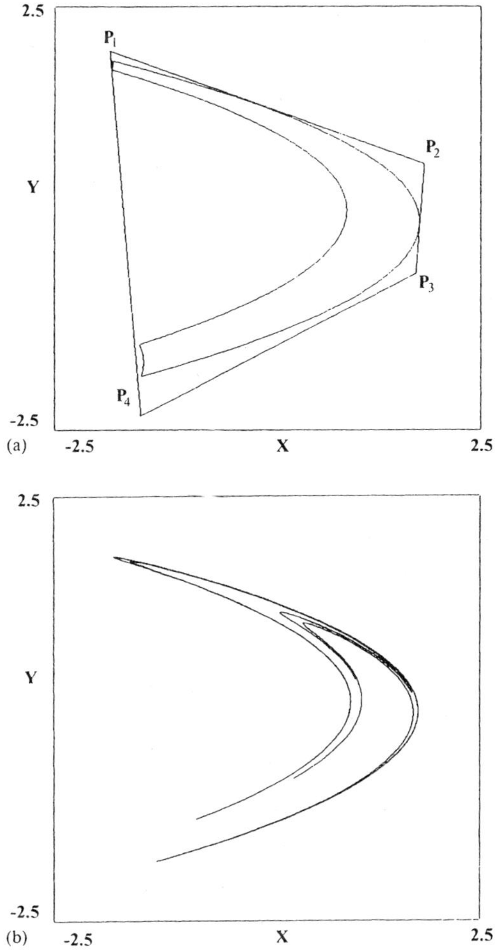

Fig. 4. (a) The superposition of the quadrilateral $Q = P_1P_2P_3P_4$ and its first iterate. (b) The Hénon attractor. Is it a chaotic attractor?

4.6. A chaotic attractor for the Lozi map, a simpler case

In 1978 Lozi [32] introduced a piecewise-linear map as an analog of the Hénon map. The dynamics of the Lozi map may be easier to understand than that of the Hénon map. Consider the map $L_{a,b} : \mathbb{R}^2 \to \mathbb{R}^2$ defined by

$$L_{a,b}(x, y) = (1 - a|x| + by, x). \tag{10}$$

For every $a, b \in \mathbb{R}$ with $-1 \le b \le 1$, the map $L_{a,b}$ is a continuous, piecewise C^2-diffeomorphism (it is a C^2-diffeomorphism on $\mathbb{R}^2 \setminus \{(x, y) \in \mathbb{R}^2 : x = 0\}$). For $a = 1.7$, $b = 0.5$, numerical simulations suggested the existence of a strange attractor (we will discuss this term later, but you may think of a chaotic attractor). Choosing initial point $z_0 = (1, 1)$, ignoring the initial 100 iterates and plotting the next 10^6 iterates, the resulting picture is

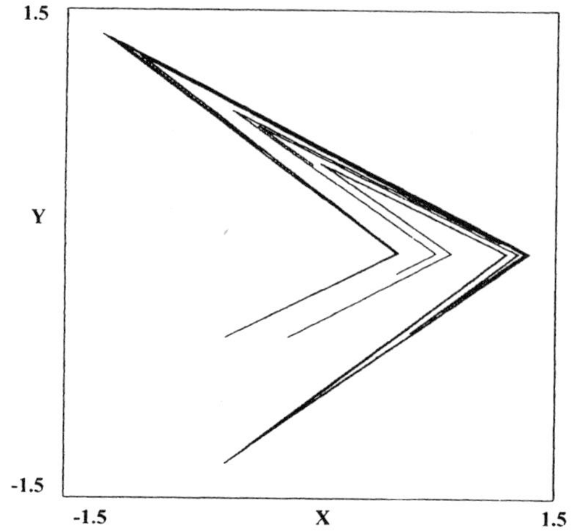

Fig. 5. The Lozi attractor.

shown in Fig. 5. If $0 < b < 1$, $a > 0$ and $a + b > 1$, the map $L_{a,b}$ is a homeomorphism. For an open set of parameter values (which does not include $a = 1.7$, $b = 0.5$), Misiurewicz [33] showed there is a trapping region, i.e. he constructed a compact region Q such that Q is mapped into the interior of Q when the map $L_{a,b}$ is applied once. The intersection of all forward iterates $L^* = \cap_{n=0}^{\infty} L_{a,b}^n(Q)$ of the region Q is a compact set of Lebesgue measure zero. Hence, L^* is a Milnor attractor. Let P be the fixed point of $L_{a,b}$ that has two positive coordinates. Let $W^u(P)$ be the unstable "manifold" of P, i.e. $W^u(P)$ is the collection of points whose orbits converge to P as the map is iterated backwards. Misiurewicz [33] proved that L^* is a strange attractor by showing that the map had a hyperbolic structure and that L^* equals the closure of the unstable "manifold" $W^u(P)$. (Since the map is not differentiable at $z = (0, y)$, its hyperbolic structure must be understood only as the existence of a hyperbolic splitting at those points at which it may exist.) It appears that the Lozi map is an intermediate stage between the Axiom A systems and more complicated systems like the Hénon map.

4.7. An invariant measure for the Lozi map

Bowen and Ruelle [7] assumed the maps were smooth in the expanding directions. The measure they described however was typically not smooth because the map had the property of having contracting directions. In contrast, Lasota and Yorke [10] studied maps which were expanding. Their measure was typically not smooth because the maps had the property of being only piecewise smooth, and the successive iterates of the non-C^2 points (points at which the map is not C^2) could be dense in a collection of intervals. The first papers to deal with maps having both properties were the papers by Rychlik [34,35], Young [18] and by Collet and Levy [36]. They dealt with maps that had the properties of having a contracting direction and being non-smooth in the expanding direction.

Collet and Levy [36] investigated the Lozi maps for the same values of a and b as Misiurewicz but assumed in addition that b is small. Recall that $L_{a,b}$ is a homeomorphism and that $W^u(P) = \{z \in \mathbb{R}^2 : \lim_{n\to\infty} L_{a,b}^{-n}(z) = P\}$ is the unstable "manifold" of P. The maximal smooth component $W_{\text{loc}}^u(P)$ of $W^u(P)$ containing P is called the *local unstable manifold* of P. Collet and Levy [36] construct an invariant measure as follows. They start off with the local unstable manifold of P. Let $B_0 = W_{\text{loc}}^u(P)$. From Misiurewicz results one knows that L^* is the closure of $\cup_{n=0}^{\infty} L_{a,b}^n(B_0) = W^u(P)$. Therefore, it is natural to try to obtain an invariant measure by iterating the Lebesgue measure supported by B_0. They define a sequence $\{\mu'_n\}_{n\in\mathbb{Z}}$ of probability measures on $\mathcal{M}_{L^*}$ by $\mu'_n(A) = \ell(L^{-n}_{a,b}(A) \cap B_0)/\ell(B_0)$ for $A \in \mathcal{M}_{L^*}$, where $\mathcal{M}_{L^*}$ denotes the Borel sub σ-algebra of $\mathbb{R}^2$ corresponding to L^* and ℓ is the one-dimensional Lebesgue measure. Note that for every n, μ'_n has support in $L^n_{a.b}(B_0)$. The sequence $\{\mu_n\}_{n\in\mathbb{N}}$ defined by $\mu_n = (1/n)\sum_{i=0}^{n-1} \mu'_i$ is a sequence of probability measures on $\mathcal{M}_{L^*}$. Since L^* is compact, one can extract a subsequence which converges in the weak star topology to an invariant Borel probability measure μ. They show that μ is unique, that it is absolutely continuous with respect to the Lebesgue measure in the unstable direction, and that it is in fact a Bowen–Ruelle measure (i.e. a measure generated by the orbits of almost all points). This approach resembles the Lasota and Yorke's pushing forward of the Lebesgue measure construction.

The existence of invariant measures for the Lozi maps for parameter values studied by Misiurewicz, was shown independently by Rychlik [34,35] by a method independent of Young [18] and Collet and Levy [36]. The results were reported in [34]. The paper [34] contains only sketches of the arguments and the full proofs can be found in Rychlik's dissertation [35], which is available on the author's website. The paper [34] and the dissertation [35] present a hybrid approach to the "Lasota–Yorke" measures and the "Sinai–Ruelle–Bowen" measures. In fact, they contain an abstract theorem which implies both the existence of Lasota–Yorke measures for $C^{1+\varepsilon}$ piecewise expanding maps of an interval and Sinai–Ruelle–Bowen measures for Lozi mappings.

4.8. Generalized Lozi maps and their invariant measures

Young [18] introduced a much more general class of piecewise C^2-diffeomorphisms on the region $R = [0, 1] \times [0, 1]$ which are continuous one-to-one maps and which map the region R into itself. These maps are hyperbolic piecewise smooth maps. She based them on the geometric properties of Lozi maps and calls them "generalized Lozi maps". Recall that the above result of Lasota and Yorke states that piecewise expanding maps of the interval have absolutely continuous invariant measures (Lasota–Yorke invariant measure). Young proves an analogous statement for her generalized Lozi maps. A part of the proof follows a combination of the methods used by Sinai, and by Lasota and Yorke. Young's result is that the generalized Lozi maps admit certain invariant measures which imply that the maps have "Bowen–Ruelle measures", i.e. measures generated by the orbits of almost all initial points. More precisely, she defines a Borel probability measure μ on R to be a *Bowen–Ruelle measure* if there is a set $U \subset R$ of positive Lebesgue measure such that for every continuous function φ: $R \to R$, $(1/n)\sum_{i=0}^{n-1} \varphi(f^i(x)) \to \int \varphi\, d\mu$ for Lebesgue-almost every $x \in U$.

Note. Jakobson and Newhouse [37] also studied invariant measures for hyperbolic piecewise smooth mappings of a rectangle.

4.9. Invariant measures for the Hénon map

We now return to the Hénon map $H_{a,b}$. For the parameter values $a = 1.4$, $b = 0.3$, numerical simulations suggest that for almost every initial condition in a certain region, the orbit has a well-defined distribution as $n \to \infty$, and this distribution is independent of the choice of initial condition. However, in this case as well as for other parameter values for which there seems to be a strange attractor, the situation is very delicate. The question is whether there exists an invariant measure that reflects the statistical behavior of Lebesgue-almost every point in a much larger set. But why should such an invariant measure exist? On the other hand, if (one can show that) such a measure exists, it is of great significance.

In the proofs of Theorem SRB, Markov partitions were used to connect the dynamics of F to certain one-dimensional lattice systems in statistical mechanics, and μ was realized as a Gibbs state or equilibrium state. For an introduction to the applications in nonequilibrium statistical mechanics of chaotic dynamics, see [38]. However, Markov partitions are not available for systems more general than Axiom A, like the Hénon map. To see why it is so difficult to prove the existence of a natural invariant measure for the Hénon map, consider the following. Let P be the fixed point of $H_{a,b}$ that has two positive coordinates. Let $W^u(P)$ be the unstable manifold of P. Recall that $W^u(P)$ is the collection of points whose orbits converge to P as the map is iterated backwards. Similar to the construction of an invariant measure by Collet and Levy [36] for the Lozi map, suppose we start with the local unstable manifold $W^u_{\text{loc}}(P) = B_0$ of the fixed point P and push forward Lebesgue measure on B_0. The definition of a local unstable manifold says that if $x \in B_0$ then $H^{-n}_{a,b}(x) \to P$ as $n \to \infty$, but generally, one has no idea what happens to points on B_0 under forward iteration of the map. For example, it is easy to imagine that parts of $W^u(P) = \cup_{n=0}^{\infty} H^n_{a,b}(B_0)$ belong to basins of attraction of periodic orbits, so the orbits of certain points of B_0 converge to a periodic orbit attractor when the map $H_{a,b}$ is applied repeatedly. Hence, the question of existence of natural invariant (SRB) measures is very delicate. (The Lozi map question is simpler since it has a uniformly expanding direction.) It is in some sense analogous to the problem of existence of absolutely continuous invariant measures for nonuniformly expanding maps (Jakobson theorem) only that it is more complicated because one has to consider not only rates of expansion but their directions as well.

Despite these difficulties, there have been some major breakthroughs. Benedicks and Carleson [39] developed elaborate machinery for analyzing the dynamics of $H_{a,b}$ for a positive measure set of parameters near $a = 2$, $b = 0$. Subsequently, Benedicks and Young [20] constructed SRB measures for these attractors. Here we state their important results.

Theorem 2 ([39]). *Let $H_{a,b}$ be the Hénon map. Then there is a rectangle $\Delta = (a_0, a_1) \times (0, b_1)$ in parameter space such that for all $(a, b) \in \Delta$, $H_{a,b}$ has an attractor A with a dense orbit.*

Theorem 3 ([20]). *Let $H_{a,b}$ be the Hénon map. Then for all sufficiently small $b > 0$, there exists a positive Lebesgue measure set $\Delta_b \subset \mathbb{R}^2$ such that for all $a \in \Delta_b \subset \Delta$, $H_{a,b}$ admits an SRB measure λ^*. Morevoer, λ^* is unique and it is supported on the entire attractor.*

Corollary ([20]). *For $b > 0$ small and $a \in \Delta_b$, the map $H = H_{a,b}$ has the following property: Let U be a neighborhood of the attractor. Then there exists a set $\tilde{U} \subset U$ with positive Lebesgue measure such that for all continuous functions φ: $U \to \mathbb{R}$, $(1/n)\sum_{i=0}^{n-1} \varphi \circ H^i(x) \to \int \varphi \, d\lambda^*$ for every $x \in \tilde{U}$.*

We quote from [20]: "The property expressed in this corollary is often assumed to be true in physics; it is also taken for granted in numerical experiments".

Note. More recently, Young [40] and Alves et al. [41] investigated diffeomorphisms which are not hyperbolic, but partially hyperbolic. These papers contain important results that are beyond the scope of this brief overview. Under quite general assumptions, Young [40] proves that correlations for hyperbolic systems with singularities decay exponentially. If Lyapunov exponents are bounded away from zero, Alves et al.[41] prove that there are only finitely many natural invariant (Sinai–Ruelle–Bowen) measures for such maps and Lebesgue almost every point is in one of the corresponding basins.

5. SLYRB chaos and SLYRB measures

5.1. Strange attractors

In the 1980s a lot of attention was spent on attractors in dynamical systems. Conferences devoted to this single topic were held. The notion of "strange attractor" was originally introduced by Ruelle and Takens [42]. It meant an attractor which is not a fixed point attractor, a periodic orbit or a smooth manifold without boundary such as a circle or torus. Nowadays the terms "chaotic attractor" and "strange attractor" are quite common. Generally, a *chaotic attractor* is a Milnor attractor on which the dynamics is chaotic, i.e. the system has at least one positive Lyapunov exponent. Frequently, the two notions are used as synonyms. Indeed, a chaotic attractor is always a strange attractor in the sense of Ruelle and Takens. However, an example of a strange attractor which is not a chaotic attractor is the so-called Feigenbaum attractor which occurs in the one-parameter family of quadratic maps $f_\alpha(x) = \alpha x(1-x)$. (The *Feigenbaum attractor* is a Milnor attractor which is a Cantor set and the Lyapunov exponent of Lebesgue almost every trajectory is zero.) On the other hand, one encounters in the literature the following. A *strange attractor* is a Milnor attractor that has a Cantor-like structure. Of course, under this definition a chaotic attractor need not be strange. The chaotic attractor consisting of three intervals in the example of skew tent map above demonstrates this. There is also a large literature on "strange non-chaotic attractors" in systems that have terms that are quasi-periodic in time (see, e.g. [43]). We concentrate on chaotic attractors.

We recall that the Lasota–Yorke measures are absolutely continuous invariant measures, while the Sinai–Ruelle–Bowen (or SRB) measures are not. Hence, from the point of view of measure theory applied to dynamics, those invariant measures are considered to be different. However, from the point of view of dynamics using measure theory, the existence of a chaotic attractor implies in certain systems the existence of the special invariant measures as discussed in the previous sections. We propose the following: a dynamical system displays "SLYRB chaos" whenever there is a chaotic attractor which admits a certain natural invariant measure (e.g. a Lasota–Yorke measure or a Sinai–Ruelle–Bowen measure). The notion SRB varies somewhat from paper to paper. In a recent paper, Young [40] limits the definition of SRB measures to the context of no zero Lyapunov exponents and states a definition of Sinai–Ruelle–Bowen measure. She notes that this definition is a generalization of the ideas in Sinai [5] and first appeared in a paper by Ledrappier [44].

5.2. SLYRB (we pronounce it "slurb" as a single word)

After comparing the construction of SRB measures and its properties to the construction and properties of Lasota–Yorke measure, we suggest that the natural invariant measure supported on any chaotic attractor might be called a "SLYRB measure". (The ordering of the letters S-LY-RB correspond to the dates of the papers 1972–1973–1975.) We propose the following. A measure μ is an *SLYRB measure* if and only if μ is the weighted sum $\mu = \sum_{i\geq 1} \alpha_i \mu_i$ of invariant measures μ_i such that each μ_i is the asymptotic distribution of a trajectory for a set of initial points of positive measure, where $a_i \geq 0$.

The SRB concept of measure can be motivated by asking how a trajectory from a typical initial point is distributed asymptotically. Similarly the SLYRB concept of measure can be motivated by asking what the average distribution is for trajectories of a large collection of initial points in some region not necessarily restricted to a single basin. (The coefficient a_i is the fraction of initial conditions in the ith basin B_i.) The latter is analogous to asking where all the rain drops from a rain storm go and the former asks about where a single rain drop goes, perhaps winding up distributed throughout a particular lake. Consider a system that has exactly one attractor A and almost all initial points chosen near the attractor A generate trajectories that are asymptotically distributed according to a single measure μ. Then μ would be both an SRB measure (by some standards) and a SLYRB measure. Also, if the SLYRB measure μ is unique, then the system has exactly one attractor A, and a Lebesgue typical initial point chosen near the attractor A generates a trajectory that is asymptotically distributed according to μ. Recall that an invariant measure with this latter property for certain two-dimensional systems has been called a Bowen–Ruelle measure by Young [18] and for certain one-dimensional systems a Bowen–Ruelle–Sinai (or BRS) measure by Tsujii [19]. Therefore, the Bowen–Ruelle measure and Tsujii's BRS measure are special cases of SLYRB measures. Furthermore, the SBR measure defined by Benedicks and Young [20] being an invariant measure with a positive Lyapunov exponent and absolutely continuous conditional measures along unstable manifolds [20] is also an example of a SLYRB measure. It is not hard to show that the Lasota–Yorke measure is an example of a SLYRB measure too.

To see that a Lasota–Yorke measure must also be a SLYRB measure, recall that according to their hypotheses the map τ^K is piecewise C^2 and expanding wherever its derivative is defined. Each attractor for τ^K must contain at least one point where τ^K is not expanding; since there are only finitely many such points, τ^K has only finitely many attractors. Thus τ itself has only finitely many attractors, and Lebesgue almost every initial condition is in the basin of one of the attractors, see also [17]. Applying Theorem LaY to an initial density function whose support is in the basin of a particular attractor, we conclude that each attractor must support an absolutely continuous invariant measure μ_i. By the Birkhoff ergodic theorem, μ_i-almost every initial condition, and hence Lebesgue almost every initial condition within the attractor, generates a trajectory that is asymptotically distributed according to μ_i. We can then make the same statement about Lebesgue almost every initial condition in the basin of the attractor because, since τ is piecewise expanding, Lebesgue almost every point in the basin has a neighborhood that maps diffeomorphically onto an interval in the attractor under a finite number of iterations of τ. Finally, an arbitrary initial density function f can be decomposed into a sum of functions f_i, each of which is supported in the basin of a single attractor; the Lasota–Yorke measure associated to f by Theorem LaY is then a linear combination of the measures μ_i with coefficients determined by the integrals of the f_i (see also [17]).

5.3. The relationship of the LaY approach and the SRB approach

Assume that the measures exist and processes converge for the system in question. If an initial point x is in a basin, iterating it forward will in certain circumstances yield a trajectory distributed according to some measure μ_x according to the SRB approach and almost all x in that basin will yield the same measure. The Lasota–Yorke approach is to choose a density f that need not be restricted to one basin. Pushing it forward and averaging yields a measure m_f^* that depends on f. (Theorem LaY guarantees that the sequence $(1/n)\sum_{i=0}^{n}\mathcal{P}^i f$ converges in the L^1-norm to a function $f^* \in L^1(I,m)$ as $n \to \infty$ and the measure m_f^* is defined by $\mathrm{d}m_f^* = f^*\,\mathrm{d}m$.)

We now want to relate the Lasota–Yorke and SRB measures. Set aside the differences in the settings of the Lasota–Yorke and SRB measures and compare the similarities. Theorem LaY deals with functions that can have several (disjoint) chaotic attractors, while Theorem SRB deals with maps that have one attractor. How could one change the SRB theorem to handle that more general kind of situation when the location of the attractors and their basins are not known? Depending on the precise hypotheses used, there might still be an invariant measure μ_x for almost every x but it would depend on x. Therefore one could choose a probability density function f and define a measure which is an average of μ_x weighted by f, i.e. $m_f = \int f(x)\mu_x\,\mathrm{d}x$. Then the (generalized) SRB theorem would say that for every integrable f and bounded continuous φ,

$$\lim_{n\to\infty}\int f(x)\frac{1}{n}\sum_{i=0}^{n-1}\varphi(F^i(x))\,\mathrm{d}x = \int \varphi\,\mathrm{d}m_f \tag{11a}$$

for some measure m_f (where m_f depends linearly on f). Eq. (11a) is equivalent to

$$\lim_{n\to\infty}\frac{1}{n}\sum_{i=0}^{n-1}\int f(x)\varphi(F^i(x))\,\mathrm{d}x = \int \varphi\,\mathrm{d}m_f. \tag{11b}$$

Since F is one-to-one, the Frobenius–Perron operator for F is simply $(\mathcal{P}f)(x) = f(F^{-1}(x))/|\det DF(x)|$. Using the change of variables $y = F^{-i}(x)$ of Eq. (6), we have that $\int f(x)\varphi(F^i(x))\,\mathrm{d}x = \int(\mathcal{P}^i f)(y)\varphi(y)\,\mathrm{d}y$. Hence,

Eq. (11b) can be rewritten as

$$\lim_{n\to\infty}\frac{1}{n}\sum_{i=0}^{n-1}\int \mathcal{P}^i f(y)\varphi(y)\,\mathrm{d}y = \int \varphi\,\mathrm{d}m_f. \tag{12a}$$

Eq. (12a) is equivalent to

$$\lim_{n\to\infty}\int \frac{1}{n}\sum_{i=0}^{n-1}\mathcal{P}^i f(y)\varphi(y)\,\mathrm{d}y = \int \varphi\,\mathrm{d}m_f, \tag{12b}$$

which is the formula used by [10] in one dimension.

Finally, we propose: a dynamical system displays *SLYRB chaos* whenever there exists a chaotic attractor that supports a SLYRB measure.

All the computer assisted pictures were made by using the program Dynamics [2].

Acknowledgements

This work was supported in part by the National Science Foundation (Grant No. 0104087), and by the W.M. Keck Foundation.

References

[1] J. Milnor, On the concept of attractor, Commun. Math. Phys. 99 (1985) 177–195;
J. Milnor, Comments on the concept of attractor: corrections and remarks, Commun. Math. Phys. 102 (1985) 517–519.
[2] H.E. Nusse, J.A. Yorke, Dynamics: Numerical Explorations, 2nd Edition, Accompanied by Software Program Dynamics 2, An Interactive Program for IBM Compatible PCs and Unix Computers (the Unix version of the program is by B.R. Hunt, E.J. Kostelich), Applied Mathematical Sciences, Vol. 101, Springer, New York, 1998.
[3] D.J. Farmer, E. Ott, J.A. Yorke, The dimension of chaotic attractors, Physica D 7 (1983) 153–180.
[4] A. Lasota, M. Mackey, Chaos, Fractals and Noise, Stochastic Aspects of Dynamics, 2nd Edition, Applied Mathematical Sciences, Vol. 97, Springer, New York, 1994.
[5] Ya.G. Sinai, Gibbs measures in ergodic theory, Russ. Math. Surv. 27 (1972) 21–69.
[6] R. Bowen, Equilibrium states and the ergodic theory of Anosov diffeomorphisms, Springer Lecture Notes in Mathematics, Vol. 470, Springer, New York, 1975.
[7] R. Bowen, D. Ruelle, The ergodic theory of Axiom A flows, Invent. Math. 29 (1975) 181–202.
[8] D. Ruelle, Statistical mechanics on a compact set with Z^v action satisfying expansiveness and specification, Trans. Am. Math. Soc. 187 (1973) 237–251.
[9] D. Ruelle, A measure associated with Axiom A attractors, Am. J. Math. 98 (1976) 619–654.
[10] A. Lasota, J.A. Yorke, On the existence of invariant measures for piecewise monotonic transformations, Trans. Am. Math. Soc. 186 (1973) 481–488.
[11] M. Dellnitz, O. Junge, On the approximation of complicated dynamical behavior, SIAM J. Numer. Anal. 36 (1999) 491–515.
[12] S.M. Ulam, J. von Neumann, On combination of stochastic and deterministic processes, Abstract 403, Bull. Am. Math. Soc. 53 (1947) 1120.
[13] O.W. Rechard, Invariant measures for many-one transformations, Duke Math. J. 23 (1956) 477–488.
[14] K. Krzyzewski, W. Szlenk, On invariant measures for expanding differentiable mappings, Studia Math. 33 (1969) 83–92.
[15] T.-Y. Li, J.A. Yorke, Period three implies chaos, Am. Math. Monthly 82 (1975) 985–992.
[16] J. Guckenheimer, Sensitive dependence to initial conditions for one-dimensional maps, Commun. Math. Phys. 70 (1979) 133–160.
[17] T.-Y. Li, J.A. Yorke, Ergodic transformations from an interval into itself, Trans. Am. Math. Soc. 235 (1978) 183–192.
[18] L.-S. Young, Bowen–Ruelle measures for certain piecewise hyperbolic maps, Trans. Am. Math. Soc. 287 (1985) 41–48.
[19] M. Tsujii, On continuity of Bowen–Ruelle–Sinai measures in families of one-dimensional maps, Commun. Math. Phys. 177 (1996) 1–11.
[20] M. Benedicks, L.-S. Young, Sinai–Bowen–Ruelle measures for certain Hénon maps, Invent. Math. 112 (1993) 541–576.
[21] Y.B. Pesin, Dimension Theory in Dynamical Systems, Contemporary Views and Applications, University of Chicago Press, Chicago, 1997.
[22] D. Ruelle, Applications conservant une measure absolument continue par raport a dx sur [0,1], Commun. Math. Phys. 55 (1977) 47–51.

[23] M. Misiurewicz, Absolutely continuous measures for certain maps of the interval, IHES Publ. Math. 53 (1981) 17–51.
[24] P. Collet, J.-P. Eckmann, Positive Lyapunov exponents and absolutely continuity, Ergod. Th. Dynam. Syst. 3 (1983) 13–46.
[25] M. Jakobson, Absolutely continuous invariant measures for one-parameter families of one-dimensional maps, Commun. Math. Phys. 81 (1981) 39–88.
[26] M.R. Rychlik, Another proof of Jakobson's theorem and related results, Ergod. Th. Dynam. Syst. 8 (1988) 93–109.
[27] J. Graczyk, G. Swiatek, Generic hyperbolicity in the logistic family, Ann. Math. 146 (1997) 1–52.
[28] J. Graczyk, G. Swiatek, The Real Fatou Conjecture, Annals of Mathematical Studies, Vol. 144, Princeton University Press, Princeton, NJ, 1998.
[29] M. Lyubich, Dynamics of quadratic polynomials, Parts I and II, Acta Math. 178 (1997) 185–297.
[30] M. Lyubich, Almost every real quadratic map is either regular or stochastic, Preprint IMS at Stony Brook, No. 1997/1998, Ann. Math, in press. Available at http://www.math.sunysb.edu/~mlyubich/.
[31] M. Hénon, A two-dimensional mapping with a strange attractor, Commun. Math. Phys. 50 (1976) 69–77.
[32] R. Lozi, Un attracteur étrange (?) du type attracteur de Hénon, J. Phys. (Paris) 39 (Coll. C5) (1978) 9–10.
[33] M. Misiurewicz, Strange attractors for Lozi mappings, in: R.H.G. Helleman (Ed.), Nonlinear Dynamics, Annals of New York Academy of Sciences, Vol. 357, New York Academy of Sciences, New York, 1980, pp. 348–358.
[34] M. Rychlik, Measures invariantes et principe variationel pour les applications de Lozi, CR Acad. Sci. (Paris) 296 (1983) 19–22.
[35] M. Rychlik, Invariant measures and variational principle for Lozi applications, Ph.D. Dissertation, University of California, Berkeley, 1983. http://alamos.math.arizona.edu/~rychlik/publications.html.
[36] P. Collet, Y. Levy, Ergodic properties of the Lozi mappings, Commun. Math. Phys. 93 (1984) 461–481.
[37] M.V. Jakobson, S.E. Newhouse, A two-dimensional version of the folklore theorem, Am. Math. Soc. Trans., Ser. 2 171 (1996) 89–105; M.V. Jakobson, S.E. Newhouse, Asymptotic measures for hyperbolic piecewise smooth mappings of a rectangle, Astérisque 261 (2000) 103–159.
[38] J.R. Dorfman, An introduction to chaos in nonequilibrium statistical mechanics, Cambridge Lecture Notes in Physics, Vol. 14, Cambridge University Press, Cambridge, 1999.
[39] M. Benedicks, L. Carleson, The dynamics of the Hénon map, Ann. Math. 133 (1991) 73–169.
[40] L.-S. Young, Statistical properties of dynamical systems with some hyperbolicity, Ann. Math. 147 (2) (1998) 585–650.
[41] J.F. Alves, C. Bonatti, M. Viana, SRB measures for partially hyperbolic systems whose central direction is mostly expanding, Invent. Math. 140 (2000) 351–398.
[42] D. Ruelle, F. Takens, On the nature of turbulence, Commun. Math. Phys. 20 (1971) 167–192.
[43] E. Ott, Chaos in Dynamical Systems, Cambridge University Press, Cambridge, 1993.
[44] F. Ledrappier, Propriétés ergodiques des measures de Sinai, IHES Publ. Math. 59 (1984) 163–188.

Credits

We would like to extend our thanks to the following Authors and Publishers for granting permission to reprint the following articles:

E.N. Lorenz, Deterministic nonperiodic flow, J. Atm. Sc. 20 (1963), 130–141. Reprinted with permission from American Meteorological Society.

K. Krzyzewski and W. Szlenk On invariant measures for expanding differentiable mappings, Studia Math. 33 (1969), 83–92. Reprinted with the permission of the Institute of Mathematics, Polish Academy of Sciences.

A. Lasota and J.A. Yorke, On the existence of invariant measures for piecewise monotonic transformations Trans. Amer. Math. Soc. 186 (1973), 481–488. Reprinted with permission from the American Mathematical Society.

R. Bowen and D. Ruelle, The ergodic theory of Axiom A flows, Invent. Math. 29 (1975), 181–202. Reprinted with permission from Springer-Verlag.

T.-Y. Li and J.A. Yorke, Period three implies chaos. Amer. Math. Monthly 82 (1975), 985–992. Reprinted with permission from Mathematical Association of America.

R.M. May, Simple mathematical models with very complicated dynamics, Nature 261 (1976), 459–467. Reprinted with permission from Macmillan Publishers Limited.

M. Henon, A two-dimensional mapping with a strange attractor, Commun. Math. Phys. 50 (1976), 69–77. Reprinted with permission from Springer-Verlag.

E. Ott, Strange attractors and chaotic motions of dynamical systems, Rev. Modern Phys. 53 (1981), 655–671. Reprinted with permission from American Physical Society.

F. Hofbauer and G. Keller, Ergodic properties of invariant measures for piecewise monotonic transformations Math. Z. 180 (1982), 119–140. Reprinted with permission from Springer-Verlag.

D. J. Farmer, E. Ott and J.A. Yorke, The dimension of chaotic attractors, Physica D 7 (1983), 153–180. Reprinted with permission from Elsevier Science.

P. Grassberger and I. Procaccia, Measuring the strangeness of strange attractors, Physica D 9 (1983), 189–208. Reprinted with permission from Elsevier Science.

M. Rychlik, Invariant measures and Variational Principle for Lozi Applications, Ph.D. dissertation, University of California at Berkeley (1983). Unpublished, reprinted with permission from the author.

P. Collet and Y. Levy, Ergodic properties of the Lozi mappings, Commun. Math. Phys. 93 (1984), 461–481. Reprinted with permission from Springer-Verlag.

J. Milnor, On the concept of attractor, Commun. Math. Phys. 99 (1985), 177–195; Comments "On the concept of attractor": corrections and remarks, Commun. Math. Phys. 102 (1985), 517–519. Reprinted with permission from Springer-Verlag.

L.-S. Young, Bowen-Ruelle measures for certain piecewise hyperbolic maps, Trans. Amer. Math. Soc. 287 (1985), 41–48. Reprinted with permission from American Mathematical Society.

J.-P. Eckmann and D. Ruelle Ergodic theory of chaos and strange attractors, Rev. Modern Phys. 57 (1985), 617–656. Reprinted from permission from American Physical Society.

M.R. Rychlik, Another proof of Jakobson's theorem and related results, Ergodic Theory Dynamical Systems 8 (1988), 93–109. Reprinted with permission from Cambrige University Press.

C. Grebogi, E. Ott, and J.A. Yorke, Unstable periodic orbits and the dimensions of multifractal chaotic attractors, Phys. Rev. A 37 (1988), 1711–1724. Reprinted with permission from American Physical Society.

P. Gora and A. Boyarsky, Absolutely continuous invariant measures for piecewise expanding C^2 transformation in R^N, Israel J. Math. 67 (1989), 272–286. Reprinted with permission from The Hebrew University, Magness Press.

M. Benedicks and L.-S. Young, Sinai-Bowen-Ruelle measures for certain Hwnon maps, Invent. Math. 112 (1993), 541–576. Reprinted with permission from Springer-Verlag.

M. Dellnitz and O. Junge, On the approximation of complicated dynamical behavior, SIAM J. Numer. Anal. 36 (1999), 491–515. Reprinted with permission from SIAM.

M. Tsujii, Absolutely continuous invariant measures for piecewise real-analytic expanding maps on the plane, Comm. Math. Phys. 208 (2000), 605–622. Reprinted with permission from Springer-Verlag.

J.F. Alves, C. Bonatti and M. Viana, SRB measures for partially hyperbolic systems whose central direction is mostly expanding, Invent. Math. 140 (2000), 351–398. Reprinted with permission from Springer-Verlag.

B.R. Hunt, J.A. Kennedy, T.-Y. Li and H.E. Nusse, SLYRB measures: natural invariant measures for chaotic systems, Physica D 170 (2002), 50–71. Reprinted with permission from Elsevier Science.